T0189240

Lecture Notes in Computer Science 11538

Commenced Publication in 1973
Founding and Former Series Editors:
Gerhard Goos, Juris Hartmanis, and Jan van Leeuwen

More information about this series at http://www.springer.com/series/7407

João M. F. Rodrigues · Pedro J. S. Cardoso ·
Jânio Monteiro · Roberto Lam ·
Valeria V. Krzhizhanovskaya ·
Michael H. Lees · Jack J. Dongarra ·
Peter M. A. Sloot (Eds.)

Computational Science – ICCS 2019

19th International Conference
Faro, Portugal, June 12–14, 2019
Proceedings, Part III

 Springer

Editors
João M. F. Rodrigues ⓘ
University of Algarve
Faro, Portugal

Pedro J. S. Cardoso ⓘ
University of Algarve
Faro, Portugal

Jânio Monteiro ⓘ
University of Algarve
Faro, Portugal

Roberto Lam ⓘ
University of Algarve
Faro, Portugal

Valeria V. Krzhizhanovskaya ⓘ
University of Amsterdam
Amsterdam, The Netherlands

Michael H. Lees
University of Amsterdam
Amsterdam, The Netherlands

Jack J. Dongarra ⓘ
University of Tennessee at Knoxville
Knoxville, TN, USA

Peter M. A. Sloot ⓘ
University of Amsterdam
Amsterdam, The Netherlands

ISSN 0302-9743   ISSN 1611-3349 (electronic)
Lecture Notes in Computer Science
ISBN 978-3-030-22743-2   ISBN 978-3-030-22744-9 (eBook)
https://doi.org/10.1007/978-3-030-22744-9

LNCS Sublibrary: SL1 – Theoretical Computer Science and General Issues

This Springer imprint is published by the registered company Springer Nature Switzerland AG
The registered company address is: Gewerbestrasse 11, 6330 Cham, Switzerland

Preface

Welcome to the 19th Annual International Conference on Computational Science (ICCS - https://www.iccs-meeting.org/iccs2019/), held during June 12–14, 2019, in Faro, Algarve, Portugal. Located at the southern end of Portugal, Algarve is a well-known touristic haven. Besides some of the best and most beautiful beaches in the entire world, with fine sand and crystal-clear water, Algarve also offers amazing natural landscapes, a rich folk heritage, and a healthy gastronomy that can be enjoyed throughout the whole year, attracting millions of foreign and national tourists. ICCS 2019 was jointly organized by the University of Algarve, the University of Amsterdam, NTU Singapore, and the University of Tennessee.

The International Conference on Computational Science is an annual conference that brings together researchers and scientists from mathematics and computer science as basic computing disciplines, as well as researchers from various application areas who are pioneering computational methods in sciences such as physics, chemistry, life sciences, engineering, arts, and humanitarian fields, to discuss problems and solutions in the area, to identify new issues, and to shape future directions for research.

Since its inception in 2001, ICCS has attracted an increasingly higher quality and numbers of attendees and papers, and this year was no exception, with over 350 participants. The proceedings series have become a major intellectual resource for computational science researchers, defining and advancing the state of the art in this field.

ICCS 2019 in Faro was the 19th in this series of highly successful conferences. For the previous 18 meetings, see: http://www.iccs-meeting.org/iccs2019/previous-iccs/.

The theme for ICCS 2019 was "Computational Science in the Interconnected World," to highlight the role of computational science in an increasingly interconnected world. This conference was a unique event focusing on recent developments in: scalable scientific algorithms; advanced software tools; computational grids; advanced numerical methods; and novel application areas. These innovative novel models, algorithms, and tools drive new science through efficient application in areas such as physical systems, computational and systems biology, environmental systems, finance, and others.

ICCS is well known for its excellent line-up of keynote speakers. The keynotes for 2019 were:

- Tiziana Di Matteo, King's College London, UK
- Teresa Galvão, University of Porto/INESC TEC, Portugal
- Douglas Kothe, Exascale Computing Project, USA
- James Moore, Imperial College London, UK
- Robert Panoff, The Shodor Education Foundation, USA
- Xiaoxiang Zhu, Technical University of Munich, Germany

This year we had 573 submissions (228 submissions to the main track and 345 to the workshops). In the main track, 65 full papers were accepted (28%); in the workshops, 168 full papers (49%). The high acceptance rate in the workshops is explained by the nature of these thematic sessions, where many experts in a particular field are personally invited by workshop organizers to participate in their sessions.

ICCS relies strongly on the vital contributions of our workshop organizers to attract high-quality papers in many subject areas. We would like to thank all committee members for the main track and workshops for their contribution to ensure a high standard for the accepted papers. We would also like to thank Springer, Elsevier, and Intellegibilis for their support. Finally, we very much appreciate all the local Organizing Committee members for their hard work to prepare this conference.

We are proud to note that ICCS is an A-rank conference in the CORE classification.

June 2019

João M. F. Rodrigues
Pedro J. S. Cardoso
Jânio Monteiro
Roberto Lam
Valeria V. Krzhizhanovskaya
Michael Lees
Jack J. Dongarra
Peter M. A. Sloot

Organization

Workshops and Organizers

Advanced Modelling Techniques for Environmental Sciences – AMES

Jens Weismüller
Dieter Kranzlmüller
Maximilian Hoeb
Jan Schmidt

Advances in High-Performance Computational Earth Sciences: Applications and Frameworks – IHPCES

Takashi Shimokawabe
Kohei Fujita
Dominik Bartuschat

Agent-Based Simulations, Adaptive Algorithms, and Solvers – ABS-AAS

Maciej Paszynski
Quanling Deng
David Pardo
Robert Schaefer
Victor Calo

Applications of Matrix Methods in Artificial Intelligence and Machine Learning – AMAIML

Kourosh Modarresi

Architecture, Languages, Compilation, and Hardware Support for Emerging and Heterogeneous Systems – ALCHEMY

Stéphane Louise
Löic Cudennec
Camille Coti
Vianney Lapotre
José Flich Cardo
Henri-Pierre Charles

Biomedical and Bioinformatics Challenges for Computer Science – BBC

Mario Cannataro
Giuseppe Agapito
Mauro Castelli

Riccardo Dondi
Rodrigo Weber dos Santos
Italo Zoppis

Classifier Learning from Difficult Data – CLD^2

Michał Woźniak
Bartosz Krawczyk
Paweł Ksieniewicz

Computational Finance and Business Intelligence – CFBI

Yong Shi
Yingjie Tian

Computational Methods in Smart Agriculture – CMSA

Andrew Lewis

Computational Optimization, Modelling, and Simulation – COMS

Xin-She Yang
Slawomir Koziel
Leifur Leifsson

Computational Science in IoT and Smart Systems – IoTSS

Vaidy Sunderam

Data-Driven Computational Sciences – DDCS

Craig Douglas

Machine Learning and Data Assimilation for Dynamical Systems – MLDADS

Rossella Arcucci
Boumediene Hamzi
Yi-Ke Guo

Marine Computing in the Interconnected World for the Benefit of Society – MarineComp

Flávio Martins
Ioana Popescu
João Janeiro
Ramiro Neves
Marcos Mateus

Multiscale Modelling and Simulation – MMS

Derek Groen
Lin Gan
Stefano Casarin
Alfons Hoekstra
Bartosz Bosak

**Simulations of Flow and Transport: Modeling, Algorithms,
and Computation – SOFTMAC**

Shuyu Sun
Jingfa Li
James Liu

**Smart Systems: Bringing Together Computer Vision, Sensor Networks,
and Machine Learning – SmartSys**

João M. F. Rodrigues
Pedro J. S. Cardoso
Jânio Monteiro
Roberto Lam

Solving Problems with Uncertainties – SPU

Vassil Alexandrov

Teaching Computational Science – WTCS

Angela Shiflet
Evguenia Alexandrova
Alfredo Tirado-Ramos

**Tools for Program Development and Analysis
in Computational Science – TOOLS**

Andreas Knüpfer
Karl Fürlinger

Programme Committee and Reviewers

Ahmad Abdelfattah
Eyad Abed
Markus Abel
Laith Abualigah
Giuseppe Agapito
Giovanni Agosta
Ram Akella

Elisabete Alberdi
Marco Aldinucci
Luis Alexandre
Vassil Alexandrov
Evguenia Alexandrova
Victor Allombert
Saad Alowayyed

Stanislaw
 Ambroszkiewicz
Ioannis Anagnostou
Philipp Andelfinger
Michael Antolovich
Hartwig Anzt
Hideo Aochi

Rossella Arcucci
Tomasz Arodz
Kamesh Arumugam
Luiz Assad
Victor Azizi Tarksalooyeh
Bartosz Balis
Krzysztof Banas
João Barroso
Dominik Bartuschat
Daniel Becker
Jörn Behrens
Adrian Bekasiewicz
Gebrail Bekdas
Stefano Beretta
Daniel Berrar
John Betts
Sanjukta Bhowmick
Bartosz Bosak
Isabel Sofia Brito
Kris Bubendorfer
Jérémy Buisson
Aleksander Byrski
Cristiano Cabrita
Xing Cai
Barbara Calabrese
Carlos Calafate
Carlos Cambra
Mario Cannataro
Alberto Cano
Paul M. Carpenter
Stefano Casarin
Manuel Castañón-Puga
Mauro Castelli
Jeronimo Castrillon
Eduardo Cesar
Patrikakis Charalampos
Henri-Pierre Charles
Zhensong Chen
Siew Ann Cheong
Andrei Chernykh
Lock-Yue Chew
Su Fong Chien
Sung-Bae Cho
Bastien Chopard
Stephane Chretien
Svetlana Chuprina

Florina M. Ciorba
Noelia Correia
Adriano Cortes
Ana Cortes
Jose Alfredo F. Costa
Enrique
    Costa-Montenegro
David Coster
Camille Coti
Carlos Cotta
Helene Coullon
Daan Crommelin
Attila Csikasz-Nagy
Loïc Cudennec
Javier Cuenca
Yifeng Cui
António Cunha
Ben Czaja
Pawel Czarnul
Bhaskar Dasgupta
Susumu Date
Quanling Deng
Nilanjan Dey
Ergin Dinc
Minh Ngoc Dinh
Sam Dobbs
Riccardo Dondi
Ruggero Donida Labati
Goncalo dos-Reis
Craig Douglas
Aleksandar Dragojevic
Rafal Drezewski
Niels Drost
Hans du Buf
Vitor Duarte
Richard Duro
Pritha Dutta
Sean Elliot
Nahid Emad
Christian Engelmann
Qinwei Fan
Fangxin Fang
Antonino Fiannaca
Christos
    Filelis-Papadopoulos
José Flich Cardo

Yves Fomekong Nanfack
Vincent Fortuin
Ruy Freitas Reis
Karl Frinkle
Karl Fuerlinger
Kohei Fujita
Wlodzimierz Funika
Takashi Furumura
Mohamed Medhat Gaber
Jan Gairing
David Gal
Marco Gallieri
Teresa Galvão
Lin Gan
Luis Garcia-Castillo
Delia Garijo
Frédéric Gava
Don Gaydon
Zong-Woo Geem
Alex Gerbessiotis
Konstantinos
    Giannoutakis
Judit Gimenez
Domingo Gimenez
Guy Gogniat
Ivo Gonçalves
Yuriy Gorbachev
Pawel Gorecki
Michael Gowanlock
Manuel Graña
George Gravvanis
Marilaure Gregoire
Derek Groen
Lutz Gross
Sophia
    Grundner-Culemann
Pedro Guerreiro
Kun Guo
Xiaohu Guo
Piotr Gurgul
Pietro Hiram Guzzi
Panagiotis Hadjidoukas
Mohamed Hamada
Boumediene Hamzi
Masatoshi Hanai
Quillon Harpham

William Haslett
Yiwei He
Alexander Heinecke
Jurjen Rienk Helmus
Alvaro Herrero
Bogumila Hnatkowska
Maximilian Hoeb
Paul Hofmann
Sascha Hunold
Juan Carlos Infante
Hideya Iwasaki
Takeshi Iwashita
Alfredo Izquierdo
Heike Jagode
Vytautas Jancauskas
Joao Janeiro
Jiří Jaroš
Shantenu Jha
Shalu Jhanwar
Chao Jin
Hai Jin
Zhong Jin
David Johnson
Anshul Joshi
Manuela Juliano
George Kallos
George Kampis
Drona Kandhai
Aneta Karaivanova
Takahiro Katagiri
Ergina Kavallieratou
Wayne Kelly
Christoph Kessler
Dhou Khaldoon
Andreas Knuepfer
Harald Koestler
Dimitrios Kogias
Ivana Kolingerova
Vladimir Korkhov
Ilias Kotsireas
Ioannis Koutis
Sergey Kovalchuk
Michał Koziarski
Slawomir Koziel
Jarosław Koźlak
Dieter Kranzlmüller

Bartosz Krawczyk
Valeria Krzhizhanovskaya
Paweł Ksieniewicz
Michael Kuhn
Jaeyoung Kwak
Massimo La Rosa
Roberto Lam
Anna-Lena Lamprecht
Johannes Langguth
Vianney Lapotre
Jysoo Lee
Michael Lees
Leifur Leifsson
Kenneth Leiter
Roy Lettieri
Andrew Lewis
Jingfa Li
Yanfang Li
James Liu
Hong Liu
Hui Liu
Zhao Liu
Weiguo Liu
Weifeng Liu
Marcelo Lobosco
Veronika Locherer
Robert Lodder
Stephane Louise
Frederic Loulergue
Huimin Lu
Paul Lu
Stefan Luding
Scott MacLachlan
Luca Magri
Maciej Malawski
Livia Marcellino
Tomas Margalef
Tiziana Margaria
Svetozar Margenov
Osni Marques
Alberto Marquez
Paula Martins
Flavio Martins
Jaime A. Martins
Marcos Mateus
Marco Mattavelli

Pawel Matuszyk
Valerie Maxville
Roderick Melnik
Valentin Melnikov
Ivan Merelli
Jianyu Miao
Kourosh Modarresi
Miguel Molina-Solana
Fernando Monteiro
Jânio Monteiro
Pedro Montero
James Montgomery
Andrew Moore
Irene Moser
Paulo Moura Oliveira
Ignacio Muga
Philip Nadler
Hiromichi Nagao
Kengo Nakajima
Raymond Namyst
Philippe Navaux
Michael Navon
Philipp Neumann
Ramiro Neves
Mai Nguyen
Hoang Nguyen
Nancy Nichols
Sinan Melih Nigdeli
Anna Nikishova
Kenji Ono
Juan-Pablo Ortega
Raymond Padmos
J. P. Papa
Marcin Paprzycki
David Pardo
Héctor Quintián Pardo
Panos Parpas
Anna Paszynska
Maciej Paszynski
Jaideep Pathak
Abani Patra
Pedro J. S. Cardoso
Dana Petcu
Eric Petit
Serge Petiton
Bernhard Pfahringer

Daniela Piccioni
Juan C. Pichel
Anna
  Pietrenko-Dabrowska
Laércio L. Pilla
Armando Pinto
Tomasz Piontek
Erwan Piriou
Yuri Pirola
Nadia Pisanti
Antoniu Pop
Ioana Popescu
Mario Porrmann
Cristina Portales
Roland Potthast
Ela Pustulka-Hunt
Vladimir Puzyrev
Alexander Pyayt
Zhiquan Qi
Rick Quax
Barbara Quintela
Waldemar Rachowicz
Célia Ramos
Marcus Randall
Lukasz Rauch
Andrea Reimuth
Alistair Rendell
Pedro Ribeiro
Bernardete Ribeiro
Robin Richardson
Jason Riedy
Celine Robardet
Sophie Robert
João M. F. Rodrigues
Daniel Rodriguez
Albert Romkes
Debraj Roy
Philip Rutten
Katarzyna Rycerz
Augusto S. Neves
Apaar Sadhwani
Alberto Sanchez
Gabriele Santin

Robert Schaefer
Olaf Schenk
Ulf Schiller
Bertil Schmidt
Jan Schmidt
Martin Schreiber
Martin Schulz
Marinella Sciortino
Johanna Sepulveda
Ovidiu Serban
Vivek Sheraton
Yong Shi
Angela Shiflet
Takashi Shimokawabe
Tan Singyee
Robert Sinkovits
Vishnu Sivadasan
Peter Sloot
Renata Slota
Grażyna Ślusarczyk
Sucha Smanchat
Maciej Smołka
Bartlomiej Sniezynski
Sumit Sourabh
Hoda Soussa
Steve Stevenson
Achim Streit
Barbara Strug
E. Dante Suarez
Bongwon Suh
Shuyu Sun
Vaidy Sunderam
James Suter
Martin Swain
Grzegorz Swisrcz
Ryszard Tadeusiewicz
Lotfi Tadj
Daniele Tafani
Daisuke Takahashi
Jingjing Tang
Osamu Tatebe
Cedric Tedeschi
Kasim Tersic

Yonatan Afework
  Tesfahunegn
Jannis Teunissen
Andrew Thelen
Yingjie Tian
Nestor Tiglao
Francis Ting
Alfredo Tirado-Ramos
Arkadiusz Tomczyk
Stanimire Tomov
Marko Tosic
Jan Treibig
Leonardo Trujillo
Benjamin Uekermann
Pierangelo Veltri
Raja Velu
Alexander von Ramm
David Walker
Peng Wang
Lizhe Wang
Jianwu Wang
Gregory Watson
Rodrigo Weber dos
  Santos
Kevin Webster
Josef Weidendorfer
Josef Weinbub
Tobias Weinzierl
Jens Weismüller
Lars Wienbrandt
Mark Wijzenbroek
Roland Wismüller
Eric Wolanski
Michał Woźniak
Maciej Woźniak
Qing Wu
Bo Wu
Guoqiang Wu
Dunhui Xiao
Huilin Xing
Miguel Xochicale
Wei Xue
Xin-She Yang

Dongwei Ye
Jon Yosi
Ce Yu
Xiaodan Yu
Reza Zafarani
Gábor Závodszky

H. Zhang
Zepu Zhang
Jingqing Zhang
Yi-Fan Zhang
Yao Zhang
Wenlai Zhao

Jinghui Zhong
Xiaofei Zhou
Sotirios Ziavras
Peter Zinterhof
Italo Zoppis
Chiara Zucco

Contents – Part III

Track of Computational Finance and Business Intelligence

Track of Computational Optimization, Modelling and Simulation

Track of Computational Science in IoT and Smart Systems

Track of Biomedical and Bioinformatics Challenges for Computer Science

Parallelization of an Algorithm for Automatic Classification of Medical Data

Victor M. Garcia-Molla[1]([⊠]), Addisson Salazar[2], Gonzalo Safont[2], Antonio M. Vidal[1], and Luis Vergara[2]

[1] Department of Information Systems and Computing (DSIC), Universitat Politècnica de València, 46022 Valencia, Spain
{vmgarcia, avidal}@dsic.upv.es
[2] Institute of Telecommunications and Multimedia Applications (ITEAM), Universitat Politècnica de València, 46022 Valencia, Spain
{asalazar, lvergara}@dcom.upv.es,
gonsaar@upvnet.upv.es

Abstract. In this paper, we present the optimization and parallelization of a state-of-the-art algorithm for automatic classification, in order to perform real-time classification of clinical data. The parallelization has been carried out so that the algorithm can be used in real time in standard computers, or in high performance computing servers. The fastest versions have been obtained carrying out most of the computations in Graphics Processing Units (GPUs). The algorithms obtained have been tested in a case of automatic classification of electroencephalographic signals from patients.

Keywords: High performance computing · Bioinformatics · Automatic classification · ICA (independent component analysis) · SICAMM · GPU computing

1 Introduction

Modern clinical monitoring systems usually deliver large amounts of data, which in some cases must be processed in real time. Typical examples of such medical procedures are electroencephalography (EEG), functional magnetic resonance imaging (fMRI), etc. Furthermore, sometimes the medical data must be heavily processed, for example, the processing of medical images or the spectral analysis of EEG data. Very often the amount of data cannot be handled appropriately (in a reasonable time) by standard computers and parallel computing becomes necessary for achieving acceptable computing times [1–3]. This is especially true when real time computing is needed.

In this paper, we use the SICAMM (Sequential Independent Component Analysis Mixture Modeling) algorithm, a Bayesian classification procedure [4]. The case study tackled in this paper is the analysis of EEG signals in epileptic patients taken during memory and learning neuropsychological tests. Since this case study provides large amounts of output data, initial versions of the SICAMM algorithm (written in Matlab language [12]) would need large, unacceptable computing times.

© Springer Nature Switzerland AG 2019
J. M. F. Rodrigues et al. (Eds.): ICCS 2019, LNCS 11538, pp. 3–16, 2019.
https://doi.org/10.1007/978-3-030-22744-9_1

This paper describes several parallelization techniques that are applied to the SICAMM algorithm. The parallel algorithms proposed have been implemented in different hardware platforms so that the system can be adapted to different requirements. The best results have been obtained using GPUs for the heaviest computations. Moreover, the GPU utilization has allowed an increase in the performance of SICAMM algorithm, making it possible its use in real-time applications.

The rest of the paper is structured as follows. Section 2 describes the SICAMM algorithm. Section 3 discusses the different optimizations and parallelization applied to the SICAMM algorithm. Section 4 presents a case study of EEG signal processing where the different implementations were applied. Section 5 includes an analysis of the results in terms of efficiency. Conclusions and future work are presented in Sect. 6.

2 State of the Art: Automatic Classification, Sequential ICAMM

The automatic classification problem can be described as follows: let $x_i \in \mathbb{R}^M$, $i = 1, 2, \ldots . Nx$ be the observed data, M the number of data sources and Nx the number of signal to be processed. Each data vector x_i may belong to one of K different classes. A procedure is sought that classifies the data, i.e. a procedure that indicates if the vector x_i belongs to one of the K existing classes. The output of the classification is given in a vector of class assignments, $Ce \in \mathbb{R}^{Nx}$, such that if the i-th data vector x_i is found to belong to the j-th class ($j = 1, \ldots, K$), then $Ce(i)$ is given the value j. There are many automatic classification methods (neural networks, support vector machines, k-nearest neighbors, etc. see for example [7]). Here, we only consider and optimize the SICAMM method [4].

We will start by describing the ICAMM (Independent Component Analyzer Mixture Modeling) method. ICAMM is an automatic classification method where several classes are considered and each class is modeled using an independent component analyzer (ICA). ICAMM has been applied to a number of real-world applications, for instance, non-destructive testing, image processing, and change detection (see [8–10] and their references). The SICAMM algorithm is the extension of ICAMM to the case where data has sequential dependence i.e., when x_i depends in some way on the values of $x_j, j < i$.

Let us review the main concepts of the SICAMM algorithm. An ICA is formulated as a linear model of the observed data, the vectors $x_i \in \mathbb{R}^M$. These vectors are assumed to be a linear transformation of a vector of sources $s_i \in \mathbb{R}^M$ given by a mixing matrix $\in \mathbb{R}^{M \times M}$, as $x_i = As_i$.

If the mixing matrix A is invertible (with $W = A^{-1} \in \mathbb{R}^{M \times M}$ being the demixing matrix), we can express the joint probability density $p(x_i)$ in terms of the product of the marginal densities of the elements of $p(s_i)$, as $p(x_i) = |\det W| p(s_i)$, where $s_i = Wx_i$. The general expression of ICAMM requires some bias vectors to separate the components of the mixture of K classes or data groups. To do this, each class will have its own mixing matrix (A_k) and its own bias vector (b_k). Therefore, if x_i belongs to class k ($Ce(i) = k$), then $x_i = A_k s_{k,i} + b_k$, where it is assumed that x_i belongs to class k, denoted by $Ce(i)$. A_k and s_k are, respectively, the mixing matrix and the source vector of the ICA model of class k, and b_k is the corresponding bias vector.

SICAMM extended the mixture model to consider time dependencies. This is modelled through transition probabilities, π_{kj}, which give the probability that the i-th data vector belongs to class k given that the $i-1$-th data vector belongs to class j: $\pi_{kj} = P(Ce(i) = k|Ce(i-1) = j)$. An initial description of the SICAMM algorithm in its more general form (K classes) is shown as Algorithm 1.

---

1 Algorithm 1: SICAMM with K classes:

2 Input: demixing matrices $W_1, W_2, \ldots W_K$; bias vectors $b_1, b_2, \ldots, b_K$ and signals $x_i, i = 1 \ldots Nx$

3 Output: vector Ce

4 /* Initialization; $i = 1$, Classification of x_1 */

5 Calculate sources for each class, $s_k = W_k(x_1 - b_k), k = 1 \ldots K$

6 Calculate posterior probabilities $P(Ce(1) = k|x_1)$ using an ICAMM algorithm (e.g., [11]):

$$P(Ce(1) = k|x_1) = \frac{|\det W_k|\, p(s_k)\, P(Ce(1) = k)}{\sum_{j=1}^{K} |\det W_j|\, p(s_j)\, P(Ce(1) = j)}$$

7 Initial signal x_1 is assigned to the class with maximum posterior probability

8 /* classification of $x_i, i > 1$ */

9 for $i = 2 \ldots Nx$, with Nx being the number of signals:

10   Select current signal, x_i, and build $X_i = [x_1, \ldots, x_i]$

11   Calculate sources for each class, $s_k = W_k(x_i - b_k), k = 1 \ldots K$

12   Calculate conditional class probabilities using:

$$P(Ce(i) = k|X_{i-1}) = \sum_{j=1}^{K} \pi_{kj}\, P(Ce(i-1) = j|X_{i-1})$$

13   Calculate posterior probabilities using:

$$P(Ce(i) = k|X_i) = \frac{|\det W_k|\, p(s_k)\, P(Ce(i) = k|X_{i-1})}{\sum_{j=1}^{K} |\det W_j|\, p(s_j)\, P(Ce(i) = j|X_{i-1})}$$

14   Current signal x_i is assigned to the class with maximum $P(Ce(i) = k|X_i)$

15 end for

---

3 Optimization and Parallelization

In this section, we discuss the implementation of the SICAMM algorithm plus the different optimizations. We start with an initial implementation of the SICAMM algorithm, given in pseudo code.

We restrict ourselves to the case where two classes ($K = 2$) are considered. The source probabilities $p(s_1)$ and $p(s_2)$ ($ps1$ and $ps2$ in Algorithm 2) are computed using two sets of (correctly) pre-classified signals, $S_1 \in \mathbb{R}^{M \times Nf1}$ and $S_2 \in \mathbb{R}^{M \times Nf2}$. Therefore, the input data is the following: the two pre-classified sets S_1 and S_2, the demixing matrices $W_1 \in \mathbb{R}^{M \times M}$, $W_2 \in \mathbb{R}^{M \times M}$; the bias vectors $b_1 \in \mathbb{R}^M$, $b_2 \in \mathbb{R}^M$; and the probability transition parameter $r \in \mathbb{R}$, where $r = \pi_{11} = \pi_{22}$. Accordingly, $\pi_{21} = \pi_{12} = 1 - r$. The signals to be classified are the vectors $x_i \in \mathbb{R}^M$; in a real situation, these vectors should be processed at a fast rate. A total of Nx signals are analyzed. The first version of this algorithm was developed in Matlab (R2016b), using the Statistics Toolbox function "ksdensity" to obtain the probability densities $ps1$ and $ps2$. Although the results were correct, the code was quite inefficient due to (among other circumstances) repeated pre-computations of medians and typical deviations of the pre-classified signals, S_1 and S_2. A basic (but more efficient) version of the SICAMM algorithm is presented here as Algorithm 2, where these pre-computations are made only once, outside of the main loop.

```
 1 Algorithm 2: Optimized SICAMM for 2 classes:
 2 Input: W₁, W₂, b₁, b₂, S1, S2, Nx, r , and signals xᵢ, i = 1..Nx
 3 Output: vector Ce
 4 /*Precomputation of sig1, sig2 */
 5 for j=1:M
 6     med1=median(S1(j,:)); med2=median(S2(j,:))
 7     sig1[j]= median(abs(S1(j,:)-med1)/0.6745)
 8     sig2[j]= median(abs(S2(j,:)-med2)/0.6745)
 9 end for
10 dW1=det(W1); dW2=det(W2)
11 /*Main Loop */
12 for i=1:Nx
13     s₁ = W₁ * (xᵢ − b₁); s₂ = W₂ * (xᵢ − b₂)
14     ps1=density(S1, s₁,sig1); ps2=density(S2, s₂,sig2)
15     p1=abs(dW1)*ps1;  p2=abs(dW2)*ps2
16     if (i==1)
17          p=p1+p2
18          PC1=p1/p; PC2=p2/p
19     else
20          PC1X=r*PC1+(1-r)*PC2;  PC2X=(1-r)*PC1+r*PC2
21          p=p1*PC1X+p2*PC2X
22          PC1=(p1*PC1X)/p;   PC2=(p2*PC2X)/p
23     end if
24     if PC1>PC2
25          Ce(i)=1
26     else
27          Ce(i)=2
28      end if
29 end for
```

For each incoming signal x_i, the output is the value of the vector $Ce(i)$, which will be 1 if the signal x_i belongs to class 1 and 2 if the signal x_i belongs to class 2. The part of algorithm with the larger computational cost is the computation of the source probabilities $ps1$ and $ps2$. The "density" routine (Algorithm 3) is a simplified and adapted version of the Matlab "ksdensity" function. In order to obtain reasonable execution times, this algorithm was coded in C language.

```
1 Algorithm 3: density
2 Input: S, s, sig; Output: ps
3 [M,Nf]=size(S)
4 weight=1.0/Nf
5 Pi_constant=1/sqrt(2*PI)
6 /* computation of the density for each component s(j)
7 for each s(j), the density probability is out(j)*/
8 for j=1:M
9     u=sig(j) * pow(4/(3*Nf),(1.0/5.0))
10    aux=0
11    for i=1:Nf
12          z=(S(j,i)-s(j))/u
13          aux=aux + exp(-0.5*z*z)*Pi_constant
14    end for
15    out(j)=aux*weight/u
16 end for
17 /* computation of the joint probability for all the vector s */
18 ps=1.0
19 for j=1:M
20    ps=ps*out(j)
21 end
```

Some interesting aspects of Algorithms 2 and 3 are commented on below:

- The initial iteration of the algorithm (line 20, for the first signal) is slightly different from the rest of the iterations in order to correctly initialize the probabilities $PC1$ and $PC2$.
- For each incoming signal, the most computationally expensive part is the execution of the "density" function (Algorithm 3). The computational cost of Algorithm 3 for each signal (discarding lower order terms) can be established by examining lines 12 and 13 in Algorithm 3. These two lines are executed $M \cdot Nf$ times, and there are six floating point operations and one call to the exponential function in these lines. Therefore, the cost per signal is $6 \cdot M \cdot Nf$ flops and $M \cdot Nf$ evaluations of the exponential function. Algorithm 3 is called twice with each signal, one with

$S1 \in \mathbb{R}^{M \times Nf1}$ and the other with $S2 \in \mathbb{R}^{M \times Nf2}$. Therefore, the theoretical cost of the whole "density" algorithm for a single entry signal x_i can be established as $6 \cdot M \cdot (Nf1 + Nf2)$ flops and $M \cdot (Nf1 + Nf2)$ exponentials.

- Algorithm 3 is easily parallelizable in different ways. The inner loop ("For i=1:Nf", line 11) can be parallelized using, for example, standard OpenMP directives [13]. The outer loop ("For j=1:M", line 8) can also be executed trivially in parallel. In such situations, usually the best strategy is to parallelize the outer code using OpenMP (so that each "j" index is processed by a single core), and to use compiler directives to force the compiler to "vectorize" the inner loop. This means that, in each core, the computation is accelerated by using vector registers and vector instructions, such as the AVX instruction set, which can carry out several operations in a single clock cycle [14]. The parallel version tested in Sect. 5 was parallelized in this way.

3.1   Block Version

Algorithm 2 must process incoming signals sequentially because of the dependence of the probabilities $PC1X$ and $PC2X$ on the accumulated probabilities $PC1$ and $PC2$, which come from former iterations. However, the computational cost of the code where these probabilities are used (lines 16 to 28) is minimal; the largest computational cost is in the calls to the "density" function and, to a lesser extent, in the matrix-vector products in line 13. On the other hand, the computations carried out in "density" for an incoming signal, and the matrix-vector products, are completely independent from the computations for other signals. Therefore, it is possible to group several incoming signals in a block and perform several calls to "density" (possibly in parallel), and later on, process the output of the "density" function for all of the signals of the block, obtaining the corresponding values of the Ce vector.

A block version of Algorithms 2 and 3 was developed where the signals are processed in blocks of B_{size} signals. The use of blocks improves the memory traffic, and allows the matrix-vector products in line 13 to be embodied as matrix-matrix products, which are significantly more efficient [14]. They are carried out using calls to BLAS routines [15].

This version obtained good performance, but it cannot be presented here due to lack of space. Nevertheless, the best results were obtained with GPU-accelerated versions, which are described in the next section.

3.2 GPU Version

Nowadays GPUs are used in many science fields where heavy computations are required. These devices, which were originally designed to handle the graphic interface with the computer user, have had strong development driven by the videogame market. Therefore, GPUs have actually become small supercomputers. The main feature of GPUs (considered as computing devices) is that they have a large number of small, relatively slow cores (slow compared with CPU cores). The number of cores in a NVIDIA GPU ranges today from less than 100 in cheap GeForce cards to around 4000 cores in a Tesla K40.

The use of GPUs for general-purpose computation received an enormous push with the launch of CUDA, a parallel programming environment for GPUs created by NVIDIA. CUDA allows for relatively easy programming of the GPUs by using extensions of the C language (it is also possible to program in other languages). CUDA programs look like standard C programs, but they have special routines to send/receive data from the GPU, and allow special pieces of code called "kernels", which contain the code to be executed in the GPU. A detailed description of CUDA can be found in many recent papers; we will assume a basic knowledge of CUDA from the reader in order to avoid another lengthy description of the basics of CUDA programming [6].

3.2.1 Description of the Main GPU Algorithm

The idea behind our GPU implementation is: (1) to process the signals in blocks, processing them as matrices with B_{size} columns, (2) to leave the sequential part of the main loop of Algorithm 2 (lines 16 to 28) untouched, and (3) to use the GPUs for the bulk of the computation (lines 13 to 14).

The main GPU algorithm is Algorithm 4. The first change is that the data must be sent to the GPU; the data that must remain in the GPU all of the time (variables $W1$, $W2$, $b1$, $b2$, $S1$, $S2$, $sig1$, $sig2$) is sent to the GPU before the start of the main loop (line 9). Then the main loop starts like Algorithm 2, but the signals are grouped in blocks of B_{size} columns which are sent to the GPU. The matrix-matrix products are carried out using the CUBLAS "cublas_dgemm" routine [17]. The calls to "density" are rewritten as CUDA kernels that process a block with B_{size} signals in each call. A value of $B_{size} = 100$ has shown to be adequate to obtain good performance with this routine. These CUDA kernels are described in the next section. The goal is the same, to compute the probabilities $ps1$ and $ps2$. Once these probabilities are computed, they are sent back to the CPU, where the rest of the algorithm is identical to Algorithm 2.

```
 1  Algorithm 4: GPU SICAMM for 2 classes.
 2  Input: W1, W2, b1, b2, S1, S2, x, Nx, B_size
 3  Output: vector Ce
 4  /*Precomputation of sig1, sig2 like in Algorithm 2 */
 5  Initial_iteration(PC1,PC2, x_1)
 6  indce=2
 7  /*Main Loop */
 8  Send Data to GPU: W1, W2, b1, b2, S1, S2, sig1, sig2
 9  for i=2:B_size: Nx
10      xB_i = x(:, i: i + B_size − 1)          /*select B_size signals */
11      Send xB_i to GPU
12      compute    in    GPU    sB_1 = W_1 * (xB_i − repmat(b_1, 1, B_size))
/*CUBLAS call*/
13      compute    in    GPU    sB_2 = W_2 * (xB_i − repmat(b_2, 1, B_size))
/*CUBLAS call*/
14      out1=density_block_GPU(S1, sB_1,sig1)  /*kernel with with M*B_size
blocks */
15      out2=density_block_GPU(S2, sB_2,sig2)    /*kernel with M*B_size
blocks */
16      ps1=product_reduction_GPU(out1)        /*kernel with B_size blocks
*/
17      ps2=product_reduction_GPU(out2)        /*kernel with B_size blocks
*/
18      Send ps1,ps2 to CPU.
19      for j=1:B_size
20          p1=abs(det(W1))*ps1(j); p2=abs(det(W2))*ps2(j)
21          PC1X=M(1,1)*PC1+M(2,1)*PC2
22          PC2X=M(1,2)*PC1+M(2,2)*PC2
23          p=p1*PC1X+p2*PC2X
24          PC1=(p1*PC1X)/p; PC2=(p2*PC2X)/p
25          if PC1>PC2
26            Ce(indce)=1
27          else
28            Ce(indce)=2
29          end if
30          indce=indce+1
31      end for
32  end for
```

3.2.2   Description of the "Density" Algorithm for GPU

The "density" algorithm must be rewritten as a CUDA kernel. The work unit in the GPU is the thread. The cores of the GPU can execute these threads concurrently so there can be thousands of threads running in a GPU at the same time. GPU threads can be organized in thread blocks that can be one-, two-, or three-dimensional. The thread blocks can be visualized as "teams" of blocks working together with some shared information. The kernels are executed in parallel by all of the threads in all of the blocks, but they execute their tasks on different data. The blocks can also be organized in grids, which can also be one-, two-, or three-dimensional. Clearly, there is plenty of flexibility.

We have chosen the following scheme to obtain a kernel called "density_block_GPU" equivalent to "density", but computes the probabilities of B_{size} signals in a single call to the kernel. We have chosen one-dimensional blocks, with 512 threads each. Each block of threads is responsible for the computation of an $out(j, k)$ value. In other words, each block performs computations analogous to the inner loop ("For i=1:Nf") in "density". This loop is a reduction, where all the "aux" values must be added to obtain the final $out(j, k)$ value. This reduction has been programmed using a simplified version of the algorithm for reductions in CUDA proposed by Mark Harris [18]. All of the threads of the block cooperate using shared memory.

```
1 Algorithm 5: density_block_GPU (CUDA kernel)
2 Input: S, sB, sig, Bsize; Output: out
3 [M,Nf]=size(S)
4 weight=1.0/Nf
5 Pi_constant=1/sqrt(2*PI)
6 j=BlockIdx.x;  /* BlockIdx.x is a CUDA built-in function giving the Block
row in the grid of blocks (of threads)*/
7 k=BlockIdx.y /* BlockIdx.y is a CUDA built-in function giving the Block
column in the grid of blocks (of threads)*/
8 u=sig(j) * pow(4/(3*Nf),(1.0/5.0))
9 aux=0
10 /*All threads of block (j,k) cooperate in the "parallel for" to compute aux */
11 parallel_for  i=1:Nf
12          z=(S(j,i)-sB(j,k))/u
13          aux=aux + exp(-0.5*z*z)*Pi_constant
14 end_parallel_For
15 out(j,k)=aux*weight/u
```

Then we use a two-dimensional grid of blocks of threads of size $M \cdot B_{size}$. The number of rows of the grid is M, and each row (the blocks of threads of that row) performs computations that are analogous to the "for j=1:M" loop (line 8, Algorithm 3). Finally, the number of columns of the grid is B_{Size}, and each column (the blocks of threads of that column) of the grid deals with a column of the B_{size} blocks of signals being processed. Then, the block of the grid with index (j, k) computes the value $out(j, k)$. A simplified version of the kernel is shown in Algorithm 5.

Lines 18 to 21 of the "density" function involve a further reduction, which is needed to compute the final ps values. In the GPU version, this computation must be carried out for B_{size} vectors. For the sake of simplicity, it was more convenient to carry out this reduction in a different kernel. Thus, the "density_Block_GPU" returns the matrix *"out"*, of size $M \cdot B_{Size}$, and a different kernel (Algorithm 6) carries out the final reduction through two kernel calls (lines 16 and 17 of Algorithm 4).

```
1 Algorithm 6: product_reduction_GPU (CUDA kernel)
2 Input: out; Output: ps
3 [M,B_size]=size(out)
4 k=BlockIdx.x
5 ps(k)=1.0
6 /*All threads of block k cooperate in the "parallel for" to compute ps(k)*/
7 parallel_for  j=1:M
8                 ps(k)=ps(k)*out(j,k)
9 end_parallel_For
```

4   Case Study

As mentioned above, we considered the analysis of EEG signals taken from six epileptic patients that were performing a memory and learning neuropsychological test. This test was performed as part of the clinical evaluation of the patients.

The signals were recorded using an EEG device of 19 channels with a sampling frequency of 500 Hz positioned according to the 10–20 system. The signals were filtered and split into epochs of 0.25 s length (i.e., window size of 125 with 124 samples of overlap). This short length was selected in order to ensure that all stages of the test (some of which were very short) were spread over multiple epochs, thus improving parameter estimation. The theta-slow-wave-index (TSI) was estimated for each epoch as a feature for classification. (e.g. see [19], for its definition and computation). The neuropsychological test was drawn from the Wechsler Adult Intelligence Scale (WAIS, [20]) suite. WAIS is designed to measure intelligence in adults and older adolescents.

The specific sub-test of WAIS used was the Digit Span, which is divided into two stages. In the first stage, called Digit Span Forward, the participant is read a sequence of numbers and then is asked to recall the numbers in the same order. In the second stage, Digit Span Backward, the participant is read a sequence of numbers and is asked to recall the numbers in reverse order. In both stages, the objective is to correctly recall as many sequences as possible. The test is repeated a number of trials until the subject fails two consecutive trials. In the case of Subject #5, the total number of trials performed was 30, of which the data of the first 14 trials was used for training and the data of the last 16 trials was used for testing the developed algorithms. The TSI index was recorded with the goal of classifying the phases of stimuli presentation (phase 1) and subject response (phase 2).

The classification accuracy obtained for the six subjects is shown in Table 1. The results obtained have reasonable accuracy.

The case selected for testing the algorithms was the data from subject #5. In this case, the classification accuracy was 77.85% and the total number of EEG signals was 115886 (231.77 s). Fifty percent of this data (57943 signals) was used for the training

Table 1. Classification accuracy results

Subject #	1	2	3	4	5	6
Classification accuracy %	61.05	82.44	78.01	81.60	77.85	80.41

stage of the algorithm and the rest (57943 signals) was used for testing. Thus, the dimensions of the parameters were the following: $k = 1, 2$ (1: stimuli presentation; 2: subject response); matrices $W1$ and $W2$ of size 19×19; bias vectors $b1$ and $b2$ of size 19×1; transition probability parameter r; signals of the first class $S1$ of size 19×30654; and signals of the second class S, of size 19×27289.

5 Analysis of Results in Terms of Efficiency

The data of the experiment described above for Subject #5 was processed with the initial, sequential, and parallel versions. The results of all of the versions were identical (in terms of classification) in all of the cases. We used two machines with different characteristics. The first one was a standard desktop computer with a core-i7 Intel CPU, with 4 cores, 10 GB of RAM, and a GeForce GT 530 GPU with 96 cores. The second computer is a relatively expensive machine, with a Xeon CPU with 24 cores, and a k40 T NVIDIA GPU with 2880 cores.

In the standard desktop, we tested the following versions of the code: single core (to evaluate the benefits of parallelization); version parallelized with OpenMP, running with 4 cores and GPU version running in a GT530 card.

The results are summarized in Table 2:

Table 2. Execution time for the experiment on a standard desktop computer.

Version	Single core	4 cores	GPU GT530
Time (seconds)	951	320	126

This table shows the effect of the different parallelization/optimizations. Since the total duration of the experiment was 231.77 s, the last version (with a cheap GPU) would allow real-time processing of the data.

We tested two versions on the second machine: v1: version parallelized with OpenMP, running with 24 cores, and v2: GPU version running in a k40 T card. The results are shown in Table 3.

In order to obtain a better evaluation of the computational cost, we generated different subsets of the EEG data of Subject #5 and tested the code by varying the "size" of the problem. Taking into account that the computational cost per signal

Table 3. Execution time for the experiment on a high performance server.

Version	v1: 24 cores	v2: GPU k40
Time (seconds)	104	14,5

received was $6 * M \cdot (Nf1 + Nf2)$ flops and $M \cdot (Nf1 + Nf2)$ exponentials, we defined the "size" of the problem as $M \cdot (Nf1 + Nf2)$. We fixed the number of incoming signals (20000) and executed the different versions of the code for increasing values of problem size: 100000, 300000, 600000, and 800000 corresponding to $2, 5, 10,$ and 14 EEG channels from the total of the 19 EEG channels measured. The results are summarized in Figs. 1 and 2.

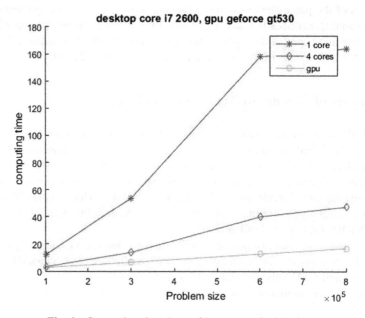

Fig. 1. Computing time (seconds) on a standard desktop

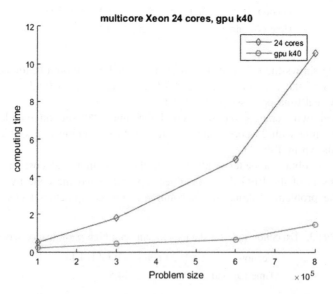

Fig. 2. Computing time (seconds) on a high performance computing server

6   Conclusion

The results show the large reduction of computational time due to the proposed parallelization. GPU versions are especially appropriate in problems of this kind. Specifically, a small laptop with a GPU with some data-acquisition equipment may be enough to register and process large amounts of biomedical data in an affordable period of time.

The approach has been illustrated in the automatic classification of electroencephalographic signals from epileptic patients that were performing a neuropsychological test. The state-of-the-art SICAMM algorithm has been considered due to its computational complexity and good results. Moreover, it is especially suited to parallelization. This parallelization opens up the use of SICAMM in real time on a realistic clinical setting. There are many possible applications for this setting, for instance, the real-time detection of epileptic seizures would allow the activation of certain clinical procedures or analyses.

As a future work, the proposed approach might be implemented in wearable devices for medical applications, thus facilitating fast, real-time monitoring and diagnosis.

Acknowledgements. This work was supported by Spanish Administration (Ministerio de Economía y Competitividad) and European Union (FEDER) under grants TEC2014-58438-R and TEC2017-84743-P; and Generalitat Valenciana under grants PROMETEO II/2014/032 and PROMETEO II/2014/003.

References

1. Shi, L., Liu, W., Zhang, H., Xie, Y., Wang, D.: A survey of GPU-based medical image computing techniques. Quant. Imag. Med. Surg. 2(3), 188–206 (2012). https://doi.org/10.3978/j.issn.2223-4292.2012.08.02
2. Montagnat, J., et al.: Medical images simulation, storage, and processing on the European DataGrid testbed. J. Grid Comput. 2(4), 387 (2005). https://doi.org/10.1007/s10723-004-5744-y
3. Saxena, S., Sharma, S., Sharma, N.: Image registration techniques using parallel computing in multicore environment and its applications in medical imaging: an overview, In: International Conference on Computer and Communication Technology (ICCCT) (2014)
4. Salazar, A., Vergara, L., Miralles, R.: On including sequential dependence in ICA mixture models. Signal Process. **90**, 2314–2318 (2010)
5. Niedermeyer, E., da Silva, F.L.: Electroencephalography: Basic Principles, Clinical Applications, and Related Fields, 5th edn. Lippincott Williams & Wilkins, Philadelphia (2005)
6. CUDA C Programming Guide. http://docs.nvidia.com/cuda/cuda-c-programming-guide/index.html
7. Common, P., Jutten, C.: Handbook of Blind Source Separation: Independent Component Analysis and Applications. Academic Press, Oxford (2010)
8. Salazar, A., Igual, J., Vergara, L., Serrano, A.: Learning hierarchies from ICA mixtures. In: IEEE International Conference on Neural Networks, IJCNN 2007, pp. 2271–2276 (2007)

9. Salazar, A.: On Statistical Pattern Recognition in Independent Component Analysis Mixture Modelling. Springer, New York (2013). https://doi.org/10.1007/978-3-642-30752-2
10. Safont, G., Salazar, A., Vergara, L., Gomez, E., Villanueva, V.: Probabilistic distance for mixtures of independent component analyzers. IEEE Trans. Neural Netw. Learn. Syst. 29(4), 1161–1173 (2018)
11. Salazar, A., Vergara, L., Serrano, A., Igual, J.: A general procedure for learning mixtures of independent component analyzers. Pattern Recogn. 43(1), 69–85 (2010)
12. The Mathworks Inc.: MATLAB R14 Natick MA (2004)
13. http://www.openmp.org/
14. Intel(R) AVX - Intel(R) Software Network. http://software.intel.com/en-us/avx/
15. Golub, G.H., Van Loan, C.F.: Matrix Computations, 3rd edn. The Johns Hopkins University Press, Baltimore (1996)
16. Anderson, E.: LAPACK Users' Guide. SIAM, Philadelphia (1999)
17. CUBLAS. http://docs.nvidia.com/cuda/cublas/index.html
18. Harris, M.: Optimizing parallel reduction in CUDA. https://developer.download.nvidia.com/assets/cuda/files/reduction.pdf
19. Motamedi-Fakhr, S., Moshrefi-Torbati, M., Hill, M., Hill, C.M., White, P.R.: Signal processing techniques applied to human sleep EEG signals-a review. Biomed. Signal Process. Control 10, 21–33 (2014)
20. Strauss, E.: A Compendium of Neuropsychological Tests. Oxford University Press, Oxford (2006)

The Chain Alignment Problem

Leandro Lima[1,2]([✉])[iD] and Said Sadique Adi[3][iD]

[1] Laboratoire de Biométrie et Biologie Evolutive UMR5558,
Univ Lyon, Université Lyon 1, CNRS, 69622 Villeurbanne, France
[2] EPI ERABLE - Inria Grenoble, Rhône-Alpes, France
leandro.ishi-soares-de-lima@inria.fr
[3] Faculdade de Computação, Universidade Federal de Mato Grosso do Sul,
Campo Grande 79070-900, Brazil
said@facom.ufms.br

Abstract. This paper introduces two new combinatorial optimization problems involving strings, namely, the Chain Alignment Problem, and a multiple version of it, the Multiple Chain Alignment Problem. For the first problem, a polynomial-time algorithm using dynamic programming is presented, and for the second one, a proof of its $\mathcal{NP}$-hardness is provided and some heuristics are proposed for it. The applicability of both problems here introduced is attested by their good results when modeling the Gene Identification Problem.

Keywords: Chain Alignment Problem · Dynamic programming · $\mathcal{NP}$-hardness · Gene prediction

1 Introduction

Problems involving strings can be found in many theoretical and practical areas, such as molecular biology, pattern search, text editing, data compression, etc. The present paper proposes two new combinatorial optimization problems involving strings: the Chain Alignment Problem and a multiple version of it, called the Multiple Chain Alignment Problem. Both can serve as models for other problems in several research areas. In Bioinformatics, for example, they can be used as a combinatorial optimization formulation for the gene prediction task [9] and for the problem of protein comparison by local structure segments [18]. Other applications include automated text categorization [16] and sequence data clustering [7]. The Chain Alignment Problem is closely related to a well-known problem in Bioinformatics, namely the Spliced Alignment Problem [4].

This paper explores the theoretical properties and algorithmic solutions for the Chain Alignment Problem and the Multiple Chain Alignment Problem. More specifically, for the first problem, a polynomial-time algorithm using dynamic programming is presented. For the second problem, it is proved that it is very unlikely that there exists a polynomial-time algorithm for it, i.e., we show it is $\mathcal{NP}$-hard through a reduction from the

J. M. F. Rodrigues et al. (Eds.): ICCS 2019, LNCS 11538, pp. 17–30, 2019.
https://doi.org/10.1007/978-3-030-22744-9_2

Longest Common Subsequence Problem. Given the $\mathcal{NP}$-hardness of the Multiple Chain Alignment Problem, we propose three different heuristics for it.

Despite our focus being a theoretical study of the problems introduced here, we also show that they can be applied to practical problems. More specifically, we model a variant of the Gene Identification Problem as the Chain Alignment Problem and as the Multiple Chain Alignment Problem, and present some results which attest their applicability.

This paper is structured as follows. Section 2 formally defines the Chain Alignment Problem and the Multiple Chain Alignment Problem. Section 3 describes an efficient dynamic programming algorithm for the Chain Alignment Problem. Section 4 is devoted to the proof of the $\mathcal{NP}$-hardness of the Multiple Chain Alignment Problem and Sect. 5 to a brief description of three heuristics for this problem. Section 6 discusses the application of both problems in the Gene Identification Problem. Finally, some concluding remarks and future directions for research are discussed in Sect. 7.

2   The Chain Alignment Problem

For a better understanding of the Chain Alignment Problem, consider the following definitions. An *alphabet* is a finite set of *symbols*[1]. Let $s = s[1]s[2]\ldots s[n]$ be a finite string over an alphabet Σ, whose *length* is denoted by $|s| = n$. The *empty string*, whose length equals to zero, is denoted by ϵ. We say that a string $s' = s'[1]s'[2]\ldots s'[m]$ is a *subsequence* of s if there exists a strictly increasing sequence $\mathcal{I} = i_1, i_2, \ldots, i_m$ of indices of s such that $s[i_j] = s'[j]$ for all $1 \leq j \leq m$. If $i_{k+1} - i_k = 1$ for all $1 \leq k < m$, we call s' a *segment* or a *substring* of s.

Further, let $b = s[i]\ldots s[j]$ be a segment of a string s. The position of the first (last) symbol of b in s is denoted by $first(b) = i$ ($last(b) = j$). Let $\mathcal{B} = \{b_1, b_2, \ldots, b_u\}$ be a set of u segments of s. $\mathcal{B}$ is defined as an *ordered set of segments* if: 1) $first(b_i) < first(b_{i+1})$ or 2) $first(b_i) = first(b_{i+1})$ and $last(b_i) < last(b_{i+1})$, for $1 \leq i \leq u - 1$. Moreover, a segment $b' = s[i]\ldots s[j]$ of s *overlaps* another segment $b'' = s[k]\ldots s[l]$ of s if $k \leq i \leq l$, or $k \leq j \leq l$, or $i \leq k \leq j$, or $i \leq l \leq j$. On the other hand, if $j < k$, we say that b' *precedes* b'', and this relation is denoted by $b' \prec b''$. Using the previous definitions, a *chain* $\Gamma_{\mathcal{B}}$ of an ordered set of segments $\mathcal{B}$ is defined as a subset $\Gamma_{\mathcal{B}} = \{b_i, b_j, \ldots, b_p\}$ of $\mathcal{B}$ such that $b_i \prec b_j \prec \ldots \prec b_p$. In addition, the string resulting from the concatenation of the segments of a chain $\Gamma_{\mathcal{B}}$ is denoted by $\Gamma_{\mathcal{B}}^{\bullet}$, i.e., $\Gamma_{\mathcal{B}}^{\bullet} = b_i \bullet b_j \bullet \ldots \bullet b_p$, where $\bullet$ is the string concatenation operator.

Finally, given two strings s and t, $sim_{\omega}(s, t)$ denotes the *similarity* (or the score of an *optimal alignment*) between s and t under a scoring function $\omega: \bar{\Sigma} \times \bar{\Sigma} \to \mathbb{R}$, where $\bar{\Sigma} = \Sigma \cup \{-\}$ [12][2]. With all the previous definitions in mind, the Chain Alignment Problem can be formally stated as follows:

---

[1] In this paper, we will only consider alphabets that do not include the *space* (denoted by the symbol $-$) as one of its elements.

[2] For the sake of simplicity, we will also use the terms *similarity* and *optimal alignment* with chains. This means that when we refer to the similarity (optimal alignment) between two chains $\Gamma_{\mathcal{B}}$ and $\Gamma_{\mathcal{C}}$, we are referring to the similarity (optimal alignment) between $\Gamma_{\mathcal{B}}^{\bullet}$ and $\Gamma_{\mathcal{C}}^{\bullet}$.

Chain Alignment Problem (CAP): given two strings s and t, an ordered set of segments $\mathcal{B} = \{b_1, b_2, \ldots, b_u\}$ of s, an ordered set of segments $\mathcal{C} = \{c_1, c_2, \ldots, c_v\}$ of t, and a scoring function ω, find a chain $\Gamma_\mathcal{B} = \{b_p, b_q, \ldots, b_r\}$ of $\mathcal{B}$ and a chain $\Gamma_\mathcal{C} = \{c_w, c_x, \ldots, c_y\}$ of $\mathcal{C}$ such that $sim_\omega(\Gamma_\mathcal{B}^\bullet, \Gamma_\mathcal{C}^\bullet)$ is maximum among all chains of $\mathcal{B}$ and $\mathcal{C}$.

Figure 1 illustrates an instance of the CAP.

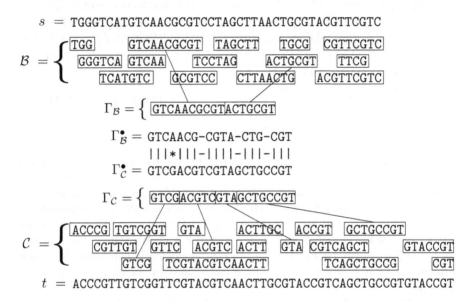

Fig. 1. An instance of the CAP and its solution. In this example we are considering the scoring function $\omega(a, b) = \{1,$ if $a = b; -1,$ if $a \neq b; -2,$ if $a = -$ or $b = -\}$. In the center of the figure there is depicted an optimal solution to this example, the chains $\Gamma_\mathcal{B}$ and $\Gamma_\mathcal{C}$, and an optimal alignment between $\Gamma_\mathcal{B}^\bullet$ and $\Gamma_\mathcal{C}^\bullet$.

The **Multiple Chain Alignment Problem**, in turn, is a multiple version of the CAP and can be defined as follows:

Multiple Chain Alignment Problem (MCAP): given $n > 2$ strings $s_1, s_2, \ldots, s_n$, n ordered sets of segments $\mathcal{B}_1, \mathcal{B}_2, \ldots, \mathcal{B}_n$, where $\mathcal{B}_i$ is an ordered set of segments of s_i, and a scoring function ω, find n chains $\Gamma_1, \Gamma_2, \ldots, \Gamma_n$, where Γ_i is a chain of $\mathcal{B}_i$, such that $\sum_{i=2}^n \sum_{j=1}^{i-1} sim_\omega(\Gamma_i^\bullet, \Gamma_j^\bullet)$ is maximum.

3   An Algorithmic Solution for the CAP

It is easy to check that a brute-force algorithm is not suitable to solve the CAP. However, a careful analysis of this problem shows that it exhibits the *optimal substructure* and *overlapping subproblems* properties, and as such, it is possible to develop an efficient algorithm for it using dynamic programming.

In order to understand the recurrence that solves the **CAP**, consider the following definitions. Let $b_k[1..i] = b_k[1] \bullet b_k[2] \bullet \ldots \bullet b_k[i]$ and let $\Gamma_{\mathcal{B}}(k[i]) = \{b_q, b_r, \ldots, b_k[1..i]\}$ be a chain of an ordered set of segments $\mathcal{B}$ ending with $b_k[1..i]$. For the sake of simplicity, we will sometimes denote $\Gamma_{\mathcal{B}}(k[\|b_k\|])$ simply by $\Gamma_{\mathcal{B}}(k)$. Given an instance $I = \langle s, t, \mathcal{B} = \{b_1, \ldots, b_u\}, \mathcal{C} = \{c_1, \ldots, c_v\}, \omega \rangle$ of the **CAP**, let $\Gamma^{\star}_{k[i]l[j]}$ be a pair of chains $(\Gamma^{\star}_{\mathcal{B}}(k[i]), \Gamma^{\star}_{\mathcal{C}}(l[j]))$, where $\Gamma^{\star}_{\mathcal{B}}(k[i])$ is a chain of $\mathcal{B}$ ending with $b_k[1..i]$ and $\Gamma^{\star}_{\mathcal{C}}(l[j])$ is a chain of $\mathcal{C}$ ending with $c_l[1..j]$ such that $v(\Gamma^{\star}_{k[i]l[j]}) = sim_{\omega}(\Gamma^{\star}_{\mathcal{B}}(k[i])^{\bullet}, \Gamma^{\star}_{\mathcal{C}}(l[j])^{\bullet})$ is maximum, i.e., $sim_{\omega}(\Gamma^{\star}_{\mathcal{B}}(k[i])^{\bullet}, \Gamma^{\star}_{\mathcal{C}}(l[j])^{\bullet}) \geq sim_{\omega}(\Gamma_{\mathcal{B}}(k[i])^{\bullet}, \Gamma_{\mathcal{C}}(l[j])^{\bullet})$ for all pairs of chains $\Gamma_{\mathcal{B}}(k[i])$ of $\mathcal{B}$ ending with $b_k[1..i]$ and $\Gamma_{\mathcal{C}}(l[j])$ of $\mathcal{C}$ ending with $c_l[1..j]$. By definition, $\Gamma^{\star}_{k[i]l[j]}$ is defined for all $1 \leq k \leq |\mathcal{B}|$, $1 \leq l \leq |\mathcal{C}|$, $1 \leq i \leq |b_k|$, and $1 \leq j \leq |c_l|$. Lastly, let $\mathcal{M}$ be a four-dimensional matrix such that

$$\mathcal{M}[i, j, k, l] = v(\Gamma^{\star}_{k[i]l[j]}) = \max_{\substack{\text{all pairs of chains} \\ (\Gamma_{\mathcal{B}}(k[i]), \Gamma_{\mathcal{C}}(l[j]))}} sim_{\omega}(\Gamma_{\mathcal{B}}(k[i])^{\bullet}, \Gamma_{\mathcal{C}}(l[j])^{\bullet}).$$

As defined, $\mathcal{M}$ can be efficiently calculated by means of a recurrence using dynamic programming.

For a better understanding of the base cases of this recurrence, consider that the sets of segments $\mathcal{B}$ and $\mathcal{C}$ will each include a *virtual segment* (a segment not given in the input) b_0 and c_0, respectively, such that $b_0 = c_0 = \epsilon$ (both are empty strings), $b_0 \prec b$ for all $b \in \mathcal{B}$ and $c_0 \prec c$ for all $c \in \mathcal{C}$. We thus define $\Gamma^{\star}_{\mathcal{B}}(0[0])$ and $\Gamma^{\star}_{\mathcal{C}}(0[0])$ as empty chains. Finally, $\mathcal{M}$ can be efficiently calculated by means of Recurrence 1 if $k = 0$ or $l = 0$ (base cases) and by means of Recurrence 2 otherwise[3].

As to Recurrence 1, its first line corresponds to the score of an optimal alignment between the empty chains $\Gamma^{\star}_{\mathcal{B}}(0[0])$ and $\Gamma^{\star}_{\mathcal{C}}(0[0])$. Its second line, to the score of an optimal alignment between a chain of $\mathcal{B}$ ending with $b_k[1..i]$ and the empty chain $\Gamma^{\star}_{\mathcal{C}}(0[0])$ of $\mathcal{C}$. The best alignment we can get here is aligning $b_k[1..i]$ with i spaces. The third line is analogous to the second one.

The general idea behind Recurrence 2 is to consider all possible extensions of previously computed subproblem solutions and choose the best one. The options the recurrence considers for this extension, called the *set of candidate alignments*, and how this extension is carried out, change according to the position of the matrix being calculated. When calculating $\mathcal{M}[i > 1, j > 1, k, l]$, i.e., finding $v(\Gamma^{\star}_{k[i]l[j]}) = sim_{\omega}(\Gamma^{\star}_{\mathcal{B}}(k[i])^{\bullet}, \Gamma^{\star}_{\mathcal{C}}(l[j])^{\bullet})$, the recurrence can extend the optimal alignment between $\Gamma^{\star}_{\mathcal{B}}(k[i-1])$ and $\Gamma^{\star}_{\mathcal{C}}(l[j-1])$ (stored in $\mathcal{M}[i-1, j-1, k, l]$) by adding $b_k[i]$ paired with $c_l[j]$ to this alignment. It also considers the options of extending the optimal alignment between $\Gamma^{\star}_{\mathcal{B}}(k[i-1])$ and $\Gamma^{\star}_{\mathcal{C}}(l[j])$ (stored in $\mathcal{M}[i-1, j, k, l]$) by adding $b_k[i]$ paired with a space and the optimal alignment between $\Gamma^{\star}_{\mathcal{B}}(k[i])$ and $\Gamma^{\star}_{\mathcal{C}}(l[j-1])$ (stored in $\mathcal{M}[i, j-1, k, l]$) by adding a space

---

[3] To simplify the calculation of the base cases, we are assuming in Recurrence 1 that $\omega(a, b) < 0$, if $a = -$ or $b = -$, where ω is the scoring function given in the input. Nonetheless, it can be easily modified to cope with any scoring function.

paired with $c_l[j]$. Evidently, in this case, the set of candidate alignments consists of $\mathcal{M}[i-1,j-1,k,l]$, $\mathcal{M}[i-1,j,k,l]$, and $\mathcal{M}[i,j-1,k,l]$. Given these three options, it chooses to extend the one that maximizes the score of the optimal alignment being calculated.

$$\mathcal{M}[0,0,0,0] = 0;$$

$$\mathcal{M}[i,0,k,0] = \sum_{t=1}^{i} \omega(b_k[t], -), \text{ for } 1 \leq i \leq |b_k| \text{ and } 1 \leq k \leq |\mathcal{B}|; \tag{1}$$

$$\mathcal{M}[0,j,0,l] = \sum_{t=1}^{j} \omega(-, c_l[t]), \text{ for } 1 \leq j \leq |c_l| \text{ and } 1 \leq l \leq |\mathcal{C}|.$$

$$\mathcal{M}[i,j,k,l] = \begin{cases} \text{if } i > 1 \text{ and } j > 1: \\ \quad \max \begin{cases} \mathcal{M}[i-1,j-1,k,l] + \omega(b_k[i], c_l[j]), \\ \mathcal{M}[i,j-1,k,l] + \omega(-, c_l[j]), \\ \mathcal{M}[i-1,j,k,l] + \omega(b_k[i], -) \end{cases} \\ \text{if } i = 1 \text{ and } j = 1: \\ \quad \max \begin{cases} \max_{b_{k'} \prec b_k,\ c_{l'} \prec c_l} \{\mathcal{M}[|b_{k'}|, |c_{l'}|, k', l'] \\ \qquad\qquad\qquad\qquad\qquad + \omega(b_k[1], c_l[1])\}, \\ \max_{c_{l'} \prec c_l} \{\mathcal{M}[1, |c_{l'}|, k, l'] + \omega(-, c_l[1])\}, \\ \max_{b_{k'} \prec b_k} \{\mathcal{M}[|b_{k'}|, 1, k', l] + \omega(b_k[1], -)\} \end{cases} \\ \text{if } i = 1 \text{ and } j > 1: \\ \quad \max \begin{cases} \mathcal{M}[1, j-1, k, l] + \omega(-, c_l[j]), \\ \max_{b_{k'} \prec b_k} \begin{cases} \mathcal{M}[|b_{k'}|, j-1, k', l] + \omega(b_k[1], c_l[j]), \\ \mathcal{M}[|b_{k'}|, j, k', l] + \omega(b_k[1], -) \end{cases} \end{cases} \\ \text{if } i > 1 \text{ and } j = 1: \\ \quad \max \begin{cases} \mathcal{M}[i-1, 1, k, l] + \omega(b_k[i], -), \\ \max_{c_{l'} \prec c_l} \begin{cases} \mathcal{M}[i-1, |c_{l'}|, k, l'] + \omega(b_k[i], c_l[1]), \\ \mathcal{M}[i, |c_{l'}|, k, l'] + \omega(-, c_l[1]) \end{cases} \end{cases} \end{cases} \tag{2}$$

The same idea explained in the previous paragraph is used to calculate all the remaining cells of $\mathcal{M}$, changing only the set of candidate alignments. When calculating $\mathcal{M}[i=1, j=1, k, l]$, for instance, the recurrence takes into account more candidates. Firstly, it considers all previously computed optimal alignments ending with any segment $b_{k'} \prec b_k$ ($\Gamma_{\mathcal{B}}^{\star}(k')$) and any segment $c_{l'} \prec c_l$ ($\Gamma_{\mathcal{C}}^{\star}(l')$), which are all included in the set $\{\mathcal{M}[|b_{k'}|, |c_{l'}|, k', l']\}$, and extends them by adding $b_k[1]$ paired with $c_l[1]$. It also considers extending all optimal alignments between $\Gamma_{\mathcal{B}}^{\star}(k[1])$ and $\Gamma_{\mathcal{C}}^{\star}(l')$, $c_{l'} \prec c_l$ ($\{\mathcal{M}[1, |c_{l'}|, k, l']\}$), by adding a space paired with $c_l[1]$ to them. Analogously, it further considers extending all optimal alignments between $\Gamma_{\mathcal{B}}^{\star}(k')$, $b_{k'} \prec b_k$, and $\Gamma_{\mathcal{C}}^{\star}(l[1])$ ($\{\mathcal{M}[|b_{k'}|, 1, k', l]\}$) by adding $b_k[1]$ paired with a space to them. Finally, it chooses the best extension of all of these.

In the third case, Recurrence 2 calculates $\mathcal{M}[i=1, j>1, k, l]$. First off, it considers extending the optimal alignment between $\Gamma_{\mathcal{B}}^{\star}(k[1])$ and $\Gamma_{\mathcal{C}}^{\star}(l[j-1])$ (stored in $\mathcal{M}[1, j-1, k, l]$) by adding a space paired with $c_l[j]$ to this alignment. But it also has to look at several other candidate alignments. It further considers extending all optimal alignments between $\Gamma_{\mathcal{B}}^{\star}(k')$, $b_{k'} \prec b_k$, and $\Gamma_{\mathcal{C}}^{\star}(l[j-1])$ ($\{\mathcal{M}[|b_{k'}|, j-1, k', l]\}$) by adding $b_k[1]$ paired with $c_l[j]$ to them. The last set of candidates consists of all optimal alignments between $\Gamma_{\mathcal{B}}^{\star}(k')$, $b_{k'} \prec b_k$, and $\Gamma_{\mathcal{C}}^{\star}(l[j])$ ($\{\mathcal{M}[|b_{k'}|, j, k', l]\}$). The recurrence extends them by adding $b_k[1]$ paired with a space. After computing all these extensions, it chooses the best one. The fourth case is analogous to the third one.

Figure 2 shows a way of viewing $\mathcal{M}$ as a $(u+1) \times (v+1)$ matrix, where each cell of $\mathcal{M}$ is an *alignment submatrix* representing the best alignment ending with two specific segments.

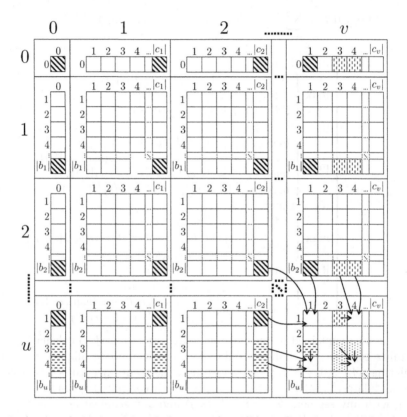

Fig. 2. A way of viewing $\mathcal{M}$ and the sets of candidate alignments considered when Recurrence 2 calculates the positions $\mathcal{M}[i > 1, j > 1, u, v]$, $\mathcal{M}[i = 1, j = 1, u, v]$, $\mathcal{M}[i = 1, j > 1, u, v]$ and $\mathcal{M}[i > 1, j = 1, u, v]$. Some arrows were hidden for the purpose of clarity, but cells with the same hatching pattern are in the same set of candidate alignments. As we can see in this specific example, it is necessary that $b_0 \prec b_1 \prec \ldots \prec b_u$ and $c_0 \prec c_1 \prec \ldots \prec c_v$ (i.e., no overlapping segments in $\mathcal{B}$ and $\mathcal{C}$), for the sets of candidate alignments to be like those depicted above.

Upon completion of $\mathcal{M}$, the value of the optimal solution can be found as

$$\max_{b_k \in \mathcal{B}, c_l \in \mathcal{C}} \{\mathcal{M}[|b_k|, |c_l|, k, l]\}.$$

The solution for the CAP, i.e., a chain $\Gamma_\mathcal{B}$ of $\mathcal{B}$ and a chain $\Gamma_\mathcal{C}$ of $\mathcal{C}$ such that $sim_\omega(\Gamma_\mathcal{B}^\bullet, \Gamma_\mathcal{C}^\bullet)$ is maximum, can be constructed by a traceback procedure. This procedure builds a path that begins at the position in $\mathcal{M}$ that stores the value of the optimal solution, and goes back to the cell used to calculate this value. It keeps building this path backwards until $\mathcal{M}[0,0,0,0]$ is reached. To determine $\Gamma_\mathcal{B}$ ($\Gamma_\mathcal{C}$), the procedure logs every segment of $\mathcal{B}$ ($\mathcal{C}$), except b_0 (c_0), in this path in LIFO order.

The proposed algorithm runs in $\mathcal{O}(|\mathcal{B}| * |\mathcal{C}| * b_{max} * c_{max})$ time and it needs $\mathcal{O}(|\mathcal{B}| * |\mathcal{C}| * b_{max} * c_{max})$ space to store $\mathcal{M}$, where b_{max} (resp. c_{max}) is the size of the longest segment in $\mathcal{B}$ (resp. $\mathcal{C}$).

4   The MCAP Is $\mathcal{NP}$-hard

In this section it is shown that the MCAP is $\mathcal{NP}$-hard, that is, it has a polynomial-time algorithm if and only if $\mathcal{P} = \mathcal{NP}$. To this end, consider the following decision version of the MCAP, called MCAPD:

> **Multiple Chain Alignment Problem - decision version (MCAPD):** given $n > 2$ strings $s_1, s_2, \ldots, s_n$, n ordered sets of segments $\mathcal{B}_1, \mathcal{B}_2, \ldots, \mathcal{B}_n$, where $\mathcal{B}_i$ is an ordered set of segments of s_i, a scoring function ω, and a positive integer l, are there n chains $\Gamma_1, \Gamma_2, \ldots, \Gamma_n$, where Γ_i is a chain of $\mathcal{B}_i$, such that $\sum_{i=2}^{n} \sum_{j=1}^{i-1} sim_\omega(\Gamma_i^\bullet, \Gamma_j^\bullet) = l$?

To prove the $\mathcal{NP}$-completeness of the MCAPD, we will reduce from the following decision version of the **Longest Common Subsequence Problem**:

> **Longest Common Subsequence Problem - decision version (LCSPD):** given $n > 2$ strings $s_1, s_2, \ldots, s_n$ and a positive integer k, is there a string t with $|t| = k$ such that t is a subsequence of $s_1, s_2, \ldots, s_n$?

A $\mathcal{NP}$-completeness proof of the LCSPD can be found in [10].

Theorem 1. *The MCAPD is $\mathcal{NP}$-complete.*

Proof 1. It is not hard to see that the MCAPD is in $\mathcal{NP}$. For this matter, consider a simple verification algorithm that receives as input an instance $\langle\{s_1, s_2, \ldots, s_n\}, \{\mathcal{B}_1, \mathcal{B}_2, \ldots, \mathcal{B}_n\}, \omega, l\rangle$ of the MCAPD and a certificate $\langle \Gamma_1, \Gamma_2, \ldots, \Gamma_n \rangle$, and then checks if each Γ_i is a chain of $\mathcal{B}_i$ and if $\sum_{i=2}^{n} \sum_{j=1}^{i-1} sim_\omega(\Gamma_i^\bullet, \Gamma_j^\bullet) = l$. It is trivial to see that this naive algorithm runs in polynomial time.

For the reduction step, suppose there is a polynomial-time decision algorithm for the MCAPD called *AlgMCAPD*. We will thus prove there is a polynomial-time algorithm for the LCSPD by transforming an arbitrary instance $I = \langle\{s_1, s_2, \ldots, s_n\}, k\rangle$ of the LCSPD into an instance $I' = \langle\{s'_1, s'_2, \ldots, s'_n\}, \{\mathcal{B}_1, \mathcal{B}_2, \ldots, \mathcal{B}_n\}, \omega, l\rangle$ of the MCAPD such that I is positive if and only if I' is positive.

In what follows, four steps are described to build I' from I. In the first step, $\{s_1, s_2, \ldots, s_n\}$ is assigned to $\{s'_1, s'_2, \ldots, s'_n\}$, i.e., the sets of strings are the same for both instances.

The second step defines the n ordered sets of segments $\mathcal{B}_1, \mathcal{B}_2, \ldots, \mathcal{B}_n$. An ordered set of segments $\mathcal{B}_i$ is constructed by adding, in order, each symbol of s_i as a segment of $\mathcal{B}_i$.

The third step defines the scoring function ω as

$$\omega(a,b) = \begin{cases} 1, & \text{if } a = b \\ -(Opt^+ + 1), & \text{if } a \neq b \\ -(Opt^+ + 1), & \text{if } a = - \text{ or } b = -, \end{cases}$$

where $Opt^+ = \frac{n*(n-1)}{2} * |s_{max}|$ and s_{max} is the longest input string. In other words, Opt^+ is an upper bound for $\sum_{i=2}^{n} \sum_{j=1}^{i-1} sim_\omega(\Gamma_i^\bullet, \Gamma_j^\bullet)$.

Finally, the fourth step defines

$$l = \frac{n * (n-1) * k}{2}.$$

It is not hard to check that the transformation of I into I', using the four steps previously explained, takes polynomial time. It is thus possible to build an algorithm, called $AlgLCSPD$, to decide the LCSPD. This algorithm receives an arbitrary instance I of the LCSPD, transforms I into an instance I' of the MCAPD, and then calls $AlgMCAPD$ taking I' as argument. $AlgLCSPD$ then decides I as positive (negative) if $AlgMCAPD$ decides I' as positive (negative). Since both the transformation of I into I' and the execution of $AlgMCAPD$ take polynomial time, it is easy to see that $AlgLCSPD$ also runs in polynomial time. To complete the proof, we need to show that I is positive if and only if I' is positive.

(Proof of "only if.") Suppose I is positive. In this case, there is a string t with $|t| = k$ such that t is a subsequence of $s_1, s_2, \ldots, s_n$. We will prove that I' is also positive. As I' is defined, it is always possible to build n chains $\Gamma_1, \Gamma_2, \ldots, \Gamma_n$ such that $\Gamma_1^\bullet = \Gamma_2^\bullet = \ldots = \Gamma_n^\bullet = t$. Therefore,

$$\sum_{i=2}^{n} \sum_{j=1}^{i-1} sim_\omega(\Gamma_i^\bullet, \Gamma_j^\bullet) = \sum_{i=2}^{n} \sum_{j=1}^{i-1} sim_\omega(t, t)$$

$$= \sum_{i=2}^{n} \sum_{j=1}^{i-1} |t|$$

$$= \sum_{i=2}^{n} \sum_{j=1}^{i-1} k$$

$$= \frac{n*(n-1)*k}{2}$$

$$= l.$$

Hence I' is positive.

(Proof of "if.") Suppose I' is positive. Thus, there are n chains $\Gamma_1, \Gamma_2, \ldots, \Gamma_n$, where Γ_i is a chain of $\mathcal{B}_i$, such that $\sum_{i=2}^{n} \sum_{j=1}^{i-1} sim_\omega(\Gamma_i^\bullet, \Gamma_j^\bullet) = l$. We will show that I is also positive.

Since I' is positive, we can show that $\Gamma_1 = \Gamma_2 = \ldots = \Gamma_n$. Suppose, by contradiction, that $\Gamma_1 = \Gamma_2 = \ldots = \Gamma_n$ is not true, i.e., there is at least one pair of chains (Γ_i, Γ_j) such that $\Gamma_i \neq \Gamma_j$. In this case, in at least one of the $\frac{n*(n-1)}{2}$ alignments between $\Gamma_1^\bullet, \Gamma_2^\bullet, \ldots, \Gamma_n^\bullet$, there will be a mismatch or a space. As the score of a mismatch or a space is negative enough to make $\sum_{i=2}^{n} \sum_{j=1}^{i-1} sim_\omega(\Gamma_i^\bullet, \Gamma_j^\bullet) < 0$, we have a contradiction of our hypothesis that $\sum_{i=2}^{n} \sum_{j=1}^{i-1} sim_\omega(\Gamma_i^\bullet, \Gamma_j^\bullet) = l$, since l is a positive integer.

As $\Gamma_1 = \Gamma_2 = \ldots = \Gamma_n$, we have that $\Gamma_1^\bullet = \Gamma_2^\bullet = \ldots = \Gamma_n^\bullet$. Besides, due to the way we transformed I into I', it is easy to check that each $\Gamma_i^\bullet$ is a subsequence of $s_1, s_2, \ldots s_n$. With the previous observations in mind, we have

$$\sum_{i=2}^{n} \sum_{j=1}^{i-1} sim_\omega(\Gamma_i^\bullet, \Gamma_j^\bullet) = \sum_{i=2}^{n} \sum_{j=1}^{i-1} sim_\omega(\Gamma_x^\bullet, \Gamma_x^\bullet)$$
$$= \sum_{i=2}^{n} \sum_{j=1}^{i-1} |\Gamma_x^\bullet|$$
$$= \frac{n*(n-1)*|\Gamma_x^\bullet|}{2}.$$

Since I' is positive, $\sum_{i=2}^{n} \sum_{j=1}^{i-1} sim_\omega(\Gamma_i^\bullet, \Gamma_j^\bullet) = l$. Thus, we have

$$l = \frac{n*(n-1)*|\Gamma_x^\bullet|}{2}$$

$$|\Gamma_x^\bullet| = \frac{2*l}{n*(n-1)} = k.$$

Therefore, if $t = \Gamma_x^\bullet$, for all $1 \leq x \leq n$, we have that $|t| = k$ so that t is a subsequence of $s_1, s_2, \ldots, s_n$. Hence, I is positive. ∎

5   Heuristics for the Multiple Chain Alignment Problem

Given the $\mathcal{NP}$-hardness of the MCAP, we propose here three heuristics for it. Two of them are based on the solution proposed by us for the CAP and the third heuristic is based on the algorithm described in [4] to solve the Spliced Alignment Problem (SAP). Informally, this problem consists in, given as input two strings s and t, and an ordered set of segments $\mathcal{B} = \{b_1, b_2, \ldots, b_u\}$ of s, finding an ordered subset $\Gamma = \{b_p, b_q, \ldots, b_r\}$ of non-overlapping segments of $\mathcal{B}$ such that the string resulting from the concatenation of all segments in Γ is very similar to t.

For a better understanding of the proposed heuristics, consider the following. The input to the Multiple Chain Alignment Problem is given by the triple $(\{s_1, s_2, \ldots, s_n\}, \{\mathcal{B}_1, \mathcal{B}_2, \ldots, \mathcal{B}_n\}, \omega)$. We can invoke the algorithm that solves the Chain Alignment Problem as $CAP(s_i, s_j, \mathcal{B}_i, \mathcal{B}_j, \omega)$, which in turn will return a pair of chains $\mathcal{P}_{i,j} = (\Gamma_i, \Gamma_j)$ such that $sim_\omega(\Gamma_i^\bullet, \Gamma_j^\bullet)$ is maximum among all chains of $\mathcal{B}_i$ and $\mathcal{B}_j$. Further, given a pair of chains $\mathcal{P}_{i,j}$, we can reference its first chain by $\mathcal{P}_{i,j}.\Gamma_i$, its second chain by $\mathcal{P}_{i,j}.\Gamma_j$, and its value by $val(\mathcal{P}_{i,j}) = sim_\omega(\mathcal{P}_{i,j}.\Gamma_i^\bullet, \mathcal{P}_{i,j}.\Gamma_j^\bullet)$.

All three heuristics here described work on a common framework. They compute, at each iteration i, the chain Γ_i to be included in the final solution. Thus, it is enough to describe how only one Γ_i is determined. In addition, their first step in order to calculate Γ_i is to always invoke $\mathcal{P}_{i,j} = CAP(s_i, s_j, \mathcal{B}_i, \mathcal{B}_j, \omega)$ for all $1 \leq j \leq n, j \neq i$. Thus, in the description of the heuristics, we will always suppose that all pairs $\mathcal{P}_{i,j}$ are already precomputed.

5.1 $\mathcal{H}1$ - Heuristic of the Consensus Chain

The consensus chain $\mathcal{C}(\Gamma_i)$ of an ordered set of segments $\mathcal{B}_i$ of a sequence s_i is obtained as follows. We count how many times each segment of $\mathcal{B}_i$ is present in each pair $\mathcal{P}_{i,j}$ for $1 \leq j \leq n, j \neq i$. $\mathcal{C}(\Gamma_i)$ is then defined as the segments of $\mathcal{B}_i$ that are present more than $\lfloor \frac{n-1}{2} \rfloor$ times. This heuristic thus defines Γ_i as $\mathcal{C}(\Gamma_i)$.

5.2 $\mathcal{H}2$ - Greedy Heuristic

The Greedy Heuristic determines each Γ_i as

$$\Gamma_i = \mathcal{P}_{i,j}.\Gamma_i | val(\mathcal{P}_{i,j}) = \max_{1 \leq j \leq n, j \neq i} \{val(\mathcal{P}_{i,j})\}.$$

5.3 $\mathcal{H}3$ - Heuristic of the Central Chain

For the Heuristic of the Central Chain, consider that we can invoke the algorithm described in [4] to solve the **Spliced Alignment Problem** as $SAP(s, t, \mathcal{B}, \omega)$, and that this invocation returns the value of optimal solution. This heuristic let

$$\Gamma_i = \mathcal{P}_{i,j}.\Gamma_i | \sum_{k=1,\ k \neq i}^{n} SAP(s_k, \mathcal{P}_{i,j}.\Gamma_i^{\bullet}, \mathcal{B}_k, \omega) \text{ is maximum } \forall\ 1 \leq j \leq n, j \neq i.$$

6   Application to the Gene Identification Problem

In this section we model a variant of the **Gene Identification Problem** as the CAP and the MCAP, and present some results which attest one of the applications of these latter problems, justifying their previous formalization and study.

In short, the **Gene Identification Problem** (GIP) consists in, given a DNA sequence, determine the exons of the genes encoded by the given sequence [11]. The GIP is one of the most important problems in Molecular Biology, since it is directly related to disease prevention and treatment, pest control, etc. However, even nowadays, it still remains a challenging and complicated problem, which justifies additional research on this topic.

We propose to tackle a variant of the GIP by modelling it as the CAP. To formally define this variant, we briefly recall the definition of some biological concepts. Firstly, consider the fact that an organism's DNA may change permanently during time. These changes are called *mutations* and they can be either beneficial, neutral or negative, according to their effect on the organism. It is also important to note that mutations are observed more frequently on non-coding regions of the DNA, which impact less on the protein biosynthesis process, than on coding regions. Beneficial mutations may be transferred to the organism's descendants and, in some cases, even originate new species. In this context, given two different organisms $\mathcal{O}_1$ and $\mathcal{O}_2$, we say that a gene $\mathcal{G}_1$ of $\mathcal{O}_1$ is homologous to a gene $\mathcal{G}_2$ of $\mathcal{O}_2$ if $\mathcal{G}_1$ and $\mathcal{G}_2$ evolved from a same gene present in a common ancestor of $\mathcal{O}_1$ and $\mathcal{O}_2$. Finally, we can define the **Homologous Gene Identification Problem** as follows.

Homologous Gene Identification Problem (HGIP): given two DNA sequences $\mathcal{D}_1$ and $\mathcal{D}_2$, which contain two unknown homologous genes $\mathcal{G}_1$ and $\mathcal{G}_2$, respectively, a set of putative exons $\mathcal{E}_1$ of $\mathcal{G}_1$, and a set of putative exons $\mathcal{E}_2$ of $\mathcal{G}_2$, find a set $\mathcal{R}_1$ of $\mathcal{E}_1$ and a set $\mathcal{R}_2$ of $\mathcal{E}_2$ such that $\mathcal{R}_1$ contains the real exons of $\mathcal{G}_1$ and $\mathcal{R}_2$ contains the real exons of $\mathcal{G}_2$.

To model the HGIP as the CAP, let's consider $\Sigma = \{A, C, G, T\}$, map the input of the HGIP to an input of the CAP, and the output of the CAP to an output of the HGIP. To map the inputs, let s, t, $\mathcal{B}$ and $\mathcal{C}$ be $\mathcal{D}_1$, $\mathcal{D}_2$, $\mathcal{E}_1$ and $\mathcal{E}_2$, respectively, and let $\omega(a, b) = \{1, \text{ if } a = b; -1, \text{ if } a \neq b; -2, \text{ if } a = -\text{ or } b = -\}$.

To map the outputs, let $\mathcal{R}_1$ and $\mathcal{R}_2$ be $\Gamma_{\mathcal{B}}$ and $\Gamma_{\mathcal{C}}$, respectively. The main supposition behind this modeling is that the coding regions of the two unknown homologous genes $\mathcal{G}_1$ and $\mathcal{G}_2$ will be similar, even if they have been affected by independent mutations. In this sense, the probability that the segments in $\Gamma_{\mathcal{B}}$ and $\Gamma_{\mathcal{C}}$ correspond to the real exons of $\mathcal{G}_1$ and $\mathcal{G}_2$ is high.

Similarly as we extended the CAP to the MCAP, we can extend the HGIP to a multiple version, where we receive n DNA sequences containing n unknown homologous genes, and a set of putative exons of each gene, and we aim to find the set of real exons of each gene. In this way, it is also possible to apply the MCAP to the HGIP. The strength of doing so stands out when several homologous genes are available, instead of only two. In these situations, we apply the heuristics proposed for the MCAP to make an effective use of this additional data, in order to get better solutions.

The dataset considered to evaluate the accuracy of the proposed model was built using data from the ENCODE project [3] and from the HomoloGene database [15]. In short, the ENCODE project aimed at identifying all functional elements in the human genome sequence, and the HomoloGene database allowed for the automatic detection of homologous genes from several eukaryotic, completely sequenced DNAs. To build the dataset, we firstly selected all genes from the 44 regions studied in the pilot phase of ENCODE project, such that (1) coded for only one protein; (2) their size were a multiple of 3; (3) presented the canonical exon-intron structure. For each of these selected human genes, we searched for homologous genes from other species by using HomoloGene. We removed all homologous sequences that: (1) did not code for a protein; (2) coded for hypothetical proteins; (3) not all exons were completely identified; (4) did not present the canonical exon-intron structure; (5) had exons composed by other bases besides A, C, G or T. Finally, we added to the dataset the DNA sequences encoding all genes selected so far (these sequences also included 1000 bases before the start of the gene and after its end).

To finish the construction of the dataset, a simple tool was implemented to retrieve the set of putative exons for each DNA sequence. To do so, we used the method GAP3 described in [6] coupled with some metrics. For each pair of human and homologous sequences, GAP3 is applied to find the similar regions of both sequences, which we consider as the putative exons. As we might have a large number of putative exons in this step, we decided to filter the 50 best, by using a combination of four metrics: the WAMs defined in [14], the codon usage

[5], the optimal general alignment score [6], and the PWMs defined in [2]. We would like to note that the sequences in the dataset presented an average of 6.8 exons. However, the sets of putative exons generated by this tool included less than 50% of the real exons. Thus, we ultimately decided to add the missing real exons to each set. Finally, the dataset included 206 test instances.

The results of the application of the three heuristics for the MCAP and the algorithm that solves the CAP to the HGIP can be found in Table 1. The accuracy of each solution was evaluated by using the measures introduced by Burset and Guigó in [1] to evaluate the performance of gene prediction tools on three distinct levels: nucleotide, exon and border. Briefly speaking, at the nucleotide level, a true positive (TP) is a coding nucleotide that was correctly predicted as coding, a true negative (TN) is a non-coding nucleotide that was correctly predicted as non-coding, a false positive (FP) is a non-coding nucleotide that was incorrectly predicted as coding, and a false negative (FN) is a coding nucleotide that was incorrectly predicted as non-coding. Considering this, the specificity (resp. sensitivity) at the nucleotide level is defined as $Sp_n = \frac{TP}{TP+FP}$ (resp. $Sn_n = \frac{TP}{TP+FN}$), and the approximate correlation, which summarize both measures, as $Ac = \frac{1}{2}*(\frac{TP}{TP+FN} + \frac{TP}{TP+FP} + \frac{TN}{TN+FP} + \frac{TN}{TN+FN}) - 1$. At the exon level, let $\mathcal{R}_e$ be the set of real exons and let $\mathcal{P}_e$ be the set of predicted exons. The specificity (resp. sensitivity) at the exon level is thus defined as $Sp_e = \frac{|\mathcal{R}_e \cap \mathcal{P}_e|}{|\mathcal{P}_e|}$ (resp. $Sn_e = \frac{|\mathcal{R}_e \cap \mathcal{P}_e|}{|\mathcal{R}_e|}$) and both measures are summarized as $Av_e = \frac{Sp_e+Sn_e}{2}$. For the border level, suppose that the exon borders denote its start and end positions, and let $\mathcal{PR}_b$ be the set of exon borders that were correctly predicted, let $\mathcal{R}_e$ be the set of real exons, and let $\mathcal{P}_e$ be the set of predicted exons. The specificity (resp. sensitivity) at the border level is thus defined as $Sp_b = \frac{|\mathcal{PR}_b|}{2*|\mathcal{P}_e|}$ (resp. $Sn_b = \frac{|\mathcal{PR}_b|}{2*|\mathcal{R}_e|}$) and both measures are summarized as $Av_b = \frac{Sp_b+Sn_b}{2}$.

In this work these measures were calculated considering only the predictions on the human sequence. The algorithm for the CAP was applied to the human sequence and to each of its homologous, and the best solution was selected.

Table 1. The results of the application of the three heuristics for the MCAP and the algorithm that solves the CAP.

Algorithm	Nucleotide			Exon			Border		
	Sp_n	Sn_n	Ac	Sp_e	Sn_e	Av_e	Sp_b	Sn_b	Av_b
$\mathcal{H}1$	**0,945**	0,935	0,923	**0,897**	0,881	**0,889**	**0,919**	0,901	0,910
$\mathcal{H}2$	0,910	**0,994**	0,940	0,803	0,880	0,841	0,838	0,925	0,882
$\mathcal{H}3$	0,942	0,988	**0,958**	0,865	**0,909**	0,887	0,894	**0,943**	**0,919**
CAP	0,916	0,969	0,933	0,826	0,881	0,854	0,856	0,916	0,886

As we can see in Table 1, the **CAP** got the worst performance, followed by $\mathcal{H}2$. The best results were obtained by $\mathcal{H}3$, followed by $\mathcal{H}1$, as $\mathcal{H}3$ got the best performance on Ac and Av_b, and lost on Av_e to $\mathcal{H}1$ by a very small margin. As expected, these results show that if we increase the number of evidences in the HGIP, we can get better results. In this specific test, by obtaining several homologous genes to a human gene, and by processing these data collectively, using the heuristics for the MCAP, it was possible to produce better results than processing these homologous sequences individually, using the algorithm for the **CAP**. We can conclude then, even though we do not have an exact solution for the MCAP, in some applications, the extra data and evidence given by multiple inputs are too valuable to be ignored. Thus, future works can focus on a deeper study of better heuristics or approximate solutions for the MCAP.

7   Concluding Remarks

This paper presented two new string related combinatorial optimization problems, namely the **Chain Alignment Problem** and the **Multiple Chain Alignment Problem**, and some results for them. More specifically, the first problem was solved via a polynomial-time dynamic programming algorithm. For the second one, it was proved that it is hard to develop a polynomial-time exact algorithm, unless $\mathcal{P} = \mathcal{NP}$. We also attested that these problems can model practical problems, by applying them to a variant of the **Gene Identification Problem**, and by getting appropriate results.

A particularly interesting point to be investigated concerns the cases where each ordered set of segments β_i of s_i has linear size on s_i. In such cases, algorithms for the classical **Spliced Alignment Problem** were improved from cubic to almost quadratic time [8,13,17]. Surely, such restriction may not be applicable depending on the modeled problem, but regarding the main application of both the **Chain Alignment Problem** and the **Multiple Chain Alignment Problem**, namely the **Gene Identification Problem**, such constraint is appropriate. Therefore, the techniques described in those papers could lead to faster solutions to the problems presented in this work. It is also worth noting that the results shown in this paper for the **Multiple Chain Alignment Problem** assumed a specific scoring function. As future works, it would be of interest to explore the complexity of this problem under other scoring functions, especially those usually employed in practical problems. Still regarding this problem, the study of probabilistic algorithms, linear programming models, and heuristics for it could reveal interesting results. The application of both problems in other practical tasks also remains as a future work.

Acknowledgements. The authors acknowledge Coordenação de Aperfeiçoamento de Pessoal de Nível Superior (CAPES), Brazil, for the support of this work.

References

1. Burset, M., Guigó, R.: Evaluation of gene structure prediction programs. Genomics **34**, 353–367 (1996)
2. Cavener, D.R., Ray, S.C.: Eukaryotic start and stop translation sites. Nucleic Acids Res. **19**(12), 3185–3192 (1991)
3. The ENCODE Project Consortium: The ENCODE (ENCyclopedia Of DNA elements) project. Science **306**(5696), 636–640 (2004)
4. Gelfand, M.S., Mironov, A.A., Pevzner, P.A.: Gene recognition via spliced sequence alignment. Proc. Natl. Acad. Sci. U.S.A. **93**, 9061–9066 (1996)
5. Guigo, R.: DNA composition, codon usage and exon prediction, Chap. 4. In: Genetic Databases, pp. 53–80. Academic Press (1999)
6. Huang, X., Chao, K.M.: A generalized global alignment algorithm. Bioinformatics **19**(2), 228–233 (2003)
7. Jain, A.K., Dubes, R.C.: Algorithms for Clustering Data. Prentice-Hall, Upper Saddle River (1988)
8. Kent, C., Landau, G.M., Ziv-Ukelson, M.: On the complexity of sparse exon assembly. J. Comput. Biol. **13**(5), 1013–1027 (2006)
9. Lewin, B., Krebs, J., Goldstein, E., Kilpatrick, S.: Lewin's Genes X. Jones and Bartlett (2009)
10. Maier, D.: The complexity of some problems on subsequences and supersequences. J. ACM **25**(2), 322–336 (1978)
11. Majoros, W.: Methods for Computational Gene Prediction. Cambridge University Press, Cambridge (2007)
12. Needleman, S.B., Wunsch, C.D.: A general method applicable to the search for similarities in the amino acid sequence of two proteins. J. Mol. Biol. **48**, 443–453 (1970)
13. Sakai, Y.: An almost quadratic time algorithm for sparse spliced alignment. Theory Comput. Syst. **48**(1), 189–210 (2011)
14. Salzberg, S.: A method for identifying splice sites and translational start sites in eukaryotic mRNA. Bioinformatics **13**(4), 365–376 (1997)
15. Sayers, E.W., et al.: Database resources of the National Center for Biotechnology Information. Nucleic Acids Res. **38**(Database-Issue), 5–16 (2010)
16. Sebastiani, F.: Machine learning in automated text categorization. ACM Comput. Surv. **34**(1), 1–47 (2002)
17. Tiskin, A.: Semi-local string comparison: algorithmic techniques and applications. Math. Comput. Sci. **1**(4), 571–603 (2008)
18. Ye, Y., Jaroszewski, L., Li, W., Godzik, A.: A segment alignment approach to protein comparison. Bioinformatics **19**, 742–749 (2003)

Comparing Deep and Machine Learning Approaches in Bioinformatics: A miRNA-Target Prediction Case Study

Valentina Giansanti[1]([✉]), Mauro Castelli[2], Stefano Beretta[1], and Ivan Merelli[1]

[1] Institute for Biomedical Technologies, National Research Council of Italy, Segrate, MI, Italy
{valentina.giansanti,stefano.beretta,ivan.merelli}@itb.cnr.it
[2] Information Management School (NOVA IMS), Universidade Nova de Lisboa, Campus de Campolide, 1070-312 Lisbon, Portugal
mcastelli@novaims.unl.pt

Abstract. MicroRNAs (miRNAs) are small non-coding RNAs with a key role in the post-transcriptional gene expression regularization, thanks to their ability to link with the target mRNA through the complementary base pairing mechanism. Given their role, it is important to identify their targets and, to this purpose, different tools were proposed to solve this problem. However, their results can be very different, so the community is now moving toward the deployment of integration tools, which should be able to perform better than the single ones.

As Machine and Deep Learning algorithms are now in their popular years, we developed different classifiers from both areas to verify their ability to recognize possible miRNA-mRNA interactions and evaluated their performance, showing the potentialities and the limits that those algorithms have in this field.

Here, we apply two deep learning classifiers and three different machine learning models to two different miRNA-mRNA datasets, of predictions from 3 different tools: TargetScan, miRanda, and RNAhybrid. Although an experimental validation of the results is needed to better confirm the predictions, deep learning techniques achieved the best performance when the evaluation scores are taken into account.

Keywords: miRNA · Deep learning · Machine learning · miRNA-target prediction

1 Introduction

MicroRNAs are a particular family of RNAs characterized by a length of about 22 nucleotides originated from the non-coding RNAs [1]. Their ability to link with the target leads to mRNA degradation and the translation process's block [2], preventing the production of proteins. These small molecules have an important role also in the control of many biological processes, of which homeostasis is one of the possible examples [3].

J. M. F. Rodrigues et al. (Eds.): ICCS 2019, LNCS 11538, pp. 31–44, 2019.
https://doi.org/10.1007/978-3-030-22744-9_3

Since their first discovery in 1993 [4,5], miRNAs are at the center of the scientific community's interest. The computational analysis states that at last 30% of human genes are regulated by miRNA [6], and it has been shown that their dysfunction can lead to the development and progression of different kind of diseases, from cancer to cardiovascular dysfunction [7], and also neurological disorders such as Alzheimer's [8]. Indeed, some of the works done so far describe the role of miR-21 in different kinds of diabetes mellitus [9] and refer to the miRNAs of the has-let-7 family as associated with metabolic disease [10]. Regarding cancer, miRNAs can act as tumor inducers or suppressors, therefore they are the core of the development of new drugs and therapeutic methods [11–13].

Given their importance, understanding the complex mechanisms behind the interactions between miRNAs and their targets is one of the challenges of these years. As a matter of fact, predicting which are the molecules that link together is of vital importance to produce a specific solution to their misbehaviour. That said, the use of biological approaches alone is not a sufficient strategy to resolve the problem because, as they only partially match, each miRNA has multiple mRNA targets and vice versa. Since there are more than 2000 miRNAs in the human genome [14], it is impossible to experimentally test the enormous number of possible combinations between miRNA and mRNA. It is time-consuming and costly at the same time. There is, therefore, a need to identify in advance miRNA-target interactions so to apply experimental approaches that can provide their functional characterization and, thus, the understand of their effects.

The bioinformatics community proposed many computational prediction tools which scope is to provide putative miRNA-target interactions to be evaluated in laboratory. The problem lies in the fact that those tool predictions are often inconsistent with each other. Indeed, the biological properties and parameters used in the algorithms are often different and not complete, so the scientist has difficulties in understanding which tool provides the best prediction and choosing the appropriate miRNA to validate. Up to now, only a limited number of targets have been experimentally validated as the tools suffer from a high level of false positives [15].

The tools available so far include different computational approaches mainly based on the modelling of physical interactions. However, new tools based on *Machine Learning* (ML) and *Deep Learning* (DL) are starting to emerge. These tools are able to automatically extract information from sequences: deepTarget [16] relies on autoencoders and Recurrent Neural Networks (RNN) to model RNA sequences and learn miRNA-mRNA interactions; miRAW [17] uses DL to analyse the mature miRNA transcript without making assumptions on which physical characteristics are the best suitable to impute targets. Other DL algorithms make use of selected features to predict the genes targeted by miRNAs; for example in [18] the authors used three conservative features, eight accessible features, and nine complementary ones to train a Convolutional Neural Network (CNN) classifier.

Many works have been done to compare the performance of the available tools, and for this we refer the reads to [15,19–25]. Almost all of them compare the most famous tools: PITA, microRNA, miRSystem, miRmap, microT-CDS, CoMir, mirWalk, TargetScan, PicTar, miRU, RNAhybrid, miRanda etc. All these methods use specific features and parameters, but one of the main difference among them is that some generate original scores by which the interaction is evaluated (e.g. TargetScan), while others are based on the development of older tools, like microRNA, which is based on a development of the miRanda algorithm.

Despite their variety, only few tools are effectively used in standard procedures. One reason is the confidence of their results, and another one is the easiness that characterizes their use: the web-based service is the preferred platform, followed by the downloaded, and last are the packages. This and the fact that the targets identified by more than one tool are supposed to have a higher probability to be validated in the lab [22] made us choose to test integrated tools for the prediction of miRNA-mRNA interactions based only on selected features.

From the above-mentioned ones, we selected three tools: TargetScan, miRanda, and RNAhybrid. These are among the most used ones and generate scores which describe different interaction mechanisms that we believe are to be considered simultaneously. This selection gave us the possibility to use the knowledge available on the processes behind the miRNA-target interactions without redundancy. That is not true for those tools using multiple scores to describe the same feature or relying on a wide number of other tools since, if the scores are produced viewing the same characteristic, this could lead to a bias.

We tested miRNA-mRNA positive and negative interactions on the selected tools, their scores were collected, and a dataset was built. As reported in a previous study, the approach combining the results of different prediction tools achieves better performance than those obtained by the single ones [22,35].

In the recent years, DL gained a huge success in many classification problems, outperforming ML models [26], so we decided to verify if a case study as the one we described could be resolved by DL architectures in a better way compared to ML methods.

In this work, we employ two datasets of predictions from 3 different miRNA-mRNA interaction tools, namely, TargetScan, miRanda, and RNAhybrid, to train two DL classifiers and three different ML models: support vector machine (SVM), logistic regression (LR), and random forest (RF). More precisely, we use a dataset of positive and negative miRNA-target interactions of small dimension (572 examples), and a larger one (13505 examples) obtained by generating new negative examples. As expected, from the results we observed that DL models need more time to train, but they achieve the best performance when the evaluation scores are taken into account. Given these results, we can confirm a limit of machine and deep learning: they both need a considerable amount of data to train on.

In Sect. 2 we briefly describe the main biological properties of the miRNA-mRNA interactions, the three selected tools used for the sequence-based prediction, the scores they produce, and we introduce the ML and DL algorithms we chose to test. In Sect. 3 we describe the data and the specifics of the DL and ML methods we trained. In Sect. 4 we compare their performance on the dataset, and in Sect. 5 we draw the conclusion of our study.

2   Background

2.1   miRNA-mRNA Interactions and Prediction Tools

As mentioned before, miRNAs have a key role in various biological processes, especially in gene expression regularization, by binding to mRNA molecules. The miRNA-mRNA interactions are predicted by computational tools that commonly evaluate four main features:

Seed region. The seed region of a miRNA is defined as the first 2 to 8 nucleotides starting from the 5'-end to the 3'-end which is the chemical orientation of a single strand of nucleic acid [27]. This is the small sequence where miRNAs link to their targets in a Watson-Crick (WC) match: adenosine (A) pairs with uracil (U), and guanine (G) pairs with cytosine (C). It is considered a perfect seed match if there are no gaps in the whole seed region (8 nucleotides) but other seed matches are also possible, like the 6mer where the WC pairing between the miRNA seed and mRNA is up to 6 nucleotides.

Site accessibility. The miRNA-mRNA interaction is possible only if the mRNA can be unfolded after its binding to the miRNA. The mRNA secondary structure can obstruct the hybridization, therefore the energy necessary to provide the target accessibility can be considered to evaluate the possibility that the mRNA is the real target of a miRNA.

Evolutionary conservation. A sequence conservation across species may provide evidence that a miRNA-target interaction is functional because it is being selected by positive natural selection. Conserved sequences are mainly the seed regions [28].

Free energy. Since the Gibbs free energy is a measure of the stability of a biological system, the complex miRNA-mRNA energy variation can be evaluated to predict which is the most likely interaction [29]: a low free energy corresponds to a high probability that the interaction will occur.

From the miRNA-target prediction tools that use the aforementioned features, three are particularly popular in the scientific community:

TargetScan. It was the first algorithm able to predict miRNA-target interactions in vertebrates. It has been upgraded several times in the years and now it estimates the cumulative weighted context score (CWCS) for each miRNA submitted. CWCS is the sum of the contribution of six features of site context to confirm the site efficacy [30], and it can vary from −1 to 1. The lowest score is representative of a higher probability for the given mRNA to be targeted.

miRanda. It was also one of the earlier prediction tools. The inputs are sequences, and it searches for potential target regions in a sequence dataset. It outputs two different scores, one evaluating the alignment and the other the free energy: to describe a possible target, accordingly to the scores meaning, the former has to be positive and high, while the latter must be negative.

RNAhybrid. Given two sequences (miRNA and target), RNAhybrid determines the most energy favourable hybridization site. Thus, its main output is a score evaluating the free energy for a given seed region. The tool provides also a *p*-value score for the miRNA-mRNA interaction that is an abundance measure of the target site.

2.2   Machine Learning and Deep Learning

Given a miRNA and a mRNA we wanted to test if the combination of scores produced by different tools could give a more accurate indication of their likely to be linked (positive outcome) or not (negative outcome). This can be seen as a binary classification problem (true or false linkage) and, thus, ML and DL techniques are nowadays the most suitable choice to deal with this kind of question. They are automatic techniques for learning to make accurate predictions based on past observations [31]. The data used to learn the model represent the so-called *training set*, while the ones used to assess the generalization ability of the model is the *test set*. The learning performance is evaluated observing a chosen score, like the *accuracy*.

In the latest applications, DL approaches outperformed ML given their ability to learn complex features. This is the reason why we implemented DL architectures and compared their performance with suitable ML techniques like SVM, RF, and LR [33].

These methods have a very different characterization:

Deep Network. DL architectures are essentially neural networks with multiple layers which perform non-linear inputs elaboration [32]. A deep network is characterized by a large number of hidden layers, which relates to the depth. The number of layers is specific to the net because it indicates the complexity of the relationships it is able to learn. Another important parameter is the number of nodes in the layer. It is possible to choose between different kind of layers (e.g., dense, convolutional, probabilistic, or memory), each able to combine the input with a set of weights. A network with only dense layers is a standard DNN, a RNN is instead characterized by memory cells, like the LSTM. Non-linear functions like sigmoid and rectified linear unit are then used to compute the output. Deep networks are suitable to analyse high-dimensional data.

Support Vector Machines. SVMs are one of the most famous ML algorithms, capable of performing linear and nonlinear classification. They aim to select the coefficients for the optimal hyperplane able to separate the input variables into two classes (e.g., 0 and 1). SVMs perform better on complex but small or medium datasets.

Random Forest. A RF is an ensemble of decision trees. Decision trees are created to select suboptimal split points by introducing randomness. Each tree makes a prediction on the proposed data, and all the predictions are averaged to give a more accurate result. The ensemble has a similar bias but a lower variance than a single tree.

Logistic Regression. LR is the go-to method for binary classification in the ML area, commonly used to estimate the probability that an instance belongs to a particular class. The goal is to find the right coefficients that weight each input variable. The prediction for the output is transformed using a non-linear logistic function that transforms any value into the range 0 to 1. LR is commonly used for datasets that do not contain correlated attributes.

3   Methods

3.1   Data

The classifiers were trained with data from a reference dataset of 48121946 miRNA-target predictions. This dataset was obtained starting from the sequences of the miRNA families and of the untranslated regions (UTRs, the genomic loci targeted by miRNA) from 23-way alignment, filtering the information relative to the Homo sapiens species. More precisely, obtained a total of 30887 UTR (mRNA genes) and 1558 miRNA sequences, which were used as the starting point of our analysis. Then, we run the 3 tools (namely, miRanda, TargetScan, and RNAhybrid) and we combined their results in a matrix by looking at the positions on the UTR. Finally, to deal with the missing values of some tools, we assigned penalizing scores to them, which have been chosen after some experiments we conducted to assess how these penalizing scores influence the final classification results. As anticipated, the input matrix was composed of 48121946 rows (30887 UTRs × 1558 miRNAs), in which the first five columns contain scores provided by TargetScan (Tscan-score), miRanda (miRanda-score and miRanda-energy) and RNAhybrid (RNAhybrid-mfe and RNAhybrid-pvalue). The last column contains the classification of the instances used to partition the dataset into five classes: negative examples (a), positive and experimentally validated examples (b), only experimentally validated examples (c), only positive examples (d) and unknown examples (e), as described in Table 1.

This latter classification is obtained by considering the predictions coming from [36] in which the authors generated two sets of "positive" and "negative" miRNA-target examples. The former set (positive examples) was obtained by biologically verified experiments, while the latter examples (negative) were identified from a pooled dataset of predicted miRNA-target pairs. Moreover, we downloaded the data from miRTarBase [37], a database of experimentally validated miRNA-target interactions, to additionally classify the positive examples into positive and experimentally validated examples, only experimentally validated examples, and only positive examples. More specifically, we crossed miRTarBase interactions with the positive examples from [36], in order to make the dataset more robust.

The unknown examples were not useful to train the classifiers, thus we put aside these examples, while we merged together b, c, and d data to obtain a unique *positive* class. The rearranged dataset was composed of 6841 positive and 286 negative interactions. A great class imbalance like the one present in

Table 1. Dataset description.

Original		New	
Classification	# Examples	Classification	# Examples
Negative (a)	286	Negative	286
Positive and exp. validated (b)	179	Positive	6841
Only exp. validated (c)	286		
Only positive (d)	6376		
Unknown (e)	48114996	Unknown	48114996
Total	**48121946**		**48121946**

this dataset is a huge problem for any ML or DL classification algorithm. In the training phase, the classifiers receive much more information of one class than of the other, and are not able to learn equally: the classifier tends to infer new data as part of the majority class. A possible way to deal with this problem is to make a *balanced* dataset by reducing the items of the majority class or by increasing the minority class examples. We tried both methods and compared the results obtained by the classifiers.

In the first case the dataset, that we called *small* dataset, was composed of the 286 available negative examples and 286 positive examples, sampled from the positive class: 179 from the original *b* classification, 54 from *c*, and 53 from *d*. In the second case, we constructed a dataset (called *large* dataset) of 13505 examples: 6664 positive and 6841 negative interactions. The negative examples were comprehensive of the 286 already available ones and 6555 new generated examples. The generation of the artificial negative examples was made through *k*-mer exchange between key and non-key regions of miRNA, as suggested in [34]. After their production, the miRNA were processed by *HappyMirna* [35], a tool for the integration of miRNA-target predictions and comparison. Thanks to HappyMirna we were able to obtain possible target for the new-generated miRNAs and the scores provided by TargetScan, miRanda, and RNAhybrid to construct a matrix equal to the previous one.

All datasets had missing values (NaN) whenever the prediction tools were not able to produce a result for the input. Since classifiers can not deal with NaN, we replaced all of them with penalizing scores chosen according to the range of the tool score, e.g. NaN for TargetScan were replaced with 1000. RNAhybrid was able to assign a score to 43744510 miRNA-mRNA interactions, while TargetScan only to 5341653 and miRanda to 4370618.

We observed that almost all the times a NaN was in the TargetScan record a corresponding NaN was also in miRanda. Instead, RNAhybrid does not fail to give a predictive value when the others meet a NaN.

3.2 Classifiers

We developed five classifiers (SVM, RF, LR, DNN, and RNN) using the python libraries Keras and Scikit-learn built on top of TensorFlow. All the classifiers were fine-tuned by selecting the optimal hyperparameters with GridSearchCV. When possible, we tried to maintain the same characteristics between models, as the number of folds for the cross-validation used during the training.

On the *small* dataset we trained the SVM, RF, LR, and DNN. The SVM best parameters were gamma = 1, C = 100, degree = 3, kernel = rbf, and random state = 100. The parameters tuned for the RF classifier were the number of estimators and the number of leaves for each estimator, obtaining 400 and 50, respectively. For the LR we investigated the solver, C, and the number of iterations, obtaining the best performance with the lbfgs solver, C = 3, and 50 iterations. For the DNN, we built a network with 5 fully connected layers, the first characterized by 80 nodes, from the second to the fourth with 40 nodes, and the last with 2 (necessary for a binary classification). The network was trained with 3000 epochs, a batch size of 100 and the ADAM optimizer. The score used to evaluate the performance was in all cases the accuracy. The training was with a 5-fold cross-validation.

On the *large* dataset, we used instead a 10-fold cross-validation. The best parameters for the SVM were gamma = 2, C = 100, degree = 3, kernel = rbf, and random state = 100. The RF had 100 estimators with a number of leaves for each estimator of 60. For the LR, C was set to 0.5, the iterations to 200, and the best solver was lbfgs. As for the deep architectures, we trained the same DNN used on the *small* dataset and a RNN characterized by three layers: the first and second were LSTM layers, with 150 and 100 units, respectively, the third was a dense layer like in the DNN. During the cross-validation, each fold was processed for 1000 epochs with a batch size of 1000.

4 Results

A way to compare the performance of different classifiers is to evaluate the accuracy, the ROC curve, and compute the area under the curve. Figure 1 shows the results obtained on the *small* dataset.

More precisely, the DNN and the RF had the best performance compared to the other classifiers, achieving an area under the curve of 0.66. Other measures were calculated and compared (see Table 2): the DNN had the best accuracy, precision, and f1 score over all the classifiers, while the recall was comparable. Based on this and the AUC we can state that the DNN was the model that best learned to recognize the miRNA-mRNA interactions.

All the positive data not used for the training were stored in separated datasets, called *two* and *three* dataset, so to remember the classification of origin. We used these datasets to check the percentage of the positive (and thus correct) predictions made by the trained classifiers, which was a way to further assess their performance. The same was done on the unknown data (the *zero* dataset) but in this case we could not say if the performance of the classifiers

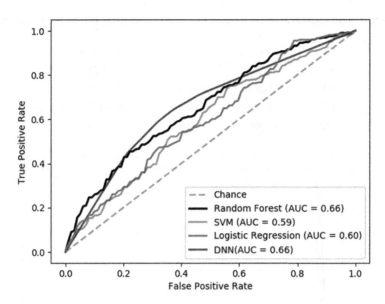

Fig. 1. Receiver Operating Characteristic (ROC) curve: performance obtained on the *small* dataset by Random Forest, SVM, Logistic Regression, and DNN. The models learned how to classify the items, but made a considerable percentage of mistakes.

Table 2. Evaluation scores for the implemented classifiers on the *small* dataset.

Classifier	Accuracy	Precision	Recall	f1
SVM	0.5717	0.5622	0.6847	0.6115
RF	0.5753	0.5862	0.6288	0.5910
LR	0.5280	0.5345	0.6451	0.5710
DNN	0.6417	0.6620	0.6250	0.6232

was good or not. In fact, since we did not have prior knowledge, we could only observe how different the predictions were. As we can see in Table 3, the SVM and DNN models were the ones able to recognize the higher number of interactions, especially on the *three* dataset. The RF model had the worst performance on all the datasets: we could not be sure of the results of this model as it gave only the 50% or less of correct evaluation on the unseen data that correspond to the *two* and *three* dataset. On the *zero* dataset the SVM and DNN were again the ones that proposed as possible the larger portion of interactions.

What we obtained training the models on the *large* dataset was instead very different. All the models were able to correctly classify over the 95% of the items, and none performed considerably better than the others (see Fig. 2 and Table 4). We believe that the scores improvement was thanks to the supplementary information the models were able to learn in the *large* dataset: miRNA-target inter-

Table 3. Percentage of positive classification obtained from the trained classifiers.

Classifier	Two dataset	Three dataset	Zero dataset
SVM	62.5593	80.3571	46.8071
RF	51.2970	42.8571	25.3370
LR	52.0563	58.9285	29.5477
DNN	59.9968	76.7857	50.3155

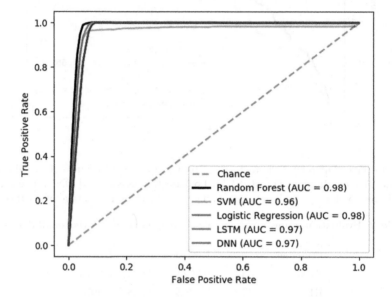

Fig. 2. Receiver Operating Characteristic (ROC) curve: performance obtained on the *large* dataset by Random Forest, SVM, Logistic Regression, deep net with lstm layer (LSTM), and deep net with only dense layers (DNN). All the models learned to correctly recognize the items with minor mistakes.

action is a complex problem which needs a considerable amount of data to be addressed.

We trained the recurrent network to compare the results of the two deep architectures. Finding good parameters for the RNN was easier compared to the DNN, and also the number of necessary layers was smaller. Consequently, the training time was considerably reduced. As we used the *three* data and *two* data to build the *large* dataset, we could only observe how all the trained models behaved on the *zero* dataset (see Table 5): almost all the miRNA-target interactions were recognized as possibly true in all models.

Table 4. Evaluation scores for the implemented classifiers on the *large* dataset.

Classifier	Accuracy	Precision	Recall	f1
SVM	0.9145	0.9582	0.8646	0.9087
RF	0.9745	0.9532	0.9975	0.9748
LR	0.9658	0.9428	0.9910	0.9662
DNN	0.9679	0.9390	1.0	0.9685
RNN	0.9671	0.9389	0.9982	0.9676

Table 5. Percentage of positive classifications obtained from the trained classifiers on the *zero* dataset.

Dataset	SVM	RF	LR	DNN	LSTM
Zero	96.0268	99.8839	99.7372	99.9781	99.9654

5   Conclusions

miRNA-target interactions are predicted by a variety of tools that frequently give divergent results, thus are becoming diffuse solutions that integrate their outputs to give a unique and (possibly) reliable decision on the couples validity. Machine learning and deep learning methods are the ones preferably used to integrate different outcomes, with deep learning methods usually surpassing machine learning ones in terms of performance.

In our work, we trained five models from machine and deep learning area to test the possibility to identify a miRNA-mRNA interactions based on the scores provided by TargetScan, miRanda, and RNAhybrid. We used two different datasets: the first was a dataset of positive and negative miRNA-target interactions of small dimension (572 examples), the other was a larger dataset (13505 examples) obtained generating new negative examples. The performance of all the models was comparable in both cases: they performed poorly on the *small* dataset and very well on the *large* one.

Given these results, we can confirm a limit of machine and deep learning: they both need a considerable amount of data to train on. On the *small* dataset, comparing more scores, we said that the DNN performed fairly better than the other models, while on the *large* dataset we obtained from all the models very good and comparable results.

In the latter case we have to say that, as all the performance were very good (and as all the models gave the same results on the *zero* dataset), we do not suggest to use deep network solutions. In fact they have a difficult nature and they require a lot of time for the training and, moreover, on our problem the efforts did not give results upon the mean. It is much easier and faster to use standard machine learning implementations. On the *small* dataset the DNN results were better, but we can not recommend using this method over the others as the data were not enough to efficiently train the models.

In conclusion, we recognized the possibility to implement integrated tools based on machine and deep learning, the goodness of which can be finally evaluated only when the miRNA-target interactions they propose will be experimentally validated. The dimension of the dataset used to train is one of the main problems for a good integration. We showed that with a reduced dataset it's hard to find a model which can easily recognized miRNA-target interaction (even if more DL architectures should be evaluated). If instead lots of examples are available, the choice of model type is irrelevant from the accuracy point of view.

Acknowledgments. This work was partially supported by national funds through FCT (Fundação para a Ciência e a Tecnologia) under project DSAIPA/DS/0022/2018 (GADgET).

References

1. Bartel, D.P.: MicroRNAs: genomics, biogenesis, mechanism, and function. Cell **116**(2), 281–297 (2004)
2. He, L., Hannon, G.J.: MicroRNAs: small RNAs with a big role in gene regulation. Nat. Rev. Genet. **5**(7), 522–531 (2004)
3. Liu, B., Li, J., Cairns, M.J.: Identifying miRNAs, targets and functions. Brief. Bioinform. **15**(1), 1–19 (2012)
4. Lee, R.C., Feinbaum, R.L., Ambros, V.: The C. elegans heterochronic gene lin-4 encodes small RNAs with antisense complementarity to lin-14. Cell **75**(5), 843–854 (1993)
5. Wightman, B., Ha, I., Ruvkun, G.: Posttranscriptional regulation of the heterochronic gene lin-14 by lin-4 mediates temporal pattern formation in C. elegans. Cell **75**(5), 855–862 (1993)
6. Ross, J.S., Carlson, J.A., Brock, G.: miRNA: the new gene silencer. Am. J. Clin. Pathol. **128**(5), 830–836 (2007)
7. Hackfort, B.T., Mishra, P.K.: Emerging role of hydrogen sulfide-microRNA crosstalk in cardiovascular diseases. Am. J. Physiol.-Heart Circ. Physiol. **310**(7), H802–H812 (2016)
8. Hebert, S.S.: MicroRNA regulation of Alzheimer's Amyloid precursor protein expression. Neurobiol. Dis. **33**(3), 422–428 (2009)
9. Sekar, D., Venugopal, B., Sekar, P., Ramalingam, K.: Role of microRNA 21 in diabetes and associated/related diseases. Gene **582**(1), 14–18 (2016)
10. Shi, C., et al.: Adipogenic miRNA and meta-signature miRNAs involved in human adipocyte differentiation and obesity. Oncotarget **7**(26), 40830–40845 (2016)
11. Ling, H., Fabbri, M., Calin, G.A.: MicroRNAs and other non-coding RNAs as targets for anticancer drug development. Nat. Rev. Drug Discov. **11**, 847–865 (2013)
12. Samanta, S., et al.: MicroRNA: a new therapeutic strategy for cardiovascular diseases. Trends Cardiovasc. Med. **26**(5), 407–419 (2016)
13. Riquelme, I., Letelier, P., Riffo-Campos, A.L., Brebi, P., Roa, J.: Emerging role of miRNAs in the drug resistance of gastric cancer. Int. J. Mol. Sci. **17**(3), 424 (2016)
14. Hammond, S.M.: An overview of microRNAs. Adv. Drug Delivery Rev. **87**, 3–14 (2015)
15. Akhtar, M.M., Micolucci, L., Islam, M.S., Olivieri, F., Procopio, A.D.: Bioinformatic tools for microRNA dissection. Nucleic Acids Res. **44**(1), 24–44 (2015)

16. Lee, B., Baek, J., Park, S., Yoon, S.: deepTarget: end-to-end learning framework for microRNA target prediction using deep recurrent neural networks. In: Proceedings of the 7th ACM International Conference on Bioinformatics, Computational Biology, and Health Informatics, pp. 434–442. ACM, October 2016

17. Planas, A.P., Zhong, X., Rayner, S.: miRAW: a deep learning-based approach to predict microRNA targets by analyzing whole microRNA transcripts. PLoS Comput. Biol. **14**(7), e1006185 (2017)

18. Cheng, S., Guo, M., Wang, C., Liu, X., Liu, Y., Wu, X.: MiRTDL: a deep learning approach for miRNA target prediction. IEEE/ACM Trans. Comput. Biol. Bioinform. **13**(6), 1161–1169 (2016)

19. Bartel, D.P.: MicroRNAs: target recognition and regulatory functions. Cell **136**(2), 215–233 (2009)

20. Mendes, N.D., Freitas, A.T., Sagot, M.F.: Current tools for the identification of miRNA genes and their targets. Nucleic Acids Res. **37**(8), 2419–2433 (2009)

21. Ruby, J.G., Stark, A., Johnston, W.K., Kellis, M., Bartel, D.P., Lai, E.C.: Evolution, biogenesis, expression, and target predictions of a substantially expanded set of Drosophila microRNAs. Genome Res. **17**(12), 1850–1864 (2007)

22. Alexiou, P., Maragkakis, M., Papadopoulos, G.L., Reczko, M., Hatzigeorgiou, A.G.: Lost in translation: an assessment and perspective for computational microRNA target identification. Bioinformatics **25**(23), 3049–3055 (2009)

23. Fan, X., Kurgan, L.: Comprehensive overview and assessment of computational prediction of microRNA targets in animals. Brief. Bioinform. **16**(5), 780–794 (2014)

24. Srivastava, P.K., Moturu, T.R., Pandey, P., Baldwin, I.T., Pandey, S.P.: A comparison of performance of plant miRNA target prediction tools and the characterization of features for genome-wide target prediction. BMC Genomics **15**(1), 348 (2014)

25. Faiza, M., Tanveer, K., Fatihi, S., Wang, Y., Raza, K.: Comprehensive overview and assessment of miRNA target prediction tools in human and drosophila melanogaster (2017). arXiv:1711.01632

26. Chen, H., Engkvist, O., Wang, Y., Olivecrona, M., Blaschke, T.: The rise of deep learning in drug discovery. Drug Discov. Today **23**(6), 1241–1250 (2018)

27. Lewis, B.P., Burge, C.B., Bartel, D.P.: Conserved seed pairing, often flanked by adenosines, indicates that thousands of human genes are MicroRNA targets. Cell **120**(1), 15–20 (2005)

28. Lewis, B.P., Shih, I.H., Jones-Rhoades, M.W., Bartel, D.P., Burge, C.B.: Prediction of mammalian MicroRNA targets. Cell **115**(7), 787–798 (2003)

29. Yue, D., Liu, H., Huang, Y.: Survey of computational algorithms for MicroRNA target prediction. Curr. Genomics **10**(7), 478–492 (2009)

30. Garcia, D.M., Baek, D., Shin, C., Bell, G.W., Grimson, A., Bartel, D.P.: Weak seed-pairing stability and high target-site abundance decrease the proficiency of lsy-6 and other microRNAs. Nat. Struct. Mol. Biol. **18**(10), 1139–1146 (2011)

31. Schapire, R.E.: The boosting approach to machine learning: an overview. In: Denison, D.D., Hansen, M.H., Holmes, C.C., Mallick, B., Yu.B. (eds.) Nonlinear Estimation and Classification. LNS, vol. 171, pp. 149–171. Springer, New York (2003). https://doi.org/10.1007/978-0-387-21579-2_9

32. Min, S., Lee, B., Yoon, S.: Deep learning in bioinformatics. Brief. Bioinform. **18**(5), 851–869 (2017)

33. Goodfellow, I., Bengio, Y., Courville, A., Bengio, Y.: Deep Learning, vol. 1. MIT Press, Cambridge (2016)

34. Mitra, R., Bandyopadhyay, S.: Improvement of microRNA target prediction using an enhanced feature set: a machine learning approach. In: IEEE International Advance Computing Conference, pp. 428–433. IEEE, March 2009

35. Beretta, S., Giansanti, V., Maj, C., Castelli, M., Goncalves, I., Merelli, I.: HappyMirna: a library to integrate miRNA-target predictions using machine learning techniques. In: Proceedings of Intelligent Systems in Molecular Biology, July 2018

36. Bandyopadhyay, S., Mitra, R.: TargetMiner: microRNA target prediction with systematic identification of tissue-specific negative examples. Bioinformatics $25(20)$, 2625–2631 (2009)

37. Hsu, S.D., et al.: miRTarBase update 2014: an information resource for experimentally validated miRNA-target interactions. Nucleic Acids Res. 42, D78–D85 (2014)

Automated Epileptic Seizure Detection Method Based on the Multi-attribute EEG Feature Pool and mRMR Feature Selection Method

Bo Miao[1], Junling Guan[2], Liangliang Zhang[1], Qingfang Meng[1], and Yulin Zhang[1(✉)]

[1] University of Jinan, Jinan 250000, Shandong, China
ise_zhangyl@ujn.edu.cn
[2] University of New South Wales, Sydney, NSW 1466, Australia

Abstract. Electroencephalogram (EEG) signals reveal many crucial hidden attributes of the human brain. Classification based on EEG-related features can be used to detect brain-related diseases, especially epilepsy. The quality of EEG-related features is directly related to the performance of automated epileptic seizure detection. Therefore, finding prominent features bears importance in the study of automated epileptic seizure detection. In this paper, a novel method is proposed to automatically detect epileptic seizure. This work proposes a novel time-frequency-domain feature named global volatility index (GVIX) to measure holistic signal fluctuation in wavelet coefficients and original time-series signals. Afterwards, the multi-attribute EEG feature pool is constructed by combining time-frequency-domain features, time-domain features, nonlinear features, and entropy-based features. Minimum redundancy maximum relevance (mRMR) is then introduced to select the most prominent features. Results in this study indicate that this method performs better than others for epileptic seizure detection using an identical dataset, and that our proposed GVIX is a prominent feature in automated epileptic seizure detection.

Keywords: Medical signal processing · Epileptic seizure detection · Minimum redundancy maximum relevance · Global volatility index

1 Introduction

Epilepsy is a common chronic brain disorder characterized by convulsions from epileptic seizures [1]. More than 50 million people suffer from epilepsy all over the world, with the vast majority living in developing countries. The clinical manifestations of epileptic people are mainly sudden loss of consciousness, general convulsions, and abnormality of mind [2]. As a disease with complex and diverse causes, epilepsy severely affects patients' physical and psychological

© Springer Nature Switzerland AG 2019
J. M. F. Rodrigues et al. (Eds.): ICCS 2019, LNCS 11538, pp. 45–59, 2019.
https://doi.org/10.1007/978-3-030-22744-9_4

health. At the physical level, the violent convulsion of the body during epileptic seizures can result in a fracture. At the psychological level, the uncertainty of epileptic seizures can be very disturbing. Moreover, epileptic people are often stigmatized at school or at work and are thus mentally traumatized [3]. Therefore, finding prominent features and correctly detecting epileptic seizure bears importance in diagnosing and curing epilepsy.

Electroencephalography is a method of recording brain activity through electrophysiological indicators without inflicting trauma to the subject. Electroencephalogram (EEG) signals can be detected when many neurons in the same brain region are activated simultaneously. EEGs are widely used to detect brain-related disorders due to its noninvasiveness, low cost and high temporal resolution. However, artificially interpreting EEG recording is costly and subjective. Machine learning methods can be used for classification and detection and have shown good performance [4]. Therefore, machine learning combined with EEG signals are used to detect brain-related disorders [5]. Selecting effective features is crucial to epileptic seizure detection and pathological discovery. This means that more comprehensive features need to be used and more discriminating feature subsets need to be selected. To describe the information contained in EEG signals more comprehensively, many time-domain, frequency-domain and entropy-based features are used to automatically identify epilepsy [6–8]. However, a dearth of studies currently investigate effective features to measure holistic signal fluctuation in EEG signals. This work attempts to fill this gap.

This study primarily aims to improve the accuracy of automated epileptic seizure detection by establishing a novel method for the automated detection of epileptic seizure. First, a novel time-frequency-domain feature named global volatility index (GVIX) is proposed to measure holistic signal fluctuation in wavelet coefficients and original time-series signals. Second, to generalize information in a signal more comprehensively, time-frequency-domain features, time-domain features, nonlinear features, and entropy-based features are used to construct the multi-attribute EEG feature pool. Third, minimum redundancy maximum relevance (mRMR) feature selection algorithm is introduced to identify the most prominent features based on the multi-attribute EEG feature pool. Finally, 10-fold cross validation is achieved using a support vector machine (SVM) classifier. The key contributions and novelties of this work are as follows.

- Developing a novel time-frequency-domain feature GVIX to measure holistic signal fluctuation in the wavelet coefficients and original time-series signals.
- Constructing the multi-attribute EEG feature pool based on time-frequency-domain features, time-domain features, nonlinear features, and entropy-based features.
- Detecting epileptic seizure using a novel framework that applying the multi-attribute EEG feature pool combined with mRMR feature selection method and SVM.

The remainder of this paper is summarized as follows. Section 2 presents the different features and methods used in the prior studies, which has focused on

EEG classification. The details of the data used in this work and its processing framework are presented in Sect. 3. Section 4 presents our experimental results and compares them with previous work. Finally, this work is concluded in Sect. 5.

2   Related Work

Many previous studies have used machine learning methods to classify EEG signals. Many features are also used to display information in EEG signals.

Song et al. [9] utilized sample entropy (SampEn) as a feature extraction method to extract the features of EEG signals. Based on these features, back-propagation neural networks and extreme learning machines are used to achieve epilepsy detection. Results of their study show that SampEn is an outstanding feature in the automated epilepsy detection.

Acharya et al. [10] utilized nonlinear higher order spectra (HOS) and wavelet packet decomposition to construct a feature set. Finally, epileptic EEG signals are detected using SVM. Their results show that the automated epilepsy detection using HOS-based features is a promising approach.

Some nonlinear features such as fractal dimension (FD) and Hurst exponent are also used in the automated epilepsy detection. Acharya et al. [7] achieved the automated identification of epileptic EEG signals using some nonlinear features such as FD and Hurst exponent. Six classifiers have been used in their study, and the fuzzy classifier achieves the highest classification accuracy.

Wavelet transform (WT) is considered as a powerful tool for time-frequency analysis. EEG signals can reflect spontaneous and rhythmic neural activity of the brain, so WT has been introduced to obtain neural activity in different bands in some previous studies. Ibrahim et al. [11] obtained features using discrete wavelet transform (DWT) and cross-correlation. Afterwards, epilepsy and autism spectrum disorder diagnosis are achieved using four classifiers, whereas the k-nearest neighbor algorithm obtains the highest classification accuracy. Guo et al. [12] used multiwavelet transform and approximate entropy (ApEn) to constructed features, and epileptic seizure detection is achieved using these features and artificial neural network.

Principal component analysis (PCA) is a classical method for feature dimension reduction, which can effectively deal with the curse of dimensionality. Zarei et al. [13] combined PCA and cross-covariance to extract features. They subsequently used multilayer perceptron neural networks, least-square SVM, and logistic regression to achieve the classification of EEG signals.

The above-mentioned methods have realized the classification of EEG signals based on different methods. In the present study, we utilize a novel method for automated epileptic seizure detection. The method used in this paper is described in detail in the next section.

3   Data and Methods

In this study, a novel framework is proposed for the automated epileptic seizure detection. Figure 1 illustrates the steps of this work. First, wavelet coefficients of

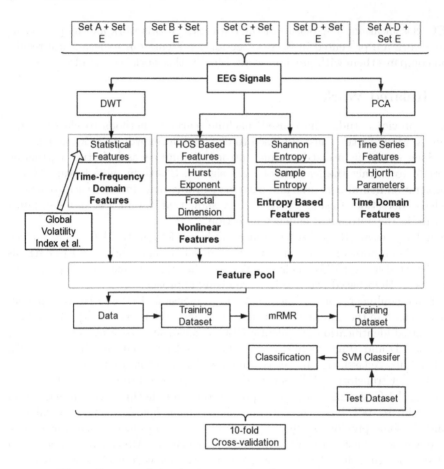

Fig. 1. Framework for the automated epileptic seizure detection.

different EEG frequency bands are analyzed using DWT, and related statistical features like GVIX are used to obtain features from both wavelet coefficients and EEG signals. Afterwards, PCA is used to extract time-series features, and many other features such as entropy-based features and nonlinear features are calculated. The multi-attribute EEG feature pool is then constructed based on these features and used for mRMR. Finally, 10-fold cross validation is achieved based on SVM.

3.1 Dataset

The EEG database (Set A–E) we used is obtained from a publicly available EEG database developed by University of Bonn [14]. The entire EEG database includes five sets each containing 100 segments with a duration of 23.6 s (4097 time points per segment). All these EEG signals are recorded using a 128-channel amplifier system, digitized with a sampling rate of 173.61 Hz and 12-bit A/D

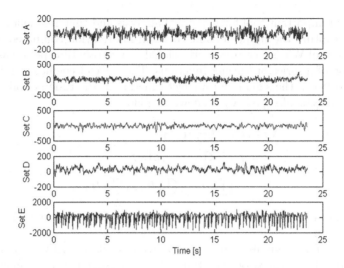

Fig. 2. Examples of EEG signals in the five different sets.

resolution and filtered using a 0.53–40 Hz (12 dB/octave) band pass filter. Manual and eye movement disturbances are removed from all EEG signals. The specific information of EEG data is summarized in Table 1. Figure 2 also shows some examples of Set A to E.

Table 1. Description of EEG data in five sets

	Set A	Set B	Set C	Set D	Set E
Subjects	Healthy	Healthy	Epileptic	Epileptic	Epileptic
State	Eyes opened	Eyes closed	Interictal	Interictal	Ictal
Electrode type	External	External	Intracranial	Intracranial	Intracranial

3.2 Feature Extraction

To establish a better classification model, the multi-attribute EEG feature pool is used. These features can be subdivided into four categories: (1) time-frequency-domain features, (2) time-domain features, (3) nonlinear features, and (4) entropy-based features.

Time-Frequency-Domain Features. The time-frequency-domain features used in this work include wavelet coefficients and related statistical features which can describe wavelet coefficients in sub-bands and original time-series signals.

WT is an important tool for numerical analysis and time-frequency analysis, and it can capture the frequency and location information compared with FT. The basic idea of WT is to represent the signal in a certain time period as a linear combination of a series of wavelet functions. The wavelet coefficient reflects the similarity between the signal and wavelet function in the time period. A multi-resolution analysis of signal X_J is shown as follows:

$$
\begin{aligned}
X_J &= L_{J-1} \oplus H_{J-1} \\
&= L_{J-2} \oplus H_{J-2} \oplus H_{J-1} \\
&= ... \oplus H_{J-3} \oplus H_{J-2} \oplus H_{J-1}
\end{aligned}
\tag{1}
$$

where H represents high frequency, L represents low frequency, and $\oplus$ represents the intersection.

Picking different numbers of decomposition levels for EEG signals should be based on the purpose of the study. An EEG signal usually shows different rhythms in different frequency ranges. Most useful frequency components contained in EEG signals are found to be below 30 Hz [15, 16]. Therefore, the decomposition levels used in this study are set to 5, and the signals are decomposed into details D1-D5 and final approximation A5. Some previous studies have compared the effects of several wavelets and found that the Daubechies wavelet of order 4 is the most suitable one for automated epileptic seizure detection [17], so the wavelet coefficients are computed using the db4 in this work. DWT is used for each data set, approximation and details are thus obtained and are shown in Fig. 3. Table 2 shows the frequency range of different decomposition levels for db4 with a sampling frequency of 173.6 Hz. These wavelet coefficients, calculated for A5 and D3-D5 are used for automated epileptic seizure detection.

Table 2. Frequency range of different decomposition levels

Decomposed signal	Frequency range (Hz)
D1	43.4–86.8
D2	21.7–43.4
D3	10.8–21.7
D4	5.4–10.8
D5	2.7–5.4
A5	0–2.7

To represent the information in the wavelet coefficients and original EEG signals from multiple perspectives, statistical features are used to achieve this goal. In this paper, a novel feature GVIX is proposed to measure the holistic signal fluctuation in wavelet coefficients and original time-series signals, since it considers the signal fluctuation at any time interval. GVIX is calculated based on

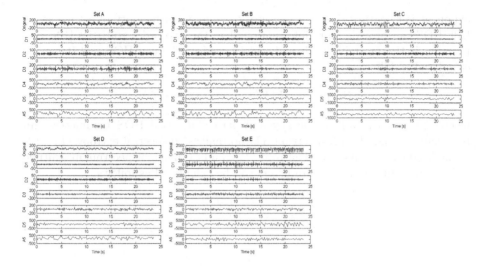

Fig. 3. DWT coefficients of EEG signals taken from the subject of each set.

Manhattan Distance. Calculation of the GVIX of a sequence data $X(1)$, $X(2)$,..., $X(n)$ can be obtained as follows:

$$GVIX(X) = \frac{2}{n^2 - n} \sum_{i=2}^{n} \sum_{j=1}^{i-1} |X(i) - X(j)| \tag{2}$$

where n is the number of points in the sequence data. The other statistical features used to describe wavelet coefficients in sub-bands and original time-series signals in this study are mean (M), mean square (MS), standard deviation (SD), skewness (Ske), kurtosis (Kur), interquartile range (IQR) and volatility index (VIX).

Time-Domain Features. The time-domain features used in this work include principal components and Hjorth parameters (HP).

PCA transforms the original data into a set of linearly independent representations between dimensions through linear transformation, and the linearly independent variables are named principal components [18]. In this paper, PCA is used to reduce the dimension of time-series data, and the standard of dimension reduction is to save 99% of the original information. HP are measurements used to study epileptic lateralization [19]. Mobility (Mobi) and Complexity (Comp) are used in this work.

Non-linear Features. The nonlinear features used in this work include FD, Hurst exponent and HOS parameters.

FD is a statistic that measures the dimensional complexity of signals [7]. The FD calculation algorithm proposed by Higuchi is used in this study. Hurst exponent is a feature that can measure the long-term memory of a time series [20]. HOS can provide more information than the two order statistics [10]. Bispectrum is the most in-depth and widely used method in HOS and it can be calculated as follows:

$$B(f_1, f_2) = E[X(f_1)X(f_2)X(f_1 + f_2)] \tag{3}$$

where $X(f)$ is the FT of the signal $X(nT)$. The used HOS parameters can be calculated based on bispectrum: (1) normalized bispectral entropy (P_1), (2) normalized bispectral squared entropy (P_2), and (3) mean bispectrum magnitude (M_{ave}). In addition, these parameters can be calculated as follows:

$$p_n = \frac{|B(f_1, f_2)|}{\sum_\Omega |B(f_1, f_2)|}$$
$$P_1 = \sum_n p_n \log p_n \tag{4}$$

$$q_n = \frac{|B(f_1, f_2)|^2}{\sum_\Omega |B(f_1, f_2)|^2}$$
$$P_2 = \sum_n q_n \log q_n \tag{5}$$

$$M_{ave} = \frac{1}{L} \sum_\Omega |B(f_1, f_2)| \tag{6}$$

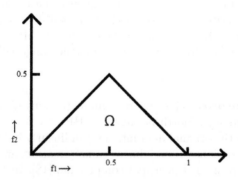

Fig. 4. Region of points that can avoid redundant computation.

where Ω is the region that avoids redundant computation, and its range is shown in Fig. 4. L is the number of points in Ω.

Entropy-Based Features. The entropy-based features used in this work include Shannon entropy (ShEn) and SampEn.

Shannon entropy (ShEn) can be used to measure the uncertainty of EEG signals [21]. As an improved version of ApEn, SampEn is a measure of time-series complexity [22]. SampEn is proportional to time series complexity, while it is inversely proportional to self-similarity. The SampEn of a given time series $X(1)$, $X(2)$,..., $X(n)$ is calculated as follows:

(1) Constructing an m dimensional vector based on the embedding dimension;
(2) Defining the distance function $d[X_m(i), X_m(j)]$ based on Chebyshev Distance;
(3) Calculating SampEn based on the similar tolerance r:

$$SampEn(X) = -\log \frac{A}{B} \qquad (7)$$

where A is the number of template vector pairs having $d[X_{m+1}(i), X_{m+1}(j)] < r$, and B is the number of template vector pairs having $d[X_m(i), X_m(j)] < r$. In this paper, embedding dimension m is set to 2, and similar tolerance r is set to 0.2 based on a previous study [23].

Minimum Redundancy Maximum Relevance. mRMR algorithm is an effective feature selection algorithm, and widely used in the study of bioinformatics [24]. mRMR algorithm uses incremental search to select features and rates them based on mutual information. The selected features are added to the selected feature set in an incremental manner until the number of selected features achieves the termination condition. The average mutual information between a feature and its category is considered relevance (R), and the average mutual information between unselected features is considered redundancy (D). Each feature will get a score when using mRMR algorithm, and the evaluation criterion for each feature is calculated as follows:

$$max R(F, C), Relevance = \frac{1}{|F|} \sum_{f_r \in F} I(f_r, c)$$

$$min D(F, C), Redundancy = \frac{1}{|F|^2} \sum_{f_r, f_o \in F} I(f_r, f_o) \qquad (8)$$

$$max \Phi(R, D), \Phi = R - D \qquad (9)$$

where F represents the feature set, and C represents the target category.

Performance Evaluation Methods. To evaluate the performance of SVM, a 10-fold cross validation method is used. The data are divided into 10 parts. In each cross-validation process, one piece of data is used as the test set and nine pieces of data are used as the training set. In this paper, many criteria such as classification accuracy, sensitivity, specificity and F1 score are used to measure the performance of our method.

Table 3. Five cases for automated epileptic seizure detection

Case	Class 1	Class 2
1	A	E
2	B	E
3	C	E
4	D	E
5	ABCD	E

4   Results and Discussion

4.1   Results

Effective feature is one of the most important factors determining classification performance. In this paper, multi-attribute EEG feature pool is constructed for automated epileptic seizure detection. In this work, a total of five cases are considered for automated epileptic seizure detection, as shown in Table 3.

mRMR algorithm is used to select features for each case, and the classification performance of each case is shown in Table 4. Wilcoxon rank sum test is used to detect whether there were significant difference in GVIX of wavelet coefficients and original time-series signals between set E and control groups (all $P < 0.001$). Moreover, Fig. 5 shows average performance comparison between the multi-attribute EEG feature pool with and without GVIX in each case. It can be found that adding GVIX into the multi-attribute EEG feature pool can achieve better performance in automated epileptic seizure detection.

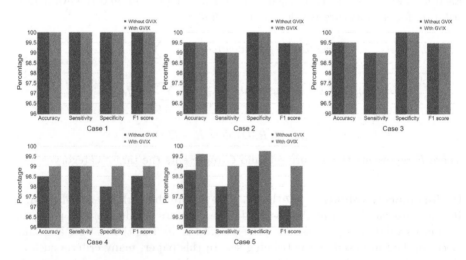

Fig. 5. Average performance comparison with and without adding GVIX into the multi-attribute EEG feature pool.

Table 4. Classification performance for each case

Case	Accuracy (%)	Sensitivity (%)	Specificity (%)	F1 score (%)
1	100	100	100	100
2	99.5	99.0	100	99.47
3	99.5	99.0	100	99.47
4	99.0	99.0	99.0	99.0
5	99.6	99.0	99.75	99.0

Results of Fig. 5 also show that case 1 has the best classification performance, whereas case 4 has the worst classification performance among all the cases. This finding is due to the data in case 1 being either healthy or epileptic; conversely, the data in case 4 is either interictal or ictal. Results of our study indicate that our method performs well in automated epileptic seizure detection. Moreover, a comparison of weight of each feature in different cases is shown in Fig. 6. It can be found that there were more significant differences in GVIX, ShEn, SampEn, SD, IQR and VIX between the two groups due to the calculation of the weight is based on mutual information. Results of our study obviously show that our proposed GVIX is a prominent time-frequency-domain feature in automated epileptic seizure detection. In addition, the accuracy comparison with other previous methods is discussed in the next subsection.

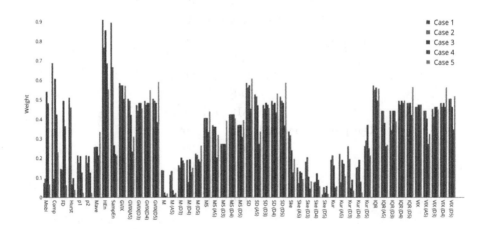

Fig. 6. Comparison of weight of each feature in different cases.

4.2   Comparison with Other State-of-the-Art Results

Table 5 shows a summary of studies that other current methods to epileptic seizure detection using the same data used in this work. It can be clearly seen

Table 5. Comparison between our method and other current methods

Authors	Cases	Accuracy (%)
Nicolaou [26]	A-E	93.55
	B-E	82.88
	C-E	88
	D-E	79.94
Tawfik [27]	A-E	98.5
	B-E	85
	C-E	93.5
	D-E	96.5
Fu [28]	A-E	99.85
	ABCD-E	98.8
Swami [25]	A-E	100
	B-E	98.89
	C-E	98.72
	D-E	93.33
	ABCD-E	95.24
Jaiswal [29]	A-E	99.3
	B-E	95.65
	C-E	97.79
	D-E	94.77
	ABCD-E	96.57
Mursalin [5]	A-E	100
	B-E	98
	C-E	99
	D-E	98.5
	ABCD-E	97.4
Our proposed method	A-E	100
	B-E	99.5
	C-E	99.5
	D-E	99
	ABCD-E	99.6

from Table 5 that it is the most difficult to identify epileptic seizure in case 4 and case 5. Moreover, results show that our method is comparable.

In case 1, this method shows the best performance and completely separates the normal EEG from the ictal EEG. The same results were also obtained in the works of [5] and [25]. Mursalin et al. [5] achieved automated epileptic seizure detection using improved correlation feature selection (ICFS) algorithm and random forest. Swami et al. [25] achieved epileptic seizure detection using dual-tree complex wavelet transform (DTCWT), entropy-based features and a general regression neural network (GRNN).

In case 2, this method achieves a classification accuracy of 99.5%, and it is the best compared with other studies. In case 3, classification accuracy of 99.5% is obtained in this work. In case 4, this method achieves 99% classification accuracy, which is the best compared with other current methods. Classification has also been achieved in the works of [26] and [27]. Nicolaou et al. [26] achieved epileptic seizure detection using Permutation Entropy (PeEn) and SVM. Tawfik et al. [27] achieved epileptic seizure detection using weighted PeEn and SVM.

In case 5, the classification accuracy obtained in this study is 99.6% which is higher than those obtained with other state-of-art methods. The best performance of these previous studies has been obtained in the work of [28], in which 98.8% classification accuracy is achieved using Hilbert marginal spectrum (HMS) and SVM. Similar works have been done in [29]. In [29], local neighbor descriptive pattern (LNDP) and SVM are used for epileptic seizure detection.

5 Conclusion

Finding prominent features and correctly detecting epileptic seizure is crucial in diagnosing and curing epilepsy. The main contribution of this work is automated epileptic seizure detection using a novel method. In this paper, a novel time-frequency-domain feature GVIX that can measure the holistic signal fluctuation in wavelet coefficients and original time-series signals is proposed. Time-frequency-domain features, time-domain features, nonlinear features, and entropy-based features are used to construct the multi-attribute EEG feature pool and feed them to mRMR algorithm. Results of this study show that GVIX is a prominent feature in automated epileptic seizure detection. We further compare our findings to other studies, and comparison results show that our proposed method is comparable. It can be concluded that using this proposed method would help clinicians make more efficient and reasonable decisions in the automatic detection of epileptic seizure.

Acknowledgements. This work was supported by the National Natural Science Foundation of China (Grant No. 61671220).

References

1. Fisher, R.S., et al.: ILAE official report: a practical clinical definition of epilepsy. Epilepsia **55**(4), 475–482 (2014)
2. Chang, B.S., Lowenstein, D.H.: Epilepsy. N. Engl. J. Med. **349**(13), 1257–1266 (2003)
3. de Boer, H.M., et al.: The global burden and stigma of epilepsy. Epilepsy Behav. **12**(4), 540–546 (2008)
4. Anand, V., et al.: Pediatric decision support using adapted Arden Syntax. Artif. Intell. Med. **92**, 15–23 (2018)

5. Mursalin, R.S., et al.: Automated epileptic seizure detection using improved correlation-based feature selection with random forest classifier. Neurocomputing **241**(C), 204–214 (2017)
6. Martis, R.J., et al.: Application of intrinsic time-scale decomposition ITD to EEG signals for automated seizure prediction. Int. J. Neural Syst. **23**(5), 1350023 (2013)
7. Acharya, U.R., et al.: Application of non-linear and wavelet based features for the automated identification of epileptic EEG signals. Int. J. Neural Syst. **22**(2), 1250002 (2012)
8. Acharya, U.R., et al.: Automated diagnosis of epileptic EEG using entropies. Biomed. Signal Process. Control **7**(4), 401–408 (2012)
9. Song, Y., et al.: A new approach for epileptic seizure detection: sample entropy based feature extraction and extreme learning machine. J. Biomed. Sci. Eng. **3**(6), 556–567 (2010)
10. Acharya, U.R., et al.: Automatic detection of epileptic EEG signals using higher order cumulant features. Int. J. Neural Syst. **21**(5), 403–411 (2011)
11. Ibrahim, S., et al.: Electroencephalography (EEG) signal processing for epilepsy and autism spectrum disorder diagnosis. Biocybern. Biomed. Eng. **38**(1), 16–26 (2017)
12. Guo, L., et al.: Epileptic seizure detection using multiwavelet transform based approximate entropy and artificial neural networks. J. Neurosci. Methods **193**(1), 156–163 (2010)
13. Zarei, R., et al.: A PCA aided cross-covariance scheme for discriminative feature extraction from EEG signals. Comput. Methods Programs Biomed. **146**, 47–57 (2017)
14. Andrzejak, R.S., et al.: Indications of nonlinear deterministic and finite dimensional structures in time series of brain electrical activity: dependence on recording region and brain state. Phys. Rev. E Stat. Nonlinear Soft Mater. Phys. **64**, 061907 (2001)
15. Jahankhani, P., et al.: Signal classification using wavelet feature extraction and neural networks. In: IEEE John Vincent Atanasoff 2006 International Symposium on Modern Computing, pp. 120–124 (2006)
16. Gerrard, P., Malcolm, R.: Mechanisms of modafinil: a review of current research. Neuropsychiatr. Dis. Treat. **3**(3), 349–364 (2007)
17. Subasi, A., et al.: EEG signal classification using wavelet feature extraction and a mixture of expert model. Expert Syst. Appl. **32**(4), 1084–1093 (2007)
18. Abdi, H., Williams, L.J.: Principal component analysis. Wiley Interdisc. Rev. Comput. Stat. **2**(4), 433–459 (2010)
19. Cecchin, T., et al.: Seizure lateralization in scalp EEG using Hjorth parameters. Clin. Neurophysiol. **121**(3), 290–300 (2010)
20. Hurst, H.E.: Long term storage capacity of reservoirs. Am. Soc. Civ. Eng. **116**(12), 770–808 (1951)
21. Shannon, C.E.: A mathematical theory of communication. Bell Syst. Tech. J. **27**(3), 379–423 (1997)
22. Richman, J.S., Moorman, J.R.: Physiological time-series analysis using approximate entropy and sample entropy. Am. J. Physiol. Heart Circ. Physiol. **278**(6), 2039–2049 (2000)
23. Yentes, J.M., et al.: The appropriate use of approximate entropy and sample entropy with short data sets. Ann. Biomed. Eng. **41**(2), 349–365 (2013)
24. Peng, L., et al.: Feature selection based on mutual information criteria of max-dependency, max-relevance, and min-redundancy. IEEE Trans. Pattern Anal. Mach. Intell. **27**(8), 1226–1238 (2005)

25. Swami, P., et al.: A novel robust diagnostic model to detect seizures in electroencephalography. Expert Syst. Appl. **56**(C), 116–130 (2016)
26. Nicolaou, N., Georgiou, J.: Detection of epileptic electroencephalogram based on permutation entropy and support vector machines. Expert Syst. Appl. **39**(1), 202–209 (2012)
27. Tawfik, N., et al.: A hybrid automated detection of epileptic seizures in EEG records. Comput. Electr. Eng. **53**, 177–190 (2015)
28. Fu, K., et al.: Hilbert marginal spectrum analysis for automatic seizure detection in EEG signals. Biomed. Signal Process. Control **18**, 179–185 (2015)
29. Jaiswal, A.K., Banka, H.: Local pattern transformation based feature extraction techniques for classification of epileptic EEG signals. Biomed. Signal Process. Control **34**, 81–92 (2017)

An Approach for Semantic Data Integration in Cancer Studies

Iliyan Mihaylov, Maria Nisheva-Pavlova$^{(\boxtimes)}$, and Dimitar Vassilev

Faculty of Mathematics and Informatics, Sofia University St. Kliment Ohridski,
5 James Bourchier Blvd., 1164 Sofia, Bulgaria
{mihaylov,marian,dimitar.vassilev}@fmi.uni-sofia.bg

Abstract. Contemporary development in personalized medicine based both on extended clinical records and implementation of different high-throughput "omics" technologies has generated large amounts of data. To make use of these data, new approaches need to be developed for their search, storage, analysis, integration and processing. In this paper we suggest an approach for integration of data from diverse domains and various information sources enabling extraction of novel knowledge in cancer studies. Its application can contribute to the early detection and diagnosis of cancer as well as to its proper personalized treatment.

The data used in our research consist of clinical records from two particular cancer studies with different factors and different origin, and also include gene expression datasets from different high-throughput technologies – microarray and next generation sequencing. An especially developed workflow, able to deal effectively with the heterogeneity of data and the enormous number of relations between patients and proteins, is used to automate the data integration process. During this process, our software tool performs advanced search for additional expressed protein relationships in a set of available knowledge sources and generates semantic links to them. As a result, a set of hidden common expressed protein mutations and their subsequent relations with patients is generated in the form of new knowledge about the studied cancer cases.

Keywords: Data integration · Ontology · Linked data ·
Knowledge extraction · Cancer studies

1 Introduction

Data integration is one of the challenges of contemporary data science with great impact on different practical and research domains. Data generated in medicine from both clinical and omics high-throughput sources contribute to changes in data storage and analytics and to a number of related bioinformatics approaches aiming at better diagnostics, therapy and implementation of personalized medicine [1, 2].

These integrative efforts for research and therapy generate a huge amount of raw data that could be used to discover new knowledge in the studied domains. The extraction of new knowledge is a complex and labor-intensive task with many components – different data sources may have the same attributes but with different

J. M. F. Rodrigues et al. (Eds.): ICCS 2019, LNCS 11538, pp. 60–73, 2019.
https://doi.org/10.1007/978-3-030-22744-9_5

semantics. The used data sources are also heterogeneous: each of them has its own structure and its own data format.

In this paper we present a study of neuroblastoma and breast cancer. These cancers are a great threat for children and women respectively. Breast cancer concerns approximately one in eight women over their lifetime [3], whilst neuroblastoma is the most common cancer in less than one year old children. It accounts for about 6% of all cancers in children [4]. Both cancer datasets vary strongly for each particular case study.

It is essential to discover relations between these two cancers, based on an effective mapping of common mutated proteins. Such an approach can present classes of proteins related to both diseases, which can serve as a basis for more accurate and well annotated discovery of potential molecular markers in cancer studies. From a data science point of view this set of problems can be tackled by an approach based on semantic data integration [5, 6].

Data integration is understood as a mean to combining data from different sources, creating a unified view and improving their accessibility to a potential user [3, 4]. Data integration and biomedical analysis are separate disciplines and have evolved in relative isolation. There is a general agreement that uniting both these disciplines in order to develop more sustainable methods for analysis is necessary [7, 8]. Data integration fundamentally involves querying across different data sources. These data sources could be, but are not limited to, separate relational databases or semi-structured data sources distributed across a network. Data integration facilitates dividing the whole data space into two major dimensions, referring to where data or metadata or knowledge reside and to the representation of data and data models. Biomedical experiments [9] take advantage of a vast number of different analytical methods that facilitate mining relevant data from the dispersed information. Some of the most frequent experiments are related to gene expression profiling, clinical data analytics [10], rational drug design [6], which attempt to use all available biological and clinical knowledge to make informed development decisions. Moreover, machine learning-based approaches for finding and highlighting the useful knowledge in the vast space of abundant and heterogeneous data are applied for improving these analytics. Metadata, in particular, is gaining importance, being captured explicitly or inferred with the aid of machine learning models. Some examples include the use of machine learning methods in the inference of data structure, data distribution, and common value patterns.

The heterogeneity of data makes any integrative analysis highly challenging. Data generated with different technologies include different sets of attributes. Where data is highly heterogeneous and weakly related, two interconnected integrative approaches are applied: horizontal and vertical integration (Fig. 1). The horizontal data integration unites information of the same type, but from different sources and in different formats. It helps to unite heterogeneous data from many different sources in one data model. The vertical integration has a potential to relate different kinds of information, helping for example to manage links between the patient, gene expression, clinical information, chemical knowledge and existing ontologies, e.g., via web technologies [11–13]. Most existing approaches for data integration are focused on one type of data or one disease and cannot facilitate cross-type or disease integration [1, 2].

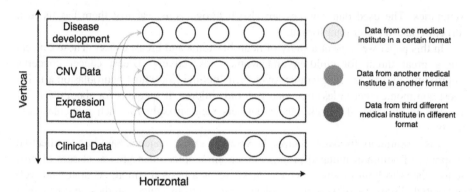

Fig. 1. Horizontal and vertical data integration. Grey arrows show the relations between the used types of data (clinical, expression, CNV and disease development). The horizontal arrow shows a flow of integration of all provided data sources, like medical institutes. The vertical arrow shows a potential to link all existing different types of data.

The main objective of this paper is to present a novel efficient data integration model for the studied cancer cases by data mining and knowledge extraction approaches which can find relationships between certain data patterns, related to mutated proteins, expression and copy number variation (CNV). Our approach utilizes NoSQL databases, where we combine clinical and expression profile data, using both raw data records and external knowledge sources.

2   Problem Description

Intelligent exploitation of large amounts of data from different sources, with different formats, and different semantics is among the most important challenges especially in the biomedical area [14]. Life sciences and in particular medicine and medical research generate a lot of such data due to recent developments of high-throughput molecular technologies and large clinical studies [2]. The major challenge here is to integrate, analyze and interpret this data in the scope of contemporary personalized medicine. Personalized medicine (PM) has the potential to tailor therapy and to ensure adequate patient care. By enabling each patient to receive early diagnoses, risk assessments, and optimal treatments, PM holds promise for improving health care while also lowering costs. The information background of the PM is a key element for its successful implementation. This information background is based on semantically reach, accurate and precisely analyzed bio-medical data [15].

The circle of problems in our study can be described in the scope of a successful analysis and integration of the massive datasets, now prevalent in the medical sciences – more precisely, integration of these datasets with linked data sources to find additional relations between proteins and clinical attributes for the two studied diseases. The challenge is to provide a method based on Linked Data and open source technologies [16, 17] to combine knowledge from many existing open sources for efficient integration of raw libratory data. Raw libratory data that can eventually be integrated for the comprehensive elucidation of complex phenotypes include functional gene annotations, gene expression profiles, proteomic profiles, DNA polymorphisms, DNA copy number variations, epigenetic modifications, etc. [18].

The general challenges in our study are based on the use of different data: clinical data, RNA-Seq and microarray gene expression data, and CNV from comparative genome hybridization (aCGH) [19]. The solution of these challenges is to demonstrate and examine the power of semantic data integration and knowledge extraction for the purposes of real world clinical settings in breast cancer (BC) and neuroblastoma (NB).

A specific challenge in the study is to integrate and analyze sets of unbalanced and non-structured data. The molecular data is in raw format with all fields and attributes generated from the sequencing or microarray technology. Before starting the integration process, it is necessary to perform some preprocessing operations on the raw formats and to generate an appropriate new data structure. We are working with datasets, rich in relations, and it is essential to be able to find many annotations for the existing relations which will help one to enhance the set of relations by proper resources from the available knowledge bases.

Solutions based on semantic integration of data have already been successfully applied to cancer datasets to find driver proteins and pathways [20]. We chose an approach based on semantic integration because most features of our data have different semantics for each patient, which is an essential background for personalized medicine. The expected results of such type of approach include identification of hidden protein subtypes distinguished by common patterns of network alteration and a predictive model for cancer development based on the knowledge about joined proteins.

3 Data Description

The raw data in each studied dataset are in a specific format and have specific semantics. A field (an attribute) in each dataset has different meanings due to the technologies and the subsequent recording. The provided data by itself also contain information for mutated proteins, expression and CNV.

The initial point for transformation, grouping and integration are the patients/ sample files. The generated record for each particular patient contains attributes like age, gender, nationality, etc. Two datasets – neuroblastoma (NB) and breast cancer (BC), are used in this study. The neuroblastoma set contains RNA-Seq gene expression profiles of 498 patients as well as Agilent microarray expression and aCGH data for a matched subset of 145 patients and corresponding clinical information. The breast cancer set contains profiles for microarray and copy number data, and clinical information (survival time, multiple prognostic markers, therapy data) for about 2,000 patients.

Each patient record contains many files with expression, mutation and CNV data shown on Fig. 2. For example, each record from expression files refers to another file with detailed information. It includes 100–200 sample records for one expression and contains also information about the mutation type, expressed chromosome, expression start position in the DNA, expression end position. All properties were generated by the used sequencing technology. A mutation file contains proteins, their attributes and reference to the expression file with detailed information. The relationships patient – protein expression and patient – protein mutation are fundamentally different. A patient who has an expressed protein may not have the same protein mutated.

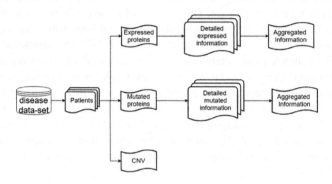

Fig. 2. General structure of raw data.

Both datasets contain some additional files for medical patient stability, meta clinical information, meta expression/mutation information, etc.

4   Related Work

Data integration is a real challenge for querying across multiple autonomous and heterogeneous data sources [5, 9]. It is crucial in the medical and health sector that use multiple sources for producing datasets and is of great importance for their subsequent analysis for study, diagnostics, and therapy purposes [11–13].

A major objective of data integration is to enable the use of data with implicit knowledge and to aid the use of the integrated sources to answer queries [6, 21].

In medical studies and in particular in cancer studies, there are several quite different and heterogeneous kinds of data that need to be integrated: clinical data, medical check data, various types of molecular information [2]. All these massively emerging amounts of data, accompanied with the new requirements for the purposes of preventive and personalized medicine, emphasize the great importance of data integration in cancer studies [10].

The semantic side of data integration in cancer studies gives a large horizon for using related data with different meaning and structure and makes possible to extract unknown knowledge about various aspects of the studied cancer case(s) [15]. The semantic integration of data from different types of cancer studies challenges the problem how to use all provided data aside with the opportunity to make proper associations of clinically controlled parameters and information based on "omics" data profiles [19, 22].

Technically the integration of cancer studies data can be supported by development of workflows and systems, capable for fast and accurate integration of disparate cancer data and enhanced models for cancer data analytics. Such systems allow constructing and executing queries at a conceptual level and in a way utilizing the available rich in semantics cancer information. Some existing tools for semantic data integration in cancer studies are based on utilization of external domain knowledge sources like Gene Ontology (GO). There are several popular RDF-based data integration platforms: Reactome [23], BioModels [24], BioSamples [25], Expression Atlas [5], ChEMBL [9], UniProt RDF [11]. These platforms can be used to search across datasets. For example, a query for gene expression data will integrate results from Expression Atlas with relevant pathway information from Reactome and compound-target information from ChEMBL. The structured data are available for download or can be queried directly.

This long list of platforms and resources can be used as good examples of linking various sources of domain knowledge. Our methodology goes much further aiming at the extraction of new knowledge from the results of integration of raw, unstructured and heterogeneous data.

5 Main Characteristics and Novelty of the Proposed Approach

The approach to data integration we suggest in this paper was originally oriented to a particular application – to unite data from real studies and treatments of neuroblastoma and breast cancer – but its design characteristics make it sufficiently generic and applicable in a wide range of subject areas. As a result of its application, different datasets are joined and the semantic integrity of the data is kept and enriched. In our particular case, through combining data from multiple sources (Fig. 3) we create a new network of data where entities, like proteins, clinical features and expression features, are linked with each other [26]. In this network, nodes represent patients and edges represent similarities between the patients' profiles, consisting of clinical data, expression profiles and CNV data. Such a network can be used to group patients and to associate these groups with distinct features. The main challenges here are: (1) building an appropriate linked data network, discovering a data model semi-structure [14] and mapping assertions by the applied model for data integration [27]; and (2) data cleaning, combined into a formal workflow for data integration.

We focus on two aspects of data integration: horizontal and vertical. As explained, horizontal data integration means combining data from different sources for the same entity. Entities can be identified in our particular datasets as clinical data, patients, expression profiles and CNV. Each kind of data is measured by a specific technology and available in various data formats. Groups of entities are semantically similar. Vertical data integration, on the other hand, is applied to creating relations between all horizontally integrated objects. This vertical data integration provides a connection between all different types of entities. This connection, in our case, covers relations between patients, expression profiles, clinical data and CNV data. Based on these relations we can easily detect all patients closely related to each other by protein mutations, diagnosis and therapy. We used different databases for horizontal and for vertical data integration. These different databases are required because horizontal and vertical data integration address different aspects of the integration problem. Data for the horizontal data integration are unstructured and heterogeneous. Thus, we use a document-oriented database, which can handle different data types and formats. For vertical data integration a graph database is used, as it is suitable for representing relations – crucial in this case. In this study, all relations are established between existing records for each entity, and represented by a semi-structure.

An integration model over a NoSQL database can potentially unite medical studies data, alternatively to the most frequently used statistical/machine learning methods. Most NoSQL database systems share common characteristics, supporting scalability, availability, flexibility and ensuring fast access times for storage, data retrieval and analysis [9, 11]. Very often when applying cluster analysis methods for grouping or joining data issues, small classes occur – mainly with outliers and mostly with data dynamically changing their relatedness. Applying our integration model we do believe that all these problems can be overcome. Moreover, we can extend the potential of the model by using multiple datasets, regardless of the level of heterogeneity, particular formats, types of data, etc. – all very specific for cancer studies.

6 Suggested Methodology

Our methodology for integration of unstructured data from the studied samples is based on the proper use of schema-less databases and domain ontologies like the Gene Ontology (GO) [28]. The data we are manipulating contain hidden relationships between the proteins provided from different patients within studies of both diseases (BC and NB). We use all available information about already built relationships in our data sources and try to find additional information in some third party sources to attain semantic integration of the data. Thus, step-by-step, we develop a network, which combines protein relations between patients and diseases. The challenge here is to store all relationships with their cycle dependencies. The latter are possible because one patient has relationships with mutated protein(s), mutated protein(s) has/have reference (s) to expression and other patients have references to the same protein(s).

In each disease (BC and NB) every patient has a different set of mutated or expressed proteins. Only small sets of mutated proteins are equal and exist in each patient. All proteins belong to families, which contain many related proteins. By application of semantic annotation and search techniques we aim to find and combine all proteins which are semantically related to the studied diseases. So, we can aggregate and discover all needed information for an enhanced number of related proteins.

The role of GO in our methodology is to provide a controlled vocabulary for annotating homologous gene and protein sequences in the studied cancers. GO classifies genes and gene products on the base of three hierarchical structures that describe a given entry's biological processes, cellular components and molecular functions, and organizes them into a parent-child relationship [22]. For the purposes of our study, an essential key are protein families and relationships between them. We associate all this information with the inferred relationships between patients and proteins to provide a complete schema of mutated and related proteins for each studied patient. In this way we demonstrate the possibilities of using appropriate subject knowledge bases (like GO) for the purpose of semantic integration of data from different sources. As a result we build a base of semantically linked data from biomedical research and create a framework supporting the practical extraction of new knowledge in life sciences and other significant areas (Fig. 3).

We developed a workflow for data integration in order to overcome the heterogeneity of the data and the enormous number of relations between the studied patients and proteins. First of all we are aiming to integrate datasets from the two cancer studies. In this line we are trying to find some relations within the given data as well as to generalize some information about commonly related proteins. This process invokes the semantic integration of the data, which is a key part of our study.

During the analysis of raw data we create a sort of "semi-structure" of the data – a structure containing only attributes, existing in each record. In semi-structured data, the entities belonging to the same class (protein mutated, expressed and CNV) may have different attributes even though they grouped together, and the attributes' order is not important. Semi-structured data is becoming more and more prevalent, e.g. in structured documents and when performing simple integration of data from multiple sources. Traditional data models and query languages are inappropriate, since semi-structured data often is irregular: some data is missing, similar concepts are represented using different types, heterogeneous sets are present, or object structure is not entirely known [13].

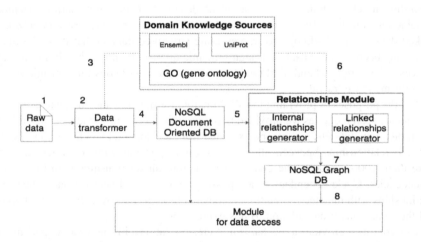

Fig. 3. Workflow of semantic data integration.

Our workflow covers eight stages (Fig. 3). The first two stages correspond to data generated from different high-throughput technologies – microarray and RNA-Seq data (1), preprocessed (2) for databases generation. In this step we aim to complete the missing data (3), to define data for a patient, to generate the "semi-structured data" and to store it. Raw and "semi-structured" data are stored in a document-oriented NoSQL database (MongoDB) with all attributes (4). At the next stage (5) we try to find relationships between proteins, patients and diseases. For the enrichment of our dataset we try to find additional protein relationships (6) in a set of external domain knowledge sources and to generate semantic links to them. These enriched protein relations are traced with other proteins and the discovered relations are explored. We use in this case the Neo4J database. In this graph database we store only relationships between entities (7). Each relation contains a "semi-structure" with ID from MongoDB. Module 8 is intended for user access to data.

Starting point (1) in our workflow on Fig. 3 is the raw data entry, read in different formats (CSV, TXT, XML, etc.). We analyze each record and create a general data structure ("semi-structure"). The meaning of this semi- structure plays a key role in the semantic integration of data. The process of generating the semi-structure is dynamically changed after reading of each record. This parallelized process finishes by reading of all records. Data from different sequencing technologies has different attributes for proteins. In order to unify these attributes, we use several external domain knowledge sources (EDKS): Ensembl [21], UniProt [29], and GO, which provide additional knowledge about the existing annotated proteins. We search for proteins and their annotations in these EDKS by generating a request by URL. This URL is stored in the protein records in the document-oriented NoSQL database, which is schema-less and can store both structured and unstructured data of the same record. Each record contains different number of attributes.

The next step is to find relationships and to connect our data to other linked data from the explored EDKS in order to generate more relationships between the proteins. According to [16], "linked data is a set of design principles for sharing machine-readable data on the WWW for use by public administrations, business and citizens". Linked data is a methodology for representing structured data so that it can be inter-linked and become more useful through semantic queries. It allows data from different sources to be combined and used together when executing queries for semantic search and information retrieval.

For the purposes of providing tools for flexible formulation and adequate execution of semantic search queries we provided two types of relationships: ones, extracted from the studied raw datasets, and others, extracted from the domain knowledge sources like GO. All relationships between all entities (proteins) in the semi-structured data are more than 1,000,000,000. The goal here is to store all relationships with their cycle dependencies. Cycle dependencies are possible and expected because one patient has relationships with mutated protein(s), the mutated proteins have reference to expression and the same patient has relation to this expression.

Our approach provides a mechanism to create relationships between all patients on the basis of the mutations, expressions and CNV. We use a graph database (Neo4j) to manage and store all known relationships. In Neo4j we insert only the "semi-structure" for each related entity. This "semi-structure" contains a 128-bit identifier, and by using it we create and manage the relations between MongoDB and Neo4j. This is necessary also because full data attributes are stored in MongoDB which cannot provide such relationship management.

After relationships are generated, we generate the new relations via Ensembl and UniProt using GO, because GO by itself does not provide suitable programmable access. These sources contain information about proteins, protein families and relationships between them. We take this information and store the links of similar proteins as relationships in our database (Neo4j). This approach provides a method for creation of co-relationships between proteins. The relations generated by dint of the used external sources are assigned as unreliable. They are considered at this stage as unreliable because in deep search all proteins by one or another way are related. In the working process, if one unreliable relationship is "accessed" by the same proteins multiple (more than 10) times, such relationship is transformed as a normal (or trusted) one [30]. Therefore, we dynamically create reliable relationships between proteins in the workflow. For example, we use the "similar proteins" section of UniProt to extract the related proteins. This section provides links to proteins that are similar to the protein sequence(s) described in the search query at different levels of sequence identity thresholds (100%, 90% and 50%) based on their membership in UniProt Reference Clusters.

At the final step, we create a workflow module for sending queries to the involved databases. So the workflow can produce pathways, predict relationships between proteins, patients and diseases and thus discover new knowledge about the protein relationships based on the semantic integration of data and external domain knowledge sources. Our approach enables users to formulate and execute a wide spectrum of dynamically constructed semantic search queries.

7   Results and Discussion

An essential part of our methodology is based on the use of the validation mechanisms provided by MongoDB. The latter are applied to create a specific "validation filter" for all attributes in the semi-structure. In our case these filters are set with minimal level which guarantees that there will be a value for each attribute and thus our semi-structure cannot contain empty attribute fields. On the other hand, our approach is consistent with the fact that document-oriented databases are not appropriate for storing such related data. For storage and management of all relationships we use a graph database – Neo4j. Graph databases (GDB) provide a suitable environment for developing and managing relations between the entities (proteins, patients). GDB have native solutions for management of complex cycling dependencies (relations). As already mentioned, our data imply many cycling references between patients and proteins and it is necessary to have a trusted path for each relation between patient and protein. Methodologically this problem is solved in graph theory by trusted relational trees [31], and we use this solution with Neo4j.

It is possible for each relation in Neo4j to insert additional information, containing the relation type, IDs etc. We use this to connect our semi-structured data to the linked data via dynamically generated URLs, which refer to external domain knowledge sources for more detailed annotations (protein, protein family).

The mentioned EDKS are accessed via specific APIs based on HTTP/S protocol realized by RESTFul methodology – a software architectural style that defines a set of constraints to be used for creating Web services. For example, for RERE protein (a protein which is related to apoptosis triggering) we generate the following URL [12]: https://www.ebi.ac.uk/proteins/api/proteins?offset=0&size=100&gene=RERE.

Based on the answer from the EDKS we dynamically create new relationships between all references of the searched protein. This enhances the network where indirectly all proteins are related. Such type of indirectly generated relationships has a small score in the query answer. Initially they are accounted as untrusted. Their score is generated dynamically and depends on the number of requests. Based on this score we rank the respond to the user who has posed the request. Automatically these relationships become trusted after a certain level of score value (Fig. 4).

Obviously the used EDKS have some limitations. GO provides no interface for programming access to it and the only way to use directly the ontology is to download it. That is why we use Ensembl to access GO as an internal resource. Initially, we create a request to EBI. The returned information from EBI contains reference IDs to other knowledge sources. We use the particular reference ID for GO and create a request to Ensembl for getting ontology-based relationships for the requested protein. For tracing the relationships, we build the following request:

https://rest.ensembl.org/ontology/ancestors/GO:0003677?content-type=application/json.

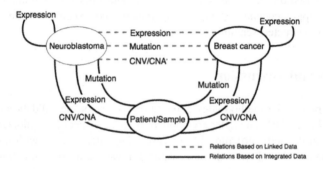

Fig. 4. Relationships between the proteins.

Here GO:0003677 is the reference ID from the first request to EBI. If information about this protein does not exist in GO, the same procedure is repeated to find it in another database (InterPro), etc. As a consequence of such intelligent integration of the two studied cancer datasets, we discover new knowledge about the mutated proteins, related to those two types of cancer. We suppose that following this approach, based on semantic data integration, it is possible to discover a unique set of proteins and their functional annotations for a particular cancer.

As an example for the provided opportunities, we can demonstrate how to find relations between the proteins RERE, EMP2 and KLF12. These three proteins are related to neuroblastoma and breast cancer. In both diseases they are mutated. To find other related proteins we create a request to our system using GraphQL – a query language for APIs, and a server-side runtime for managing and executing typified queries. The result is illustrated on Fig. 5 where proteins are shown as circles (graph nodes). The size of each node is based on its resulting score. Proteins are connected by different types of relationships shown on the figure in different shades of grey and different style.

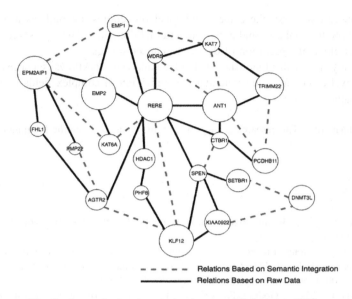

Fig. 5. Semantic data integration by use of linked data URLs.

Found proteins have two types of relations. If the relation is generated from linked data (as URLs), it is classified as non-reliable. Many relations are generated from linked data in external sources and marked by a score. This score is generated dynamically and depends on the number of requests to our system. Based on this score we rank the response to the user who posed the request. If a found relation is built on the raw dataset then it has a higher score than the ones built from linked data. The score of such relation is increasing and after a number of requests it can be classified as a trusted one. In such manner our workflow enables automated enrichment of protein relationships and extraction of new knowledge related to both cancers.

8 Conclusions

In this paper we discuss an original methodology for extraction of hidden relations in an integrated dataset by combining data from disparate sources and consolidating them into a meaningful and valuable information pool by the use of semantic technologies. The use of linked and overplayed NoSQL database technologies allowed us to aggregate the non-structured, heterogeneous cancer data with their various relationships. The applied semantic integration of different cancer datasets has an obvious merit concerning the enrichment of the studied data by discovery of mutual internal relations and relations with external domain knowledge sources.

A novel approach for automated semantic data integration has been proposed and analyzed. It provides means, supporting augmented and precise enough search for hidden common (protein) relations. The discovery of these hidden common proteins and joining their functionality is in fact an extraction of new knowledge about the studied cancer cases. All these methodological procedures are built as a workflow, based on NoSQL databases and exploring external domain knowledge sources for the purposes of efficient integration of data from cancer studies.

This study shows also that using customized analysis workflows is a necessary step towards novel discoveries and potential generalization in complex fields like personalized therapy.

Acknowledgements. This research is supported by the National Scientific Program eHealth in Bulgaria.

References

1. Stein, L.: Creating a bioinformatics nation. Nature **417**, 119–120 (2002). https://doi.org/10.1038/417119a
2. Kashyap, V., Hongsermeier, T.: Can semantic web technologies enable translational medicine? In: Baker, C.J.O., Cheung, K.H. (eds.) Semantic Web, pp. 249–279. Springer, Boston (2007). https://doi.org/10.1007/978-0-387-48438-9_13
3. DeSantis, C., Ma, J., Goding, S., Newman, L., Jemal, A.: Breast cancer statistics, 2017, racial disparity in mortality by state. Cancer J. Clin. **67**(6), 439–448 (2017)
4. American Cancer Society: Cancer Statistics Center. http://cancerstatisticscenter.cancer.org. Accessed 23 Mar 2019
5. Kapushesky, M., et al.: Gene expression atlas update—a value-added database of microarray and sequencing-based functional genomics experiments. Nucleic Acids Res. **40**(D1), D1077–D1081 (2012)
6. Lenzerini, M.: Data integration: a theoretical perspective. In: PODS 2002, pp. 233–246 (2002)
7. Sioutos, N., et al.: A semantic model integrating cancer-related clinical and molecular information. J. Biomed. Inform. **40**(1), 30–43 (2007)
8. Louie, B., et al.: Data integration and genomic medicine. J. Biomed. Inform. **40**(1), 5–16 (2007)
9. Gaulton, A., et al.: ChEMBL: a large-scale bioactivity database for drug discovery. Nucleic Acids Res. **40**(D1), D1100–D1107 (2012)
10. Ruttenberg, A., et al.: Advancing translational research with the semantic web. BMC Bioinform. **8**(Suppl. 3), S2 (2007)
11. Nicole, R.: UniProt in RDF: Tackling Data Integration and Distributed Annotation with the Semantic Web. Nature Precedings (2009). http://dx.doi.org/10.1038/npre.2009.3193.1
12. Jupp, S., et al.: The EBI RDF platform: linked open data for the life sciences. Bioinformatics **30**(9), 1338–1339 (2014)
13. Beeri, C., Milo, T.: Schemas for integration and translation of structured and semi-structured data. In: Beeri, C., Buneman, P. (eds.) ICDT 1999. LNCS, vol. 1540, pp. 296–313. Springer, Heidelberg (1999). https://doi.org/10.1007/3-540-49257-7_19
14. Zhang, H., Guo, Y., et al.: Data integration through ontology-based data access to support integrative data analysis: a case study of cancer survival. In: Proceedings of IEEE International Conference on Bioinformatics Biomed, pp. 1300–1303 (2017)
15. Pittman, J., et al.: Integrated modeling of clinical and gene expression information for personalized prediction of disease outcomes. PNAS **10**(22), 8431–8436 (2004)
16. Berners-Lee, T.: Linked Data (2006). http://www.w3.org/DesignIssues/LinkedData.html. Accessed 23 Mar 2019

17. Oren, E., et al.: Sindice.com: a document-oriented lookup index for open linked data. J. Metadata Semant. Ontol. **3**(1), 37–52 (2008)
18. Famili, F., Phan, S., Fauteux, F., Liu, Z., Pan, Y.: Data integration and knowledge discovery in life sciences. In: García-Pedrajas, N., Herrera, F., Fyfe, C., Benítez, J.M., Ali, M. (eds.) IEA/AIE 2010. LNCS (LNAI), vol. 6098, pp. 102–111. Springer, Heidelberg (2010). https://doi.org/10.1007/978-3-642-13033-5_11
19. Holford, M., et al.: A semantic web framework to integrate cancer omics data with biological knowledge. BMC Bioinform. **13**(Suppl. 1), S10 (2012)
20. Glaab, E., et al.: Extending pathways and processes using molecular interaction networks to analyse cancer genome data. BMC Bioinform. **11**, 579 (2010)
21. Yates, A., et al.: Ensembl 2016. Nucleic Acids Res. **44**(Database issue), D710–D716 (2016)
22. Zhang, H., et al.: Data integration through ontology-based data access to support integrative data analysis: a case study of cancer survival. In: IEEE International Conference on Bioinformatics and Biomedicine (BIBM), Kansas City, MO, pp. 1300–1303 (2017)
23. Fabregat, A., et al.: The reactome pathway knowledgebase. Nucleic Acids Res. **46**(Database issue), D649–D655 (2018)
24. Chen, L., et al.: BioModels database: an enhanced, curated and annotated resource for published quantitative kinetic models. BMC Syst. Biol. (2010). https://doi.org/10.1186/1752-0509-4-92
25. Gostev, M., et al.: The BioSample Database (BioSD) at the European bioinformatics institute. Nucleic Acids Res. **40**(D1), D64–D70 (2012)
26. Semantic Web Health Care and Life Sciences (HCLS) Interest Group. https://www.w3.org/blog/hcls/. Accessed 23 Mar 2019
27. Heath, T., Bizer, C.: Linked Data: Evolving the Web into a Global Data Space. Morgan & Claypool (2011)
28. The gene ontology consortium: gene ontology consortium: going forward. Nucleic Acids Res. **43**(D1), D1049–D1056 (2015)
29. The UniProt Consortium, Bateman, A., et al.: UniProt: the universal protein knowledgebase. Nucleic Acids Res. **45**(Database issue), D158–D169 (2017)
30. Golbeck, J., Parsia, B.: Trust network-based filtering of aggregated claims. Int. J. Metadata Semant. Ontol. **1**(1), 58–65 (2006)
31. He, P.: A new approach of trust relationship measurement based on graph theory. Int. J. Adv. Comput. Sci. Appl. **3**(2), 19–22 (2012)

A Study of the Electrical Propagation in Purkinje Fibers

Lucas Arantes Berg[1]([✉])[ID], Rodrigo Weber dos Santos[1][ID],
and Elizabeth M. Cherry[2]

[1] Federal University of Juiz de Fora, Rua José Lourenço Kelmer,
s/n - Campus Universitário, Juiz de Fora, MG 36036-900, Brazil
berg@ice.ufjf.br
[2] Rochester Institute of Technology,
One Lomb Memorial Drive, Rochester, NY 14623-5603, USA

Abstract. Purkinje fibers are fundamental structures in the process of the electrical stimulation of the heart. To allow the contraction of the ventricle muscle, these fibers need to stimulate the myocardium in a synchronized manner. However, certain changes in the properties of these fibers may provide a desynchronization of the heart rate. This can occur through failures in the propagation of the electrical stimulus due to conduction blocks occurring at the junctions that attach the Purkinje fibers to the ventricle muscle. This condition is considered a risk state for cardiac arrhythmias. The aim of this work is to investigate and analyze which properties may affect the propagation velocity and activation time of the Purkinje fibers, such as cell geometry, conductivity, coupling of the fibers with ventricular tissue and number of bifurcations in the network. In order to reach this goal, several Purkinje networks were generated by varying these parameters to perform a sensibility analysis. For the implementation of the computational model, the monodomain equation was used to describe mathematically the phenomenon and the numerical solution was calculated using the Finite Volume Method. The results of the present work were in accordance with those obtained in the literature: the model was able to reproduce certain behaviors that occur in the propagation velocity and activation time of the Purkinje fibers. In addition, the model was able to reproduce the characteristic delay in propagation that occurs at the Purkinje-muscle junctions.

Keywords: Computational electrophysiology · Purkinje fibers · Finite Volume Method

Supported by Coordenação de Aperfeiçoamento de Pessoal de Nível Superior (CAPES), Fundação de Amparo à Pesquisa do Estado de Minas Gerais (FAPEMIG), Conselho Nacional de Desenvolvimento Científico e Tecnológico (CNPq), Universidade Federal de Juiz de Fora (UFJF) and National Science Foundation under grant no. CNS-1446312.

J. M. F. Rodrigues et al. (Eds.): ICCS 2019, LNCS 11538, pp. 74–86, 2019.
https://doi.org/10.1007/978-3-030-22744-9_6

1   Introduction

Ischaemic heart diseases and strokes are still the main cause of deaths worldwide, counting up to approximately 15.2 million of deaths in 2016 [1]. Within this critical scenario it is expected that the cost in cardiac diseases surpass 1044 billions of dollares by 2030 [2].

Even with the recent advances in the treatment of these diseases, like arrhythmias and ischemias, research in this area is still needed. The whole process that leads to this risk state is not completely understood. There are open questions that needs to be further investigated. In this context, computational models that reproduce the electrophysiology of the heart began to be a valuable tool over the years by providing more knowledge about the complex phenomena that are responsible for causing these diseases. Most of these studies aims to analyze the cardiac conduction system.

The cardiac conduction system is a group of specialized cardiac cells of the heart which sends electrical signals to the muscles, enabling them to contract in order to pump the blood from the ventricles to all parts of the body.

It is mainly composed by the sinoatrial node (SA node), the atrioventricular node (AV node), the bundle of His and the Purkinje fibers. The SA node is the natural pacemaker of the heart, on its normal state, it deliveries stimulus to the system. After the stimulus pass through the AV node and the bundle of His, the signal reaches the Purkinje fibers, which are responsible for stimulating the ventricle tissue leading to a contraction of the muscle [3].

Recently, there are several studies that relates problems in the cardiac conduction system and cardiac arrhythmias [4–8]. Furthermore, other studies that mapped the electrical activity of dog and pig hearts, suggest that the Purkinje fibers play an important role in generating ventricular fibrillations at the junctions sites that link the fibers with the ventricular tissue [9–12]. At these junctions, known as Purkinje-muscle junctions (PMJ's), reentry currents could occur, which are a cyclic stimulation that happens over the tissue and is normally triggered by a unidirectional block that occurs at some point of the cardiac conduction system [13].

Also, it was demonstated that the behaviour of the Purkinje system changes when the fibers are couple to a large ventricular mass. There are electrotonic interactions between the fiber and the myocardium which makes the passage of the stimulus from one domain to the another more difficult, leading to delays and even a complete propagation block [14]. These delays normally range from 5 to 15 ms [15] and the main reason for them to happen is a combination of high resistance that the stimulus encounter at the PMJ's together with a low current which unables the ventricular tissue to despolarized. This condition is consider a source-sink mismatch [8, 16].

In this work, it is presented a study which analyzes what factors could affect the electrical stimulus over the Purkinje fibers by evaluating which parameters change the propagation velocity and the activation time of the cells. The main reason for this work is that in previous works the studies only focused in specific features, like the evaluation only of the activation of the PMJ's [17] or by considering how the network geometry affects the activation map [7] or even studying only how the properties of the cells that form the fiber changes the propagation [19].

The parameters that were analyzed in the present work were the geometric properties of the cells constituting the fibers, like the diameter and length. Also, the ionic properties were evaluated by using two different cellular models in order to model the dynamics of the Purkinje cells. The first model was proposed by Noble [20], and not only captures the essential features that happens in these cells, but it also provides an opportunity for direct comparison of our results with previous studies. The second model was proposed by Li and Rudy [21] and is a more recent model that is able to capture more complex behaviours, like the calcium cycle.

The implemented model also study the delays that could occur at the PMJ's sites by varying both the electrical resistance and the volume to be stimulated in these regions. Thus, the model was able to reproduce scenarios of conduction block when certain factors were attended.

In addition, the geometry of the Purkinje network could be very irregular, as it was verified in the recent studies from [22]. Because of this, we also consider how the number of bifurcations presented in the network could affect both the propagation velocity and activation time in our simulations.

2   Modeling of the Heart Electrophysiology

The contraction of the cardiac cells is initiated by an electrical activation from an action potential (AP), which is a despolarization current that rises the transmembrane potential of an excitable cell from its resting value, normally between -90 and -80 mV to slightly positive values.

The propagation of the AP from one cell to another can only happen because of gap junctions, which are specialized proteins that enables the flux of ions between neighboring cells as represented by Fig. 1.

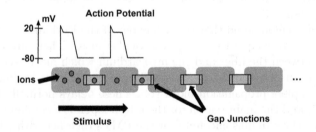

Fig. 1. Representation of the electrical propagation through a Purkinje fiber. The difference in ionic concentration of the cell membrane generates a difference in the potential, which is responsible for triggering an AP, despolarizing the cell when it reaches a certain threshold. In addition, ions can pass from one cell to another by gap junctions, activating the adjacent cells in a wave-like form.

2.1 Monodomain Model

In order to mathematically reproduce the electrical propagation over the Purkinje fibers we use the unidimensional monodomain model, which is a reaction-diffusion equation.

$$\sigma_x \frac{\partial^2 V_m}{\partial x^2} = \beta \left(C_m \frac{\partial V_m}{\partial t} + I_{ion} + I_{stim} \right), \tag{1}$$

where V_m is the transmembrane potential of the cell given in mV, σ_x is the conductivity from the cells given in mS/cm, β is the surface per volume ratio of the cell given in cm^{-1}, C_m is the capacitance of the membrane given in μF/cm^2, I_{ion} is an ionic current that depends of the cellular model being used, I_{stim} is a stimulus current.

2.2 Propagation Velocity

One important feature of the electrical propagation through the heart is the velocity in which the stimulus travels. There are differences in the propagation velocity between the Purkinje cells and the ventricular ones. The human cardiac Purkinje cells are characterized by a fast propagation velocity, ranging from 2 to 4 m/s [23]. On the other hand, ventricular cells have relative slower velocity with values between 0.3 to 1.0 m/s [24].

The propagation velocity along a Purkinje fiber could be calculated using the cable equation [27] and by considering that the cable is composed by a set of cells with length h and a diameter d.

$$v = \frac{c}{2C_m} \sqrt{\frac{d}{R_m R_c}}, \tag{2}$$

where v is the propagation velocity of the fiber given in cm/ms, R_c is the citoplasmatic resistivity given in Ω.cm, R_m is the membrane resistivity given in Ω.cm^2 and c is a parameter which depends on the cellular model.

Following the work of [17] the propagation velocity is considered to be constant along the fibers. Because of that, given the distance between two points of the network and the velocity that was measured in this region it is possible to calculate the activation time of a particular cell in the network.

2.3 Numerical Solution

The numerical solution of the monodomain equation was done by applying the Finite Volume Method (FVM). The main reason for using the FVM was that is a method based on conservative principals, so that dealing with bifurcations would not be a problem. Also, the method can be applied in complex geometries, such as the Purkinje fibers [18].

For the time discretization of the FVM we divided the diffusion part from the reaction one in Eq. (1) using the Godunov splitting operator [28]. From this, at each timestep we need to solve two distinct problems, a non-linear system of ODE's given by:

$$\begin{cases} \dfrac{\partial V_m}{\partial t} = \dfrac{1}{C_m}[-I_{ion}(V_m, \eta) + I_{stim}], \\ \dfrac{\partial \eta}{\partial t} = f(V_m, \eta), \end{cases} \tag{3}$$

and a parabolic PDE

$$I_v = \beta \left(C_m \frac{\partial V_m}{\partial t} \right) = \nabla \cdot (\sigma_x \nabla V_m). \tag{4}$$

where η represents the set of variables which controls the ionic channels related to the ionic current I_{ion} of the cellular model.

To approximate the time derivative from Eq. (4) a implicit Euler scheme was used in order to avoid instabilities problems.

$$\frac{\partial V_m}{\partial t} = \frac{V_m^{n+1} - V_m^n}{\Delta t}, \tag{5}$$

where V^n is the transmembrane potential at timestep n and Δt is the size of the time discretization, which is necessary to advance the PDE in time.

For the time discretization of the non-linear system of ODE's, given by Eq. (3) an explicit Euler scheme was used with the same timestep of the PDE. For the spacial discretization, the diffusive term from Eq. (4) was approximated using the following flux:

$$J = -\sigma \nabla V, \tag{6}$$

where J ($\mu A/cm^2$) is the intracelular flux of current density

$$\nabla \cdot J = -I_v, \tag{7}$$

where I_v ($\mu A/cm^3$) is a volumetric current which corresponds to the left side of Eq. (4).

In addition, we will consider an unidimensional cable, which is composed by cylinders and will represent the Purkinje cells, as it is illustrated by Fig. 2. These will be the control volumes of FVM, in which at the center of each one will be a node containing the amount of interest V_m.

After that, we define the FVM equations by integrating (7) over an individual control volume V_i with a value equal to $\pi d^2 h/4$.

$$\int_\Omega \nabla \cdot J dv = - \int_\Omega I_v dv. \tag{8}$$

Then, applying the divergence theorem on (8).

$$\int_\Omega \nabla \cdot J dv = \int_{\partial \Omega} J \cdot \xi ds, \tag{9}$$

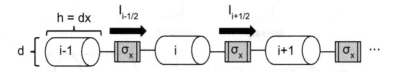

Fig. 2. Space discretization used for modeling the Purkinje fibers. Each Purkinje cell has diameter d and a length h. The length of the cells have the same value of the space discretization size dx. Between the cells, there are current fluxes through the faces of the control volumes.

where ξ is an unitary vector which points to the boundary $\partial\Omega$.

$$\int_{\partial\Omega} J_i \cdot \xi ds = -\int_{\Omega} I_v dv. \tag{10}$$

Finally, assuming that I_v represents a mean value for each particular volume and substituting (4) into (10).

$$\beta C_m \frac{\partial V_m}{\partial t}\bigg|_i = \frac{-\int_{\partial\Omega} J_i \cdot \xi ds}{\pi . d^2 . h / 4}. \tag{11}$$

Calculating J_i for each control volume by dividing the flux into a sum of fluxes over each face and remembering that we have only an input and a output flux at each control volume.

$$\int_{\partial\Omega} J_i \cdot \xi ds = (I_{out} - I_{in}), \tag{12}$$

where the current between two control volumes i and j are calculated as follows

$$I_{i,j} = -\sigma_x \frac{\partial V_m}{\partial x}\bigg|_{i,j} \frac{\pi . d^2}{4}, \tag{13}$$

where a central finite difference approximation was used for the space derivative on Eq. (13):

$$\frac{\partial V_m}{\partial x}\bigg|_{i,j} = \frac{V_j - V_i}{h}. \tag{14}$$

In order to model the PMJ's we consider that the last Purkinje cell of the fiber will be linked to a large myocardium cell, which has a volume equivalent to the region to be stimulated over the ventricular tissue. This connection was implemented by adding a discrete resistor R_{PMJ} between the last Purkinje cell and the myocardium cell as shown in Fig. 3.

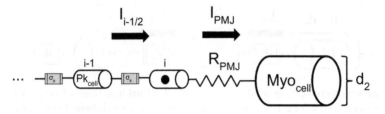

Fig. 3. Space discretization used for modeling the Purkinje-muscle junctions. The myocardium cell has a larger diameter d_2 than the Purkinje cells.

Furthermore, the value of the current I_{PMJ} that passes through the resistor R_{PMJ} is given by

$$I_{PMJ} = \frac{(V_{pk} - V_{myo})}{R_{PMJ}},$$ (15)

where V_{pk} is the tramsmembrane potential from the Purkinje cell and V_{myo} is the transmembrane potential of the myocardium cell.

To be able to analyze the behaviour of the characteristic delay that occur at the PMJ's sites we define a γ parameter, which is the product between the resistance of the PMJ and the volume of the myocardium cell.

$$\gamma = \frac{\pi.d_2^2.h}{4}.R_{PMJ}$$ (16)

3   Results and Discussions

3.1   Stimulus Protocol and Cellular Models

For all the experiments of this work we use the following stimulus protocol.

Initially, we perform a simulation so that the system could reach its steady-state, where $\Delta t = 0.01$ ms and a stimulus current I_{stim} is applied in periods of 500 ms in the first 5 cells of the fiber. At the end of this first simulation we save the state of every cell in the network and load this state as the initial condition for a second simulation, in which the propagation velocity and activation times will be calculated. This was done in order to get a more reliable result since the system will be at equilibrium.

Regarding the cellular models for the ionic current I_{ion} that appears on Eq. (3) we use two models that are specifically for cardiac Purkinje cells, the Noble model [20] and the Li and Rudy model [21].

3.2   Model for the Propagation Velocity

The first experiment of this work has the objective of calibrating the parameters related to the monodomain model until it reaches the same behaviour of the analytical expression given by (2), which describes the propagation velocity over the fiber.

In order to achieve this the parameter c from Eq. (2) needs to be calculated based on the cell properties we were using. On [27] there is a table with typical values for the parameters from Eq. (2) for different types of cells, for this work we use the values related to the cardiac mammal muscle in this table.

Considering some works found in the literature [29–31] an average value for the propagation velocity in a dog Purkinje fiber with a diameter in the range of $d = 35\,\mu\mathrm{m}$ is approximately 2 to 3.5 m/s. By using the values $R_m = 7\,\mathrm{k\Omega.cm^2}$, $R_c = 150\,\Omega.\mathrm{cm}$, $C_m = 1.2\,\mu\mathrm{F/cm^2}$, $d = 35\,\mu\mathrm{m}$ and $v = 2.6\,\mathrm{m/s}$ on Eq. (2) we calculate $c = 10.808$. This result will be consider the analytical solution for the model in this experiment.

For the numerical approximation we changed the value of the conductivity σ_x in order to get a propagation velocity close to the one given by the analytical solution. Then, we measured the velocity on a 2 cm cable composed of Purkinje cells with a diameter and length equal to $d = 35\,\mu\mathrm{m}$ and $dx = 164\,\mu\mathrm{m}$, respectively. After some tests we found the values $\sigma_x = 0.004\,\mathrm{mS/cm}$ for the Noble model and $\sigma_x = 0.0019\,\mathrm{mS/cm}$ for the Li and Rudy model, which resulted in a propagation velocity along the cable equals to $v = 2.645\,\mathrm{m/s}$, which was close to the value from the analytical model.

By varying the value of the diameter d of the Purkinje cells we could them verify how the propagation velocity behaves along the cable. In Fig. 4 we measured the propagation velocity in a 2 cm cable by using an initial diameter of $d = 10\,\mu\mathrm{m}$ and we incremented this value by $5\,\mu\mathrm{m}$ until it reaches $d = 50\,\mu\mathrm{m}$.

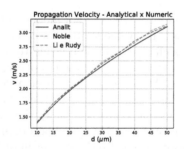

Fig. 4. Comparison between the analytical solution given by (2) and the numerical one. The propagation velocity was calculated for a Purkinje cell set at the middle of the 2 cm cable.

As can be verify, the results from our numerical model were really close to the analytical one. Furthermore, the behaviour of the propagation velocity was in accordance with the observations found in the literature [30], which says that the velocity along a cable decays with a factor proportional to the $\sqrt{d}$.

3.3 Model with Purkinje-Muscle Junctions

In the next experiment we explore how the introduction of a PMJ would affect the activation time along the fiber. The simulations were done using again a

2 cm cable, which will be representing the Purkinje fiber. At the end of the cable a PMJ with resistance $R_{PMJ} = 11000\,k\Omega$ was added and then linked to a myocardium cell. The characteristic delay was calculated by taking the difference between the activation time of the last Purkinje cell and the myocardium cell.

To analyze the characteristic delay that happens at the PMJ's sites we vary the value of the parameter γ and the diameter d of the Purkinje cells, so that we could study in which condition there will be propagation blocks as can be seen in Fig. 5.

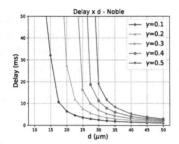 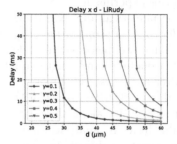

Fig. 5. Comparison between the characteristic delay given in ms and the diameter of the Purkinje cells when the γ parameter changes for both the Noble and Li and Rudy model.

As can be visualized in Fig. 5, both cellular models obtain similar results. It is important to note that there is a value for the diameter where the characteristic delay begins to assume high values until it leads to a complete block of the stimulus, unabling the myocardium cell to be despolarized, characterized by the sudden peaks in the results. Also, with this simulation we could check that there is a relation between the characteristic delay and the γ parameter. As we increase the value γ the delay time increases proportionally, since a high value of γ means either a large ventricular mass to be stimulated or a high resistance of the PMJ. In both cases we need more current from the fiber in order to trigger an AP at the myocardium cell.

Regarding the diameter of the Purkinje cells the relation between this parameter and the characteristic delay is inversely proportional. The reason is that with a larger diameter more current could travel through the fiber since its transversal area will be wider, which makes the despolarization of the myocardium cell to happen more easily.

3.4   Model with Bifurcations

The last experiment of this work studies the effects of bifurcations over the Purkinje fibers. Different types of networks were generated with different levels of bifurcations as shown in Fig. 6.

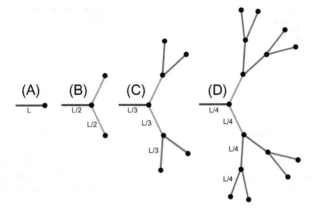

Fig. 6. Networks that were used in the experiments that evaluate the influence of bifurcations over the activation time. Each one of the four networks represents a structure with a different level of bifurcations. The distance between the source of the stimulus and the terminal points of the networks are always equals to L. (A) Level 0. (B) Level 1. (C) Level 2. (D) Level 3.

In addition, we decrease the diameter of the Purkinje cells from one level of the network to another. This decision was made by observing images of Purkinje fibers of a calf [26]. The value of the diameter at a certain level k of the network it now be given by (17).

$$\begin{cases} d_k = d_0, \; if \; k = 0, \\ d_k = \delta.d_{k-1}, \; if \; k \neq 0, \end{cases} \tag{17}$$

where d_k is the diameter of a Purkinje cell at a level k in the network and δ is the ratio in which the diameter is decreased.

The simulations were done using the same two cellular models and by adding a PMJ with $\gamma = 0.2$ at the end of each fiber. For the space discretization we use $dx = 164\,\mu m$, a initial diameter $d_0 = 30\,\mu m$ for the Noble model and $d_0 = 55\,\mu m$ for the Li and Rudy model. The decrease ratio for the diameter was tested within the range $\delta = \{90\%, 80\%, 70\%\}$ and the distance from the source of the stimulus to each terminal was $L = 2\,cm$.

In Fig. 7 it can be verified that in both celullar models the activation time of the myocardium cell increases proportionally to the number of bifurcations. This happens because after a bifurcation the diameter decreases by following Eq. (17), which makes the activation much harder, since the networks that had more bifurcations will have a smaller diameter at the last level of the fibers.

Based in this observation we perform another experiment that test the effects of also decreasing the myocardium volume to be stimulated by the fibers. Mainly because, the Purkinje fibers seems to be well distributed over the ventricular tissue in order to proper activate the muscles and to reduce propagation blocks [26]. This was done by calculating the value of γ based on how many levels of bifurcation a certain network had.

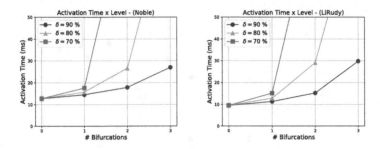

Fig. 7. Results showing the activation time of the myocardium cell considering the number of bifurcations the stimulus had to pass through until it reaches the end of the fiber as the diameter decreases by a ratio δ, for both the Noble and Li and Rudy model.

$$\gamma_k = \frac{\gamma_0}{2^k}, \tag{18}$$

Analyzing the results from Fig. 8 the decreasing of the γ parameter could avoid some of the propagation blocks that happen in the previous test. This behaviour can be justify by the fact that with less resistance at the PMJ sites, the fibers which had a small diameter could now despolarize the myocardium cell.

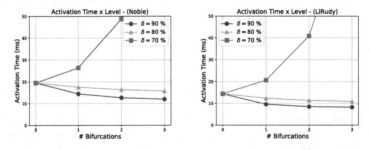

Fig. 8. Results showing the activation time of the myocardium cell considering the decreasing of the parameter γ using (18) as the diameter of cells decreases by a ratio δ.

4   Conclusions

In this work it was studied which features from the Purkinje fibers change the propagation velocity and the activation time within these structures. From the results of the experiments, geometric parameters, like the diameter of the Purkinje cells, were the ones that had the most influence among the others. Meaning that they could play an important role in sustaining arrhythmias.

From the results of the experiments presented in this work we could retrieve some relevant information about the factors that could affect the propagation velocity and the activation over the Purkinje fibers.

In the first experiment, it was possible to validate certain behaviours found in the literature. The implemented model was able to reproduce the proportionality between the diameter of the Purkinje cells and propagation velocity along the fiber, which according to some works is proportional to $\sqrt{d}$ [25,27]. Thus, both cellular models reproduced the same behaviour as it was expected.

From the results of the second experiment it was shown that Purkinje fibers with a small diameter were not able to stimulate a large region of the myocardium, generating propagation blocks over the PMJ's sites. These blocks mainly occur because there are cells that could not achieve the current threshold necessary to activate the myocardium cell, which is characterized as a source-sink mismatch.

The last experiment fortifies the idea that the activation of the ventricular tissue depends of how well distributed the fibers are over this domain, which is actually observed in nature. The ramifications of the Purkinje network seems to be a way to homogeneously cover the myocardium tissue in order to increase the superficial area of the network and minimize the energetic cost for the stimulation of the ventricles.

Furthermore, the implemented model was able to reproduce the characteristic delay that occurs at the PMJ's sites. The implementation using the Finite Volume Method and the coupling using a resistor linked to a larger control volume has proved to be very effective, since it was able to generate the same results that were observed in the literature.

References

1. World Health Organization. https://www.who.int. Accessed 28 Dec 2018
2. American Heart Association et al.: Heart Disease and Stroke Statistics: 2017 At-a-Glance (2017)
3. Clayton, R.H., et al.: Models of cardiac tissue electrophysiology: progress, challenges and open questions. Progress Biophys. Mol. Biol. **104**(1–3), 22–48 (2011)
4. Deo, M., Boyle, P.M., Kim, A.M., Vigmond, E.J.: Arrhythmogenesis by single ectopic beats originating in the Purkinje system. Am. J. Physiol.-Heart Circulatory Physiol. **299**(4), H1002–H1011 (2010)
5. Behradfar, E., Nygren, A., Vigmond, E.J.: The role of Purkinje-myocardial coupling during ventricular arrhythmia: a modeling study. PloS One **9**(2), 1–9 (2014)
6. Quan, W., Rudy, Y.: Unidirectional block and reentry of cardiac excitation: a model study. Circ. Res. **66**(2), 367–382 (1990)
7. Boyle, P.M., Deo, M., Plank, G., Vigmond, E.J.: Purkinje-mediated effects in the response of quiescent ventricles to defibrillation shocks. Ann. Biomed. Eng. **38**(2), 456–468 (2010)
8. Oliveira, R.S., et al.: Ectopic beats arise from micro-reentries near infarct regions in simulations of a patient-specific heart model. Sci. Rep. **8**(1), 16392 (2018)
9. Berenfeld, O., Jalife, J.: Purkinje-muscle reentry as a mechanism of polymorphic ventricular arrhythmias in a 3-dimensional model of the ventricles. Circ. Res. **82**(10), 1063–1077 (1998)
10. Allison, J.S., et al.: The transmural activation sequence in porcine and canine left ventricle is markedly different during long-duration ventricular fibrillation. J. Cardiovasc. Electrophysiol. **18**(12), 1306–1312 (2007)

11. Tabereaux, P.B., et al.: Activation patterns of Purkinje fibers during long-duration ventricular fibrillation in an isolated canine heart model. Circ. Res. **116**(10), 1113–1119 (2007)
12. Li, L., Jin, Q., Huang, J., Cheng, K.A., Ideker, R.E.: Intramural foci during long duration fibrillation in the pig ventricle. Circ. Res. **102**(10), 1256–1264 (2008)
13. Sigg, D.C., Iaizzo, P.A., Xiao, Y.-F., He, B.: Cardiac Electrophysiology Methods and Models. Springer, Heidelberg (2010)
14. dos Santos, R.W., Campos, F.O., Ciuffo, L.N., Nygren, A., Giles, W., Koch, H.: ATX II effects on the apparent location of M cells in a computational model of a human left ventricular wedge. J. Cardiovasc. Electrophysiol. **17**, S86–S95 (2006)
15. Palamara, S., Vergara, C., Faggiano, E., Nobile, F.: An effective algorithm for the generation of patient-specific Purkinje networks in computational electrocardiology. J. Comput. Phys. **283**, 495–517 (2015)
16. Xie, Y., Sato, D., Garfinkel, A., Qu, Z., Weiss, J.N.: So little source, so much sink: requirements for afterdepolarizations to propagate in tissue. Biophys. J. **99**(5), 1408–1415 (2010)
17. Vergara, C., et al.: Patient-specific computational generation of the Purkinje network driven by clinical measuraments. MOX Report (9) (2013)
18. Oliveira, R.S., Rocha, M.B., Burgarelli, D., Meira Jr., W., Constantinides, C., dos Santos, R.W.: Performance evaluation of GPU parallelization, space-time adaptive algorithms, and their combination for simulating cardiac electrophysiology. Int. J. Numer. Methods Biomed. Eng. **34**(2), e2913 (2018)
19. Campos, F.O., Shiferaw, Y., Prassl, A.J., Boyle, P.M., Vigmond, E.J., Plank, G.: Stochastic spontaneous calcium release events trigger premature ventricular complexes by overcoming electrotonic load. Cardiovasc. Res. **107**(1), 175–183 (2015)
20. Noble, D.: A modification of the Hodgkin Huxley equations applicable to Purkinje fibre action and pacemaker potentials. J. Physiol. **160**(2), 317–352 (1962)
21. Li, P., Rudy, Y.: A model of canine Purkinje cell electrophysiology and Ca2+ cycling. Circ. Res. **109**, 71–79 (2011). CIRCRESAHA-111
22. Duan, D., Yu, S., Cui, Y., Li, C.: Morphological study of the atrioventricular conduction system and Purkinje fibers in yak. J. Morphol. **278**(7), 975–986 (2017)
23. Carnefield, P.F., Wit, A.L., Hoffman, B.F.: Conduction of the cardiac impulse: III. Characteristics of very slow conduction. J. Gener. Physiol. **59**(2), 227–246 (1972)
24. Bers, D.: Excitation-Contraction Coupling and Cardiac Contractile Force. Springer, Heidelberg (2001). https://doi.org/10.1007/978-94-010-0658-3
25. Sperelakis, N.: Cell Physiology Source Book: Essentials of Membrane Biophysics. Elsevier, Amsterdam (2012)
26. Sebastian, R., Zimmerman, V., Romero, D., Sanchez-Quintana, D., Frangi, A.F.: Characterization and modeling of the peripheral cardiac conduction system. IEEE Trans. Med. Imaging **32**(1), 45–55 (2013)
27. Keener, J., Sneyd, J.: Mathematical Physiology. Springer, Heidelberg (1998)
28. Sundnes, J., Lines, G.T., Cai, X., Nielsen, B.F., Mardal, K.A., Tveito, A.: Computing the Electrical Activity in the Heart. Springer, Heidelberg (2007). https://doi.org/10.1007/3-540-33437-8
29. Ten Tusscher, K., Panfilov, A.: Modelling of the ventricular conduction system. Progress Biophys. Mol. Biol. **96**(1), 152–170 (2008)
30. Bigger, J.T., Mandel, W.J.: Effect of lidocaine on conduction in canine Purkinje fibers and at the ventricular-muscle-Purkinje fiber junction. J. Pharmacol. Exp. Ther. **172**(2), 239–254 (1970)
31. Kassebaum, D.G., Van Dyke, A.R.: Electrophysiological effects of isoproterenol on Purkinje fibers of the heart. Circ. Res. **19**(5), 940–946 (1966)

A Knowledge Based Self-Adaptive Differential Evolution Algorithm for Protein Structure Prediction

Pedro H. Narloch📧 and Márcio Dorn$^{(\boxtimes)}$📧

Institute of Informatics, Federal University of Rio Grande do Sul,
Porto Alegre, Brazil
mdorn@inf.ufrgs.br

Abstract. Tertiary protein structure prediction is one of the most challenging problems in Structural Bioinformatics, and it is a NP-Complete problem in computational complexity theory. The complexity is related to the significant number of possible conformations a single protein can assume. Metaheuristics became useful algorithms to find feasible solutions in viable computational time since exact algorithms are not capable. However, these stochastic methods are highly-dependent from parameter tuning for finding the balance between exploitation (local search refinement) and exploration (global exploratory search) capabilities. Thus, self-adaptive techniques were created to handle the parameter definition task, since it is time-consuming. In this paper, we enhance the Self-Adaptive Differential Evolution with problem-domain knowledge provided by the angle probability list approach, comparing it with every single mutation we used to compose our set of mutation operators. Moreover, a population diversity metric is used to analyze the behavior of each one of them. The proposed method was tested with ten protein sequences with different folding patterns. Results obtained showed that the self-adaptive mechanism has a better balance between the search capabilities, providing better results in regarding root mean square deviation and potential energy than the non-adaptive single-mutation methods.

Keywords: Protein structure prediction ·
Self-Adaptive Differential Evolution · Structural Bioinformatics ·
Knowledge-based methods

1 Introduction

Proteins are macro-molecules composed by a sequence of amino acids, assuming different shapes accordingly to this sequence and environment conditions [1]. The three-dimensional structural conformation of a protein is related to its biological function, where any modification might influence the protein's biological function [26]. Thus, the determination of these structures is significant to understanding proteins role performed inside a cell [9]. Nowadays, the determination of three-dimensional structures is through experimental methods such as

© Springer Nature Switzerland AG 2019
J. M. F. Rodrigues et al. (Eds.): ICCS 2019, LNCS 11538, pp. 87–100, 2019.
https://doi.org/10.1007/978-3-030-22744-9_7

X-ray crystallography and Nuclear Magnetic Resonance. However, these experimental strategies are time-consuming and expensive [12]. In light of the importance of these molecules and limitations of experimental methods, computational strategies became interesting approaches to reduce costs and the difference between sequenced and determined structures. However, the determination
of three-dimensional protein structures is classified, in computational complexity theory, as an NP-hard problem [15] due to the explosive of possible shapes
a protein can assume, making impossible the use of exact methods to solve the
problem. In light of the complexity of the Protein Structure Prediction (PSP)
problem, metaheuristics became attractive to finding feasible solutions for one
of the most challenging problems in Structural Bioinformatics [9], although these
techniques do not guarantee the finding of optimal solution [13]. There are three
steps needed to build a possible solver for the protein structure prediction, (i)
the computational representation of proteins; (ii) a scoring method to measure
the molecule's free energy; and (iii) a search method to explore the conformational search space [12]. Different metaheuristics have been used in many
NP-hard problems but, the Differential Evolution [24] (DE) is one of the most
effective search strategy for complex problems [11] in a vast type of problems,
including PSP [18–20].

Besides the capacity of finding good solutions for NP-Complete problems that
different metaheuristics have, they are very dependent on the balance between
two search characteristics: the *exploitation* and the *exploration* [10]. This balance
helps the algorithm to avoid local optima, prevent the premature convergence,
and ensure the neighborhood exploitation for better final solutions. This balance
can be affected by tuning parameters and modifying different operators, but this
is not a trivial task. In this way, we propose the use of a Self-Adaptive Differential Evolution (SaDE) [22] in the PSP problem, since its adaptive mechanisms
tend to preserve the balance between exploration and exploitation capabilities
during the search process. As the PSP be a complex problem, we use the Angle
Probability List [5] (APL), a valuable source of problem-domain data, to enhance
the algorithm. Moreover, we also use a populational diversity metric [10] to
monitor the SaDE behavior during the search process, comparing it with four
mutation operators that compose the set used in the self-adaptive version. Some
interesting convergence behaviors were observed as well as good results for the
problem. The next sections in this paper are organized as follows. Section 2
presents the concepts used in this works such as the problem formulation, the
SaDE algorithm, APL construction, and related works. The proposed method is
described in Sect. 3. In Sect. 4 the results obtained by the different approaches
are discussed. Conclusions and future works are given in Sect. 5.

2   Preliminaries

2.1   Three-Dimensional Protein Structure Prediction

A protein molecule is formed by a linear sequence of amino acids (primary structure). The thermodynamic hypothesis of Anfinsen [1] states that protein's folding

depends on its primary structure. The native functional conformation of a protein molecule coincides with its lowest free energy conformation. Over the years, different computational efforts were made in the PSP problem, creating energy functions, proteins representation, and search mechanisms to simulate the folding process [12]. However, as proteins are complex molecules, the definition of each of these three components is not a trivial task. The computational representation of proteins can vary, from the most simple ones such as two-dimensional *lattice* models [4] to the full-atom model in a three-dimensional space. The trade-off among these different representations is related to the computational cost versus real protein representation. One adequate way to computationally represent these complex molecules is by their rotational angles, known as dihedral angles, maintaining the closeness of real systems and reducing the computational complexity of its representation.

The dihedral angles of proteins are present in chemical bonds among the atoms that compose the molecule. The amino acids present in the protein's primary structure are chained together by a chemical bond known as a peptide bond. In general, all amino acids found in proteins have the same basic structure, with an amino-group N, the central carbon atom C_α, a carboxyl-group C, and four hydrogens. The difference among the 20 known amino acids is in their side-chain atoms. When bonding two amino acids, the peptide bond is formed by the C-N interaction, forming a planar angle known as ω. The ϕ angle represents the rotation around the N-C_α and ψ the rotation angle that rotates around C_α-C. These two angles (ϕ, ψ) are free to rotate in the space, varying from $-180°$ to $+180°$. Due to this fact, there is an explosion of possible conformation a protein can assume since each amino acid's backbone is composed of two free rotational and one planar angle. Beyond that, there are the side-chain angles noted as χ-angles, and their number varies from 0 to 4 accordingly to the amino acid type. The values of these rotation angles modify the position of different atoms along the whole protein structure, forming different structural patterns (secondary structure). The most stable and important secondary structures present in protein's structures are the α-helix and β-sheets. Another type of secondary structure is the β-turn, composed of short segments and generally responsible for connecting two β-strands. There are structures responsible for connecting different secondary structures, known as coils. In this way, ones can computationally represent a protein as a sequence of dihedral angles, where each set of angles serve as an amino acid. It is possible to imagine that as the size of a protein (quantity of amino acids in its primary structure) increases, the problem dimension grows as well.

The physicochemical interactions among the atoms should be considered to determine the correct orientation of them. In this way, different energy functions were proposed to simulate proteins molecular mechanics [3]. Prediction methods use a potential energy function to describe the search space, which the minimum global energy represents the native conformation of the protein. The *Rosetta energy function* [23] is one of the popular scoring tools for all-atom energy determination, and it is used in this paper. The Eq. 1 presents the different components this energy function considers.

$$E_{Rosetta} = \begin{cases} E_{physics-based} + E_{inter-electrostatic} \\ +E_{H-bonds} + E_{knowledge-based} + E_{AA} \end{cases} \tag{1}$$

where $E_{physics-based}$ calculates the 6–12 Lennard-Jones interactions and Solvatation potential approximation, $E_{inter-electrostatic}$ stands for inter-atomic electrostatic interactions and $E_{H-bonds}$ hydrogen-bond potentials. In $E_{knowledge-based}$ the terms are combined with knowledge-based potentials while the free energy of amino acids in the unfolded state is in E_{AA} term.

Angle Probability List: Over the years different methods have been proposed for the PSP problem. These methods can be classified in four classes [12]. The fold recognition and comparative modeling are two classes of methods that strictly depends on existing structures to predict the structure of another protein. Besides their efficiency, they can not find new folds of proteins. In *first principles prediction without database information*, known as *ab initio*, the folding process uses only the amino acid as information for finding the lowest energy in the energy space, making possible the prediction of new folding patterns. However, methods purely *ab initio* have some limitations due to the size of the conformational search space [12]. In this work, we use a variation of the *ab initio* class, the *first principles with database information*. In this way, adding problem-domain knowledge to enhance the search mechanism, better structures are found, and it does not preclude the finding of new folding patterns.

As amino acids can assume different torsion angle values depending on their secondary structure [17], it is worth to consider these occurrences as information to reduce the search space while enhancing algorithms with better search capabilities. In light of these facts, the *Angle Probability List*, APL, was proposed in [5] based on the conformational preferences of amino acids based on their secondary structures. The data was retrieved from the Protein Data Bank (PDB) [2], considering only high-quality information. To compose this database, a set of 11,130 structures with resolution ≤ 2.5 Å was used. The APL was built based in a histogram matrix of $[-180, 180] \times [-180, 180]$ for each amino acid and secondary structure. To generate the APLs, a web tool known as NIAS[1] (*Neighbors Influence of Amino acids and Secondary structures*) was used [6].

Self-Adaptive Differential Evolution: The Differential Evolution (DE) algorithm was proposed initially by Storn and Price [24] and since then it has been one of the most efficient metaheuristics in different areas [11]. The DE is a populational-based evolutionary algorithm which depends on three parameters, the crossover rate (CR), a mutation factor (F) and the size of the population (NP). In the SaDE [22] version, parameters CR and F are modified by the algorithm instead of pre-fixed values for the whole optimization process. This strategy is interesting since the parameter fine-tuning is a time-consuming task. Another important fact is that there is not a global parameter value that might be the optimum parameter for all problems.

As the F factor be related to the convergence speed, in SaDE algorithm the F parameter assume random values in the range of $[0, 2]$, with a normal distribu-

---

[1] http://sbcb.inf.ufrgs.br/nias.

tion of mean 0.5 and standard deviation of 0.3. In this way, the global (large F values) and local (low F values) search abilities are maintained during the whole optimization process. The CR parameter is changed along the evolutionary process, starting with a random mean value of 0.5 (CRm) and a standard deviation of 0.1. The CRm is adapted during the optimization process based on its success rate. Furthermore, the SaDE also adapts the mutation mechanism used for creating new individuals. In classical DE algorithm, only one mutation mechanism is employed during the whole optimization process. The first SaDE approach proposed the usage of two different mutation mechanisms, with different exploration and exploitation capabilities. To chose the method to be employed, a learning stage is applied during some generations before the real optimization process. In this way, a probability of occurrence is associated with a mutation mechanism accordingly its success and failure rate.

Related Works: Some of the most well-known search algorithms used in the PSP problem are Genetic Algorithms (GA), Differential Evolution (DE), Artificial Bee Colony (ABC), Particle Swarm Optimization (PSO) and many others. The DE behavior was previously analyzed in [18] and [19], where different mutation strategies were employed to increase the diversity capabilities of the algorithm. As the author used the diversity metric, it is possible to notice that the diversity maintenance is a key factor to avoid local optima solutions and, consequently, the premature convergence. A self-adaptive multi-objective DE was proposed in [25], showing the importance of how self-adaptive strategies could be interesting to the PSP problem. The Self-Adaptive Differential Evolution was employed by [20] with two sources of knowledge: the APL and the Structure Pattern List (SPL). In this version, authors demonstrated how important it is to combine problem-domain knowledge with the SaDE algorithm. Besides the contribution of using APL and SPL as a source of structural information, the authors have not analyzed each mutation operator separately, either the algorithm behavior regarding convergence and diversity maintenance, creating a gap in the application. Besides some works have already used APL as a source of information [5,7,9], none of them have used some self-adaptive mechanism or are concerned about the behavior of the algorithms regarding diversity maintenance. Thus, in our approach, we close this gap using a diversity index to monitor and analyze the behavior of each mutation operator and a self-adaptive version of the DE algorithm combined with information provided by the APL. Also, our application uses different mutation operators from [20] based on the exploration and exploitation capabilities of each mutation strategy.

3   Materials and Methods

There are three essential components needed to create a PSP predictor: (i) a way to computationally represents the protein structure; (ii) a scoring function to evaluate the protein's potential energy; and (iii) a search strategy to explore the protein's conformational search space and find feasible structures.

The main contribution of this work is related to the (iii) search strategy, providing a populational convergence analysis of each mutation mechanism used in a knowledge-based SaDE algorithm for the PSP problem.

Protein's Representation and Scoring Function: In this work, we represented a protein molecule as a set of torsion angles. Each possible solution assumes $2N$ dimensions, where N is the length of the protein's primary structure. Therefore, this set of angles modifies the cartesian coordinates of protein's atoms to do the energy evaluation of the molecule. As we use the PyRosetta [8], a well-known interface to Python-based Rosetta energy function interface [23], we opted to reduce the search space optimizing only the protein backbone torsion angles (ϕ and ψ) without losing the molecule's characteristics. In light of preserving well-formed secondary structures, we used the PyRosetta to identify secondary structures using DSSP implementation [16] and considering it as an additional term in the *score3* energy function as shown by Eq. 2.

$$E_{total} = E_{score3} + E_{SS} \tag{2}$$

Another important metric to evaluate a possible solution is the *Root Mean Square Deviation* (RMSD), which compares the distance, in angstroms, among the atoms in two structures. In this work, the RMSD is used to compare the final solution with the already known experimental structure. Equation 3 displays the RMSD_α metric, which compares the backbone between two structures.

$$\text{RMSD}(a, b) = \sqrt{\frac{\sum_{i=1}^{n} |\, r_{ai} - r_{bi} \,|^2}{n}} \tag{3}$$

where r_{ai} and r_{bi} are the ith atoms in a group of n atoms from structures a and b. The closer RMSD is from 0 Å more similar are the structures.

Search Strategy: In any metaheuristic, the adjustment of parameters is important but not a trivial task since they affect the quality of possible solutions [14]. In order to sidestep the time-consuming task of parameter tuning, different self-adaptive strategies were proposed [21]. In this work we combine the SaDE [22] approach with the APL knowledge-database considering the high-quality information it provides [5,7,9]. As far as we know, the only SaDE application that used some kind of structural information was proposed in [20]. Differently from [22], we have used four DE mutation mechanisms (Table 1), which are also different from the used in [20]. We took in consideration the exploratory ($\text{DE}_{rand/1/bin}$ and $\text{DE}_{curr-to-rand}$) and exploitative ($\text{DE}_{best/1/bin}$ and $\text{DE}_{curr-to-bes}$) capabilities they provide to compose the set of mutation mechanisms that SaDE can choose. The Algorithm 1 shows the how we have structured our approach. The "learning stage" uses the same structure but with few numbers of generations to set the initial probability rates of each mutation strategy and CRm.

Moreover, we use a diversity measure (Eq. 4) to monitor the algorithm behavior during the optimization process. This metric takes into consideration the individual dimensions instead of the fitness, making possible to verify if the population has lost its diversity. This index was proposed in [10] for continuous-domain

Table 1. Classical mutation strategies in DE.

Approach	Equation
$DE_{best/1/bin}$	$v_i^{g+1} = x_{best}^g + F \cdot (x_{r2}^g - x_{r3}^g)$
$DE_{rand/1/bin}$	$v_i^{g+1} = x_{r1}^g + F \cdot (x_{r2}^g - x_{r3}^g)$
$DE_{curr-to-rand}$	$v_i^{g+1} = x_i^g + F1 \cdot (x_{r1}^g - x_i^g) + F2 \cdot (x_{r2}^g - x_{r3}^g)$
$DE_{curr-to-best}$	$v_i^{g+1} = x_i^g + F1 \cdot (x_{best}^g - x_i^g) + F2 \cdot (x_{r2}^g - x_{r3}^g)$

problems. The index ranges from $[0, 1]$, where 1 is the maximum diversity in the population and 0 the full convergence of the population to a single solution.

$$GDM = \frac{\sum_{i=1}^{N-1} ln \left(1 + \min_{j[i+1,N]} \frac{1}{D} \sqrt{\sum_{k=1}^{D} (x_{i,k} - x_{j,k})^2} \right)}{NMDF} \tag{4}$$

where D represents the dimensionality of the solution vector, N is the population size and x the individual (the solution vector). The NMDF is a normalization factor which corresponds to the maximum diversity value so far.

Algorithm 1. Self-Adaptive Differential Evolution with APL

Data: NP
Result: The best individual in population
Generate initial population with NP individuals based on APL
while $g \leq$ *number of generations* **do**
  F ← norm(0.5, 0.3)
  if *past 25 generations* **then**
    | CRm ← update based on the success rate of previous CR values.
  end
  for *each i individual in population* **do**
      mStrategy ← random(0,1) //Probability to choose the mutation strategy
      modifies the individual $u_{i,g}$ with the mutation strategy accordingly to mStrategy
      if $u_{i,fitness} \leq x_{i,fitness}$ **then**
      | add u_i in the offspring
      else
      | add x_i in the offspring
      end
  end
  update the mutation probabilities based on their success rate
  population ← offspring
  | g ← g +1
end

---

4   Experiments and Analysis

The algorithms was ran 30 times in five different DE configurations: $DE_{rand/1/bin}$, $DE_{best/1/bin}$, $DE_{curr-to-rand}$, $DE_{curr-to-best}$ and $DE_{Self-Adaptive}$. For the four non-adaptive versions of DE, we used the parameter CR as 1 and F as 0.5, while $DE_{Self-Adaptive}$ the parameters are initialized with 0.5 for CRm and 0.5 for F. One million of fitness evaluations were done in each run, corresponding to 10 thousand generations in total. To keep a fair comparison among all different versions, the initial population for each DE configuration is the same, avoiding that one mechanism starts with a better population than others. Achieved results are present in Table 2, by protein and DE version. Tests were performed in an *Intel Xeon E5-2650V4 30 MB, 4 CPUs, 2.2 Ghz, 96 cores/threads, 128G of RAM, and 4 TB* in disk space. To test our approach we used 9 proteins based on literature works that can be foun at the PDB. The PDB ID are: 1AB1 (46 amino acids), 1ACW (29 amino acids), 1CRN (46 amino acids), 1ENH (54 amino acids), 1ROP (63 amino acids), 1UTG (70 amino acids), 1ZDD (35 amino acids), 2MR9 (44 amino acids), and 2MTW (20 amino acids).

In order to compare the SaDE approach with the other 4 DE variations, we have applied the *Wilcoxon Signed Rank Test* (Table 2 - 3rd column), where *p-values* lower than 0.05 indicates that there is statistical relevance. It is possible to notice that $DE_{Self-Adaptive}$ got relevant results in 5 of 9 cases, and better average energy in 8 of them. In the other 3 cases, SaDE showed equivalence with $DE_{curr-to-rand}$ (1ENH and 1UTG) and $DE_{rand/1/bin}$ (1ROP), getting worst results only for 2MTW, meaning that the Self-Adaptive approach is better, or at least equivalent, to the non-adaptive version of DE using only one mutation mechanism in 8 of 9 cases. The convergence analysis are presented by two proteins (1ROP and 1UTG) in Figs. 1 and 2. For other proteins, the patterns are quite similar, changing accordingly with the dimensionality each protein presents. For 1ROP protein (Fig. 1) it is possible to notice that $DE_{Self-Adaptive}$ better explore the search space during the optimization process, leading to better energy values, and avoiding premature convergence as observed by $DE_{best/1/bin}$. A similar analysis can be done in the 1UTG protein's optimization process (Fig. 2), where the diversity index from $DE_{Self-Adaptive}$ is significant even in the end of the optimization process. This behavior shows that it is possible to keep optimizing the search space for even better solutions.

This analysis shows that the combination of different mutation mechanisms during the optimization process can be beneficial to the balance between exploration and exploitation capabilities. It is possible to observe that elitist approaches ($DE_{best/1/bin}$ and $DE_{curr-to-best}$) are not so good when used alone during the whole process, but they are useful in small portions of generations. Since the determination of **when** apply this type of technique, the Self-Adaptive mechanism can decide by itself when to use each mutation operator. Also, the self-adaptive mechanism used adapts the mutation and crossover factors (F and CR), which might contribute to better search space exploration. It is noteworthy that each protein configures a different search space. Hence, the parameter setting for one protein might not be better for every other protein. The same

Table 2. Results obtained by the 5 DE approaches. Bolded lines presents the approaches with best energy results accordingly to *Wilcoxon Signed Rank Test*.

PDB	Strategy	Energy	p-value
1AB1	$DE_{rand/1/bin}$	$-98.00(-75.48 \pm 9.54)$	0.00
	$DE_{best/1/bin}$	$-152.24(-95.32 \pm 18.48)$	0.00
	$DE_{curr-to-rand}$	$-169.14(-109.07 \pm 17.29)$	0.00
	$DE_{curr-to-best}$	$-158.14(-122.57 \pm 15.64)$	0.00
	$DE_{Self-Adaptive}$	**$-184.62(-157.08 \pm 20.37)$**	–
1ACW	$DE_{rand/1/bin}$	$-148.22(-25.17 \pm 41.97)$	0.00
	$DE_{best/1/bin}$	$-133.85(-88.22 \pm 39.33)$	0.00
	$DE_{curr-to-rand}$	$-135.75(-63.13 \pm 24.92)$	0.00
	$DE_{curr-to-best}$	$-160.84(-111.85 \pm 26.69)$	0.00
	$DE_{Self-Adaptive}$	**$-203.31(-161.35 \pm 20.04)$**	–
1CRN	$DE_{rand/1/bin}$	$-95.03(-72.76 \pm 6.13)$	0.00
	$DE_{best/1/bin}$	$-136.18(-93.92 \pm 16.06)$	0.00
	$DE_{curr-to-rand}$	$-188.41(-113.55 \pm 23.59)$	0.00
	$DE_{curr-to-best}$	$-173.95(-129.20 \pm 23.17)$	0.00
	$DE_{Self-Adaptive}$	**$-185.67(-154.13 \pm 18.55)$**	–
1ENH	$DE_{rand/1/bin}$	$-343.13(-334.83 \pm 3.08)$	0.00
	$DE_{best/1/bin}$	$-364.38(-348.84 \pm 7.92)$	0.00
	$DE_{curr-to-rand}$	**$-376.11(-363.21 \pm 10.90)$**	0.20
	$DE_{curr-to-best}$	$-368.94(-359.37 \pm 5.06)$	0.00
	$DE_{Self-Adaptive}$	**$-375.94(-367.26 \pm 4.35)$**	–
1ROP	**$DE_{rand/1/bin}$**	**$-498.18(-485.32 \pm 6.59)$**	0.28
	$DE_{best/1/bin}$	$-471.52(-458.66 \pm 6.13)$	0.00
	$DE_{curr-to-rand}$	$-484.88(-475.80 \pm 3.14)$	0.00
	$DE_{curr-to-best}$	$-477.11(-468.65 \pm 4.64)$	0.00
	$DE_{Self-Adaptive}$	**$-507.13(-488.14 \pm 8.78)$**	–
1UTG	$DE_{rand/1/bin}$	$-514.55(-487.69 \pm 10.24)$	0.00
	$DE_{best/1/bin}$	$-516.13(-497.01 \pm 9.29)$	0.00
	$DE_{curr-to-rand}$	**$-545.70(-533.13 \pm 8.03)$**	0.42
	$DE_{curr-to-best}$	$-536.09(-515.88 \pm 9.49)$	0.00
	$DE_{Self-Adaptive}$	**$-544.34(-534.29 \pm 5.72)$**	–
1ZDD	$DE_{rand/1/bin}$	$-233.00(-225.00 \pm 3.78)$	0.00
	$DE_{best/1/bin}$	$-232.28(-225.54 \pm 3.66)$	0.00
	$DE_{curr-to-rand}$	$-245.71(-236.38 \pm 4.22)$	0.00
	$DE_{curr-to-best}$	$-240.61(-231.89 \pm 4.05)$	0.00
	$DE_{Self-Adaptive}$	**$-245.49(-240.38 \pm 3.26)$**	–
2MR9	$DE_{rand/1/bin}$	$-287.20(-264.20 \pm 11.33)$	0.00
	$DE_{best/1/bin}$	$-282.84(-270.72 \pm 6.96)$	0.00
	$DE_{curr-to-rand}$	$-296.22(-289.38 \pm 3.28)$	0.02
	$DE_{curr-to-best}$	$-290.33(-283.44 \pm 4.76)$	0.00
	$DE_{Self-Adaptive}$	**$-299.87(-290.89 \pm 3.86)$**	–
2MTW	**$DE_{rand/1/bin}$**	**$-109.56(-102.87 \pm 3.45)$**	0.01
	$DE_{best/1/bin}$	$-95.02(-90.62 \pm 2.12)$	0.00
	$DE_{curr-to-rand}$	$-104.58(-98.74 \pm 2.88)$	0.01
	$DE_{curr-to-best}$	$-101.91(-94.70 \pm 2.53)$	0.00
	$DE_{Self-Adaptive}$	$-105.3(-100.66 \pm 2.13)$	–

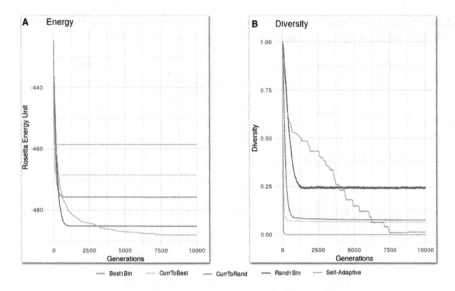

Fig. 1. PDB ID 1ROP convergence of energy and diversity for all five Differential Evolution versions. Both plots consider the average among all runs.

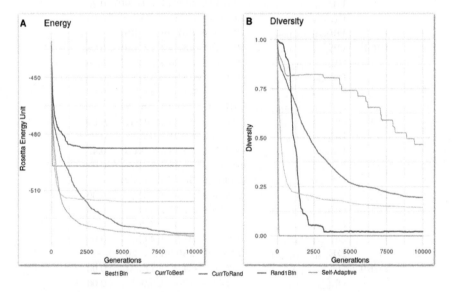

Fig. 2. PDB ID 1UTG convergence of energy and diversity for all five Differential Evolution versions. Both plots consider the average among all runs.

assumption can be used for mutation operators. Final conformations are compared with the experimental ones and reported in Fig. 3.

It is possible to notice that $DE_{Self-Adaptive}$ achieved the better results in terms of Energy and RMSD when compared with the other approaches. Of course,

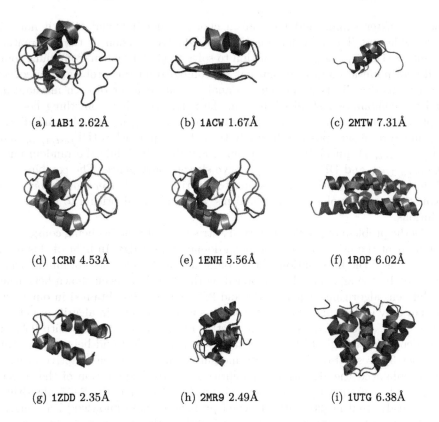

(a) 1AB1 2.62Å   (b) 1ACW 1.67Å   (c) 2MTW 7.31Å

(d) 1CRN 4.53Å   (e) 1ENH 5.56Å   (f) 1ROP 6.02Å

(g) 1ZDD 2.35Å   (h) 2MR9 2.49Å   (i) 1UTG 6.38Å

Fig. 3. Cartoon representation of experimental structures (red) compared with lowest energy solutions (blue) found by SaDE version. (Color figure online)

it is needed further investigations to improve the conformational search method, helping the algorithm to reach more similar structures, with lower energy and RMSD. It is important to realize that the RMSD values should decrease within the energy values, but as the energy functions are computational approximations, and the search space has multimodal characteristics, it is possible to have conformations with higher energy values but with lower RMSD values. However, it is expected that if the minimum global energy is found, the RMSD might be 0, finding the correct structure.

5   Conclusion

As shown in many works in literature, the PSP problem still an open issue in Bioinformatics that can contribute for life-sciences. Besides the significant advances in the problem, it is still needed advances in better search methods for prediction of proteins. As proteins have different characteristics among them, and metaheuristics are very sensitive in the use of parameters and operators, the parameter tuning and mechanism choice is not a trivial task, if not an impossible one. Thus, we have used a self-adaptive version of the differential evolution ($DE_{Self-Adaptive}$) algorithm to solve the PSP problem, combining four well-known mutation mechanisms not yet combined for this problem. Moreover, we have used a diversity measure to analyze the behavior of each mechanism and the combination of all of them in the SaDE algorithm, something not yet explored in the literature. Accordingly to the convergence graphs and diversity measure, it was possible to verify that elitist approaches ($DE_{best/1/bin}$ and $DE_{curr-to-best}$) quickly loss the populational diversity while the random ones ($DE_{rand/1/bin}$ and $DE_{curr-to-rand}$) have slower convergence. The combination of them in a self-adaptive model seems to contribute to a better balance between exploitation and exploration mechanisms, allowing the algorithm to find better solutions.

As the problem of predicting tertiary structures of proteins being complex, it is imminent the usage of some problem-domain knowledge. In light of this fact, we have used the information of the conformational preferences of amino acids provided by the APL. The data supplied by the APL have been shown beneficial in different algorithms, such as GAs and PSO. The results obtained in our work are not only interesting regarding problem-solving, but also in algorithm behavior analysis. The $DE_{Self-Adaptive}$ got better results in 5 of 9 cases with 95% of confidence accordingly to the *Wilcoxon Signed Rank Test* and being equivalent in other 3 cases. Also, the diversity measure showed that self-adaptive mechanisms enhanced the algorithm capabilities for better exploration of the search space and, consequently, better energy results. Although the SaDE algorithm was already used in [20] with attached problem-domain knowledge, an analysis of each mutation mechanism was not found, neither the energy values or any type of convergence trace was done. In this way, the present work closed this gap, providing the opportunity to do further investigations of self-adaptive algorithms using APL as a knowledge database to enhance the algorithm capabilities.

For future works, it is intended to expand the usage of APL in different ways, not only in the initial population. Also, it would be interesting to add more DE mechanisms, comparing their behavior with specific metrics (such as the diversity measurement), and how they contribute to the self-adaptive algorithm for better search capabilities. It is important to do better investigations about the energy functions, verifying the possibility of multiobjective problem formulation as already seen in other PSP predictors that used different energy functions to guide the search mechanism.

Acknowledgements. This work was supported by grants from FAPERGS [*16/2551-0000520-6*], MCT/CNPq [*311022/2015-4; 311611/2018-4*], CAPES-STIC AMSUD [*88887.135130/2017-01*] - Brazil, Alexander von Humboldt-Stiftung (AvH) [*BRA 1190826 HFST* CAPES-P] - Germany. This study was financed in part by the Coordenacão de Aperfeiçoamento de Pessoal de Nível Superior - Brazil (CAPES) - Finance Code 001.

References

1. Anfinsen, C.B.: Principles that govern the folding of protein chains. Science **181**(4096), 223–230 (1973)
2. Berman, H.M., et al.: The protein data bank. Nucleic Acids Res. **28**, 235–242 (2000)
3. Boas, F.E., Harbury, P.B.: Potential energy functions for protein design. Curr. Opin. Struct. Biol. **17**(2), 199–204 (2007)
4. Bonneau, R., Baker, D.: Ab initio protein structure prediction: progress and prospects. Ann. Rev. Biophys. Biomol. Struct. **30**(1), 173–189 (2001)
5. Borguesan, B., e Silva, M.B., Grisci, B., Inostroza-Ponta, M., Dorn, M.: APL: an angle probability list to improve knowledge-based metaheuristics for the three-dimensional protein structure prediction. Comput. Biol. Chem. **59**, 142–157 (2015)
6. Borguesan, B., Inostroza-Ponta, M., Dorn, M.: NIAS-server: neighbors influence of amino acids and secondary structures in proteins. J. Comput. Biol. **24**(3), 255–265 (2017)
7. Borguesan, B., Narloch, P.H., Inostroza-Ponta, M., Dorn, M.: A genetic algorithm based on restricted tournament selection for the 3D-PSP problem. In: 2018 IEEE Congress on Evolutionary Computation (CEC), pp. 1–8. IEEE, July 2018
8. Chaudhury, S., Lyskov, S., Gray, J.J.: PyRosetta: a script-based interface for implementing molecular modeling algorithms using Rosetta. Bioinformatics **26**(5), 689–691 (2010)
9. de Lima Corrêa, L., Borguesan, B., Krause, M.J., Dorn, M.: Three-dimensional protein structure prediction based on memetic algorithms. Comput. Oper. Res. **91**, 160–177 (2018)
10. Corriveau, G., Guilbault, R., Tahan, A., Sabourin, R.: Review of phenotypic diversity formulations for diagnostic tool. Appl. Soft Comput. J. **13**(1), 9–26 (2013)
11. Das, S., Mullick, S.S., Suganthan, P.N.: Recent advances in differential evolution - an updated survey. Swarm Evol. Comput. **27**, 1–30 (2016)
12. Dorn, M., e Silva, M.B., Buriol, L.S., Lamb, L.C.: Three-dimensional protein structure prediction: methods and computational strategies. Comput. Biol. Chem. **53**(PB), 251–276 (2014)
13. Du, K.L.: Search and Optimization by Metaheuristics Techniques and Algorithms Inspired by Nature (2016)
14. Eiben, E., Hinterding, R., Michalewicz, Z.: Parameter control in evolutionary algorithms - evolutionary computation. IEEE Trans. Evol. Comput. **3**(2), 124–141 (1999)
15. Guyeux, C., Côté, N.M.L., Bahi, J.M., Bienie, W.: Is protein folding problem really a NP-complete one? First investigations. J. Bioinformat. Comput. Biol. **12**(01), 1350017(1)–1350017(24) (2014)
16. Kabsch, W., Sander, C.: Dictionary of protein secondary structure: pattern recognition of hydrogen-bonded and geometrical features. Biopolymers **22**(12), 2577–2637 (1983)
17. Ligabue-Braun, R., Borguesan, B., Verli, H., Krause, M.J., Dorn, M.: Everyone is a protagonist: residue conformational preferences in high-resolution protein structures. J. Comput. Biol. **25**(4), 451–465 (2017)
18. Narloch, P.H., Parpinelli, R.S.: Diversification strategies in differential evolution algorithm to solve the protein structure prediction problem. In: Madureira, A.M., Abraham, A., Gamboa, D., Novais, P. (eds.) ISDA 2016. AISC, vol. 557, pp. 125–134. Springer, Cham (2017). https://doi.org/10.1007/978-3-319-53480-0_13

19. Narloch, P., Parpinelli, R.: The protein structure prediction problem approached by a cascade differential evolution algorithm using ROSETTA. In: Proceedings - 2017 Brazilian Conference on Intelligent Systems, BRACIS 2017 (2018)
20. Oliveira, M., Borguesan, B., Dorn, M.: SADE-SPL: a self-adapting differential evolution algorithm with a loop structure pattern library for the PSP problem. In: 2017 IEEE Congress on Evolutionary Computation (CEC), pp. 1095–1102 (2017)
21. Parpinelli, R.S., Plichoski, G.F., Samuel, R., Narloch, P.H.: A review of techniques for on-line control of parameters in swarm intelligence and evolutionary computation algorithms. Int. J. Bio-inspired Comput. **13**, 1–20 (2018)
22. Qin, A., Suganthan, P.: Self-adaptive differential evolution algorithm for numerical optimization. In: 2005 IEEE Congress on Evolutionary Computation, vol. 2, 1785–1791 (2005)
23. Rohl, C.A., Strauss, C.E., Misura, K.M., Baker, D.: Protein Structure Prediction Using Rosetta, pp. 66–93 (2004)
24. Storn, R., Price, K.: Differential evolution - a simple and efficient heuristic for global optimization over continuous spaces. J. Glob. Optim. **11**(4), 341–359 (1997)
25. Venske, S.M., Gonçalves, R.A., Benelli, E.M., Delgado, M.R.: ADEMO/D: an adaptive differential evolution for protein structure prediction problem. Expert Syst. Appl. **56**, 209–226 (2016)
26. Walsh, G.: Proteins: Biochemistry and Biotechnology. Wiley, Hoboken (2014)

A Multi-objective Swarm-Based Algorithm for the Prediction of Protein Structures

Leonardo de Lima Corrêa and Márcio Dorn[✉]

Institute of Informatics, Federal University of Rio Grande do Sul,
Porto Alegre, Rio Grande do Sul, Brazil
{llcorrea,mdorn}@inf.ufrgs.br

Abstract. The protein structure prediction is one of the most challenging problems in Structural Bioinformatics. In this paper, we present some variations of the artificial bee colony algorithm to deal with the problem's multimodality and high-dimensionality by introducing multi-objective optimization and knowledge from experimental proteins through the use of protein contact maps. Obtained results regarding measures of structural similarity indicate that our approaches surpassed their previous ones, showing the real need to adapt the method to tackle the problem's complexities.

Keywords: Swarm intelligence · Multi-objective optimization · PSP

1 Introduction

The protein structure prediction (PSP) remains as one of the most challenging problems in Bioinformatics. Proteins are in all living systems and are responsible for a massive set of functions, participating in almost all cellular processes. Knowing the protein structure allows one to study biological processes more thoroughly. The PSP is classified as NP-hard problem in accord with the computational complexity theory [19], due to the multi-modal search space and high dimensionality, presenting an exponential growth of difficulty as the protein's size increases. Problem complexity relies on protein conformations' explosion, where a long amino acid (aa) chain can give rise to few conformations around a native state among numerous existing possibilities.

An extensive range of computational methods has been presented for the PSP problem. The existing methods can be classified into two major categories in accordance with the target protein characteristics [7,15]: (i) template-based modeling (TBM); and (ii) free-modeling (FM). So the first one encompasses aa sequences that have detectable evolutionary similarities to the experimentally determined ones, making it possible to identify similar structural models and ease prediction process. Differently, FM represents aa sequences which do not exhibit similarities to the experimentally determined proteins. Difficulty relies on

© Springer Nature Switzerland AG 2019
J. M. F. Rodrigues et al. (Eds.): ICCS 2019, LNCS 11538, pp. 101–115, 2019.
https://doi.org/10.1007/978-3-030-22744-9_8

the target modeling through *ab initio* methods which may incorporate protein structural information from databases. Methods under this classification generally represent hybrid approaches that use *aa* fragments combined to a purely *ab initio* strategy. *Ab initio* methods are based only on thermodynamic concepts and physicochemical properties of the folding process of proteins in nature.

It is well known that the energy function inaccuracies and the multi-modal search space are enough factors in expanding efforts to develop new strategies to obtain not only better structural results but insights about intrinsic and hidden problem properties. Multi-objective (MO) strategies aim to deal with optimization problems from different perspectives. Generally, complex problems present objective functions with several terms, many of them conflicting with each other, which, in turn, makes it hard to simultaneously optimize them properly [13]. Also, such problems may have specific properties that are not often considered in optimization processes, for reasons of simplicity or even inability to integrate them into the evaluation function when single-objective optimization [9]. In this sense, we adapted the Mod-ABC algorithm [5] to deal with the PSP by introducing MO strategies [9,13], in order to minimize the existing conflicts between energy function terms and reach an acceptable balance among them, and evaluate the MO algorithms in the face of a quite difficult problem. These new algorithms incorporated another experimentally determined protein structures' knowledge strategy besides the ones already integrated into the Mod-ABC. Encouraged by the latest CASP results [18], we modeled the information of contact maps (CMs) [12,18] as a term added to the energy function. CMs are predicted from analysis of correlated evolutionary mutations achieved from multiple sequence alignments. In this work, it was used as constraints in the algorithm calculation to support the heuristic, deal with the search space roughness and reduce its size. An assessment of CMs contribution to the solution quality was carried out regarding single and multi-objective optimization. Our major contribution in this work is the development and assessment of the ABC algorithm adaptation to work with MO strategies and also handle the information of CMs as constraints in optimization to reach better prediction results.

2   Problem Background

The methods described in this work are variants of the Mod-ABC algorithm [5]. All of them adopt the same computational protein representation and the Angle Probability List technique. Methods accept as input parameters the protein primary structure, its expected secondary structure (SS) and the generated CMs.

A. Protein Representation: From a structural perspective, a peptide is formed by two or more amino acids joined by a chemical bond known as a peptide bond. Larger peptides are known as polypeptides or proteins. So the proteins are represented by linear *aa* sequences, responsible for determining their conformations. The protein folding gives the protein-specific properties, which dictate its role in the cell. The amino acids found in proteins present all the same main structure, the backbone, and differ in the side chain structure. In an *aa* chain, the peptide bond, known as Omega angle (C-N, ω), has a partial double bond character which does not allow the free molecule rotation around it. Conversely, the free molecule rotation is allowed over the bonds known as Phi (N-C$_\alpha$) and Psi (C$_\alpha$-C) dihedral angles, ranging under a continuous domain from $-180°$ to $+180°$. Such free rotation is mostly responsible for the 3-D structure assumed by the protein, whereas the amino acids' stable local arrangements define the SS. As the polypeptide backbone, side chains present dihedral angles too, known as Chi angles (χ). Their conformations contribute to the stabilization and packing of the protein structure. The Chi angles number in an *aa* is concerned to its type, varying from 0 to 4, ranging under a continuous domain from $-180°$ to $+180°$. Thereby, the protein's set of dihedral angles form its 3-D structure. In this paper, the protein structure was computationally represented by its dihedral angles as a way to reduce the use complexity of all-atom representation of the protein.

B. Objective Function: To assess the quality of a modeled protein structure, we adopted as fitness function the Rosetta energy function (all-atom high-resolution and minimization function) [17] provided by the PyRosetta toolkit https://www.rosettacommons.org. The Rosetta energy function considers more than 18 energy terms, most of them derived from knowledge-based potentials [17]. The function has terms based on Newtonian physics, inter-atomic electrostatic interactions and hydrogen bonding energies dependent on the orientation. According to the CASP experiments, Rosetta methods have reached one of the best results in the competition [15]. The final energy value of the Rosetta function ($E_{rosetta}$) is given by the sum of all weighted terms considered in the calculation. The terms' weights are defined based on the energy function *Talaris2014*, that is the standard Rosetta function used to assess *all-atom* protein structures. Additionally to the Rosetta terms, the solvent accessible surface area from the PyRosetta was included as a term ($SASA_{term}$) into the final energy function [5] with an atomic radius of 1.4 Å, to assist the 3-D structures packing given the difficulties presented by *Talaris2014* in such task. Also, to support the secondary structures formation, the SS term (Eq. 1) was added to the fitness function. The procedure gives: (i) a positive reinforcement to the energy function, adding a negative constant (-1000) to the sum of amino acids of the protein structure P, if the SS (zp_i) corresponding to the i-th amino acid (aa_i) is equal to the SS (zi_i) of the same *aa* informed as input to the method; or (ii) gives a negative reinforcement to the sum, adding a positive constant ($+1000$), when the SS of the corresponding amino acids are not the same. All protein amino acids are compared throughout the model evaluation. We used the DSSP method (https://swift.cmbi.umcn.nl/gv/dssp/) to assign the secondary structures. Finally, the terms previously described were integrated to the Rosetta function composing the evaluation function (E_{final}) (Eq. 3) used in this work.

$$SS_{term} = \sum_{aa \in P} V(aa_i, zp_i, zi_i) \tag{1}$$

$$V(aa, zp, zi) = \begin{cases} -const, & zp = zi \\ +const, & zp \neq zi \end{cases} \tag{2}$$

$$E_{final} = E_{rosetta} + SASA_{term} + SS_{term} \tag{3}$$

C. Amino Acids Conformational Preferences: The methods in this paper use the knowledge of experimental protein structures in the Protein Data Bank (PDB) (https://www.rcsb.org). The main benefit of using this information is to reduce the search space size and increase the effectiveness of the method. In the Mod-ABC [5], the authors incorporated the structural information of known protein templates to determine the conformational preferences of a target protein using the Angle Probability List (APL) strategy [4]. Such technique assigns the dihedral angles to the target amino acids by the conformational preferences analysis of such amino acids in experimental structures, regarding the secondary structures and the neighboring amino acids. To use it, according to the authors, they built histograms of $[-180°, 180°] \times [-180°, 180°]$ cells for each amino acid and SS, generating combinations up to 9 amino acids (1–9) and their secondary structures, and taking into account the reference aa's neighborhood for combinations larger than 1. We note that the angle values are attributed only to the reference aa. Each histogram cell (i,j) has the number of times that a given aa (or combination of amino acids) presents a torsion angles pair ($i \leq \phi < i+1$, $j \leq \psi < j+1$) concerning a SS. The angle probability list was calculated for each histogram, representing the normalized frequency of each cell. APL was incorporated in the methods to create short combinations of amino acids aiming the use of high-quality individuals as a starting point and after a restarting function. A weighted random selection was employed to select the angle values from APL. It gives greater chances to the histograms' cells that present a higher relative frequency of occurrence. Furthermore, for a full APL description, we point out our web server NIAS-Server [4] created to investigate the amino acids conformational preferences.

D. Protein Contact Maps: The prediction of protein contact maps is based on the knowledge discovery from experimental protein structures data and tries to probabilistically determine which residues are in contact. There are several proposed contact map predictors in the literature [18]. Most of them explore strategies of machine learning, such as deep learning networks and support vector machines with classical biological features, like SS, solvent accessibility and sequence profile [2]. Ultimately, the incorporation of contact predictions from coevolution-based methods as additional features also significantly improved their performance [2,18]. In the last years, contact predictions were shown to be a valuable addition to the PSP methods [18]. As reported, improved contact methods can lead to improved FM model accuracy [1]. However, despite the improvement in the residue-residue contact prediction, its use in an efficient way into the PSP algorithms configures the major challenge [18]. Various factors determine the methods' performance, such as the number of contacts considered and how they are incorporated into the modeling. Hence, the most suitable contact prediction technique and the number of contacts to consider are dependent on the PSP algorithm. As pointed out by the last CASP report [18], the use of size lists of $L/2$ contacts can improve the performance, reducing the false positives and taking into account the predicted residue contacts with higher probabilities of being in contact. L represents the target aa sequence length. By the prediction results carried out in the experiments, list sizes of $L/2$ seemed to be one of the best choices. In this paper, we used a reduced list of $L/2$ predicted contacts. The CMs were predicted by the MetaPSICOV predictor [10].

In a CM, two amino acids are close enough or in contact, if the distance between their $C\beta$ side chain atoms, or $C\alpha$ of backbone for Glycine, is less than or equal to a distance threshold, generally $8\,\text{Å}$. A term of distance constraint is generally used to get the information from CMs and to overcome some inaccuracies of the energy function [12]. In this paper, besides the terms of the fitness function described in Eq. 3, we proposed a scheme to employ the information of CMs in the problem as a new term in the energy function. This term was idealized based on an atom distance constraint function presented in the work of Kim et al. [12]. It was modified to follow the same idea of weighting used in the SS term (Eq. 1). The CM term is a function of the distances between the aa contained into the CMs, and it aims to positively reinforce the aa pairs that are within the contact bounds or to penalize the ones that are out of the threshold, according to Eq. 4.

$$CM_{term} = \sum_{i,j}^{CM_{pairsL/2}} = \begin{cases} p \times -c, & d(i,j) \leq ub \\ p \times -c \div 2, & ub < d(i,j) \leq ub + 2 \\ p \times +c, & d(i,j) > ub + 2 \end{cases} \tag{4}$$

where p denotes the probability of the residues are in contact, c is a constant, ub is a residue contact upper bound and $d(i,j)$ represents the Euclidean distance between a pair of amino acids in the predicted contact list. The MetaPSICOV considers the ub contact threshold of $8\,\text{Å}$, so in this paper, we adopted the same threshold of distance. For the constant c, we adopted $c = 1000$ to follow the reinforcement values defined in the SS term (Eq. 1). So for a target protein, the procedure goes through the $L/2$ aa pairs in the predicted CM, measuring the distances between these pairs regarding a given protein model. It gives (i) a positive reinforcement to the term summation, adding a negative constant $(-c)$ multiplied by the probability of the residues are being in contact, if the distance between them is less than or equal to the ub threshold; (ii) also a positive reinforcement to the term summation but considering the negative constant divided by 2 $(-c \div 2)$, if the distance between the amino acids is greater than the ub but does not exceed $ub + 2$ (tolerance threshold); and (iii) a negative reinforcement to the term summation, adding a positive constant $(+c)$ multiplied by the probability of the residues are being in contact, if the distance between the residues is greater than the threshold $ub + 2$.

E. Multi-objectivization: The multi-objectivization avoids conflicting terms compete to reach the best solutions and also favors the regarding of new properties about the problem, to aggregate information to the evaluation terms already considered in the optimization and better guide the search through new visions of the state space. In such approaches, the final optimization result encompasses a set of good solutions, called Pareto front (PF) [13]. PF represents a set of so-called non-dominated solutions. It comprises solutions where there is no possibility to improve one objective without disfavor another. Switching from one non-dominated solution to another will always result in a trade-off between objectives. The method's solutions can still be evaluated under different aspects, such as emerging features and unknown properties about the problem and input data.

In PSP, there are often conflicts between different terms of the energy function, as demonstrated by Cutello et al. [6]. The modeling of the energy function terms as independent objectives can provide a new exploration of the search space. Another interesting point is the possibility of inserting additional objectives containing information and constraints on the problem more naturally, avoiding the use of weighting coefficients, as it happens when inserted in traditional single objective approaches. Thus, the multi-objectivization of the prediction methods tends to ease the process of knowledge incorporation about the problem. An unconstrained MO optimization problem can be mathematically formulated as follows. Let $x = [x_1, x_2, ..., x_n]$ be a n-dimensional vector of decision variables, X be the search space (decision space) and Z be the objective space:

$$Minimize\ z = f(x) = [f_1(x), f_2(x), ..., f_m(x)], x \in X, z \in Z \tag{5}$$

where $m \geq 2$ is the number of objectives. Considering that during the optimization exists more than one single solution, the solutions are compared based on Pareto dominance, and the final answer is a set of non-dominate solutions (Pareto set). Let $M = [1, 2, ..., m]$ be the set of objectives, the Pareto set is defined according to the Eq. 6. A solution $x \in X$ dominates $y \in X$ ($x < y$) if and only if:

$$\forall i \in M : f_i(x) \leq f_i(y) \land \exists i \in M : f_i(x) < f_i(y), f_i(.) \in Z \tag{6}$$

To incorporate MO optimization in our algorithms and sort the solutions based on multiple objectives, we used the Pareto rank definition integrated into the evaluation function [16]. The Pareto rank of a solution measures the number of solutions that dominate it overall considered optimization objectives, regarding strict comparison ($<$), as shown in Eq. 6. So less the Pareto rank, less dominated is the solution. To order a set of solutions by Pareto rank as a minimization function, first the solutions are ordered from low to high Pareto rank and within this sorted order, those with the same Pareto rank are further ordered from low to high based on their energy values scored by an energy function.

3   Proposed Strategies

In this paper, we presented some ABC algorithm variations to tackle the PSP problem. We started from a previously proposed work, presented by Corrêa et al. [5], which has shown an ABC algorithm variation [11] implemented from suggested improvements in literature for the original ABC but never tested for the problem under study. It is called Mod-ABC and was designed to explore the specific properties of the problem. So the proposed algorithm variations were designed from the Mod-ABC based on an incremental development approach, in an attempt to improve the previously reached results. It was done by exploring additional features about the problem and adapting it to MO optimization to restrict the conformational space and overcome some energy function inaccuracies. In the following sections, Mod-ABC and the designed variations of it are presented.

A. Artificial Bee Colony Algorithm: ABC consists of a swarm intelligence based metaheuristic. It mimes the foraging process of honeybee swarms and is suitable for multi-numerical and multi-modal optimization [3,11]. Various works and ABC variations have been proposed indicating the algorithm competitiveness concerning other metaheuristics, such as genetic and differential evolution algorithms, particle swarm optimization and swarm-based algorithms [11]. It is said the key advantage of the heuristic is the use of a few control parameters [8]. In the ABC, the solution exploration and exploitation (refinement) are crucial optimization components. But the method has some inefficiencies, such as to perform well at the exploration but not so much at the solution refinement step [8]. This causes the heuristic's convergence slower and can be a problem on some occasions. To overcome it, improved ABC versions have been proposed in the literature. It was shown that these modified variations could be able to perform better than the original ABC [14]. Thus, the Mod-ABC assembles two proposed strategies for the algorithm. The first component, introduced in the work of Akay and Karaboga [3], concerns changes in the mechanisms that control the mutation frequency of variables of an individual and at the use of the most reasonable parameterization in the exploration ABC stage. The second one, presented by Zhu and Kwong [20], is related to the gbest-guided ABC (GABC). It uses the information regarding the best population's solution in the individual's mutation equation to improve the exploitation step. Authors of both methods pointed out that the ABC could be considered a promising metaheuristic regarding global and local optimization.

B. Mod-ABC Algorithm: In the ABC [3,11], each food source is a problem solution, and the solution quality is defined by the fitness value. Concerning the PSP, the food source means a possible solution for the protein under study and the quality of it is given by the energy value. The food sources are exploited by employed bees. Thus, the number of employed bees is the same number of food sources, i.e., the size of the population. The onlooker bees amount in the swarm is the same employed bees amount. Suppose that SN is the food sources amount (population's solutions), eb and ob the number of employed and onlooker bees, respectively. So $SN = eb = ob$. The algorithm mimics the foraging behavior of honeybees regarding three steps: (i) in the employed bees' step (Algorithm 1, lines 3 to 10) each algorithm's solution represents a food source that is *updated* by a mutation procedure; (ii) in the onlooker bees' step (Algorithm 1, lines 18 to 27), *ob* individuals are randomly selected through the rank-based selection and the *update* procedure of the preceding stage is performed in the selected individuals; and (iii) in the scout bees' step (Algorithm 1, line 28) the most inactive population's individual is discarded and a new one is generated. An inactive individual is a solution that did not suffer improvements (fitness value) for a given number of generations. The update procedure (Algorithm 1, lines 5 and 21) used in the first two stages is responsible for generate a new individual from an existing one. So the generation of an individual $v_i = [v_{i1}, v_{i2}, ..., v_{in}]$ from the i-th individual $x_i = [x_{i1}, x_{i2}, ..., x_{in}]$, such that $x_i = v_i$, is described by (7).

$$v_{ij} = x_{ij} + \delta_{ij}(x_{ij} - x_{kj}) + \gamma_{ij}(y_j - x_{ij}), \tag{7}$$

where $i = [1, ..., SN]$, $j = [1, ..., n]$. SN represents the population size and n is the problem dimensionality. x_{ij} represents the j-th variable of individual x_i, v_{ij} is the new x_{ij} value, x_{kj} represents the j-th variable of the k-th population's individual ($k = [1, ..., SN]$) randomly chosen, and δ_{ij} means a random value in the continuous range $[-1, 1]$. The last term of 7 considers the population's best solution in the mutation operation. y_j denotes the j-th variable of the best individual and γ_{ij} represents a random value in the continuous range $[0, 1.5]$. Thus, the term presented by Zhu and Kwong [20] tries to guide the individual towards the population's best solution, increasing the algorithm convergence. Each variable j of the individual x_i is mutated regarding the control parameter MR (Algorithm 1, lines 4 and 20). Mod-ABC was set with $MR = 0.4$, according to the work of Akay and Karaboga [3]. So the update of a variable is done under the probability of 40%. The updating procedure concludes with a greedy selection between v_i and x_i (Algorithm 1, lines 8 and 24). Following the representation adopted in the paper (Sect. 2-A), each variable is an *aa* of the protein which has up to seven angles. Thus, the dihedral angles of the same variable are mutated in the same manner. To adjust the algorithm to the specific problem's characteristics, the Mod-ABC incorporates the function of *angle verification* (Algorithm 1, lines 6 and 22) into the updating procedure concerning the new generated values. The function verifies, at each angle mutation of the variable v_{ij}, if the newly generated value is in APL-1. It defines the *aa* conformational preferences regarding the variable v_{ij} and is used to avoid unfavorable state space regions or out of interval $[-180, 180]$. If the procedure verifies that the new value is not in the APL-1 or is out of the allowed interval, this value is discarded and the previous value is maintained. Lastly, in the scout bees' stage, if some population's individual did not suffer improvements over l generations, it is discarded and a new solution is included in the population (Algorithm 1, line 28). Suppose that l is the discarding threshold. We have used $l = 200$ according to Akay and Karaboga [3] and $SN = 300$ as population size, according to Corrêa et al. [5].

Irregular regions of proteins, such as coils and turns, are the hardest ones to predict because of the solvent exposure, configuring then structures with high flexibility level and low stability. Regarding the Mod-ABC, the algorithm focuses its search effort solely in such protein regions, excluding the more stable secondary structures, as β-sheets and α-helices, from the refinement process. Thus, the *updating* function (Algorithm 1, lines 5 and 21) is performed just in variables concerned the amino acids which present irregular secondary structures. To enhance the exploration aspect of the algorithm and increase the solutions diversity, as the updating of variables (Algorithm 1, lines 5 and 21) is constrained to the protein irregular secondary structures, the algorithm incorporates a crossover operation between two solutions of the population (Algorithm 1, line 14). The crossover was included between the first two Mod-ABC stages. The parents are selected through the rank-based strategy of selection (Algorithm 1, lines 12 and 13) and the operation is performed over the SS uniform crossover. The crossover concludes with a greedy selection between the generated solution and its parents (Algorithm 1, line 16). It is noteworthy that the Mod-ABC was implemented to assess in which way the knowledge-based strategies contribute to the algorithm performance facing a complex problem. The authors have shown by the obtained results that the method was able to outperform the ABC algorithm, corroborating the necessity of adapting the method to tackle the problem.

SS Uniform Crossover: From the proteins' structural preferences, it was created to support the secondary structures formation. The operator gives priority to the solutions that formed the appropriate arrangement concerning the SS input parameter. The crossover aims to maintain the similarity found so far between the solutions' secondary structures that are being optimized and the previously informed SS to create offspring with suitable secondary arrangements. Analogous to the uniform crossover, for each *aa* (specific positions of the angles in the vector solution), all the angles related to it are considered either from parent 1 or 2. The probability of 0.5 is used if both the secondary structures regarding the individuals' amino acids are equal or different from the previously informed SS. If only one of them is equal to the SS sequence parameter, the dihedral angles related to this amino acid are attributed to the offspring.

C. First Variation of the Mod-ABC Algorithm: The first variation of the Mod-ABC encompass modification just in the energy function used to assess the quality of a given protein structure. This version is called Mod-ABC-CM and incorporates the CM term (Eq. 4), already described in Sect. 2-D, into the final evaluation function. The CM term was designed to consider the information of protein contact maps in the PSP. The term was idealized in a way that penalizes violation of a predefined contact threshold regarding the distance of *aa* pairs in the CMs. In this sense, the CM term is added to the summation of all the terms already considered in the energy function (Rosetta energy function, SASA term, and SS term) (Sect. 2-B), forming then the final scoring function (Eq. 8) for the Mod-ABC-CM.

D. MO Versions of the Mod-ABC Algorithm

MO-ABC-1 Algorithm: The first MO version adapted from the Mod-ABC algorithm, called as MO-ABC-1, considers two objectives in the optimization process (bi-objective optimization). As first objective, the algorithm uses the final evaluation function (E_{final}) (Eq. 3) defined in Sect. 2-B. This scoring function is the summation result of three different terms, that is, Rosetta energy, SASA, and SS term. It is the fitness function used in the Mod-ABC algorithm. The second objective used in the MO-ABC-1 is the CM term (Eq. 4).

$$E_{finalCM} = E_{rosetta} + SASA_{term} + SS_{term} + CM_{term} \qquad (8)$$

Algorithm 1. MO-ABC-1 algorithm's pseudocode.

Require: number of energy evaluations, primary and secondary *aa* sequence
Ensure: best individual found
 1: **initialize** population using APL
 2: **while** stop criteria not satisfied **do**
 3:　　**for** each *individual* in *population* **do**   *//Employed bees's step*
 4:　　　**if** $rand(0, 1) \leq MR$ **then**
 5:　　　　**update** *individual* by (7)
 6:　　　　apply the **angle verification function**
 7:　　　　**calculate** the Pareto rank of *individual*
 8:　　　　apply a **greedy selection** between the new and old *individual*
 9:　　　**end if**
10:　　**end for**
11:　　**Sort** *population* by Pareto rank and energy value (tiebreaker criterion)
12:　　$bee_1 \leftarrow$ **select** an *individual* through rank-based selection   *//Crossover step*
13:　　$bee_2 \leftarrow$ **select** an *individual* through rank-based selection
14:　　$bee_{offspring} \leftarrow$ **SSUniformCrossover**(bee_1, bee_2)
15:　　**calculate** the Pareto rank of $bee_{offspring}$
16:　　apply a **greedy selection** between $bee_{offspring}$ and its parents
17:　　**Sort** *population* by Pareto rank and energy value (tiebreaker criterion)
18:　　**for** $i \leftarrow 1 : ob$ **do**   *//Onlooker bees's step*
19:　　　**select** an *individual* through rank-based selection
20:　　　**if** $rand(0, 1) \leq MR$ **then**
21:　　　　**update** *individual* by (7)
22:　　　　apply the **angle verification function**
23:　　　　**calculate** the Pareto rank of *individual*
24:　　　　apply a **greedy selection** between the new and old *individual*
25:　　　　**Sort** *population* by Pareto rank and energy value (tiebreaker criterion)
26:　　　**end if**
27:　　**end for**
28:　　**Discard** the most inactive individual   *//Scout bees's step*
29:　　**Sort** *population* by Pareto rank and energy value (tiebreaker criterion)
30: **end while**

One of the main reasons to consider the scoring function E_{final} as a unique objective besides the CM term is that SASA and SS terms tend to stabilize during the optimization, as the population reaches some degree of convergence. Final solutions at the end of the process tend to present similar values for these terms, as can be seen in Table 1, regarding the average and standard deviation values for eight executions of the Mod-ABC algorithm for each listed target protein [5]. So it indicates that both of the terms are more necessary at the beginning of the optimization when the population is quite diversified. Both terms improve the search space exploration providing well-formed SS and more packing protein models. On the other hand, CMs were treated as a different objective as the contacts consider punctual atom distances in a more locally point of view, based on experimental protein knowledge, which can guide the search during the entire process making finer adjustments even when the algorithm reach some diversity degree. Another reason to categorize the objectives in this fashion was to assess the potential of the MO-Mod-ABC face a complex problem but including known and promising scoring potential. It is not so obvious how to organize terms of an energy function or include new ones into MO optimization for the PSP. However, it is indicated to keep the number of objectives small [16].

Algorithm 1 shows the MO-ABC-1 algorithm's pseudocode. The main difference of the MO-ABC-1 concerning its previous versions consists of the use of the Pareto rank strategy to compare and sort solutions during the optimization. The Pareto rank strategy, as well as how it is applied to sort the population's solution was already described in Sect. 2-E. The energy function employed as tiebreaker criterion when solutions present the same Pareto rank value was the final scoring function ($E_{finalCM}$) (Eq. 8) used in the Mod-ABC-CM.

MO-ABC-2 Algorithm: The MO-ABC-2 is the second MO version idealized from the Mod-ABC algorithm. It is basically the same MO-ABC-1 algorithm. However, it considers four objectives in the optimization process. The algorithm models each term of the final evaluation function (E_{final}) (Eq. 3), defined in Sect. 2-B, as different objectives. Thus, the first objective is the Rosetta energy function, the second is the SASA term, and the third is the SS term. The MO-ABC-2 also considers the CM term as a fourth objective during the optimization process. The energy function employed as tiebreaker criterion is the same used in the MO-ABC-1.

4   Computational Experiments

The described algorithms in this paper were run 8 times with a stop criterion of 10^6 calculations of energy per run on each target protein. We have used as case studies in our tests the *aa* sequences of 8 target proteins (Table 1) obtained from the PDB. To classify our algorithms concerning the most significant methods in the area, we have compared them to the Rosetta *ab initio* protocol [17]. Following the last CASP reports, Rosetta is one of the most relevant algorithms used to tackle the PSP problem [1,15]. Obtained results are presented in the next section.

Results and Discussion: For each case study, we have analyzed the best solutions among the performed executions, regarding the root-mean-square deviation (RMSD, minimization measure) and the global distance total score test (GDT_TS, maximization measure) of the predicted structures in comparison with their corresponding experimental ones. Table 2 summarizes the obtained results of the Mod-ABC, Mod-ABC-CM, both MO Mod-ABC variations, and method of Rosetta applied to the target proteins.

Table 1. Target *aa* sequences. Average and standard deviation values for SASA and SS terms considering the best solutions of eight runs of the Mod-ABC algorithm [5] for each target protein.

Protein	Length	SS Content	SASA term		SS term	
			Avg.	Std.	Avg.	Std.
1AB1 (Fig. 1a)	46	1 β-sheet/2 α-helices	3022.81	184.91	−42500.0	1936.49
1ACW (Fig. 1b)	29	1 β-sheet/1 α-helix	2168.96	79.74	−27000.0	0.0
1AIL (Fig. 1c)	70	3 α-helices	4512.33	116.53	−70000.0	0.0
1DFN (Fig. 1d)	30	1 β-sheet	2610.51	100.2	−24500.0	1322.88
2MR9 (Fig. 1e)	44	3 α-helices	2698.88	93.02	−44000.0	0.0
2P5K (Fig. 1f)	64	1 β-sheet/3 α-helices	4581.71	282.31	−63000.0	0.0
3V1A (Fig. 1g)	48	2 α-helices	3329.42	97.05	−48000.0	0.0
T0820-D1 (Fig. 1h)	90	3 α-helices	6304.73	291.76	−89750.0	661.44

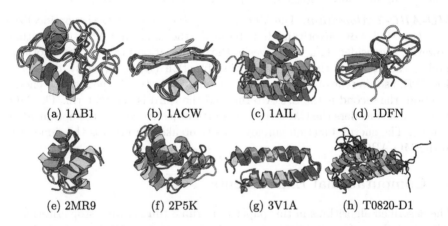

| (a) 1AB1 | (b) 1ACW | (c) 1AIL | (d) 1DFN |
| (e) 2MR9 | (f) 2P5K | (g) 3V1A | (h) T0820-D1 |

Fig. 1. Graphic representation of the experimental (red) and the predicted structures (lowest RMSD) for the Mod-ABC (green), MO-ABC-1 (blue) and Rosetta (yellow) (Color figure online).

According to the results summarized in the Table 2, we observe that the Mod-ABC-CM outperformed its previous version in almost all cases regarding lowest and average RMSD values, except for the 1ACW and T0820-D1. Similar results are noticeable analyzing the average and highest GDT_TS values, where Mod-ABC-CM performed better than Mod-ABC in 5 of the eight targets. We strongly believe that Mod-ABC-CM surpassed Mod-ABC due to the use of experimental protein knowledge through the protein CMs incorporated into the fitness function. It reduced the size and complexity of the conformational space and eased the search process. These results reinforce the need to incorporate previous knowledge about the problem in the metaheuristics.

Regarding Table 2, we observe that the MO-ABC-1 reached better average RMSD values in 5 targets in comparison with the Mod-ABC-CM, and in 4 cases regarding lowest RMSD values. Related to the GDT_TS values, MO-ABC-1 outperformed Mod-ABC-CM in 6 targets for average results and 4 cases for highest ones. We should note the MO algorithm did not show great improvement when compared to its previous version. However, these results indicate that the MO strategy has great potential to be improved. It is observable that in this work we did not explore more sophisticated strategies to improve the multi-objectivation, and even though the algorithm was able to perform better in some cases. One of the reasons for that is the MO strategies capability to keep a set of non-dominated solutions over the PF. This sort of idea can increase the solutions' diversity by exploring different perspectives of the problem. It is observable that MO-ABC-1 in average presented better results than MO-ABC-2, corroborating that the arrangement of objectives also influences the search process.

Table 2. Methods simulation results. The **boldface** numbers represent the best results concerning RMSD and GDT_TS. The (*) denotes the best results between only Mod-ABC and its variations.

Method	RMSD (Å)							
	Lowest	Avg. (std.)	Lowest	Avg. (std.)	Lowest	Avg. (std.)	Lowest	Avg. (std.)
	1AB1		1ACW		1AIL		1DFN	
Mod-ABC	4.96	6.15 ± (1.43)	1.65	2.43* ± (0.69)	6.85	7.67 ± (0.55)	4.35	5.31 ± (0.55)
Mod-ABC-CM	4.31	**5.0*** ± (0.61)	1.85	3.01 ± (0.8)	3.9	6.77 ± (1.46)	3.55	4.5 ± (0.71)
MO-ABC-1	3.83*	5.09 ± (0.7)	**1.54***	2.8 ± (0.74)	3.8*	**5.24*** ± (1.57)	3.05*	**3.91*** ± (0.62)
MO-ABC-2	4.0	5.06 ± (0.67)	2.05	2.89 ± (0.82)	3.82	5.3 ± (1.33)	3.07	4.12 ± (0.7)
Rosetta	**3.45**	5.55 ± (1.02)	1.66	**2.11** ± (0.38)	6.85	9.45 ± (1.05)	3.63	5.29 ± (0.86)
Method	2MR9		2P5K		3V1A		T0820-D1	
Mod-ABC	2.32	4.12 ± (1.83)	6.81	10.3 ± (1.72)	1.74	2.53 ± (0.81)	**6.06***	11.82* ± (2.47)
Mod-ABC-CM	1.71*	2.29 ± (0.48)	2.69*	3.75* ± (0.94)	1.28*	2.2* ± (0.56)	9.66	14.39 ± (3.43)
MO-ABC-1	2.0	2.27* ± (0.28)	3.18	4.09 ± (0.65)	1.64	2.27 ± (0.47)	10.79	13.74 ± (2.63)
MO-ABC-2	1.84	2.44 ± (0.35)	3.76	4.42 ± (0.52)	1.73	2.35 ± (0.44)	9.86	13.3 ± (1.86)
Rosetta	**1.43**	**2.22** ± (0.69)	**1.57**	**2.29** ± (1.0)	**0.7**	2.51 ± (1.9)	7.34	**9.19** ± (1.7)
Method	GDT_TS							
	Highest	Avg. (std.)	Highest	Avg. (std.)	Highest	Avg. (std.)	Highest	Avg. (std.)
	1AB1		1ACW		1AIL		1DFN	
Mod-ABC	57.07	52.17 ± (2.94)	**77.72***	69.93* ± (5.34)	50.71	44.96 ± (2.52)	46.67	41.88 ± (3.43)
Mod-ABC-CM	63.04	59.1 ± (2.46)	72.41	65.41 ± (5.07)	60.36	47.9 ± (5.72)	**53.33***	46.98 ± (3.46)
MO-ABC-1	65.76	61.21 ± (2.41)	77.59	67.03 ± (5.97)	56.79	52.46 ± (5.95)	50.83	**48.13*** ± (2.11)
MO-ABC-2	**69.02***	**61.62*** ± (4.75)	74.14	67.24 ± (4.8)	**61.43***	**53.44*** ± (6.09)	50.0	46.35 ± (2.53)
Rosetta	62.5	56.45 ± (4.27)	77.59	**73.49** ± (3.33)	48.93	39.33 ± (5.36)	49.17	44.69 ± (2.6)
Method	2MR9		2P5K		3V1A		T0820-D1	
Mod-ABC	71.02	59.66 ± (6.86)	40.48	33.93 ± (3.02)	**55.73***	**53.97*** ± (1.1)	40.0	35.59 ± (2.6)
Mod-ABC-CM	79.55*	73.22* ± (4.72)	51.19*	47.02* ± (2.7)	55.2	52.66 ± (1.6)	36.94	34.13 ± (2.13)
MO-ABC-1	75.57	71.09 ± (3.13)	49.6	45.68 ± (2.18)	55.21	53.45 ± (1.04)	40.0	35.1 ± (4.04)
MO-ABC-2	76.7	71.45 ± (2.87)	45.63	44.44 ± (0.97)	54.17	52.67 ± (1.12)	40.28*	35.9* ± (2.94)
Rosetta	**83.52**	**73.79** ± (6.59)	**53.97**	**51.54** ± (1.85)	55.21	51.44 ± (4.63)	**45.28**	**39.62** ± (3.71)

Figure 1 shows the comparison between the 3-D topology of the models predicted by Mod-ABC (green), MO-ABC-1 (blue) and Rosetta (yellow) superimposed upon the experimentally determined structures (red). Analyzing the Table 2, we notice that Rosetta surpassed all of the other algorithms regarding the lowest and average RMSD values in 4 targets and related to the highest and average GDT_TS values in 3 and 4 cases, respectively. Although it is observable by visual inspection of Fig. 1 that the MO-ABC-1 and Rosetta reached overall target folding very similar to each other and comparable to the experimentally determined structures. Finally, such results denote the importance of adapting the metaheuristic to handle the specific complexities of the PSP problem.

5 Conclusion

In this paper, we proposed some variations of the artificial bee colony algorithm to deal with the protein structure prediction problem by introducing multiobjective strategies and exploration of knowledge from experimental proteins by the use of protein contact maps. The obtained results showed that our algorithms were able to find acceptable solutions concerning RMSD and GDT_TS structural measures and outperform their previous version in most of the cases, and also reached comparable solutions to the state of the art method of Rosetta regarding experimental protein structures. Besides that the obtained results are topologically similar to the experimentally determined structures, thus corroborating the proposed strategies' promising performance for the problem.

Acknowledgements. This work was supported by grants from FAPERGS [16/2551-0000520-6], MCT/CNPq [311022/2015-4; 311611/2018-4], CAPES-STIC AMSUD [88887.135130/2017-01] - Brazil, Alexander von Humboldt-Stiftung (AvH) [BRA 1190826 HFST CAPES-P] - Germany. This study was financed in part by CAPES - Finance Code 001.

References

1. Abriata, L.A., Tamò, G.E., Monastyrskyy, B., Kryshtafovych, A., Dal Peraro, M.: Assessment of hard target modeling in CASP12 reveals an emerging role of alignment-based contact prediction methods. Proteins: Struct. Funct. Bioinf. **86**, 97–112 (2018)
2. Adhikari, B., Hou, J., Cheng, J.: Protein contact prediction by integrating deep multiple sequence alignments, coevolution and machine learning. Proteins: Struct. Funct. Bioinf. **86**, 84–96 (2018)
3. Akay, B., Karaboga, D.: A modified artificial bee colony algorithm for real-parameter optimization. Inf. Sci. **192**, 120–142 (2012)
4. Borguesan, B., Inostroza, M., Dorn, M.: NIAS-server: neighbors influence of amino acids and secondary structures in proteins. J. Comput. Biol. **24**, 255–265 (2016)
5. Corrêa, L.D.L., Dorn, M.: A knowledge-based artificial bee colony algorithm for the 3-D protein structure prediction problem. In: 2018 IEEE Congress on Evolutionary Computation (CEC), pp. 1–8, July 2018

6. Cutello, V., Narzisi, G., Nicosia, G.: A multi-objective evolutionary approach to the protein structure prediction problem. J. R. Soc. Interface **3**(6), 139–151 (2006)
7. Dorn, M., e Silva, M.B., Buriol, L.S., Lamb, L.C.: Three-dimensional protein structure prediction: methods and computational strategies. Comput. Biol. Chem. **53**, 251–276 (2014)
8. Gao, W., Liu, S., Huang, L.: A global best artificial bee colony algorithm for global optimization. J. Comput. Appl. Math. **236**(11), 2741–2753 (2012)
9. Handl, J., Lovell, S.C., Knowles, J.: Investigations into the effect of multiobjectivization in protein structure prediction. In: Rudolph, G., Jansen, T., Beume, N., Lucas, S., Poloni, C. (eds.) PPSN 2008. LNCS, vol. 5199, pp. 702–711. Springer, Heidelberg (2008). https://doi.org/10.1007/978-3-540-87700-4_70
10. Jones, D.T., Singh, T., Kosciolek, T., Tetchner, S.: MetaPSICOV: combining coevolution methods for accurate prediction of contacts and long range hydrogen bonding in proteins. Bioinformatics **31**(7), 999–1006 (2014)
11. Karaboga, D., Basturk, B.: A powerful and efficient algorithm for numerical function optimization: artificial bee colony (ABC) algorithm. J. Glob. Optim. **39**(3), 459–471 (2007)
12. Kim, D.E., DiMaio, F., Yu-Ruei Wang, R., Song, Y., Baker, D.: One contact for every twelve residues allows robust and accurate topology-level protein structure modeling. Proteins: Struct. Funct. Bioinf. **82**, 208–218 (2014)
13. Konak, A., Coit, D.W., Smith, A.E.: Multi-objective optimization using genetic algorithms: a tutorial. Reliab. Eng. Syst. Saf. **91**(9), 992–1007 (2006)
14. Li, G., Niu, P., Xiao, X.: Development and investigation of efficient artificial bee colony algorithm for numerical function optimization. Appl. Soft Comput. **12**(1), 320–332 (2012)
15. Moult, J., Fidelis, K., Kryshtafovych, A., Schwede, T., Tramontano, A.: Critical assessment of methods of protein structure prediction (CASP)-Round XII. Proteins: Struct. Funct. Bioinf. **86**, 7–15 (2018)
16. Olson, B., Shehu, A.: Multi-objective optimization techniques for conformational sampling in template-free protein structure prediction. In: International Conference on Bioinformatics and Computational Biology (2014)
17. Rohl, C.A., Strauss, C.E., Misura, K.M., Baker, D.: Protein structure prediction using Rosetta. Methods Enzymol. **383**, 66–93 (2004)
18. Schaarschmidt, J., Monastyrskyy, B., Kryshtafovych, A., Bonvin, A.M.: Assessment of contact predictions in CASP12: co-evolution and deep learning coming of age. Proteins: Struct. Funct. Bioinf. **86**, 51–66 (2018)
19. Unger, R., Moult, J.: Finding the lowest free energy conformation of a protein is an NP-hard problem. Bull. Math. Biol. **55**(6), 1183–1198 (1993)
20. Zhu, G., Kwong, S.: Gbest-guided artificial bee colony algorithm for numerical function optimization. Appl. Math. Comput. **217**(7), 3166–3173 (2010)

Combining Polynomial Chaos Expansions and Genetic Algorithm for the Coupling of Electrophysiological Models

Gustavo Montes Novaes[1,3](✉) (ID), Joventino Oliveira Campos[1,3] (ID),
Enrique Alvarez-Lacalle[2] (ID), Sergio Alonso Muñoz[2] (ID),
Bernardo Martins Rocha[1] (ID), and Rodrigo Weber dos Santos[1] (ID)

[1] Post-graduated Program in Computational Modeling,
Federal University of Juiz de Fora, Juiz de Fora, MG, Brazil
gtvmontes@gmail.com, joventinoo@gmail.com
[2] Department of Physics, Universitat Politècnica de Catalunya-BarcelonaTech,
08028 Barcelona, Spain
{enric.alvarez,s.alonso}@upc.edu
[3] Department of Computation and Mechanics, Federal Center of Technological
Education of Minas Gerais, Leopoldina, MG, Brazil
{bernardo.rocha,rodrigo.weber}@ufjf.edu.br

Abstract. The number of computational models in cardiac research
has grown over the last decades. Every year new models with different
assumptions appear in the literature dealing with differences in inter-
species cardiac properties. Generally, these new models update the physiological knowledge using new equations which reflect better the molecular basis of process. New equations require the fitting of parameters
to previously known experimental data or even, in some cases, simulated data. This work studies and proposes a new method of parameter
adjustment based on Polynomial Chaos and Genetic Algorithm to find
the best values for the parameters upon changes in the formulation of
ionic channels. It minimizes the search space and the computational cost
combining it with a Sensitivity Analysis. We use the analysis of different
models of L-type calcium channels to see that by reducing the number of
parameters, the quality of the Genetic Algorithm dramatically improves.
In addition, we test whether the use of the Polynomial Chaos Expansions
improves the process of the Genetic Algorithm search. We find that it
reduces the Genetic Algorithm execution in an order of 10^3 times in the
case studied here, maintaining the quality of the results. We conclude
that polynomial chaos expansions can improve and reduce the cost of
parameter adjustment in the development of new models.

Supported by organizations Coordenação de Aperfeiçoamento de Pessoas de Nível
Superior (CAPES), Fundação de Amparo à Pesquisa do Estado de Minas Gerais
(FAPEMIG), Conselho Nacional de Desenvolvimento Científico e Tecnológico (CNPq),
Universidade Federal de Juiz de Fora (UFJF), Centro Federal de Educação Tecnológica
de Minas Gerais (CEFET-MG) and E. Alvarez-Lacalle acknowledges funding from
Spanish Ministry for Science, Innovation and Universities under grant SAF2017-88019-
C3-2-R.

J. M. F. Rodrigues et al. (Eds.): ICCS 2019, LNCS 11538, pp. 116–129, 2019.
https://doi.org/10.1007/978-3-030-22744-9_9

Keywords: Dimensional Reduction · Emulation · Genetic Algorithm

1   Introduction

The first mathematical models that simulate the electrical activity of heart cells use Hodgkin-Huxley [9] type equations to describe the ions channels present in the cell membrane [2, 20]. This type of equations is well consolidated in whole-cell models since it is the cornerstone of full-heart models which tries to reproduce organ-scale behavior.

However, these models fail when cardiac cells do not present homogeneous properties, such as differences in calcium concentrations due to intracellular calcium waves, across the cell. Indeed, recent intracellular models have introduced stochastic equations based on Markov Chains [13]. These Markov Chains formulations reflect the structure of the ion channels and are critical to any intracellular model. However, precisely because of its better physiological relation with reality, this new type of formulation has also found its way as average non-stochastic equations useful for homogeneous whole-cell models [16]. For instance, studies involving different mathematical models of cardiomyocytes have shown that the I_{CaL} current is better modeled when it is formulated using Markov Chains since it may reproduce the different states that a single channel can assume [1, 3].

The use of the older models still remain useful but some studies [7, 18] show an effort to update the old formulations in order to reproduce new experiments and improve the biological meaning of the equations. In this update process, it is common to merge different models due to the fact that each one might complement the other. A very common issue that arises from this coupling process is that they may generate inconsistent results since each model has different assumptions in its conception. To correct these problems a parameter adjustment has been an effective tool. However, the way how this process is done has not been studied properly and presents a wide range of different scenarios.

One of the analyses that have been used in recent studies [8, 10, 11] to see how the parameters are associated with the biological behavior is Uncertainty Quantification (UQ). Since this analysis considers the existence of uncertain measures associated with the studied object, it provides some methods to quantify the impacts of uncertainty in these parameters values upon model outputs. A technique used to perform this analysis is the polynomial chaos expansion (PCE), which approximate model outputs through orthogonal polynomials in terms of the uncertain model inputs [19].

Therefore, the main objective of this paper is to study and propose a systematic process to merge different ionic models involving parameter adjustment based on polynomial chaos expansion and genetic algorithm.

2   Methods

2.1   Modeling the L-Type Calcium Current. The Pandit-Mahajan Baseline Model

We take the rat ventricular electrophysiological model developed by Pandit et al. [14] as the structural model we are going to use. The original Pandit formulations models calcium, sodium and potassium channels to reproduce the action potential of the rat's ventricle. The calcium current entering the cell via thousand of L-type calcium channels I_{CaL} is written in terms of the Eq. (1).

$$I_{CaL} = g_{CaL}d\left[\left(0.9 + \frac{Ca_{inact}}{10.0}\right)f_{11} + \left(0.1 - \frac{Ca_{inact}}{10.0}\right)f_{12}\right](V - E_{CaL}) \quad (1)$$

where the parameter g_{CaL} is the maximum conductivity, d is associated with gating activation, f_{11}, f_{12} and Ca_{inact} are parameters associated with the gating inactivation, V is the transmembrane potential and E_{CaL} is the Nernst potential associated with the Ca^{2+} ions[1].

The above equation fits the experimental average calcium entering into the cell due to thousands of L-type calcium channels. It does not take into account, however, the typical structure of the L-type calcium channel. We take as a typical benchmark of any model development the ability to replace the Pandit formulation based on Hodgkin-Huxley approach with a more detailed averaged formulation of Markovian states. We take the formulation for the L-type Ca^{2+} as a Markov Chain developed by Mahajan et al. for rabbit which reads

$$I_{CaL} = g_{CaL}O_M(V - E_{CaL}) \quad (2)$$

where O_M is the state of the Mahajan Markov Chain that is associated with the channel open state. The model of the LCC has two relevant closed states and two inactivated, besides a single open state. As can be seen, the new formulation represents a single replacement in the usage of the gating formulations to the Markov Chain formulations[2].

Given that the original models were developed to simulate the electrophysiology of different animals, a parameter adjustment is necessary to make the two parts of the coupled model compatible. Figure 1 shows the difference between the original Pandit I_{CaL} and Action Potential (AP) curves and the same curves where the Mahajan model of LCC has been introduced without any parameter adjustment (we will call this model the Pandit-Mahajan baseline). As can be seen in Fig. 1, the new I_{CaL} current was too small with respect to the Pandit original values. Thereby, this difference ends up influencing the main model values, as shown by the AP curves.

In order to describe the I_{CaL} curve we will use the "Time to Peak", which is the time that the current take from the beginning of the channels opening

---

[1] For more details about the original Pandit I_{CaL} equation, see Pandit et al. [14].
[2] For more details about the transition rates between LCC rates and its dependence with voltage and calcium, see Mahajan et al. [13].

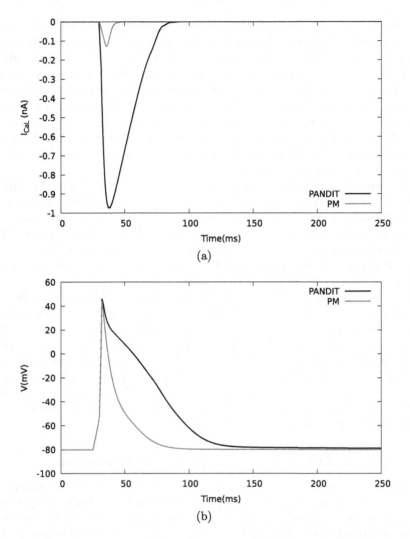

Fig. 1. Comparison of the I_{CaL} (a) and AP (b) curves of the original Pandit model (PANDIT) with the I_{CaL} and AP curves generated after Pandit-Mahajan (PM) coupling.

up to the moment when it reaches the minimum value; "Peak Value", which is the minimum value that the current reaches; and "Time to Decay", which is the time that the current take from the minimum value up to the channels close. The Pandit model has values of "Time to Peak", "Peak Value" and "Time to Decay" of 6.86 ms, −0.97 nA and 38.77 ms, respectively. While Pandit-Mahajan baseline model (without any adjustment) has 3.04 ms, −0.13 nA and 3.55 ms. So, analyzing these features of the two model, Pandit and the baseline Pandit-Mahajan, it is possible to conclude that a parameter adjustment is necessary.

It is clear now the aim of our work: we want to obtain a reliable Pandit-Mahajan model which reproduces the results of the rat ventricle (Pandit) using the formulation of the LCC obtained in the study of the physiology of the rabbit.

The LCC Markov Chain from Mahajan has a total of seven states and twenty transition rates. However, a considerable number of these rates is a combination of others. So, in the parameter fitting process, those composed rates was not considered. After that, eight transition rates - k_1, k_2, k_1', k_2', r_1, r_2, s_1 and s_1' - and the parameter τ_{po} (time constant of activation associated with the α and β transition rates) together with the Pandit's maximum conductance of I_{CaL} current - g_{CaL} - were considered in this study.

The parameter adjustment process may be done using a Genetic Algorithm. However, a minimization problem solved by a genetic algorithm may become computationally expensive depending on the dimension of the search space. In the case of parameter adjustment from the baseline Pandit-Mahajan model, the GA would have to find a set of ten values for the parameters so that those values should reproduce the original Pandit I_{CaL} current values. In this search space, it might be difficult to find the optimal values.

To reduce the GA search space, Sensitivity Analysis and Dimensional Reduction techniques were used to determinate how the parameters of the coupled model influence the main I_{CaL} characteristics. With this analysis, it is possible to choose the most relevant parameters in order to adjust the new I_{CaL} current. Furthermore, the search process has to simulate the mathematical model several times and, due to it, the computational cost associated with this process becomes expensive. Trying to minimize it, the use of Emulations [10,11] based on Polynomial Chaos Expansion was tested in order to replace the mathematical model in the evaluation process of the GA.

2.2 Polynomial Chaos Expansion (PCE) and Emulations

Emulations, also known as Metamodels in the literature, has a very important role in the study involving complex systems models [15]. This technique aims to decrease the complexity of the mathematical models caused by the over-parametrization but maintaining the results as expected. With this process, the new model loses a few phenomenological information but, on a high scale, it is still able to reproduce the expected results and to help on the decision-making process.

In studies involving the Uncertainty Quantification, a common technique to analyze the insertion of uncertainty in deterministic models is the Polynomial Chaos Expansion (PCE). In this technique, an orthogonal polynomial is used to approximate some outputs of a forward model [19]. As a polynomial evaluation is computationally fast, this approach becomes very useful when the original model has a high computational cost to be simulated. Then, some process where a lot of simulations are required, such as Sensitivity Analysis and Parameter fitting, become more practical to be performed.

Considering a vector $\boldsymbol{\xi} = (\xi_1, \xi_2, \ldots, \xi_N)^T$ of model inputs composed by independent random variables and assume that the quantity of interest y is written in terms of these variables, it is possible to express this quantity through an infinite polynomial chaos expansion [12]. In practical applications, this quantity of interest can be approximated by a finite expansion obtained through a linear combination of the elements from the polynomial chaos basis:

$$\bar{y}(\boldsymbol{\xi}) = \sum_{i=1}^{P} b_i \Phi_i(\boldsymbol{\xi}), \tag{3}$$

where b_i are the unknown coefficients and Φ_i are orthogonal polynomial functions in terms of the inputs. This polynomial chaos expansion with N inputs and order d has P terms, where $P = \frac{(N+d)!}{N!d!}$.

Defined the approximation for the quantity of interest, as in Eq. (3), it is necessary to determine the coefficients b_i that define the polynomial in terms of the inputs. To this end, a non-intrusive approach named Probabilistic Collocation method is used, where a weighted residual formulation in the random space is defined [17] and the polynomial expansion needs to be equal to the model evaluation in a number of collocation points, which are samples of the random inputs. The result is a linear system in terms of the b_i coefficients, then, the polynomial coefficients that approximate the quantity of interest are found by solving this system.

2.3 Sensitivity Analysis

In complex systems modeled as differential equations, like a Markov Chain with a large number of states and transitions between them, it is very common to appear a high number of parameters. In these cases, it is important to understand how each parameter influences the entire system.

A Sensitivity Analysis of the ten parameters associated in the I_{CaL} current from the Pandit-Mahajan model was done in order to determinate which parameters have more influence on the main characteristics of the I_{CaL} curve. To do this analysis, the ChaosPy library [6] implemented in Python was used.

The ChaosPy library has methods to quantify the uncertainty combining the use of the Monte Carlo method and also the PCE. To determine the PCE coefficients, it provides the Probabilistic Collocation method and the Pseudospectral Projection method.

The ten parameters, τ_{po}, k_1, k_2, k_1', k_2', r_1, r_2, s_1, s_2 and g_{CaL}, were analyzed and the results showed that three of them have more influence on the studied I_{CaL} curve. Figure 2 shows these results.

As can be seen in Fig. 2, the parameters r_1, r_2 and g_{CaL} have high influence on the main I_{CaL} current from the Pandit-Mahajan model. So, considering these results, the search space of the Genetic Algorithm might be reduced from ten parameters to only three.

2.4 Parameter Adjustment Using Genetic Algorithm

A parameter adjustment using a Genetic Algorithm was used to adjust the LCC Markov Chain parameters so that the I_{CaL} curve presented in Pandit

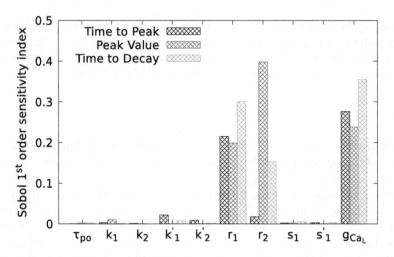

Fig. 2. The Sobol 1st order sensitivity index (The 1st order sensitivity index quantifies the portion that an input parameter contributes directly to the total variance of the quantity of interest, for more details see Eck *et al.* [5].) of the parameters associated with the main I_{CaL} curve characteristics: the Time to Peak, the Peak Value and the Time to Decay.

original model remains unaltered in the new Pandit-Mahajan model. The GA was implemented using the "Simple Genetic Algorithm" provided by Pagmo C++ library [4].

Two important processes when defining a Genetic Algorithm are the choices of the variable of search (Genes) and the objective function (Fitness function). In order to evaluate the effectiveness of the Sensitivity Analysis done, the variables of the search were defined in two ways: the first with the ten parameters and, the second, with only the three parameters highlighted by the Sensitivity Analysis. The second process, the definition of the Fitness function, used the main characteristics of the I_{CaL} curve. In this process, two methods were tested to evaluate the objective function: the first one was considering the Pandit-Mahajan mathematical model simulations and the second was considering the emulations done by the PCE generated through the ChaosPy library. Mathematically, this process can be seen as a minimization problem with objective functions presented by the Eqs. (4) and (5) for the approach using the simulations of the model and the emulations using the PCE, respectively.

$$F(v, I_{CaL_p}) = \frac{1}{|TimeToPeak(I_{CaL_p})|}|TimeToPeak(v) - TimeToPeak(I_{CaL_p})|$$

$$+ \frac{1}{|PeakValue(I_{CaL_p})|}|PeakValue(v) - PeakValue(I_{CaL_p})|$$

$$+ \frac{1}{|TimeToDecay(I_{CaL_p})|}|TimeToDecay(v) - TimeToDecay(I_{CaL_p})|$$

$$(4)$$

where v is the I_{CaL} curve generated by the Pandit-Mahajan simulation to be evaluated by the function and I_{CaL_p} is the original curve of the Pandit model.

$$F(\xi, I_{CaL_p}) = \frac{1}{|TimeToPeak(I_{CaL_p})|}|P_{TTP}(\xi) - TimeToPeak(I_{CaL_p})|$$

$$+ \frac{1}{|PeakValue(I_{CaL_p})|}|P_{PV}(\xi) - PeakValue(I_{CaL_p})| \qquad (5)$$

$$+ \frac{1}{|TimeToDecay(I_{CaL_p})|}|P_{TTD}(\xi) - TimeToDecay(I_{CaL_p})|$$

where ξ are the set o parameters to be evaluated by the function, $P_\bullet(\xi)$ are the 4th order PCE that approximate the "Time To Peak" ($\bullet_{TTP}$), "Peak Value" ($\bullet_{PV}$) and the "Time To Decay" ($\bullet_{TTD}$) features and I_{CaL_p} is the original curve of the Pandit model.

The chromosome C of the genetic algorithm is defined as $C = \{c_p \in \mathbb{R} | 0.1 < c_p < 5.0\}$ where c_p is the multiplier for the specific parameter p. All Genetic Algorithm executions were done with the standard setup from the Pagmo library except the use of the crossover strategy. In this study was used the Simulated Binary Crossover ("sbx") strategy, also provided by Pagmo library. All executions used 15 generations and the population was composed of 150 individuals.

3   Results

The main objective of this study is to evaluate a new process of parameter adjustment based on uncertainty quantification techniques and genetic algorithm.

After the use of the methods described in Sect. 2, two classes of results were generated. The first one using the simulations of the mathematical model into the GA evaluation and, the second one, using the emulations done by the Polynomial Chaos Expansions obtained using the ChaosPy library. Figure 3 shows the results obtained in the different tests.

The best results obtained by the GA using the simulation of Pandit-Mahajan model and the emulation by the PCE are presented in Tables 1 and 2, respectively.

Figure 4 shows the I_{CaL} (4(a)), AP (4(b)) and $[Ca]_i$ (4(c)) curves of the proposed Pandit-Mahajan model after the parameter adjustment using the 3 parameter results considering both simulations of the model (PM-SIM) and emulation of the PCE (PM-EMU) in comparison with the original Pandit model.

As can be seen, the GA adjustment was able to fit the new model formulations in order to maintain the main characteristics of the I_{CaL} curve as presented in the original model using both simulation and emulation techniques.

Since the I_{CaL} curve was well reproduced, the main electrophysiology variables, the Action Potential (AP) and the Intracellular Calcium Concentration ($[Ca]_i$), also were well reproduced.

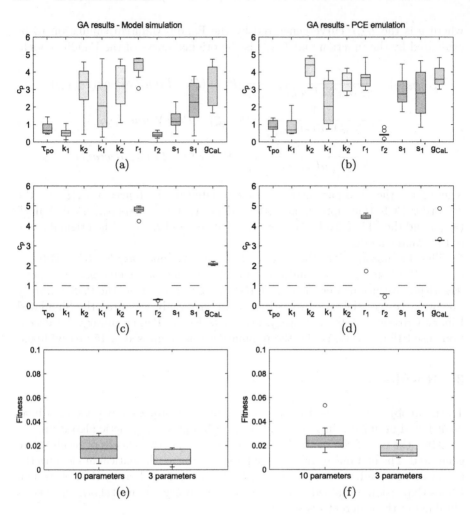

Fig. 3. Results obtained by the GA in the cases of study. The descriptive measures (minimum, first quartile, median, third quartile and maximum values) of the multiplier c_p result adjusting 10 parameters using the mathematical model simulations (a) and the PCE emulations (b); the descriptive measures of the multiplier c_p result adjusting 3 parameters using the mathematical model simulations (c) and PCE emulations (d); the descriptive measures for the best Fitness found considering the simulations of the mathematical model (e) and emulations using the PCE (f) for 10 and 3 parameters.

3.1 Dimensional Reduction

The first method studied was the Dimensional Reduction. Based on the Sensitivity Analysis done, it was possible to notice that three parameters have more influence on the main features of the I_{CaL} curve.

The GA was executed 10 times with the search space of 10 parameters and 10 times with search space of 3 parameters. The values found by the GA in both

Table 1. The parameter values from the Original Model presented with the values encountered by the best execution of the Genetic Algorithm considering the mathematical model simulation using both the 10 parameters and the 3 parameters.

Parameter	Mahajan et al. [13]	10 parameters	3 parameters
τ_{po}	1	**1.434**	1
k_1	0.03	**0.015**	0.03
k_2	$1.04\ 10^{-4}$	**3.88 10^{-4}**	$1.036\ 10^{-4}$
k_1'	$4.13\ 10^{-3}$	**4.361 10^{-3}**	$4.13\ 10^{-3}$
k_2'	$2.24\ 10^{-3}$	**7.092 10^{-3}**	$2.24\ 10^{-3}$
r_1	0.3	**1.247**	**1.482**
r_2	3	**0.667**	**0.914**
s_1	0.02	**0.023**	0.02
s_1'	$1.95\ 10^{-3}$	**6.671 10^{-4}**	$1.95\ 10^{-3}$
g_{CaL}	0.031	**0.052**	**0.064**

Table 2. The parameter values from the Original Model presented with the values encountered by the best execution of the Genetic Algorithm considering the mathematical model emulation done with the PCE using both the 10 parameters and the 3 parameters.

Parameter	Mahajan et al. [13]	10 parameters	3 parameters
τ_{po}	1	**1.232**	1
k_1	0.03	**0.023**	0.03
k_2	$1.036\ 10^{-4}$	**4.09 10^{-4}**	$1.036\ 10^{-4}$
k_1'	$4.13\ 10^{-3}$	**0.014**	$4.13\ 10^{-3}$
k_2'	$2.24\ 10^{-3}$	**9.448 10^{-3}**	$2.24\ 10^{-3}$
r_1	0.3	**0.988**	**1.309**
r_2	3	**1.279**	**1.731**
s_1	0.02	**0.043**	0.02
s_1'	$1.95\ 10^{-3}$	**7.939 10^{-4}**	$1.95\ 10^{-3}$
g_{CaL}	0.031	**0.115**	**0.101**

cases, using the simulation of the model or the emulation by the PCE, are shown in the Figs. 3(a), (c) and in Figs. 3(b) and (d), respectively.

As can be seen in Figs. 3(e) and (f), the reduction in the search space dimension of the GA generated an increase in the quality of the results obtained.

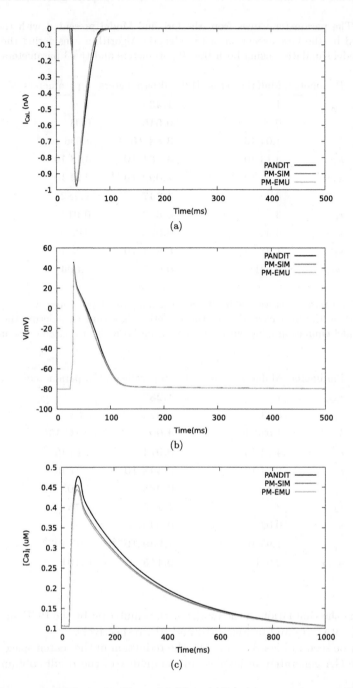

Fig. 4. I_{CaL} current (a), Action Potential (b) and $[Ca]_i$ (c) curves generated by Pandit original model (PANDIT) compared with the new Pandit-Mahajan model after the best parameter adjustment using the simulation of the model (PM-SIM) or the emulation of the PCE (PM-EMU) into the Genetic Algorithm search process.

3.2   Emulations

An Emulator based on PCE generated by ChaosPy library was tested instead of the mathematical model in the GA Fitness evaluation. Figures 3(b) and (d) show the results obtained by the GA using the Emulation technique.

When compared with the GA results using the mathematical model (Figs. 3(b) and (d)), the use of the Emulation obtained satisfactory results. The values found by GA were similar using both simulations of the mathematical model or using emulations. Furthermore, even changing the evaluation method, the obtained Fitness values were similar using both methods as can be seen in Figs. 3(e) and (f).

The great advantage of using the Emulation technique is associated with the computational cost. The same results considering both techniques was obtained by the GA but the search process using the simulations as evaluation function took approximately 90 min to find the results, whereas using the emulation it took less than 1 s. Considering that ChaosPy took around of 30 min to generate the PCE, it is possible to conclude that the use of the Emulation technique speeds up the problem solution at least in 3 fold.

4   Conclusions and Discussions

The coupling of different models or even an update of an existing model is an important tool to develop new electrophysiological studies. However, this process involves different procedures and techniques that imply a set of parameter adjustments, which may determinate the quality of the results.

In this paper, we presented a new method based on the combination of Polynomial Chaos expansion and a Genetic Algorithm to optimize the efficiency and the efficacy of the process of coupling two different models. Furthermore, a dimensional reduction in the Genetic Algorithm search space was done based on the Sensitivity Analysis, which showed that for the Pandit-Mahajan specific coupling process, only three parameters have the biggest influence in the parameter estimations needed to model I_{CaL} curve. This result motivated an adjustment involving only these three parameters instead of the ten previously adjusted. This reduction showed to be a good option since it generated smaller Fitness values when adjusted with the Genetic Algorithm. This reduction may be explained due to the fact that this dimensional reduction in the Genetic Algorithm search space, considering now only the more important parameters, facilitated the process to find the optimal values.

Another important point to highlight is that, considering all the ten parameters, the Genetic Algorithm results did not converge to a small range of values. The results of the parameters multipliers c_p assumed different values in each algorithm execution. This behavior is not interesting since the results are associated with parameters that, in some cases, are related to the biological characteristics of the model. So this variation may cause a decrease in confidence of its value. This unwanted behavior did not occur in the Genetic Algorithm considering the

Sensitivity Analysis. With the combined techniques, the results had converged to a small range of values and, with this behavior, they may be more trustworthy.

The Fitness evaluation process in a Genetic Algorithm can be a computationally expensive process, mainly when this process involves a simulation of a mathematical model. In this paper, a method using an Emulator based in Polynomial Chaos Expansions was tested instead of the mathematical model in order to replace a complex model simulation for a single polynomial evaluation and, with this, reduce the cost associated with this process.

The Polynomial Chaos Expansions method was more effective than the Genetic Algorithm since it was able to obtain the roughly the same results, however, reducing the Genetic Algorithm search process from 90 min to less than 1 s in the configuration used in this paper. If we consider a larger number of generations or a bigger population size of the Genetic Algorithm, the speed up will be highly increased.

In conclusion, in this paper, we employ a parameter adjustment process combining Polynomial Chaos Expansions and Genetic Algorithm to couple two models: Pandit and Mahajan into a new combined Pandit-Mahajan model. The procedure showed to be effective to facilitate the coupling of the two models since it arrives in excellent agreement with the original model.

References

1. Armstrong, C.M., Bezanilla, F.: Inactivation of the sodium channel. II. Gating current experiments. J. Gener. Physiol. **70**(5), 567–590 (1977)
2. Beeler, G.W., Reuter, H.: Reconstruction of the action potential of ventricular myocardial fibres. J. Physiol. **268**(1), 177–210 (1977)
3. Bezanilla, F., Armstrong, C.M.: Inactivation of the sodium channel. I. Sodium current experiments. J. Gener. Physiol. **70**(5), 549 (1977)
4. Biscani, F., Izzo, D.: esa/pagmo2: pagmo 2.9, August 2018
5. Eck, V.G., et al.: A guide to uncertainty quantification and sensitivity analysis for cardiovascular applications. Int. J. Numer. Methods Biomed. Eng. **32**(8), e02755 (2016)
6. Feinberg, J., Langtangen, H.P.: Chaospy: an open source tool for designing methods of uncertainty quantification. J. Comput. Sci. **11**, 46–57 (2015)
7. Gattoni, S., Røe, Å.T., Frisk, M., Louch, W.E., Niederer, S.A., Smith, N.P.: The calcium-frequency response in the rat ventricular myocyte: an experimental and modelling study. J. Physiol. **594**(15), 4193–4224 (2016)
8. Hauseux, P., Hale, J.S., Cotin, S., Bordas, S.P.: Quantifying the uncertainty in a hyperelastic soft tissue model with stochastic parameters. Appl. Math. Model. **62**, 86–102 (2018)
9. Hodgkin, A.L., Huxley, A.F.: A quantitative description of membrane current and its application to conduction and excitation in nerve. J. Physiol. **117**(4), 500–544 (1952)
10. Lawson, B.A., Burrage, K., Burrage, P., Drovandi, C.C., Bueno-Orovio, A.: Slow recovery of excitability increases ventricular fibrillation risk as identified by emulation. Front. Physiol. **9**, 1114 (2018)

11. Lawson, B.A., Drovandi, C.C., Cusimano, N., Burrage, P., Rodriguez, B., Burrage, K.: Unlocking data sets by calibrating populations of models to data density: a study in atrial electrophysiology. Sci. Adv. **4**(1), e1701676 (2018)
12. Li, H., Zhang, D.: Probabilistic collocation method for flow in porous media: comparisons with other stochastic methods. Water Resour. Res. **43**(9), W09409 (2007)
13. Mahajan, A., et al.: A rabbit ventricular action potential model replicating cardiac dynamics at rapid heart rates. Biophys. J. **94**(2), 392–410 (2008)
14. Pandit, S.V., Clark, R.B., Giles, W.R., Demir, S.S.: A mathematical model of action potential heterogeneity in adult rat left ventricular myocytes. Biophys. J. **81**(6), 3029–3051 (2001)
15. Ratto, M., Castelletti, A., Pagano, A.: Emulation techniques for the reduction and sensitivity analysis of complex environmental models (2012)
16. Stary, T.: Mathematical and computational study of Markovian models of ion channels in cardiac excitation (2016)
17. Tatang, M.A., Pan, W., Prinn, R.G., McRae, G.J.: An efficient method for parametric uncertainty analysis of numerical geophysical models. J. Geophys. Res.: Atmos. **102**(D18), 21925–21932 (1997)
18. Terkildsen, J.R., Niederer, S., Crampin, E.J., Hunter, P., Smith, N.P.: Using physiome standards to couple cellular functions for rat cardiac excitation-contraction. Exp. Physiol. **93**(7), 919–929 (2008)
19. Xiu, D.: Numerical Methods for Stochastic Computations: A Spectral Method Approach. Princeton University Press, Princeton (2010)
20. Yanagihara, K., Noma, A., Irisawa, H.: Reconstruction of sino-atrial node pacemaker potential based on the voltage clamp experiments. Japan. J. Physiol. **30**(6), 841–857 (1980)

A Cloud Architecture for the Execution of Medical Imaging Biomarkers

Sergio López-Huguet[1]([✉]), Fabio García-Castro[3], Angel Alberich-Bayarri[2,3], and Ignacio Blanquer[1]

[1] Instituto de Instrumentación para Imagen Molecular (I3M) Centro mixto CSIC, Universitat Politècnica de València, Camino de Vera s/n, 46022 Valencia, Spain
serlohu@upv.es, iblanque@dsic.upv.es
[2] GIBI 230 (Biomed. Imag. Res. Group), La Fe Health Research Institute, Valencia, Spain
[3] QUIBIM (Quantitative Imaging Biomarkers in Medicine) SL, Valencia, Spain
{fabiogarcia,angel}@quibim.com

Abstract. Digital Medical Imaging is increasingly being used in clinical routine and research. As a consequence, the workload in medical imaging departments in hospitals has multiplied by over 20 in the last decade. Medical Image processing requires intensive computing resources not available at hospitals, but which could be provided by public clouds. The article analyses the requirements of processing digital medical images and introduces a cloud-based architecture centred on a DevOps approach to deploying resources on demand, adjusting them based on the request of resources and the expected execution time to deal with an unplanned workload. Results presented show a low overhead and high flexibility executing a lung disease biomarker on a public cloud.

Keywords: Cloud computing · Medical imaging · DevOps

1 Introduction

Traditionally, medical images have been analysed qualitatively. This type of analysis relies on the experience and knowledge of specialised radiologists in charge of carrying out the report. This entails a high temporal and economic cost. The rise

The work in this article has been co-funded by project SME Instrument Phase II - 778064, QUIBIM Precision, funded by the European Commission under the INDUSTRIAL LEADERSHIP - Leadership in enabling and industrial technologies - Information and Communication Technologies (ICT), Horizon 2020, project ATMOSPHERE, funded jointly by the European Commission under the Cooperation Programme, Horizon 2020 grant agreement No 777154 and the Brazilian Ministério de Ciência, Tecnologia e Inovação (MCTI), number 51119. The authors would like also to thank the Spanish "Ministerio de Economía, Industria y Competitividad" for the project "BigCLOE" with reference number TIN2016-79951-R.

© Springer Nature Switzerland AG 2019
J. M. F. Rodrigues et al. (Eds.): ICCS 2019, LNCS 11538, pp. 130–144, 2019.
https://doi.org/10.1007/978-3-030-22744-9_10

of computer image analysis techniques and the improvement of computer systems lead to the advent of quantitative analysis. Contrary to qualitative analysis, the quantitative analysis aims to measure different characteristics of a medical image (for example, the size, texture, or function of a tissue or organ and the evolution of these features in time) to provide radiologists and physicians with additional, objective information as a diagnostic aid. In turn, the quantitative analysis requires image acquisitions with the highest possible quality to ensure the accuracy of the measurements.

An imaging biomarker is a characteristic extracted from medical images, regardless of the acquisition modality. These characteristics must be measured objectively and should depict changes caused by pathologies, biological processes or surgical interventions [21,27]. The availability of large population sets of medical images, the increase of their quality and the access to affordable intensive computing resources has enabled the rapid extraction of a huge amount of imaging biomarkers from medical images. This process allows to transform medical images into mineable data and to analyze the extracted data for decision support. This practice, known as radiomics, provides information that cannot be visually assessed by qualitative radiological reading and reflects underlying pathophysiology. This methodology is designed to be applied at a population level to extract relationships with the clinical endpoints of the disease that can be directed to manage the disease of an individual patient.

The execution of medical image processing tasks, such as biomarkers, is a process that sometimes requires high-performance computing infrastructures and, in some cases, specific hardware (GPUs) that is not available in most medical institutions. Cloud service providers make it possible to access specific and powerful hardware that fits the needs of the workload [28]. Another interesting advantage of the Cloud platforms is the capability of fitting the infrastructure capacity to the dynamic workload, thus improving cost contention. The seamless transition from local image processing and data analytic development environments to cloud-based production-quality processing services implies a Continuous Integration and Deployment DevOps problem that is not properly addressed in current platforms.

1.1 Motivation and Objectives

The objective of this work is to design and implement a cloud-based platform that could address the needs of developing and exploiting medical image processing tools. In this sense, the work focuses on the development of an architecture focusing on the following principles:

- Agnostic to the platform, so the same solution can be deployed on different public and on-premise cloud offerings, adapting to the different needs and requirements of the users and avoiding lock-in.
- Capable of integrating High-performance computing and storage back-ends to deal with the processing of massive sets of medical images.
- Seamlessly integrating development, pre-processing, validation and production from the same platform and automatically.
- Open, reusable, extendable, secure and traceable platform.

1.2 Requirements

A requirement elicitation and analysis process was performed, leading to the identification of 13 requirements, classified into 9 mandatory requirements, 3 recommendable requirements and 3 desirable requirements. The requirements are described in Tables 1, 2 and 3.

Table 1. Requirements of the infrastructure (**M**andatory, **R**ecommended, **D**esirable)

RiD	Name	Description	Level
RI1	Resource provisioning	Resources should be automatically configured in the deployment. Provisioning should be performed with minimal intervention by system administrators and should be able to work in multiple IaaS platforms	R
RI2	Resource isolation	Jobs should run on the system at the maximum level of isolation. Workload may have different and even incompatible software dependencies, and the failure of the execution of a job should not affect the rest of the executions	M
RI3	Resource scalability	Virtual infrastructures should be automatically reconfigured when adding or removing nodes (limited by a minimum and maximum number of nodes for each type of resource). Elasticity could be triggered externally	R
RI4	Manag. of releases	Software should be easy to update and releases should be easy to deploy. This implies automation, minimal customer intervention, progressive rollouts, roll backs, and version freezing	M
RI5	User authentication	Users should be able to log-in the system using ad-hoc credentials or an external Identity Provider (IDP) such as Google or Microsoft LiveID	D
RI6	User authorisation	Access to the services should be granted only to authorised users. This implies access to data, services and resources. Only special users would be able to access resources	D
RI7	High availability (HA)	Services must be deployed in HA to guarantee Quality of Service	M

2 State of the Art

Since the appearance of Cloud services, a large number of applications have been adapted to facilitate the access of the application users to Cloud infrastructures. In the field of biomedicine we find examples in [22,32]. On the other hand, there

Table 2. Requirements for job execution.

RiD	Name	Description	Level
RE1	Batch execution	The system should run batch jobs. A job will comprise a set of files, software dependencies, hardware requirements, execution arguments, input and output sandbox, job type, memory and CPU requirements	M
RE2	Workflow execution	A job may include several linked steps that need to be executed according to a data flow. The workflow will imply the automatic execution of the different stages as dependencies are solved	M
RE3	Job customization	Linked to requirements RI2 and RI4, this requirement poses the need of jobs to run on a customizable environment requiring special hardware, specific software configuration, operating system, and licenses	M
RE4	Execution triggers	Jobs could also be initiated by means of events. Uploading a file or messages in a queue can spawn the execution of jobs. These reactive jobs will be defined through rules	D
RE5	Efficient execution	Jobs should be efficiently executed in the platform. This performance is defined at two levels: (a) minimum overhead with respect to the execution on an equivalent pre-installed physical node; (b) capability of integrating high-performance resources as GPUs and multicore CPUs	M

Table 3. Requirements with respect to the data.

RiD	Name	Description	Level
RD1	POSIX access	Jobs expect to find the data to be processed in a POSIX file system in a specific directory route	M
RD2	ACLs access	Storage access authorization based on a coarse granularity (access granted/denied for both read & write)	M
RD3	Provenance and traceability	Traceability for the derived data is key to bound to the GDPR regulations (e.g. a trained model should be invalidated if the permissions for any part of the data used in the training is revoked)	R

are works that offer pre-configured platforms with a large number of tools for bioinformatic analysis. An example is the Galaxy Project [7], a web platform that can be deployed on public and on-premises Cloud offerings (e.g. using CloudMan [5] for Amazon EC2 [1] and OpenStack [15]).

Another example is Cloud BioLinux [4], a project that offers a series of pre-configured virtual machine images for Amazon EC2, VirtualBox and Eucalyptus [6]. Finally, the solution proposed in [30] is specially designed for medical image analysis using ImageJ [8] in an on-premise infrastructure using Eucalyptus.

Before taking a decision on the architecture, an analysis has been done at three levels: Container technologies, Resource Managers and Job Scheduling.

Containers are a set of methodologies and technologies that aim at isolating execution environments at the level of processes, network namespaces, disk areas and resource limitations. Containers are isolated with respect to: the host filesystem (as they can only see a limited section of it, using techniques such as chroot or FreeBSD Jails), the processes running on the host (only the processes derived executed within the container are visible, for example, using namespaces), and the resources the container can use (as the processes in a container can be bound to a CPU, memory or I/O share, for example, using cgroups).

Containers are used in application delivery, isolation and light encapsulation of resources in the same tenant, execution of processes with incompatible dependencies and improved system administration. Three of the most prominent technologies in the market supporting Containers are Docker [23], Linux Containers [10] and Singularity [24]. Docker has reached the maximum popularity for application delivery due to its convenient and rich ecosystem of tools. However, Docker containers run under the root user space and do not provide multi-tenancy. On the other side, Singularity run containers on the user-space, but access to specific devices is complex. LxC/D is better in terms of isolation but have limited support (e.g. LxD only works in ubuntu [11]). The solution for container isolation selected will be Docker on top of isolated virtual machines.

Resources should be provisioned and allocated for deploying containers. Resource Management Systems (RMSs) deal with the principles of managing a pool of resources and splitting them across different workloads. RMSs manage the resources of physical and virtual machines reserving a fraction of them for a specific workload. RMSs deal with different functionalities, such as: Resource discovery, Resource Monitoring, Resource allocation and release and Coordination with Job Schedulers. We identify 3 technologies to orchestrate resources are: Kubernetes [9], Mesos [2] and EC3 [17,18].

Finally, Job schedulers manage the remote execution of a job on the resources provided by the RMS. Job Schedulers retrieve job specifications from different interfaces, run them on the remote nodes using a dedicated input and output sandbox for the job, monitor its status and retrieve the results.

Job Schedulers may provide other features, such as fault tolerance, complex jobs (bag of tasks, parallel or workflow jobs) and deeply interact with the RMS to access and release the needed resources. We consider in this analysis Marathon [12] Chronos [3], Kubernetes [9] and Nomad [13]. Marathon and Chronos require a Mesos Resource Management System and can deploy containers as long-term fault-tolerant services (Marathon) or periodic jobs (Chronos). Kubernetes has the capacity of deploying containers (mainly Docker but not limited to it) as services, or running batch jobs as containers. However, any of them deal seamlessly with non-containerised and container-based jobs. In this sense, Nomad can deal with multi-platform hybrid workloads with minimal installation requirements.

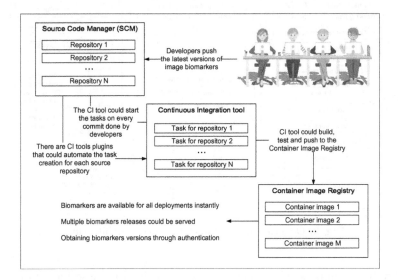

Fig. 1. Overview of Container Delivery architecture

3 Architecture

The service-oriented architecture is described and implemented in a modular manner, so components could easily be replaced. In Sect. 3.1, the architecture is described in a technology-agnostic way, so different solutions could fit into the architecture. Section 3.2 describes all architecture components and how they fulfill the use case requirements identified in Sect. 1.2. Finally, Sect. 3.3 shows the final version of the architecture including the technologies selected and how each one addresses the requirements.

3.1 Overview of the Architecture

The system architecture addresses the requirements described in Sect. 1.2. The architecture can be divided into two parts: Container Delivery (CD), and Container Execution (CE). It should be noted that although all the components of the architecture can be installed in different nodes, in some cases they could be installed in the same node to reduce costs. As it can been seen in Fig. 1, the CD architecture is composed of three components: the Container Image Registry, the Source Code Manager (SCM), and the Continuous Integration (CI) tool.

Figure 2 depicts the CE architecture. The system consists of four types of logical nodes: Front-end, job schedulers nodes, working nodes and Container Image

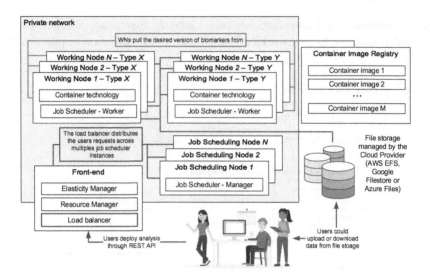

Fig. 2. Overview of Container Execution architecture

Registry (this component also appear in CD architecture describe above). The Front-end and job schedulers nodes are interconnected using a private network. The Front-end logical node exposes a REST API, which allows load-balanced communication with the REST API of the job scheduler nodes. Furthermore, it contains the Resource Management Service, which is composed of the Cloud Orchestrator and the horizontal elasticity manager. Job schedulers nodes comprise the master services of the Job Scheduler (JS). Furthermore, as the Front-end is the gateway between users and the job scheduler service, a service for providing authorization and load balancing (between job scheduler nodes) is required. Different working nodes will run the Job scheduler executors. Working nodes mount locally a volume available from a global external storage. Is should be noted that the set of working nodes can be heterogeneous.

3.2 Components

Resource Management service (RMS): It is in charge of deploying the resources, configuring them and to reconfigure them according to the changes on the workload. The requirements stated in Sect. 1.2 focus on facilitating deployment, higher isolation, scalability, application releases management and generic authentication and authorisation mechanisms. RMS may require to interact with the infrastructure provider to deploy new resources (or undeploy them), and to configure the infrastructure accordingly. Furthermore, the deployment should be maintainable, reliable, extendable and platform agnostic.

Job Scheduling service: It will perform the execution of containerised jobs requested remotely by a client application through the REST API. It is required that the job scheduler service includes a monitoring system to provide up-to-date information on the status of the jobs running in the system. Clients will submit jobs through the load balancer service providing a job description formed by any additional information required by the Biomarkers platform, the information related to the container image, the input and output sandbox, and the software and hardware requirements (such GPUs, memory, etc.).

Horizontal Elasticity service: It is necessary to fulfill the Resource scalability requirement, which is strongly related to the Job scheduler and the Resource Manager. Horizontal elasticity tool has to be able to monitor the job scheduler queue and running jobs, and current working node resources in order to scale in or scale out the infrastructure. It is desirable that the horizontal elasticity manager could maintain a certain number of nodes always idle or stopped to reduce the time that the jobs are queued.

Source Code Manager (SCM): It is required to manage the coding source for developers. Due to the release management requirement and the development complexity, it is mandatory to lean on this kind of tools.

Container Image Registry: In order to store and delivery the biomarker applications, it is necessary to use a Container Image Registry. Biomarker applications could be bound to Intellectual Property Rights (IPR) restrictions so Container Image Registry must be private. For this reason, authentication mechanisms are required for obtaining images from the registry. Working nodes will pull the application images when the container image not exists or recent version exists.

Continuous Integration (CI) tool: CI eases the development cycle because it automates building, testing and pushing to the Container Image Resgistry the biomarker application (with a certain version or tag). Developers or (the CI experts) define this workflow to do it. Furthermore, some of CI tools could trigger these tasks for each SCM commits.

Storage: Biomarker applications make use of legacy code and standard libraries which expect data to be provided in a POSIX filesystem. For this reason, the storage technology must mount the filesystem in the container.

3.3   Detailed Architecture

The previous sections describe the architecture and its components. After the technology study done in Sect. 2 and the feature identification for each architecture component in Sect. 3.2, the technologies selected for addressing the requirements are the following:

- Jenkins as the CI tool due to the wide variety of available plugins (such us SCM plugins). Furthermore, it allows you to easily define workflows for each task using a file named Jenkinsfile. Jenkins provides means to satisfy the requirements RI4, RI5 and RI6.
- GitHub as SCM because it supports both private (commercial license) and public repositories, which are linked to the CI tool. Furthermore, there is

a Jenkins plugin that could scan a GitHub organization and create Jenkins tasks for each repository (and also for each branch) that contains a Jenkinsfile, addressing to requirement RI4. This component meets the requirements RI5, RI6 and RI7.

- Hashicorp Nomad is the Job scheduler. We selected Nomad instead of Kubernetes due to Kubernetes can only run Docker containers. Furthermore, Nomad incorporates job monitoring that can be consulted by users. Additionally, it is designed to work in High Availability mode, addressing requirement RI7. Hashicorp Consul is used to resources service discovery as Nomad could use it natively. By using this job scheduler, the architecture meets the requirements RI2, RI4 (job versioning), RE1, RE2, RE3 and satisfyies RI5 and RI6 using its Access Control List (ACL) feature.
- Docker is the container platform selected because it is the most popular container technology and it is supported by wide variety of Job schedulers. It provides the resources isolation required by RI2 and support version management (by tagging the different images) of RI4.
- As Docker is the container platform used in this work, Docker Hub and Docker Registry are used as, respectively, public and private container image registry. Requirements RI4 and RI6 are address by using Docker.
- Infrastructure Manager (IM) [17] is the orchestrator chosen because it is open source, cloud agnostic and provides the required functionality to fulfill the use case requirements RI1, RI3, RI5 and RI6. In [25], IM is used for deploy 50 simultaneous nodes.
- CLUES [19] has been chosen for addressing RI3 because it is open source and can scale up or down infrastructures using IM by monitoring the Nomad jobs queue. CLUES can auto-scale the infrastructure according to different types of workloads [16,25].
- The RMS selected is EC3, which is a tool for system administrators that combines IM and CLUES to configure, create, suspend, restart and remove infrastructures. By using EC3 the system could address the requirements RI, RI3, RI5 and RI6.
- Due to the experimentation will be done in Azure and the current storage solution of QUIBIM is Azure Files, it has been selected as storage. It allows to mount (entirely o partially) the data as a POSIX filesystem, which is the requirement RD1. Also, it provides the mechanisms required to fulfill RI5, RI6 and RD2. Additionally, it allows to mount the same filesystem concurrently.
- HAProxy is used for load balancing because it is reliable, open source and support LDAP or OAuth for authentication.

It should be remarked that the proposed architecture (which is depicted in Fig. 3) is a simplification of Fig. 2. For the experiment performed, the job scheduling services are deployed in the Front-end node but users connect with the job schedulers using the load balancer service. So, this simplification does not affect to the users-services communication. Furthermore, in order to avoid costs, the CI tool (Jenkins) and the Docker Private Registry are in the same resource.

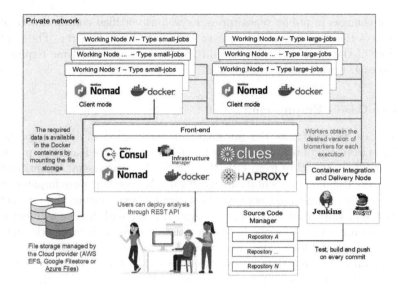

Fig. 3. Proposed architecture with selected technologies.

4   Results

The experiments have been performed on the public Cloud Provider Microsoft Azure. The infrastructure is composed by three type of nodes. The *front-end* node corresponds with the A2 v2 instance, which has two Intel(R) Xeon(R) CPU E5-2660 2.20 GHz, 3.5 GB of RAM memory, 20 GB of Hard Drive disk and two network interfaces. IM version 1.7.6 and CLUES version 2.2.0b has been installed in this node. HAProxy 1.8.14-52e4d43 and Consul 1.3.0 are running on Docker containers also in that node. The second type of node, *smallwn*, corresponds with the NC6 instance with six Intel(R) Xeon(R) CPU E5-2690 v3 2.60 GHz, 56 of memory RAM, 340 GB of Hard Drive disk, one NVIDIA Tesla K80 and one network interface. Finally, the *largewn* node type corresponds with the D13 v2 instance, which has eight Intel(R) Xeon(R) CPU E5-2690 v3 @ 2.60 GHz, 56 GB of RAM memory, 400 GiB of Hard Drive disk and one network interface. The operating system is CentOS Linux release 7.5.1804. Nomad version 0.8.6 and Docker version 18.09.0 build 4d60db4 are installed in all nodes.

4.1   Deployment

The infrastructure configuration is coded into Ansible roles[1] and RADL recipes, and they include parameters to differentiate among the deployment. The roles will reference a local repository of packages or specific versions to minimize the

---

[1] All Ansible Roles used in this work are available in a GitHub repository https:// github.com/grycap/.

impact of changes in public repositories, as well as certified containers. Deployment time is the time required to create and configure a resource. The deployment time of the front-end takes 29 min 28 s on average. The time required to configure each worker node is 9 min 30 s.

4.2   Use Case - Emphysema

The use case selected was the automatic quantification of lung emphysema biomarkers from computed tomography images. This pipeline features a patented air thresholding algorithm from QUIBIM [26,29] for emphysema quantification and an automatic lung segmentation algorithm. Two versions were implemented. A fast one with rough lung segmentation can be used during the interactive inspection and validation of parameters. Another one with higher segmentation accuracy is implemented for the batch, production case. This brings the need of supporting short and long cases, which take respectively, 4 and 20 min.

The small cases are related with executions that take minutes to be completed. In order to provide QoS, CLUES is configured to provide always more resources than required (one node always free). As these type of jobs are very fast, the small-jobs nodes that are IDLE too much time are suspended for avoiding the deployment time. The large cases of QUIBIM biomakers could take many hours and use huge amount of resources, so the deployment time is negligible. For this reason, although the large case of this work takes the same amount of time that the deployment time and the application does not need the all resources of the VM, large-jobs nodes are not suspended and restarted, and only one large-job can run concurrently on the same VM. The "small" Emphysema used in this work consumes 15 GB of memory RAM and 2 vCPUs, so three Emphysema small-jobs could run simultaneously on the same node.

The main goal of the experiment is to demonstrate the capabilities of the proposed architecture. The experiment consists of submitting 35 small-jobs and 5 large-jobs in order to ensure that there are workload peaks that require starting up new VMs and idle periods long enough to remove (in case of large-jobs nodes) or suspend VMs (in case of small-jobs nodes). Table 4 shown the time frames were jobs are submitted.

Figure 4 shows the number of jobs (vertical axis) along time (horizontal axis) in the different status: SUBMITTED, STARTED, FINISHED, QUEUED and RUNNING. The first three metrics denote the cumulative number of jobs that have been submitted, have actually started and have been completed over time, respectively. The remaining metrics denote the number of jobs that are queued or concurrently running at a given time.

As depicted in Fig. 4, the length of the queue does not grow above ten jobs. The delay between the submission and the start of a job (the difference in the horizontal axis between submitted and started lines) is negligible during the first twenty minutes of the experiment. After a period without new submissions

Table 4. Scheduling of the jobs to be executed.

Name	Time	Name	Time	Name	Time	Name	Time
small-1	0:01:19	small-10	0:47:25	small-18	1:00:31	small-27	1:10:42
large-1	0:01:20	small-11	0:49:25	small-19	1:01:32	large-5	1:11:44
small-2	0:01:50	small-12	0:50:25	small-20	1:01:33	small-28	1:11:46
small-3	0:05:50	large-2	0:51:27	small-21	1:02:33	small-29	1:14:47
small-4	0:08:51	small-13	0:51:27	small-22	1:03:34	small-30	1:15:48
small-5	0:11:51	small-14	0:52:28	small-23	1:03:36	small-31	1:16:48
small-6	0:16:23	small-15	0:55:28	small-24	1:08:38	small-32	1:17:49
small-7	0:16:24	small-16	0:55:29	small-25	1:08:38	small-33	1:18:49
small-8	0:19:24	small-17	0:58:30	small-26	1:08:39	small-34	1:18:50
small-9	0:22:25	large-3	0:59:30	large-4	1:09:40	small-35	1:21:50

(between 00:24:00 and 00:47:00 min), the workload grows again due to the submission of new jobs, triggering the deployment of four new nodes (as it can be seen in Fig. 5).

The largest number of queued jobs (10) is reached during this second deployment of new resources at 01:15:09, although it decreases to four only three minutes later (and to two at 1:23:00). Besides, the four last jobs queued were large-jobs. It should be pointed out that large-jobs have greater delays than short-jobs between submission and starting as they use dedicated nodes and this type of nodes are eliminated after 1 min without jobs allocated (a higher value will be used in production).

Figure 5 depicts the status of the nodes along the experiment, which could be USED (executing jobs), IDLE (powered-on and without jobs allocated), POWON (being started, restarted or configured), POWOFF (being suspended or removed), OFF (not deployed or suspended) or FAILED. CLUES is configured to ensure that a small-node is always active (in status IDLE or USED). It can be seen in Fig. 5 that two nodes are deployed at the start of the experiment. Then, during the period without new submissions, the running jobs end their executions, so these new nodes are powered off after one minute in the IDLE status. As the workload grows, the system deployed four new nodes between 00:50:00 and 01:13:00. Besides, CLUES tried to deploy one more node (a large-node) but, as the quota limit of the cloud provider's account has reached, the deployment was failed. It should be noted that the system is resilient to this type of problems and successfully ended the experiment. After two hours, all jobs are completed, so CLUES suspends or removes all nodes (except one small-node that has to be active always).

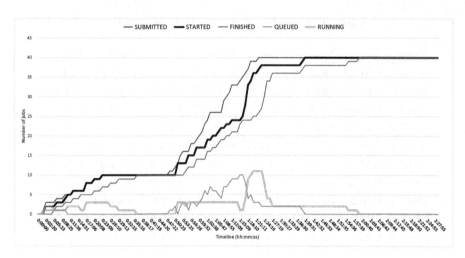

Fig. 4. Status of jobs during the experiment.

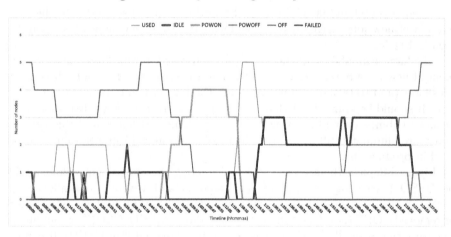

Fig. 5. Status of working nodes.

5   Conclusions and Future Work

This paper has presented a agnostic and elastic architecture and a set of open-source tools for the execution of medical imaging biomarkers. Regarding the technical requirements defined in Sect. 1.2, the experiment of a real use case and the results exposed, it can be concluded that all the requirements proposed were fulfilled by the architecture presented. In Sect. 4.2, a combination of 40 batch jobs was scheduled to be executed in a specific time and the cluster achieve to execute all of them by adjusting the resources available. Furthermore, when there are wasted resources too much time, the nodes are suspended (or eliminated).

The architecture uses IM and CLUES which has proven a good scalability [25] and the capability to work with unplanned workloads [16].

The proposed architecture are not only related to the execution of batch jobs, it provides to developers a workflow to ease the building, testing, delivery and version management of their application.

Future work includes implementing the proposed architecture on QUIBIM ecosystem, testing other solutions for distributed storage as Ceph [33] or One-Data [20], the study of Function As a Services (FaaS) frameworks for executing batch jobs (SCAR [31] or OpenFaas [14]) or using Kubernetes for ensuring that services (Nomad, Consul and HAProxy) are always up.

References

1. Amazon EC2 web site. https://aws.amazon.com/es/ec2/. Accessed 29 Dec 2018
2. Apache Mesos web site. http://mesos.apache.org/. Accessed 29 Dec 2018
3. Chronos web site. https://mesos.github.io/chronos/. Accessed 29 Dec 2018
4. Cloudbiolinux web site. http://cloudbiolinux.org/. Accessed 29 Dec 2018
5. Cloudman web site. https://galaxyproject.org/cloudman. Accessed 29 Dec 2018
6. Eucalyptus web site. https://www.eucalyptus.cloud/. Accessed 29 Dec 2018
7. Galaxy Platform web site. https://galaxyproject.org. Accessed 29 Dec 2018
8. Imagej web site. https://imagej.nih.gov/ij/. Accessed 29 Dec 2018
9. Kubernetes web site. https://kubernetes.io. Accessed 29 Dec 2018
10. Linux containers. https://linuxcontainers.org/. Accessed 29 Dec 2018
11. LXD documentation. https://lxd.readthedocs.io/. Accessed 29 Dec 2018
12. Marathon. https://mesosphere.github.io/marathon/. Accessed 29 Dec 2018
13. Nomad web site. https://www.nomadproject.io/. Accessed 29 Dec 2018
14. OpenFaas web site. https://www.openfaas.com/. Accessed 29 Dec 2018
15. OpenStack web site. https://www.openstack.org/. Accessed 29 Dec 2018
16. de Alfonso, C., Caballer, M., Calatrava, A., Moltó, G., Blanquer, I.: Multi-elastic Datacenters: auto-scaled virtual clusters on energy-aware physical infrastructures. J. Grid Comput. (2018). https://doi.org/10.1007/s10723-018-9449-z
17. Caballer, M., Blanquer, I., Moltó, G., de Alfonso, C.: Dynamic management of virtual infrastructures. J. Grid Comput. **13**(1), 53–70 (2015). https://doi.org/10.1007/s10723-014-9296-5
18. Calatrava, A., Romero, E., Moltó, G., Caballer, M., Alonso, J.M.: Self-managed cost-efficient virtual elastic clusters on hybrid Cloud infrastructures. Future Gener. Comput. Syst. **61**, 13–25 (2016). https://doi.org/10.1016/j.future.2016.01.018
19. De Alfonso, C., Caballer, M., Alvarruiz, F., Hernández, V.: An energy management system for cluster infrastructures. Comput. Electr. Eng. **39**, 2579–2590 (2013). https://doi.org/10.1016/j.compeleceng.2013.05.004
20. Dutka, Ł., et al.: Onedata - a step forward towards globalization of data access for computing infrastructures. Procedia Comput. Sci. **51**, 2843–2847 (2015). https://doi.org/10.1016/j.procs.2015.05.445. International Conference On Computational Science, ICCS 2015
21. European Society of Radiology (ESR): White paper on imaging biomarkers. Insights Imaging **1**(2), 42–45 (2010). https://doi.org/10.1007/s13244-010-0025-8
22. Lee, H.: Using Bioinformatics Applications on the Cloud (2013). http://dsc.soic.indiana.edu/publications/bioinformatics.pdf. Accessed 29 Dec 2018

144 S. López-Huguet et al.

23. Docker Inc.: Docker. https://www.docker.com/. Accessed 29 Dec 2018
24. Kurtzer, G.M., Sochat, V., Bauer, M.W.: Singularity: scientific containers for mobility of compute. PLOS ONE **12**(5), 1–20 (2017). https://doi.org/10.1371/journal.pone.0177459
25. López-Huguet, S., et al.: A self-managed Mesos cluster for data analytics with QoS guarantees. Future Gener. Comput. Syst. **96**, 449–461 (2019). https://doi.org/10.1016/j.future.2019.02.047
26. Martí-Bonmatí, L., García-Martí, G., Alberich-Bayarri, A., Sanz-Requena, R.: QUIBIM SL.: Método de segmentación por umbral adaptativo variable para la obtención de valores de referencia del aire corte a corte en estudios de imagen por tomografía computarizada, ES 2530424B1, 02 September 2013
27. Martí-Bonmatí, L., Alberich-Bayarri, A.: Imaging Biomarkers: Development and Clinical Integration. Springer, Cham (2017). https://doi.org/10.1007/978-3-319-43504-6
28. Marwan, M., Kartit, A., Ouahmane, H.: Using cloud solution for medical image processing: issues and implementation efforts. In: 2017 3rd International Conference of Cloud Computing Technologies and Applications (CloudTech). IEEE, October 2017. https://doi.org/10.1109/cloudtech.2017.8284703
29. Mayorga-Ruiz, I., García-Juan, D., Alberich-Bayarri, A., García-Castro, F., Martí-Bonmatí, L.: Fully automated method for lung emphysema quantification for Multidetector CT images. http://quibim.com/wp-content/uploads/2018/02/ECR_Fully-automated-quantification-of-lung-emphysema-using-CT-images.pdf. Accessed 22 Mar 2019
30. Mirarab, A., Fard, N.G., Shamsi, M.: A cloud solution for medical image processing. Int. J. Eng. Res. Appl. **4**(7), 74–82 (2014)
31. Pérez, A., Moltó, G., Caballer, M., Calatrava, A.: Serverless computing for container-based architectures. Future Gener. Comput. Syst. **83**, 50–59 (2018). https://doi.org/10.1016/j.future.2018.01.022
32. Shakil, K.A., Alam, M.: Cloud computing in bioinformatics and big data analytics: current status and future research. In: Aggarwal, V.B., Bhatnagar, V., Mishra, D.K. (eds.) Big Data Analytics. AISC, vol. 654, pp. 629–640. Springer, Singapore (2018). https://doi.org/10.1007/978-981-10-6620-7_60
33. Weil, S.A., Brandt, S.A., Miller, E.L., Long, D.D.E., Maltzahn, C.: Ceph: a scalable, high-performance distributed file system. In: Proceedings of the 7th Symposium on Operating Systems Design and Implementation, OSDI 2006, pp. 307–320. USENIX Association, Berkeley (2006). http://dl.acm.org/citation.cfm?id=1298455.1298485

A Self-adaptive Local Search Coordination in Multimeme Memetic Algorithm for Molecular Docking

Pablo Felipe Leonhart[ID], Pedro Henrique Narloch[ID], and Márcio Dorn[✉][ID]

Institute of Informatics, Federal University of Rio Grande do Sul,
Porto Alegre, Brazil
mdorn@inf.ufrgs.br

Abstract. Molecular Docking is a methodology that deals with the problem of predicting the non-covalent binding of a receptor and a ligand at an atomic level to form a stable complex. Because the search space of possible conformations is vast, molecular docking is classified in computational complexity theory as a NP-hard problem. Because of the high complexity, exact methods are not efficient and several metaheuristics have been proposed. However, these methods are very dependent on parameter settings and search mechanism definitions, which requires approaches able to self-adapt these configurations along the optimization process. We proposed and developed a novel self-adaptive coordination of local search operators in a *Multimeme Memetic Algorithm*. The approach is based on the *Biased Random Key Genetic Algorithm* enhanced with four local search algorithms. The self-adaptation of methods and radius perturbation in local improvements works under a proposed probability function, which measures their performance to best guide the search process. The methods have been tested on a test set based on HIV-protease and compared to existing tools. Statistical test performed on the results shows that this approach reaches better results than a non-adaptive algorithm and is competitive with traditional methods.

Keywords: Self-adaptation · Multimeme Memetic Algorithms ·
Molecular docking

1 Introduction

Molecular Docking (MD) is a computer-aided approach used in Drug Discovery to predict the conformation of a small molecule (ligand) inside a larger molecule binding site (receptor), measuring the binding affinity between these molecules. Due to the massive number of possible conformations that a molecule can assume, the MD problem is considered as a NP-Hard one by computational complexity theory. Thus, the development of search methods with the capability to explore the conformational search space is essential. Due to the complexity related to the conformational search space, exact methods are not efficient since

© Springer Nature Switzerland AG 2019
J. M. F. Rodrigues et al. (Eds.): ICCS 2019, LNCS 11538, pp. 145–159, 2019.
https://doi.org/10.1007/978-3-030-22744-9_11

they can not retrieve a solution in a viable time. In this manner, metaheuristics, like *Memetic Algorithms* (MA), employing algorithms as global and local search procedures, became interesting approaches for finding good solutions in a possible computational time. Regardless of the possibility of using metaheuristics to explore the search space, they are very dependent on parameter setting and search mechanism definition. In this way, self-adaptive mechanisms are reliable techniques to give the algorithm the option to self-decide which parameter, or search mechanism, should be used along the optimization process [1].

Besides the search mechanism definition, another important component of MD methods is the energy function responsible for describing the interaction between receptor-ligand, evaluating different physicochemical aspects related to the binding process. Also, the scoring function is important for the search method, since it is used to distinguish and rank different solutions regarding energy. Different scoring functions are available in the literature [2], but the RosettaLigand scoring function had achieved interesting results in the last years [3,4] in MD predictions. In light of these facts, we propose an MD method based on RosettaLigand scoring function. To explore the conformational search space, a novel self-adaptive multimeme algorithm based on *Biased Random Key Genetic Algorithm* (BRKGA) enhanced with four local-search algorithms is evaluated and compared with state-of-art methods. Our contributions are related to the self-adaptive mechanism used to select the local search strategy to be applied, and the radius perturbation for each docking case. Results obtained showed that our approach reaches better energy results in comparison with the non-adaptive BRKGA version and traditional methods such as AutoDock Vina [5], and jMetal [6]. The remaining of this paper is organized as follows: Sect. 2 describe the main concepts related to the molecular docking problem and related works. Section 3 presents the proposed self-adaptive approach. In Sect. 4 are analyzed the achieved results. Finally, Sect. 5 presents the conclusions and future works from this research.

2   Preliminaries

Molecule Flexibility: An important aspect related to the MD problem is the flexibility associated with each molecule. It is possible to classify the MD approaches in rigid-ligand and rigid-protein; flexible-ligand and rigid-protein; and flexible-ligand and flexible-protein. These approaches dictate how the problem is computationally encoded. Therefore, the molecule flexibility is directly related to the problem's dimensionality, where more flexible molecules have more degrees of freedom. Thus, leading to a more complex optimization search space. For the flexible-ligand with rigid-receptor scheme (used in this work), a possible solution for the problem can be described by $7 + n$ variables, where three of them are the ligand translation values (T_x, T_y, and T_z), followed by four values corresponding the ligand's orientation (Q_x, Q_y, Q_z, and T_w), and n varies according to the dihedral angles present in the ligand.

Scoring Function: In order to compute the binding affinity between a ligand and a receptor, scoring functions are used to score and rank the steric and electrostatic interactions of both molecules. As the energy measurement is a computationally expensive operation [7], approximation methods have been used. It is possible to categorize these functions in force-field based, empirical and semi-empirical, and knowledge-based functions [8]. In this paper, we have used a knowledge-based energy function known as RosettaLigand, which considers statistical analysis of the *Protein Data Bank* (PDB) in its composition [3] in addition to physicochemical terms [4]. The RosettaLigand score function was tested in different comparative studies, showing good results when compared with well-known scoring functions such as AutoDock [3].

Related Works: The multidimensionality and complexity of the conformational search space avoid that any known computational technique optimally solves the MD problem. Many methods based on metaheuristics have been applied to attempt to get optimal solutions to these problems. The major of algorithms utilized in MD are *Genetic Algorithms* (GA), *Differential Evolution* (DE), and *Particle Swarm Optimization* (PSO). Hybrid strategies also have been applied, like in Rosin et al. [9] where was implemented a GA hybridized with *Simulated Annealing* (SA) and *Solis-Wets* as LS strategies to explore the binding conformational search space. In Tagle et al. [10] three variations of SA were employed as local search procedure in a *Memetic Algorithm*. Recently, Leonhart et al. [11] proposed a hybridization of BRKGA with three variations of *Hill-climbing* and SA algorithms working as LS chained operators applied in test cases of HIV-protease. The approach proposed by Krasnogor et al. [12] and applied to bioinformatics problems (prediction of structures) is also relevant to highlight. The authors developed a simple, but efficient, inheritance mechanism (SIM) for a discrete combinatorial search. The strategy consists of encoding the memetic material in the individual's representation, where this material indicates the LS to be applied. During the evolution, the crossover is responsible for choosing what method attach to the offspring, according to the parent with better fitness. Jakob [13] proposed another self-adaptive approach to schedule local search methods over a function probability. In this idea, all LS have equal chance to be selected at the beginning of the process. During the evolutionary process, the probabilities of applying each algorithm are updated, according to the relative fitness gain and the required evaluations considered. In Domínguez-Isidro et al. [14] is proposed an adaptive local search coordination, based on a cost-benefit scheme, for a multimeme DE for constrained numerical optimization problems. Also, in Jin et al. [1] is proposed a MA, combining GA, DE, and *Hill-climbing*, working under a strategy that use weights to measure the contribution of the algorithms according to their improvement over the individuals of the population and in the stage of evolution. In this study, we investigate the possibility of applying a self-adaptive memetic algorithm, based in a proposal probability function and inspired on the BRKGA to coordinate the use of different local search methods, and exploring different neighborhoods by adapting the radius perturbation.

Memetic Algorithms: A Memetic Algorithm (MA) is an evolutionary algorithm which is composed of global and local procedures [15]. Global search algorithms can explore the whole search space, while, the exploitation of a neighborhood solution is attributed to a local search method, which can obtain good precision [16]. They are inspired by Darwinian's principles of natural evolution and Dawkins' notion of a *meme*, defined as a unit of cultural evolution that can perform local refinements.

Biased Random Key Genetic Algorithm: In this study, our global search strategy is inspired on the *Biased Random Key Genetic Algorithm* (BRKGA) initially proposed by Gonçalves et al. [17]. The algorithm combines aspects of GA and random keys to encode the solutions by real values. In our approach, however, we adopt real-coded values to represent each gene, instead of the keys. In BRKGA, the set of solutions is ordered by fitness value and divided in *castes*, called *elite* and *non-elite* groups. The initial population is randomly generated and the individuals ordered in the castes. The crossover operator gets one parent from each set to create the offspring following a probabilistic parameter, that prefers more genes from the elite group. The mutation procedure is responsible for generating new individuals instead of modifying them. Thus, the next population will be composed by the caste 'A' (elite set) from the current population; the caste 'B', formed by all offsprings; and the caste 'C', which are the mutated solutions. Then, the new set is ordered to update the individuals in the castes.

Search Space Discretization: The MD problem needs a definition of the search space around the binding site of the receptor. In this work, we adopt the idea proposed by Leonhart et al. [11] representing this area as a cube formed by smaller cubes, where the central atom of the ligand defines the center of this box. Thus, with this point and the area's volume is possible to set the size of smaller cubes. This division has the objective to explore the whole ligand-receptor binding search space better. At the beginning of the memetic execution, the population is generated and equally distributed between the subareas of the cube. To keep the diversity, at least one individual must be in each area. This feature builds a local competition, where the better solutions dispute location between them. Procedures like crossover, mutation and local search operators could move solutions between regions, following the evolutionary process, but respecting the percentual of individuals non-migrants. The crossover can create offsprings from different positions of their parents according to the cube of each one. In mutation process, eventually, are filled empty cubes, and generated solutions in any area of the search space. Also, the local search methods can modify solutions of a subcube, depending on the approach to visit the neighborhood solutions.

Local Search Methods: Local search is a heuristic method which moves from solution to solution in the candidate's space by performing local changes until getting a local optimum or reach another limit. However, this movement is only possible if a neighborhood relation is defined in the search space. The neighborhood generation is made taking into account three aspects: (i) order of gene visitation, i.e., which position would be first considered to explore; (ii) radius perturbation, what means how much each gene must be modified; and (iii) the direction of search, that indicates if the radius value would be added or subtracted from the value encoded in the individual.

The *Hill Climbing algorithm* (HC) [18], also known as descent improvement, is an old and simple local search that starts with an initial solution and replace it, at each iteration, by the best neighbor found that improves the value of objective function. The procedure ends when there is no solution better than the current one, reaching a local optimum. This algorithm presents variations in the way that the neighborhood is explored. The order of solution's generation (deterministic or stochastic) is one of them, as well as the strategy selection of the next solution, known as a pivoting rule. We highlight three variants of HC (pivoting rules) to select the best neighbor: (i) *Best Improvement* (BI), which performs a fully and deterministic search in the whole neighborhood. This strategy could be time-consuming because it evaluates all the movements, although ensuring the selection of the best solution in every iteration. (ii) *First Improvement* (FI) this variation swap the current solution by the first best neighbor found. The neighbors are evaluated in a deterministic way following a pre-defined order of candidates' generation. The approach is faster than BI because it does not visit the whole neighborhood, just in the worst case, meaning the end of improvements. (iii) *Stochastic Hill Descent* (SHD) is similar to FI. The swap rule of solutions is the same, but the order of neighbors' visitation is randomized at each iteration of the process. This characteristic ensures an equal selection of candidates from all regions of the search space.

Another local search method is the *Simulated Annealing* (SA), a stochastic algorithm which accepts, in some moments, worst solutions [19]. The objective is to avoid local optimums and delay the search convergence. The process starts with an initial solution and at each iteration generates a random neighbor. The candidate is accepted if better than the current solution, and if it is worst, there is a rule with a decreasing probability to allow this solution. Thus, the procedure begins running like a random walk accepting many solutions, but when the probability decreases, the method is more similar to the HC algorithm.

A recent feature in LS is the concept of *Local Search Chains* proposed by Molina et al. [20]. The idea is to share the state of search between different local search's applications. The LS call may not explore all neighborhood of a solution, so the final strategy parameter values achieved by the candidate will be the initial state of a possible subsequent local search application in this individual. This strategy allows the LS operators to be extended in some promising search zones, avoiding different algorithms to evaluate the same candidates by chaining the searches. According to [20] there are some aspects to consider in the LS chains' management. A fixed intensity search is one of them, LS intensity stretch (I_{str}), which ensures that every local search algorithm has the same computational effort applied. Another one is to save the configurations (visitation order) that guided the search and the current state at the end of the call. In the HC methods, for instance, the last neighbor generated is saved in all variants, in the FI and SHD approaches are also kept the visitation's order of the neighborhood.

Adaptive Memetic Algorithms (MAs): Recent studies have applied hybridization with adaptation in MAs. The adjustment of parameters and operators has presented a promising area of computation in this evolutionary algorithm. This approach can self-adjust to a given problem without previous knowledge by utilizing acquired information to adapt itself with the search progress. The challenge to design a robust and efficient MA has some questions to be considered, like: (i) where and when the local search should be applied; (ii) which individuals should be improved and how must be chosen; (iii) the computational effort required in each LS call; (iv) the equilibrium of ratio between global and local search. Several adaptive memetic algorithms have addressed these questions [1,12,13].

Multimeme Memetic Algorithm (MMA): The *Multimeme Memetic Algorithm* (MMA) was originally proposed by Krasnogor and Smith [21]. In a MA only one LS method is used, while in the MMA a set of local searchers is employed. The idea of the algorithm is self-adaptively choose from this set which method to use for different instances, phases of the search or individuals in the population. In their approach, the individual was represented by its genetic and memetic material, where the last one specifies which meme will be used to perform local search.

3   The Proposed Method

In this paper, we propose and test a MMA algorithm for the MD problem. Our algorithm self-adaptively choose which local search operators to apply in different individuals. We represent a solution only by its genetic material, i.e., encoding values to perform conformational operations with the ligand structure, which includes the translation and rotation of the molecule, and internal angle rotations. Thus, from the definition of our solution representation, we have a neighborhood size equal to $2n$, where n represents the genes' number in an individual. We have adapted and implemented the three variations of the HC algorithm and the Simulated Annealing (for more details see Sect. 2). Also, we incorporate the LS chaining to enjoy best the benefits of each algorithm applied. The main contribution of this study is the self-adaptive coordination of the local search methods and the radius perturbation, which works under a probabilistic rule composed of two main elements, inspired in the studies of Jakob [13,22] and Domínguez-Isidro et al. [14]. Equation 1 show these two terms: the ratio fitness gain (rfg), and the ratio of success application (rs_{app}). The first element it is a measurement of how much each LS operator improves the fitness of solutions, considering the history of values reached and the effort applied in each one. The idea is to ponder LS operators (LSO), performing few iterations and considering improvements in the individuals. The second term is simple and represents the benefit of the LS, by counting how many times solutions were improved.

$$rfg = \left| \frac{f_{original} - f_{LS}}{f_{original} - f_{mean}} \right| \qquad rs_{app} = \frac{n_{improvements}}{n_{app}} \qquad (1)$$

The rfg is obtained from the LS application to one individual, where $f_{original}$ is the individual fitness before the local improvement, f_{LS} is the fitness reached by the algorithm, and f_{mean} is the fitness average from all solutions of the caste 'A' in the moment of LS call. The objective is to measure the effectiveness of the local operator over a solution comparing the improvement with the best individuals from the population at that moment. The rs_{app} is the ratio between the number of improvements ($n_{improvements}$) and the number of applications (n_{app}) of a given LSO. To keep the methods competitive and cooperative, inspired in the study of Jin et al. [1], we adopt *weights* to measure the contribution of each algorithm, the Eq. 2 shows the proposal:

$$weight_{rfg} = \left(\frac{\sum_{i=1}^{n_{app}} rfg_i}{n_{app}} \right) * w_{rfg} \qquad weight_{rs_{app}} = rs_{app} * w_{rs_{app}} \tag{2}$$

$$weight_X = weight_{rfg} * weight_{rs_{app}}$$

To measure the effectiveness of a LS we consider the history of improvements by calculating the average rfg and multiplying by a fixed value, represented by w_{rfg}. In the same way, at the success ratio is attributed another weight ($weight_{rs_{app}}$). These weight values added is equal to zero. Finally, with the multiplication of both, we get $weight_X$, where X is any LS operator. After a local search application, the respective weight is calculated, and then the probabilities of each one are obtained by Eq. 3 shows:

$$prob_X = \frac{weight_X}{\sum_{j=1}^{n_{LS}} weigth_j} \tag{3}$$

The probability of a given local search is obtained by calculating its contribution percentage in the weights of all LS methods (n_{LS}). Thus, each operator has a value between 0 and 1, allowing to mount a roulette wheel to sort and obtain an algorithm to be applied. It is important to highlight that at the beginning of the MMA execution all LSO has the same evaluations number to fairly calibrate the initial weights. Along the execution, the best method will get a higher probability, so having more opportunity to be applied. Algorithm 1 shows the pseudo-code of our approach. The input values P, P_e, P_m, n, I_{ls} represent the population size, the elite size, the mutant population size, the number of individual genes, and the minimum number of individuals to run a LS call, respectively. During the crossover and mutation processes, if the intensity stretch of global search is reached, the local search function is called. The individuals to be improved are selected from the most populated cubes (see space discretization in Sect. 2). After each LS application, its contribution is calculated and the probabilities adjusted for the next call. The execution's output is the best solution found until satisfies the maximum of energy evaluations.

4 Experimental Results

To test our MA approach, we use a test set with 16 complexes containing receptor and ligand molecules. We perform the tests into the following stages: (i) comparison of the proposed MMA with other two self-adaptive algorithms considering

Algorithm 1. Pseudocode of the developed MMA algorithm.

Input: P, P_e, P_m,n, I_{ls}
Output: Best Solution X
1:   $P \leftarrow$ initialize with n vectors of real-coded values
2:   **while** not reach the maximum of energy evaluations **do**
3:      Evaluate solutions of P and divide in P_e and $P_{\bar{e}}$
4:      $P^+ \leftarrow P_e \bigcup crossover() \bigcup mutation()$ ▷ Making the next population
5:      **if** currentEvaluations is multiple of LS intensity **then**
6:         *prepareLocalSearch()*   ▷ Individuals improved are introduced in P^+
7:      **end if**
8:   **end while**
9: **return** best solution X

Local search preparation

10: Make a list of potential candidates to apply LS   ▷ According to I_{ls} value
11: **while** not enough evaluations in the I_{str} **do**
12:     individual = getNextIndividual()   ▷ Iterating over the previous list
13:     method, radius = chooseOperators()   ▷ Sort values to determine which one
14:     *runLocalSearch(individual, method, radius)*
15:     rfg and rs_{app} are calculated ▷ According to method and radius perturbation
16:     updateProbabilities()   ▷ Probabilities of LSO are updated under the weights
17: **end while**

a fixed radius perturbation; (ii) the MMA approach with three variations of radius search; (iii) finally, the self-adaptive algorithm was compared with other methods in the state-of-art.

Benchmark: The 16 selected structures are based on the HIV-protease receptor and was previously classified in [23]. The set was obtained from the PDB database and divided into four groups, following the ligand structure size. The PDB code and the range of crystallographic resolution in Ångströms (Å) are: *small* (1AAQ, 1B6L, 1HEG, 1KZK, 1.09–2.50); *medium* (1B6J, 1HEF, 1K6P, 1MUI, 1.85–2.80); *large* (1HIV, 1HPX, 1VIK, 9HVP, 2.00–2.80) size inhibitors, as well as *cyclic urea* (1AJX, 1BV9, 1G2K, 1HVH, 1.80–2.00) inhibitors. PYMOL was used to remove molecules such as solvent, non interacting ions, and water from the target structures.

The *AutoDockTools* was used to add partial charges and hydrogens to structure, as well, to define the maximum number of torsional angles in the ligand. This maximum number of torsions was set to 10, but can be less according to the ligand structure/size, and their selection considers the angles that fewer move atoms, keeping freeze the ligand center [23]. After, the *Open Babel tool* [24] was used to convert and generate necessary files to be manipulated by our algorithm and then used by the *PyRosetta* to load the complexes and evaluate the energy score. As already mentioned, the search space was represented by a cube including the binding site of the receptor, the size of this box was set in 11 Å for each axis, and the grid spacing defined as 0.375 Å. From this definition, the initial ligand conformation and position in the cube are randomized for each algorithm run, to take no advantage of the known crystal structure.

Parametrization: Considering the random values for the ligand in each run is fair that every algorithm starts from the same point. Thus, for each instance in each run, the initial population and the ligand structure was the same. In BRKGA we adopted the recommended values to the parameters according to [17]. The population size is 150 individuals, where 20% is the elite group, 30% is the mutation set, and the remaining is crossover offsprings. The elite allele inheritance probability on the random choice is 0.5–0.7 to crossover operation and the percentual of individuals non-migrants is set to 30% also. The LS parameters were defined as 0.5 to radius perturbation, in the first experiments, after this value varied in 0.1 and 0.25 also. The intensity stretch was equal to 500, and the global/local search ratio equal to 0.5, following the values adopted in [20]. Finally, the MMA weights were defined as 0.6 to the historical ratio fitness gain and 0.4 to the ratio of LS success application. Since the use of a stochastic method, we execute 11 (first and second stages) and 31 (third stage) independent runs per instance with a stop criterion of 1,000,000 energy evaluations. So, we are acquiring statistical confidence in the present results.

Achieved Results: We divide our tests into three stages described as following. Firstly, we ran the MMA approach, working under our proposal probability function, and compared with another two simple self-adaptation methods. The SIM technique [21] is one of them, and the other one is a random selection of the LS operators at the moment to be used. In the SIM approach we start the population equally distributing the LS methods to each solution, and during the iterations, the crossover and mutation procedures changes this memetic material. The random approach just sorts a method when the local search must be applied. All algorithms were applied over half of the instances with 11 runs each one. In this first test, we are interested in verifying the behavior of the approach based on probabilities against the inheritance and random methods. It is important to note that the radius perturbation is fixed in 0.5 to every gene. This value was chosen empirically in preliminaries tests with the generation and evaluation of random individuals. We evaluated the results by its energy in *kcal/mol* and with the *Root Mean Square Deviation* (RMSD), which is a structural measure of the ligand get by the algorithm with the crystallographic experimental structure of the molecule. Table 1 presents the best energy found and the corresponding RMSD, as well as the average and standard deviation of each one, based on the best solutions generated in each configuration.

The SIM method has a slight advantage over our proposal, considering just the best energies found, but looking for the average, we get the best values in half instances. Also, regarding RMSD, the probability approach gets the best solutions in five test cases, and about the average, the best results were divided between the algorithms. We applied a nonparametric test for multiple comparisons procedure, the *Dunn test* [25], to analyze the statistical significance in the achieved results. The test indicated that there is no difference between the approaches compared. In the second stage, we have applied our MMA algorithm shifting the radius perturbation of the LS with the values: 0.10, 0.25, and 0.50. The idea of this stage is to verify if different modifications in the genes might improve the

Table 1. Achieved results from the MMA, based on the probability function, with an inheritance (SIM) and random (RAND) approaches. Cells highlighted in gray shows the best solution obtained, and cells in blue, the best solutions average, for each instance.

ID	Method	Best solution		11 runs average		ID	Best solution		11 runs average	
		Energy	RMSD	Energy	RMSD		Energy	RMSD	Energy	RMSD
1AJX	RAND	-250.16	1.02	-242.06 ± 3.61	10.49 ± 4.04	1HPX	-356.80	2.88	-349.66 ± 3.61	4.86 ± 1.73
	SIM	-250.64	1.40	-243.21 ± 4.17	9.11 ± 4.47		-355.18	2.25	-349.30 ± 2.77	5.10 ± 1.36
	MMA	-250.20	3.21	-241.34 ± 2.86	11.48 ± 2.68		-357.00	2.73	-350.15 ± 3.90	4.89 ± 1.19
1B6J	RAND	-327.38	2.05	-322.45 ± 5.03	4.69 ± 2.83	1K6P	-383.38	5.26	-380.48 ± 2.56	6.42 ± 1.60
	SIM	-326.64	2.13	-321.75 ± 2.82	5.58 ± 2.41		-383.63	4.32	-381.08 ± 1.58	5.48 ± 1.33
	MMA	-329.11	1.11	-322.75 ± 3.12	3.87 ± 2.71		-383.37	3.95	-380.91 ± 2.57	4.95 ± 1.52
1G2K	RAND	-456.64	1.06	-449.38 ± 6.43	4.40 ± 2.90	1KZK	-440.35	1.58	-435.47 ± 4.49	4.17 ± 2.10
	SIM	-456.13	1.30	-451.27 ± 6.67	3.32 ± 2.53		-441.05	1.54	-435.40 ± 9.85	3.25 ± 2.12
	MMA	-456.70	0.83	-452.69 ± 5.78	2.68 ± 2.06		-441.03	1.53	-435.30 ± 9.21	3.29 ± 1.72
1HEG	RAND	358.31	8.12	359.47 ± 1.25	8.21 ± 1.32	1VIK	175.39	3.59	182.95 ± 11.34	3.74 ± 1.70
	SIM	358.04	5.74	359.52 ± 1.36	7.96 ± 2.30		173.33	2.29	191.48 ± 21.64	5.29 ± 3.65
	MMA	358.05	8.75	359.20 ± 0.70	8.05 ± 1.24		173.94	2.15	191.45 ± 21.94	4.32 ± 2.75

potential of the applied local search method. Table 2 summarizes the achieved results over the same benchmark utilized in the first test. In the same way, we evaluate the energy and RMSD from the best solutions, as well as the averages and standard deviations. Results show that we have an almost equal distribution of the best values found between the three variations, comparing just the best energy reached or the average. Thus, we conclude that all variations in the radius search could improve the LS process, for example, a bigger value can be more adequate at the beginning of the search, while a smaller one would be better in the algorithm refinement stage.

Table 2. Achieved results from the MMA approach with three variations of the LS radius perturbation. Third and fourth columns contains the lowest energy (*kcal/mol*) and its RMSD values for each version. Cells shaded in gray highlight the best solution obtained, and cells shaded in blue, the best solutions average, for each instance.

ID	Radius Search	Best solution		11 runs average		ID	Best solution		11 runs average	
		Energy	RMSD	Energy	RMSD		Energy	RMSD	Energy	RMSD
1AJX	0.10	-251.18	1.39	-243.06 ± 3.89	10.00 ± 4.05	1HPX	-355.42	1.24	-348.88 ± 2.10	5.50 ± 1.69
	0.25	-250.54	3.09	-242.69 ± 4.71	9.68 ± 4.07		-356.61	3.43	-350.85 ± 3.30	4.80 ± 1.34
	0.50	-250.20	3.21	-241.34 ± 2.86	11.48 ± 2.68		-357.00	2.73	-350.15 ± 3.90	4.89 ± 1.19
1B6J	0.10	-328.58	2.10	-325.19 ± 2.47	3.50 ± 2.45	1K6P	-383.59	4.96	-380.35 ± 2.57	5.70 ± 1.66
	0.25	-328.13	2.21	-321.98 ± 3.60	5.84 ± 2.57		-383.52	0.92	-380.80 ± 2.43	6.04 ± 2.45
	0.50	-329.11	1.11	-322.75 ± 3.12	3.87 ± 2.71		-383.37	3.95	-380.91 ± 2.57	4.95 ± 1.52
1G2K	0.10	-456.69	0.78	-443.59 ± 9.81	5.04 ± 2.51	1KZK	-440.93	1.56	-434.06 ± 9.59	4.51 ± 1.92
	0.25	-456.58	1.28	-452.81 ± 5.60	2.88 ± 2.04		-441.12	0.84	-438.83 ± 2.28	3.15 ± 2.07
	0.50	-456.70	0.83	-452.69 ± 5.78	2.68 ± 2.06		-441.03	1.53	-435.30 ± 9.21	3.29 ± 1.72
1HEG	0.10	357.60	9.53	359.46 ± 1.62	8.96 ± 2.41	1VIK	173.89	2.28	185.06 ± 13.93	4.68 ± 3.62
	0.25	358.10	5.57	359.41 ± 1.03	8.02 ± 1.92		172.99	2.16	191.22 ± 18.04	4.95 ± 2.39
	0.50	358.05	8.75	359.20 ± 0.70	8.05 ± 1.24		173.94	2.15	191.45 ± 21.94	4.32 ± 2.75

Also, the application of test Dunn showed that only in three instances (1B6L, 1G2K and 1HEF) there is a statistical difference between the three variations of the MMA algorithm. With these preliminary results, we have decided to extend

Table 3. Comparison results of MMA with BRKGA, two MA methods, AUTODOCK VINA, DOCKTHOR, and jMETAL. The lowest energy (*kcal/mol*) and its RMSD Å values are shown (best values highlighted in gray), as well as their respective averages and standard deviation (cells shaded in blue for the best values) for 31 runs of each algorithm.

ID	Radius Search	Best Energy	Best RMSD	11 runs avg Energy	11 runs avg RMSD	ID	Best Energy	Best RMSD	11 runs avg Energy	11 runs avg RMSD
1AAQ	BRKGA	36.70	1.14	42.75 ± 5.89	3.12 ± 2.47	1HIV	-164.16	2.92	-150.52 ± 3.10	6.65 ± 1.11
	SHD	36.88	1.13	37.94 ± 0.47	1.09 ± 0.14		-164.94	3.41	-163.67 ± 0.82	2.28 ± 0.55
	SA	37.04	1.14	39.12 ± 2.68	1.59 ± 1.51		-165.15	3.39	-163.19 ± 2.67	2.65 ± 1.04
	MMA	36.72	1.24	38.76 ± 3.73	1.64 ± 1.70		-165.12	3.38	-163.35 ± 3.58	2.85 ± 1.39
	VINA	3.93	9.75	9.07 ± 5.05	9.06 ± 0.86		-0.29	7.49	12.01 ± 18.70	8.09 ± 0.92
	DOCK	3.52	0.94	5.12 ± 1.41	9.25 ± 4.92		55.13	0.29	55.32 ± 0.14	0.29 ± 0.04
	JMETAL	-16.00	2.90	-12.56 ± 3.48	2.11 ± 1.15		-32.00	7.93	-18.14 ± 3.68	2.58 ± 1.45
1AJX	BRKGA	-246.85	4.40	-239.68 ± 1.87	12.24 ± 1.50	1HPX	-348.40	5.28	-346.81 ± 1.59	5.87 ± 1.04
	SHD	-250.51	0.98	-243.68 ± 4.39	8.59 ± 4.75		-357.34	3.15	-352.56 ± 3.99	4.04 ± 1.36
	SA	-250.62	1.22	-242.21 ± 3.84	10.32 ± 4.22		-356.79	2.97	-351.63 ± 3.73	4.66 ± 1.49
	MMA	-250.89	3.08	-243.00 ± 4.56	9.42 ± 4.57		-357.49	3.52	-349.83 ± 3.40	5.08 ± 1.17
	VINA	-10.74	1.52	-9.82 ± 0.42	6.21 ± 2.82		-9.08	5.74	-6.52 ± 2.18	6.18 ± 1.03
	DOCK	48.68	0.85	49.33 ± 2.30	0.88 ± 0.09		88.64	7.81	96.05 ± 9.69	7.07 ± 2.11
	JMETAL	-18.00	4.40	-12.84 ± 4.35	3.88 ± 1.64		-18.00	4.86	-11.73 ± 4.77	3.84 ± 0.94
1B6J	BRKGA	-328.69	1.11	-316.83 ± 7.49	5.85 ± 2.33	1HVH	428.01	8.94	434.16 ± 2.60	10.33 ± 2.22
	SHD	-328.70	1.09	-322.42 ± 3.26	5.06 ± 2.76		427.29	8.38	429.98 ± 2.45	8.65 ± 1.62
	SA	-328.27	1.18	-320.81 ± 3.99	5.94 ± 2.22		427.00	8.42	430.81 ± 2.26	8.66 ± 1.80
	MMA	-329.00	1.13	-322.17 ± 3.89	5.63 ± 2.84		427.95	8.41	431.74 ± 2.50	9.06 ± 2.79
	VINA	-9.12	2.89	-2.39 ± 3.64	6.92 ± 3.08		-8.65	7.34	-7.04 ± 1.12	5.98 ± 1.71
	DOCK	38.78	0.61	38.96 ± 0.11	0.61 ± 0.06		127.18	8.34	130.22 ± 2.76	6.10 ± 1.58
	JMETAL	-18.00	3.47	-11.45 ± 6.79	2.49 ± 0.71		-18.00	2.44	-12.45 ± 4.42	2.89 ± 0.52
1B6L	BRKGA	-317.46	3.09	-312.62 ± 3.11	6.64 ± 1.94	1K6P	-382.38	4.42	-375.89 ± 4.85	6.88 ± 1.76
	SHD	-319.21	2.80	-316.64 ± 1.83	3.54 ± 1.99		-383.78	0.83	-368.36 ± 67.28	5.39 ± 1.96
	SA	-318.57	2.96	-315.72 ± 2.34	4.42 ± 2.41		-383.08	5.36	-380.13 ± 1.85	6.29 ± 1.56
	MMA	-319.56	2.48	-315.66 ± 2.25	5.10 ± 2.43		-384.23	4.07	-381.27 ± 1.60	5.73 ± 1.93
	VINA	-12.71	0.89	-12.02 ± 1.29	2.22 ± 3.07		-5.16	5.22	-0.53 ± 3.20	7.45 ± 2.09
	DOCK	30.51	0.46	30.72 ± 0.13	0.43 ± 0.02		143.30	1.73	150.40 ± 6.76	2.10 ± 1.88
	JMETAL	-16.00	2.17	-13.14 ± 3.68	2.17 ± 0.63		-20.00	2.38	-14.69 ± 4.77	3.20 ± 0.85
1BV9	BRKGA	-80.07	0.74	-16.33 ± 44.55	7.27 ± 2.86	1KZK	-440.32	1.91	-422.16 ± 17.60	5.68 ± 2.14
	SHD	-80.64	0.67	-57.17 ± 18.63	6.83 ± 4.84		-441.16	1.54	-438.45 ± 2.99	2.83 ± 1.97
	SA	-80.41	0.70	-53.44 ± 22.76	6.90 ± 4.59		-441.03	1.55	-435.52 ± 5.89	3.94 ± 2.35
	MMA	-80.81	0.71	-53.52 ± 16.20	7.48 ± 4.36		-441.19	1.54	-437.01 ± 6.58	3.47 ± 2.31
	VINA	14.56	5.78	20.65 ± 2.64	8.41 ± 1.49		-9.85	2.37	-8.05 ± 0.83	5.73 ± 2.57
	DOCK	55.69	0.89	56.59 ± 1.04	0.94 ± 0.05		27.61	0.79	27.79 ± 0.16	0.75 ± 0.04
	JMETAL	-20.00	2.20	-14.61 ± 5.64	2.45 ± 1.31		-24.00	9.02	-10.44 ± 12.79	7.95 ± 1.63
1G2K	BRKGA	-455.77	1.34	-436.89 ± 9.44	6.04 ± 2.05	1MUI	-36.39	1.83	-26.57 ± 6.68	5.85 ± 3.02
	SHD	-456.71	0.88	-452.56 ± 5.60	2.83 ± 2.28		-36.79	1.82	-34.99 ± 1.86	1.72 ± 0.68
	SA	-456.62	0.78	-450.82 ± 6.38	3.43 ± 2.36		-36.57	1.93	-33.67 ± 2.64	2.53 ± 2.21
	MMA	-456.78	1.10	-451.14 ± 7.35	3.21 ± 2.52		-36.85	1.95	-33.83 ± 3.38	3.35 ± 2.75
	VINA	-10.01	4.41	-9.34 ± 0.86	6.05 ± 1.77		-8.22	6.68	-6.29 ± 1.15	7.48 ± 1.09
	DOCK	15.46	0.36	16.10 ± 1.49	0.43 ± 0.17		17.06	0.34	17.42 ± 0.15	0.46 ± 0.20
	JMETAL	-20.00	2.60	-12.84 ± 5.31	3.01 ± 1.69		-20.00	4.65	-13.04 ± 4.16	2.57 ± 0.98
1HEF	BRKGA	175.78	13.48	176.69 ± 0.57	12.25 ± 0.56	1VIK	174.70	2.18	228.25 ± 50.90	6.85 ± 2.82
	SHD	173.69	11.43	175.26 ± 1.00	11.79 ± 0.67		173.46	2.20	186.06 ± 16.52	3.82 ± 2.81
	SA	173.69	11.36	175.46 ± 0.87	12.00 ± 0.80		173.35	2.16	187.99 ± 18.04	4.59 ± 2.93
	MMA	173.58	11.40	174.81 ± 1.19	11.66 ± 0.33		173.34	2.15	182.83 ± 13.13	4.20 ± 2.66
	VINA	-1.79	9.36	1.42 ± 2.26	8.77 ± 0.56		96.97	11.19	140.08 ± 29.70	9.18 ± 2.02
	DOCK	67.68	1.75	68.07 ± 0.36	1.83 ± 0.15		165.54	1.38	267.46 ± 123.25	2.97 ± 2.60
	JMETAL	-22.00	6.60	-10.74 ± 6.25	5.81 ± 1.67		-40.00	4.90	-19.12 ± 14.44	4.01 ± 0.98
1HEG	BRKGA	357.76	6.91	361.20 ± 1.45	8.67 ± 1.68	9HVP	359.99	1.09	372.04 ± 14.29	4.98 ± 3.39
	SHD	357.18	6.45	359.00 ± 0.59	7.53 ± 1.36		360.26	1.46	362.91 ± 4.20	3.11 ± 3.44
	SA	356.76	5.56	359.11 ± 1.32	7.77 ± 1.82		360.09	1.05	362.94 ± 4.75	2.50 ± 2.46
	MMA	356.91	9.52	358.99 ± 1.11	8.19 ± 1.49		359.97	1.11	363.17 ± 4.78	3.36 ± 3.40
	VINA	-5.85	5.50	-5.51 ± 0.26	6.00 ± 0.98		-3.74	10.09	0.50 ± 4.03	9.39 ± 0.95
	DOCK	58.37	2.75	60.50 ± 1.91	4.02 ± 1.49		25.96	1.19	26.04 ± 0.05	1.21 ± 0.05
	JMETAL	-14.50	3.16	-9.15 ± 3.90	5.24 ± 1.73		-22.00	4.17	-14.89 ± 4.13	2.72 ± 1.01

and apply our proposal probability function to the choice of which radius perturbation to use during the optimization process, combining it with the previous variation in the LS methods. In this third stage, the weights and probabilities of the radius pool are updated after a local search execution. Similarly, the application and evaluation of the LS algorithms are made with any radius perturbation applied. Thus, we compare the MMA approach against a BRKGA version without LS, and another two MA versions with SHD and SA. Additionally, we compare these results with AUTODOCK VINA [5], DOCKTHOR [26], and jMETAL [6], a multi-objective docking approach. Table 3 shows the achieved results. The comparison of energy values was made only between our implemented methods because of the energy function utilized, which differs from other methods. The RMSD comparison was made between all algorithms. The values show that the MMA is better than BRKGA, when we compare the average of energy and RMSD. Comparing the SHD and SA approaches with the self-adaptive, it is possible to notice that values are very similar. Considering the average RMSD values, the MMA got better results only in the 1AAQ complex in comparison with the state-of-art algorithms. The best method overall was the DOCKTHOR with better RMSD values in 81% of the instances used in this work.

In the same way, we applied the Dunn test in this stage. Table 4 shows the significant levels of energy and RMSD, adopting a significance of $\alpha < 0.05$, guaranteeing a confidence of 95% in the analysis. Cells above the main diagonal (highlighted) shows the p-values comparing the energy results, and the remaining cells show the RMSD comparisons. Comparing MMA with BRKGA we find significant difference in all instances regarding energy and 50% in RMSD. Also, looking for the differences between the self-adaptive approach and SHD and SA, there are significant energy values only for complexes 9HVP and 1AAQ, respectively. Thus, the MMA shows to be better than a BRKGA method, but is equivalent with MA versions.

Table 4. Analysis of results with a significant level equal to $p < 0.05$, cells above the main diagonal show the p-values for energy, and the entries below the RMSD values.

ID	Method	BRKGA	SHD	SA	MMA	ID	BRKGA	SHD	SA	MMA	ID	BRKGA	SHD	SA	MMA	ID	BRKGA	SHD	SA	MMA
1AAQ	BRKGA	---	0.00	0.43	0.00	1BV9	---	0.00	0.00	0.00	1HIV	---	0.00	0.00	0.00	1KZK	---	0.00	0.00	0.00
	SHD	0.00	---	0.10	0.72		1.00	---	1.00	1.00		0.00	---	1.00	0.44		0.00	---	0.39	1.00
	SA	0.00	1.00	---	0.00		1.00	1.00	---	1.00		0.00	1.00	---	0.67		0.05	0.31	---	0.67
	MMA	0.02	1.00	1.00	---		1.00	1.00	1.00	---		0.00	0.64	1.00	---		0.00	1.00	1.00	---
1AJX	BRKGA	---	0.00	0.01	0.00	1G2K	---	0.00	0.00	0.00	1HPX	---	0.00	0.00	0.00	1MUI	---	0.00	0.00	0.00
	SHD	0.00	---	1.00	1.00		0.00	---	0.86	1.00		0.00	---	1.00	1.00		0.00	---	0.45	1.00
	SA	0.34	0.88	---	0.64		0.00	1.00	---	0.32		0.03	0.73	---	1.00		0.00	1.00	---	1.00
	MMA	0.00	1.00	0.70	---		0.00	1.00	1.00	---		0.27	0.10	1.00	---		0.02	0.18	1.00	---
1B6J	BRKGA	---	0.00	0.20	0.00	1HEF	---	0.00	0.00	0.00	1HVH	---	0.00	0.00	0.00	1VIK	---	0.00	0.00	0.01
	SHD	1.00	---	1.00	1.00		0.01	---	1.00	1.00		0.03	---	1.00	0.10		0.00	---	1.00	0.68
	SA	1.00	1.00	---	0.71		0.19	1.00	---	0.44		0.01	1.00	---	1.00		0.08	0.78	---	0.18
	MMA	1.00	1.00	1.00	---		0.00	1.00	0.44	---		0.11	1.00	1.00	---		0.01	1.00	1.00	---
1B6L	BRKGA	---	0.00	0.00	0.00	1HEG	---	0.00	0.00	0.00	1K6P	---	0.00	0.00	0.00	9HVP	---	0.34	0.08	0.00
	SHD	0.00	---	1.00	0.83		0.12	---	1.00	1.00		0.09	---	1.00	1.00		0.54	---	1.00	0.02
	SA	0.00	1.00	---	1.00		0.28	1.00	---	1.00		1.00	1.00	---	0.21		0.18	1.00	---	0.12
	MMA	0.05	0.13	1.00	---		1.00	0.83	1.00	---		0.06	1.00	1.00	---		0.39	1.00	1.00	---

5   Conclusion and Future Work

The statistical test performed shows that the multimeme approach improves the results for the MD problem instead use only an evolutionary algorithm. Although, the comparisons with memetic implementations showed to be equivalent techniques. This is explained because in the MMA version is utilized, in a distributed way, four LS algorithms instead only one in the full search process. In this case, we may be losing computational effort with methods that do not contribute much to find local optimums. Also, in the MA is adopted a fixed radius whereas in the multimeme is utilized three values to be self-adapted. This feature is interesting because the methods can explore different neighborhoods in different stages of the evolution, and then compensate for a possible loss of processing. So, this pool of local search methods combined with the pool of radius perturbation confirms to be an interesting initial variation to reach reasonable solutions to this problem, but still, need improvements.

This work brings contributions to the use of a self-adaptive memetic computational technique. Also, the proposal and application of a function to evaluate, in execution, the local search impacts, to best adapt the parameters and guide the search process. Further investigations might consider different global search algorithms, e.g., *Differential Evolution* and *Particle Swarm Optimization*, as well as methods to run as local searches, such as *Solis and Wets* and *Nelder-Mead*. Besides that, the use of another objective function to evaluate the ligation energy, such as AutoDock Vina, it is interesting to confirm the performance of the algorithm. The results of minimum energy would be different but the ability of MMA will be the same.

About the self-adaptive approach is possible to change the pool of LSO, using only the SHD and SA methods, for instance. The values for radius perturbation can also be modified, as well as different values applied to each gene, i.e., the translation operation may have a specific value different of the rotation and dihedral angles. Also, in the our probability function there are issues to be explored since the weights utilized in both terms until other details. The option by considering the average fitness of caste 'A' can be replaced by the fitness value of the best solution only, or the average of all solutions, or the average of the best individuals from each subcube. Another possibility is to implement a time window to evaluate the LS contributions, considering the last one hundred thousand evaluations, for example. Finally, future works can explore many features to turn this approach better to find solutions to molecular docking problem.

Acknowledgements. This work was supported by grants from FAPERGS [*16/2551-0000520-6*], MCT/CNPq [*311022/2015-4; 311611/2018-4*], CAPES-STIC AMSUD [*88887.135130/2017-01*] - Brazil, Alexander von Humboldt-Stiftung (AvH) [*BRA 1190826 HFST* CAPES-P] - Germany. This study was financed in part by CAPES - Finance Code 001.

References

1. Jin, X., Zhihua, C., Wenyin, G.: An adaptive strategy to adjust the components of memetic algorithms. In: 2014 IEEE 26th International Conference on Tools with Artificial Intelligence, pp. 55–62, November 2014
2. Chen, Y.C.: Beware of docking!. Trends Pharmacol. Sci. **36**, 78–95 (2015)
3. Combs, S.A., et al.: Small-molecule ligand docking into comparative models with Rosetta. Nature Protoc. **8**, 1277–1298 (2013)
4. Davis, I.W., Baker, D.: Rosettaligand docking with full ligand and receptor flexibility. J. Mol. Biol. **385**, 381–392 (2009)
5. Trott, O., Olson, A.J.: Autodock vina: improving the speed and accuracy of docking with a new scoring function, efficient optimization, and multithreading. J. Comput. Chem. **31**(2), 455–461 (2010)
6. López-Camacho, E., Godoy, M.J.G., Nebro, A.J., Aldana-Montes, J.F.: jMetal-Cpp: optimizing molecular docking problems with a C++ metaheuristic framework. Bioinformatics **30**, 437–438 (2013)
7. Dar, A.M., Mir, S.: Molecular docking: approaches, types, applications and basic challenges. J. Anal. Bioanal. Tech. **08**, 8–10 (2017)
8. Kitchen, D.B., Decornez, H., Furr, J.R., Bajorath, J.: Docking and scoring in virtual screening for drug discovery: methods and applications. Nat. Rev. Drug Discov. **3**(11), 935–949 (2004)
9. Rosin, C.D., Halliday, R.S., Hart, W.E., Belew, R.K.: A comparison of global and local search methods in drug docking. In: In Proceedings of the Seventh International Conference on Genetic Algorithms, pp. 221–228. Morgan Kaufmann (1997)
10. Ruiz-Tagle, B., Villalobos-Cid, M., Dorn, M., Inostroza-Ponta, M.: Evaluating the use of local search strategies for a memetic algorithm for the protein-ligand docking problem. In: 2017 36th International Conference of the Chilean Computer Science Society (SCCC), pp. 1–12, October 2017
11. Leonhart, P.F., Spieler, E., Braun, R., Dorn, M.: A biased random key genetic algorithm for the proteinligand docking problem. Soft Comput. **23**, 1–22 (2018)
12. Krasnogor, N.: Studies on the theory and design space of memetic algorithms. Ph.D. thesis, University of the West of England (2002)
13. Jakob, W.: A general cost-benefit-based adaptation framework for multimeme algorithms. Memetic Comput. **2**(3), 201–218 (2010)
14. Domínguez-Isidro, S., Mezura-Montes, E.: A cost-benefit local search coordination in multimeme differential evolution for constrained numerical optimization problems. Swarm Evol. Comput. **39**, 249–266 (2018)
15. Moscato, P.: On evolution, search, optimization, genetic algorithms and martial arts - towards memetic algorithms (1989)
16. Krasnogor, N., Smith, J.: A tutorial for competent memetic algorithms: model, taxonomy, and design issues. EEE Trans. Evol. Comput. **9**(5), 474–488 (2005)
17. Gonçalves, J.F., Resende, M.G.C.: Biased random-key genetic algorithms for combinatorial optimization. J. Heuristics **17**(5), 487–525 (2011)
18. Aarts, E., Lenstra, J.K. (eds.): Local Search in Combinatorial Optimization, 1st edn. Wiley, New York (1997)
19. Kirkpatrick, S., Gelatt, C.D., Vecchi, M.P.: Optimization by simulated annealing. Science **220**(4598), 671–680 (1983)
20. Molina, D., Lozano, M., Sánchez, A.M., Herrera, F.: Memetic algorithms based on local search chains for large scale continuous optimisation problems: MA-SSW-Chains. Soft Comput. **15**(11), 2201–2220 (2011)

21. Krasnogor, N., Smith, J.: Emergence of profitable search strategies based on a simple inheritance mechanism (2001)
22. Jakob, W.: Towards an adaptive multimeme algorithm for parameter optimisation suiting the engineers' needs. In: Runarsson, T.P., Beyer, H.-G., Burke, E., Merelo-Guervós, J.J., Whitley, L.D., Yao, X. (eds.) PPSN 2006. LNCS, vol. 4193, pp. 132–141. Springer, Heidelberg (2006). https://doi.org/10.1007/11844297_14
23. Morris, G.M., et al.: AutoDock4 and AutoDockTools4: automated docking with selective receptor flexibility. J. Comput. Chem. **30**(16), 2785–2791 (2009)
24. O'Boyle, N.M., Banck, M., James, C.A., Morley, C., Vandermeersch, T., Hutchison, G.R.: Open babel: an open chemical toolbox. J. Cheminform. **3**(1), 1–14 (2011)
25. Dunn, O.J.: Multiple comparisons using rank sums. Technometrics **6**(3), 241–252 (1964)
26. de Magalhães, C.S., Almeida, D.M., Barbosa, H.J.C., Dardenne, L.E.: A dynamic niching genetic algorithm strategy for docking highly flexible ligands. Inf. Sci. **289**, 206–224 (2014)

Parallel CT Reconstruction for Multiple Slices Studies with SuiteSparseQR Factorization Package

Mónica Chillarón[1]([⊠]) [ID], Vicente Vidal[1] [ID], and Gumersindo Verdú[2] [ID]

[1] Departamento de Sistemas Informáticos y Computación (DSIC),
Universitat Politècnica de València, València, Spain
mnichipr@inf.upv.es, vvidal@dsic.upv.es
[2] Instituto de Seguridad Industrial, Radiofísica y Medioambiental (ISIRYM),
Universitat Politècnica de València, València, Spain
gverdu@iqn.upv.es

Abstract. Algebraic factorization methods applied to the discipline of Computerized Tomography (CT) Medical Imaging Reconstruction involve a high computational cost. Since these techniques are significantly slower than the traditional analytical ones and time is critical in this field, we need to employ parallel implementations in order to exploit the machine resources and obtain efficient reconstructions.

In this paper, we analyze the performance of the sparse QR decomposition implemented on SuiteSparseQR factorization package applied to the CT reconstruction problem. We explore both the parallelism provided by BLAS threads and the use of the Householder reflections to reconstruct multiple slices at once efficiently. Combining both strategies, we can boost the performance of the reconstructions and implement a reliable and competitive method that gets high-quality CT images.

Keywords: CT · Medical Imaging · Reconstruction ·
Matrix factorization · QR · Few projections · Parallel QR ·
SuiteSparseQR

1 Introduction and Background

In recent years, medical tests such as Magnetic Resonance Imaging (MRI) [1] have gained prominence in clinical practice. MRIs are not harmful to the patient, since the image is produced from the application of magnetic fields on a body. In contrast, Computerized Tomographies (CT) [2] project X-rays, which induce a dose of radiation that can be harmful to the patient. However, despite being harmful, CT scans are still necessary.

On the one hand, they obtain better images than the MRI for certain types of objects of interest (bones and tumors) while the magnetic resonance is mostly applied to soft tissues since it achieves greater contrast between different tissues.

© Springer Nature Switzerland AG 2019
J. M. F. Rodrigues et al. (Eds.): ICCS 2019, LNCS 11538, pp. 160–169, 2019.
https://doi.org/10.1007/978-3-030-22744-9_12

On the other hand, not all people are suitable for both tests. MRI is not recommended for patients who have a pacemaker or metal implants in their body, while they can undergo a CT scan. But CT is contraindicated for pregnant patients, infants and children, due to the dose of radiation induced with the test.

For all the above, although the current perception is that the CTs are losing ground to the MRI, it is not entirely true, so we believe that it is necessary to continue improving CT scanners and also the techniques of image reconstruction they use.

In our previous works [3–7], we have studied the option of working with algebraic methods instead of the traditional analytical methods to reconstruct CT images. In this way, we can solve the problem mathematically by either iterative or direct algebraic algorithms. While other works focus on reducing radiation by applying X-rays with lower voltage [8], we focus on taking fewer shots. With this approach, we can work with a low number of projections and still get high-quality images. This means that we could reduce the radiation dose to which the patient is exposed, which is our main objective.

Using iterative methods, we can reduce the number of views or projections that we use to a really small number if we make a good selection of the projection angles [6,9,10]. However, with the direct methods [7,11], we have reached the conclusion that it is necessary that the matrix of the CT system has full rank, which will determine the number of projections required according to the image resolution that you want to achieve. Regardless, both approaches need a lower number of X-ray projections than other methods.

However, the types of methods we use require more computational resources. In addition, the time needed to reconstruct the images is much higher than with the analytical methods (minutes or hours versus milliseconds). Therefore, if we want the algebraic approach to be employed, we have to reduce the reconstruction time to the maximum. For this we can use High-Performance Computing (HPC) techniques, exploiting the hardware resources of the machine through parallel implementations of the algorithms.

In this paper, we focus on the analysis of the performance of the QR [12] factorization employed to reconstruct CT images. To do this, we make use of the SuiteSparseQR factorization package [13], analyzing the effect of using a different number of BLAS threads in operations. In addition, we compare the time efficiency when we reconstruct a single slice, or when we have a volume formed by multiple slices, which fits best a real situation.

In Sect. 2.1, we will describe the CT image reconstruction problem. We also make a brief description of the scanner we simulate and the dataset DeepLesion, used to take as reference. In Sects. 2.2 and 2.3 we explain how to perform the CT reconstructions using the explicit QR factorization or the Householder form. The results of the study are discussed in Sect. 3, analyzing on the one hand the performance of the QR factorization using a different number of threads (Sect. 3.1) as well as the performance of the reconstruction step (Sect. 3.2). Additionally, in Sect. 3.3 we show the obtained images measuring their quality. Finally, in Sect. 4 we summarize the work done and discuss the future lines of work.

2 Materials and Methods

2.1 Algebraic CT Image Reconstruction

When dealing with the reconstruction of CT images in an algebraic way, it is necessary to model the associated problem. Initially, we only have the data obtained through the scanner using X-rays. Therefore, it will be necessary to transform this data into an image that represents the projected object or body part.

We pose the problem as a system of linear equations as proposed in Eq. (1). Here, g is the data acquired by the scanner. It is usually called projections vector or sinogram for fanbeam CTs. As we see in (2), g is a vector of size M, which depends on the physical parameters of the scanner. We calculate M as the product of the number of detectors of the CT and the number of projections taken. It is worth mentioning than one projection (one X-ray shot) obtains as many data as detectors we have.

The system matrix A is defined in Eq. (4). This is a sparse weight matrix that represents the influence a of each ray beam traced (i) on each image pixel (j). As we can see, the number of rows of A is M, and the number of columns is N, which is the resolution in pixels of the reconstructed image we want to get.

In our case, both the matrix and the projections vector is simulated. We calculate both using the forward projection ray-tracing algorithm proposed by Joseph [14], simulating a CT scanner with 1025 detectors and taking equiangular projections around the 360 degrees of rotation. The images we projected are a selection of the DeepLesion dataset [15], which contains thousands of CT images of numerous patients for the study of different types of lesions.

Finally, u is the solution of our equations system. It is the reconstructed image. If we store in vector form, as we see in (3), its size is the total number of pixels on the image. For instance, for a final image of resolution 256×256 pixels, u is a vector of size 1×256^2. It is very easy to go from vector form to image form and vice versa.

$$A * u = g \tag{1}$$

$$g = [g_1, g_2, ..., g_M]^T \in \mathbb{R}^M \tag{2}$$

$$u = [u_1, u_2, ..., u_N]^T \in \mathbb{R}^N \tag{3}$$

$$A = a_{i,j} \in \mathbb{R}^{MxN} \tag{4}$$

2.2 QR Factorization Applied to the CT Problem

Ideally, the previously modeled problem could be solved very simply by obtaining the inverse of the system matrix as shown in the Eq. (5), provided it had full rank and the matrix was square. But in our application, this matrix dimensions can be really large for the highest resolutions and rectangular (more rows than columns). It is not feasible to explicitly compute the inverse since it requires a high computational cost and really advanced hardware. Besides, it is a highly unstable computation, so the errors could spoil the resulting image.

For this reason, we need to solve the problem with an iterative method as we did in [3,6] or apply a factorization to the matrix so we can solve it directly. The factorization we propose is the QR with pivoting [16], as described in (6). Here, Q is an orthonormal matrix (its columns are orthogonal unit vectors so $Q^t Q = I$). R is an upper triangular matrix and P is the permutation matrix used to reduce the filling.

With this decomposition done, we can emulate the inverse of the matrix A as shown in (7), or the pseudoinverse if the matrix is non-invertible [17,18]. Now we can solve the problem as (8). And since the factorization can be performed only once and stored for future use, every time we need to reconstruct we will only have to perform a matrix-matrix product, a permutation and solving one upper triangular equations system. This means faster reconstructions.

In addition, here g can be a matrix $\mathbb{R}^{MxS}$ with S columns (the number of slices we want to reconstruct), so we can get multiple images within the same operation, which also reduces the time per slice.

$$u = A^{-1} * g \tag{5}$$

$$A * P = Q * R \tag{6}$$

$$A^{-1} = PR^{-1}Q^T \tag{7}$$

$$u = P * (R^{-1}(Q^T * g)) \tag{8}$$

2.3   Q-Less Factorization Using Householder Reflections

As we mentioned, our matrices can get relatively big depending on the desired image resolution, so it is possible the computational resources needed to compute the decomposition are extensive. Even if the system matrix A is sparse, Q can be dense. As a way to spare main memory resources, we could decide to not calculate the Q matrix explicitly. Instead, we could perform the factorization using Householder reflections, and store only the set of Householder vectors, which describe transforms to be applied as shown in (9). Here H_i are the successive reflection matrices to apply to our right-hand side g. Apart from main memory, this technique will also reduce the computation time.

$$Q = H_1 H_2 \cdots H_{N-2} H_{N-1} \tag{9}$$

2.4   SuiteSparseQR Factorization Package

SuiteSparseQR [13] is an implementation of the multifrontal sparse QR factorization method. It uses both BLAS and Intel's Threading Building Blocks, a shared-memory programming model for modern multicore architectures to exploit parallelism. The package is written in C++ with user interfaces for MATLAB, C, and C++. It works for real and complex sparse matrices.

In our case, we are using the MATLAB interface, making use of BLAS parallelism and working with real sparse matrices.

3   Results and Discussion

In order to test the performance of the method we run both the matrix factorization and the reconstructions in a server with four Intel Xeon E5-4620 8c/16T processors (8 cores/processor, 32 cores/node) and 256 GB DDR3 RAM memory (ratio 8 GB/core).

We use a matrix corresponding to resolution 256 × 256 pixels and 90 projections. Thus, the size of the matrix is 90200 × 65536, with 40871478 non-zero elements and 0.0069 density. In the next subsections we show the experimental time results when using a different number of BLAS threads for the two alternatives of the factorization method, using one physical processor per thread.

3.1   Factorization Step

In the fist step we need to compute the factorization shown in Eq. (6). As explained in Sect. 2.2, the factorization of the matrix can be done once before the reconstruction of the images. Therefore, the computational speed is not going to be that important in this case. Even so, it is desirable to reduce it to the maximum, both to save computing time and to avoid possible system failures that can abort our process and spoil hours of work.

As we can see in Table 1, it is much faster to store the factorization in Householder form than to form the Q matrix explicitly. Regardless of the number of BLAS threads employed, it is around 3 times faster, which is a significant difference.

In Figs. 1 and 2 we can see the Speedup and Efficiency of both of the methods. We compute the Speedup for p processors as $S_p = T_1/T_p$, being T_1 the time with 1 processor and T_p the time with p processors. The Efficiency is $E_p = S_p/p$. In a perfect parallel algorithm, the Speedup is equal to the number of processors and the Efficiency is 1, which means we are taking advantage of the 100% of the resources.

However, we can observe that we get lower results. The Householder factorization has slightly better performance except when using 32 processors, with a

Table 1. Factorization time

BLAS threads	Factorization method		Improvement factor
	Explicit Q	Householder	
	Time (secs.)		
1	172784	57628	3.00
2	119407	31955	3.74
4	72186	21105	3.42
8	56325	18481	3.05
16	36451	10422	3.50
32	16805	7191	2.34

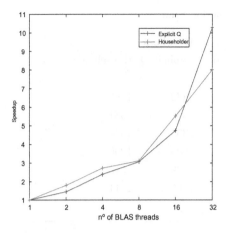

Fig. 1. Factorization Speedup

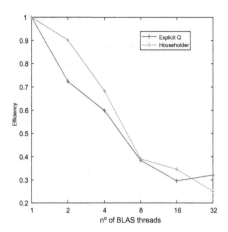

Fig. 2. Factorization Efficiency

Speedup of 8 versus the 10.2 of the explicit Q factorization. In Fig. 2 we can see that with more than 4 processors we are using less than a 50% of the computational resources. Since we are working with sparse matrices with a low density it is usual to get lower efficiency that with dense matrices.

3.2 Reconstruction Step

We also need to verify which method is better in the reconstruction phase and if we get good performance when using more resources. In Table 2 we observe the results for reconstructing just 1 slice (1 right-hand side vector) or 128 slices, as per Eq. (8). As we can see, when we want to reconstruct only 1 slice, the Explicit Q factorization gets lower computational times. Besides, they improve slightly when using more processors, up to 16. With the Householder form, we only get better performance with up to 8 processors. However, the performance here is worse than in the factorization step, as we can see in Figs. 3 and 4. The Speedup is very low regardless of the number of threads used, and the highest efficiency we get is only 0.5 using 2 threads.

When we are dealing with multiple right-hand sides, in this case 128, the performance is different. We can see in the table that in this case, it is faster to perform the reconstruction using the Householder reflections. We get the reconstructions around twice as fast (2.8 times faster for 16 threads). On the other hand, the Speedup and Efficiency for each of the methods is not better than in the previous case, as we can see in Figs. 5 and 6. We get very low Efficiency, wasting resources.

Table 2. Reconstruction time

BLAS threads	1 Slice		128 Slices	
	Explicit Q	Householder	Explicit Q	Householder
	Time (secs.)			
1	239	281	2432	1392
2	226	272	2370	1381
4	221	256	2351	1370
8	210	232	2638	1296
16	177	253	2539	910
32	222	269	2607	1111

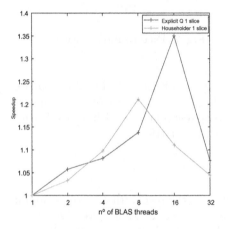

Fig. 3. 1 slice rec. Speedup   **Fig. 4.** 1 slice rec. Efficiency

3.3 Image Quality

In our preliminary work [7] we concluded that if we work with a full-rank system matrix, the images reconstructed by the QR factorization have really high quality. In order to verify this for the matrix we are studying here, we show the quality results, measuring with Mean Absolute Error (MAE), PSNR (Peak Signal-to-Noise Ratio) [19]. We also measure the SSIM (Structural Similarity Index). Since we are reconstructing 128 slices, we show the minimum, maximum and average results of both reconstruction techniques.

In Figs. 7 and 8 we can see the PSNR and MAE results respectively. We don't show the SSIM results, since all the images get a SSIM equal to 1. This means that all the reconstructions have the same structure as the reference image (we are not losing significant internal structures). As we observe in the Figures, the reconstructions obtained through the Householder matrix have better quality, which reflects the better numerical stability of the factorization. It gets around 10 units of PSNR higher and lower error. But both of the methods get really high quality. Notice that we are getting MAEs of order 10^{-12}, which is almost insignificant.

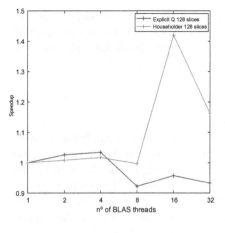

Fig. 5. 128 slices rec. Speedup

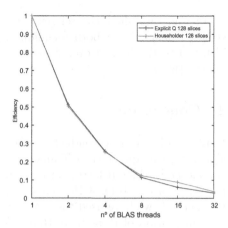

Fig. 6. 128 slices rec. Efficiency

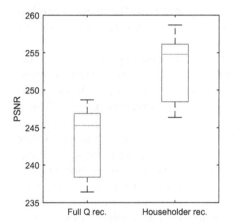

Fig. 7. PSNR results

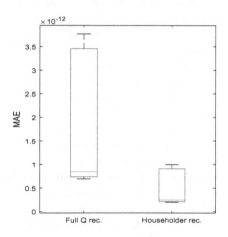

Fig. 8. MAE results

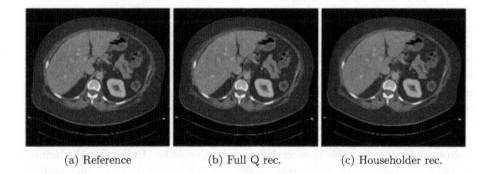

(a) Reference   (b) Full Q rec.   (c) Householder rec.

Fig. 9. Abdomen CT reconstructions

In Fig. 9 we show the reference CT image of an abdomen and the reconstructed images with both techniques. As we can see, the images are almost identical to the human eye, since we can not discern significant differences or information loss.

4 Conclusion

In this work, we have conducted a study of the efficiency of applying a direct algebraic technique to the CT image reconstruction problem. We have compared two methods, the QR factorization forming the matrix Q explicitly, or using the matrix Q in the form of Householder reflections. To verify the performance of the methods we have used the SuiteSparseQR library, which includes a parallel implementation of these algorithms. We have used a server to get the experimental time performance, using up to 32 processors.

Regarding the quality obtained by both methods, we have determined that the Householder factorization is more numerically stable, so the images have less error. However, we speak of very small magnitudes that are not perceptible to the human eye.

On the other hand, we have verified that the time used to perform the factorization of the system matrix is much higher if we form the matrix Q in an explicit manner. This can lead to errors as we expose ourselves to system failures, power outages, etc. In addition, we have verified that the reconstructions are faster when we reconstruct several images simultaneously. We get less time per slice in general, and less time using the Householder form in particular. Since in reality we are going to reconstruct more than one slice at a time (usually a CT study has at least 100 slices), we determine that it is more efficient to use Householder reflections for our application. In this way we can reconstruct volumes with high quality in less than 30 min.

Finally, we have determined that the efficiency that is achieved using fine-grained parallelism in many-core servers is not good. We observe that we do not take advantage of all the allocated resources, obtaining very low Speedups. Still, the time used to solve 128 slices with 32 threads is slightly lower than using the 32 threads to solve each slice independently using coarse-grain parallelism.

At this point, it is still necessary to employ HPC computers since this implementation requires a high amount of main memory. That is the reason we are not able to compute the problem for higher image resolutions. As future work, we plan to work with out-of-core techniques that read the data stored in blocks from the hard drive when the particular block is needed for the computation, instead of having it always loaded in main memory. In this way, we could achieve to reconstruct bigger problems in workstations with lower amount of RAM memory and thus lower cost.

Acknowledgements. This research has been supported by "Universitat Politècnica de València", "Generalitat Valenciana" under PROMETEO/2018/035 co-financed by FEDER funds, as well as ACIF/2017/075 predoctoral grant, and the "Spanish Ministry of Economy and Competitiveness" under Grant TIN2015-66972-C5-4-R and TIAMHA co-financed by FEDER funds.

References

1. Brown, R.W., Haacke, E.M., Cheng, Y.C.N., Thompson, M.R., Venkatesan, R.: Magnetic Resonance Imaging: Physical Principles and Sequence Design. Wiley, Hoboken (2014)
2. Brooks, R., Chiro, G.D.: Principles of computer assisted tomography (CAT) in radiographic and radioisotopic imaging. Phys. Med. Biol. **21**(5), 689–732 (1976)
3. Flores, L., Vidal, V., Verdú, G.: Iterative reconstruction from few-view projections. Procedia Comput. Sci. **51**, 703–712 (2015)
4. Parcero, E., Flores, L., Sánchez, M., Vidal, V., Verdú, G.: Impact of view reduction in CT on radiation dose for patients. Radiat. Phys. Chem. **137**, 173–175 (2017)
5. Flores, L.A., Vidal, V., Mayo, P., Rodenas, F., Verdú, G.: Parallel CT image reconstruction based on GPUs. Radiat. Phys. Chem. **95**, 247–250 (2014)
6. Chillarón, M., Vidal, V., Segrelles, D., Blanquer, I., Verdú, G.: Combining grid computing and Docker containers for the study and parametrization of CT image reconstruction methods. Procedia Comput. Sci. **108**, 1195–1204 (2017)
7. Chillarón, M., Vidal, V., Verdú, G., Arnal, J.: CT medical imaging reconstruction using direct algebraic methods with few projections. In: Shi, Y., et al. (eds.) ICCS 2018. LNCS, vol. 10861, pp. 334–346. Springer, Cham (2018). https://doi.org/10. 1007/978-3-319-93701-4_25
8. Padole, A., Ali Khawaja, R.D., Kalra, M.K., Singh, S.: CT radiation dose and iterative reconstruction techniques. Am. J. Roentgenol. **204**(4), W384–W392 (2015)
9. Andersen, A.H., Kak, A.C.: Simultaneous algebraic reconstruction technique (SART): a superior implementation of the ART algorithm. Ultrason. Imaging **6**(1), 81–94 (1984)
10. Yu, W., Zeng, L.: A novel weighted total difference based image reconstruction algorithm for few-view computed tomography. PLoS ONE **9**(10), 1–10 (2014). https:// doi.org/10.1371/journal.pone.0109345
11. Rodríguez-Alvarez, M.J., Sanchez, F., Soriano, A., Moliner, L., Sanchez, S., Benlloch, J.M.: QR-factorization algorithm for computed tomography (CT): comparison with FDK and conjugate gradient (CG) algorithms. IEEE Trans. Radiat. Plasma Med. Sci. **2**(5), 459–469 (2018)
12. Golub, G.H., Van Loan, C.F.: Matrix Computations, vol. 3. Johns Hopkins University Press, Baltimore (2013)
13. Davis, T.A.: Algorithm 915, suitesparseQR: multifrontal multithreaded rank-revealing sparse QR factorization. ACM Trans. Math. Softw. **38**(1), 8:1–8:22 (2011)
14. Joseph, P.: An improved algorithm for reprojecting rays through pixel images. IEEE Trans. Med. Imaging **1**(3), 192–196 (1982)
15. Yan, K., Wang, X., Lu, L., Summers, R.M.: DeepLesion: automated mining of large-scale lesion annotations and universal lesion detection with deep learning. J. Med. Imaging **5**(3), 036501 (2018)
16. Golub, G.H., Ortega, J.M.: Scientific Computing: An Introduction with Parallel Computing. Academic Press Professional Inc., Cambridge (1993)
17. Arridge, S., Betcke, M., Harhanen, L.: Iterated preconditioned LSQR method for inverse problems on unstructured grids. Inverse Prob. **30**(7), 075009 (2014)
18. Hansen, P.C.: The L-curve and its use in the numerical treatment of inverse problems (1999)
19. Hore, A., Ziou, D.: Image quality metrics: PSNR vs. SSIM. In: 2010 20th International Conference on Pattern Recognition, pp. 2366–2369. IEEE (2010)

Track of Classifier Learning from Difficult Data

ARFF Data Source Library for Distributed Single/Multiple Instance, Single/Multiple Output Learning on Apache Spark

Jorge Gonzalez-Lopez[1], Sebastian Ventura[2], and Alberto Cano[1(✉)]

[1] Department of Computer Science, Virginia Commonwealth University,
Richmond, USA
{gonzalezlopej,acano}@vcu.edu
[2] Department of Computer Science and Numerical Analysis,
University of Cordoba, Córdoba, Spain
sventura@uco.es

Abstract. Apache Spark has become a popular framework for distributed machine learning and data mining. However, it lacks support for operating with Attribute-Relation File Format (ARFF) files in a native, convenient, transparent, efficient, and distributed way. Moreover, Spark does not support advanced learning paradigms represented in the ARFF definition including learning from data comprising single/multiple instances and/or single/multiple outputs. This paper presents an ARFF data source library to provide native support for ARFF files, single/multiple instance, and/or single/multiple output learning on Apache Spark. This data source extends seamlessly the Apache Spark machine learning library allowing to load all the ARFF file varieties, attribute types, and learning paradigms. The ARFF data source allows researchers to incorporate a large number of diverse datasets, and develop scalable solutions for learning problems with increased complexity. The data source is implemented on Scala, just like the Apache Spark source code, however, it can be used from Java, Scala, and Python. The ARFF data source is free and open source, available on GitHub under the Apache License 2.0.

Keywords: ARFF · Apache Spark · Multi-instance · Multi-output

1 Introduction

The exponential growth of data, both in size and complexity, has resulted in a pressing need to develop scalable solutions to learn models from large-scale data. As ever-increasing data sizes have outpaced the capabilities of single machines, distributing computations to multiple nodes has gained more relevance [5]. Several frameworks have been developed based on the MapReduce programming model [7]. This model offers a simple and robust paradigm to handle large-scale datasets in a cluster. One of the first implementations of the MapReduce model

© Springer Nature Switzerland AG 2019
J. M. F. Rodrigues et al. (Eds.): ICCS 2019, LNCS 11538, pp. 173–179, 2019.
https://doi.org/10.1007/978-3-030-22744-9_13

was Hadoop [18], yet one of its critical disadvantages is that it processes the data from the distributed file system, which introduces a high latency. On the contrary, Spark [19] provides in-memory computation which results in a big performance improvement, especially on iterative jobs. Using in-memory data has been proved to be of the utmost importance for speeding up machine learning algorithms [2,8].

The WEKA [10] machine learning suite unifies access to well-known and state-of-the-art techniques for traditional classification, regression, clustering, and feature selection. This tool has been extended to accommodate numerous types of advanced learning paradigms including MULAN [17] and MEKA [14] for multi-output learning, MILK [10] for multi-instance learning, MOA [3] for data stream mining, JCLEC [6] for all previous, among others. These frameworks are built on top of WEKA or provide wrappers to its methods, and despite considering different learning paradigms all of them are based on the Attribute-Relation File Format (ARFF). Most of the benchmark datasets used in supervised learning are provided in this relational format, forcing the researchers to use one of the mentioned frameworks or to implement parsers to load the data in their codes. There are wrapper approaches to facilitate the use of Weka methods on Spark through the Weka interface [12] but the Spark machine learning library still lacks native support for ARFF files and the advanced learning paradigms.

This paper presents a native data source library to support ARFF files, single/multiple instance, and single/multiple output learning on Apache Spark[1]. The functionality of the data source extends the ones included in Apache Spark machine learning library but incorporating all the advanced learning paradigms considered in the definition of the ARFF. Therefore, researchers in these areas are provided with a powerful tool to facilitate the implementation and usage of their methods on Apache Spark, taking advantage of the distributed performance capabilities. This is expected to attract even more attention and users to the Apache Spark framework and help in its future development.

The rest of the manuscript is organized as follows: Sect. 2 introduces the learning paradigms supported by the data source, Sect. 3 presents the library framework, Sect. 4 shows some illustrative examples, and finally Sect. 6 presents some conclusions and future work.

2   Learning Paradigms

Learning paradigms can be categorized according to the task goal into supervised learning (classification for predicting a discrete label, regression for predicting a continuous output) and unsupervised learning (clustering, pattern mining, among others). However, these paradigms can be also broken down according to the induction process from a single/multiple instances and according to the prediction of a single/multiple outputs. This way, new paradigms such as multi-instance learning, multi-label classification, multi-target regression, multi-instance multi-label learning have become popular in recent years.

---

[1] https://github.com/jorgeglezlopez/spark-arff-data-source.

Let $\mathcal{X}$ denote the domain of instances in a dataset, a single instance $\boldsymbol{x} \in \mathcal{X}$ is represented as a set of features $\boldsymbol{x} = \{\boldsymbol{x}_1, \boldsymbol{x}_2, \ldots, x_d\}$, where $\boldsymbol{d}$ is the number of features. Traditional supervised learning, single-instance single-output, finds a mapping function that associates a single instance $\boldsymbol{x}_i$ to a single output $\boldsymbol{y}_i$, where classification considers $y_i \in \mathbb{Z}$ and regression $y_i \in \mathbb{R}$.

Multi-instance learning associates a set of instances $\boldsymbol{b}_i = \{\boldsymbol{x}_1, \boldsymbol{x}_2, \ldots, x_{|b_i|}\}$, known as a bag, to a single output $\boldsymbol{y}_i$. Multi-instance classification [1] considers $y_i \in \mathbb{Z}$, while multi-instance regression [11] defines $y_i \in \mathbb{R}$.

Multi-output learning associates a single instance $\boldsymbol{x}_i$ to a finite set of outputs $\boldsymbol{y}_i = \{\boldsymbol{y}_1, \boldsymbol{y}_2, \ldots, \boldsymbol{y}_t\}$. In multi-output classification [9,16], also known as multi-label classification, the output is represented as a binary vector $\boldsymbol{y}_j \in \{0, 1\}^t$ where each value can be 1 if the label is present and 0 otherwise. In multi-output regression [4,13,15], also known as multi-target regression, each output is a vector of real values $\boldsymbol{y}_j \in \mathbb{R}^t$.

Multi-instance multi-output learning combines both paradigms to represent an example as a bag (set of instances) associated with multiple outputs. Most of the learning paradigms presented are recent problems, therefore the availability of algorithms and datasets is fairly restricted. Our data source aims to bring these new paradigms to the distributed environment on Apache Spark.

3   Library Framework

The data source has been implemented in Scala, just like the original Apache Spark source code. This source implements the *FileFormat* trait in the original *datasources* package. This trait allows Spark to read data from external storage systems, like HDFS or a local file system, through *DataFrameReader* interface. Figure 1 shows a class diagram with the structure of the data source.

- *ARFFFileFormat*: It represents the entry point of the data source from the *DataFrameReader* interface. This class inherits from a series of interfaces in order to ensure the correct communication from the *DataFrameReader*. The first interface is the *DataSourceRegister* which can register the data source under an alias. The second set of interfaces is the *TextBasedFileFormat* and *FileFormat*, which define the methods that will be called from the *DataFrameReader* in order to create the proper DataFrame. The creation of the DataFrame is split into creating a suitable *schema* for the attributes and parsing all the instances. Both processes are isolated to each other because of the *FileFormat* interface.
- *ARFFInferSchema*: This class receives the header and the options defined by the user. The header comes either from the beginning of the file or from an independent file. The class uses the *ARFFAttributeParser* class to extract the information of each attribute using regular expressions. This information is used to create the required *schema* that matches the learning paradigm, as well as to store the information of the attributes in the *metadata*. Each column uses the *ARFFAttributeParser* and the *ExtendedAttributeGroup* to

transform the corresponding information of the header into *metadata*. The *ExtendedAttributeGroup* adds support for new types of attributes, such as *String* and *Date*.

- *ARFFInstanceParser*: It parses each of the lines of data in the file into *Rows* for the DataFrame. It reconstructs the *ARFFAttributeParser* of each attribute from the *metadata* received in the *schema*. Once all the parsers have been constructed, it reads each line of data allowing to read both *dense* and *sparse* instances. In every instance, the original values are transformed into a *numeric* format and stored in the corresponding fields of a *Row* following the same order they present in the *header*.
- *ARFFOptions*: This class handles the options set by the user, which can only be set from there at the beginning and will be final during the execution of the data source. The supported options are explained in Sect. 4.

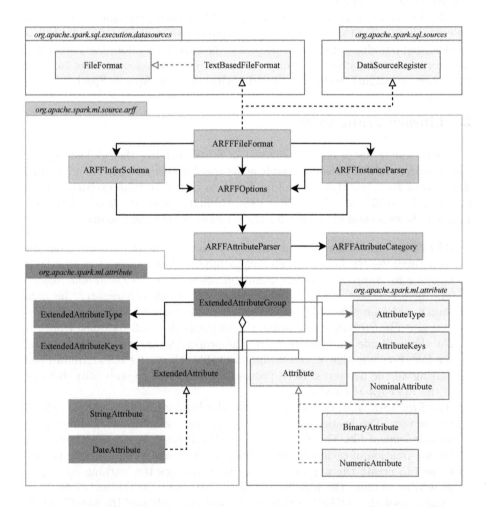

Fig. 1. ARFF data source class diagram for Apache Spark.

This data source has a series of advantages over the *libSVM* data source, which is the data source used by the machine learning library. The main functionalities and advantages are:

- Support for different types of learning paradigms, each with a different *schema*.
- Automatic conversion of all the features to numeric types. It transforms types such as *Date*, *String*, or *Nominal* to a Double. This allows using the Dataframe directly with the machine learning methods.
- Storage of the information of each attribute in the metadata of the *schema*. This information can be used in different algorithms such as finding the best splits in decision trees over *nominal* data.
- Dynamic and automatic conversion to either *dense* or *sparse* instances, whichever uses less storage space.

4   Illustrative Examples

The process to create a DataFrame out of an ARFF file is the same as for the built-in sources. Algorithm 1.1 illustrates the simplest form to call the data source. It also supports to manually specify extra options (through *ARFFOptions*) to define the learning paradigm, formed by a key and a value. The available options are:

- *("schemaFile", "path")*: Specifies a file with the definition of the ARFF header. By default, it uses the header at the top for the files being loaded.
- *("xmlMultilabelFile", "path")*: Specifies the XML file that defines the attributes that are considered outputs in a multi-label paradigm. By default, it is empty.
- *("numOutputs", number)*: Specifies the number of attributes at the end of the header that are considered outputs. However, if "xmlMultilabelFile" is defined there is no need to specify a number of outputs. By default, only the last attribute is an output.
- *("multiInstance, boolean)*: Indicates if the file defines a multi-instance paradigm. By default it is *false*.

```
val dataFrame = sparkSession
  .read
  .format(''org.apache.spark.ml.source.arff'')
  .option(key, value) // Add the options required
  .load(path)
```

Algorithm 1.1. DataFrame creation from an ARFF file

5   Current Software Version

Table 1 presents the information about the current version of the software. This table indicates the address where the source code can be found, as well as the documentation where the details of the implementation can be reviewed.

Table 1. Software metadata

Nr.	Software metadata description	Information
S1	Current software version	1.1
S2	Permanent link to executables of this version	https://github.com/jorgeglezlopez/spark-arff-data-source/
S3	Legal Software License	Apache License 2.0
S4	Computing platform	Apache Spark
S5	Installation requirements & dependencies	Apache Spark 2.0 or newer, and Scala 2.10 or 2.11
S6	User manual	https://github.com/jorgeglezlopez/spark-arff-data-source/blob/master/README.md
S7	Support email for questions	`gonzalezlopej@vcu.edu` `acano@vcu.edu`

6 Conclusions

This paper presented a new data source library for Apache Spark, which extends seamlessly the built-in data sources to support the ARFF file format and advanced learning paradigms including single/multiple instances and single/-multiple outputs learning. This new data source aims to significantly increase the number of machine learning datasets and algorithms ready-to-use for the Spark community. It provides the tools to help researchers to develop new algorithms in a distributed environment to address the increasing complexity.

This software has defined the schema structure for different paradigms, besides incorporating the data format. As future work, it is expected that new frameworks with non-traditional learning paradigms will extend this work by building novel distributed learning algorithms.

Acknowledgments. This research was partially supported by the 2018 VCU Presidential Research Quest Fund and an Amazon AWS Machine Learning Research award.

References

1. Amores, J.: Multiple instance classification: review, taxonomy and comparative study. Artif. Intell. **201**, 81–105 (2013)
2. Bei, Z., Yu, Z., Luo, N., Jiang, C., Xu, C., Feng, S.: Configuring in-memory cluster computing using random forest. Future Gener. Comput. Syst. **79**, 1–15 (2018)
3. Bifet, A., Holmes, G., Kirkby, R., Pfahringer, B.: MOA: massive online analysis. J. Mach. Learn. Res. **11**, 1601–1604 (2010)
4. Borchani, H., Varando, G., Bielza, C., Larrañaga, P.: A survey on multi-output regression. Wiley Interdisc. Rev.: Data Min. Knowl. Discov. **5**(5), 216–233 (2015)
5. Cano, A.: A survey on graphic processing unit computing for large-scale data mining. Wiley Interdisc. Rev.: Data Min. Knowl. Discov. **8**(1), e1232 (2018)
6. Cano, A., Luna, J.M., Zafra, A., Ventura, S.: A classification module for genetic programming algorithms in JCLEC. J. Mach. Learn. Res. **16**, 491–494 (2015)
7. Dean, J., Ghemawat, S.: MapReduce: simplified data processing on large clusters. Commun. ACM **51**(1), 107–113 (2008)
8. Gonzalez-Lopez, J., Cano, A., Ventura, S.: Large-scale multi-label ensemble learning on Spark. In: IEEE Trustcom/BigDataSE/ICESS, pp. 893–900 (2017)
9. Gonzalez-Lopez, J., Ventura, S., Cano, A.: Distributed nearest neighbor classification for large-scale multi-label data on Spark. Future Gener. Comput. Syst. **87**, 66–82 (2018)
10. Hall, M., Frank, E., Holmes, G., Pfahringer, B., Reutemann, P., Witten, I.H.: The WEKA data mining software: an update. ACM SIGKDD **11**(1), 10–18 (2009)
11. Herrera, F., et al.: Multi-instance regression. In: Multiple Instance Learning, pp. 127–140. Springer, Cham (2016). https://doi.org/10.1007/978-3-319-47759-6_6
12. Koliopoulos, A.K., Yiapanis, P., Tekiner, F., Nenadic, G., Keane, J.: A parallel distributed weka framework for big data mining using Spark. In: IEEE International Congress on Big Data, pp. 9–16 (2015)
13. Melki, G., Cano, A., Kecman, V., Ventura, S.: Multi-target support vector regression via correlation regressor chains. Inf. Sci. **415–416**, 53–69 (2017)
14. Read, J., Reutemann, P., Pfahringer, B., Holmes, G.: MEKA: a multi-label/multi-target extension to WEKA. J. Mach. Learn. Res. **17**(21), 1–5 (2016)
15. Reyes, O., Cano, A., Fardoun, H., Ventura, S.: A locally weighted learning method based on a data gravitation model for multi-target regression. Int. J. Comput. Intell. Syst. **11**(1), 282–295 (2018)
16. Tsoumakas, G., Katakis, I.: Multi-label classification: an overview. Int. J. Data Warehous. Min. **3**(3), 1–13 (2007)
17. Tsoumakas, G., Spyromitros-Xioufis, E., Vilcek, J., Vlahavas, I.: MULAN: a Java library for multi-label learning. J. Mach. Learn. Res. **12**, 2411–2414 (2011)
18. White, T.: Hadoop: The Definitive Guide. O'Reilly Media, Inc., Sebastopol (2012)
19. Zaharia, M., et al.: Apache Spark: a unified engine for big data processing. Commun. ACM **59**(11), 56–65 (2016)

On the Role of Cost-Sensitive Learning in Imbalanced Data Oversampling

Bartosz Krawczyk[1] and Michal Wozniak[2(✉)]

[1] Department of Computer Science, Virginia Commonwealth University,
Richmond, VA, USA
bkrawczyk@vcu.edu
[2] Department of Systems and Computer Networks,
Wrocław University of Science and Technology, Wrocław, Poland
michal.wozniak@pwr.edu.pl

Abstract. Learning from imbalanced data is still considered as one of the most challenging areas of machine learning. Among plethora of methods dedicated to alleviating the challenge of skewed distributions, two most distinct ones are data-level sampling and cost-sensitive learning. The former modifies the training set by either removing majority instances or generating additional minority ones. The latter associates a penalty cost with the minority class, in order to mitigate the classifiers' bias towards the better represented class. While these two approaches have been extensively studied on their own, no works so far have tried to combine their properties. Such a direction seems as highly promising, as in many real-life imbalanced problems we may obtain the actual misclassification cost and thus it should be embedded in the classification framework, regardless of the selected algorithm. This work aims to open a new direction for learning from imbalanced data, by investigating an interplay between the oversampling and cost-sensitive approaches. We show that there is a direct relationship between the misclassification cost imposed on the minority class and the oversampling ratios that aim to balance both classes. This becomes vivid when popular skew-insensitive metrics are modified to incorporate the cost-sensitive element. Our experimental study clearly shows a strong relationship between sampling and cost, indicating that this new direction should be pursued in the future in order to develop new and effective algorithms for imbalanced data.

Keywords: Machine learning · Imbalanced data ·
Cost-sensitive learning · Data preprocessing · Oversampling · SMOTE

1 Introduction

Class imbalance occurs when the distribution of instances among classes in the training set is skewed [2]. As the training procedure of most classifiers is based on the predictive accuracy (or 0-1 loss function), an equal importance of all training instances is inherently assumed. Therefore, learning algorithms tend to

ⓒ Springer Nature Switzerland AG 2019
J. M. F. Rodrigues et al. (Eds.): ICCS 2019, LNCS 11538, pp. 180–191, 2019.
https://doi.org/10.1007/978-3-030-22744-9_14

get biased towards the majority class, as it leads to overall smaller error than when trying to properly model infrequent and difficult minority class. Despite more than two decades of constant progress, learning from imbalanced data still poses a challenge for machine learning community [12]. It can be contributed to the constant emergence of new real-life problems, in which instances coming from one of the classes are much less frequent than from the others. Traditional examples of such cases include medicine, where we deal with diagnosis of a rare disease, or fraud detection systems, where we have a plethora of correct transactions versus a handful of fraudulent ones. Recent advances in machine learning and data mining brought the challenge of tackling class imbalance into new fields, such as big data [15], data stream mining [21], or structured outputs [6], among others. This creates new challenges force researchers to come up with new algorithms that are able to scale-up to ever-increasing volume and velocity of data, as well as adapt to emerging difficulties embedded in the nature of analyzed datasets.

To address the problem of imbalanced data two main approaches are used: data-level [7] and algorithm-level solutions [13]. The former ones concentrate on modifying the training set by removing or generating instances, in order to achieve rebalanced distributions. The latter ones aim at gaining an insight into what causes a given classifier to fail and modify its underlying mechanisms. Data-level solutions can be seen as more general ones, as they usually do not involve a specific classifier while performing sampling. Therefore, the processed dataset can be used by any conventional machine learning technique. The algorithm-level solutions are more specialized, usually designed for a given specific type of classifier and cannot be that easily transfered to another family of learners. At the same time, they may offer a more precise solution for tackling class imbalance.

Cost-sensitive learning is arguably the most wide-spread algorithm-level solution [8]. It assumes the modification of the standard 0-1 loss function and adding a learning penalty for misclassification of the minority class [4]. This will lead to an increased importance of the minority class instances during training and alleviation of the bias towards the better-represented majority class. It can be seen either as modifying the cost matrix for a classifier [1], or as an realization of instance weighting [23]. While this approach is efficient and many existing classifiers can be easily modified to their cost-sensitive versions [5,9], its main limitation lies in a lack of well-defined techniques for estimating the optimal misclassification cost. When improperly set, the cost parameter may significantly deteriorate the performance of a classifier, which is a main cause of many researchers preferring the data-level solutions [14].

One should notice that in many real-life imbalanced problems the parameter cost may be obtained from a domain expert [12]. In case of medical diagnosis, it will be a cost of making a wrong prediction about a patient and thus following issues with incorrect medications. In fraud detection, it will be the cost of allowing for a adversarial transaction to take place. Despite this fact, many solutions to these problems ignore the underlying cost and focus on data-level solutions. We will discuss here that this is not a correct approach to such applications.

In this paper, we propose to investigate a relationship between data-level algorithms and cost-sensitive learning. We argue that one cannot simply apply a sampling technique without any regard for the associated costs. Additionally, existing cost-sensitive algorithms use the cost parameter during training, but never take it into account during the evaluation phase. This leads to incorrect error estimations that may be too optimistic. Through a thorough experimental study, we investigate the interplay between varying misclassification costs and oversampling ratios used by popular data-level techniques. We show that using cost-sensitive modifications of skew-insensitive performance metrics reveals a clear correlation between these two factors that cannot be neglected. This is a starting work on proposing a new paradigm for learning from imbalanced data that combined sampling and cost-sensitive algorithms.

The contributions of this work are as follow:

- A proposal of new direction in learning from imbalanced data that uses the information from cost-sensitive learning in data-level solutions.
- A new experimental setup for imbalanced cost-sensitive learning, where misclassification cost is taken into account both during training and testing.
- A thorough experimental study investigation relationships between cost-sensitive framework and oversampling performance.

The remaining of this manuscript presents an insight into the problem of imbalanced data classification, with special emphasis on cost-sensitive solutions, discusses the relationships between cost and oversampling, depicts and discusses results of the experimental study, as well as presents lines for future research in this topic.

2   Learning from Imbalanced Data

Imbalanced data is widely known problem in machine learning domain, where unequal distribution of possible classes occurs in datasets [2,12]. In this paper, we will focus on the imbalanced data problem with two-class problem being taken into account in which two classes can be specified and one of them is underrepresented. Imbalanced dataset provides insufficient or inadequate representation of one class known as *minority* class, while *majority* class refers to the one that is more representative or even overrepresented.

Due to its nature, imbalanced data is mostly characterized by its *Imbalance Ratio* (IR) as well as intrinsic characteristic like disjuncts of overlapping of classes. Imbalance Ratio is defined as a ratio between number of objects that corresponds to the majority class and number of instances of the majority class. In other words, the higher the value, the more imbalanced dataset is due to the minority class being highly underrepresented.

However, the IR is not the sole source of learning difficulties. Small sample size of the minority class may inhibit any generalization capabilities of a classifier, while local data characteristics make some instances harder to classify than the others [18]. Such cases as borderline or noisy instances pose additional challenge to a classifier and thus should be paid special attention during the learning phase.

As a solution to class imbalance, three main groups of techniques were developed. *Preprocessing methods* refers to group of algorithms that alters inner structure of dataset by either introducing new minority class samples (*oversampling*) or removing majority class samples (*undersampling*). Oversampling technique can be done simply by randomly duplicating minority class samples or by artificially introducing new instances of minority class as it is done in quite popular SMOTE algorithm [7]. Other group of methods that are used for dealing with such problem are *algorithm level methods* which refers to the modification of base classifier in order to make it more sensitive to the imbalanced datasets [3]. Finally, *ensemble methods* involve forming a pool of classifiers and may combine its learners either both preprocessing or algorithm level methods [22].

3   On the Role of Misclassification Cost in Data Oversampling

In this paper we aim to investigate if there is a connection between the performance of oversampling methods and the underlying cost associated with a given problem. As we mentioned in the previous section, sampling and cost-sensitive methods have been considered as separate approaches [20]. We propose to change this way of thinking and initiate a discussion on cost-sensitive sampling for imbalanced data. This section will focus on two core challenges in this new area: (i) how to tune oversamling methods when cost is involved; (ii) how to properly evaluate classifiers when cost is involved.

3.1   Cost-Sensitive Oversampling

Oversampling is one of the most efficient approaches for handling skewed data distributions, as new artificial instances are being introduced into the minority class. Regardless of the fact if a simple random oversampling or guided sampling algorithms are used, the number of introduced instances remains as an ad-hoc parameter. There are no clear rules on how to select (sub)optimal oversampling ratio, despite a crucial role of this factor [17]. Oversampling should be seen as a trade-off approach. Too small number of artificial instances will fail to adjust the class distributions properly, while too high number may lead to minority class shift and negatively impact the performance on the majority class.

It seems interesting to investigate if having an access to the cost associated with misclassification of minority instances would lead to a better control over the artificial instance generation procedure. While all data-level algorithms ignore the cost, even if it is provided by a domain expert, one may see that this leads to simply discarding useful information about the problem.

Cost may be associated to a degree in which the minority class is important for the considered problem. Higher costs of misclassification should force the classification system to concentrate more on the minority class, even if it comes at the cost of impairing performance on the majority class. On the other hand, low misclassification cost should direct the classification system towards achieving a balanced performance on both of classes.

We propose to analyze if there is a relationship between the provided misclassification cost and performance of oversampling methods, with special emphasis put on the number of generated instances. Our hypothesis is that problems characterized by a higher cost would benefit from increasing the oversampling ratio. At the same time, for problems with a low misclassification cost the role of oversampling ratio should not be that significant. If our hypothesis is verified, then it would lead to a development of new branch of hybrid algorithms for imbalanced data that are cost-sensitive, while working on data-level.

3.2   Cost-Sensitive Evaluation of Algorithms

Another issue related with existing cost-sensitive approaches lies in their evaluation [11]. The cost parameter is usually taken into an account during classifier training phase. During the testing phase, most of works in the literature use one of many skew-insensitive metrics, such as G-mean or F-measure [19]. While this is a proper approach from the class imbalance point of view, it neglects completely the presence of the cost parameter, as all of skew-insensitive measures assume 0-1 loss function.

Such an experimental framework is therefore flawed, as misclassification cost, if known for a given problem, should be considered during all steps of learning and evaluation. Furthermore, by neglecting the role of cost, one puts cost-sensitive methods in a disadvantaged position. There were only few efforts in the literature to propose evaluation metrics tackled specifically for cost-sensitive problems [10,16], however they do not explicitly take into an account imbalanced data distributions. Additionally, as for imbalanced data there is already a plethora of established metrics proposed [2], it would be more interesting in adapting these metrics to cost-sensitive data, rather than adding more metrics to the stack.

In this paper, we formulate a hypothesis that misclassification cost, if known, should be taken into account during evaluation for all types of algorithms. Such an analysis would allow to gain a deeper insight into the performance of popular data- and algorithm-level solutions, as well as formulate a more realistic evaluation framework.

For the mentioned investigation of relationship between the misclassification cost and oversampling ratio, we will adopt cost-sensitive modifications of existing metrics. This will allow for a fair evaluation of the role of cost-sensitive learning in imbalanced data oversampling.

4   Experimental Study

This experimental study was designed in order to answer the following research questions:

– Is there any relationship between the provided misclassification cost and performance of oversampling algorithms, with a special emphasis put on the oversampling ratio that returns the best performance.
– Is it worthwhile to use cost-sensitive modifications of popular skew-insensitive evaluation metrics and does such an evaluation leads to gaining an additional insight into evaluated algorithms.

For experimental purposes, a number of diverse benchmark datasets were selected from the public KEEL Imbalanced Data repository. Datasets related to the two-class problem were already prepared for the 5-Fold Cross Validation and selected with specific Imbalance Ratio (IR) in mind as shown in Sect. 4.1. Algorithms used for the evaluation purpose, as well as their implementations are covered in Sect. 4.2, where detailed information about evaluation methodology and metrics can be seen in Sect. 4.3.

4.1 Datasets

Selected datasets that were used for the experiment are shown in Table 1, sorted by the value of Imbalance Ratio. Each dataset is described by the Imbalance

Table 1. Selected datasets for evaluation

Dataset	IR	Feat.	Inst.	Maj.	Min.
glass1	1.82	10	214	138	76
wisconsin	1.86	10	683	444	239
pima	1.87	9	768	500	268
haberman	2.78	4	306	225	81
vehicle2	2.88	19	846	628	218
vehicle1	2.90	19	846	629	217
vehicle3	2.99	19	846	634	212
glass-0-1-2-3_vs_4-5-6	3.20	10	214	163	51
vehicle0	3.25	19	846	647	199
new-thyroid1	5.14	6	215	180	35
segment0	6.02	20	2308	1979	329
glass6	6.38	10	214	185	29
vowel0	9.98	14	988	898	90
cleveland-0_vs_4	12.31	14	173	160	13
abalone9-18	16.40	9	731	689	42
glass5	22.78	10	214	205	9
lymphography-normal-fibrosis	23.67	19	148	142	6
winequality-red-4	29.17	12	1599	1546	53
winequality-white-3_vs_7	44.00	12	900	880	20
kddcup-buffer_overflow_vs_back	73.43	42	2233	2203	30

Ratio, number of features, instances as well by the amount of majority and minority samples.

4.2 Set-Up

For experimental purposes, a framework written in *R language* with parts of code related to the *k-Nearest Neighbors* search written in *C++11* was introduced. In order to fairly assess performance of proposed solution, *5-Fold Cross Validation* (5-CV) was done on selected datasets. As a base classifier, C5.0 decision tree was used from the *C50* package. Experiment depends on two implemented oversampling techniques, more precisely Random Oversampling as well as SMOTE which allows to emphasize minority class either by duplicating instances or artificially introducing new samples respectively. Implemented SMOTE technique was used with Euclidean metric with parameter $k = 5$ which corresponds to the amount of neighbors taken into account in the neighborhood of computed instance.

4.3 Cost Sensitive Metrics

The basic metrics for the classifier evaluation for binary imbalanced datasets are true positive (*TP*), true negative (*TN*), false positive (*FP*) and false negative (*FN*) which can be deducted from the confusion matrix built from the predictions and reference labeling of test subset. However, aggregated measures are needed in order to compare different classifiers with or without preprocessing methods applied. For our experimental study, we will use the following ones with cost sensitivity taken into account which is applied to the false negative (*FN*) as shown in Eq. 1. Cost sensitivity depends on the cost value provided to such metric, which varies in range $cost \in \{1, 2, 8, 16, 32, 64\}$ as shown in the results of experiment done in Sect. 4.4.

$$FN_{cost} = FN * cost \tag{1}$$

Information about proper classification of minority class can be obtained by *Sensitivity* metric also known as *Recall* or *True Positive Rate*, shown in Eq. 2.

$$Senstivity_{cost} = \frac{TP}{TP + FN_{cost}} \tag{2}$$

As the above metric takes only one class into consideration, Geometric Mean shown in Eq. 3 is used as it balances between classification accuracy over the instances from both minority and majority classes at the same time.

$$GM_{cost} = \sqrt{\frac{TP}{TP + FN_{cost}} * \frac{TN}{FP + TN}} \tag{3}$$

F-Measure shown in Eq. 4 can be considered as a harmonic mean of both precision and sensitivity which can measure accuracy of the test.

$$FMeasure_{cost} = \frac{2 * TP}{2 * TP + FP + FN_{cost}} \tag{4}$$

Balanced Accuracy shown in Eq. 5 is a metric that was used for performance evaluation and can be described as an average accuracy received from both minority and majority class.

$$BAccuracy_{cost} = \frac{1}{2} \left(\frac{TP}{TP + FP} + \frac{TN}{TN + FN_{cost}} \right) \tag{5}$$

4.4 Results and Discussion

Results for both, Random Oversampling and SMOTE preprocessing methods are shown in Figs. 1, 2, 3 and 4. For each metric, averaged results on all datasets from the Sect. 4.1 are shown with different *Cost* as well as the *Oversampling percentage* which refers to the amount of minority samples to be introduced either by simply duplicating or artificially creating new one, relative to the reference amount of minority instances.

Presented figures should be analyzed from two levels. The individual analysis should focus on the impact of varying oversampling ratios on the performance of evaluated methods under a pre-set cost. The global analysis should focus on capturing the trends in performance related to increasing cost value and how does this affect the stability of oversampling methods.

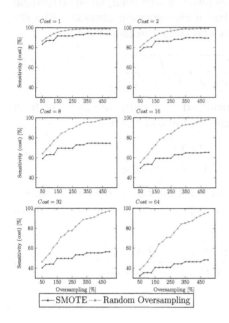

Fig. 1. Cost-sensitive sensitivity.

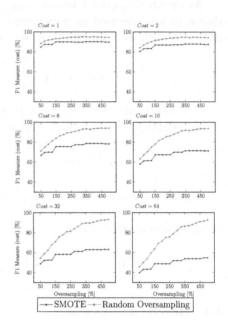

Fig. 3. Cost-sensitive F1-measure.

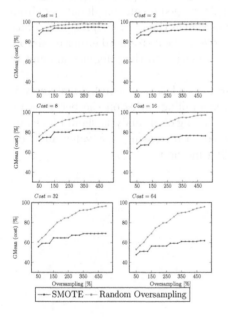

Fig. 2. Cost-sensitive G-mean.

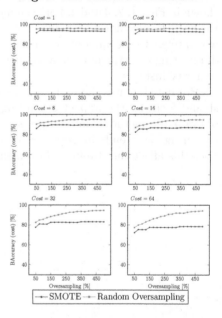

Fig. 4. Cost-sensitive balanced accuracy.

The obtained results allow us to draw a number of interesting conclusions. The most important one is the fact that there is a clear correlation between the cost and oversampling ratios. Regardless of the metric chosen, one can observe that for higher costs an increased oversampling ratio is preferred. When high values of cost are used (e.g., cost = 64) a high number of instances needs to be introduced in order to maximize the performance. On the other hand, for low cost values a good performance of oversampling methods is achieved even with <100% oversampling ratio. When cost is not taken into account (i.e., cost = 1), all oversampling methods display similar performance regardless of the number of instances introduced. These observations prove our hypothesis that the underlying cost has a crucial impact on the performance of data-level solutions. It allows to better tune the balancing process and as we can see from the trends associated with the increasing cost, it is also beneficial for avoiding pitfalls related to introducing incorrect number of instances, such as data shift or increased computational complexity of the learning process. Therefore, we may conclude that cost-sensitive imbalanced data preprocessing is a direction worth pursuing.

When comparing random oversampling and SMOTE, one can see that they display different performance when combined with cost-sensitive information. SMOTE, while still strongly affected by cost values, stabilities its performance with a lower values of oversampling ratios. This was to be expected, as SMOTE aims at introducing more meaningful instances than randomized approaches. Random oversampling is much more sensitive to cost and benefits from much higher oversampling ratios. However, especially for high cost parameter values, random oversampling easily outperforms SMOTE. This is an interesting observation, as one would expect SMOTE to be superior. It seems that by combining high misclassification costs with high oversampling ratios, random oversampling is capable for better empowering the minority class regions, thus translating to alleviated classification bias. This shows that each data-level method should be analyzed individually, in order to learn how it copes with cost-sensitive paradigm.

Finally, the results prove the usefulness of cost-sensitive metrics for gaining an insight into the nature of class imbalance learning algorithms. When no cost is taken into account (i.e., cost = 1), one cannot see significant differences between SMOTE and random oversampling. By scaling our metrics with cost value, the differences in performance between these two methods become obvious. We hope that this evaluation framework for any imbalanced algorithms will lead to better understanding which algorithms succeed and which fall under varying conditions.

5   Conclusions

In this paper, we proposed a new approach for looking at imbalanced data oversampling from a cost-sensitive perspective. We stated that when the misclassification associated with a given dataset is known, then it is beneficial to take it into an account when introducing new artificial instances to balance class distributions. Additionally, we pointed out the fact that in most works related to class imbalance the cost parameter is taken into account only during the learning phase, not during the testing phase. We argued that such an approach is incorrect, as one cannot neglect the role of associated cost when evaluating learning algorithms. Therefore, we have proposed to use cost-sensitive modifications of popular skew-insensitive metrics in scenarios where value of the cost parameter is known.

Our experimental study revealed a clear correlation between the value of cost parameter and the oversampling ratio. Higher costs, when used with cost-sensitive measures, favored higher number of artificial instances being introduced. For lower costs, the higher oversampling ratios did not contributed to the improvement of predictive power. This showed that cost-sensitive approaches may be used to tune and guide the oversampling, by allowing a more precise and automatic adaptation to a given imbalanced problem.

Obtained results encourage us to continue works in the new direction of cost-sensitive data-level solutions to class imbalance. Our next steps will be to propose an automatic way for embedding cost into oversampling methods in order to tune their parameters, and to evaluate this approach for multi-class imbalanced data scenarios.

Acknowledgement. This work was supported by the Polish National Science Centre under the grant No. 2017/27/B/ST6/01325 as well as by the statutory funds of the Department of Systems and Computer Networks, Faculty of Electronics, Wroclaw University of Science and Technology.

References

1. Bernard, S., Chatelain, C., Adam, S., Sabourin, R.: The multiclass ROC front method for cost-sensitive classification. Pattern Recognit. **52**, 46–60 (2016)
2. Branco, P., Torgo, L., Ribeiro, R.P.: A survey of predictive modeling on imbalanced domains. ACM Comput. Surv. **49**(2), 31:1–31:50 (2016)
3. Cano, A., Zafra, A., Ventura, S.: Weighted data gravitation classification for standard and imbalanced data. IEEE Trans. Cybern. **43**(6), 1672–1687 (2013)
4. Cao, P., Zhao, D., Zaiane, O.: An optimized cost-sensitive SVM for imbalanced data learning. In: Pei, J., Tseng, V.S., Cao, L., Motoda, H., Xu, G. (eds.) PAKDD 2013. LNCS (LNAI), vol. 7819, pp. 280–292. Springer, Heidelberg (2013). https://doi.org/10.1007/978-3-642-37456-2_24
5. Castro, C.L., de Pádua Braga, A.: Novel cost-sensitive approach to improve the multilayer perceptron performance on imbalanced data. IEEE Trans. Neural Netw. Learn. Syst. **24**(6), 888–899 (2013)
6. Charte, F., Rivera, A.J., del Jesús, M.J., Herrera, F.: Addressing imbalance in multilabel classification: measures and random resampling algorithms. Neurocomputing **163**, 3–16 (2015)
7. Chawla, N.V., Bowyer, K.W., Hall, L.O., Kegelmeyer, W.P.: Smote: synthetic minority over-sampling technique. J. Artif. Intell. Res. **16**(16), 321–357 (2002)
8. Domingos, P.M.: Metacost: a general method for making classifiers cost-sensitive. In: Proceedings of the Fifth ACM SIGKDD International Conference on Knowledge Discovery and Data Mining, San Diego, CA, USA, 15–18 August 1999, pp. 155–164 (1999)

9. Ducange, P., Lazzerini, B., Marcelloni, F.: Multi-objective genetic fuzzy classifiers for imbalanced and cost-sensitive datasets. Soft Comput. **14**(7), 713–728 (2010)
10. George, N.I., Lu, T., Chang, C.: Cost-sensitive performance metric for comparing multiple ordinal classifiers. Artif. Intell. Res. **5**(1), 135–143 (2016)
11. Holte, R.C., Drummond, C.: Cost-sensitive classifier evaluation using cost curves. In: Washio, T., Suzuki, E., Ting, K.M., Inokuchi, A. (eds.) PAKDD 2008. LNCS (LNAI), vol. 5012, pp. 26–29. Springer, Heidelberg (2008). https://doi.org/10.1007/978-3-540-68125-0_4
12. Krawczyk, B.: Learning from imbalanced data: open challenges and future directions. Prog. Artif. Intell. **5**(4), 221–232 (2016)
13. Ksieniewicz, P., Woźniak, M.: Dealing with the task of imbalanced, multidimensional data classification using ensembles of exposers. In: First International Workshop on Learning with Imbalanced Domains: Theory and Applications, LIDTA@PKDD/ECML 2017, 22 September 2017, Skopje, Macedonia, pp. 164–175 (2017)
14. López, V., Fernández, A., Moreno-Torres, J.G., Herrera, F.: Analysis of preprocessing vs. cost-sensitive learning for imbalanced classification. Open problems on intrinsic data characteristics. Expert Syst. Appl. **39**(7), 6585–6608 (2012)
15. López, V., del Río, S., Benítez, J.M., Herrera, F.: Cost-sensitive linguistic fuzzy rule based classification systems under the mapreduce framework for imbalanced big data. Fuzzy Sets Syst. **258**, 5–38 (2015)
16. McDonald, R.A.: The mean subjective utility score, a novel metric for cost-sensitive classifier evaluation. Pattern Recognit. Lett. **27**(13), 1472–1477 (2006)
17. del Río, S., Benítez, J.M., Herrera, F.: Analysis of data preprocessing increasing the oversampling ratio for extremely imbalanced big data classification. In: 2015 IEEE TrustCom/BigDataSE/ISPA, Helsinki, Finland, 20–22 August 2015, vol. 2, pp. 180–185 (2015)
18. Skryjomski, P., Krawczyk, B.: Influence of minority class instance types on SMOTE imbalanced data oversampling. In: First International Workshop on Learning with Imbalanced Domains: theory and applications, LIDTA@PKDD/ECML 2017, 22 September 2017, Skopje, Macedonia, pp. 7–21 (2017)
19. Thai-Nghe, N., Gantner, Z., Schmidt-Thieme, L.: Cost-sensitive learning methods for imbalanced data. In: International Joint Conference on Neural Networks, IJCNN 2010, Barcelona, Spain, 18–23 July 2010, pp. 1–8 (2010)
20. Wang, S., Li, Z., Chao, W., Cao, Q.: Applying adaptive over-sampling technique based on data density and cost-sensitive SVM to imbalanced learning. In: The 2012 International Joint Conference on Neural Networks (IJCNN), Brisbane, Australia, 10–15 June 2012, pp. 1–8 (2012)
21. Wang, S., Minku, L.L., Yao, X.: Resampling-based ensemble methods for online class imbalance learning. IEEE Trans. Knowl. Data Eng. **27**(5), 1356–1368 (2015)
22. Woźniak, M., Graña, M., Corchado, E.: A survey of multiple classifier systems as hybrid systems. Inf. Fusion **16**, 3–17 (2014)
23. Zhao, H.: Instance weighting versus threshold adjusting for cost-sensitive classification. Knowl. Inf. Syst. **15**(3), 321–334 (2008)

Characterization of Handwritten Signature Images in Dissimilarity Representation Space

Victor L. F. Souza[1(✉)], Adriano L. I. Oliveira[1], Rafael M. O. Cruz[2], and Robert Sabourin[3]

[1] Centro de Informática, Universidade Federal de Pernambuco, Recife, Pernambuco, Brazil
{vlfs,alio}@cin.ufpe.br
[2] Stradigi AI, Montreal, QC, Canada
rafaelmenelau@gmail.com
[3] École de Technologie Supérieure, Université du Québec, Montreal, QC, Canada
robert.sabourin@etsmtl.ca

Abstract. The offline Handwritten Signature Verification (HSV) problem can be considered as having difficult data since it presents imbalanced class distributions, high number of classes, high-dimensional feature space and small number of learning samples. One of the ways to deal with this problem is the writer-independent (WI) approach, which is based on the dichotomy transformation (DT). In this work, an analysis of the difficulty of the data in the space triggered by this transformation is performed based on the instance hardness (IH) measure. Also, the paper reports on how this better understanding can lead to better use of the data through a prototype selection technique.

Keywords: Offline signature verification ·
Writer-independent signature verification · Dichotomy transformation ·
Prototype selection · Instance hardness

1 Introduction

Handwritten signature is one of the oldest accepted biometric characteristics and is still widely used to verify if a person is who he/she claims to be [2]. The handwritten signatures verification (HSV) systems are used to classify query signature as genuine or forgeries. While genuine signatures are those that really belong to the indicated person, forgeries are those created by other people and can be categorized as [11]: (i) random forgeries, where the forger does not know both the name and the signature pattern of the signer; (ii) simple forgeries, in which the forger only has the access to the name of the writer but does not know the signature pattern; (iii) skilled forgeries, where the forger has the knowledge of both the name and the signature pattern of the signer (resulting in forgeries more similar to the genuine signatures).

© Springer Nature Switzerland AG 2019
J. M. F. Rodrigues et al. (Eds.): ICCS 2019, LNCS 11538, pp. 192–206, 2019.
https://doi.org/10.1007/978-3-030-22744-9_15

While being researched for a long time the HSV problem still remains challenging. Depending on how it is handled, the following challenges can be faced: imbalanced class distributions, high number of classes, high-dimensional feature space and small number of learning samples [8]. A specific concern is related to the skilled forgeries since they tend to be very similar to the genuine signatures and, in real applications, they are not available during the training phase of the classifier (which should be trained only with genuine signatures and the random forgeries) [8].

There are two approaches for building offline HSV systems. In the Writer-Dependent (WD) systems, a verification model is trained for each user. Although, in general, WD systems present good performance for the HSV task, requiring a classifier for each user increases the complexity and the cost of the system operations as more users are added. Also, the small number of genuine samples per user is a problem that often needs to be addressed. The other systems are known as Writer-Independent (WI). In WI, a single model is trained for all users from a dissimilarity space generated by the dichotomy transformation (DT). Thus, the classification inputs are dissimilarity vectors, which represent the difference between each feature of a questioned and a reference signature of the writer. WI systems are less complex but in general obtain worse results, when compared to the WD approach [7].

Since the samples in dissimilarity space generated by the dichotomy transformation are formed through the combination two by two of signatures (a questioned and a reference signature), this approach is able to increase the number of samples in the WI-HSV scenario. Thus, the small number of samples is no longer a problem. Moreover, the system can be developed to handle the class imbalance by generating a similar number of samples for the positive and negative classes. However, many samples in the WI-HSV scenario are redundant and have little influence for training the verification model. Thus, the use of prototype selection (PS) techniques in the dissimilarity space may enable the reduction of the complexity and the computational cost of training a classifier without deteriorating the final model performances [5].

The objective of this paper is (i) to understand the difficulty of the data and (ii) to analyze the use of prototype selection in the offline WI-HSV based on the dichotomy transformation. Related to (i) the instance hardness (IH) measure is used to achieve the stated objective. The IH is a metric used both to identify hard to classify instances and also to understand why they are hard to classify [12]. One of the advantages of understanding the instances misclassification is to have ideas about the best preprocessing technique or the best classifier to be used [12]. To complement this understanding, in (ii), we analyze if a prototype selection preprocessing can be applied without degrading the performance of the classifier and whether preprocessing based on a systematic prototypes selection technique is better than a random subsampling.

This paper is organized as follows: Sect. 2 presents the HSV problem and the dichotomy transformation as fundamentals related to this work. Section 3 contains the discussion and the conducted experiments for both the prototype selection and the instance hardness analysis. In the last section, the conclusion and the future works are presented.

2 Fundamentals

2.1 Handwritten Signature Verification (HSV)

The problem of automatic handwritten signature verification (HSV) is commonly modeled as classifying a given signature as genuine (i.e. belonging to the indicated writer) or forgery (created by someone else) [2,8]. Figure 1 depicts examples of signatures, from the GPDS dataset. Each row represents a different writer and for each writer the first three signatures are genuine and the last one is a skilled forgery.

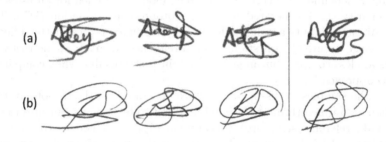

Fig. 1. Signatures.

In the skilled forgeries the forger knows both the name and the signature pattern of the signer and will attempt to imitate the genuine signature. Thus, genuine signatures and skilled forgeries tend to be very similar. From Fig. 1, one can see that skilled forgeries are more similar to the genuine signature than the random forgeries (the genuine signatures from other users).

Also, in a real scenario, the systems are trained with partial knowledge. In general, the training set of HSV systems are composed only of genuine signatures without access to skilled forgeries [8]. So, the classifier is trained without information capable of distinguishing between genuine signatures and forgeries. However, during the verification process the system will have to both reject the forgeries and accept the genuine signatures.

Furthermore, the number of genuine samples per user is often small (between 3–5 signatures) and there is great intra-class variability. This is difficult to tackle since the few available signatures are not sufficient to capture the full range of variation [8]. Figure 1 shows the variability in the genuine signatures.

In the WD systems, one classifier is trained for each user. In the WI case, a single model is trained for all users and the classification only depends on the input reference signature. The common practice for WI-HSV systems is to train using the development set D and to test using the exploitation set ε. In general, these sets have different subset of users [7].

The current state-of-the-art in feature representation for offline signatures is reported in the paper by Hafemann et al. [7], which uses Deep Convolutional Neural Networks (DCNN) for learning the signature representations in a WI way. So, it tries to learn a new feature space with the most representative properties of the handwritten signatures. As a WI approach, the learned representation space is not specific for a single set of users and is able to use data from as many users as possible. In this work, the 2048 features obtained from the FC7 layer of the DCNN called SigNet are used as feature vectors [7] (available online[1]).

2.2 Dichotomy Transformation (DT)

The Dichotomy Transformation (DT), proposed by Cha and Srihari [3], is an approach that allows to transform a multi-class problem (as the offline HSV) in a 2-class problem. The Dichotomy Transformation has already been used in various contexts, including writer identification [1] and for handwritten signature verification [4,11,14]. For the HSV context, it can be presented as follows: given a reference signature and a questioned signature, the objective is to determine whether the two signatures were produced by the same writer.

In a more formal definition, let $\mathbf{x}_q$ and $\mathbf{x}_r$ be two feature vectors in the feature space, the distance vector in the dissimilarity space resulting from the Dichotomy Transformation, $\mathbf{u}$, is computed by Eq. 1:

$$
\mathbf{u}(\mathbf{x}_q, \mathbf{x}_r) = \begin{bmatrix} |x_{q1} - x_{r1}| \\ |x_{q2} - x_{r2}| \\ \vdots \\ |x_{qn} - x_{rn}| \end{bmatrix} \tag{1}
$$

where $|\cdot|$ represents the absolute value of the difference, x_{qi} and x_{ri} are the n-th feature of the signatures $\mathbf{x}_q$ and $\mathbf{x}_r$ respectively, and n is the number of features. It is worth highlighting that each component of the $\mathbf{u}$ vector is equal to the distance between the corresponding components of the vectors $\mathbf{x}_q$ and $\mathbf{x}_r$. Thus, both the distance vector and the feature vectors have the same dimensionality.

As previously noted, in the dissimilarity space, regardless of the number of writers, there are only two classes: (i) The within/positive class w_+: composed of distance vectors computed from samples of the same writer (i.e., intraclass distances). (ii) The between/negative class w_-: composed of distances vectors computed from samples of different writers (i.e., interclass distances).

Systems based on the DT approach need datasets already transposed into the dissimilarity space to train a dichotomizer (two classes classifier), which will be used to perform the verification task. Generally, the writers that are in the training set are not part of the test set [3].

---

[1] http://en.etsmtl.ca/Unites-de-recherche/LIVIA/Recherche-et-innovation/Projets/ Signature-Verification.

When users have more than one reference signature, the dichotomy transformation is applied between the feature vector $\mathbf{x}_q$ of the questioned signature and the writer's reference set $\{\mathbf{x}_r\}_1^R$, producing a set of dissimilarity vectors $\{\mathbf{u}_r\}_1^R$, where R is the number of signatures in the reference set. For example, if a writer has 3 reference signatures ($R = 3$) and $\{\mathbf{u}_r\}_1^R = \{\mathbf{u}_1, \mathbf{u}_2, \mathbf{u}_3\}$. Then, the dichotomizer evaluates each dissimilarity vector individually and produces a set of partial decisions $\{f(\mathbf{u}_r)\}_1^R$ [11]. The final decision about the questioned signature is based on the fusion of all partial decisions by a function $g(\cdot)$ and depends on the output of the dichotomizer, e.g. (i) in a label case, then the majority vote is an appropriate function, (ii) in a probability or distance output, the sum, mean, median, max, and min functions can be used [11].

It is expected that the signatures from the same writer be close to each other in the feature space. Hence, they will be clustered close to the origin in the dissimilarity space. In turn, signatures of different writers are typically distant from each other in the feature space and away from the origin in the dissimilarity space [3]. This behavior can be seen in Fig. 2, which depicts a 2D feature space with three writers (classes 1, 2 and 3), each one with 10 signatures and the respective dichotomy transformation to the dissimilarity space.

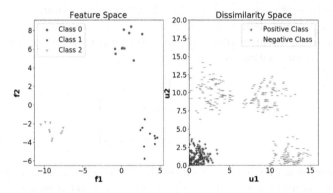

Fig. 2. On the left, samples from three different writers are represented in the feature space. On the right is the representation after the dichotomy transformation.

However, this does not always happen. A disadvantage of DT is that writers perfectly grouped in the feature space may not be perfectly separated in the dissimilarity space [3]. In other words, the more scattered the writers' samples are in the feature space, the smaller is the ability of the dichotomy transformation to separate the samples from the positive and negative classes [11]. This behavior can be seen in Fig. 3.

Other properties of the dichotomy transformation deserve to be highlighted. Firstly, DT is able to increase the number of samples in the dissimilarity space, hence it is composed of each pair of signatures. That is, if K writers provide a set of R reference signatures each, the Eq. 1 generates up to $\binom{KR}{2}$ different distances

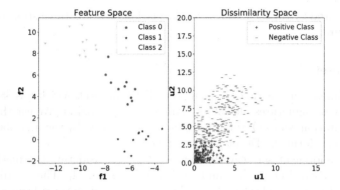

Fig. 3. On the right the dissimilarity space was not able to perfectly separate samples from the within and between classes. This was due to the sparsity of the samples in the feature space (left).

vectors. Of these, $K\binom{R}{2}$ belong to the positive class and $\binom{K}{2}R^2$ to the negative class [11]. Thus, even using a small amount of reference samples from each writer, the dichotomy transformation is able to generate a large amount of samples in the dissimilarity space. The increased number of samples can be visualized in both Figs. 2 and 3. In Fig. 2, for example, 30 samples from the feature space were transformed into 435 samples in the dissimilarity space (being 135 samples from positive class and 300 negative samples).

Also, DT affects the geometry of the data distribution. In addition, the vectors in the dissimilarity space are always non-negative, since they consist of distances transformed into absolute values [3]. Both of these properties can be seen in both Figs. 2 and 3.

To illustrate how the verification process through dichotomy transformation is independent of writer, given $\mathbf{x}_q$ and $\mathbf{x}_r$ as respectively a questioned and a reference feature vector, both of a new writer "class 4". The DT computes the distance vector $\mathbf{u}$ between $\mathbf{x}_q$ and $\mathbf{x}_r$ (Eq. 1), which must be located in the within region of the dissimilarity space, being the dichotomizer able to authenticate the questioned and reference signatures as belonging to the same writer. On the other hand, if the same scenario were used in the feature space, the model would fail to perform the classification. In fact, it is impossible for the feature domain model to properly classify signatures as belonging to the "class 4" writer, since this writer is not present in the training set. Therefore, the writer-independence of system is obtained by the use of the dichotomy transformation [11].

3   Experiments

The objective of the experiments is to both analyze the use of prototype selection applied to the offline WI-HSV based on dichotomy transformation and also obtain a better understanding on the difficulty of the data from the dissimilarity space generated by the dichotomy transformation based on the instance hardness (IH) measure.

3.1 Dataset

The experiments are carried out using the GPDS dataset, which has 881 writers and 24 genuine signatures plus 30 skilled forgeries per writer. We use the GPDS-300 segmentation, so the Exploitation set ε is composed by the first 300 writers, the other ones form the Development set D.

The Development set segmentation was done considering the methodology adopted by Rivard et al. [11] and by Eskander et al. [4]. The learning set L is generated using a subset of 14 of the 24 genuine signatures from the development dataset. So, the positive class samples are all computed using genuine signatures from every writer, as in Table 1. To generate an equivalent number of counterexamples, the negative class is formed by using 13 genuine signatures (reference signatures) against 7 random forgeries, each one selected from a genuine signature of 7 different writers (Table 1). The Exploitation set is acquired as in [7].

Table 1. Development set segmentation of the GPDS-300 dataset.

Learning set (L)	
Positive class	Negative class
Distances among the 14 signatures per writer (D_1)	Distances among 13 signatures per writer and 7 random signatures of other writers
$581 \cdot 14 \cdot 13/2 = 52,871$ samples	$581 \cdot 13 \cdot 7 = 52,871$ samples

3.2 Experimental Setup

Before feeding the classifier, the distance vectors $\mathbf{u}$ (in the dissimilarity space) are standardize by removing the mean and scaling to unit variance.

In this paper, the SVM is used as writer-independent classifier with the following settings: RBF kernel, $\gamma = 2^{-11}$ and $C = 1.0$ [14]. The predicted confidence scores for samples are used as classifiers output. The confidence score for a sample is the signed distance of that sample to the classifier's hyperplane [14].

All data were randomly selected and a different SVM was trained for each replication (ten replications for each experimental configuration were performed).

The performance evaluation of the classification methods is based on the Equal Error Rate (EER) metric with user thresholds (considering just the genuine signatures and the skilled forgeries) [7]. In the paper by Souza et al. [14], in general, for the tested dataset, the best results are obtained using the highest number of references and max as fusion function. So, only this approach is considered. To evaluate the effectiveness of the results, the Wilcoxon paired signed-rank test with a 5% level of significance was conducted to confirm if two methods are significantly different.

3.3   Using Prototype Selection (PS)

Considering the main characteristics of the WI dichotomy transformation, it is able to handle with some of the HSV problem difficulties when compared to the WD approach. (i) The DT reduced the high number of classes to a 2-class problem. (ii) The problem is no longer imbalanced as both positive and negative classes have the same number of samples (Table 1). (iii) The small number of samples is no longer a problem (Table 1). The dichotomy transformation was able to increase the number of samples in the WI-HSV scenario, yet many of them are redundant (a disadvantage). Thus, the use of prototype selection in the dissimilarity space can reduce the impact of this redundancy issue.

Prototype Selection (PS) approaches aim to obtain a representative training subset, in general, with a lower number of samples compared to the original one ($SelectedSubset \subseteq TrainingSet$) [5]. One of the main advantages of PS methods is the capacity to choose relevant training examples. So, by using the selected subset, it is expected to obtain similar or even better performance of the classifier during the generalization phase.

In the paper by Pekalska et al. [10], the authors present the prototype selection as an important preprocessing technique to be considered when dealing with dissimilarity-based classification. In their experiments they showed that by using well chosen selected prototypes, it is possible obtain a similar or higher classification performance at a lower computational cost in the classifier training process, when compared to the use of all the original training samples. To the best of our knowledge, this analysis has not yet been done specifically for the dichotomy transformation scenario.

In this work, the classical Condensed Nearest Neighbors (CNN) is used as prototypes selection technique. This approach maintains the instances that are misclassified by a 1-NN classifier, discarding them otherwise [9]. The CNN was choosen because its main property is to reduce the training set size by removing redundant instances (i.e. samples that will not affect the classification accuracy of the training set), retaining the instances closer to the decision boundaries [5]. In our experiments, the K_{CNN} is set to 1, as in the original algorithm [9].

The following experiments compare the application of prototype selection before training the SVM, considering the GPDS-300 dataset. The $\%\_SVM$ represent the models with uniformly random subsampling: 1.0%, 5.0% and 10.0% of the original training set were used. The Condensed Nearest Neighbors is called CNN_SVM in the tables.

Tables 2 and 3 present, respectively, the comparative analysis on the number of training samples and the EER metric obtained by the WI-SVMs (with and without prototype selection). It is worth noting that Table 3 contains both the comparison of the SVMs and also the results from the state of the art models.

Table 2. Comparison of WI-SVMs considering the number of training samples.

Model	#Positive samples	#Negative samples	#Retained samples (%)
SVM	52871	52871	100.00 (0.00)
$1\%\_SVM$	531.70 (17.04)	526.30 (17.04)	1.00 (0.00)
$5\%\_SVM$	2648.10 (24.78)	2639.90 (24.78)	5.00 (0.00)
$10\%\_SVM$	5289.30 (31.69)	5285.70 (31.69)	10.00 (0.00)
CNN_SVM	345.90 (15.25)	4437.80 (125.11)	4.52 (0.13)

Table 3. Comparison of EER with the state-of-the-art on the GPDS-300 dataset, WI-SVMs using Max as fusion function (errors in %).

Type	Model	#references	EER
WD	Soleimani et al. [13]	10	20.94
WD	Hafemann, Sabourin and Oliveira [6]	12	12.83
WD	Hafemann et al. [7]	5	3.92 (0.18)
WD	Hafemann et al. [7]	12	3.15 (0.18)
WI	SVM_{max}	12	3.69 (0.18)
WI	$1\%\_SVM_{max}$	12	3.54 (0.26)
WI	$5\%\_SVM_{max}$	12	3.62 (0.32)
WI	$10\%\_SVM_{max}$	12	3.48 (0.12)
WI	CNN_SVM_{max}	12	3.47 (0.15)

From Table 2, the use of the PS method allowed the SVM to be trained with a much smaller number of samples. Thus, when compared to the model trained with all the original training set, by using PS it was possible to obtain comparable performance (Table 3) with a reduction in computational cost and complexity in SVM training, considering the offline WI-HSV scenario.

As can be observed in Tables 2 and 3, a simple random subsampling with 1.0% of the training samples maintains the results when compared to the SVM trained with all the original training set. This demonstrates how redundant the samples resulting from the dichotomy transformation are for this dataset.

As presented in Table 3, even operating in a writer-independent way, both models with and without preprocessing obtained comparable results for the EER metric when compared to the WD-model from Hafemann et al. [7] for the GPDS-300 dataset. In the comparison to the other models, the proposed approach obtained better results.

By using a systematic PS, such as the CNN, more attention can be given for border samples. Thus, the prototype selection can be used without degrading the performance of the WI-classifier and still avoid to store more instances than are necessary for an accurate generalization.

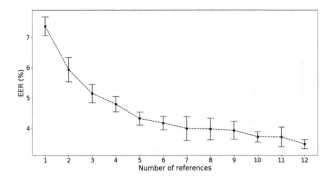

Fig. 4. Performance of the CNN_SVM varying the number of references.

Another point that should be considered when studying the difficulty of the data in the WI-HSV scenario is the available number of reference signatures. Figure 4 depicts the average EER for the CNN_SVM as a function of the number of reference signatures used for verification. As can be observed, the less signatures are used, the more difficult the problem becomes.

3.4 Instance Hardness (IH) Analysis

The instance hardness (IH) is a metric used to identify instances that are hard to classify and to understand why they are misclassified [12]. In this work, the kDisagreeing Neighbors (kDN) is used as instance hardness measure. Given an instance's neighborhood, the kDN represents the percentage of instances that do not have the same label of itself. This metric has a high correlation with the probability that a given instance is misclassified by different classification methods [12]. The kDN hardness measure is computed by Eq. 2:

$$kDN(x_q) = \frac{|x_k : x_k \in KNN(x_q) \wedge label(x_k) \neq label(x_q)|}{K} \qquad (2)$$

where $KNN(x_q)$ represents the set of K nearest neighbors of a query instance x_q and x_k represents an instance in its neighborhood. $label(x_q)$ and $label(x_k)$ represent the class labels of the instances x_q and x_k respectively [12].

In this section we analyze the data difficulty of the HSV problem, by using the IH measure, using the exploitation set to characterize the problem. This analysis considers different types of signatures (genuine signatures, random and skilled forgeries) and the evaluation of different values of the neighborhood size K in the kDN measure (K in the interval [3, 51]).

The main characteristics of the dichotomy transformation (see Figs. 2 and 3) are: (i) signatures that are close to each other in the feature space will be close to the origin in the dissimilarity space and (ii) the further away two signatures are from each other in the feature space, the farther from the origin will be the transformed vector [3].

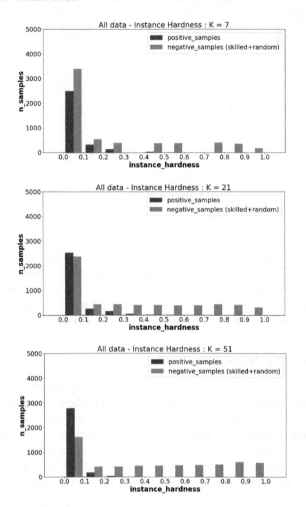

Fig. 5. Scenario (1): instance hardness considering the all selected data.

In the original feature space [7], genuine signatures from the writers form dense clusters; the skilled forgeries can present two different behaviors: (i) for some writers skilled forgeries are very separate from the genuine signatures, and (ii) in other cases they are closer to the genuine signatures.

Considering this, it is expected that the dissimilarity space generated by the dichotomy transformation presents the following characteristics: (i) positive samples will be close to the origin, forming also a dense cluster in the dissimilarity space (DS), (ii) skilled forgeries with a larger separation will generate negative samples farther away from the origin in the DS, (iii) skilled forgeries close to the genuine signatures will generate negative samples closer to the origin in the DS. As random forgeries are genuine samples from other writers and different writers

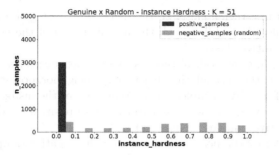

Fig. 6. Scenario (2): IH considering the positive samples and negative samples (only random forgeries).

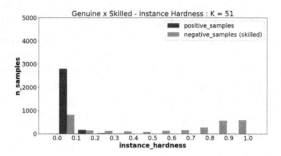

Fig. 7. Scenario (3): IH considering the positive samples and negative samples (only skilled forgeries).

occupy different regions of the feature space, negative samples from random forgeries will be located far from the DS origin.

Figures 5, 6 and 7 depict, respectively, the histograms of the instance hardness considering the scenarios: (1) all the data; (2) positive samples and negative samples (only from random forgeries); (3) positive samples and negative samples (only from skilled forgeries). With the scenarios (2) and (3), we can see the relationship between each type of negative samples and the positive data.

As depicted in Fig. 5, the vast majority of positive samples have $IH = 0.0$ and almost all them have a $IH < 0.3$. As we are considering the kDN measure, a higher values of K result in a more embracing investigation and lower K values represent a more compact investigation of space. To illustrate this, for $IH < 0.3$ and $K = 7$, at least 5 of the 7 neighbors of the positive samples are from the positive class itself. On the other hand, given $K = 51$, the $IH < 0.3$ represents that at least 36 neighbors of the positive samples are from the positive class itself. So, regardless of the neighborhood size, in the dissimilarity space the majority of neighbors of the positive samples are from the positive class. So, positive samples form a dense cluster close to the origin in the dissimilarity space.

For the negative samples, the IH values are spread along the histogram. Thus, negative samples are expected to be in a sparse region of the dissimilarity space with some them in a region closer to the dense positive region of the space, as some samples have IH = 1.0 (i.e. all the neighborhood of the negative sample belongs to the positive class).

Still from Fig. 5, as a result of the positive samples density and the negative samples sparsity in the dissimilarity space, higher neighborhood sizes (K) results in an increase of negative samples on the right side of the histogram (i.e. some negative samples are in a region closer to the dense positive region of the space than to the negative samples themselves).

When considering the positive samples and the negative samples (only random forgeries), we can observe that (Fig. 6): almost all the positive samples are in the IH = 0.0 bin, for this to happen there is no class overlap between the class distributions in the dissimilarity space.

Combining Figs. 5 and 7, we can see that almost all the positive samples with $IH \neq 0.0$ from Fig. 5 are derived from skilled forgeries. Thus, there should be class overlapping in the dissimilarity space. This behavior is expected, since the skilled forgeries are more similar to the genuine ones when compared to random forgeries. Still from Fig. 7, as the negative samples include some with higher IH, the overlap of the classes occurs in the positive region of the dissimilarity space.

Thus, in general, positive samples are located in a dense cluster close to the origin and the negative samples are scattered throughout the dissimilarity space. Also, the clusters are disjoint (based on the concentration of the IH with low values) with a small area of overlap (because of the high similarity between genuine signatures and few skilled forgerie). Considering that hard to classify samples are in the border region the use of a condensation PS technique, such as the CNN, was shown to produce good experimental results because it maintains the samples in the decision boundaries [5].

4   Conclusion

In this work we presented the Handwritten Signature Verification problem as having difficult data and tried to understand its behavior in the dissimilarity space generated from the dichotomy transformation used by a writer-independent approach. The evaluation was based on the instance hardness measure.

As presented, the WI dichotomy transformation is able to handle with some of the HSV problem difficulties, such as the imbalanced class distributions, high number of classes, high-dimensional feature space and small number of learning samples, when compared to the writer-dependent approach. Also, the WI approach presented good adaptability to the data, since after training, the classifier can verify signatures regardless of the writer has been used during the training, depending only on the input reference signature.

The reported IH analysis showed that, in general, in the transformed space the positive samples are located in a dense cluster close to the origin and the negative samples are scattered throughout the dissimilarity space generated by the dichotomy transformation. This better understanding of the transformed space, allowed us to make a better use of the samples by using a prototype selection technique, the Condensed Nearest Neighbors (CNN), that is more suited to the worked context.

The experimental results showed that, the dichotomy transformation is able to increase the number of samples in the offline WI-HSV scenario, yet many of them are redundant. Thus, by using prototype selection it is possible to speed up the classifier training and still achieve a similar or better classification performance than by using all the training samples. Even being a classic and simple technique, the Condensed Nearest Neighbors [9] applied as systematic approach was able to select fewer prototypes and still maintain performance when compared to both the SVM trained with all the original training set and the random subsampling approach.

Future works may include the study of feature selection in the dissimilarity space and the adaptation of the WI classifier over new incoming data.

Acknowledgment. This work was supported by the FACEPE (Fundação de Amparo à Ciência e Tecnologia de Pernambuco), CNPq (Conselho Nacional de Desenvolvimento Cientìfiico e Tecnológico) and the École de Technologie Supérieure (ÉTS Montréal).

References

1. Bertolini, D., Oliveira, L.S., Sabourin, R.: Multi-script writer identification using dissimilarity. In: 2016 23rd International Conference on Pattern Recognition (ICPR), pp. 3025–3030. IEEE (2016)
2. Bouamra, W., Djeddi, C., Nini, B., Diaz, M., Siddiqi, I.: Towards the design of an offline signature verifier based on a small number of genuine samples for training. Expert Syst. Appl. **107**, 182–195 (2018)
3. Cha, S.-H., Srihari, S.N.: Writer identification: statistical analysis and dichotomizer. In: Ferri, F.J., Iñesta, J.M., Amin, A., Pudil, P. (eds.) SSPR/SPR 2000. LNCS, vol. 1876, pp. 123–132. Springer, Heidelberg (2000). https://doi.org/10.1007/3-540-44522-6_13
4. Eskander, G.S., Sabourin, R., Granger, E.: Hybrid writer-independent-writer-dependent offline signature verification system. IET Biometrics **2**(4), 169–181 (2013)
5. Garcia, S., Derrac, J., Cano, J., Herrera, F.: Prototype selection for nearest neighbor classification: taxonomy and empirical study. IEEE Trans. Pattern Anal. **34**(3), 417–435 (2012)
6. Hafemann, L.G., Sabourin, R., Oliveira, L.S.: Writer-independent feature learning for offline signature verification using deep convolutional neural networks. In: IEEE IJCNN 2016, pp. 2576–2583. IEEE (2016)
7. Hafemann, L.G., Sabourin, R., Oliveira, L.S.: Learning features for offline handwritten signature verification using deep convolutional neural networks. Pattern Recogn. **70**, 163–176 (2017)
8. Hafemann, L.G., Sabourin, R., Oliveira, L.S.: Offline handwritten signature verification—literature review. In: 2017 Seventh International Conference on Image Processing Theory, Tools and Applications (IPTA), pp. 1–8. IEEE (2017)
9. Hart, P.: The condensed nearest neighbor rule (Corresp.). IEEE Trans. Inf. Theory **14**(3), 515–516 (1968)

10. Pekalska, E., Duin, R.P., Paclik, P.: Prototype selection for dissimilarity-based classifiers. Pattern Recogn. **39**(2), 189–208 (2006)
11. Rivard, D., Granger, E., Sabourin, R.: Multi-feature extraction and selection in writer-independent off-line signature verification. Int. J. Doc. Anal. Recogn. (IJDAR) **16**(1), 83–103 (2013)
12. Smith, M.R., Martinez, T., Giraud-Carrier, C.: An instance level analysis of data complexity. Mach. Learn. **95**(2), 225–256 (2014)
13. Soleimani, A., Araabi, B.N., Fouladi, K.: Deep multitask metric learning for offline signature verification. Pattern Recogn. Lett. **80**, 84–90 (2016)
14. Souza, V.L.F., Oliveira, A.L.I., Sabourin, R.: A writer-independent approach for offline signature verification using deep convolutional neural networks features. In: 2018 7th Brazilian Conference on Intelligent Systems (BRACIS), pp. 212–217. IEEE (2018)

Missing Features Reconstruction and Its Impact on Classification Accuracy

Magda Friedjungová$^{(\boxtimes)}$ (ID), Marcel Jiřina, and Daniel Vašata (ID)

Faculty of Information Technology, Czech Technical University in Prague,
Prague, Czech Republic
{magda.friedjungova,marcel.jirina,daniel.vasata}@fit.cvut.cz

Abstract. In real-world applications, we can encounter situations when a well-trained model has to be used to predict from a damaged dataset. The damage caused by missing or corrupted values can be either on the level of individual instances or on the level of entire features. Both situations have a negative impact on the usability of the model on such a dataset. This paper focuses on the scenario where entire features are missing which can be understood as a specific case of transfer learning. Our aim is to experimentally research the influence of various imputation methods on the performance of several classification models. The imputation impact is researched on a combination of traditional methods such as k-NN, linear regression, and MICE compared to modern imputation methods such as multi-layer perceptron (MLP) and gradient boosted trees (XGBT). For linear regression, MLP, and XGBT we also propose two approaches to using them for multiple features imputation. The experiments were performed on both real world and artificial datasets with continuous features where different numbers of features, varying from one feature to 50%, were missing. The results show that MICE and linear regression are generally good imputers regardless of the conditions. On the other hand, the performance of MLP and XGBT is strongly dataset dependent. Their performance is the best in some cases, but more often they perform worse than MICE or linear regression.

Keywords: Missing features · Imputation methods ·
Feature reconstruction · Transfer learning

1 Introduction

While solving a classification task one often faces demanding preprocessing of data. One of the preprocessing steps is the treatment of missing values. In practice, we struggle with randomly located single missing data in instances or with missing entire features. In real-world scenarios, e.g. [10, 25, 26], we have to deal with missing data. Missing values can also be part of a cold-start problem. Imputation treatments for missing values have been widely investigated [8, 14, 24] and plenty of methods how to reconstruct missing data were designed, but these methods are not directly designated for entire missing features reconstruction.

© Springer Nature Switzerland AG 2019
J. M. F. Rodrigues et al. (Eds.): ICCS 2019, LNCS 11538, pp. 207–220, 2019.
https://doi.org/10.1007/978-3-030-22744-9_16

This work focuses on the influence of missing entire features and possibilities of their reconstruction for usage in predictive modeling. We consider the following scenario: a classification model is trained on a dataset containing a complete set of continuous features but has to be used for prediction of classes of a dataset with some entire features missing. Entire feature reconstruction and its usage in an already learned model in order to perform with a reconstructed dataset distinguishes our work from others. Our point of interest is to find out how missing features impact the accuracy of the classification model, what possibilities of missing entire features reconstruction exist, and how the model performs with imputed data. In our work, the reconstruction of missing features, i.e. data imputation, is the very first task of transfer learning methods [17], where the identification of identical, missing, and new features is crucial.

Experimental results of this work should shed more light onto the applicability of state-of-the-art imputation methods on data and their ability to reconstruct entire missing features. We deal with traditional imputation methods: linear regression, k-nearest neighbors (k-NN), and multiple imputation by chained equations (MICE) [24], as well as with modern methods: multi-layer perceptron (MLP), and gradient boosted trees (XGBT) [6]. Experiments are performed on four real and six artificial datasets. The imputation influence is studied on six commonly used binary classification models: random forest, logistic regression, k-NN, naive Bayes, MLP, and XGBT. The amount of missing data varies between one feature and 50%.

This paper is structured as follows. In the next section we briefly review related work. Section 3 introduces imputation methods that are being analyzed in this work. Multiple features imputation is also discussed here. In Sect. 4 we describe the experiments that were carried out and present their results in Sect. 5. Finally, we conclude the paper in Sect. 6.

2   Related Work

There exist many surveys which summarize missing value imputation methods such as [5,8,10,11,16,22,25]. A lot of them are more than five years old and focus on traditional imputation methods.

A very good review of methods for imputation of missing values was provided by [8]. This study is focused on discrete values only with up to 50% missingness. They experimentally evaluated six imputation methods (hot-deck, imputation framework with hot-deck, naive Bayes, imputation framework with naive Bayes, polynomial multiple regression, and mean imputation) on 15 datasets used in 6 classifiers. Their results show that all imputation methods except for mean imputation improve classification error when missingness is more than 10%. The decision tree and naive Bayes classifiers were found to be missing data resistant, however other classifiers benefit from the imputed data.

In [25], performance of imputation methods was evaluated on datasets with varying amounts of missingness (up to 50%). Two scenarios were tested: values are missing only during the prediction phase, and values are missing during both induction and prediction phases. Three classifiers were used in this study: a decision tree, k-NN, and a Bayesian network. Imputation by mean, k-NN, regression and ensemble were used as imputation methods. The experimental results show that the presence of missing values always leads to performance reduction of the classifier, no matter which imputation method is used to deal with the missing values. However, if there are no missing data in the training phase, imputation methods are highly recommended at the time of prediction.

Finally, in [3], Arroyo et al. present imputation of missing values of Ozone in real-life datasets using various imputation methods - multiple linear and nonlinear regression, MLP and radial basis function networks, where the usefulness of artificial neural networks is presented.

3   Imputation Methods

Plenty of methods of missing data reconstruction have been designed. They perform differently on various datasets and in practice the most suitable imputation method for a given dataset is usually chosen according to the evaluation of the average performance (e.g. RMSE) of each method in the phase of training [20].

First let us briefly introduce imputation methods which we focus on within this study. The most basic methods are linear regression and the k-NN (see e.g. [9]).

The MICE [21,24] does not simply impute missing values using the most fitting single value, but it also tries to preserve some of the randomness of the original data distribution. This is being accomplished by performing multiple imputations, see [19]. The MICE comes up with very good results and is currently one of the best-performing methods [24]. In our research we use MICE in a simplified way. This means that multiple imputations are pooled using the mean before the classification model is applied. The reason is that we want to simulate the situation when the use of a classification model is restricted.

The MLP [22] with at least one hidden layer and no activation function in the output layer and the XGBT, see [6] for more details, are considered to be modern imputation methods.

3.1   Multiple Features Imputation

To impute several missing features, there are two ways of accomplishing this task using the previously mentioned methods. The first is to impute all features simultaneously which can be done using k-NN and MLP models. The second, which is usable for all other methods, is to apply the model sequentially one missing feature after another. However, to do this, it is important to choose some order in which the features will be imputed. We focus on an ordering where the most easy to impute features are treated first.

In the case of k-NN and MICE such a sequential imputation is not needed. The reason is that, in the case of k-NN, the neighbors typically do not change in subsequent steps and MICE is already prepared for multiple features imputation using an internal chained equation approach [21,24].

Linear Imputability

A simple way of measuring imputation easiness of features is to use the multiple correlation coefficient [2]. Multiple correlation coefficient $\rho_{X,\boldsymbol{X}'}$ between a random variable X and a random vector $\boldsymbol{X}' = (X_1', \ldots, X_n')^T$ is the highest correlation coefficient between X and a linear combination $\alpha_1 X_1' + \ldots + \alpha_n X_n' = \boldsymbol{\alpha}^T \boldsymbol{X}'$ of random variables $X_1', \ldots, X_n'$,

$$\rho_{X,\boldsymbol{X}'} = \max_{\boldsymbol{\alpha} \in \mathbb{R}^n} \rho_{X,\boldsymbol{\alpha}^T \boldsymbol{X}'}.$$

It takes values between 0 and 1, where $\rho_{X,\boldsymbol{X}'} = 1$ means that the prediction by linear regression of X based on $\boldsymbol{X}'$ can be done perfectly and $\rho_{X,\boldsymbol{X}'} = 0$ means that the linear regression will not be successful at all.

When $X_1, \ldots, X_p$ are the p features, we call the multiple correlation coefficient $\rho_{X_i, \boldsymbol{X}_{-(i)}}$ between X_i and a random vector of other features $\boldsymbol{X}_{-(i)} = (X_1, \ldots, X_{i-1}, X_{i+1}, \ldots, X_p)^T$ the *linear imputability* of feature X_i.

The estimation of the linear imputability is based on the following expression

$$\rho_{X_i, \boldsymbol{X}_{-(i)}}^2 = \frac{\operatorname{cov}(X_i, \boldsymbol{X}_{-(i)})^T \big(\operatorname{cov}(\boldsymbol{X}_{-(i)}) \big)^{-1} \operatorname{cov}(X_i, \boldsymbol{X}_{-(i)})}{\operatorname{var}(X_i)},$$

where $\operatorname{cov}(X_i, \boldsymbol{X}_{-(i)})$ is a vector of covariances between X_i and remaining features $X_1, \ldots, X_{i-1}, X_{i+1}, \ldots, X_p$, and $\operatorname{cov}(\boldsymbol{X}_{-(i)})$ is a $p - 1 \times p - 1$ variance-covariance matrix of covariances between remaining features.

If we want to impute multiple features, say $X_i, X_{i+1}, \ldots, X_{i+k}$, in the first step we choose $X_j, i \leq j \leq i + k$ such that $\rho_{X_j, \boldsymbol{X}_{-(i,\ldots,i+k)}}$ is the largest, where $\boldsymbol{X}_{-(i,\ldots,i+k)} = (X_1, \ldots, X_{i-1}, X_{i+k+1}, \ldots, X_p)^T$ is a vector of the remaining features. Then, in the next step, we repeat the process where X_j is taken as a known feature. Thus we choose $X_l, i \leq l \leq i + k, l \neq j$ such that its linear imputability with respect to random vector $\boldsymbol{X}_{-(i,\ldots,j-1,j+1,\ldots,i+k)}$ is the largest. We continue this way until all missing features are imputed.

Note that we are recalculating linear imputability in every step. This should not be done if the imputation is performed with linear regression since after the re-estimation (on the full training set) one obtains unachievable values.

Information Imputability

Linear imputability is a simple measure of how the linear regression imputation will perform. However, when one uses more sophisticated imputation models like MLP or XGBT that can handle non-linear dependencies, the linear imputability may not be suitable.

Hence we propose another way how to measure the imputability which is based on a particular result from Information theory. If a feature X_j is predicted by an estimator $\hat{X}_j$ based on other features represented by a vector $\boldsymbol{X}_{-(j)}$, i.e. $\hat{X}_j \equiv \hat{X}_j(\boldsymbol{X}_{-(j)})$, then it can be shown (see [7]) that

$$\mathrm{E}\left(X_j - \hat{X}_j\right)^2 \geq \frac{1}{2\pi \mathrm{e}} \mathrm{e}^{2H(X_j | \boldsymbol{X}_{-(j)})},$$

where $H(X_j|\boldsymbol{X}_{-(j)})$ is the conditional (differential) entropy of X_j given $\boldsymbol{X}_{-(j)}$.

Hence the lower bound of the expected prediction error is determined by the conditional entropy $H(X_j|\boldsymbol{X}_{-(j)})$. The greater the entropy is the worse predictions one can achieve at best when estimating X_j from other features. Thus one may measure imputability through the value of a conditional entropy multiplied by -1 in order to have larger values which correspond to better imputability. Hence we define the *information imputability* as a value of $-H(X_j|\boldsymbol{X}_{-(j)})$.

The process of multiple feature imputation is now exactly the same as it was using linear imputability. One first imputes the feature with the largest information imputability. The only difference is that in the second and all subsequent steps the recalculation does not make sense since one is not able to get any new information no matter what model will be used for the imputation. This partially simplifies the process of imputation order selection.

On the other hand, the problem that strongly limits its practical usage is the estimation of the conditional entropy. Even the most recently proposed estimators in [15,23] suffer from the curse of dimensionality. This is due to the fact that all these estimators are based on the k-NN approach introduced by Kozachenko and Leonenko in [12]. As our numerical experiments indicate, the method is limited to approximately five features depending on the underlying joint distribution.

4   Experiments

Our experiments consist of the following steps. First the original dataset is divided into a training part (70%) and a test part (30%). Several classification models as well as all imputation methods are trained on the training part. The imputation models are trained to impute in scenarios where each individual feature is missing and where randomly selected combinations of multiple features are missing. The degree of missingness varies from 10% to 50%. Finally, an evaluation of the accuracy of all classification models combined with all imputation methods is performed on the test dataset.

4.1   Settings and Parameters of Imputation Methods

Experiments were done using various settings. In order to keep the report short we present only those with satisfying results. All experiments were implemented in Python 3.

The k-NN imputation (knn) was implemented using the `fancyimpute` library[1]. A missing value is imputed by sample mean of the values of its neighbors weighted proportionally to their distances. In the case where multiple features are missing we impute all missing values at once (per row). In the presented results the hyper-parameter k is always taken as $k = 5$. This value was chosen based on preliminary experiments and with respect to computational time.

---

[1] `Fancyimpute` repository: https://github.com/iskandr/fancyimpute.

For the MICE method (*mice*) we also used the `fancyimpute` library. The parameter setup was inspired by [4] and we chose the number of imputations to be 150, the internal imputation model to be a Bayesian ridge regression, and the multiple imputed values to be pooled using the mean.

Linear regression imputation was implemented using the `scikit-learn` library[2] [18]. We tested two scenarios within the case when multiple features were missing. The first scenario was based on the linear imputability (*linreg-li*) and an iterative approach (*linreg-iter*) which corresponds to chained equations in MICE. This approach repeats two steps. First, every single missing value is imputed from the known features only. Second, all the imputed values are iteratively re-imputed from other features (all features except the one being imputed).

The MLP imputation is implemented using the `scikit-learn` library in two scenarios. The first (*mlp*) imputes all missing features at once and the second (*mlp-li*) imputes subsequently based on linear imputability. The hyperparameters of MLP (learning rate, numbers and sizes of hidden layers, activation function, number of training epochs) were tuned using randomized search. The XGBT was implemented using the `xgboost` library[3] in two scenarios. The first (*xgb-li*) is an analogy to *mlp-li* and the second (*xgb-iter*) to *linreg-iter*. The hyperparameters (learning rate, number of estimators, max depth of trees) were again tuned using randomized search.

The multiple features subsequent imputation scenario using information imputability is not presented here since in preliminary experiments it does not bring any benefits over linear imputability.

4.2 Evaluation

Imputation methods were evaluated using six binary classification models: k-NN, MLP, logistic regression (LR), XGBT, random forest (RF), and naive Bayes (NB), where LR, RF, and NB were provided by the `scikit-learn` library. We again used the randomized search algorithm to get classifier hyper-parameter configurations for each dataset.

First, we trained all classification models and measured their performance on the full test dataset (no missing features) (see Table 1 for results). Second, we combined them with imputation methods. We then measured the accuracies of all classification models on the imputed test dataset. Finally, we calculated the imputation performances as changes with respect to the accuracies on the full test dataset.

4.3 Datasets

We use both artificial and real datasets which are presented in Table 1. All datasets have continuous features and binary target labels. All datasets contain

---

[2] `Scikit-learn` repository: https://github.com/scikit-learn/scikit-learn.

[3] `XGBoost` repository: https://github.com/dmlc/xgboost.

Table 1. Details of datasets with corresponding classification model accuracies. The number of features (# feat.) does not include the target label. The name ds_a_b_c stands for an artificial dataset where a is the number of features, b is the number of informative features, and c is the number of redundant features. Bold values of accuracy correspond to the two best models for a given dataset.

Name	Type	# feat.	# records	LR	MLP	k-NN	NB	XGBT	RF
Cancer [13]	real	9	699	**0.966**	0.956	0.961	0.941	0.961	**0.966**
MAGIC [13]	real	10	19020	0.789	0.829	0.825	0.726	**0.869**	**0.857**
Wine [13]	real	11	4898	0.751	0.682	0.696	0.696	**0.786**	**0.773**
Spambase [1]	real	57	4597	0.912	**0.943**	0.783	0.834	0.930	**0.932**
Ringnorm [1]	artificial	20	7400	0.762	0.817	0.679	**0.979**	**0.945**	0.936
Twonorm [1]	artificial	20	7400	**0.980**	0.978	0.968	0.980	**0.980**	0.952
ds_10_7_3	artificial	10	4000	0.836	**0.990**	0.971	0.869	**0.980**	0.973
ds_20_14_6	artificial	20	4000	0.839	**0.992**	0.959	0.818	0.968	**0.977**
ds_50_35_15	artificial	50	4000	0.886	**0.995**	0.943	0.861	**0.975**	0.919
ds_100_70_30	artificial	100	4000	0.819	**0.975**	**0.878**	0.790	0.819	0.792

complete data without missing values. We assume all features are in a suitable form for the classification of the target label.

The real Wine Quality dataset originally contains ten target classes that were symmetrically merged in order to have a binary classification task. The artificial datasets were generated using the `make_classification` method in the `scikit-learn` library. They contain informative and redundant features. Informative features are drawn independently from the standard normal distribution. Redundant features are generated as random linear combinations of the informative features. A noise drawn from a centered normal distribution with variance 0.1 is added to each feature.

5 Results of Experiments

Results of the single feature imputation are shown in Table 2, where we present measured accuracy changes using the sample mean ± the sample standard deviation. The top 10% of imputation methods for each dataset and classification model are indicated by the value printed in bold. Two typical scenarios are shown in more details in Fig. 1.

Results of the multiple features imputation for two best models on each dataset are presented in Tables 3 and 4 for real and artificial datasets, respectively. Visualizations of typical results are given in Fig. 2 for a selected real dataset and in Fig. 3 for a selected artificial dataset. Box plots are used to show the results for different imputation methods and portions of missing features.

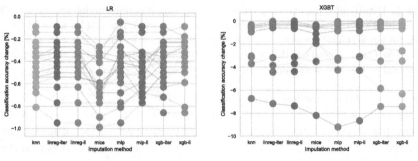

(a) LR model on Twonorm dataset   (b) XGBT model on MAGIC dataset

Fig. 1. Change in classification accuracy under all imputation methods for single missing features. Each feature is linked between different methods using a line.

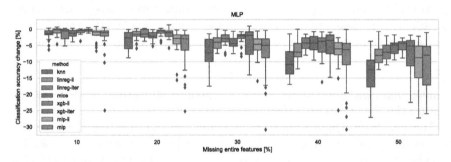

Fig. 2. Classification accuracy change of MLP model on real dataset Spambase.

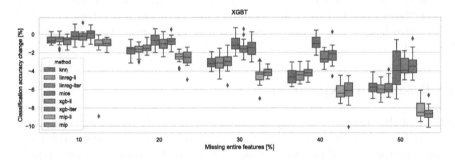

Fig. 3. Classification accuracy change of XGBT model on artificial dataset Ringnorm.

5.1 Discussion

The results are highly dataset specific. For some datasets (Cancer, all ds_....
datasets) the decreases in the classification accuracy were only minor, less than
1%, even for 50% of missing values. On the other hand for some datasets
(MAGIC, Ringnorm) the decrease is much greater, 1%–2% for 10% of missing
features and 10% for 50% of missing features.

Table 2. Mean accuracy changes in percentages (± standard deviation) for single missing feature imputation. Classification methods are shown in the last six columns and imputations methods are given in the second column.

Dataset	Method	LR	MLP	k-NN	NB	XGBT	RF
Cancer	knn	**-0.16 ± 0.35**	0.05 ± 0.29	0.05 ± 0.86	0.44 ± 0.38	0.33 ± 0.42	**0.44 ± 0.38**
Cancer	linreg-iter	**-0.16 ± 0.35**	0.05 ± 0.29	-0.0 ± 1.07	0.44 ± 0.38	0.33 ± 0.42	**0.44 ± 0.29**
Cancer	linreg-li	**-0.16 ± 0.35**	0.05 ± 0.29	0.00 ± 1.07	0.44 ± 0.38	0.33 ± 0.45	**0.44 ± 0.29**
Cancer	mice	-0.22 ± 0.61	**0.27 ± 0.5**	-0.22 ± 0.89	0.65 ± 0.42	0.11 ± 0.41	**0.44 ± 0.52**
Cancer	mlp	-0.38 ± 0.73	0.0 ± 0.6	**0.11 ± 1.03**	**0.71 ± 0.55**	0.16 ± 0.49	0.33 ± 0.42
Cancer	mlp-li	-0.38 ± 0.48	0.22 ± 0.61	**0.11 ± 0.8**	0.38 ± 0.59	-0.11 ± 0.77	0.11 ± 0.41
Cancer	xgb-iter	**-0.11 ± 0.22**	0.22 ± 0.55	**0.11 ± 0.59**	0.6 ± 0.33	**0.54 ± 0.52**	0.38 ± 0.33
Cancer	xgb-li	**-0.16 ± 0.35**	0.11 ± 0.59	**0.11 ± 0.64**	0.6 ± 0.22	0.38 ± 0.48	0.33 ± 0.35
Wine	knn	-1.38 ± 2.33	-0.07 ± 1.0	-1.19 ± 1.97	-0.37 ± 1.01	-1.86 ± 1.52	-1.1 ± 0.91
Wine	linreg-iter	**-0.45 ± 0.97**	0.2 ± 1.01	-0.87 ± 2.1	**-0.1 ± 0.59**	-1.39 ± 1.22	-0.74 ± 1.0
Wine	linreg-li	**-0.45 ± 0.97**	0.2 ± 1.01	-0.87 ± 2.1	**-0.1 ± 0.59**	-1.39 ± 1.22	-0.74 ± 1.0
Wine	mice	-0.77 ± 1.65	-0.19 ± 0.67	**-0.35 ± 0.72**	-0.48 ± 0.6	**-0.79 ± 0.98**	**-0.54 ± 0.55**
Wine	mlp	-3.17 ± 5.3	-0.28 ± 0.73	-1.46 ± 2.71	-2.23 ± 4.94	-3.32 ± 3.44	-1.84 ± 1.76
Wine	mlp-li	-2.2 ± 2.98	-0.24 ± 0.55	-1.78 ± 2.98	-2.52 ± 6.64	-2.7 ± 2.55	-1.74 ± 1.84
Wine	xgb-iter	-1.68 ± 3.92	**0.28 ± 0.7**	-0.57 ± 1.35	**-1.33 ± 3.52**	-2.43 ± 2.97	-1.02 ± 2.35
Wine	xgb-li	-1.71 ± 4.19	**0.22 ± 0.65**	-0.67 ± 1.47	-1.35 ± 3.51	-2.62 ± 3.13	-1.1 ± 2.3
MAGIC	knn	-0.74 ± 1.94	-1.18 ± 2.58	-1.44 ± 2.53	**0.01 ± 0.54**	-2.0 ± 2.11	-1.84 ± 2.49
MAGIC	linreg-iter	-0.63 ± 1.66	**-1.12 ± 2.56**	**-1.27 ± 2.2**	-0.11 ± 0.6	-2.03 ± 2.48	-1.95 ± 2.72
MAGIC	linreg-li	-0.63 ± 1.66	**-1.12 ± 2.56**	**-1.27 ± 2.2**	-0.11 ± 0.6	-2.03 ± 2.48	-1.95 ± 2.72
MAGIC	mice	-0.86 ± 2.13	-1.24 ± 2.86	-1.5 ± 2.68	-0.1 ± 0.46	**-1.88 ± 2.42**	**-1.36 ± 2.44**
MAGIC	mlp	-0.66 ± 1.71	-1.2 ± 2.73	-1.31 ± 2.34	-0.07 ± 0.64	-2.25 ± 2.89	-1.96 ± 3.07
MAGIC	mlp-li	-0.67 ± 1.72	**-1.17 ± 2.65**	**-1.26 ± 2.26**	-0.11 ± 0.59	-2.11 ± 2.77	-1.83 ± 2.83
MAGIC	xgb-iter	**-0.36 ± 1.31**	-1.49 ± 2.66	-0.95 ± 1.66	-0.33 ± 0.99	-2.1 ± 2.65	-2.37 ± 3.32
MAGIC	xgb-li	**-0.38 ± 1.3**	-1.63 ± 2.95	-0.99 ± 1.67	-0.32 ± 0.98	-2.17 ± 2.74	-2.43 ± 3.37
Spambase	knn	-0.15 ± 0.64	-0.16 ± 0.71	-0.36 ± 1.69	**-0.27 ± 1.22**	0.08 ± 4.99	-0.7 ± 2.75
Spambase	linreg-iter	-0.11 ± 0.66	-0.26 ± 1.18	-0.34 ± 1.76	-0.3 ± 1.02	-0.45 ± 1.84	-0.24 ± 0.79
Spambase	linreg-li	-0.11 ± 0.66	-0.26 ± 1.18	-0.34 ± 1.76	-0.3 ± 1.02	-0.45 ± 1.84	-0.24 ± 0.79
Spambase	mice	-0.18 ± 0.46	-0.18 ± 0.4	**-0.03 ± 0.35**	-0.1 ± 0.43	1.03 ± 1.12	**-0.2 ± 0.78**
Spambase	mlp	-0.38 ± 0.99	-0.49 ± 1.65	-0.17 ± 0.93	-3.6 ± 6.18	-0.55 ± 2.23	-0.36 ± 1.28
Spambase	mlp-li	-0.16 ± 0.41	-0.11 ± 0.49	-0.18 ± 0.87	-1.87 ± 3.25	-0.5 ± 2.1	-0.33 ± 1.16
Spambase	xgb-iter	0.01 ± 0.17	0.01 ± 0.24	-0.13 ± 0.64	-0.09 ± 0.96	-0.77 ± 4.27	**-0.19 ± 0.9**
Spambase	xgb-li	0.01 ± 0.21	0.01 ± 0.23	-0.08 ± 0.46	**-0.13 ± 0.93**	-0.78 ± 4.28	-0.24 ± 0.97
Ringnorm	knn	-0.65 ± 0.32	-0.47 ± 0.47	-1.32 ± 0.37	-0.5 ± 0.17	-0.4 ± 0.21	**-0.16 ± 0.19**
Ringnorm	linreg-iter	-0.58 ± 0.36	-0.49 ± 0.45	-1.34 ± 0.35	-0.5 ± 0.16	-0.42 ± 0.22	-0.15 ± 0.21
Ringnorm	linreg-li	-0.58 ± 0.36	-0.49 ± 0.45	-1.34 ± 0.35	-0.5 ± 0.16	-0.42 ± 0.22	-0.15 ± 0.21
Ringnorm	mice	-0.53 ± 0.43	**-0.14 ± 0.55**	**0.08 ± 0.31**	-0.46 ± 0.15	-0.43 ± 0.18	-0.16 ± 0.17
Ringnorm	mlp	-0.77 ± 0.37	-0.68 ± 0.45	-0.8 ± 0.32	**-0.41 ± 0.17**	**-0.39 ± 0.21**	**-0.14 ± 0.2**
Ringnorm	mlp-li	-0.75 ± 0.45	-0.72 ± 0.54	-0.74 ± 0.33	-0.43 ± 0.14	**-0.39 ± 0.21**	-0.14 ± 0.22
Ringnorm	xgb-iter	**-0.3 ± 0.36**	**-0.2 ± 0.5**	-1.28 ± 0.35	-0.5 ± 0.15	-0.4 ± 0.22	-0.16 ± 0.2
Ringnorm	xgb-li	**-0.3 ± 0.43**	-0.21 ± 0.51	-1.3 ± 0.35	-0.49 ± 0.15	**-0.38 ± 0.23**	-0.16 ± 0.21
Twonorm	knn	**-0.42 ± 0.18**	-0.43 ± 0.21	**-0.1 ± 0.19**	**-0.38 ± 0.19**	-0.43 ± 0.18	-0.12 ± 0.28
Twonorm	linreg-iter	-0.42 ± 0.21	**-0.42 ± 0.18**	-0.11 ± 0.18	-0.37 ± 0.2	-0.41 ± 0.19	**-0.08 ± 0.33**
Twonorm	linreg-li	-0.42 ± 0.21	**-0.42 ± 0.18**	-0.11 ± 0.18	-0.37 ± 0.2	**-0.41 ± 0.19**	**-0.08 ± 0.33**
Twonorm	mice	-0.58 ± 0.21	-0.54 ± 0.21	-0.21 ± 0.18	-0.56 ± 0.2	-0.55 ± 0.21	-0.15 ± 0.23
Twonorm	mlp	-0.45 ± 0.23	-0.43 ± 0.22	**-0.1 ± 0.22**	-0.4 ± 0.23	**-0.42 ± 0.23**	**-0.09 ± 0.27**
Twonorm	mlp-li	-0.47 ± 0.19	-0.46 ± 0.25	**-0.11 ± 0.22**	-0.45 ± 0.19	-0.45 ± 0.19	-0.14 ± 0.24
Twonorm	xgb-iter	**-0.42 ± 0.19**	**-0.41 ± 0.19**	-0.17 ± 0.2	**-0.39 ± 0.22**	**-0.41 ± 0.21**	-0.33 ± 0.25
Twonorm	xgb-li	**-0.41 ± 0.2**	**-0.42 ± 0.2**	-0.16 ± 0.19	**-0.39 ± 0.2**	-0.43 ± 0.22	-0.28 ± 0.29
ds_10_7_3	knn	**0.03 ± 0.13**	-0.09 ± 0.12	**0.07 ± 0.18**	-0.04 ± 0.14	-0.17 ± 0.17	-0.29 ± 0.19
ds_10_7_3	linreg-iter	**0.05 ± 0.09**	-0.07 ± 0.1	**0.13 ± 0.14**	-0.03 ± 0.11	-0.14 ± 0.12	-0.29 ± 0.18
ds_10_7_3	linreg-li	**0.05 ± 0.09**	-0.07 ± 0.1	**0.13 ± 0.14**	-0.03 ± 0.11	-0.14 ± 0.12	-0.29 ± 0.18
ds_10_7_3	mice	-0.02 ± 0.17	-0.19 ± 0.23	-0.02 ± 0.18	**0.07 ± 0.13**	-0.1 ± 0.14	**0.05 ± 0.14**
ds_10_7_3	mlp	**0.03 ± 0.12**	-0.11 ± 0.11	**0.11 ± 0.16**	-0.06 ± 0.2	-0.15 ± 0.13	-0.21 ± 0.12
ds_10_7_3	mlp-li	**0.04 ± 0.15**	-0.08 ± 0.1	0.1 ± 0.16	-0.03 ± 0.15	-0.18 ± 0.17	-0.26 ± 0.17
ds_10_7_3	xgb-iter	-0.19 ± 0.39	-0.56 ± 0.6	-0.08 ± 0.38	-0.19 ± 0.67	-0.71 ± 0.91	-0.55 ± 0.53
ds_10_7_3	xgb-li	-0.15 ± 0.21	-0.55 ± 0.5	-0.11 ± 0.38	-0.07 ± 0.49	-0.62 ± 0.58	-0.43 ± 0.47
ds_20_14_6	knn	**-0.24 ± 0.21**	**-0.03 ± 0.07**	-0.03 ± 0.14	-0.02 ± 0.14	-0.12 ± 0.17	**-0.06 ± 0.09**
ds_20_14_6	linreg-iter	**-0.25 ± 0.22**	-0.01 ± 0.05	-0.01 ± 0.13	**0.03 ± 0.12**	-0.09 ± 0.15	**-0.06 ± 0.09**
ds_20_14_6	linreg-li	**-0.25 ± 0.22**	-0.01 ± 0.05	-0.01 ± 0.13	**0.03 ± 0.12**	-0.09 ± 0.15	**-0.06 ± 0.09**
ds_20_14_6	mice	-0.58 ± 0.42	-0.16 ± 0.18	-0.05 ± 0.17	-0.15 ± 0.24	**-0.01 ± 0.15**	-0.06 ± 0.14
ds_20_14_6	mlp	-0.29 ± 0.25	-0.02 ± 0.06	**0.01 ± 0.12**	**0.0 ± 0.14**	-0.07 ± 0.13	**-0.07 ± 0.09**
ds_20_14_6	mlp-li	**-0.24 ± 0.27**	-0.01 ± 0.05	-0.03 ± 0.14	-0.03 ± 0.1	-0.08 ± 0.11	**-0.07 ± 0.12**
ds_20_14_6	xgb-iter	-0.7 ± 0.64	-0.21 ± 0.25	-0.22 ± 0.22	-0.23 ± 0.37	-0.22 ± 0.21	-0.21 ± 0.27
ds_20_14_6	xgb-li	-0.76 ± 0.7	-0.23 ± 0.29	-0.18 ± 0.24	-0.26 ± 0.39	-0.27 ± 0.3	-0.22 ± 0.27
ds_50_35_15	knn	-0.12 ± 0.14	**-0.02 ± 0.04**	-0.07 ± 0.08	0.04 ± 0.11	-0.02 ± 0.05	-0.05 ± 0.11
ds_50_35_15	linreg-iter	**-0.06 ± 0.09**	**-0.01 ± 0.02**	**-0.02 ± 0.04**	0.03 ± 0.07	0.0 ± 0.04	-0.02 ± 0.05
ds_50_35_15	linreg-li	**-0.06 ± 0.09**	**-0.01 ± 0.02**	**-0.02 ± 0.04**	0.03 ± 0.07	0.0 ± 0.04	-0.02 ± 0.05
ds_50_35_15	mice	-0.09 ± 0.15	-0.15 ± 0.14	-0.26 ± 0.22	-0.06 ± 0.15	-0.03 ± 0.15	-0.05 ± 0.14
ds_50_35_15	mlp	-0.08 ± 0.12	**-0.02 ± 0.04**	-0.04 ± 0.06	**0.05 ± 0.07**	**-0.01 ± 0.07**	-0.03 ± 0.09
ds_50_15_35	mlp-li	-0.09 ± 0.13	**-0.01 ± 0.04**	-0.05 ± 0.06	**0.05 ± 0.08**	0.0 ± 0.06	-0.01 ± 0.08
ds_50_35_15	xgb-iter	-0.15 ± 0.2	-0.12 ± 0.13	-0.21 ± 0.23	-0.04 ± 0.21	-0.12 ± 0.17	-0.13 ± 0.22
ds_50_35_15	xgb-li	-0.18 ± 0.26	-0.13 ± 0.14	-0.24 ± 0.22	-0.05 ± 0.26	-0.15 ± 0.23	-0.19 ± 0.28
ds_100_70_30	knn	**-0.3 ± 0.99**	-0.05 ± 0.07	-0.07 ± 0.12	-0.04 ± 0.14	0.03 ± 0.16	-0.02 ± 0.08
ds_100_70_30	linreg-iter	0.0 ± 0.14	**0.0 ± 0.01**	-0.02 ± 0.06	0.0 ± 0.04	0.01 ± 0.05	-0.01 ± 0.06
ds_100_70_30	linreg-li	0.0 ± 0.14	**0.0 ± 0.01**	-0.02 ± 0.06	0.0 ± 0.04	0.01 ± 0.05	-0.01 ± 0.06
ds_100_70_30	mice	-1.76 ± 1.85	-0.25 ± 0.18	-0.09 ± 0.17	**0.02 ± 0.18**	-0.14 ± 0.22	-0.02 ± 0.12
ds_100_70_30	mlp	**-0.03 ± 0.35**	-0.05 ± 0.07	-0.05 ± 0.09	0.0 ± 0.09	**0.01 ± 0.08**	**-0.01 ± 0.1**
ds_100_70_30	mlp-li	**0.03 ± 0.32**	-0.05 ± 0.07	-0.05 ± 0.1	**0.01 ± 0.08**	-0.01 ± 0.1	-0.01 ± 0.06
ds_100_70_30	xgb-iter	-3.67 ± 3.86	-0.22 ± 0.18	-0.13 ± 0.29	-0.05 ± 0.23	-0.16 ± 0.31	-0.08 ± 0.18
ds_100_70_30	xgb-li	-3.58 ± 4.25	-0.22 ± 0.18	-0.15 ± 0.27	-0.06 ± 0.22	-0.15 ± 0.32	-0.08 ± 0.19

Table 3. Mean accuracy changes shown in percentages ($\pm$ standard deviation) on real datasets with missingness from 10% up to 50% where only the two best classification models for each dataset are shown.

Dataset	Model	Method	10%	20%	30%	40%	50%
Cancer	LR	knn	-0.16 ± 0.35	-0.37 ± 0.31	-0.49 ± 0.45	-0.66 ± 0.7	-0.98 ± 0.95
Cancer	LR	linreg-iter	-0.16 ± 0.35	-0.29 ± 0.4	-0.43 ± 0.56	-0.68 ± 0.7	-1.06 ± 1.01
Cancer	LR	linreg-li	-0.16 ± 0.35	$\mathbf{-0.27 \pm 0.34}$	$\mathbf{-0.37 \pm 0.47}$	-0.47 ± 0.72	-1.1 ± 0.98
Cancer	LR	mice	-0.22 ± 0.61	$\mathbf{-0.27 \pm 0.63}$	-0.61 ± 0.76	-0.98 ± 0.99	-1.32 ± 1.1
Cancer	LR	mlp	-0.38 ± 0.73	-0.44 ± 0.61	-0.63 ± 0.69	-0.81 ± 1.03	-1.14 ± 1.06
Cancer	LR	mlp-li	-0.38 ± 0.48	-0.59 ± 0.49	-0.74 ± 0.74	-1.05 ± 0.82	-1.54 ± 1.74
Cancer	LR	xgb-iter	$\mathbf{-0.11 \pm 0.22}$	-0.32 ± 0.37	-0.54 ± 0.5	-0.47 ± 0.61	-1.0 ± 1.15
Cancer	LR	xgb-li	-0.16 ± 0.35	$\mathbf{-0.25 \pm 0.46}$	-0.56 ± 0.83	$\mathbf{-0.29 \pm 0.64}$	$\mathbf{-0.78 \pm 0.99}$
Cancer	RF	knn	$\mathbf{0.44 \pm 0.38}$	$\mathbf{0.49 \pm 0.45}$	0.22 ± 0.77	0.12 ± 0.91	-0.69 ± 1.52
Cancer	RF	linreg-iter	$\mathbf{0.44 \pm 0.29}$	$\mathbf{0.5 \pm 0.4}$	$\mathbf{0.33 \pm 0.66}$	-0.03 ± 0.85	$\mathbf{-0.63 \pm 1.24}$
Cancer	RF	linreg-li	$\mathbf{0.44 \pm 0.29}$	0.39 ± 0.41	$\mathbf{0.42 \pm 0.75}$	$\mathbf{0.17 \pm 0.8}$	-0.66 ± 1.14
Cancer	RF	mice	$\mathbf{0.44 \pm 0.52}$	0.37 ± 0.67	-0.2 ± 0.86	-0.83 ± 1.35	-1.23 ± 0.95
Cancer	RF	mlp	0.33 ± 0.42	0.1 ± 0.8	0.02 ± 0.8	-0.46 ± 1.0	-1.08 ± 1.35
Cancer	RF	mlp-li	0.11 ± 0.41	-0.12 ± 0.71	-0.47 ± 1.14	-1.3 ± 1.57	-2.72 ± 4.01
Cancer	RF	xgb-iter	0.38 ± 0.33	0.29 ± 0.56	0.12 ± 0.61	-0.12 ± 0.79	$\mathbf{-0.44 \pm 1.1}$
Cancer	RF	xgb-li	0.33 ± 0.35	0.42 ± 0.43	0.22 ± 0.83	0.1 ± 0.82	$\mathbf{-0.66 \pm 0.7}$
Wine	XGBT	knn	-1.86 ± 1.52	-3.21 ± 1.25	-5.57 ± 2.36	-7.42 ± 2.4	-9.23 ± 2.81
Wine	XGBT	linreg-iter	-1.39 ± 1.22	-2.9 ± 1.53	-3.36 ± 1.66	-6.78 ± 2.44	-7.61 ± 2.59
Wine	XGBT	linreg-li	-1.39 ± 1.22	-2.94 ± 1.29	-4.54 ± 2.5	-7.5 ± 3.08	-7.5 ± 3.08
Wine	XGBT	mice	$\mathbf{-0.79 \pm 0.98}$	$\mathbf{-1.15 \pm 0.97}$	$\mathbf{-2.45 \pm 1.95}$	$\mathbf{-3.75 \pm 2.25}$	$\mathbf{-4.13 \pm 2.67}$
Wine	XGBT	mlp	-3.42 ± 2.67	-5.49 ± 3.86	-7.69 ± 4.55	-10.75 ± 5.12	-13.25 ± 5.57
Wine	XGBT	mlp-li	-3.51 ± 3.78	-5.91 ± 3.88	-8.4 ± 4.1	-10.33 ± 4.37	-12.29 ± 4.97
Wine	XGBT	xgb-iter	-2.43 ± 2.97	-5.55 ± 4.53	-6.72 ± 5.75	-9.57 ± 5.51	-14.86 ± 6.54
Wine	XGBT	xgb-li	-2.62 ± 3.13	-5.07 ± 4.38	-8.17 ± 4.82	-9.75 ± 4.31	-11.57 ± 5.83
Wine	RF	knn	-1.1 ± 0.91	-1.94 ± 0.78	-3.61 ± 1.44	-4.72 ± 1.63	-5.8 ± 1.73
Wine	RF	linreg-iter	-0.74 ± 1.0	-1.5 ± 1.19	$\mathbf{-1.73 \pm 1.12}$	-3.88 ± 1.8	-5.29 ± 1.83
Wine	RF	linreg-li	-0.74 ± 1.0	-1.52 ± 1.2	-2.7 ± 2.0	-3.84 ± 1.73	-5.49 ± 2.6
Wine	RF	mice	$\mathbf{-0.54 \pm 0.55}$	$\mathbf{-0.95 \pm 0.61}$	$\mathbf{-1.79 \pm 0.99}$	$\mathbf{-2.64 \pm 1.59}$	$\mathbf{-2.87 \pm 1.71}$
Wine	RF	mlp	-1.69 ± 1.65	-2.95 ± 2.23	-5.26 ± 2.47	-7.24 ± 2.05	-8.7 ± 1.82
Wine	RF	mlp-li	-1.74 ± 1.5	-3.32 ± 1.86	-5.12 ± 2.03	-6.12 ± 2.21	-8.44 ± 1.76
Wine	RF	xgb-iter	-1.02 ± 2.35	-3.3 ± 3.05	-4.5 ± 3.26	-6.14 ± 2.53	-8.0 ± 2.94
Wine	RF	xgb-li	-1.1 ± 2.3	-2.65 ± 3.01	-5.33 ± 3.39	-6.22 ± 2.61	-7.58 ± 3.26
MAGIC	XGBT	knn	$\mathbf{-2.0 \pm 2.11}$	-3.05 ± 2.26	$\mathbf{-5.3 \pm 3.07}$	-8.06 ± 4.47	-10.71 ± 4.19
MAGIC	XGBT	linreg-iter	$\mathbf{-2.03 \pm 2.48}$	-4.19 ± 3.52	-6.28 ± 4.82	-9.99 ± 6.0	-12.82 ± 5.76
MAGIC	XGBT	linreg-li	$\mathbf{-2.03 \pm 2.48}$	-4.68 ± 3.86	-6.77 ± 4.91	-9.2 ± 6.2	-11.25 ± 4.96
MAGIC	XGBT	mice	$\mathbf{-2.03 \pm 2.52}$	$\mathbf{-2.51 \pm 2.25}$	$\mathbf{-4.78 \pm 3.39}$	$\mathbf{-7.02 \pm 4.09}$	$\mathbf{-8.52 \pm 3.55}$
MAGIC	XGBT	mlp	-5.46 ± 7.85	-3.94 ± 2.96	$\mathbf{-4.99 \pm 3.39}$	-9.59 ± 5.69	-11.77 ± 7.9
MAGIC	XGBT	mlp-li	-2.26 ± 2.9	-4.25 ± 3.59	-6.95 ± 4.43	$\mathbf{-7.77 \pm 4.76}$	$\mathbf{-8.59 \pm 4.55}$
MAGIC	XGBT	xgb-iter	-3.99 ± 6.05	-4.94 ± 6.29	-10.53 ± 7.01	-14.97 ± 7.45	-15.49 ± 9.18
MAGIC	XGBT	xgb-li	$\mathbf{-2.34 \pm 2.85}$	-3.98 ± 3.18	-9.45 ± 6.46	-9.13 ± 9.71	-16.31 ± 13.06
MAGIC	RF	knn	-2.7 ± 2.49	-3.76 ± 2.54	-5.79 ± 3.19	-8.37 ± 4.33	-10.7 ± 3.81
MAGIC	RF	linreg-iter	$\mathbf{-1.95 \pm 2.72}$	-4.77 ± 3.4	-6.65 ± 4.32	-10.73 ± 5.42	-12.84 ± 4.58
MAGIC	RF	linreg-li	$\mathbf{-1.95 \pm 2.72}$	-5.38 ± 4.0	-7.64 ± 4.75	-9.47 ± 5.73	-11.53 ± 4.25
MAGIC	RF	mice	-2.36 ± 2.54	$\mathbf{-2.73 \pm 2.19}$	$\mathbf{-4.82 \pm 3.27}$	$\mathbf{-6.84 \pm 4.05}$	$\mathbf{-8.42 \pm 3.55}$
MAGIC	RF	mlp	-5.85 ± 7.19	-4.41 ± 3.1	$\mathbf{-5.23 \pm 3.19}$	-9.62 ± 5.14	-11.25 ± 6.48
MAGIC	RF	mlp-li	-2.89 ± 2.92	-4.71 ± 3.7	-7.55 ± 4.59	-8.23 ± 4.8	$\mathbf{-9.08 \pm 4.46}$
MAGIC	RF	xgb-iter	-5.06 ± 5.88	-5.8 ± 6.55	-11.75 ± 6.02	-16.19 ± 6.39	-16.99 ± 8.59
MAGIC	RF	xgb-li	-3.57 ± 3.44	-5.61 ± 3.94	-10.93 ± 6.34	-9.92 ± 8.83	-17.16 ± 11.48
Spambase	MLP	knn	-1.55 ± 1.75	-3.72 ± 3.34	-7.87 ± 4.55	-10.01 ± 4.39	-13.55 ± 6.36
Spambase	MLP	linreg-iter	-2.24 ± 2.56	-3.36 ± 2.43	-4.91 ± 2.72	-6.25 ± 2.61	-8.34 ± 3.12
Spambase	MLP	linreg-li	-2.24 ± 2.56	-2.89 ± 2.39	-5.36 ± 2.7	-8.22 ± 4.09	-8.76 ± 3.28
Spambase	MLP	mice	-1.34 ± 1.24	-2.33 ± 1.39	-4.67 ± 1.78	$\mathbf{-4.96 \pm 2.36}$	$\mathbf{-6.56 \pm 2.31}$
Spambase	MLP	mlp	-3.44 ± 5.06	-7.17 ± 5.49	-9.64 ± 6.78	-11.38 ± 7.33	-13.9 ± 8.77
Spambase	MLP	mlp-li	-3.54 ± 3.47	-6.04 ± 6.09	-7.71 ± 4.71	-10.96 ± 5.02	-13.75 ± 7.69
Spambase	MLP	xgb-iter	$\mathbf{-0.65 \pm 0.62}$	-1.95 ± 1.31	-6.61 ± 6.25	-7.33 ± 5.25	-11.99 ± 7.68
Spambase	MLP	xgb-li	-0.89 ± 0.87	$\mathbf{-1.54 \pm 1.57}$	$\mathbf{-4.06 \pm 3.57}$	-6.41 ± 4.59	-7.5 ± 3.02
Spambase	RF	knn	-5.43 ± 6.99	-10.52 ± 9.7	-15.72 ± 8.41	-19.97 ± 8.63	-22.17 ± 9.06
Spambase	RF	linreg-iter	-2.05 ± 2.29	$\mathbf{-3.07 \pm 2.57}$	$\mathbf{-4.97 \pm 3.51}$	-7.46 ± 4.47	-11.45 ± 3.65
Spambase	RF	linreg-li	-2.05 ± 2.29	-3.57 ± 3.32	$\mathbf{-5.01 \pm 2.55}$	-8.91 ± 5.39	-9.09 ± 3.73
Spambase	RF	mice	$\mathbf{-1.64 \pm 2.15}$	$\mathbf{-2.52 \pm 2.98}$	$\mathbf{-5.05 \pm 2.54}$	$\mathbf{-4.95 \pm 3.19}$	$\mathbf{-5.96 \pm 3.0}$
Spambase	RF	mlp	-4.95 ± 5.32	-7.27 ± 5.49	-12.03 ± 8.3	-13.83 ± 6.76	-18.68 ± 8.24
Spambase	RF	mlp-li	-3.58 ± 3.87	-6.24 ± 6.33	-10.54 ± 7.14	-13.37 ± 8.97	-18.26 ± 8.02
Spambase	RF	xgb-iter	$\mathbf{-1.37 \pm 2.08}$	-4.54 ± 7.02	-10.54 ± 11.71	-17.62 ± 10.25	-21.63 ± 9.67
Spambase	RF	xgb-li	-3.71 ± 5.42	-3.89 ± 5.3	-8.38 ± 8.34	-14.68 ± 11.33	-14.78 ± 6.69

From the imputation methods point of view MICE usually performs the best on real datasets. On artificial datasets it places among the best methods only for the Ringnorm and ds_10_7_3 datasets. Its results often have smaller variance than results of other methods.

Results comparative to MICE were often reached using linear regression imputation (specifically *linreg-li*). Especially on artificial datasets it usually per-

Table 4. Mean accuracy changes shown in percentages ($\pm$ standard deviation) on artificial datasets with missingness from 10% up to 50% where only the two best classification models for each dataset are shown.

Dataset	Model	Method	10%	20%	30%	40%	50%
Ringnorm	NB	knn	-1.16 ± 0.25	-3.56 ± 0.3	-7.31 ± 0.53	-13.01 ± 0.58	-20.59 ± 0.65
Ringnorm	NB	linreg-iter	-1.81 ± 2.66	-3.51 ± 0.36	-7.41 ± 0.36	-13.25 ± 0.42	-20.92 ± 0.49
Ringnorm	NB	linreg-li	-1.23 ± 0.27	-3.46 ± 0.39	-7.75 ± 1.45	-13.0 ± 1.55	-20.96 ± 0.45
Ringnorm	NB	mice	**-1.03 ± 0.24**	-3.04 ± 0.25	**-5.8 ± 0.45**	**-9.3 ± 0.4**	**-16.98 ± 3.92**
Ringnorm	NB	mlp	-1.16 ± 0.16	-3.34 ± 0.94	-6.83 ± 0.46	-12.51 ± 1.67	-19.58 ± 0.61
Ringnorm	NB	mlp-li	-1.87 ± 3.69	**-2.95 ± 0.33**	-6.2 ± 1.34	-11.0 ± 0.4	-18.01 ± 0.68
Ringnorm	NB	xgb-iter	-1.24 ± 0.27	-3.58 ± 0.41	-7.51 ± 0.37	-12.68 ± 2.7	-19.43 ± 4.62
Ringnorm	NB	xgb-li	-1.21 ± 0.26	-3.42 ± 0.19	-7.18 ± 0.52	-13.05 ± 0.45	-20.54 ± 0.53
Ringnorm	XGBT	knn	-1.0 ± 0.29	-2.38 ± 0.37	-4.91 ± 0.62	-8.32 ± 0.82	-12.05 ± 1.15
Ringnorm	XGBT	linreg-iter	-1.34 ± 1.5	-2.54 ± 0.31	-4.91 ± 0.52	-8.12 ± 0.91	-12.42 ± 1.46
Ringnorm	XGBT	linreg-li	-0.95 ± 0.28	-2.62 ± 0.36	-5.37 ± 0.88	-8.22 ± 1.44	-12.37 ± 1.05
Ringnorm	XGBT	mice	-0.96 ± 0.29	**-2.3 ± 0.33**	**-4.32 ± 0.53**	**-6.81 ± 0.59**	**-10.76 ± 1.9**
Ringnorm	XGBT	mlp	**-0.92 ± 0.27**	-2.48 ± 0.58	-4.57 ± 0.53	-7.71 ± 1.28	-11.4 ± 1.14
Ringnorm	XGBT	mlp-li	-1.34 ± 1.97	-2.35 ± 0.41	-4.75 ± 0.9	-7.37 ± 0.83	-10.96 ± 1.1
Ringnorm	XGBT	xgb-iter	**-0.89 ± 0.23**	**-2.29 ± 0.38**	-4.76 ± 0.51	-7.67 ± 1.77	-11.12 ± 2.74
Ringnorm	XGBT	xgb-li	**-0.89 ± 0.28**	-2.38 ± 0.36	-4.85 ± 0.62	-7.88 ± 0.76	-12.28 ± 1.38
Twonorm	LR	knn	-0.85 ± 0.28	-1.69 ± 0.35	-2.88 ± 0.39	-4.13 ± 0.35	**-5.94 ± 0.58**
Twonorm	LR	linreg-iter	**-0.71 ± 0.27**	**-1.43 ± 0.28**	**-2.57 ± 0.27**	**-4.08 ± 0.41**	**-5.84 ± 0.57**
Twonorm	LR	linreg-li	-0.78 ± 0.3	-1.69 ± 0.31	**-2.61 ± 0.47**	**-3.97 ± 0.28**	-6.03 ± 0.48
Twonorm	LR	mice	-1.09 ± 0.27	-2.19 ± 0.4	-3.75 ± 0.46	-5.27 ± 0.43	-7.53 ± 0.65
Twonorm	LR	mlp	-0.91 ± 0.15	-1.79 ± 0.26	-2.7 ± 0.27	**-4.05 ± 0.56**	**-5.93 ± 0.52**
Twonorm	LR	mlp-li	-0.86 ± 0.25	-1.73 ± 0.26	-2.8 ± 0.34	-4.27 ± 0.42	-6.22 ± 0.53
Twonorm	LR	xgb-iter	-0.78 ± 0.19	-1.66 ± 0.32	-3.0 ± 0.33	-4.4 ± 0.4	-6.92 ± 0.52
Twonorm	LR	xgb-li	-0.93 ± 0.24	-1.74 ± 0.31	-2.75 ± 0.36	-4.25 ± 0.36	-6.49 ± 0.56
Twonorm	XGBT	knn	-0.82 ± 0.29	-1.68 ± 0.37	-2.85 ± 0.36	-4.15 ± 0.36	**-5.91 ± 0.6**
Twonorm	XGBT	linreg-iter	**-0.73 ± 0.25**	**-1.45 ± 0.28**	**-2.58 ± 0.26**	**-4.05 ± 0.42**	**-5.82 ± 0.56**
Twonorm	XGBT	linreg-li	**-0.75 ± 0.32**	-1.65 ± 0.31	-2.6 ± 0.44	**-3.96 ± 0.29**	-6.02 ± 0.49
Twonorm	XGBT	mice	-1.1 ± 0.27	-2.2 ± 0.37	-3.73 ± 0.46	-5.22 ± 0.45	-7.56 ± 0.65
Twonorm	XGBT	mlp	-0.87 ± 0.17	-1.72 ± 0.26	-2.71 ± 0.23	-4.04 ± 0.54	**-5.95 ± 0.52**
Twonorm	XGBT	mlp-li	-0.82 ± 0.21	-1.73 ± 0.26	-2.78 ± 0.34	-4.27 ± 0.42	-6.21 ± 0.53
Twonorm	XGBT	xgb-iter	**-0.8 ± 0.19**	-1.66 ± 0.31	-2.98 ± 0.34	-4.36 ± 0.4	-6.95 ± 0.51
Twonorm	XGBT	xgb-li	-0.92 ± 0.2	-1.71 ± 0.3	-2.73 ± 0.37	-4.25 ± 0.34	-6.53 ± 0.57
ds_10_7_3	MLP	knn	**0.25 ± 0.12**	**0.06 ± 0.26**	-1.63 ± 1.78	-9.15 ± 3.16	-13.8 ± 2.29
ds_10_7_3	MLP	linreg-iter	-0.07 ± 0.1	-0.03 ± 0.53	-0.47 ± 1.02	-5.83 ± 2.68	-11.12 ± 3.84
ds_10_7_3	MLP	linreg-li	-0.07 ± 0.1	**0.01 ± 0.5**	-1.3 ± 2.39	-6.02 ± 2.48	-11.16 ± 4.06
ds_10_7_3	MLP	mice	0.15 ± 0.23	-0.05 ± 0.29	-0.4 ± 0.45	**-1.55 ± 0.69**	**-2.51 ± 0.71**
ds_10_7_3	MLP	mlp	**0.22 ± 0.11**	0.1 ± 0.23	-0.83 ± 1.03	-3.4 ± 1.54	-7.69 ± 2.71
ds_10_7_3	MLP	mlp-li	**0.25 ± 0.1**	**0.01 ± 0.23**	-0.9 ± 1.47	-4.71 ± 2.12	-8.48 ± 3.68
ds_10_7_3	MLP	xgb-iter	-0.23 ± 0.6	-1.41 ± 1.46	-2.28 ± 1.69	-4.73 ± 2.24	-9.07 ± 2.55
ds_10_7_3	MLP	xgb-li	-0.22 ± 0.5	-0.91 ± 0.67	-2.29 ± 1.8	-3.88 ± 1.86	-7.14 ± 2.71
ds_10_7_3	XGBT	knn	-0.37 ± 0.17	-0.67 ± 0.31	-2.78 ± 2.11	-8.77 ± 4.0	-13.14 ± 3.26
ds_10_7_3	XGBT	linreg-iter	**-0.14 ± 0.12**	-0.78 ± 0.5	-1.24 ± 0.78	-6.01 ± 2.71	-10.82 ± 3.51
ds_10_7_3	XGBT	linreg-li	**-0.14 ± 0.12**	-0.72 ± 0.52	-1.87 ± 2.12	-6.53 ± 2.5	-11.45 ± 4.13
ds_10_7_3	XGBT	mice	-0.3 ± 0.14	**-0.49 ± 0.28**	**-0.83 ± 0.54**	**-1.53 ± 0.86**	**-2.61 ± 0.65**
ds_10_7_3	XGBT	mlp	-0.35 ± 0.13	**-0.61 ± 0.28**	-1.61 ± 0.94	-4.0 ± 1.74	-8.31 ± 2.92
ds_10_7_3	XGBT	mlp-li	-0.38 ± 0.17	-0.71 ± 0.31	-1.81 ± 1.51	-5.03 ± 2.21	-9.2 ± 4.22
ds_10_7_3	XGBT	xgb-iter	-0.91 ± 0.91	-2.2 ± 1.26	-3.22 ± 2.01	-5.16 ± 2.48	-10.34 ± 2.8
ds_10_7_3	XGBT	xgb-li	-0.82 ± 0.58	-1.67 ± 1.16	-3.19 ± 1.74	-5.19 ± 1.95	-8.28 ± 2.5
ds_20_14_6	MLP	knn	0.02 ± 0.26	-0.42 ± 0.33	-0.71 ± 0.39	**-1.53 ± 0.51**	**-2.16 ± 0.69**
ds_20_14_6	MLP	linreg-iter	**0.24 ± 0.07**	**0.18 ± 0.1**	-0.19 ± 0.52	-4.21 ± 1.46	-10.0 ± 1.75
ds_20_14_6	MLP	linreg-li	**0.24 ± 0.05**	**0.16 ± 0.07**	-0.18 ± 0.87	-3.83 ± 1.23	-9.83 ± 3.27
ds_20_14_6	MLP	mice	0.05 ± 0.17	-0.6 ± 0.63	-2.72 ± 1.24	-5.71 ± 1.09	-9.98 ± 2.2
ds_20_14_6	MLP	mlp	**0.24 ± 0.08**	0.11 ± 0.28	-0.15 ± 0.36	-2.56 ± 1.19	-5.99 ± 0.97
ds_20_14_6	MLP	mlp-li	0.23 ± 0.08	**0.16 ± 0.1**	-0.18 ± 0.3	-2.58 ± 0.95	-6.05 ± 1.38
ds_20_14_6	MLP	xgb-iter	-0.3 ± 0.43	-1.17 ± 0.73	-3.0 ± 0.9	-5.11 ± 1.21	-8.01 ± 1.82
ds_20_14_6	MLP	xgb-li	-0.31 ± 0.39	-1.05 ± 0.79	-1.96 ± 0.96	-3.79 ± 1.06	-6.26 ± 1.06
ds_20_14_6	RF	knn	**0.83 ± 0.16**	**0.65 ± 0.3**	**0.44 ± 0.34**	**0.04 ± 0.41**	**-0.6 ± 0.78**
ds_20_14_6	RF	linreg-iter	**0.77 ± 0.1**	0.49 ± 0.3	-0.17 ± 0.59	-4.96 ± 2.12	-11.36 ± 3.15
ds_20_14_6	RF	linreg-li	**0.73 ± 0.1**	0.45 ± 0.37	-0.61 ± 1.49	-4.22 ± 2.26	-10.34 ± 4.64
ds_20_14_6	RF	mice	0.43 ± 0.35	-0.4 ± 0.66	-2.83 ± 1.93	-6.08 ± 2.31	-10.03 ± 3.22
ds_20_14_6	RF	mlp	0.69 ± 0.15	0.4 ± 0.3	0.06 ± 0.38	-2.82 ± 1.72	-5.35 ± 1.13
ds_20_14_6	RF	mlp-li	**0.71 ± 0.19**	0.54 ± 0.28	-0.03 ± 0.42	-2.85 ± 0.89	-5.48 ± 1.29
ds_20_14_6	RF	xgb-iter	0.32 ± 0.32	-0.54 ± 0.97	-2.22 ± 1.2	-4.67 ± 1.63	-7.67 ± 2.06
ds_20_14_6	RF	xgb-li	0.36 ± 0.4	-0.26 ± 0.65	-1.62 ± 1.05	-3.68 ± 1.58	-6.53 ± 1.49
ds_50_35_15	MLP	knn	**-0.05 ± 0.06**	-0.25 ± 0.13	-0.95 ± 0.23	-2.86 ± 0.98	-7.32 ± 1.64
ds_50_35_15	MLP	linreg-iter	**-0.03 ± 0.03**	-0.06 ± 0.08	-0.27 ± 0.25	-2.77 ± 0.96	-6.4 ± 1.22
ds_50_35_15	MLP	linreg-li	**-0.02 ± 0.03**	**-0.05 ± 0.06**	**-0.19 ± 0.19**	-2.62 ± 0.57	-6.77 ± 0.9
ds_50_35_15	MLP	mice	-0.66 ± 0.35	-1.38 ± 0.47	-1.57 ± 0.33	-2.27 ± 0.31	**-3.51 ± 0.62**
ds_50_35_15	MLP	mlp	**-0.05 ± 0.06**	**-0.07 ± 0.06**	-0.2 ± 0.12	-1.98 ± 0.6	-5.17 ± 0.92
ds_50_35_15	MLP	mlp-li	-0.06 ± 0.06	-0.1 ± 0.09	-0.26 ± 0.22	**-1.34 ± 0.47**	**-3.82 ± 0.78**
ds_50_35_15	MLP	xgb-iter	-0.7 ± 0.26	-1.75 ± 0.71	-3.41 ± 0.9	-5.63 ± 0.92	-8.16 ± 1.43
ds_50_35_15	MLP	xgb-li	-0.69 ± 0.39	-1.4 ± 0.54	-2.7 ± 0.87	-4.2 ± 0.83	-6.12 ± 1.3
ds_50_35_15	XGBT	knn	-0.09 ± 0.17	-0.89 ± 0.45	-2.27 ± 1.02	-5.02 ± 1.32	-9.9 ± 2.96
ds_50_35_15	XGBT	linreg-iter	**-0.05 ± 0.1**	-0.1 ± 0.15	-0.72 ± 0.46	-4.98 ± 1.44	-8.71 ± 2.36
ds_50_35_15	XGBT	linreg-li	**-0.02 ± 0.08**	**-0.08 ± 0.19**	-0.46 ± 0.39	-3.62 ± 1.32	-9.95 ± 1.76
ds_50_35_15	XGBT	mice	-0.26 ± 0.23	-0.69 ± 0.3	-0.95 ± 0.68	**-1.43 ± 0.47**	**-2.65 ± 0.88**
ds_50_35_15	XGBT	mlp	**-0.02 ± 0.09**	-0.14 ± 0.16	-0.57 ± 0.37	-3.19 ± 0.86	-6.78 ± 1.36
ds_50_35_15	XGBT	mlp-li	**-0.02 ± 0.16**	**-0.09 ± 0.16**	**-0.39 ± 0.27**	-2.41 ± 0.66	-5.04 ± 1.04
ds_50_35_15	XGBT	xgb-iter	-1.06 ± 0.77	-2.18 ± 0.96	-4.52 ± 1.66	-7.63 ± 1.81	-9.55 ± 2.45
ds_50_35_15	XGBT	xgb-li	-0.71 ± 0.63	-1.96 ± 0.71	-3.86 ± 1.12	-5.66 ± 1.35	-8.23 ± 2.23
ds_100_70_30	MLP	knn	-0.3 ± 0.16	-0.55 ± 0.19	-1.33 ± 0.2	-3.46 ± 0.9	-7.39 ± 1.21
ds_100_70_30	MLP	linreg-iter	-0.04 ± 0.04	-0.09 ± 0.08	-0.28 ± 0.23	-2.41 ± 0.58	-6.83 ± 1.16
ds_100_70_30	MLP	linreg-li	**-0.02 ± 0.04**	**-0.05 ± 0.06**	**-0.1 ± 0.1**	**-0.41 ± 0.14**	**-1.08 ± 0.3**
ds_100_70_30	MLP	mice	-0.97 ± 0.38	-1.04 ± 0.36	-2.19 ± 0.44	-3.13 ± 0.75	-3.88 ± 0.4
ds_100_70_30	MLP	mlp	**-0.09 ± 0.07**	**-0.16 ± 0.13**	**-0.43 ± 0.21**	-2.42 ± 0.52	-6.04 ± 0.86
ds_100_70_30	MLP	mlp-li	-0.19 ± 0.18	-0.31 ± 0.17	-0.53 ± 0.27	**-1.11 ± 0.3**	-2.58 ± 0.77
ds_100_70_30	MLP	xgb-iter	-0.88 ± 0.26	-2.28 ± 0.63	-4.79 ± 0.91	-7.86 ± 1.23	-11.64 ± 1.13
ds_100_70_30	MLP	xgb-li	-0.88 ± 0.17	-1.8 ± 0.51	-3.4 ± 0.73	-5.43 ± 0.89	-8.22 ± 0.73
ds_100_70_30	k-NN	knn	-0.27 ± 0.25	-0.47 ± 0.29	-1.43 ± 0.59	-2.94 ± 0.67	-4.05 ± 1.21
ds_100_70_30	k-NN	linreg-iter	**-0.03 ± 0.13**	**0.03 ± 0.2**	**-0.02 ± 0.41**	-1.52 ± 0.57	-4.11 ± 0.8
ds_100_70_30	k-NN	linreg-li	**-0.05 ± 0.16**	**0.04 ± 0.21**	**0.01 ± 0.24**	**-0.5 ± 0.45**	**-2.45 ± 0.76**
ds_100_70_30	k-NN	mice	-0.75 ± 0.47	-1.23 ± 0.54	-2.15 ± 0.47	-3.2 ± 0.89	-4.07 ± 0.74
ds_100_70_30	k-NN	mlp	**0.01 ± 0.21**	**0.01 ± 0.2**	-0.24 ± 0.46	-1.24 ± 0.64	-3.12 ± 0.95
ds_100_70_30	k-NN	mlp-li	**-0.03 ± 0.27**	**0.0 ± 0.32**	**-0.15 ± 0.3**	**-0.69 ± 0.5**	**-2.0 ± 0.58**
ds_100_70_30	k-NN	xgb-iter	-0.93 ± 0.67	-2.13 ± 0.93	-4.16 ± 1.11	-6.28 ± 1.68	-8.71 ± 1.28
ds_100_70_30	k-NN	xgb-li	-1.21 ± 0.56	-2.06 ± 0.7	-3.55 ± 1.27	-5.25 ± 0.89	-7.23 ± 1.29

forms the best. In most cases either the MICE or linear regression are the best methods.

XGBT and MLP performances are much more dataset dependent. However, their performance is usually not comparable to the best method and it also strongly depends on what classification model is used and how many features are missing. See e.g. MAGIC dataset where MLP is performing well for 30% of missing features and performing badly for 10% of missing features or the Spambase where a similar discrepancy holds for XGBT. Finally, the k-NN almost always performs worse than other methods. The only exception is the ds_20_14_6 dataset with random forest classification model.

Considering the amount of missing features it seems that results depend on the portion of missing features and not on the absolute number of missing features.

When we restrict ourselves to one missing feature reconstruction, the results are again highly dataset specific. For Cancer, Spambase, and ds_... datasets the accuracy after imputation actually increases. This is probably due to the fact that original classification models were overfitted and the proper imputation enables them to generalize better. On the other hand for the MAGIC dataset the performance decrease was around 1%–2%.

One can summarize that the best imputation methods were MICE, which performs well on real datasets and linear regression, which performs well also on artificial datasets. In some cases comparable results were reached by XGBT and MLP imputation. Again, only the k-NN imputation is not performing well enough.

If one analyzes all classification models (not just the two best), then classification models with higher accuracy perform worse with imputed datasets than less accurate models, as can be expected. The classification accuracy decreases only slightly while using imputed data in a model with low accuracy.

6   Conclusion

We focused on missing entire features reconstruction and its impact on the classification accuracy of an already learned model. We deal with traditional imputation methods: linear regression, k-NN, and MICE, as well as modern methods: MLP, and XGBT. We also proposed two methods, linear and information imputability, for the ordering of missing features when more of them are imputed sequentially. However, in practice information imputability is hard to estimate and does not provide satisfying results.

Comprehensive experiments are presented on four real and six artificial datasets. The imputation influence is studied on six commonly used binary classification models: random forest, logistic regression, k-NN, naive Bayes, MLP, and XGBT. The amount of missing data varies between 10% and 50%.

As our results indicate MICE and linear regression are generally good imputers regardless of the amount of missingness or the classification model used. This can be seen as some kind of generality when the used classification model is unknown.

As was also shown modern imputation methods MLP and XGBT did not perform as well as expected. They rarely perform among the top methods. Their performance is often one of the lowest. This result is surprising since in many current machine learning tasks these methods are one of the best.

The experimental results of this work shed more light on the applicability of state-of-the-art imputation methods on data and their ability to reconstruct missing entire features. The study is also important thanks to its scope of datasets, methods and portions of missing data (up to 50%).

Acknowledgements. This research was supported by SGS grant No. SGS17/210/OHK3/3T/18 and by GACR grant No. GA18-18080S.

References

1. Alcalá-Fdez, J., et al.: KEEL data-mining software tool: data set repository, integration of algorithms and experimental analysis framework. J. Mult.-Valued Log. Soft Comput. **17**, 255–287 (2011)
2. Anderson, T.W.: An Introduction to Multivariate Statistical Analysis. Wiley Series in Probability and Statistics, 3rd edn. Wiley, Hoboken (2003)
3. Arroyo, Á., Herrero, Á., Tricio, V., Corchado, E., Woźniak, M.: Neural models for imputation of missing ozone data in air-quality datasets. Complexity **2018**, 14 (2018)
4. Azur, M.J., Stuart, E.A., Frangakis, C., Leaf, P.J.: Multiple imputation by chained equations: what is it and how does it work? Int. J. Methods Psychiatr. Res. **20**(1), 40–49 (2011)
5. Baitharu, T.R., Pani, S.K.: Effect of missing values on data classification. J. Emerg. Trends Eng. Appl. Sci. (JETEAS) **4**(2), 311–316 (2013)
6. Chen, T., Guestrin, C.: XGBoost: a scalable tree boosting system. In: Proceedings of the 22nd ACM SIGKDD International Conference on Knowledge Discovery and Data Mining, KDD 2016, pp. 785–794. ACM, New York (2016)
7. Cover, T.M., Thomas, J.A.: Elements of Information Theory, 2nd edn. Wiley-Interscience, New York (2006)
8. Farhangfar, A., Kurgan, L.A., Dy, J.G.: Impact of imputation of missing values on classification error for discrete data. Pattern Recogn. **41**, 3692–3705 (2008)
9. Jonsson, P., Wohlin, C.: An evaluation of k-nearest neighbour imputation using Likert data. In: Proceedings of the 10th International Symposium on Software Metrics, pp. 108–118, September 2004
10. Jordanov, I., Petrov, N., Petrozziello, A.: Classifiers accuracy improvement based on missing data imputation. J. Artif. Intell. Soft Comput. Res. **8**(1), 31–48 (2018)
11. Junninen, H., Niska, H., Tuppurainen, K., Ruuskanen, J., Kolehmainen, M.: Methods for imputation of missing values in air quality data sets. Atmos. Environ. **38**(18), 2895–2907 (2004)
12. Kozachenko, L.F., Leonenko, N.N.: Sample estimate of the entropy of a random vector. Probl. Peredachi Inf. **23**, 9–16 (1987)
13. Lichman, M.: UCI machine learning repository (2013)
14. Little, R.J.A., Rubin, D.B.: Statistical Analysis with Missing Data, vol. 333. Wiley, Hoboken (2014)

15. Lombardi, D., Pant, S.: Nonparametric k-nearest-neighbor entropy estimator. Phys. Rev. E **93**, 013310 (2016)
16. Murray, J.S., et al.: Multiple imputation: a review of practical and theoretical findings. Stat. Sci. **33**(2), 142–159 (2018)
17. Pan, S.J., Yang, Q.: A survey on transfer learning. IEEE Trans. Knowl. Data Eng. **22**(10) (2010)
18. Pedregosa, F., et al.: Scikit-learn: machine learning in Python. J. Mach. Learn. Res. **12**, 2825–2830 (2011)
19. Rubin, D.B.: Multiple Imputation for Nonresponse in Surveys. Wiley, Hoboken (1987)
20. Salgado, C.M., Azevedo, C., Proença, H., Vieira, S.M.: Missing data. In: Secondary Analysis of Electronic Health Records, pp. 143–162. Springer, Cham (2016). https://doi.org/10.1007/978-3-319-43742-2_13
21. Schafer, J.L.: Analysis of Incomplete Multivariate Data. Chapman and Hall, London (1997)
22. Silva-Ramírez, E.-L., Pino-Mejías, R., López-Coello, M.: Single imputation with multilayer perceptron and multiple imputation combining multilayer perceptron and k-nearest neighbours for monotone patterns. Appl. Soft Comput. **29**, 65–74 (2015)
23. Sricharan, K., Wei, D., Hero, A.O.: Ensemble estimators for multivariate entropy estimation. IEEE Trans. Inf. Theory **59**(7), 4374–4388 (2013)
24. Van Buuren, S.: Flexible Imputation of Missing Data. Chapman and Hall/CRC, Boca Raton (2018)
25. Zhang, Q., Rahman, A., D'este, C.: Impute vs. ignore: missing values for prediction. In: The 2013 International Joint Conference on Neural Networks (IJCNN), pp. 1–8, August 2013
26. Zhu, M., Cheng, X.: Iterative KNN imputation based on GRA for missing values in TPLMS. In: 2015 4th International Conference on Computer Science and Network Technology (ICCSNT), vol. 1, pp. 94–99. IEEE (2015)

A Deep Malware Detection Method Based on General-Purpose Register Features

Fang Li[1,2], Chao Yan[1,2], Ziyuan Zhu[1,2(✉)], and Dan Meng[1]

[1] Institute of Information Engineering, Chinese Academy of Sciences,
Beijing 100093, China
{lifang,yanchao,zhuziyuan,mengdan}@iie.ac.cn
[2] School of Cyber Security, University of Chinese Academy of Sciences,
Beijing 100049, China

Abstract. Based on low-level features at micro-architecture level, the existing detection methods usually need a long sample length to detect malicious behaviours and can hardly identify non-signature malware, which will inevitably affect the detection efficiency and effectiveness. To solve the above problems, we propose to use the General-Purpose Registers (GPRs) as our features and design a novel deep learning model for malware detection. Specifically, each register has specific functions and changes of its content contain the action information which can be used to detect illegal behaviours. Also, we design a deep detection model, which can jointly fuse spatial and temporal correlations of GPRs for malware detection only requiring a short sample length. The proposed deep detection model can well learn discriminative characteristics from GPRs between normal and abnormal processes, and thus can also identify non-signature malware. Comprehensive experimental results show that our proposed method performs better than the state-of-art methods for malicious behaviours detection relying on low-level features.

Keywords: Malware detection · Malicious attack · Registers data · Neural networks · GPRs data

1 Introduction

The malicious executable file is a computer program with a destructive purpose. In recent years, malware increasingly becomes advanced and sophisticated. Meanwhile, a large number of malicious codes are opens-source and some obfuscation tools are also made freely available, which causes new variants of malware burst with an exponential growth. According to AV-Test [1], the total number of malware has reached more than 740 million. Moreover, a recent security industry analyzes three years' security data from 26 countries and reports that more than half of the attacks resulted in financial damages of more than US$500,000

© Springer Nature Switzerland AG 2019
J. M. F. Rodrigues et al. (Eds.): ICCS 2019, LNCS 11538, pp. 221–235, 2019.
https://doi.org/10.1007/978-3-030-22744-9_17

(including but not limited to: lost revenue, customers, opportunities, and out-of-pocket costs). However, nowadays, most of ordinary users defend against attacks using signature-based scanning tools which fail to detect Zero-day attack and are not able to detect complex emerging malware. Thus, defending the evolutional malware from enormous legitimate processes in real time is imperative for an security computer application environment.

Malware detection is usually conducted based on the high-level events (i.e., system call sequence, frequency of system call and string information). Recently, some researchers began to use low-level events at micro-architecture level for malware detection, and these low-level features have been proved more effective than high-level features to identify malicious actions [9,27]. One reason is that the detector employing low-level features can perceive tiny changes between normal and malicious processes, which is useful for improving the malware detection performance. The other is that using low-level features takes less processing time than using high-level features. For example, if we use high-level events as detection features, the detector cannot make a decision until the relevant low-level information is converted into the high-level one. [12,13,26] adopted performance counters as input features for malware detection, but these methods didn't achieve a high accuracy since they directly extracted low-level features from an isolate environment interfered with the normal processes. Actually, these micro-architecture features of normal processes are noise for malware detection [27]. Although [20,23,24] can identify malicious behaviours with a high accuracy, they all use temporal statistics as features without considering the interference with normal processes. In reality, interference of normal processes always exists. Also, the above methods all need a long sample length (25K or 10K instructions) and cannot detect the non-signature malicious software.

To address the above problems, we make full use of the spatial and temporal properties in low-level features for malicious behaviours detection. First, we collect the data in GPRs [3] as our low-level features. Each register has specific functions, and the changes of its content can contain action information. E.g., EAX is a scratch register, and most of the Win32 Application Programming Interface (API) functions return values in this register [17]. Thus, changes in GPRs can be used to distinguish legal behaviours from malicious ones. Also, in this paper, we design a deep learning model to extract the spatial and temporal correlations in GPRs for malware detection, which can effectively suppress the noise (normal processes) in input features. Finally, selecting GPRs as features are effective for identifying non-signature malware. Most of the non-signature malware employ obfuscation techniques to change their original malicious codes for evading defense tools. However, the obfuscated malware will retain the harmful functionality of its original code, which means that it cannot alter the final data in GPRs when acting as a malicious behaviour. Even though the malware employs register reassignment technique, it does not change the relationship between the eight GPRs. Experimental results show that our method can achieve better malware detection performance compared with other detection models using low-level features at micro-architecture level.

This paper involves the following *contributions*:

- We propose to use GPRs as features for malware detection. By jointly using the spatial and temporal properties in GPRs, our method can effectively detect malicious behaviours.
- We design a novel deep detection model (denoted as "FusionST"), which only needs a short length of GPRs (0.8K instructions, 12.5 times shorter compared than other methods) for accurate malware detection.
- By jointly fusing spatial and temporal correlations in GPRs, our model can suppress interference with normal processes, and can also learn discriminative characteristics between normal and abnormal behaviours for non-signature malware detection.

2   Related Works

There are many methods that used low-level features for malware detection. E.g., [11] pioneeringly showed that using opcode can detect malware. Using the relevance among opcodes, [25] then proposed a weighed opcode sequence frequency method to detect malicious behaviours, but it is hard to learn the true actions of a process relying on opcodes sequence. Hence, [14] further presented a control flow-based method to extract opcode behaviors. The above works based on low-level features are all static detection techniques in Operational System (OS) level, but some malicious files, such as obfuscation malware and Zero-day malware, can easily evade these techniques since they usually cannot be disassembled properly [19].

Afterwards, some researchers adopted information at micro-architecture level to dynamically identify malware. For example, [13] used performance counters to collect multi-dimensional performance counter statistics for classifying malware, but they should collect information every 25K instructions at least, which is time-consuming. Then, [26] applied a feature reduction technique to decrease the computational load. Both [13] and [26] achieve relatively low detection Accuracy, since their collected features contain a lot of noise resulting in bad detection performance with Machine Learning (ML) techniques. [17] can obtain above 95% detection accuracy, but it needs the whole action information of one sample (registers' values when important API is invoked before and after). Then, [24] and [23] proposed a hardware-supported malware detector using low-level features for malware detection. In their works, they find that neural networks can obtain good malware classification results based on the frequency of opcodes with the largest difference. However, they need to collect 10K successive instructions for only one test. Besides, they do not indicate which type of the neural networks performs the best for detecting malicious behaviours.

The methods [13, 17, 23, 24, 26] do not make fully use of the spatial and temporal properties in low-level features. Also, they do not analyze the detection performance of their methods on unknown malware. Different from the previous works, our deep learning model, with only a short sample length, extracts the spatial and temporal properties of GPRs for malware detection. Our low-level

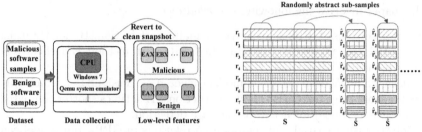

(a) Automatic data collection system   (b) Illustration of sub-sample generation

Fig. 1. Sub-sample generation using automatic data collection system. (a) shows how to collect the low-level features, and (b) illustrates the generation of sub-samples.

features are directly collected from the isolated environment based on Qemu developed in-house (shown in Sect. 3.1), which will inevitably introduce interference of normal processes to the collected data (close to the real environment). In other words, our method fetches malware features with a large amount of noise, which makes our detection task more difficult. Our designed deep learning model can effectively suppressing the noise by jointly fusing the spatial and temporal correlations in GPRs, and it can also identify non-signature malware.

3   Proposed Method

In this section, we introduce our deep malware detection model based on the spatial and temporal properties of low-level features (GPRs). Here, the spatial correlation means the relations of the eight GPRs. A malware will inevitably revises the data in GPRs, and the changes of the data may follow a particular way (i.e., correlations) which is different from that of the normal process. This difference is very helpful for malware detection.

We proceed in three parts: *low-level data collection, sub-sample generation* and *malware detection model.* The first part is used for collecting low-level features from CPU in an isolated environment. The *sub-sample generation* part introduces how to pre-process the GPRs features collected using the *low-level data collection* module, and the last part explains the architecture of the deep malware detection model.

3.1   Low-Level Data Collection

To train the deep malware detection model (FusionST) we create an automatic data collection system to collect features of low-level events (see Fig. 1(a)). Though some systems have been proposed for collecting behaviour information in recent years, they have some restrictions. For example, the techniques in [22] and [21] need to be manually set up and operated. Besides, most of them major in extracting opcode information. Here, we designed an automatic data collection system that can extract GPRs' content effectively.

We first establish a 32-bit Windows 7 operation system using virtual machine. Then, we use a system-wide emulator Qemu [10] to emulate a standard computer environment on the operation system. Here, Qemu is an open-source emulator based on dynamic binary translation and we can get the information we need by modifying the corresponding source code of Qemu. To support malicious software operations, we disable the firework and Windows Security Services on this Qemu and connected it to the network. In this isolated environment, we collect low-level information once an instruction is executed. For each executable file, we keep executing it two minutes to collect abundant behaviour information for each sample [16]. Note that collecting data of two minutes is only a necessary process for training a classification model. When performing online malware detection, we just need to run the pre-trained detection model using sub-samples with a very short length, which means the time of data collection can be ignored.

3.2 Sub-sample Generation

Based on the automatic data collection system, we can simultaneously extract multiple low-level features, such as opcode features, branch features and register features. [23, 24] proved the opcode features (existence of opcodes and frequency of opcodes with largest difference) are effective for malicious behaviour detection, but they only use the temporal statistics as input features. In this paper, we use the spatial and temporal properties of GPRs for malware detection. In the experimental part, we give comparisons of the malware detection performance using different types of low-level features.

In the preliminary experiment, we found that only using one single register can not obtain a good detection result. Hence, we utilize eight GPRs as features to train our deep malware detection model. In this section, we introduce how to transform the eight GPRs into numeric samples which can be directly input into the deep malware model for classification (see Fig. 2).

First, we combine the eight GPRs of one sample in a fixed order: EAX, EBX, ECX, EDX, EBP, ESP, ESI, EDI. As shown in Fig. 1(b), the combined data can be written as:

$$S = [r_1, r_2, \ldots, r_8], \tag{1}$$

where S is a two-dimensional matrix with height of 8, and it presents a combination of the above 8 GPRs information for one sample file. The notation r_i ($i \in \{1, \ldots, 8\}$) is a one-dimensional vector, representing one of the 8 GPRs. Since S contains two minutes' information of a malicious or normal behaviour, it is not viable to use the whole S for real time detection. Then, we thus generate sub-samples from S to train the deep malware detection model. Specifically, a sub-sample is generated by randomly cutting the combined S with a fixed length, and *we call this fixed length as **sample length** in this paper*:

$$\hat{S} = [\hat{r}_1, \hat{r}_2, \ldots, \hat{r}_8], \tag{2}$$

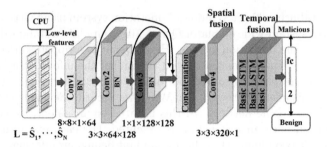

Fig. 2. Architecture of the detection model FusionST. The input of the model are the sub-samples collected from GPRs. The following CNNs and LSTMs fuse the spatial information and temporal information respectively for the final classification (output).

where $\hat{S}$ is a sub-sample from S, and $\hat{r}_i$ represents a randomly intercepted piece of one register with a certain sub-sample length. Therefore,

$$\mathbf{L} = \hat{S}_1, \cdots, \hat{S}_N, N \in R \tag{3}$$

represent all the training samples to train the deep learning model, where N indicates the number of sub-samples (see Fig. 2). Through our experiments, we find that the cut point does not affect the malware detection performance, which implicity proves the robustness of our method.

3.3   Malware Detection Model

The generated sub-sample has both spatial and temporal correlations. Based on this property, we then propose a novel deep neural network by jointly using convolutional neural networks (CNNs) and Long Short-Term Memorys (LSTMs), to spot illegal behaviours with only a short sample length.

The architecture of the proposed FusionST model is shown in Fig. 2. The input of the model is a series of sub-samples with a particular sample length. The first four CNNs abstract the spatial correlations among GPRs by fusing information from different layers, and then the following LSTM blocks encode the sequential information (temporal correlations) for the final malware detection. Here, to well learn the correlations among the eight GPRs, Conv1 adopts 8×8 filter size with 64 filters. The filters setup for Conv2 and Conv3 are $3 \times 3 \times 64 \times 128$ and $1 \times 1 \times 128 \times 128$, respectively. As shown in Fig. 2, Conv4 is a spatial fusion layer with one-dimensional convolution ($3 \times 3 \times 320 \times 1$) to incorporate information from different receptive fields, which can improve the effectiveness of the detection model. To keep the feature map from each layer in the same scale, each convolutional layer is accompanied by batch normalization (momentum of 0.99, epsilon of 0.001) followed by a rectified linear unit (ReLU) activation.

The three LSTM blocks all have 32 hidden units to extract the time feature for detecting malicious action. LSTM module employs gate mechanism to capture long-term timing dependencies, which can learn feature from time series data effectively. However, LSTM will expend a large amount of computation and hardly get effective weight update if the length of the time dimension sequence excess 0.2K [28]. In this paper, we select 0.1K as time sequence length for FusionST model.

The categorical cross entropy, together with the weight decay, is used as our loss function

$$\ell = -\frac{1}{B} \sum_{j=1}^{B} \sum_{k=1}^{K} w_k \cdot p_{jk} \log \hat{p}_{jk} + \lambda \|\theta\|_2^2, \tag{4}$$

where B is the number of batch size, K is the number of classes (2 classes for the detection task), $p_{.k}$ is the true probability of class k and $\hat{p}_{.k}$ is the predicted probability of class k. The weight w_k is used to balance class k in the training data set, and we set it to 1 for all the classes since our training data set is balanced. The weight decay term $\|\theta\|_2^2$ smoonths the parameters θ in network *FusionST* to prevent overfitting, and λ is the corresponding regularization weight [18].

4   Experiments

4.1   Experimental Setup

We implement the FusionST model using the Tensorflow [7] framework. The standard stochastic gradient descent with momentum is employed for training, where the initial learning rate, momentum and weight decay are set to 10^{-4}, 0.99 and 10^{-3}. The network converges after approximately 40 K iterations for training. All experiments are conducted in the environment: 64 Bit Ubuntu 14.04 on an Intel(R) Core (TM) i7-7800X Processor (3.50 GHz) with 16 GB of RAM. We run the networks on an Nvidia GeForce GTX 1080 Ti graphics card (GPU) and the Nvidia CUDA 8 software platform is used to accelerate the training process.

4.2   Dataset

VxHeaven [6] and VirusShare [4] are two popular datasets used for malware detection. In this paper, we randomly downloaded 1508 executables both from VxHeaven (60%) and VirusShare (40%) to make our malicious samples diverse. In contrast to [23,24], we have a larger malware dataset for evaluation (1508 malwares vs 1087 malwares). Besides, note that the executables downloaded from the VirusShare are newly released in 2017, which makes malware detection more challenging. For benign samples, they include a various of legal processes, e.g. native utilities and application executables in an operation system. These benign applications are download from the SourceForge [2] that is a popular free software source. All these malicious and benign samples are invoked by our automatic data collection system to extract the low-level features.

The collected samples are randomly divided into training set and test set as shown in Table 1. Since each program is executed for two minutes in the

Table 1. Malicious and benign dataset

Positive	# Training		# Test	
	Sample	Sub-sample	Sample	Sub-sample
Backdoor	376	50K	95	5K
Worm	187	50K	47	5K
Virus	209	50K	53	5K
Rootkit	278	50K	70	5K
Trojan	154	50K	39	5K
Total	1204	250K	304	25K
Negative	770	250K	131	25K

Table 2. Evaluation of sample lengths

Length	Accuracy	Precision	Recall	F1	FPr
0.1K	0.838	0.832	0.841	0.836	0.168
0.2K	0.888	0.883	0.890	0.886	0.117
0.4K	0.919	0.935	0.936	0.935	0.064
0.6K	0.949	0.956	0.957	0.956	0.043
0.8K	0.958	**0.962**	**0.964**	**0.963**	**0.038**
1.0K	**0.963**	0.955	0.958	0.956	0.045
1.2K	0.929	0.907	0.953	0.929	0.094
1.4K	0.922	0.961	0.964	0.962	0.064

automatic data collection environment, it will generate a large amount of data information. Hence, we generated sub-samples to train our malware detection model, which has been introduced in section: *Sub-sample Generation*.

4.3 Evaluation Metrics

In this paper, we set the labels of malicious sample and benign sample as positive and negative, respectively (see Table 1). The following are the four comprehensive metrics:

$$
\text{Accuracy} = \frac{\text{TP} + \text{TN}}{\text{TP} + \text{TN} + \text{FP} + \text{FN}},
$$
$$
\text{Precision} = \frac{\text{TP}}{\text{TP} + \text{FP}}, \text{Recall} = \frac{\text{TP}}{\text{TP} + \text{FN}} \tag{5}
$$
$$
\text{F1} = \frac{2 * \text{Precision} * \text{Recall}}{\text{Precision} + \text{Recall}}, \text{FPr} = \frac{\text{FP}}{\text{FP} + \text{TN}},
$$

where TP, TN, FP and FN represent True Positive, True Negative, False Positive and False Negative, respectively.

4.4 Experimental Results and Analysis

Evaluation of Sample Length. Table 2 shows the effect of sample length on the detection performance of the FusionST model using GPRs as features. Here, we train 8 detection models using FusionST with the different sample lengths: 0.1K, 0.2K, 0.4K, 0.6K, 0.8K, 0.1K, 1.2K, 1.4K. A short sample length leads to a low Accuracy which will increase significantly with sample length increasing. Since illegal behaviors are more likely to be missed with larger sample length, FP_r value will grow when the sample length become too long, which is consistent with the conclusion in [23, 24]. Thus, we finally choose sample length 0.8K (namely 0.8 K instructions) for malware detection to balance efficiency (sample length) and effectiveness (FP_r).

Evaluation of Low-Level Features. Table 3 shows the comparison results using three types of low-level features: *the existence of opcodes, the largest difference frequencies of opcodes* and *GPRs*. Here, we just compare our proposed

Table 3. Comparisons with other low-level malware detection methods

Methods	[23]		[24]		Proposed work
	Opcodes 10K	Opcodes 0.8K	Frequency of opcodes 10K	Frequency of opcodes 0.8K	Registers 0.8K
Accuracy	0.909	0.762	0.813	0.649	**0.953**
Precision	0.877	0.735	0.828	0.601	**0.933**
Recall	0.956	0.824	0.798	0.877	**0.976**
F1	0.914	0.777	0.813	0.713	**0.955**
FPr	0.139	0.299	0.197	0.399	**0.068**
Kappa	0.818	0.524	0.626	0.300	**0.906**
G-mean	0.907	0.763	0.813	0.609	**0.953**
BAC	0.908	0.763	0.813	0.651	**0.953**

features with opcode features proposed by [23,24], since their papers have proved that the opcode features can achieve better results than those of other various of low-level features, such as memory address distance, direction categories and branch categories. In order to ensure the fairness of the comparison, we extract these three features concurrently in the isolation environment. Furthermore, each of their samples are collected at the same moment. According to the suggestion in [23,24] (they obtain the best classification results when sample length is 10K), we train two models on opcode features using different sample lengths (10K and 0.8K), while our model are only trained on the registers' features of 0.8K instruction length. As shown in Table 3, though [23,24] can effectively identify malicious behaviours using opcode features in their experimental environment, it fails to obtain the same excellent results in our experiments. E.g., the Accuracy is only 0.813 with the largest difference frequencies of 10 K opcodes, while the Accuracy with the existence of opcodes is 0.909 with the 0.139 FPr when sample length is 10K. The reason may be that the opcode features contain noise when collected using our isolation environment. Our features with GPRs can achieve the Accuracy of 0.953 with the 0.068 FPr when the sample length is 0.8K. Since this experiment uses imbalanced data set, we also evaluate our method relying on Kappa, G-mean and BAC (Balanced Accuracy) Metrics [15]. Overall, our proposed GPRs perform better than opcodes features in terms of Accuracy, Precision, Recall, F1, FP_r, Kappa, G-mean and BAC.

Evaluation of Different Models. To evaluate the effectiveness of our proposed model, we compare FusionST with other popular classification models using the Receiver-Operating Characteristics (ROC) curves. As shown in Fig. 3, the detection algorithms can be generally categorized into Machine Learning (ML) ones and Neural Network (NN) ones. Here, the ML algorithms contain the basic linear classification algorithm: Support Vector Machine (SVM) (linear kernel) and the non-linear classification algorithms: K-NearestNeighbor (KNN),

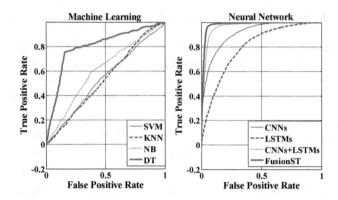

Fig. 3. Detection performance of different models.

Naive Bayes (NB) and Decision Tree (DT). From Fig. 3, the ML approaches can not differentiate malware from normal processes in the low-level space effectively. E.g., for ML algorithms, the highest detection Accuracy is 0.796 achieved by DB method while the linear SVM method hardly has the classification ability using GPRs as input features. The main reason may be that GPRs are micro-architecture level features, and they cannot directly reflect the action information before being converted into high dimensional spaces. Hence, the simple ML algorithms are difficult to detect malicious behaviors in this paper. For NN algorithms, we compare FusionST with its three variants: CNNs, LSTMs, CNNs+LSTMs. The CNNs model only has the first three convolutional layers of FusionST, while LSTMs only keeps the last three LSTM blocks of FusionST. The CNNs+LSTMs is also a combination of CNNs and LSTMs but removes the fusion layer Conv4 in FusionST (Fig. 2). As seen from the results, FusionST achieves the best detection performance, and CNNs+LSTMs is a little inferior. The reason may be that the CNNs+LSTMs model is a simple combination of CNNs and LSTMs without the fusion layer (Conv4 in Fig. 2) which is crucial to fuse multi-scale features from different layers. Meanwhile, we also find that the CNNs model performs better than LSTMs, which implies that CNNs working as filters can suppress noise in the raw GPRs. These above results also show that the combination of CNNs (spatial fusion) and LSTMs (temporal fusion) is necessary, since GPRs have both spatial and temporal correlations.

Table 4. Non-signature test sets

Non-signature set	# Positive set		# Negative set
	Download set	Generate set	
Sample	150	150	150
Sub-sample	5K	5K	5K

Table 5. Evaluation of detecting non-signature malware

Method			FusionST	AVware	ESET-NOD32	Kaspersky
Downloaded set	Accuracy		0.921	0.975	0.925	0.975
	Precision		0.881	1.00	1.00	1.00
	Recall		0.977	0.952	0.869	0.952
	F1		0.926	0.975	0.930	0.975
	FPr		0.132	0	0	0
Generated set	Accuracy		0.940	0.475	0.425	0.475
	Precision		0.917	0	0	0
	Recall		0.969	0	0	0
	F1		0.942	0	0	0
	FPr		0.087	0.513	0.541	0.513

Evaluation of Detecting Non-signature Malware. In this paper, a non-signature malware means that it does not belong to any malware families reported in Table 1. To test how the FusionST model performs when confronting with a completely new malware, we collected another two non-signature test sets for evaluation (see Table 4). The downloaded non-signature malware (positive samples) were newly released in 2018 by VirusShare. Also, we generated 150 unknown malware using Virus Maker Pack Ultimate Collection 2017 as another non-signature positive test set. This tool set contains multiple malware makers, such as DarkHorse, Poison and Necro. Then, we download another 150 benign samples from the SourceForge malware as the negative test set. Using the two new test sets, we compare the FusionST model with other three popular anti-virus softwares (AVware, ESET-NOD32 and Kaspersky) (see Table 5). From Table 5, we can find that three anti-virus softwares achieve good malware detection performance for the downloaded test set which has been signatured for these three anti-virus softwares. However, three anti-virus softwares hardly identify the generated test samples, since the generated samples are Zero-day malware for these three anti-virus softwares. We also test these generated executable files on the VirusTotal [5] website. Each unknown malware is detected using 65 latest anti-virus engines. The report results indicate that none of anti-virus engines believe these new generated files are malicious. However, these softwares are all malicious. The FusionST model obtains satisfactory detection performance on the two non-signature test sets. E.g., the FusionST can achieve Accuracy of 0.940 with 0.087 FPr on the generated test set. In conclusion, the FusionST model has a good adaptive ability for identifying non-signature malware.

4.5 Overhead Evaluation

In this paper, we use PCMark 7 [8] to measure overhead performance of our malware detection model. PCMark 7 is a performance testing tool developed by

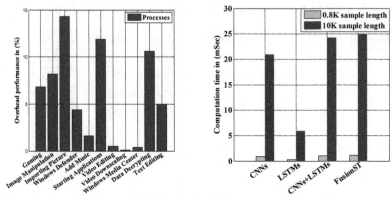

(a) Overhead performance on PCMark 7   (b) Average times of different models

Fig. 4. Overhead performance and running times. (a) shows the overhead performance of FusionST model on PCMark 7, and (b) shows the times taken by our designed four detection models (CNNs, LSTMs, CNNs+LSTMs and FusionST) to detect one sample.

Futuremark, a well-know global graphics and system test software development company.

To evaluate the overhead performance effectively, we first need to select the compared items from PCMark 7. We then iteratively test the performance of our computer five times with and without enabling our malware detection model. Finally, we compare the performance results of the two settings to compute the overhead performance of our detection model. Figure 4(a) shows the overhead performance using our model to spot malware in real time. Overall, our model has a low overhead. Though Gaming overhead is 14.3% and Starting Application overhead is 11.9%, which takes up a little more resources than with other online malware detection methods, the rest of the test items has lower overhead performance, i.e. Video Downloading overhead is only 0.5%. Besides, as long as the detection model can be implemented in hardware, nearly zero consumption can be achieved.

Evaluation of Detection Times. Figure 4(b) shows the test times taken by our designed four detection models (CNNs, LSTMs, CNNs+LSTMs and FusionST) when sample length is 0.8K and 10K. As observed, we can find that sample length can significantly affect the detection efficiency. E.g., for FusionST model, using sample length 10K consumes about 23 times longer time than using sample length 0.8K. In terms of effectiveness and efficiency, we choose the detection window as 0.8K in this paper. Overall, all the four models take less than 1ms to complete one test when the sample length is 0.8K, which is important for our detection model to monitor malicious behaviours in real time. As we mentioned before, we choose the FusionST model as our basic classification model in this work. Though FusionST model needs slightly longer test time compared with other models, it performs better in terms of FP_r (see Fig. 3).

5   Conclusions and Future Work

With micro-architecture level features, the existing methods usually use performance counters, existence of opcodes or frequency of opcodes with largest difference for malware detection. Although these methods can achieve satisfactory malware detection performance, they need collecting a long length of low-level features (about 10K instructions). Meanwhile, these detection methods are prone to be affected by noise from the low-level features and can hardly detect non-signature malware. In this paper, we use GPRs as features for malware detection. Experimental results show that only using a short length of GPRs (0.8K) can achieve high Accuracy for malware detection. Using CNNs and LSTMs, we also propose a deep detection model by jointly fusing spatial and temporal correlations of GPRs to improve the effectiveness and robustness of our detection method. Our deep detection model can well learn the correlations among GPRs for normal system processes, and thus can also identify non-signature malware due to their different characteristics from benign.

Currently, our method mainly focuses on binary classification, i.e., malware detection, and it can not identify the specific types of malwares. We will extend our work to a multi-class problem by incorporating more low-level and high-level features. Meanwhile, we also tend to apply our method to other architectures, such as $X86\_64$, ARM and MIPS.

Acknowledgement. This work was supported by the National Science and Technology Major Project of China (2018ZX01028201).

References

1. AV-TEST—The Independent IT-Security Institute. https://www.av-test.org/en/statistics/malware/. Accessed Mar 2018
2. The Complete Open-Source and Business Software Platform. https://SourceForge.net. Accessed Sept 2018
3. Software Developer Manuals for Intel 64 and IA-32 Architectures. https://www.intel.com/content/www/us/en/support/articles/000006715/processors.html?wapkw=developer/. Accessed Sept 2018
4. VirusShare. https://virusshare.com/. Accessed Sept 2018
5. VirusTotal. https://www.virustotal.com/en/. Accessed Sept 2018
6. VX Heaven. http://vxheaven.org/vl.php. Accessed Sept 2018
7. Abadi, M., et al.: TensorFlow: large-scale machine learning on heterogeneous distributed systems (2015)
8. Kumara, M.A.A., Jaidhar, C.D.: Automated multi-level malware detection system based on reconstructed semantic view of executables using machine learning techniques at VMM. Future Gener. Comput. Syst. **79**, 431–446 (2017)
9. Banin, S., Dyrkolbotn, G.O.: Multinomial malware classification via low-level features. Digit. Invest. **26**, S107–S117 (2018)

10. Bellard, F.: QEMU, a fast and portable dynamic translator. In: Conference on USENIX Technical Conference, p. 41 (2005)
11. Bilar, D.: Opcodes as predictor for malware. Int. J. Electron. Secur. Digit. Forensics **1**(2), 156–168 (2007)
12. Cronin, P., Yang, C.: Lowering the barrier to online malware detection through low frequency sampling of HPCs. In: IEEE International Symposium on Hardware Oriented Security and Trust, pp. 177–180 (2018)
13. Demme, J., et al.: On the feasibility of online malware detection with performance counters. In: International Symposium on Computer Architecture, pp. 559–570 (2013)
14. Ding, Y., Dai, W., Yan, S., Zhang, Y.: Control flow-based opcode behavior analysis for malware detection. Comput. Secur. **44**(2), 65–74 (2014)
15. Fernández, A., García, S., Galar, M., Prati, R.C., Krawczyk, B., Herrera, F.: Performance measures. In: Learning from Imbalanced Data Sets, pp. 47–61. Springer, Cham (2018). https://doi.org/10.1007/978-3-319-98074-4_3
16. Fredrikson, M., Jha, S., Christodorescu, M., Sailer, R., Yan, X.: Synthesizing near-optimal malware specifications from suspicious behaviors. In: Security and Privacy, pp. 45–60 (2010)
17. Ghiasi, M., Sami, A., Salehi, Z.: Dynamic malware detection using registers values set analysis. In: International ISC Conference on Information Security and Cryptology, pp. 54–59 (2012)
18. Hoffer, E., Banner, R., Golan, I., Soudry, D.: Norm matters: efficient and accurate normalization schemes in deep networks (2018)
19. Islam, R., Tian, R., Versteeg, S., Versteeg, S.: Review: classification of malware based on integrated static and dynamic features. J. Netw. Comput. Appl. **36**(2), 646–656 (2013)
20. Khasawneh, K.N., Ozsoy, M., Donovick, C., Abu-Ghazaleh, N., Ponomarev, D.: Ensemble learning for low-level hardware-supported malware detection. In: Bos, H., Monrose, F., Blanc, G. (eds.) RAID 2015. LNCS, vol. 9404, pp. 3–25. Springer, Cham (2015). https://doi.org/10.1007/978-3-319-26362-5_1
21. Okane, P., Sezer, S., Mclaughlin, K., Im, E.G.: Malware detection: program run length against detection rate. IET Softw. **8**(8), 42–51 (2016)
22. O'Kane, P., Sezer, S., Mclaughlin, K., Im, E.G.: SVM training phase reduction using dataset feature filtering for malware detection. IEEE Trans. Inf. Forensics Secur. **8**(3), 500–509 (2013)
23. Ozsoy, M., Donovick, C., Gorelik, I., Abughazaleh, N., Ponomarev, D.: Malware-aware processors: a framework for efficient online malware detection. In: IEEE International Symposium on High Performance Computer Architecture, pp. 651–661 (2015)
24. Ozsoy, M., Khasawneh, K.N., Donovick, C., Gorelik, I., Abughazaleh, N., Ponomarev, D.V.: Hardware-based malware detection using low level architectural features. IEEE Trans. Comput. **65**(11), 3332–3344 (2016)
25. Santos, I., Penya, Y.K., Devesa, J., Bringas, P.G.: N-grams-based file signatures for malware detection. In: International Conference on Enterprise Information Systems, pp. 317–320 (2009)
26. Sayadi, H., Makrani, H.M., Randive, O., Sai Manoj, P.D., Rafatirad, S., Homayoun, H.: Customized machine learning-based hardware-assisted malware detection in embedded devices. In: 2018 17th IEEE International Conference on Trust, Security and Privacy in Computing and Communications, 12th IEEE International Conference on Big Data Science and Engineering (TrustCom/BigDataSE), pp. 1685–1688 (2018)

27. Tang, A., Sethumadhavan, S., Stolfo, S.J.: Unsupervised anomaly-based malware detection using hardware features. In: Stavrou, A., Bos, H., Portokalidis, G. (eds.) RAID 2014. LNCS, vol. 8688, pp. 109–129. Springer, Cham (2014). https://doi.org/10.1007/978-3-319-11379-1_6
28. Yan, J., Qi, Y., Rao, Q.: LSTM-based hierarchical denoising network for Android malware detection. Secur. Commun. Netw. **2018**, 1–18 (2018)

A Novel Distribution Analysis
for SMOTE Oversampling Method
in Handling Class Imbalance

Dina Elreedy[(✉)] and Amir F. Atiya

Faculty of Engineering, Cairo University, Giza, Egypt
dinaelreedy@email.wustl.edu, amir@alumni.caltech.edu

Abstract. Class Imbalance problems are often encountered in many applications. Such problems occur whenever a class is under-represented, has a few data points, compared to other classes. However, this minority class is usually a significant one. One approach for handling imbalance is to generate new minority class instances to balance the data distribution. The Synthetic Minority Oversampling TEchnique (SMOTE) is one of the dominant oversampling methods in the literature. SMOTE generates data using linear interpolation between minority class data point and one its K-nearest neighbors. In this paper, we present a theoretical and an experimental analysis of the SMOTE method. We explore the accuracy of how faithful SMOTE method emulates the underlying density. To our knowledge, this is the first mathematical analysis of the SMOTE method. Moreover, we study the impacts of the different factors on generation accuracy, such as the dimension of data, the number of examples, and the considered number of neighbors K on both artificial, and real datasets.

Keywords: Class imbalance · Minority class · Over-sampling ·
SMOTE

1 Introduction

Imbalanced learning is encountered when one of the classes is represented fewer than others. Datasets may be naturally unbalanced such as medical diagnosis [15] and fraud detection [2], or data collection process may be too expensive such as detection of system failures. Yang et al. [22] have declared imbalanced learning as one of the ten most challenging problems in data mining. Handling class imbalance is challenging since there is a trade-off between the overwhelming influence of the majority class patterns, and an overemphasis on just a few minority class patterns.

Standard classifiers are biased towards the majority class examples while sacrificing minority class accuracy since such classifiers aim to maximize the overall classification accuracy without considering class distributions. The three main approaches for handling data imbalance problem in literature are: cost sensitive approach, algorithm level approach, and data level approach.

© Springer Nature Switzerland AG 2019
J. M. F. Rodrigues et al. (Eds.): ICCS 2019, LNCS 11538, pp. 236–248, 2019.
https://doi.org/10.1007/978-3-030-22744-9_18

The cost sensitive approach uses cost matrices to set misclassification costs according to the importance of the class and degree of imbalance. Examples of work on the cost sensitive approach include AdaCost [7], and the work done by Chawla et al. [4].

The algorithm level approach adapts the classification algorithm's to handle the class imbalance problem. For example there is work on modifying the K nearest neighbor classifier (KNN) [23], other work on adapting decision trees [18], some approaches that modify support vector machines (SVM) [16], all these methods seeks to focus on minority class.

Finally, the data level approach is based on modifying the data distribution in order to balance minority and majority classes. Data level approach is the most popular approach for handling class imbalance, since it is a simple approach that can be applied independently of the classifier being used. Data level methods balance distributions by either removing some of the majority class data points (under-sampling), or adding more of minority class instances (over-sampling).

Under-sampling can be done randomly or using some heuristics such as: the condensed nearest neighbor rule [8] and one-sided selection [1]. However, under-sampling can be considered precarious since potential important information could be lost when removing majority class examples. On the other hand, over-sampling can be done by randomly replicating minority class patterns, or by generating new minority class patterns [1]. One of the most popular over-sampling methods is "Synthetic Minority Over-sampling Technique", or SMOTE [3]. SMOTE generates patterns from the minority class by performing a linear interpolation between a minority class pattern, and a randomly chosen one of its K-nearest neighbors. A detailed description of the SMOTE method is presented in Sect. 2.

Although there is much work in literature studying sampling methods handling class imbalance problem (see the reviews [15, 19], and [12], most of this work provides empirical analysis only, and there is little work, if any, that provides a theoretical analysis of data sampling methods.

One of the empirical studies is done by Luengo et al. [20]. In this work, the authors analyze the behavior of different sampling methods including: SMOTE, its extension, SMOTE-ENN, and an evolutionary under-sampling method EUSCHC [11], by measuring the degree of feature overlapping of the different classes, and class separability and its geometrical properties. However, these measures do not consider distributional issues of the generated data.

Another empirical analysis is performed in [6], the authors analyze different under-sampling, over-sampling methods and hybrid methods using both over-sampling and under-sampling for Alzheimers disease dataset. Their experimental analysis includes: random over-sampling, SMOTE, random under-sampling and K-Medoids under-sampling, a proposed clustering-based under-sampling method. Their results show that the subtle methods such as SMOTE and K-Medoids outperform random over-sampling and random under-sampling.

A lot of methods have extended SMOTE technique [3] due to its simplicity and performance. For example, two variations of Borderline SMOTE are presented in [13], Borderline-SMOTE1 and Borderline-SMOTE2. In these methods only the minority examples near the classification boundary are over-sampled, since the near-boundary examples tend to be more informative.

Another model is the so-called Adaptive Synthetic Sampling Approach for Imbalanced Learning (ADASYN) [14]. It uses a weighted distribution for different minority class examples according to their level of difficulty in learning, where more synthetic data are generated for minority class points that are harder to learn. Difficulty of a minority example learning is determined by the class composition of the K nearest neighbors.

A recently developed over-sampling method named Sampling WIth the Majority (SWIM) handles extreme class imbalance [21]. The authors of that paper utilize the distribution of majority class to generate synthetic minority class samples in new under-represented regions of minority class. Their proposed method (SWIM) achieves that by generating synthetic data at the same Mahalanbois distance from the majority class as the minority class sample.

Although the SMOTE generation mechanism is extensively used in literature [13,14,21], and [17], the SMOTE method has a major drawback that it is not grounded on a solid mathematical theory [3]. Consequently, in this work, we aim to provide a comprehensive analysis of the SMOTE method. Specifically, our goals are the following:

- Develop a mathematical analysis of SMOTE, and test the degree of its emulation to the underlying distribution (by checking its moments).
- Provide a detailed experimental study of SMOTE, exploring the factors that affect its accuracy (in mimicking the distribution).

The paper is organized as follows: Sect. 2 introduces SMOTE method stating its advantages and potential drawbacks. Section 3 presents a mathematical analysis to derive the distribution of the patterns generated by SMOTE. Then, the experimental analysis of SMOTE is presented in Sect. 4. Finally, Sect. 5 concludes the paper and presents potential future work.

2   SMOTE Method

The SMOTE over-sampling procedure consists of the following simple steps:

- For each pattern X_0 from the minority class do the following:
  - Pick one of its K nearest neighbors X (belonging to the minority class also).
  - Create a new pattern Z on a random point on the line segment connecting the pattern and the selected neighbor, as follows:

$$Z = X_0 + w(X - X_0) \tag{1}$$

where w is a uniform random variable in the range $[0, 1]$.

Figure 1 shows an example of patterns generated by SMOTE. In contrast, this figure shows extra patterns generated from the original distribution. It can be observed that the SMOTE generated patterns are more contracted than the patterns generated from the true distribution. This is because the SMOTE generation process by linear interpolation causes them to be inward-placed. In addition, SMOTE generated patterns are allocated only on the line segments connecting the K neighbors, creating an unrealistic graph shape, where edges are studded with data points and internal portions are void of them. This problem is accentuated even more in higher dimensions. Figure 1 shows how SMOTE generated patterns cluster around some paths, with some empty spaces around them. However, means of original distribution and SMOTE generated examples' distribution are very close as shown in Fig. 2.

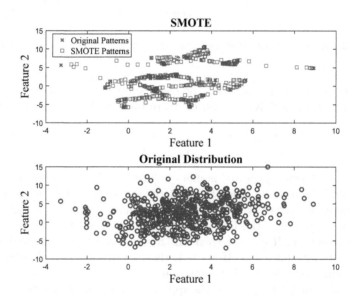

Fig. 1. SMOTE generated vs. original patterns

Another problem is that SMOTE could generate patterns in the decision regions of the majority class, this is more likely to occur in case of overlapping classes.

3   Theoretical Analysis of SMOTE

In this section, we present a theoretical analysis for SMOTE method in order to provide some mathematical basis. The success of SMOTE as a valid sampling algorithm hinges on its ability to generate patterns obeying a distribution close to the true one. We will investigate this issue here. Since the mean vector and the covariance matrix are the two major parameters characterizing any distribution,

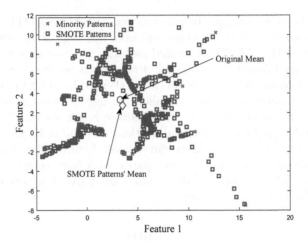

Fig. 2. Original distribution mean vs. SMOTE generated patterns' mean

we derive approximate formulas for the mean and the covariance matrix of patterns generated using SMOTE, and compare them with the true distribution's parameters.

Let $\Delta = X - X_0$, then:

$$Z = X_0 + w\Delta \tag{2}$$

where w is a uniformly generated number in $[0, w^*]$. When w^* equals to zero, distances would be zero since generated patterns are identical to original patterns. The parameter w^*, typically greater than or equal one, allows us to both extrapolate and interpolate on the line connecting the pattern X_0 and its randomly selected neighbor X. If $w^* = 1$ then this reverts back to the original SMOTE (applying only interpolation). If $w^* > 1$, then we can go beyond point X_0, i.e. we are allowing some level of extrapolation.

The basic idea for the analysis is approximating the probability density of minority class $p(X)$ using Taylor series around the point X_0 as proposed in [10]. The final approximation of the mean and covariance matrix of the generated pattern vector Z are given by Eqs. (3) and (4) respectively.

$$E[Z] \approx \mu_{X_0} + \frac{Cw^{*2}}{2} \int_{X_0} p(X_0)^{\frac{-2}{d}} \frac{\partial p(X_0)}{\partial X} \, dX_0 \tag{3}$$

where $[\frac{\partial p(X)}{\partial X}]^T = (\frac{\partial p(X)}{\partial x_1}, ..., \frac{\partial p(X)}{\partial x_d})$.

$$\Sigma_Z = \Sigma_{X_0} + \frac{Cw^{*2}}{3} \int_{X_0} p(X_0)^{1-\frac{2}{d}} \, dX_0 I$$

$$+ \frac{C^2 w^{*2}}{3} \int_{X_0} p(X_0)^{\frac{-2}{d}} \frac{\partial p(X_0)}{\partial X} \, dX_0 \int_{X_0} p(X_0)^{\frac{-2}{d}} \frac{\partial p(X_0)}{\partial X}^T \, dX_0$$

$$+ \frac{Cw^*}{2} \Big[\int_{X_0} p(X_0)^{-\frac{2}{d}} \frac{\partial p(X_0)}{\partial X} [(X_0 - \mu_{X_0})^T] \, dX_0$$

$$+ \int_{X_0} p(X_0)^{-\frac{2}{d}} (X_0 - \mu_{X_0}) \frac{\partial p(X_0)}{\partial X}^T \, dX_0 \Big] \quad (4)$$

where d is the dimension of the pattern vector, μ_{X_0} is the true mean vector of the minority class, Σ_{X_0} is the true covariance function, $p(X_0)$ is the class-conditionl density at point X_0, I is the identity matrix, and C is calculated as follows:

$$C = \frac{N! \Gamma \left(1 + \frac{2}{d}\right)^{\frac{2}{d}} \Gamma \left(K + \frac{2}{d} + 1\right)}{\pi K! (d+2) \Gamma \left(N + \frac{2}{d} + 1\right)} \quad (5)$$

If the true probability density is multivariate Gaussian, then the approximations can be simplified further to the following:

$$E[Z] \approx \mu_{X_0} \quad (6)$$

$$\Sigma_Z = \Sigma_{X_0} + \Big[(2\pi)^{\frac{1-d}{2}} \frac{Cw^{*2}}{3} \det^{\frac{1-d}{2d}} (\Sigma_{X_0}) \left(\frac{d}{2d-1}\right)^{\frac{d}{2}}$$

$$- 2\pi C w^* \det^{\frac{1}{d}} (\Sigma_{X_0}) \left(\frac{d}{d-2}\right)^{\frac{d+2}{2}} \Big] I \quad (7)$$

From Eq. (7), since the fraction $\frac{d}{d-2}$ is greater than one for any $d > 0$ and $d \neq 2$, hence the second term of the generated examples' covariance matrix Σ_Z would be negative and accordingly, the covariance matrix of SMOTE generated examples Σ_Z would be more contracted (since diagonal elements are smaller) than that of original minority class examples Σ_{X_0}.

From the above formulas one can observe the following:

- The mean vector of SMOTE-generated patterns is very close to the true one.
- The covariance matrix has some discrepancy. It is more contractive than the true one, because of the identity matrix times constant that is subtracted from the true covariance matrix (see Eq. 7). This agrees with the intuitive argument discussed last section, which argues that the SMOTE generation mechanism locates the patterns more inwards.

In order to measure how the covariance matrix of SMOTE-generated patterns diverges from the original covariance matrix, we define Total Variances Difference (TVD) measure. This measure helps us learn the amount and the polarity of the difference between synthetic and original covariance matrices. It is defined as the difference between the traces of the two covariance matrices. We normalize TVD by dividing by trace of original covariance matrix.

$$TVD = \frac{trace(\Sigma_Z) - trace(\Sigma_{X_0})}{trace(\Sigma_{X_0})} \qquad (8)$$

where the trace of the covariance matrix represents the summation of individual features' variances.

4   Experimental Analysis of SMOTE

4.1   Experiments

To have a more detailed understanding of the quality of SMOTE sampling, and its influencing factors, we set out a simulation study. In these experiments, we generate artificial datasets from multivariate Gaussian distributions, apply SMOTE over-sampling, then estimate the SMOTE-sampled examples' distribution, and compare it to the original distribution.

To have the analysis general enough, we consider 20 different distributions with different parameters. In all cases we consider the zero mean case, because the mean constitutes a shift in the center of operations, and will therefore be insignificant. However, we consider a variety of 20 different covariance matrices Σ_{X_0} varying between diagonal and off-diagonal ones. For the diagonal matrices, we sample the diagonal elements (eigenvalues) of the covariance matrix sampled from uniform distribution ranging from above zero to 40. Similarly, for the off-diagonal matrices, we first generate a diagonal matrix, named D, where its diagonal elements are randomly sampled. Then, we compute the covariance matrix Σ_{X_0} using the following equation:

$$\Sigma_{X_0} = RDR^T \qquad (9)$$

where R is an orthonormal matrix that is uniformly sampled.

We studied the effect of the same influencing parameters considered in the previous section, namely the number of original minority examples N, the dimension d, and the K parameter of the KNN. We have separately varied each of the influencing factors, while fixing the others, and in each case we documented the accuracy in the distribution of the generated points. While varying each parameter, the others are set at their "default values", which are as follows: $N = 100, \quad d = 10, \quad K = 5$. We have used over-sampling rate $R = 1$, the over-sampling rate can be defined as the amount of data points generated for each minority pattern.

Additionally, in these experiments, we have set $w^* = 1$ as used in standard SMOTE method [3] since we are interested in analyzing SMOTE method. However, for w^* can be set greater than 1, so that we can allow some extrapolation which could compensate the contraction of covariance matrix caused by SMOTE.

Algorithm 1. The experimental procedure for the SMOTE empirical analysis

> **repeat**
>> Randomly generate N patterns from the original distribution.
>> **repeat**
>>> Apply SMOTE to the generated N original patterns.
>>> Obtain the sample mean vector and covariance matrix of the SMOTE generated examples.
>> **until** L runs executed.
>> Average the sample mean vector and covariance matrix over the L inner runs.
> **until** M runs executed.
> Average mean and covariance estimates over the M outer runs to get final estimates of generated patterns' distribution.
> Compare the final estimated SMOTE generated patterns' distribution to the original distribution using the TVD distribution distance metric.

In order to estimate expectation and covariance of the SMOTE generated patterns, we apply the following procedure:

To measure how close the distribution of the SMOTE-generated patterns to the true distribution, we use the total variances difference (TVD) described and used last section. In our experiments, we set the outer number of runs M to 1000, and the inner number of runs L is set to 1000.

The following figures present the divergence of both empirical and theoretical estimates from the true distribution measured in terms of TVD metric described in last section. Figure 3 shows TVD when exploring the effect of the dimension d. As mentioned before, we fix all other factors at their default values, while varying the dimension. Similarly, Fig. 4 shows the TVD metric for the case of varying the number of minority samples N. Also, Fig. 5 show the TVD metric for the case of varying the K "number of neighbors".

It can be observed from the presented results that SMOTE behavior when varying different factors is similar in case of evaluating this behavior using our mathematical analysis and experimentally.

4.2   Experiments Using Real Data

In the other set of experiments we have applied a similar set-up as discussed on three real world UCI datasets. This provides a test for situations where the distribution is not necessarily Gaussian, and to justify that the derived conclusions apply to more complex situations, since real datasets could be noisy, and they could have sub-concepts for the minority class patterns.

We considered datasets that are originally large. This is in order to have an accurate estimate of the mean and covariance matrix. However, since SMOTE is used primarily for smaller datasets [3], we consider only a small subset (like 50 or 100) of the data, and perform the sampling using these. For example, assume that the dataset has about 10,000 points. We compute the mean and covariance matrix from the 10,000 points and assume these to be approximately the true

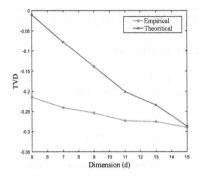

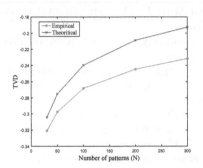

Fig. 3. TVD between empirical and theoretical estimates, and the true distribution versus dimension d

Fig. 4. TVD between empirical and Theoretical estimates, and the true distribution versus number of patterns N

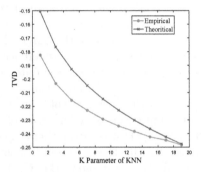

Fig. 5. TVD between empirical and theoretical estimates, and the true distribution versus K parameter of KNN in SMOTE

ones (due to the large number of points). Consider that we test the case of number of patterns $N = 100$. In such a situation we select 100 patterns randomly from the 10,000 original data points. We perform the SMOTE generation experiments on these 100 selected points. Then we repeat with a different selection of the $N = 100$ data points M times, thus implementing the outer loop of the simulation experiment along the lines discussed above for the artificial data sets.

Table 1 shows the sizes and the dimensions of the considered datasets. Adult and Default datasets are UCI datasets [9] and the third dataset, credit card, is a Kaggle dataset developed by [5]. Table 2 shows the empirical estimates of the total variance difference (TVD) metric for varying dimensionality d, where N_f indicates the total number of features for every dataset as indicated in Table 1. It can be observed from Table 2 that as dimensionality increases, the distribution distance in terms of the TVD metric is enlarged, which supports the theoretical, and empirical results on artificial data presented in Fig. 3.

In addition, Table 3 demonstrates the empirical estimates of the total variance difference (TVD) metric for varying number of patterns N. It could be noted from Table 3 that for the three considered datasets, increasing number of minority class patterns generates samples closer to the original distribution, which agrees with the theoretical, and empirical results on artificial datasets shown in Fig. 4.

Finally, Table 4 presents the empirical estimates of the total variance difference (TVD) metric for varying the K parameter of KNN in SMOTE. It can be observed that increasing K results in increasing the TVD, which means that the generated patterns incur more divergence away from the original distribution. These results agree with the theoretical and empirical results represented in Fig. 5. A further discussion on the impact of the K parameter of the KNN used in the SMOTE method is provided in Sect. 4.3.

For Tables 2, 3 and 4, only empirically estimated TVD values have been computed. The theoretical estimates as defined in Eq. (4) are hard to compute because the underlying density function $p(X_0)$ is unknown and probability densities are very hard to estimate with a reasonable error, especially for high dimensions, even in case of large data sets.

Table 1. Real world datasets description

Dataset	Number of minority class patterns	Dimension
Adult	22,654	14
Credit card	284,315	28
Client default	23,364	23

Table 2. TVD for SMOTE versus dimensionality d for the real world datasets

Dataset	$d = 3$	$d = 5$	$d = 10$	$d = N_f$
Adult	−0.111	−0.172	−0.234	−0.252
Credit card	−0.107	−0.162	−0.226	−0.275
Client default	−0.106	−0.165	−0.178	−0.227

4.3 Commentary on the Results

From the presented results, we can observe that different variables affect the accuracy in similar directions, whether based on the theoretical or the experimental results. This validates and makes these findings more general. In summary, we observe the following:

Table 3. TVD for SMOTE versus number of patterns N for the real world datasets

Dataset	$N = 50$	$N = 100$	$N = 200$	$N = 500$
Adult	−0.098	−0.064	−0.031	−0.01
Credit card	−0.3192	−0.298	−0.284	−0.265
Client default	−0.197	−0.141	−0.0950	−0.0824

Table 4. TVD for SMOTE versus K parameter of KNN in SMOTE for the real world datasets

Dataset	$K = 1$	$K = 5$	$K = 10$	$K = 20$
Adult	−0.012	−0.023	−0.036	−0.068
Credit card	−0.253	−0.275	−0.286	−0.298
Client default	−0.066	−0.099	−0.118	−0.156

- We find the TVD always negative, indicating the contractive nature of SMOTE method.
- The faithfulness of SMOTE-sampling in emulating the true density deteriorates with higher dimension d. As mentioned, whether generating from a density or estimating parameters, handling higher dimension becomes more challenging.
- The accuracy improves as the number of minority examples N is higher, exhibiting a steep decline as N becomes very small. The reason is that for higher N the K-nearest neighbor patterns become closer to each other. This has us dealing with a region of similar density function value. Going too far means going to regions of markedly different density values, and hence less "representative" generated patterns.
- The faithfulness improves with smaller K (of the KNN), becoming the best at having a single neighbor $K = 1$. But, as we mentioned, a drawback of very small K, such as $K = 1$ is that the generated examples will generally be very close to the original examples, making them highly correlated with the original examples, and lessening their contribution in improving classification performance and other estimation tasks. As a general guide, selecting K in the range of 4 to 6 seems to be a sensible choice. This would be a trade-off to avoid the high errors of large K, and the correlation issue for very small K.

5   Conclusion

In this paper, we provide a theoretical and experimental analysis of the Synthetic Minority over-sampling TEchnique (SMOTE) method. SMOTE is an effective over-sampling method that generates extra examples from the minority class in order to combat class imbalance. In this work, we investigate the distribution of the SMOTE generated patterns and analyze how it deviates from the true distribution. In addition, we study how the different factors, such as: dimensions, the number of minority patterns and the number of neighbors affect the divergence from the original distribution. We apply our experiments on both synthetic, and real datasets. The theoretical and the empirical results generally agree, and they should be a useful guide for using the SMOTE generation. As a disclaimer, this work considers only faithfulness in generating according to the true density. We do not consider how this affects classification, as this is out of scope of this work. However, an important first step in classification is to have accurate generation of patterns. A possible future work is to consider how this affects classification performance. Another possible direction to explore is to find methods or variants that would undo the contractive nature of SMOTE.

References

1. Batista, G., Prati, R., Monard, M.: A study of the behavior of several methods for balancing machine learning training data. ACM SIGKDD Explor. Newslett. **6**(1), 20–29 (2004)
2. Chan, P.K., Ave, L., York, N.: Distributed data mining in credit card fraud detection. IEEE Intell. Syst. Appl. **14**(6), 67–74 (1999)
3. Chawla, N.V., Bowyer, K.W., Hall, L.O., Kegelmeyer, W.P.: SMOTE: synthetic minority over-sampling technique. J. Artif. Intell. Res. **16**(1), 321–357 (2002)
4. Chawla, N.V., Cieslak, D.A., Hall, L.O., Joshi, A.: Automatically countering imbalance and its empirical relationship to cost. Data Min. Knowl. Discov. **17**(2), 225–252 (2008)
5. Dal Pozzolo, A., Caelen, O., Johnson, R.A., Bontempi, G.: Calibrating probability with undersampling for unbalanced classification. In: 2015 IEEE Symposium Series on Computational Intelligence, pp. 159–166. IEEE (2015)
6. Dubey, R., Zhou, J., Wang, Y., Thompson, P.M., Ye, J., Alzheimer's Disease Neuroimaging Initiative: Analysis of sampling techniques for imbalanced data: an n = 648 ADNI study. NeuroImage **87**, 220–241 (2014)
7. Fan, W., Stolfo, S.J., Zhang, J., Chan, P.K.: AdaCost: misclassification cost-sensitive boosting. ICML **99**, 97–105 (1999)
8. Fayed, H., Atiya, A.F.: A novel template reduction approach for the-nearest neighbor method. IEEE Trans. Neural Netw. **20**(5), 890–896 (2009)
9. Frank, A., Asuncion, A.: UCI machine learning repository, vol. 213. School of Information and Computer Science, University of california, Irvine (2010). http://archive.ics.uci.edu/ml
10. Fukunaga, K., Hostetler, L.: Optimization of k nearest neighbor density estimates. IEEE Trans. Inf. Theory **19**(3), 320–326 (1973)
11. García, S., Herrera, F.: Evolutionary undersampling for classification with imbalanced datasets: proposals and taxonomy. Evol. Comput. **17**(3), 275–306 (2009)
12. Haixiang, G., Yijing, L., Shang, J., Mingyun, G., Yuanyue, H., Bing, G.: Learning from class-imbalanced data: Review of methods and applications. Expert Syst. Appl. **73**, 220–239 (2016)
13. Han, H., Wang, W.-Y., Mao, B.-H.: Borderline-SMOTE: a new over-sampling method in imbalanced data sets learning. In: Huang, D.-S., Zhang, X.-P., Huang, G.-B. (eds.) ICIC 2005. LNCS, vol. 3644, pp. 878–887. Springer, Heidelberg (2005). https://doi.org/10.1007/11538059_91

14. He, H., Bai, Y., Garcia, E.A., Li, S.: ADASYN: adaptive synthetic sampling approach for imbalanced learning. In: IEEE International Joint Conference on Computational Intelligence, IJCNN 2008, pp. 1322–1328. IEEE (2008)

15. He, H., Garcia, E.: Learning from imbalanced data. IEEE Trans. Knowl. Data Eng. **21**(9), 1263–1284 (2009)

16. Imam, T., Ting, K.M., Kamruzzaman, J.: z-SVM: an SVM for improved classification of imbalanced data. In: Sattar, A., Kang, B. (eds.) AI 2006. LNCS (LNAI), vol. 4304, pp. 264–273. Springer, Heidelberg (2006). https://doi.org/10.1007/11941439_30

17. Jian, C., Gao, J., Ao, Y.: A new sampling method for classifying imbalanced data based on support vector machine ensemble. Neurocomputing **193**, 115–122 (2016)

18. Liu, W., Chawla, N.V.: A robust decision tree algorithm for imbalanced data sets. In: SDM, vol. 10, pp. 766–777. SIAM (2010)

19. Longadge, R., Dongre, S.: Class imbalance problem in data mining review. arXiv preprint arXiv:1305.1707 (2013)

20. Luengo, J., Fernández, A., García, S., Herrera, F.: Addressing data complexity for imbalanced data sets: analysis of smote-based oversampling and evolutionary undersampling. Soft Comput. **15**(10), 1909–1936 (2011)

21. Sharma, S., Bellinger, C., Krawczyk, B., Zaiane, O., Japkowicz, N.: Synthetic oversampling with the majority class: a new perspective on handling extreme imbalance. In: 2018 IEEE International Conference on Data Mining (ICDM), pp. 447–456. IEEE (2018)

22. Yang, Q., Wu, X.: 10 challenging problems in data mining research. Int. J. Inf. Technol. Decis. Making **5**(4), 597–604 (2006)

23. Zhang, X., Li, Y.: A positive-biased nearest neighbour algorithm for imbalanced classification. In: Pei, J., Tseng, V.S., Cao, L., Motoda, H., Xu, G. (eds.) PAKDD 2013. LNCS (LNAI), vol. 7819, pp. 293–304. Springer, Heidelberg (2013). https://doi.org/10.1007/978-3-642-37456-2_25

Forecasting Purchase Categories
by Transactional Data: A Comparative Study
of Classification Methods

Egor Shikov[(⊠)] and Klavdiya Bochenina

ITMO University, 49 Kronverkskiy prospect,
197101 Saint Petersburg, Russian Federation
shikovegor86@gmail.com, k.bochenina@gmail.com

Abstract. Forecasting purchase behavior of bank clients allows for development of new recommendation and personalization strategies and results in better Quality-of-Service and customer experience. In this study, we consider the problem of predicting purchase categories of a client for the next time period by the historical transactional data. We study the predictability of expenses for different Merchant Category Codes (MCCs) and compare the efficiency of different classes of machine learning models including boosting algorithms, long-short term memory networks and convolutional networks. The experimental study is performed on a massive dataset with debit card transactions for 5 years and about 1.2 M clients provided by our bank-partner. The results show that: (i) there is a set of MCC categories which are highly predictable (an exact number of categories varies with thresholds for minimal precision and recall), (ii) for most of the considered cases, convolutional neural networks perform better, and thus, may be recommended as basic choice for tackling similar problems.

Keywords: Purchase forecasting · Financial behavior · Neural networks · Machine learning · Transactional data · Machine learning

1 Introduction

Enterprise information systems incorporate different sources of information about actions of employees and clients which further can be used to create 360-degree customer view or to develop a variety of tools for predictive analytics of financial behavior. One of the problems often encountered for bank clients is expanding their debit cards payment experience, that is, increasing the number of different categories of expenses or the intensity of debit card usage. Being informed about expected future payments of a group of clients, a decision maker may provide adaptive and personalized suggestions for increasing the loyalty and improving customer debit card experience. For example, if one expects that customer A will spend amount X in category B in the next month, one may suggest to customer A to spend $1.1 \cdot X$ in category B and get a discount from a partner in this category, or to perform payment in similar category C and to get increased cashback for it.

© Springer Nature Switzerland AG 2019
J. M. F. Rodrigues et al. (Eds.): ICCS 2019, LNCS 11538, pp. 249–262, 2019.
https://doi.org/10.1007/978-3-030-22744-9_19

The transactional data for each customer are formed as a sequence of transactions with timestamp, amount and category of a transaction. Depending on a granularity of hierarchy of payment categories, one may tackle from dozens to hundreds of categories. That is, to predict payment profile for the next time period, we need to apply a binary classifier (or classifier which predicts a probability of having at least one purchase) for each of the categories. Also, one needs to mention that the frequencies of categories are highly imbalanced, and, according to the nature of the problem, different categories are of different basic predictability (e.g. almost all clients have at least one purchase in category 'food' each month, and expenses in 'petrol stations' and 'housing services' category seem to have less variance than in 'medical goods' or 'building materials').

The goals of this study are, given a massive dataset on debit card transactions: (i) to perform a comparative study of the efficiency of several modern machine learning approaches for prediction of the categories of expenses for the next time period, (ii) to study the predictability of different purchase categories and to provide the recommendations on categories which are more appropriate for planning the campaign for increasing customer involvement in debit card purchases (in terms of high precision and sufficient recall). Also, we test the consistency of predictions for different months (as patterns of purchases basically vary from month to month) and examine the quality of forecasts for the case when the last month of transactional history is unavailable for the model.

The rest of the paper is organized as follows. Section 2 presents related work on methods which were applied to solve this problem and similar problems earlier. Section 3 gives formal description of the problem, description of data processing workflow and details of implementation of different methods for our problem. Section 4 describes the dataset, the methodic of experimental study and the experimental results. Finally, Sect. 5 presents conclusion and discussion.

2 Related Work

Basically, the problem of forecasting purchase categories may be considered as multivariate time series prediction problem. Depending on the exact problem statement, these time series may be of numerical or categorical variables where each dimension is a single Merchant Category Code (MCC). In such a case, values for a single time unit comprise a vector of amount of purchases in different categories or binary flags marking the existence of at least one purchase for this MCC during given unit. Most of the categories are not presented in the payment sequences of users for a given time step, so basically these vectors are sparse. To make effective predictions, one needs to account not only the interactions between time steps, but also interactions between features. So, the difference between the models (classes of hypotheses) is determined with how these interactions are tackled.

Factorization machines (FMs) use second-order features interactions and are able to infer latent features from a highly sparse dataset using matrix factorization techniques. This method is often used for business cases such as recommender systems. For example, Lee et al. [1] use FMs for next event prediction task in business processes (namely, loan activities of bank client). There are also variants of FM which incorporate time dependencies between the events such as Factorizing Personalized Markov Chains (FPMC) [2] and Feature-Space Separated Factorization Model (FSS-FM) [3]. Dependencies between features in FM are linear. To add non-linearity in higher-order feature interactions, Neural Factorization Machine (NFM) was proposed in [4]. NFM may be considered as a generalization of classical FM and shows comparable performance with deep learning models while having simpler, shallower structure. Authors of [5] argue that matrix factorization methods systematically oversmooth distribution of user-item pairs resulting in too high probabilities of unseen items for a given user. To balance between exploration and exploitation, they introduce mixture model with two components, estimated at population and individual levels, correspondingly. The results of the mixture model are compared at seven online and offline user-item datasets and show the advantage of mixture model in terms of log-likelihood of test data and Recall@k.

To add global sequential features to the model (not only between consecutive events but also for non-adjacent cases), different architectures of recurrent neural networks (RNNs) are used. Dynamic Recurrent bAsket Model (DREAM) [6] calculates hidden state of RNN as a function of previous hidden state and latent vector representation of user's basket at time t_i. To get these latent representations, authors use operations of max pooling and average pooling over set of items in the basket (where each item, in turn, is represented as vector). Thus, the model combines representation of current interests of a user with a memory about her interests from previous baskets. Experiments have shown that DREAM outperforms simple baseline models (first-order Markov chains, non-negative matrix factorization) as well as more sophisticated models as FPMC and hierarchical representation model (HRM) [7]. As state-of-the-art model to compare with, different boosting models (as Gradient Boosting Machines in [8]) are also used.

Sequences of transactions may contain repeated patterns similar to graphical primitives and shapes in images. Thus, the next idea for predicting financial behavior is to use convolutional neural networks (CNNs) to find and use these patterns from large arrays of transactional data. This approach was used, for example, for fraud detection in [9] with excellent results on precision and recall of classifier. In [6] an author reports the results on applying CNN for a data about monthly usage of bank products to predict future usage of products.

In our study, the goal is to test existing approaches to predicting user consumption behavior for a problem of forecasting purchase categories in a next time period. For our best knowledge, there were no attempts to perform systematical comparison of predictive ability of these methods for a considered problem. We choose for the evaluation methods from different classes described above, namely recurrent neural networks, boosting and convolutional neural networks, and test them on a real-world dataset of debit card transactions provided by our bank-partner.

3 Problem Description

The problem of forecasting purchase categories may be described as follows. There is a set of clients $U = \{u_i\}, i = 1, \ldots, N$ where N is a total number of clients. Each client is characterized with a tuple $<F_i, S_i>$ where F_i is a set of static features of client i (such as gender, age), and S_i is a sequence of debit card transactions of client i. This sequence is represented as $S_i = \{ <a_{ij}, c_{ij}, y_{ij}, m_{ij}> \}, l = 1, \ldots, N_i^S$, where N_i^S – a total number of transactions of i-th client, a_{ij} – an amount of j-th transaction of i-th client, $c_{ij} \in C$ – a category of l-th transaction of i-th client, C – a set of categories (we denote a cardinality of this set, that is, a number of categories, as M).

For each client, his or her transactions may be aggregated by periods of a given length d. If total period of transactions of client u_i is equal to T_i, then the aggregated purchase matrix (APM) is defined as matrix with $K_i = \lceil T_i/d \rceil$ vectors as rows:

$$P_i = \{ <n_{i1k}, v_{i1k}, \ldots, n_{ijk}, \ldots, n_{iMk}, v_{iMk}>, z_{ik} \}_{k=1}^{K}, \qquad (1)$$

where i is an index of client, k is an index of time period, $j = 1, \ldots, M$ – an index of category, n_{ijk} – a number of transactions in category l in k-th period of client i, v_{ijk} – a total amount of transactions in category l in k-th period of client i, z_{ik} – a label which marks a 'global' index of k-th period of i-th client. For example, if we aggregate by months, earliest transaction among all of the clients was marked as Jan, 1990, the latest transaction was marked as Jun, 1991, and u_i has transactions from October, 1990 to May 1991, then local index (k) of Jan, 1991 for client u_i will be equal to 4, and global index – to 13. Also, we denote time borders of transactions for set of clients as $B = [<y_b, m_b>, <y_e, m_e>]$ (in our example $y_b = 1990, m_b = Jan, y_e = 1991$, $m_e = Jun$.

Then, the problem of forecasting purchases in a given category is formulated as follows. Given a set of tuples $U = \{u_i = <F_i, P_i>\}, i = 1, \ldots, N$ with static attributes and aggregated purchase matrices for a set of N clients (estimated by period B) predict for a given period $z^* > \max z_{ik}$ for each client u_i and category c_j if there is at least one transaction of this category in the period. That is, our goal is to get a matrix of predictions with clients as rows and categories as columns where non-zero entry with indices i and j means that a client i will spend in category j. Variants of the problem are prediction of n (number of transactions) and v (amount of transactions). In this study, we use fixed $K = K_i$ for all customers in the data set under the assumption that length of B may be larger than K. This means that a dataset may be of any length, but we use for prediction only K last months of client's transactional history (for the training sliding window technique thus should be used).

4 Methods

This section contains short descriptions of methods and some implementation details for a problem from Sect. 3. The models built for classification and regression task shared the same architecture except the fact that the counts of transactions were used for classification and expenses were used for regression.

4.1 Baseline Method (Averaging)

As a baseline method to estimate the quality of prediction, we use averaging per each client and each category over a given history of transactions. That is, a probability of purchase in j-th category by i-th client is given by:

$$p_{ij} = \frac{1}{K_i} \cdot \sum_k \mathbb{I}(n_{ijk} > 0), k = 1, \ldots, K_i. \tag{2}$$

Expression (2) provides a frequency of expenses in a given category in terms of periods of aggregation (if time unit is month, then p_{ij} is a fraction of months with expenses in category j in the history of user i).

4.2 Recurrent Neural Networks

We trained a simple LSTM with sparse vectors of monthly numbers of transactions as inputs and the hidden state of size 128. The network was trained with BPTT (Backpropagation Through Time) with cross-entropy loss for classification and MSE loss for regression.

4.3 Convolutional Neural Networks

The input layer was constructed as a concatenation of vectors of expenses. We used a simple CNN with 2 Conv2D-layers, and a pooling layer. The final layer as well as losses were the same as in LSTM. MSLE (Mean Squared Logarithmic Error) loss was also tested to reduce the influence of outliers for regression problem.

4.4 Boosting

As an inputs for the algorithm, the following features were estimated:

- minimum, maximum, average values and standard deviation for 6 last months;
- minimum, maximum, average values and standard deviation for 3 last months;
- last month expenses;
- average, minimum, maximum expenses summed over all categories;
- target month, customer age, customer gender.

All these features were concatenated to form a 759-long feature vector.

XGboost with default settings was used, a separate model for every category was trained.

5 Data

We use the dataset provided by our bank-partner (one of the largest regional banks in Russia) with $N = 180\,000$ and $z_{max} = 68$ (that is, dataset covers 5 years and 8 months). The customers were chosen to support sufficient level of transactional activity (restrictions are: $N_i^S \geq 200$ and at least 2 distinct categories of transactions in last 48 months, $N_i^S \geq 6$ in last 6 months). M (a total number of categories) is equal to 83 (an initial number of categories was equal to 86, but we did not use category 'Financial services' (it is the most frequent among the others as it consists of cash withdrawal), and categories 'Associations and organizations', 'Funeral services' according to their extremely low frequency).

Categories significantly differ in relative expenses and frequencies. Figure 1 illustrates this difference for 10 most frequent purchase categories (abbreviations for the categories are SM for 'Supermarkets', MP for 'Mobile Phones', BR for 'Bars and Restaurants', GS for 'Gas Stations', CS for 'Clothes, Shoes and Accessories', MG for 'Medical Goods', PS for 'Personal Services', C for 'Cosmetics', HS for 'Household Stores', and SS for 'Special Stores'). Figure 1a shows average monthly expenses normalized in $(0, 1]$ interval, Fig. 1b shows the average frequency of purchases for different categories measured at per-month basis, and Fig. 1c shows percentage of transactions in different categories. As we will see later, it determines different

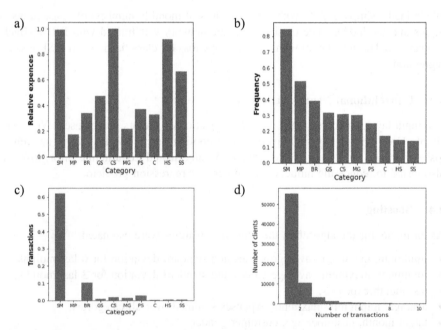

Fig. 1. Properties of expenses in a selected set of 10 most frequent categories: (a) normalized average monthly expenses, (b) frequency of payments (the average percent of months with purchase in a given category), (c) Distribution of parts of transactions in ten most popular categories, (d) Distribution of an average number of transactions of a client per month

predictive power of the algorithms for different categories. Clients also differ in the diversity of categories of expenses and frequency of payments (Fig. 1d). This suggests the existence of subpopulations with different types of purchase behavior.

As our data are time-dependent, for the test set we use all partial sequences S_i of the clients after certain timestamp z_{max}. Figure 2 illustrates the principles of formation of training and test sets for the data of a single client (columns of the table represent data for different k (months), rows of the table represent different ways of using the sequence). In this example, we know $K_i = 9$ months of transactional history of client i, and we want to train the model which predict categories using last 6 months of transactional data. Then, to form the training set, we may use sliding window for each six consecutive months of client's transactions under the condition that month to extract the label from lays within K_i months. One last thing to mention is that there can be introduced a lag between the end of the period used for prediction, and the period for making the forecast. That is, we may consider prediction for the next month or prediction for the month after the next month. It may be useful if there is delay in data collection after the end of the last month.

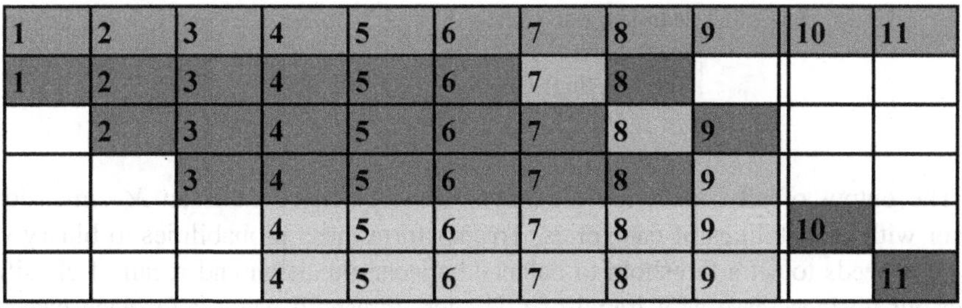

1	2	3	4	5	6	7	8	9	10	11
1	2	3	4	5	6	7	8			
	2	3	4	5	6	7	8	9		
		3	4	5	6	7	8	9		
			4	5	6	7	8	9	10	
			4	5	6	7	8	9		11

Fig. 2. Structure of train and test data for a single customer ($k = 1, \ldots, 9$ – train months, $k = 10, 11$ – test months). Blue – data for creating $[6 \times (M \cdot 2 + 1)]$ aggregated purchase matrix for training set, orange – data for extracting labels for training set (to predict month $z_{max} + 1$), dark green – data for extracting labels for training set (to predict month $z_{max} + 2$), yellow – data to create APM for evaluation of the model, red – data for extracting test labels. (Color figure online)

6 Results

The experiments were performed using a desktop PC with the following hardware configuration: Intel i7-7800X, 16 GB RAM, GeForce GTX 1060 6 GB. The dataset was divided at train (60 months) and test (11 months). We solve the problem of predicting purchase categories by last 6 months of transactional activity, so for a single client there can be several training examples for different time windows in the training set. We tried lags equal to zero and to one month (so we used both schemes from Fig. 2). As we have more than one month in our test period, all the metrics are averaged by a number of test months (to train, each time we use last year of transactional history before test month) (Table 1). Hyperparameters of classifiers were trained as it was explained in Sect. 5, and the resulting values are shown in Table 2. Further we refer different models as: Average, LSTM, CNN and Boosting.

Table 1. Parameters of different classifiers for categories prediction problem

Method	Parameters
LSTM	LSTM: hidden layer size = 128 Batch size = 64 Dropout rate = 0.2 Optimizer: Adam (learning_rate = 0.01)
CNN	Conv2D_1: nunits = 128, kernel = (2,16) ReLU MaxPooling: (2,2) Conv2D_2: nunits = 64, kernel = (2,2) ReLU Batch size = 64 Dropout rate = 0.2 Optimzer: Adam (learning_rate = 0.01)
Boosting	Nfeatures = 759 Ntrees = 100 learning_rate = 0.1 max_depth = 3 Subsampling: ON

The output of all considered classifiers for a given client is an M-dimensional vector with probabilities of categories. To transform these probabilities to binary values, one needs to set a threshold to balance between precision and recall of classifier. With our business case as a frame of reference, threshold was set for each category independently to support certain level of precision (80% or 90%). This approach was used because the results of prediction are aimed for launching campaigns for enhancement of customer debit card activities. These campaigns are planned for a restricted audience but require precise identification of target audience. So, thresholds are tuned using training set and are applied to make decisions for test set.

Figure 3 shows Precision-Recall curves for different classifiers for several categories. This curve shows achievable variants of tradeoffs between confidence in predicted values (measured by precision) and the amount of positive cases which will be captured by the model (measured by recall). The better curve is the closest to the right upper corner of the plot. We can see from Fig. 3 that for some categories (basically, the most frequent ones) curve achieves the plateau indicating both high precision and high recall (90% precision/60% recall for Gas stations, 90%/60% for Bars & Restaurants for the best classifier). For another categories, there is a region with sufficiently high value of precision up to some value of recall (80%/60% for Musical Instruments[1], 80%/40% for Medical Goods, 80%/45% for Municipal Services). For this set of categories, planning marketing campaigns based on predictions is still possible, as there is a highly

---

[1] This category mostly contains purchases in monthly paid media services like Apple Music.

predictable segment of the audience. If such algorithm operates over several hundreds of thousands of customers, then even 10% of recall may provide sufficient number of clients for setting the campaign. For the third set of categories, we cannot make a prediction with reasonable precision. Usually, these categories are less frequent.

Figure 3 also shows the difference in predictive ability of classifiers. Simple baseline in some cases outperformed more sophisticated algorithms for small values of recall. In such cases, there exists a small segment of customers with repeatable monthly behavior. Complex models like CNN seek for more sophisticated patterns in the data and then provide lower but more stable precision values.

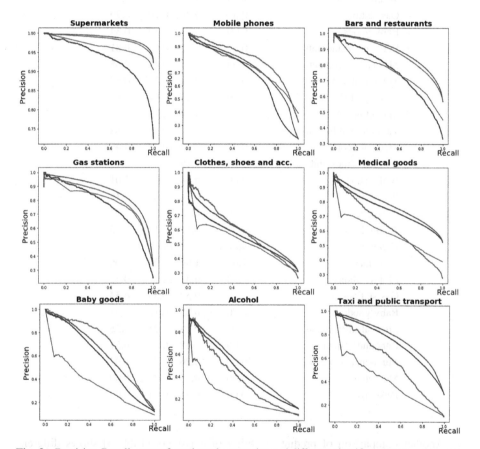

Fig. 3. Precision-Recall curves for selected categories and different classifiers (green – Average, red – CNN, blue – LSTM, magenta - Boosting) (Color figure online)

More systematic analysis of comparative precision is presented below. After setting the thresholds, we obtain values of precision and recall for a fixed precision threshold. Table 2 shows the results for different classifiers for 80% and 90% thresholds for 10 categories with highest average values of recall (over all classifiers) for 80% threshold.

For 80% threshold, CNN outperforms LSTM in all categories except three ('Musical instruments', 'Taxi and Public Transport', 'Medical Goods'). For 90% threshold, CNN outperforms LSTM in all categories.

Table 2. Comparison of recall values of classifiers for different categories with fixed precision

Category	Average	CNN	LSTM	Boosting
	80% precision threshold			
Supermarkets	**1.00**	**1.00**	**1.00**	**1.00**
Mobile phones	0.45	**0.65**	0.44	0.50
Gas stations	0.63	**0.86**	0.70	0.79
Baby goods	0.00	**0.51**	0.30	0.34
Bars and restaurants	0.41	0.76	0.67	**0.81**
Musical instruments	0.00	0.45	0.49	**0.66**
Municipal services	0.00	**0.44**	0.23	0.38
Hosting, TV	0.00	**0.26**	0.04	0.10
Marketing	0.00	**0.24**	0.02	0.05
Taxi and public transport	0.00	0.24	0.45	**0.52**
Medical goods	0.00	0.23	0.34	**0.46**
Clothes, shoes and accessories	0.00	**0.19**	0.01	0.10
Barbershop	0.00	**0.14**	0.01	0.01
Petshops	0.00	0.14	0.09	**0.15**
Cosmetics	0.00	**0.13**	0.01	0.08
Alcohol	0.00	0.11	0.12	**0.15**
Category	90% precision threshold			
Supermarkets	**1.00**	**1.00**	**1.00**	**1.00**
Mobile phones	0.00	**0.33**	0.15	0.22
Gas stations	0.00	0.52	0.38	**0.54**
Baby goods	0.00	**0.29**	0.18	0.21
Musical instruments	0.00	0.28	0.25	**0.35**
Bars and restaurants	0.00	0.28	0.52	**0.57**
Municipal services	0.00	**0.26**	0.05	0.04
Marketing	0.00	**0.15**	0.01	0.00
Petshops	0.00	**0.11**	0.01	0.05

Another comparison of predictive ability of classifiers (Table 3) shows different number of categories which are classified with precision larger than precision threshold and have recall larger than recall threshold. CNN support 1.5 times higher number of categories than LSTM for both considered values of thresholds.

For CNN classifier, we additionally investigated consistency of predictions for consequent test months and quality of predictions for instances of the problem with zero and unit lags. Figure 4 shows P-R curves for categories Baby goods, Gas stations, Restaurants and Supermarkets for several different test months. One can see that

Table 3. Number of categories which are classified with precision larger or equal to threshold and fixed recall (0.1)

	LSTM	CNN	Average	Boosting
80% precision	10	16	5	13
90% precision	6	9	1	6

although there are some fluctuations of quality for distinct months, the shape of P-R curve remains the same. Figure 5 shows P-R curves for six different categories for zero and unit lags. As expected, the absence of last month influences quality of predictions but the effect cannot be considered as drastic.

The algorithms may be used to solve the problem in a regression statement (to predict total sum of purchases in a given category for the next month). Table 4 provides results on MAPE values for different methods. One can see that CNN also provide the best percentages for all of the considered categories. The nearest competitor is twice as worse on average. Boosting algorithm does not provide any advantages compared to simply taking median value of purchases in that category. Average value gives worser percentage than median due to outliers in the sums of transactions. Finally, LSTM shows highest MAPE (almost 4 times higher than CNN). The values of MAPE for CNN vary from 77% for the pet shops to 235% for household stores with average value

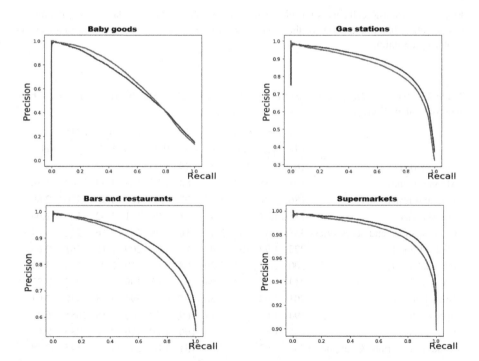

Fig. 4. Precision-Recall curves for different test months and different categories (blue – October 2018, red – November 2017) (Color figure online)

of MAPE equal to 116%. This indicates that the absolute values of MAPE are still too large to be used for planning marketing campaigns.

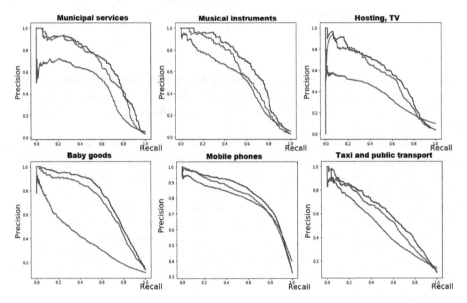

Fig. 5. Precision-Recall curves for different categories with zero and unit lags (prediction for the next month – blue, prediction for the month after the next – red; green – Average prediction) (Color figure online)

Table 4. Mean absolute percentage error for different methods ('Abs' field) and ratio to the best achieved value ('Rel' field).

	Average		Median		CNN		Boosting		RNN	
	Abs	Rel	Abs	Rel	Abs	Rel	Abs	Rel	Abs	Rel
Supermarkets	160	1.62	143	1.44	**99**	1.00	398	4.02	333	3.36
Mobile phones	108	1.35	99	1.24	**80**	1.00	107	1.34	149	1.86
Bars and restaurants	186	1.92	157	1.62	**97**	1.00	189	1.95	280	2.89
Gas stations	135	1.62	120	1.48	**81**	1.00	131	1.62	165	2.04
Clothes, shoes and accessories	333	1.35	287	2.47	**116**	1.00	251	2.16	213	1.84
Medical goods	223	1.92	193	1.99	**97**	1.00	209	2.15	261	2.69
Personal service	300	1.62	244	2.26	**108**	1.00	219	2.03	375	3.47
Cosmetics	214	1.35	192	2.04	**94**	1.00	199	2.12	213	2.27
Household stores	701	1.92	559	2.38	**235**	1.00	394	1.68	472	2.01
Special stores	594	1.62	471	2.79	**169**	1.00	369	2.18	1085	6.42
Baby goods	255	1.35	228	2.15	**106**	1.00	186	1.75	283	2.67
Building materials & supplies	708	1.92	549	3.64	**151**	1.00	257	1.70	345	2.28

(continued)

Table 4. (*continued*)

	Average		Median		CNN		Boosting		RNN	
	Abs	Rel	Abs	Rel	Abs	Rel	Abs	Rel	Abs	Rel
Hosting, TV	109	1.62	102	1.01	**101**	1.00	260	2.57	80	0.79
Sport goods	215	1.35	205	2.11	**97**	1.00	190	1.96	3103	31.99
Taxi and public transport	276	1.92	217	1.75	**124**	1.00	235	1.90	501	4.04
Alcohol	157	1.62	146	1.59	**92**	1.00	163	1.77	167	1.82
Pet shops	141	1.35	131	1.70	**77**	1.00	124	1.61	143	1.86
Municipal services	286	1.92	254	1.21	**210**	1.00	384	1.83	307	1.46
Bookstore	174	1.62	165	1.83	**90**	1.00	159	1.77	181	2.01
Medical centers	231	1.35	215	1.90	**113**	1.00	190	1.68	98	0.87
Mean	275	2.23	234	1.93	**116**	1.00	230	1.98	437	3.93

7 Conclusion

Information about consuming goods and services by a set of customers may further be used to develop different kind of personalization strategies. In this study, we consider extraction of the meaningful information to plan personalized marketing campaigns based on forecasting purchase categories for the next time period from large arrays of transactional data. For the case study, we use massive dataset provided by our industrial partner, one of the largest regional Russian banks.

To compare different machine learning algorithms, we state the problem of forecasting purchase categories as a set of binary classification problems. Data analysis shows high level of heterogeneity both in payment behavior of different clients and in different categories. In general, we observe that frequent categories are of significantly higher predictability.

We compare the results of classification and regression on a set of 83 MCC categories for recurrent neural networks, convolutional neural networks and boosting algorithm together with simple baselines as mean and median. The results of classification show that CNN outperform other competitors in terms of higher recall for a fixed precision in a majority of categories. Also, it allows to forecast larger number of categories with a minimum threshold on precision. Our study shows that there exists a set of categories which may be predicted with high accuracy and thus can be used for planning marketing campaigns. The results show their consistency on different months and for the case when data about last month before the test period is not available. As for regression problem, CNN outperforms the nearest competitor in two times. However, the resulting MAPE values are still high and may be used only as a benchmark. Further step here may be distinguishing the customer segments with more predictable expenses, stating the problem as multi-class classification and testing different classes of models.

Acknowledgements. This research is financially supported by The Russian Science Foundation, Agreement № 17-71-30029 with co-financing of Bank Saint Petersburg.

References

1. Lee, W.L.J., et al.: Predicting process behavior meets factorization machines. Expert Syst. Appl. **112**, 87–98 (2018)
2. Rendle, S., Freudenthaler, C., Schmidt-Thieme, L.: Factorizing personalized Markov chains for next-basket recommendation. In: Proceedings of the 19th International Conference on World Wide Web – WWW 2010, p. 811. ACM Press, New York (2010)
3. Cai, L., et al.: Integrating spatial and temporal contexts into a factorization model for POI recommendation. Int. J. Geogr. Inf. Sci. **32**(3), 524–546 (2018)
4. He, X., Chua, T.-S.: Neural factorization machines for sparse predictive analytics. In: Proceedings of the 40th International ACM SIGIR Conference on Research and Development in Information Retrieval - SIGIR 2017, pp. 355–364. ACM Press, New York (2017)
5. Kotzias, D., Lichman, M., Smyth, P.: Predicting consumption patterns with repeated and novel events. IEEE Trans. Knowl. Data Eng. **31**(2), 371–384 (2019)
6. Yu, F., et al.: A dynamic recurrent model for next basket recommendation. In: Proceedings of the 39th International ACM SIGIR conference on Research and Development in Information Retrieval - SIGIR 2016, pp. 729–732. ACM Press, New York (2016)
7. Wang, P., et al.: Learning hierarchical representation model for nextbasket recommendation. In: Proceedings of the 38th International ACM SIGIR Conference on Research and Development in Information Retrieval - SIGIR 2015, pp. 403–412. ACM Press, New York (2015)
8. Sheil, H., Rana, O., Reilly, R.: Predicting purchasing intent: automatic feature learning using recurrent neural networks. In: SIGIR 2018 eCom, p. 9 (2018)
9. Zhang, Z., et al.: A model based on convolutional neural network for online transaction fraud detection. Secur. Commun. Netw. **2018**, 1–9 (2018)

Recognizing Faults in Software Related Difficult Data

Michał Choraś[1,2]([✉]), Marek Pawlicki[1,2], and Rafał Kozik[1,2]

[1] ITTI Sp. z o.o., Poznań, Poland
mchoras@itti.com.pl
[2] UTP University of Science and Technology, Bydgoszcz, Poland
marek.pawlicki@utp.edu.pl

Abstract. In this paper we have investigated the use of numerous machine learning algorithms, with emphasis on multilayer artificial neural networks in the domain of software source code fault prediction. The main contribution lies in enhancing the data pre-processing step as the partial solution for handling software related difficult data. Before we put the data into an Artificial Neural Network, we are implementing PCA (Principal Component Analysis) and k-means clustering. The data-clustering step improves the quality of the whole dataset. Using the presented approach we were able to obtain 10% increase of accuracy of the fault detection. In order to ensure the most reliable results, we implement 10-fold cross-validation methodology during experiments. We have also evaluated a wide range of hyperparameter setups for the network, and compared the results to the state of the art, cost-sensitive approaches - Random Forest, AdaBoost, RepTrees and GBT.

Keywords: Pattern recognition · Faults detection · ANN ·
Data clustering

1 Introduction and Context

The development of a reliable software system, especially at a low cost, can be a significant challenge. The product also has to be market-ready in a reasonable time. Failure detection and defect proneness prediction become crucial tools for reliable software creation, helping with decision making and resource allocation. However, the analysis of software related data causes many problems and possible pitfalls due to intrinsic data difficulties. The aspects of data difficulties and motivation for this work are presented in details in Sect. 2.

To this point various metrics, such as code complexity, or number of revisions can help spot classes with high probability of bugs. Bug prediction, therefore, is a classification problem. Numerous classification methods have been employed to deal with this challenge, along with Artificial Neural Networks (ANN). While some researchers are reluctant to employ ANNs for their lack of transparency, however their prowess in modeling nonlinear functional relationships seem to make them well suited for the problem of defect prediction [1].

© Springer Nature Switzerland AG 2019
J. M. F. Rodrigues et al. (Eds.): ICCS 2019, LNCS 11538, pp. 263–272, 2019.
https://doi.org/10.1007/978-3-030-22744-9_20

Software quality is a fundamental competitive factor for the success of contemporary software houses. The Horizon 2020 Q-Rapdis project aims to augment the strategic decision-making procedures of software development by supplying strategic indicators of quality requirements. Fault-proneness is one of such metrics, as it can significantly affect the overall cost of the software. The Q-Rapids strategic indicators stem directly form the metrics and factors calculated from the software development-related data through the use of various data mining and machine learning procedures.

The paper is structured as follows: in Sect. 2 we discuss the problems and difficulties in analyzing the realistic software related data. In Sect. 5.1 the used benchmark dataset is described in details. Section 3 is devoted to Artificial Neural Network and the algorithms used in this work, while Sect. 4 addresses the problem of data imbalance. Results and the comparison with other standard machine learning approaches are presented in Sect. 5, while conclusions are given thereafter.

2 Problems and Difficulties in Real Software Related Data

In this paper we focus on pre-processing and recognizing (detecting faults/bugs) the software related data. But why is software data considered difficult anyways? There are many reasons and answers to such question, e.g. the following aspects contribute:

- Software related data from real SW companies/developers is considered sensitive commercial data. Commercial companies and SW houses are not eager to share SW related data, even if it is not directly the code. Most companies use software management and monitoring tools such as SonarQube, JIRA, GitLab, Jenkins and many others. Still the data retrieve from those tools provide information about processes, metrics, quality, testing aspects and much more, and all of those can reveal information about companies and teams as such. Moreover, such data might contain personal information (such as the names of programmers), therefore the privacy and GDPR aspects should also be taken into account [2].
- From machine learning perspective, software related data is often a one-shot learning. If you train any system/classifier on a data from one project or from one company, it is still not representing other projects and companies, so the training and adjusting algorithms have to be repeated all the time. Indeed, especially now in the era of RSD (Rapid Software Development) and agile/lean methodologies, it is difficult to observe long-term patterns in the way of working. Projects, developers and approaches change often, causing the sudden changes in the data as well, meaning that the trained models might not be relevant anymore. Therefore, lifelong learning approach to machine learning is beneficial and required.
- Software related data contains noise, and almost always the manual work on the data adaptation is needed. The good example is that, for instance each

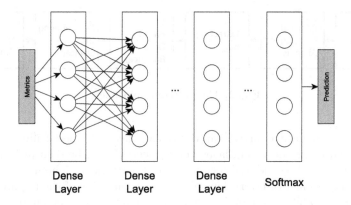

Fig. 1. A general architecture of the neural network adapted for the bug prediction (the depth of the network varies depending on the experiment).

team (even within the same company) might use different naming conventions for GitLab labels. The processes, e.g. of tickets cycles can also vary. Software data also suffer from the problem of data imbalance. The classes containing faults/bugs are under-represented, which causes the typical difficult data (d2) problem for machine learning techniques [3].

Therefore, the abovementioned problems motivate our research towards improving machine learning techniques for improving the quality of rapid software development.

3   Proposed Method

Artificial Neural Networks (ANN) constitute a functional instrument for creating machine learning models. They are a widely used tool for data mining, as they equip the user with classification, regression, clustering and time series analysis abilities. The assertion of an ANN is that it tries to imitate the learning capabilities of a biological neural network, abstractly simplified [4].

The impressive modeling capability of an ANN in fields relying on pattern recognition lies in direct proportion to its striking adaptability to data. It's extensive approximation capability is notably beneficial in handling real-world data, when there is plenty of data, but the patterns buried in the data are yet to be uncovered. Not only can the network figure out the interconnections among the variables, but it can generalize to a sufficient extent so as to provide satisfactory achievements on novel data [5]. An Artificial Neural Network is essentially like fitting a line, plane, or hyper-plane though a dataset, defining the relationships that might exist among the features [6].

A multilayer neural network is constructed with the use of multiple computational (hidden) layers. The data flows from the input layer to the following layer with adequate arithmetic along the way, and then is supplied to the following

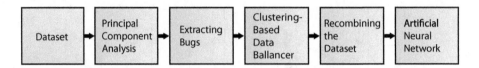

Fig. 2. The procedure pipeline.

layer and so on until it arrives at the output layer. A model illustrating the general architecture of a multilayer neural network can be seen in Fig. 1. This mechanism is dubbed the feed-forward neural network [7]. The number of neurons and the number of layers depends on the complexity of the required model and on the availability of data [5]. Using hidden layers with the number of nodes below the number of inputs creates a loss in representation, which frequently betters the network's performance. This might come as a result of eliminating the noise in data.

Designing a network with too many neurons can result in overfitting. Overfitting, or overtraining, means that the model fitted itself to extremally specific patterns of the training dataset, thus it will perform poorly on new, novel data, as it is not general enough [7].

The proposed method uses Principal Component Analysis for dimensionality reduction, singles out the bug instances in the dataset, clusters the 'clean' examples to the number of clusters that balances the number of bugs and re-merges the dataset to achieve a balanced dataset, which is then fed to the classifier. The pipeline is shown in Fig. 2.

4   Data Imbalance

A set is referred to as imbalanced when the classes are not represented in an equal manner [3]. What might initially seem like a negligible issue can cause machine learning algorithms to fail. An instance supplied in [8] explains a situation where a mammography dataset includes no more than 2% of abnormalities. In that case a classification of all the samples to the majority class would output an accuracy of 98%, strikingly missing the point of creating a machine learning algorithm to identify the minority class. Dataset imbalance exists in most of real-world research problems, so two resolutions to the challenge have been developed. Resampling - like subsampling the majority class and oversampling the minority class. Additionally, one could attach a specific cost function to the training samples [8]. Inspired by the emergence of the granular computing (GrC) paradigm, as it proved valuable in multiple scenarios [9] and after the successful application of the paradigm in [10] we are now investigating the feasibility of GrC for dataset balancing. An interesting approach using k-NN algorithm and rough sets can be observed in [11]. Our proposed method stems from the idea that one can include the characteristics of the dataset by clustering the majority class so the number of clusters matches the number of data samples in the minority class. While this method has it's drawbacks - namely the problem of the overlapping clear and bug granules, it still improves the accuracy by 10% as compared with a simple subsampling approach.

Table 1. Results of all datasets combined, using 10-fold cross validation

	Precision	Recall	F1-score	Support
False	0.61	0.57	0.59	80
True	0.64	0.67	0.66	91
Accuracy				0.6612

Table 2. Results of 4 datasets combined, new project classification

	Precision	Recall	F1-score	Support
False	0.91	0.42	0.58	1288
True	0.18	0.76	0.28	209
Accuracy				0.469

5   Results and Comparison with Classic ML Methods

5.1   Bug Prediciton Dataset

The dataset [12] which is utilised consists of an aggregation of class-level software development metrics. As mentioned in the accompanying paper, the main aim of the dataset is in providing a benchmark as an experimental field to test-run novel approaches. The set supplies characteristics derived from source code metrics in conjunction with historical and process information. A number of bugs and their impact is also supplied. Since the data is provided at a class level, the defect prediction can also be performed at the class level. The data could, however, be combined into package or subsystem level by summing class metrics.

The dataset contains metrics of 5 projects, these are: Eclipse JDT Core, Eclipse PDE UI, Equinox Framework, Lucene and Mylyn. Every project comes with a range of derived metrics, of which change log data in the form of comma separated files was used, for the use in this paper. The features were suggested by [13].

5.2   Results

In order to prove that the analysed data is difficult and to demonstrate that the proposed method allowed us to achieve superior results, we have compared various classical machine learning methods. In order to do that we have used a ROC

Table 3. Training on the 'Eclipse' project, classification on 'pde'

	Precision	Recall	F1-score	Support
False	0.89	0.90	0.90	1288
True	0.34	0.30	0.32	209
Accuracy				0.8196

curve (Fig. 3) to report the effectiveness in terms of the number of false positives (number of false alarms) and true positives (number of correctly predicted bugs). In these experiments we have considered the following classifiers:

- Random Forest
- AdaBoost
- Ensemble of RepTrees
- Gradient Boosted Trees

All of these methods have been wrapped with a metaclassifier that made the base classifier cost-sensitive. More precisely, the metaclassifier weights the training instances according to the total cost assigned to each class.

In this experiment, the Random Forest classifier is composed of 300 Random Trees that are combined together using the bagging technique. Each bag contains roughly 20% of data. During the training we control the depth of the trees. We set a hard limit to 10.

For the AdaBoost method we have used the classical approach. The ensemble is composed of one-level decision trees (decision stumps). We have noticed that increasing the ensemble size above 100 does not improve the quality.

The ensemble of RepTrees is build similarly to a Random Forest. However, instead of a Random Tree as a base classifier we have adapted a well-known Rep-Tree decision tree (Reduced Error Pruning Tree). This machine learning technique uses a pruned decision tree. First, the method generates multiple regression trees in each iteration. Afterwards, it chooses the best one. It uses regression tree adapting variance and information gain (by measuring the entropy). The algorithm prunes the tree using a back fitting method.

The GBT stands for Gradient Boosted Trees classifier. The method uses an additive learning approach. In each iteration a single tree is trained and is added to the ensemble in order to fix errors (optimise the objective function) introduced in the previous iteration. The objective function measures the loss and the complexity of the trees comprising the ensemble.

As it is shown in Fig. 3 and Table 6, we have achieved the best results for the Random Forest classifier. Although, the recall for this method is higher from that presented for our method in Table 1, the precision and f1-score remain far inferior.

When researching the ANN method multiple scenarios were evaluated all throughout the duration of the experiments. The best accuracy results along with their respective hyperparameter setups are found in Table 4, followed by the specific parameters of the comparison algorithms in Table 5. As mentioned earlier, the dataset [12] provided the metrics of 5 different coding projects. This situation differs from the one evaluated in [14] and in [15], where in order to fulfil the requirement stated by the software house's executives data from platforms like GITlab and SonarQube were proposed. One of the approaches evaluated how would the algorithm perform if it was trained on one project and tested on another. The detailed results can be seen in Table 3. A different scenario evaluated how the ANN trained on 4 of the projects would perform on a new project, as seen in Table 2. Finally a 10-fold cross validation of all the datasets combined resulted in the performance depicted in Table 1.

Table 4. Best hyperparameter setups found for the ANN's accuracy, a summary of multiple gridsearches.

4 hidden layers					
Epochs	Optimizer	Neurons	Batch_size	Activation	Accuracy
6000	adam	10	100	hard_sigmoid	0.647557
4500	rmsprop	7	500	hard_sigmoid	0.665147
4500	rmsprop	7	500	hard_sigmoid	0.665147
3 hidden layers					
Epochs	Optimizer	Neurons	Batch_size	Activation	Accuracy
4500	adam	8	500	hard_sigmoid	0.663844
4000	rmsprop	5	500	hard_sigmoid	0.663844
4000	adam	5	500	hard_sigmoid	0.663844
4000	rmsprop	10	500	hard_sigmoid	0.669055
4000	adam	10	500	hard_sigmoid	0.665798
5000	adam	10	500	hard_sigmoid	0.664495
5000	adam	5	1000	hard_sigmoid	0.663192
5000	rmsprop	10	1000	hard_sigmoid	0.665798
2 hidden layers					
Epochs	Optimizer	Neurons	Batch_size	Activation	Accuracy
10	rmsprop	5	5	sigmoid	0.644300
1 hidden layer					
Epochs	Optimizer	Neurons	Batch_size	Activation	Accuracy
4500	adam	8	500	hard_sigmoid	0.663844
4000	rmsprop	10	500	hard_sigmoid	0.669055
3000	rmsprop	10	500	hard_sigmoid	0.663844
3000	rmsprop	10	500	sigmoid	0.663844
2000	rmsprop	5	200	sigmoid	0.670358

Table 5. Comparison ML algorithms setups

Classifier name	Ensemble size	Meta-algorithm	Base classifier	Max depth
Random Forest	300	Bagging	Random Tree	10
AdaBoost	100	Boosting	Decision stump	
RepTrees	300	Bagging	Decision Tree	10
GBT	300	Boosting	Decision Tree	

Table 6. Comparison of methods on all datasets combined, using 10-fold cross validation

	Precision	Recall	F1-score
RandomForest	0.392	0.709	0.505
AdaBoost	0.294	0.765	0.425
Bag of RepTrees	0.308	0.814	0.447
GBT	0.300	0.761	0.431

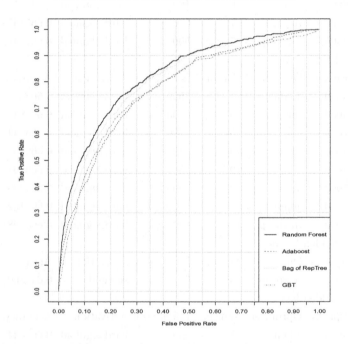

Fig. 3. ROC curve obtained for various classifiers: Random Forest, Adaboost, Bag of RepTrees, Gradient Boosted Trees (GBT).

6   Conclusions

In this paper we tackle the problem of the analysis of difficult software-related data. In general, such data can be analyzed in order to improve the software quality, detect faults and bugs or improve programming patterns. However, quite often the results are tampered by the nature of the data. Hereby, we propose to use machine learning techniques in order to detect bugs while addressing the problem of data imbalance. The presented results (Table 7) are comparable to other approaches, and we currently work to use them in practice on real commercial data from industrial software products.

Table 7. An example of ANN optimisation results - 3 hidden layers

Epochs	Optimizer	Neurons	Batch_size	Activation	ACC
2000	rmsprop	5	200	relu	0.582410
2000	adam	5	200	relu	0.595440
2000	SGD	5	200	relu	0.163518
2000	rmsprop	10	200	relu	0.578502
2000	adam	10	200	relu	0.593485
2000	SGD	10	200	relu	0.327036
2000	rmsprop	15	200	relu	0.553094

Acknowledgments. This work is funded under Q-Rapids project, which has received funding from the European Union's Horizon 2020 research and innovation programme under grant agreement No. 732253.

References

1. Lo, J.: The implementation of artificial neural networks applying to software reliability modeling. In: 2009 Chinese Control and Decision Conference, pp. 4349–4354, June 2009
2. Choraś, M., Kozik, R., Renk, R., Hołubowicz, W.: A practical framework and guidelines to enhance cyber security and privacy. In: International Joint Conference - CISIS 2015 and ICEUTE 2015, 8th International Conference on Computational Intelligence in Security for Information Systems/6th International Conference on EUropean Transnational Education, 15–17 June 2015, Burgos, Spain, pp. 485–495 (2015)
3. Kozik, R., Choraś, M.: Solution to data imbalance problem in application layer anomaly detection systems. In: Martínez-Álvarez, F., Troncoso, A., Quintián, H., Corchado, E. (eds.) HAIS 2016. LNCS (LNAI), vol. 9648, pp. 441–450. Springer, Cham (2016). https://doi.org/10.1007/978-3-319-32034-2_37
4. Maimon, O., Rokach, L.: Data Mining and Knowledge Discovery Handbook, 2nd edn. Springer, Boston (2010). https://doi.org/10.1007/978-0-387-09823-4

5. da Silva, I.N., Spatti, D.H., Flauzino, R.A., Liboni, L.H.B., dos Reis Alves, S.F.: Artificial Neural Networks. A Practical Course. Springer, Cham (2017). https://doi.org/10.1007/978-3-319-43162-8
6. Bassis, S., Esposito, A., Morabito, F.C., Pasero, E.: Advances in Neural Networks. Springer, Cham (2016). https://doi.org/10.1007/978-3-319-33747-0
7. Aggarwal, C.C.: Neural Networks and Deep Learning. A Textbook. Springer, Cham (2018). https://doi.org/10.1007/978-3-319-94463-0
8. Chawla, N.V., Bowyer, K.W., Hall, L.O., Kegelmeyer, W.P.: SMOTE synthetic minority over-sampling technique. J. Artif. Int. Res. **16**(1), 321–357 (2002)
9. Pawlicki, M., Choraś, M., Kozik, R.: Recent granular computing implementations and its feasibility in cybersecurity domain. In Proceedings of the 13th International Conference on Availability, Reliability and Security, ARES 2018, 27–30 August 2018, Hamburg, Germany, pp. 61:1–61:6 (2018)
10. Kozik, R., Pawlicki, M., Choraś, M., Pedrycz, W.: Practical employment of granular computing to complex application layer cyberattack detection. Complexity **2019**, 1–9 (2019)
11. Borowska, K., Stepaniuk, J.: Granular computing and parameters tuning in imbalanced data preprocessing. In: Saeed, K., Homenda, W. (eds.) CISIM 2018. LNCS, vol. 11127, pp. 233–245. Springer, Cham (2018). https://doi.org/10.1007/978-3-319-99954-8_20
12. D'Ambros, M., Lanza, M., Robbes, R.: An extensive comparison of bug prediction approaches. In: Proceedings of MSR 2010 (7th IEEE Working Conference on Mining Software Repositories), pp. 31–41. IEEE CS Press (2010)
13. Moser, R., Pedrycz, W., Succi, G.: A comparative analysis of the efficiency of change metrics and static code attributes for defect prediction. In: Proceedings of the 30th International Conference on Software Engineering, ICSE 2008, pp. 181–190. ACM, New York (2008)
14. Choraś, M., Kozik, R., Puchalski, D., Renk, R.: Increasing product owners' cognition and decision-making capabilities by data analysis approach. Cogn. Technol. Work **21**, 191–200 (2019)
15. Kozik, R., Choraś, M., Puchalski, D., Renk, R.: Q-rapids framework for advanced data analysis to improve rapid software development. J. Ambient Intell. Humaniz. Comput. **10**(5), 1927–1936 (2019)

Track of Computational Finance and Business Intelligence

Research on Knowledge Discovery in Database of Traffic Flow State Based on Attribute Reduction

Jia-lin Wang[1,2], Xiao-lu Li[1,2], Li Wang[1,2], Xi Zhang[3], Peng Zhang[3], and Guang-yu Zhu[1,2(✉)]

[1] MOE Key Laboratory for Transportation Complex Systems Theory and Technology, Beijing Jiaotong University, Beijing, China
gyzhu@bjtu.edu.cn
[2] Center of Cooperative Innovation for Beijing Metropolitan Transportation, Beijing, China
[3] Municipality Key Laboratory of Urban Traffic Operation Simulation and Decision Support, Beijing Transportation Research Center, Beijing, China

Abstract. Recognizing and diagnosing the state of traffic flow is an important research area, which is the basis of improving the level of traffic management and the quality of traffic information services. However, due to the increasing amount of traffic data collected, the traffic management system is facing the problem of "information surplus". After finishing several process, including data preprocessing, attribute reduction and rule acquisition, finally obtained the knowledge rules of the traffic flow's state. Using the method of knowledge discovery can reveal some hidden, unknown and valuable information from the huge amount of traffic flow information, so as to provide rules and decision-making basis for traffic management department.

Keywords: Rough set · Knowledge discovery in database ·
State of traffic flow · Attribute reduction · Rules acquisition

1 Introduction

Faced with more and more serious traffic jams, many cities in recent years have carried out targeted transformation and optimization of road network, and the overall traffic capacity of roads has been increasing. But because of the surge in the number of motor vehicles, these measures can not effective alleviate traffic congestion. The premise of making these decisions is to correctly judge the state of traffic flow. In order to get traffic flow data for judging traffic flow status, and make more effective decisions for mitigating congestion, traffic management departments have applied advanced information technology to urban traffic management. By means of advanced road detection and information processing technology, traffic management departments have been able to get a lot of road traffic data scientifically and effectively, and sort them out and store them. However, with the increasing traffic volume data collected, traffic management system is also facing the problem of "information overload", just like other information systems.

J. M. F. Rodrigues et al. (Eds.): ICCS 2019, LNCS 11538, pp. 275–285, 2019.
https://doi.org/10.1007/978-3-030-22744-9_21

Knowledge Discovery in Database (KDD) is an important research method for obtaining important rules from a large number of information. The research field of KDD is closely related to artificial intelligence and machine learning. It also contains the methods of database and pattern recognition, which has become a very important research hotspot in the academic field [1]. Rough set theory is one of the classic methods in KDD. Golan and Ziarko applied rough set theory to the analysis of stock market [2], and found the dependence between stock price and economic index. The Canadian scholar Ziarko applied rough set theory to the water resource scheduling system, predicted the rules according to the rules obtained, and achieved good prediction results [3]. In pattern recognition, Kim Dai-jin has obtained the key attributes in the handwritten character recognition system by using the attribute reduction method of rough set, and obtained the recognition rule based on the key attribute [4]. At present, there is no relevant research to apply rough set theory to traffic flow state discrimination.

In this paper, based on rough set theory, knowledge discovery technology is used to acquire valuable knowledge and rules from mass traffic data, so as to correctly judge traffic flow status and provide basis for decision-making of traffic management department. The rest contents of this paper are arranged as following.

A traffic flow data preprocessing algorithm based on filtering algorithm is designed in part 2.

A brief introduction of rough set theory is introduced in part 3.

A decision table attribute reduction algorithm for traffic flow state based on rough set theory is designed in part 4. On the basis of the attribute reduction algorithm of the decision table of the classical differential matrix, the differential matrix is improved.

A case is studied to validate the improved attribute reduction algorithm for differential matrix in part 5.

2 Data Preprocessing

The traffic flow characteristic data used in this paper is based on the coil detector. In the process of collecting traffic flow data by the coil detector, the damage of the coil, the failure of the communication equipment and the interference of the external environment will affect the accuracy and accuracy of the collected data. Therefore, the data collected by the coil detector can not be directly applied to this research, and the raw data need to be preprocessed. In view of the content of this paper, we should consider the microscopic randomness and macroscopic regularity of traffic flow data in preprocessing the data, and correct the fault data at the same time.

The sensors data record can be expressed as a 4 tuple structure $[t, q, v, h]$. Among them, t stands for time, q represents traffic volume, v stands for speed, h represents occupancy. Based on the above analysis, a traffic flow data preprocessing algorithm based on filtering algorithm is designed in this paper. The steps of the algorithm are as follows.

Step1: Define the maximum traffic volume as Q_{max}, the maximum speed as V_{max}, and the maximum occupancy as H_{max}. If $q > Q_{max}$ or $v > V_{max}$ or $h > H_{max}$, identify the data as abnormal data.

Step2: When $h > H_1(H_1 = 95\%)$, If $v > V_1(V_1 = 5\,km/h)$, identify the data as abnormal data.

Step3: If $v = 0(km/h)$ and $q \neq 0$, identify the data as abnormal data.

Step4: When $q = 0$, if $v \neq 0\&\&h \neq 0$, identify the data as abnormal data.

Step5: Calculate $AVEL$, if $AVEL \notin [1.5, 30]$, then remove it.

Step6: Replace the abnormal data with the nearest and normal data.

Step7: The method of first order differential operation is used to process the data.

If the maximum change interval of the data should be reasonable order difference scores did not fall in the mean and variance by n data difference value determined by the situation (such as $[\overline{d} - \varepsilon_o * \sigma_d, \overline{d} + \varepsilon_o * \sigma_d]$), we can define the data for data distortion abrupt change, with the former a reasonable data record with $1/\varepsilon_o$ times the difference value of $p_{n-1} + 1/\varepsilon_o * d_n$ alternative.

The specific flow of the algorithm is shown in Fig. 1 as shown.

3   Rough Set Theory

Rough set theory is a new mathematical tool for dealing with fuzzy and uncertain knowledge. Knowledge is the object and subject of research. It can be understood as the summing up and induction of knowledge and rules in a certain field. The main idea of rough set theory is to obtain the decision or classification rules of the problem by means of knowledge reduction under the premise that the ability to classify the knowledge is constant [5].

Knowledge reduction is the core of rough set theory. Some knowledge and attributes are particularly important for the purpose of decision making, while some knowledge and attributes are irrelevant or even redundant. The meaning of the concept of knowledge reduction is to find out and eliminate redundant or unimportant knowledge in knowledge base, and make the classification ability of the knowledge base unchanged.

In the definition of knowledge reduction, reduction and core are the two most important concepts. The core of knowledge is the intersection of all knowledge reduction, that is to say that the core of knowledge exists in the reduction of any knowledge, and is the most important part of the knowledge reduction. Because knowledge core is the expression part of knowledge feature, it will never be eliminated in knowledge reduction. Attribute reduction is an important part of knowledge reduction. Attribute reduction is to remove the uncorrelated or unimportant attributes under the condition of keeping the classification ability of knowledge base unchanged.

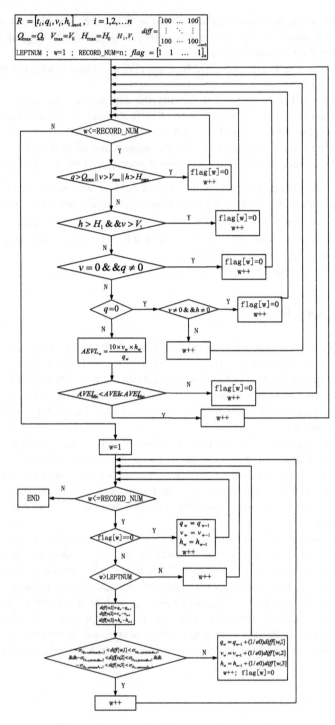

Fig. 1. Data preprocessing algorithm flow chart.

4 Attribute Reduction of Decision Table Based on Rough Set Theory

4.1 Attribute Reduction Theory of Decision Table

The decision table contains a large number of data samples, from which many decision rules can be extracted. All the decision rules are summed up, and all decision rules of the decision problem can be obtained. However, the basic decision sets obtained by this method are not practical, because many of them are not representative of the whole decision system. Therefore, it is necessary to reduce the attribute of the decision table to obtain the useful decision rules for the decision making, so that the problem of knowledge discovery can be solved in the relevant practical fields.

As the core content of rough set theory, attribute reduction based on decision table has become a branch of the rough set theory that has been focused on. For different decision systems, we can use the attribute reduction algorithm to adapt to its decision environment, so as to achieve the goal of effective decision rules. Different attribute reduction algorithms have the same goal, which is to find an optimal reduction, that is, the reduction set containing the least condition attributes.

The most commonly used attribute reduction algorithms are discernibility matrix method, attribute importance degree method, information entropy method, genetic algorithm and so on. The discernibility matrix method is intuitive and simple in dealing with general decision table problems, and its calculation process is simple and easy to understand, and its reduction result is very accurate.

4.2 Attribute Reduction Algorithm for Decision Table Based on Discernibility Matrix

In the existing algorithm of attribute reduction using rough set theory, the difference matrix method proposed by Skowron in 1992 is very classic and effective. The idea of the algorithm is that by defining a difference matrix, all the difference items of the condition attribute and the decision attribute are recorded, and then the matrix operation of the difference matrix is performed to find the kernel of the decision table and to output the set of all attribute reduction.

Give any incompatible decision table:

$DT = (U, C \cup D, V, f)$, $\forall_{xi,xj} \in U$ and $|U| = n$. We can define the improved discernibility matrix $M_{n \times n}(DT)$.

$$M_{n \times n}(DT) = \left(c_{ij}\right)_{n \times n} = \begin{bmatrix} c_{11} & c_{12} & \cdots & c_{1n} \\ c_{21} & c_{22} & \cdots & c_{2n} \\ \vdots & \vdots & \ddots & \vdots \\ c_{n1} & c_{n2} & \cdots & c_{nn} \end{bmatrix} = \begin{bmatrix} c_{11} & c_{12} & \cdots & c_{1n} \\ * & c_{22} & \cdots & c_{2n} \\ \vdots & \vdots & \ddots & \vdots \\ * & * & \cdots & c_{nn} \end{bmatrix} = \begin{bmatrix} c_{11} & * & \cdots & * \\ c_{21} & c_{22} & \cdots & * \\ \vdots & \vdots & \ddots & \vdots \\ c_{n1} & c_{n2} & \cdots & c_{nn} \end{bmatrix}$$

This includes

$$\begin{cases} c_{ij} = \{\alpha | (\alpha \in C) \wedge (f_a(x_i) \neq f_a(x_j))\}, f_D(x_i) \neq f_D(x_j) \\ \varnothing, \qquad\qquad f_D(x_i) \neq f_D(x_j) \wedge f_C(x_i) = f_C(x_j) \\ -, \qquad\qquad\qquad\qquad f_D(x_i) = f_D(x_j) \end{cases} \quad (1)$$

The algorithm of attribute reduction based on discernibility matrix for decision table is shown below.

Input: a decision table $DT = (U, C \cup D, V, f)$

Output: $RED_C(D)$ of DT

Step1: According to the specific definition of the difference matrix, the form of the lower triangulation matrix is used to generate the difference matrix $M_{n \times n}(DT)$.

Step2: Ergodic the difference matrix, searching all of the elements in detail. If $\varnothing$ does not exist, then jump to Step 3, and if $\varnothing$ exists, then quit.

Step3: Ergodic each attribute element in the difference matrix, assign the value to $CORE_C(D)$ and output $CORE_C(D) = \{\alpha | (\alpha \in C) \wedge (\exists c_{ij}, ((c_{ij} \in M_{n \times n}) \wedge (c_{ij} = \{\alpha\}))) \}$.

Step4: In all possible combinations of attributes, a combination of attributes containing relative kernel is solved and judged according to the following conditions: (1) $\forall c_{ij} \in M_{n \times n}(DT)$, if $c_{ij} \neq \varnothing$, $B \cap c_{ij} \neq \varnothing$? (ignore $c_{ij} \neq \varnothing \vee -$); (2) Is B independent?

If the two discriminant conditions are satisfied, Value to $RED_C(D)$ and ergodic a combination of attributes that contain all the relative kernel D.

Step5: Output $RED_C(D)$

The attribute reduction algorithm based on the differential matrix method above has been well applied to many practical problems. At the same time, it has a lot of advantages, especially for the relative D kernel and all the relative D reduction calculations very convenient and accurate. However, the classical algorithm based on differential matrix is only suitable for a complete and compatible decision table, and there is a problem of "nuclear explosion" caused by the large amount of calculation in the relative reduction of the decision table [6].

4.3   An Improved Attribute Reduction Algorithm for Decision Table Based on Discernibility Matrix

The traditional discernibility matrix is improved so that it can be applied to the inconsistent decision table. Give any incompatible decision table $DT = (U, C \cup D, V, f)$, $\forall x_i, x_j \in U$ and define $d(x_i) = card(\{z | z = f_D(y), \forall_y \in [x_i]c\})$. It describes the total number of all the different decision values of objects, and these decision values must be equivalent classes of the complete set of conditional attributes, so we can define the improved discernibility matrix $M_{n \times n}^*(DT)$.

$$M_{n\times n}^*(DT) = (r_{ij})_{n\times n} = \begin{bmatrix} r_{11} & r_{12} & \cdots & r_{1n} \\ r_{21} & r_{22} & \cdots & r_{2n} \\ \vdots & \vdots & \ddots & \vdots \\ r_{n1} & r_{n2} & \cdots & r_{nn} \end{bmatrix} = \begin{bmatrix} r_{11} & r_{12} & \cdots & r_{1n} \\ * & r_{22} & \cdots & r_{2n} \\ \vdots & \vdots & \ddots & \vdots \\ * & * & \cdots & r_{nn} \end{bmatrix} = \begin{bmatrix} r_{11} & * & \cdots & * \\ r_{21} & r_{22} & \cdots & * \\ \vdots & \vdots & \ddots & \vdots \\ r_{n1} & r_{n2} & \cdots & r_{nn} \end{bmatrix}$$

This includes

$$\begin{cases} c_{ij} = \{\alpha | (\alpha \in C) \wedge (f_a(x_i) \neq f_a(x_j))\}, & f_D(x_i) \neq f_D(x_j) \\ \varnothing, & f_D(x_i) \neq f_D(x_j) \wedge f_C(x_i) = f_C(x_j) \\ -, & f_D(x_i) = f_D(x_j) \end{cases}$$

$$\begin{cases} r_{ij} = c_{ij}, \min\{d(x_i), d(x_j)\} = 1 \\ \varnothing, \qquad \min\{d(x_i), d(x_j)\} > 1 \end{cases} \tag{2}$$

By adding the $\min\{d(x_i), d(x_j)\}$ condition, the traditional discernibility matrix is improved [7]. If a decision table is compatible, there will always be $\min\{d(x_i), d(x_j)\} = 1$, and the improved discernibility matrix will be degenerated into a Skowron discernibility matrix. If the decision table is incompatible, $\min\{d(x_i), d(x_j)\}$ may be more than 1. At this point, we must use the improved discernibility matrix to compute the result of attribute reduction in the correct decision table.

5 Case Study

5.1 Overview of Traffic Flow Status

Collecting traffic flow characteristic data is the first step to identify traffic flow status. Then, the traffic flow data are pre processed scientifically and effectively to analyze the relationship between traffic flow data. Traffic flow data include vehicle flow, average speed, lane occupancy, vehicle density and other parameters. The characteristics of traffic flow data can reflect the running state of traffic flow. For example, when the average speed of a section is mainly distributed between 40–60 km/h, it shows that the traffic flow condition of this section is relatively smooth [8]. Generally speaking, the traffic flow state of road sections or intersections can be divided into four modes: smooth, normal, mildly congested and congested [9].

The key step of traffic flow status recognition is to classify the traffic flow pattern, extract the characteristics of traffic flow, and identify the traffic flow status by a specific pattern recognition algorithm. There are many algorithms in the study of traffic flow status recognition, which is not the research direction of this paper. Therefore, this paper selects the traffic flow characteristic data set of known traffic flow state. This paper divides traffic flow status into three grades: smooth, mildly congested and congested.

5.2 Data Sources

The 48 h traffic section data collected by the coil detector in Beijing on May 4, 2012 (Friday) numbered 03218 is taken as the research data. There are 4 lanes in the section, Lan1, Lan2, Lan3 and Lan4 respectively. The record includes 1 sets of traffic basic data: vehicle flow, average speed of vehicles, occupancy of lanes and traffic volume of long cars. The sampling interval is 2 min.

5.3 Attribute Reduction of Traffic Flow State Decision Table

First, the data collected by the coil detector is preprocessed by using the method of part 2, and then the preprocessed traffic flow characteristic data set is preprocessed. A group of four minutes of data was extracted from the time of 6:00–10:00 in the morning. A total of 60 sets of data were extracted as the representative set of training samples, representing the form of a decision table. The 12 dimension characteristic data of Lan1, Lan2, Lan3 and Lan4 traffic volume, speed and occupancy (q_1, v_1, o_1, q_2, v_2, o_2, q_3, v_3, o_3, q_4, v_4, o_4) are taken as the conditional attributes of decision table. Among them, the flow rate of Lan1 is q_1, the speed is v_1, and the occupancy rate is o_1. Similarly, traffic flow characteristic data of Lan2, Lan3 and Lan4 can be seen. Taking the running state of the traffic flow on the basis of priori knowledge as the decision attribute of the decision table, a decision table for the running state of the traffic flow is finally constructed. Rough set method needs to deal with discretized data, but the collected traffic flow data is continuous. So first, we must simply discretize the traffic flow data and make all continuous conditional attribute values discretized. After analyzing, sorting and extracting the relevant data, the discretization interval of the specific condition attributes is set as follows.

Traffic Volume(pcu/4 min): 1/[0, 75], 2/(75, 100], 3/(100, 105], 4/(105, +∞);
Occupancy(%):1/[0, 0.2], 2/(0.2, 0.4], 3/(0.4, 0.5], 4/(0.5, 1];
Traffic states(M): 0/Smooth, 1/Mildly Congested, 2/Congested;

The conditional attribute values of all 60 sets of data are discretized into {1, 2, 3, 4}, and the decision attribute values are discretized into {0, 1, 2}. According to the discrete interval generated after discretization, the decision table of traffic flow state is constructed.

The attribute reduction algorithm of the decision table in the part 4 is realized by MATLAB programming, and the improved matrix is improved for the incompatible decision table, and the attribute reduction is completed in the decision table. After data preprocessing, data discretization, decision tables and decision tables, the final result of attribute reduction is obtained. There are 15 attribute reduction sets in the attribute reduction table, and the kernel of the decision table is {v_3}. It can be concluded from the calculation results that the number of characteristic attributes of the decision table is reduced from the initial 12 to 4. The attribute reduction can greatly reduce the computational complexity of the decision table by eliminating the redundant conditions, which provides the basis for the rule acquisition of traffic flow state.

5.4   Rule Acquisition

15 feature reduction sets are generated by arranging the attribute reduction set of traffic flow status decision table generated by MATLAB programming as shown in Table 1.

Table 1. Attribute reduction set of traffic flow state

	Reduction set
1	$\{q_1, v_1, v_3\}$
2	$\{v_1, q_2, v_3\}$
3	$\{v_1, q_3, v_3, v_4\}$
4	$\{v_1, v_3, o_3, v_4\}$
5	$\{v_1, q_3, v_4, o_4\}$
6	$\{q_1, v_2, v_3\}$
7	$\{q_2, v_2, v_3\}$
8	$\{v_2, q_3, v_3, v_4\}$
9	$\{v_2, v_3, o_3, v_4\}$
10	$\{v_3, q_4, v_4\}$
11	$\{v_2, v_3, v_4, o_4\}$
12	$\{q_1, o_1, v_3, q_4\}$
13	$\{o_1, q_2, v_3, q_4\}$
14	$\{q_1, q_3, v_3, q_4\}$
15	$\{q_2, q_3, v_3, q_4\}$

Each set of attribute reduction sets corresponds to a decision information table after reduction. For one of the reduction sets, the decision rules can be obtained by analyzing their corresponding decision information tables. A group of characteristic attribute reduction sets $\{q_1, v_1, v_3\}$ are used as an example to analyze the knowledge discovery process of the traffic flow status in this section. By deleting the feature attributes of traffic flow other than the reduction set, and merging all the same rows in the decision table, a traffic state decision table based on reduction set $\{q_1, v_1, v_3\}$ is finally generated as shown in Table 2.

Based on the above analysis, we know that knowledge discovery based on rough set theory can effectively extract all the smallest attribute reduction sets in decision tables, and we can extract corresponding traffic flow state decision rules based on deleted traffic attributes state decision tables. Based on Table 2, we can get 11 traffic flow status decision rules. Some of these rules are as follows.

Rule 1: If $\{q_1, v_1, v_3\} = \{1, 4, 4\}$, that is, Lan1's traffic volume is in the $[0, 75]$ range, the average speed of Lan1 is in the range of $(60, +\infty)$, and the average speed of Lan3 is in the range of $(60, +\infty)$. The traffic state is smooth.

Rule 2: If $\{q_1, v_1, v_3\} = \{2, 4, 3\}$, that is, Lan1's traffic volume is in the $(75, 100]$ range, the average speed of Lan1 is in the range of $(60, +\infty)$, and the average speed of Lan3 is in the range of $(45, 60]$. The traffic state is smooth.

Table 2. Decision table of deleting the redundant attributes of traffic flow state

U	q_1	v_1	v_3	M
x_1	1	4	4	0
x_2	2	4	3	0
x_3	2	3	3	0
x_4	2	3	2	0
x_5	3	2	2	0
x_6	2	2	2	1
x_7	2	2	1	1
x_8	2	1	1	1
x_9	1	1	1	1
x_{10}	2	2	4	0
x_{11}	2	3	4	0

Rule 3: If $\{q_1, v_1, v_3\} = \{2, 3, 3\}$, that is, Lan1's traffic volume is in the (75, 100] range, the average speed of Lan1 is in the range of (45, 60], and the average speed of Lan3 is in the range of (45, 60]. The traffic state is smooth.

Rule 4: If $\{q_1, v_1, v_3\} = \{2, 3, 2\}$, that is, Lan1's traffic volume is in the (75, 100] range, the average speed of Lan1 is in the range of (45, 60], and the average speed of Lan3 is in the range of (30, 45]. The traffic state is smooth.

Rule 5: If $\{q_1, v_1, v_3\} = \{3, 2, 2\}$, that is, Lan1's traffic volume is in the (100, 105] range, the average speed of Lan1 is in the range of (30, 45], and the average speed of Lan3 is in the range of (30, 45]. The traffic state is smooth.

Rule 6: If $\{q_1, v_1, v_3\} = \{2, 2, 2\}$, that is, Lan1's traffic volume is in the (75, 100] range, the average speed of Lan1 is in the range of (30, 45], and the average speed of Lan3 is in the range of (30, 45]. The traffic state is mildly congested.

Rule 7: If $\{q_1, v_1, v_3\} = \{2, 2, 1\}$, that is, Lan1's traffic volume is in the (75, 100] range, the average speed of Lan1 is in the range of (30, 45], and the average speed of Lan3 is in the range of [0, 30]. The traffic state is mildly congested.

Rule 8: If $\{q_1, v_1, v_3\} = \{2, 1, 1\}$, that is, Lan1's traffic volume is in the (75, 100] range, the average speed of Lan1 is in the range of [0, 30], and the average speed of Lan3 is in the range of [0, 30]. The traffic state is mildly congested.

Rule 9: If $\{q_1, v_1, v_3\} = \{1, 1, 1\}$, that is, Lan1's traffic volume is in the [0, 75] range, the average speed of Lan1 is in the range of [0, 30], and the average speed of Lan3 is in the range of [0, 30]. The traffic state is mildly congested.

Rule 10: If $\{q_1, v_1, v_3\} = \{2, 2, 4\}$, that is, Lan1's traffic volume is in the (75, 100] range, the average speed of Lan1 is in the range of (30, 45], and the average speed of Lan3 is in the range of (60, +∞). The traffic state is smooth.

Rule 11: If $\{q_1, v_1, v_3\} = \{2, 3, 4\}$, that is, Lan1's traffic volume is in the (75, 100] range, the average speed of Lan1 is in the range of (45, 60], and the average speed of Lan3 is in the range of (60, +∞). The traffic state is smooth.

As long as there is a set of traffic flow data at any time of this section of the road, the traffic states can be judged according to the above rules. Therefore, the recognition rule of traffic states can objectively describe the characteristics of traffic state, and provide a strong basis for traffic control guidance.

The above 11 rules are only the set of decision rules obtained from a set of traffic flow characteristic attributes reduction set $\{q_1, v_1, v_3\}$. In the same way, the other 14 groups of traffic flow characteristic reduction sets can also be used for rule acquisition. The same decision rules are merged to get the decision rule set of all traffic flow status of the section. Finally, the obtained decision rule set is stored in the corresponding rule knowledge base as the traffic flow state pattern classification knowledge of the section. In the future traffic states recognition, we only need to match the classification characteristics of the traffic flow data and the classification rules that are stored in the knowledge base, and we can identify the running the traffic states at this time. It can also predict and analyze the future traffic states based on the knowledge rule of traffic flow, so as to work out a plan for alleviating traffic jams.

6 Conclusion

This paper does not consider the influence of weather, road facilities, drivers' psychological state and other factors on traffic flow state. In the future study, these factors should also be taken into account in the identification and diagnosis of traffic state. However, in this paper, the attribute reduction of traffic flow state decision table is reduced to eliminate the redundant attributes, and the knowledge discovery of traffic flow state is realized. In the case analysis, we get the specific knowledge rules of the traffic flow status, and provide a scientific and effective decision-making basis for the traffic management department to formulate the correct traffic flow guidance measures.

References

1. Wang, L.: Knowledge Discovery of Uncertain Information Based on Rough Set and Its Application in Urban Traffic Management. Southwest Jiao Tong University, Chengdu (2011)
2. Jing, Y.G.: Research on Attribute Reduction Algorithm Based on Rough Set. Southwest Jiao Tong University Press, Chengdu (2013)
3. Sen, D., Pal, S.K.: Generalized rough sets, entropy, and image ambiguity measure. IEEE Trans. Syst. Man Cybern. **1**, 117–128 (2009)
4. Ziarko, W.: Variable precision rough set. J. Comput. Syst. Sci. **46**, 39–59 (1993)
5. Zhang, W.X., Wu, Z.W.: Rough Set Theory. Science Press, Beijing (2001)
6. Petrosino, A., Ferone, A.: Rough fuzzy set-based image compression. Math. Comput. Model. **160**(10), 1485–1506 (2009)
7. Ningler, M., Stockmanns, G., Schneider, G., et al.: Adapted variable precision rough set approach for EEG analysis. Artif. Intell. Med. **47**, 239–261 (2009)
8. Pu, S.L., Li, R.M., Shi, Q.X.: Research on traffic flow recognition algorithm based on rough set fuzzy recognition technology. J. Wuhan Univ. Technol. **34**(6), 1154–1158 (2010)
9. Yang, H., Bell, M.G.H.: Models and algorithms for road network design: a review and some new developments. Transp. Rev. **18**(3), 257–278 (1998)

Factor Integration Based on Neural Networks for Factor Investing

Zhichen Lu[1,2], Wen Long[1,2(✉)], Jiashuai Zhang[2,3], and Yingjie Tian[1,2,3]

[1] School of Economics and Management,
University of Chinese Academy of Sciences,
Beijing 100190, People's Republic of China
longwen@ucas.ac.cn
[2] Research Center on Fictitious Economy & Data Science,
Chinese Academy of Sciences, Beijing 100190, People's Republic of China
[3] School of Mathematical Sciences, University of Chinese Academy of Sciences,
Beijing 100190, People's Republic of China

Abstract. Factor investing is one kind of quantitative investing methodologies for portfolio construction based on factors. Factors with different style are extracted from multiple sources such as market data, fundamental information from financial statements, sentimental information from the Internet, etc. Numerous style factors are defined by Barra model proposed by Morgan Stanley Capital International(MSCI) to explain the return of a portfolio. Multiple factors are usually integrated linearly when being put to use, which ensures the stability of the process of integration and enhances the effectiveness of integrated factors. In this work, we integrate factors by machine learning and deep learning methodologies to explore deeper information among multiple style factors defined by MSCI Barra model. Multi-factors indexes are compiled using Smart Beta Index methodology proposed by MSCI. The results show non-linear integration by deep neural network can enhance the profitability and stability of the index compiled according to the integrated factor.

Keywords: Neural networks · Deep learning · Factor investing

1 Introduction

The definition of factors of factor investing originates from "Arbitrage pricing theory" proposed by Ross [10], which holds that the expected return of a financial asset can be modeled as a function of various macroeconomic factors or theoretical market indexes. And then researchers have tried to use specific factors to model the return of stocks. Three-factors model [4] was the primary one which modeled excess return of stock by book value, earning. Further researches verified a series of factors can be used to explain the return of investing in stocks, factors can be summarised into three main categories: macroeconomic,

© Springer Nature Switzerland AG 2019
J. M. F. Rodrigues et al. (Eds.): ICCS 2019, LNCS 11538, pp. 286–292, 2019.
https://doi.org/10.1007/978-3-030-22744-9_22

statistical, and fundamental. In risk model developed by Barra team from MSCI company, factor returns are estimated through cross-sectional regression [8]. Factor portfolios were built according to target factors to construct factor returns in Fama-French approach [1,4]. Similarly, Smart Beta Index from MSCI company [2,3] is compiled according to target factors to reflect the style and performance of specific factors under the different market situation. When being put to use, multiple factors usually need to be integrated, a common way to integrate factors is a linearly weighted sum, and weights of each factor are calculated by solving an optimization with subjectively defined target [3]. In recent years, non-linear methods such Support Vector Machine, Logistic Regression, Random Forest, Neural Networks and deep learning methodologies are well used in financial time series modeling, yet most existing works focus on stock price prediction. They learn parameters of models by fitting training samples and presume that the distribution of the training set and test set in the feature space are identical [9,13–15]. In the aspect of cross-section modeling and feature integration, only several works exist [5,6].

In our works, we introduce neural networks into the task of cross-section factor integration, and we extract factors according to the definition from Barra [8]. We use Smart Beta Index methodology to compile factor indexes to reflect performance and style of them on the Chinese market. Experimental results show the index that compiled based on factors integrated by neural networks results in better profitability and stability.

2   Factors and Factor Indexes

The changes of the stock price are not just a result of historical market behavior, but also affected by information from multiple sources such as macroeconomy and financial situation of the corresponding listed company. Indicators can be selected and defined to capture this information for usage on investment practice, and they are called factors. Factors are extracted from three main sources: technical indicators from market samples, fundamental indicators from financial statements and macroeconomic indicators.

When used in market practices, stocks are ranked and selected according to scores calculated by one or multiple factors. Factors that proven to be robust through a long time period are summarized by Barra risk model. Table 1 present the definition of factors. Original indicators are extracted from market data of stocks and financial statement of their corresponding listed companies. Factors are usually sampled in monthly frequency when being used.

To reflect performances of factors on market practices, factor indexes are compiled according to methodologies proposed by MSCI company. At beginning of each season component stocks of benchmark CSI 800 are sorted by factor score, and top 100 are selected as component of factor index and weighted according to their market value. For single factor indexes, component stocks are sorted by single target factor, for multi-factors indexes, weights of component stocks are calculated by solving optimization whose objective are maximizing multiple target factors:

Table 1. BARRA style factors

Factors	Meaning	Indicators
Size	Size of listed company	Market Value of listed company
Momentum	Degrees of trend	Risk adjusted returns of recent 20 days: $\frac{mean(r_{20})}{std(r_{20})}$
Non-linear size	Middle level of size	Residual of the regression between size and third power of size
Volatility	Uncertainty of bias from market	Standard deviation of active return
		The cumulative sum of the active return
		Standard deviation of daily return
Value (BTOP)	Book value to market value	Price earnings ratio (PE)
		Market-to-book ratio (PBR)
		Price-to-sale ratio (PS)
Liquidity	Volume and frequency of trading	Monthly logarithm turnover rate
		Mean value of monthly logarithm turnover rate in recent 3 month
		Mean value of monthly logarithm turnover rate in recent 12 month
Growth	Growth of listed company	Net profit (YoY)
		Total asset (YoY)
		Operating revenue (YoY)
Dividend (Earning Yield)	Profitability of listed company	Dividend yield
		Dividend per share
		Dividend to market value
Quality	Quality of listed company	Debt to equity
		ROE
Leverage	Leverage situation of listed company	Market leverage
		Debt to asset
		Book leverage

$$\max \quad \sum_{k=1}^{K} \sum_{i=1}^{n} \omega_i X_{ik}^{target}$$

$$s.t. \quad \sum_{i=1}^{n} \omega_i X_{ik}^{non-target} \geq \sum_{i=1}^{n} \omega_i^{benchmark} X_{ik}^{non-target} - 0.25 * std(X_k^{non-target}),$$
$$k = 1, 2, 3 \ldots, \tilde{K}$$

$$\sum_{i=1}^{n} \omega_i X_{ik}^{non-target} \leq \sum_{i=1}^{n} \omega_i^{benchmark} X_{ik}^{non-target} + 0.25 * std(X_k^{non-target}),.$$
$$k = 1, 2, 3 \ldots, \tilde{K}$$

$$max(0, \omega_i^{benchmark} - 2\%) \leq \omega_i \leq \max(10\omega_i^{benchmark}, \omega_i^{benchmark} + 2\%),$$
$$i = 1, 2, 3 \ldots, n$$

According to this methodology we compile single factor indexes and multi-factor indexes with target on Momentum, Size, Value, Dividend, which follows document from MSCI. Figure 1 is back-test results of factor indexes during 2010 to 2017. Factors present different style among different market situation. Profitability and risk of each factors are evaluated by indicators listed in Table 2, from which we can see that factor indexes reach higher returns and Sharpe ratio than benchmark, which verified the effectiveness of these factors on Chinese market. Moreover, subjectively setting the objective of optimization for factor integration may lead to unsatisfied result on profitability and risk, since factors show different performance in different market.

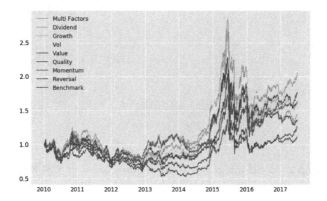

Fig. 1. Smart Beta factor indexes based on CSI 800.

Table 2. Smart Beta Index simulation results based on CSI 800

	Return	Annual Return	Volatility	Downside Beta	VaR	Alpha	Beta	Sharpe	Sortino	Loss rate	MDD	Active Return
Dividend	105.755%	10.425%	26.987%	1.0207	-2.739%	0.0860	1.0466	0.3612	0.2107	43.820%	-48.622%	94.791%
Growth	63.471%	6.988%	27.255%	1.0594	-2.779%	0.0548	1.0627	0.2437	0.0977	43.820%	-47.873%	52.507%
Vol	113.650%	10.998%	20.561%	0.7441	-2.082%	0.0838	0.7400	0.4244	0.3072	43.820%	-33.409%	102.686%
Value	26.687%	3.304%	23.491%	0.8553	-2.410%	0.0146	0.8512	0.0921	-0.0242	48.315%	-37.260%	15.722%
Quality	35.915%	4.308%	26.189%	0.9965	-2.678%	0.0275	0.9840	0.1454	0.0127	38.202%	-45.286%	24.950%
Momentum	76.557%	8.126%	26.901%	1.0443	-2.737%	0.0644	1.0230	0.2831	0.1357	46.067%	-47.984%	65.592%
Reversal	62.528%	6.903%	26.239%	1.0080	-2.676%	0.0524	0.9953	0.2397	0.1005	46.067%	-43.197%	51.564%
Multi Factors	44.600%	5.199%	28.772%	1.1226	-2.934%	0.0387	1.0962	0.1869	0.0375	41.573%	-54.206%	33.636%
Benchmark	10.964%	1.440%	24.670%	1.0011	-2.533%	0.0000	1.0000	0.0249	-0.0901	47.191%	-48.984%	0.000%

3 Neural Networks for Factor Integration

Deep learning methodology is explored on stock price prediction [7,11,12], and deep neural networks are designed to extract features from time series samples for prediction. Portfolio construction is another kind of market practice which provides cross-section level samples. In this work, we introduce Multi-layer Perceptron (MLP) to deal with cross-section factors. Traditional machine learning and linear regression are also applied in the experiment for comparison.

We use factors of each component stock of CSI 800 index from 2008 to 2017 for the experiment. Models are trained at the start of every year using monthly samples $\{\chi_t^i, y_t^i\}$ from previous 3 years, where χ_t^i denotes factors listed in Table 1 of stock i, and y_t^i denotes return of from t to $t+1$. At the start of each month, factors of each stock are integrated by models trained at the start of that year, and stocks are sorted according to integrated factors, and top 100 stocks are used for index compilation and weighted according to their market size.

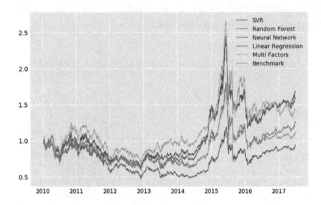

Fig. 2. Model integrated factor indexes based on CSI 800.

Results of indexes compiled based on integrated factors are performed in Fig. 2, from which we can see that the net value of most models based integrated factor indexes outperform benchmarks during most part of the back-test period. We further evaluate each index by the same performance indicators listed in Table 3. From the results of performance indicators, we can conclude that: (1) Factors integrated by neural networks and linear regression show better performance on profitability and stability than the multi-factors index. It implies that the model based integration can potentially mine the relationship between factors of stocks and their future performances. On the one hand, neural networks and linear regression based indexes show higher return than multi-factor indexes, on the other hand, volatility of multi-factor index is higher which means higher risk. Moreover, the higher Sharpe ratio still implies higher stability. (2) Neural networks show better performance than linear regression, which means the non-linear relationship between factors can be used to enhance the performance of integrated factors.

Table 3. Integrated factor indexes simulation results based on CSI 800

	Return	Annual Return	Volatility	Downside	Beta	VaR	Alpha	Beta	Sharpe	Sortino	Loss rate	MDD	Active Return
Multi Factors	44.600%	5.199%	28.772%	1.1226		-2.934%	0.0387	1.0962	0.1869	0.0375	41.573%	-54.206%	33.636%
Benchmark	10.964%	1.440%	24.670%	1.0011		-2.533%	0.0000	1.0000	0.0249	-0.0901	47.191%	-48.984%	0.000%
SVR	-5.746%	-0.810%	25.275%	0.9734		-2.606%	-0.0228	0.9867	-0.0589	-0.1696	48.315%	-52.491%	-16.710%
Random Forest	25.914%	3.218%	25.691%	1.0176		-2.630%	0.0173	0.9980	0.1020	-0.0240	44.944%	-51.071%	14.950%
Neural Network	68.440%	7.429%	25.615%	0.9938		-2.602%	0.0563	0.9537	0.2590	0.1145	44.944%	-52.198%	57.476%
Linear Regression	60.386%	6.708%	24.650%	0.9513		-2.508%	0.0487	0.9185	0.2316	0.0953	43.820%	-51.341%	49.421%

4 Conclusion

Factor indexes reflect performances of factors for factor investing so that robust factors can be filtered. Filtered factors need to be further integrated, our work introduces deep neural networks and other supervised models to integrate factors supervised by future return. And indexes are compiled according to integrated factors to evaluate their performance. Experimental results show that supervised integration by the model can enhance the effectiveness of integrated factors compared to integration by optimization with a subjectively defined objective. And Neural network is verified to be more effective since it is able to mine deep non-linear relationship between factors and future performance of stock price.

Acknowledgement. This research was partly supported by the grants from National Natural Science Foundation of China (No. 71771204, 71331005, 91546201).

References

1. Ang, A.: A five-factor asset pricing model. Fama-Miller Working Paper (2014)
2. Bender, J., Briand, R., Melas, D., Subramanian, R.: Foundations of factor investing (2013)
3. Bender, J., Briand, R., Melas, D., Subramanian, R.A., Subramanian, M.: Deploying multi-factor index allocations in institutional portfolios. In: Risk-Based and Factor Investing, pp. 339–363. Elsevier (2015)
4. Fama, E.F., French, K.R.: The cross-section of expected stock returns. J. Finance **47**(2), 427–465 (1992)
5. Gu, S., Kelly, B.T., Xiu, D.: Empirical asset pricing via machine learning. SSRN (2018). https://doi.org/10.2139/ssrn.3159577
6. Krauss, C., Do, X.A., Huck, N.: Deep neural networks, gradient-boosted trees, random forests: statistical arbitrage on the S&P 500. Eur. J. Oper. Res. **259**(2), 689–702 (2017)
7. Long, W., Lu, Z., Cui, L.: Deep learning-based feature engineering for stock price movement prediction. Knowl.-Based Syst. **164**, 163–173 (2019). http://www.sciencedirect.com/science/article/pii/S0950705118305264
8. Menchero, J., Orr, D., Wang, J.: The Barra US equity model (USE4) methodology notes. MSCI Model Insight (2011)
9. Rivest, R.L.: Learning decision lists. Mach. Learn. **2**(3), 229–246 (1987)
10. Ross, S.A.: The arbitrage theory of capital asset pricing. In: Handbook of the Fundamentals of Financial Decision Making: Part I, pp. 11–30. World Scientific (2013)

11. Shen, F., Chao, J., Zhao, J.: Forecasting exchange rate using deep belief networks and conjugate gradient method. Neurocomputing **167**, 243–253 (2015)
12. Singh, R., Srivastava, S.: Stock prediction using deep learning. Multimed. Tools Appl. **76**(18), 18569–18584 (2017)
13. Valiant, L.G.: A theory of the learnable. Commun. ACM **27**(11), 1134–1142 (1984)
14. Xiong, T., Li, C., Bao, Y., Hu, Z., Zhang, L.: A combination method for interval forecasting of agricultural commodity futures prices. Knowl.-Based Syst. **77**(C), 92–102 (2015)
15. Zhou, T., Gao, S., Wang, J., Chu, C., Todo, Y., Tang, Z.: Financial time series prediction using a dendritic neuron model. Knowl.-Based Syst. **105**(C), 214–224 (2016)

A Brief Survey of Relation Extraction Based on Distant Supervision

Yong Shi[2,3,4,5], Yang Xiao[1], and Lingfeng Niu[2,3,4(✉)]

[1] School of Computer and Control Engineering,
University of Chinese Academy of Sciences, Beijing 100190, China
`ynynny@sina.com`
[2] School of Economics and Management,
University of Chinese Academy of Sciences, Beijing 100190, China
`{yshi,niulf}@ucas.ac.cn`
[3] Key Laboratory of Big Data Mining and Knowledge Management,
Chinese Academy of Sciences, Beijing 100190, China
[4] Research Center on Fictitious Economy & Data Science,
Chinese Academy of Sciences, Beijing 100190, China
[5] College of Information Science and Technology,
University of Nebraska at Omaha, Omaha, NE 68182, USA

Abstract. As a core task and important part of Information Extraction Entity Relation Extraction can realize the identification of the semantic relation between entity pairs. And it plays an important role in semantic understanding of sentences and the construction of entity knowledge base. It has the potential of employing distant supervision method, end-to-end model and other deep learning model with the creation of large datasets. In this review, we compare the contributions and defect of the various models that have been used for the task, to help guide the path ahead.

Keywords: Relation extraction · Deep learning · Distant supervision

1 Introduction

The fundamental purpose of Information Extraction (IE), which is one of the most important task of natural language processing (NLP), is extracting structured information from primitive unstructured text. Subsequently, the structured information can be used easily by people or program. With the development of the Internet, people create and share many contents. As a result, the Internet is filled with huge amounts of data in the form of texts, and it is possible analyzing these primitive unstructured text by hand scarcely. Therefore, IE systems are extremely important. They can extract meaningful facts from texts to build Knowledge Base, which can be used for applications like search, machine reading comprehension and text generation. IE can be done in unsupervised domain, in the form of OpenIE [6]. And unsupervised approaches don't need to predefine

© Springer Nature Switzerland AG 2019
J. M. F. Rodrigues et al. (Eds.): ICCS 2019, LNCS 11538, pp. 293–303, 2019.
https://doi.org/10.1007/978-3-030-22744-9_23

any ontology or relation classes and the IE system should extract facts from the texts along with the relation phrases. Conversely, the supervised information extraction and classification methods specifically refer to the classification of an entity pair to a set of predefined relations or filling the predefined slots, which is trained by using documents containing mentions of the entity pair or structured data.

As one of the most important part of IE, the Relation Extraction (RE) is mainly responsible for identifying entities from text and extracting semantic relationships between entities [3,18,21]. RE system is able to predict whether a given document contains a relation or not for the pair. Further more, relation extraction system should predict which relation class out of a given ontology does that document point to, given that it does contain a relation, which can be regarded as a multi-class classification problem with an extra NoRelation class.

Supervised methods for relation extraction require large amount of training data for learning an desired model. Using hand annotated datasets for relation extraction takes tremendous time and effort to construct the datasets. However, there are already many knowledge bases built out such as DBpedia [1], Freebase [2], YAGO and Google Knowledge Graph. A large number of Entity-Relation-entity triplet has existed in these knowledge base, which contains useful semantic information can be used to promote the performance of relation extraction system. It needs to label the triples to the corresponding sentences in the primitive text only. Therefore, Mintz [12] proposed an assumption: if a sentence contains a pair of entities involved in a relation, then the sentence describes the relation of this pair. For example, all the sentences in the corpus that contain China and Beijing would be presumed have mentioned the relation that Beijing is the capital of China, as shown as Fig. 1. Then all these sentences are annotated as the training corpus data of the relation of the capital, and the pairs of entities are labeled simultaneously. Then put all the sentences corresponding to a relationship into a package, which is called a bag and all sentences in a bag have the same label. This work has been done later and is called multi-example learning. Such large datasets allow for learning more complex deep learning models for relation extraction. However, there are some false-positive sentences in the positive bags, which brings noise. The noise present in datasets generated through distant supervision also require special ways of modeling the problem like Multi-Instance Learning as discussed in the subsequent sections.

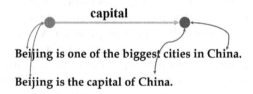

Fig. 1. A example of distant supervision.

In this review, we specifically focus on some different perspective of deep learning methods used for relation extraction.

2 Basic Concepts

In this section, we will introduce some basic concepts that are common across the models for relation extraction proposed recently.

2.1 Word Embeddings

Word Embeddings [10,11] are continuous distributional representations for the words in corpus, where each word is expressed as a continuous vector in a low dimensional latent space contrary to the high dimension of the one-hot vector. Word embeddings can capture the syntactic and semantic information about the word by predicting the context of words with unsupervised methods over large unlabeled corpus. Some work has been proposed to improve the word embeddings, such as Glove [13] and BERT [5]. After pre-training, all words are projected to an embedding matrix $E \in \mathbb{R}^{|V| \times d_w}$. Where d_w is the dimensionality of the embedding space and $|V|$ is the size of the vocabulary.

2.2 Position Embeddings

In natural language processing, the relative order of linguistic symbols has very important semantic information. Therefore, NLP system need to introduce this information into the model. In the relation extraction task, along with word embeddings, the input to the model also usually encodes the relative distance of each word from the entities in the sentence, In practice, the same continuous vectors as word embeddings are used instead of discrete form [20]. Position embeddings make neural network enable to keep track of the relative distance between words or entities in a sentence, which reserves the order information. The motivation is that words closer to the target entities probably imply more useful information reflecting the category of the relation between entities pair. The position embeddings comprise of the relative distance from current word to the entities. For example, in the sentence in Fig. 1 "Beijing is the capital of China." The relative distance between the word "capital" and head entity "Beijing" is 3 and tail entity "China" is -2. The distance are then encoded in a d_p dimensional embedding.

After obtaining word and position embeddings, the two embedding usually are Concatenated as the final representation of words and input the neural network. As the Fig. 2 shows.

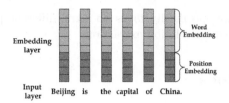

Fig. 2. A example of word representations with word and position embedding.

3 Datasets

In this section, We will introduce the commonly used data sets and evaluation metrics for relation extraction.

3.1 Supervised Dataset

The early works on relation extraction usually employed supervised training datasets. These datasets required intensive human annotation which meant that the data contained high quality tuples with little noise. But human annotation can be time-consuming, as a result of which these datasets were generally small. Both of the datasets mentioned below contain data samples in which the document sentence is already labeled with named entities of interest and the relation class expressed between the entity pair is to be predicted.

ACE 2005 Dataset: The Automatic Content Extraction dataset contains 599 documents related to news and emails. And the relations in the dataset are divided into 7 major types. 6 of the major relation types contain enough instances due to training and testing (average of 700 instances per relation type).

SemEval-2010 Task 8 Dataset: This dataset is a public dataset donated by Hendrickx et al. [7]. It contains 10717 samples which are divided as 8000 for training and 2717 for testing. It contains 9 relation types which are ordered relations. The directionality of the relations effectively doubles the number of relations, since an entity pair is believed to be correctly labeled only if the order is also correct. The final dataset thus has 19 relation classes (one for *Others* class).

3.2 Distant Supervision Datasets

To avoid the laborious task of manually building datasets for relation extraction, Mintz et al. [12] proposed the distant supervision to generate large number of relation extraction data automatically. They aligned sentences with known KBs, using the assumption that if a relation exists between an entity pair in the KB, then every sentence containing the mention of the entities pair would describe that relation. This distant supervision assumption is so strong that every sentence containing the entity pair mention may not express the relation between the pair and would bring some noise into the generated data. For example, For the tuple (*Beijing*, **Capital-of**, *China*) in the database and the sentence "Beijing is one of the biggest cities in China." At the distant supervision assumption, even this sentence do not describe the relation **Capital-of** between entity *Beijing* and entity *China*, it would also be labeled as the positive sample because of that it contains both the entities.

Riedel Dataset (also Called NYT): Riedel et al. [16] relaxed the distant supervision assumption by modeling the problem as a multi-instance learning problem, which can alleviate the problem mentioned above and reduce the noise. They released the Riedel dataset is the most popular dataset used in subsequent works building on distant supervision for relation extraction. This dataset was formed by aligning Freebase relations with the New York Times corpus (NYT). Entity mentions were found in the documents using the Stanford named entity tagger, and are further matched to the names of Freebase entities. There are 53 possible relation classes including a special relation NA which indicates there is no relation between the entity pair. The training data contains 522611 sentences, 281270 entity pairs and 18252 relational facts. The testing set contains 172448 sentences, 96678 entity pairs and 1950 relational facts.

GIDS: Jat et al. [8] created Google Distant Supervision (GIDS) dataset by extending the Google relation extraction corpus with additional instances for each entity pair. The dataset assures that the at-least-one assumption of multi-instance learning holds. This makes automatic evaluation more reliable and thus removes the need for manual verification. There are 5 possible relation classes between the entity pair. The training data contains 11297 sentences and 6498 entity pairs. The development set contains 1864 sentences and 1082 entity pairs. The testing set contains 5662 sentences and 3247 entity pairs.

4 Multi-instances Learning Models with Distant Supervision

Riedel et al. [16] regard the distant supervision relation extraction task as a multi-instances learning problem to relax the assumption, so that they could exploit the large training data created by distant supervision while being robust to the noise in the labels. Multi-instances learning is a form of weakly supervised learning problem where a label is given to a bag of instances, rather than a single instance. In multi-instances learning for relation extraction, Each entity pair in KB labels a bag of sentences. All the sentences in the bag contain the mention of the entity pair, but they do not contain the direct relation necessarily. Instead of giving a relation class label to every sentence, a label is instead given to each bag of the related entities. It assumes that if a relation exists between an entity pair, there is one document in the bag at least reflecting that relation of the given entity pair.

4.1   Piecewise Convolutional Neural Networks (PCNN)

The PCNN [19] model uses the multi-instance learning paradigm, with a neural network model to build a relation extractor using distant supervision data. The architecture of this model is similar to the model by Zeng et al. [20] proposed previously, with one important contribution of piecewise max-pooling over the sentence. The authors claim that the max-pooling layer can reduces the size of the latent feature remarkably. However, it is also losses important structure information between the entities in the sentence. This can be avoided by max-pooling in different segments of the sentence instead of the whole sentence. It is claimed that every sentence can naturally be divided into there segments based on the positions of the two entities in focus. By doing a piecewise max-pooling within each of the segments after convolution, the original sentence would be a more informative representation while still maintaining a vector that is independent of the input sentences length, which can alleviate the impact of long length sentences on relation extraction.

The disadvantage of this model is how multi-instance problem was set in the loss function. The paper defined the loss for training of the model as follows. Given T bags of sentences with each bag containing q_i sentences and having the label $y_i, i = 1, 2, ..., T$, the neural network gives the probability of extracting relation r from sentence j of bag i, d_i^j denoted as follows:

$$p(r|d_i^j, \theta); j = 1, 2, ..., q_i \tag{1}$$

where θ is the weight parameters of the neural network. Then the loss is given as follows:

$$J(\theta) = \sum_{i=1}^{T} \log \left(y_i | d_i^j, \theta \right) \tag{2}$$

$$j^* = \arg\max_j \ p \left(y_i | d_i^j, \theta \right); j = 1, 2, ..., q_i \tag{3}$$

PCNN uses the one most-likely positive sentence only for the entity pair to reduce the noise during the training and prediction stage with Eq. 2. It means that the model ignore almost all other sentences in the bag. Even though not all the sentences in the bag express the true positive relation between the entity pair, information expressed by these sentences in the bag are useful.

The PCNN with Multi-instance learning for relation extraction is shown to outperform the traditional non-deep models such as the distant supervision based model proposed by Mintz et al. [12]. As a result, it is always chosen as the baseline model.

4.2 Selective Attention over Instances

To address the drawbacks of the PCNN model which only used the one most-relevant sentence in a bag as the positive sample. Lin et al. [9] used the attention mechanism over all the instances in the bag to handle noise problem. In this model, each sentence d_i^j of bag i is first encoded into a vector representation, r_i^j as PCNN did. Then the representation of the bag is gotten by taking attention-weighted average of all the sentence vectors $(r_i^j, j = 1, 2, ..., q_i)$ in the bag. The model computes a weight α_j for each instance d_i^j of bag i. These values are dynamic in the sense that they are different for each bag and depend on the relation category and the given sentence. The final representation of the bag is given as follows:

$$r_i = \sum_{j=1}^{q_i} \alpha_j r_i^j \tag{4}$$

With the attention weighted representation of all the instances in the bag, the model is able to identify the importance of sentences from the noisy bag and all the information in the bag is utilized to predict the class of the relation.

4.3 Denoising Approach

Although the distant supervision relation extraction method can use the knowledge base to obtain a large amount of labeled data by annotating the texts automatically, much noise was introduce into the dataset, which would decline the performance of relation extraction system. Multi-instance learning models can be affected less, but it still fails to overcome the problem that all sentences in the bag is mislabeled. In order to reduce the noise of the bags, Qin et al. [14] proposed a method based on the Generative Adversarial Training to remove noise from the annotated-automatically data. The generator networks (G for short) estimates the probability distribution of the positive samples over a distant supervision bag. And then sampling positive sentences from noisy bag according to this probability distribution. The high-confidence samples generated by G are regarded as true positive samples. However, The discriminator (D for short) regards them as negative samples; conversely, the low-confidence samples are still treated as positive samples. For the generated samples, G maximizes the probability of being true positive; on the contrary, D minimizes this probability. The optimal G is obtained until the D has been greatest confused. As a result, the G is able to filter distant supervision training dataset and redistribute the false positive instances into the negative set, in which way to provide a cleaned dataset for relation classification.

Qin et al. [15] also proposed a denoise approach based on Deep reinforcement learning framework. The agent tries to remove the false positive samples from the distant supervision positive dataset P^{ori}. In order to get the reward, P^{ori} is split into the training set P_t^{ori} and the validation set P_v^{ori}; their corresponding negative part are represented as N_t^{ori} and N_v^{ori}. In each epoch i, the agent performs a series of actions to recognize the false positive samples from P_t^{ori} and treat them as negative samples. Then, a new relation classifier is trained under the new dataset $\{P_t^i, N_t^i\}$. With this relation classifier, F1 score is figured out from the new validation set $\{P_v^i, N_v^i\}$, where P_v^i is also filtered by the current agent. After that, the current reward is measured as the difference of F1 between the adjacent epochs. The above two algorithms are independent of the relation extraction model and are a plug-and-play technique that can be applied to any existing distant supervision relation extraction model.

The proposed methods on NYT dataset, the results are shown as Fig. 3. The experimental results show that the two methods can effectively remove noise and improve the extraction performance of distant supervision methods. Both method pipelines are independent of the relation prediction of entity pairs, so these models can be adopted as the true-positive indicator to filter the noisy distant supervision dataset before relation extraction. And the filter is more effective for selecting useful samples than the soft approach like [9, 19] proposed.

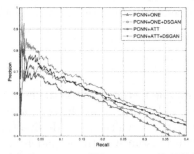 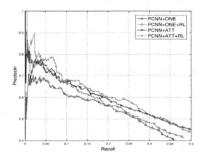

(a) Aggregate PR curves of DS-GAN(source from [14]).

(b) Aggregate PR curves of Reinforcement Learning for Distant Supervision(source from [15]).

Fig. 3. The performance of the denoising approach based relation extraction methods.

4.4 Graph-Based Model

To the best of our knowledge, it is surprising to note that only a few works for relation extraction has tried to replace Convolutional Neural Networks with Recurrent Neural Networks for encoding the sentences. One important reason is that these methods hope to use the convolutional neural network to extract the combined semantic features between words or entities independent of position. Though RNNs intuitively fits more naturally to natural language tasks and can persevere the context information of the sentence.

In the distant supervision domain, Vashishth et al. [17] proposed RESIDE, a graph model based approach, which uses Bi-GRU over the concatenated positional and word embedding for encoding the local context of each word. For capturing long-range dependencies, the Graph Convolution Networks (GCN) over dependency tree is employed to get the syntactic information representation of each word. Then, attention over tokens is used to subdue irrelevant tokens and get an embedding for the entire sentence. Finally, This model use attention over sentences to obtain a representation for the entire bag, which is fed to a softmax classifier to get the probability distribution over the relations. The performance of RESIDE is evaluated in NYT dataset and the result is shown as Fig. 4. The results validate that GCNs are effective at encoding syntactic information, which is complementary to the context information captured by RNNs. This model could improve relation extraction with these information.

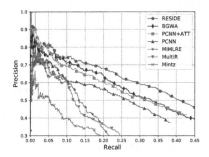

Fig. 4. The PR curves of RESIDE (source from [17])

In the supervised domain, Fenia Christopoulou et al. [4] proposed a walk based model for relation extraction. All the entities in a sentence are regarded as nodes in a fully-connected graph structure. The edges are on behalf of the position-aware contexts around the entity pairs. In order to capture different relation paths between two entities, The model construct up to a given length walks between each pair. The resulting walks are merged and iteratively used to update the edge representations. The model is evaluated on ACE 2005 for the task of relation extraction. And the model can achieve comparable performance compared with the state-of-the-art supervised relation extraction system without external syntactic tools. It shows that the dependencies between relations of entities can help extracting the final useful relations.

5   Conclusion

In general, the Entity relation extraction method with supervised learning has high accuracy, but it depends on the large labeled corpus and the construction of corpus is difficult. The unsupervised entity relation extraction does not need to define the entity relation type system in advance, which has domain independence. Scale open domain data has advantages that other methods can't match, but its clustering threshold is difficult to determine in advance. However, the distant supervision entity relation extraction only needs to label a small number of relation instances manually, which is suitable for entity relation extraction without labeling corpus, but its implementation process introduced noise into datasets, which makes the recall rate of the method lower. Many Successive works have tried to handle the noise and distant supervision assumption with mechanisms like selective attention and instance filter to improve the performance further by denoising. And the Graph-Based Model shows the huge potential to improve the relation extraction task by handling the dependency of entities. Future works for relation extraction can thus definitely try these approach to promote the RE system.

Acknowledgments. This work was supported by the National Natural Science Foundation of China [No. 11331012, Grant No. 91546201, No. 71331005, No. 71110107026, No. 11671379], UCAS Grant [No. Y55202LY00].

References

1. Auer, S., Bizer, C., Kobilarov, G., Lehmann, J., Cyganiak, R., Ives, Z.: DBpedia: a nucleus for a web of open data. In: Aberer, K., et al. (eds.) ASWC/ISWC -2007. LNCS, vol. 4825, pp. 722–735. Springer, Heidelberg (2007). https://doi.org/10.1007/978-3-540-76298-0_52

2. Bollacker, K., Evans, C., Paritosh, P., Sturge, T., Taylor, J.: Freebase: a collaboratively created graph database for structuring human knowledge. In: Proceedings of the 2008 ACM SIGMOD International Conference on Management of Data, pp. 1247–1250. ACM (2008)

3. Bunescu, R.C., Mooney, R.J.: Subsequence Kernels for relation extraction. In: NIPS (2005)

4. Christopoulou, F., Miwa, M., Ananiadou, S.: A walk-based model on entity graphs for relation extraction. In: ACL (2018)

5. Devlin, J., Chang, M.W., Lee, K., Toutanova, K.: BERT: pre-training of deep bidirectional transformers for language understanding. CoRR abs/1810.04805 (2018)

6. Etzioni, O., Fader, A., Christensen, J., Soderland, S., Mausam: Open information extraction: the second generation. In: IJCAI (2011)

7. Hendrickx, I., et al.: SemEval-2010 task 8: multi-way classification of semantic relations between pairs of nominals. In: SemEval@ACL (2010)

8. Jat, S., Khandelwal, S., Talukdar, P.: Improving distantly supervised relation extraction using word and entity based attention. CoRR abs/1804.06987 (2017)

9. Lin, Y., Shen, S., Liu, Z., Luan, H., Sun, M.: Neural relation extraction with selective attention over instances. In: ACL (2016)

10. Mikolov, T., Chen, K., Corrado, G.S., Dean, J.: Efficient estimation of word representations in vector space. CoRR abs/1301.3781 (2013)

11. Mikolov, T., Sutskever, I., Chen, K., Corrado, G.S., Dean, J.: Distributed representations of words and phrases and their compositionality. In: NIPS (2013)

12. Mintz, M., Bills, S., Snow, R., Jurafsky, D.: Distant supervision for relation extraction without labeled data. In: Proceedings of the Joint Conference of the 47th Annual Meeting of the ACL and the 4th International Joint Conference on Natural Language Processing of the AFNLP, pp. 1003–1011. Association for Computational Linguistics (2009)

13. Pennington, J., Socher, R., Manning, C.D.: Glove: global vectors for word representation. In: EMNLP (2014)

14. Qin, P., Xu, W., Wang, W.Y.: DSGAN: generative adversarial training for distant supervision relation extraction. In: ACL (2018)
15. Qin, P., Xu, W., Wang, W.Y.: Robust distant supervision relation extraction via deep reinforcement learning. In: ACL (2018)
16. Riedel, S., Yao, L., McCallum, A.: Modeling relations and their mentions without labeled text. In: Balcázar, J.L., Bonchi, F., Gionis, A., Sebag, M. (eds.) ECML PKDD 2010. LNCS (LNAI), vol. 6323, pp. 148–163. Springer, Heidelberg (2010). https://doi.org/10.1007/978-3-642-15939-8_10
17. Vashishth, S., Yao, X., Gawade, S., Bhattacharyya, C., Talukdar, P.: Reside: improving distantly-supervised neural relation extraction using side information. In: EMNLP (2018)
18. Zelenko, D., Aone, C., Richardella, A.: Kernel methods for relation extraction. J. Mach. Learn. Res. **3**, 1083–1106 (2002)
19. Zeng, D., Liu, K., Chen, Y., Zhao, J.: Distant supervision for relation extraction via piecewise convolutional neural networks. In: EMNLP (2015)
20. Zeng, D., Liu, K., Lai, S., Zhou, G., Zhao, J.: Relation classification via convolutional deep neural network. In: COLING (2014)
21. Zhou, G., Su, J., Zhang, J., Zhang, M.: Exploring various knowledge in relation extraction. In: ACL (2005)

Short-Term Traffic Congestion Forecasting Using Attention-Based Long Short-Term Memory Recurrent Neural Network

Tianlin Zhang[1,2](✉) 📧, Ying Liu[1,2](✉) 📧, Zhenyu Cui[1,2],
Jiaxu Leng[1,2], Weihong Xie[3], and Liang Zhang[4]

[1] School of Computer Science and Technology,
University of Chinese Academy of Sciences, Beijing 100190, China
zhangtianlin172@mails.ucas.ac.cn
[2] Key Lab of Big Data Mining and Knowledge Management,
Chinese Academy of Sciences, Beijing 100190, China
[3] School of Economics and Commerce, Guangdong University of Technology,
Guangzhou 510006, China
[4] School of Applied Mathematics, Guangdong University of Technology,
Guangzhou 510006, China

Abstract. Traffic congestion seriously affect citizens' life quality. Many researchers have paid much attention to the task of short-term traffic congestion forecasting. However, the performance of the traditional traffic congestion forecasting approaches is not satisfactory. Moreover, most neural network models cannot capture the features at different moments effectively. In this paper, we propose an Attention-based long short-term memory (LSTM) recurrent neural network. We evaluate the prediction architecture on a real-time traffic data from Gray-Chicago-Milwaukee (GCM) Transportation Corridor in Chicagoland. The experimental results demonstrate that our method outperforms the baselines for the task of congestion prediction.

Keywords: Traffic congestion prediction · LSTM · Attention mechanism

1 Introduction

As the population grows and the mobility increase in cities, traffic has received important concern from citizens and urban planners. Traffic congestion is one of the major problems to be solved in traffic management. For this reason, traffic congestion prediction has become a crucial issue in many intelligent transport systems (ITS) applications [1]. Short-Term traffic forecasting have beneficial impact that could increase the effectiveness of modern transportation systems. Therefore, in the past decade, many research activities have been conducted in predicting traffic congestion.

To get better prediction effect, more and more studies use real-time data, which is collected via different devices such as loop detectors, fixed position traffic sensors, or GPS. Compared with loop detectors, fixed position traffic sensors are more cost-effective and equally reliable [2]. Therefore, we use real-time data collected by these sensors to forecast the traffic congestion in our research.

© Springer Nature Switzerland AG 2019
J. M. F. Rodrigues et al. (Eds.): ICCS 2019, LNCS 11538, pp. 304–314, 2019.
https://doi.org/10.1007/978-3-030-22744-9_24

The existing traffic prediction methods can be classified into two groups [3], parametric approach and nonparametric approach. The parametric models are predetermined by some specific theoretical assumptions, such as logistic regression whose parameters can be computed from empirical data. As a commonly used parametric time series method, autoregressive integrated moving average (ARIMA) [4] is suitable for Short-Term traffic congestion prediction. Due to its non-linear complexity characteristic of traffic flow, many researchers tried to employ non-parametric method for prediction. For example, Support Vector Machine (SVM) and Support Vector Regression (SVR) [5] are considered as efficient algorithms. K-nearest neighbors (KNN) [6] is also applied to finding common features in traffic data.

In recent years, as deep learning receiving extensive attention, many neural network-based (NN-based) methods have been proposed. Since the deep learning method has flexible model structure and strong learning ability, it could provide automatic representation learning from high-dimensional data. Huang et al. [7] used Deep Belief Network (DBN) and Lv et al. [8] proposed stacked autoencoder (SAE) method. On this basis, Chen et al. [9] attempted stacked de-noising autoencoder. Due to the dynamic time-serial nature of traffic flow, Recurrent Neural Networks (RNN) that has a chain-like structure may well deal with this sequence data. However, RNN may have the vanishing or blowing up gradient problems during the back-propagation process. In order to overcome this issue, Tian et al. [10] used long short-term memory recurrent neural network (LSTM), which is a type of RNN with gated structure to learn long-term dependencies and automatically determines the time lags. Other researchers have also made some corresponding improvements, like BDLSTM [11], DBLSTM [12]. But the current LSTM models are insensitive to time-aware traffic data, which cannot distinguish the importance of different traffic states at different moments.

In order to deal with the issue and improve traffic congestion prediction accuracy, in this paper, we propose a model called Attention-based Long Short-Term Memory Recurrent Neural Network, which can capture the features of different moments more effectively. We evaluated the performance of our proposed Attention-based LSTM model with other basic traffic prediction algorithms. In the experiment, our method is clearly superior to the baselines. The remainder of this paper is organized as follows. Section 2 presents our proposed attention-based LSTM model for traffic congestion prediction in detail. Experiments design and results analysis are given in Sect. 3. Finally, we conclude our work in Sect. 4.

2 Methodology

To capture the features of traffic flow and take full advantage of time-aware flow data, we propose an Attention-based LSTM method. In Sect. 2.1, we will present the Attention-based LSTM model. In Sect. 2.2, the traffic congestion prediction architecture will be explained in detail.

2.1   Attention-Based LSTM Model

2.1.1   LSTM

Long short-term memory (LSTM) [13] is an effective approach to predict traffic congestion by capturing dependency features. It solves the vanishing gradient problem based on the gate mechanism. The structure is composed of input layer, output layer and recurrent hidden layer that has artificial designed memory cell. This cell can remove and keep the information of the cell state, which consists of three gates, including the input gate, the output gate and the forget gate. The architecture of the LSTM is illustrate in Fig. 1.

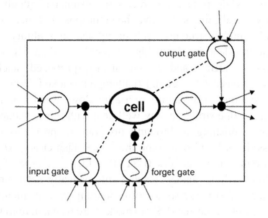

Fig. 1. The architecture of LSTM

In this model, the following equations explain the process and the notations as follow (Table 1):

$$i_t = \sigma\left(W_j x_t + U_i h_{t-1} + b_i\right) \tag{1}$$

$$f_t = \sigma\left(W_f x_t + U_f h_{t-1} + b_f\right) \tag{2}$$

$$o_t = \sigma\left(W_o x_t + U_o h_{t-1} + b_o\right) \tag{3}$$

$$g_t = \tanh\left(W_g x_t + U_g h_{t-1} + b_g\right) \tag{4}$$

$$c_t = f_t \odot c_{t-1} + f_t \odot g_t \tag{5}$$

$$h_t = o_t \odot \tanh(c_{t-1}) \tag{6}$$

Table 1. Notations for LSTM model

Notation	Definition
h_t	Hidden state
c_t	Memory cell
x_t	The input historical traffic flow
i_t	Input gate
f_t	Forget gate
o_t	Output gate
g_t	The extracted feature
W_*/U_*	Weight matrices
b_*	Bias vectors
$\odot$	Element-wise multiplication

2.1.2 Attention Mechanism

The attention mechanism in the neural networks imitates the attention of the human brain. It was proposed in the field of image recognition originally [14]. When people observe images, they often focus on some important information of the image selectively. Recently, many researchers applied the attention mechanism to natural language processing (NLP) [15, 16], because the conventional neural networks assume the weight of each word in the input is equal. Thus, they fail to distinguish the importance of different words. Therefore, attention mechanism is added to the basic model to calculate the correlation between the input and output.

Similar to natural language, traffic flow data is sequence data too. The importance of different traffic states in the flow data is not the same either. Nevertheless, since the existing methods did not solve the problem well yet, we propose an attention-based LSTM model.

2.1.3 Attention-Based LSTM

As shown in Fig. 2, the structure of Attention-based LSTM can be divided into four layers: the input layer, LSTM layer, the attention mechanism layer, and the output layer.

Attention mechanism layer [17] can highlight the importance of a particular traffic state to the entire traffic flow and consider more contextual association.

The state importance vector u_t is calculated by Eq. 8. The normalized state weight α_t is obtained through the function (Eq. 9). The aggregated of information in the traffic flow $v \leftarrow$ is the weighted sum of each h_t with α_t as the corresponding weights.

$$h_t = LSTM(vec_t) \tag{7}$$

$$u_t = \tan h(Wh_t + b) \tag{8}$$

$$\alpha_t = \frac{\exp(u_t^T a)}{\Sigma_t \exp(u_t^T a)} \tag{9}$$

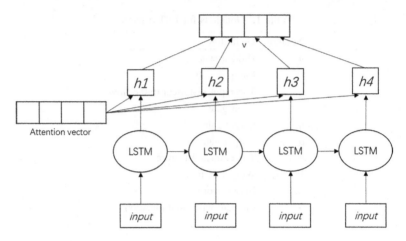

Fig. 2. The structure of the Attention-based LSTM

$$v = \sum_t a_t h_t \qquad (10)$$

Then the vector v is fed to the output layer to perform the final prediction.

2.2   The Traffic Congestion Prediction Architecture

As shown in Fig. 3, the prediction architecture mainly consists of four parts: the embedding layer, the LSTM, the attention mechanism layer and the prediction layer. The input is a sequence $\{i(0), i(1), \ldots, i(n-1)\}$ which represents the traffic flow data, and each $i(t)$ is a piece of data at a time interval encoded by one-hot representation. After the embedding layer, the data is mapped into a same dimensional vector space. Then, the LSTM network will process time-aware embedding vector and produce a hidden sequence $\{h(0), h(1), \ldots, h(n-1)\}$. An attention mechanism is used to extract traffic embedding features through the output attention probability matrix that is produced by the process in Sect. 2.1.3. Then, the prediction layer extracts mean values of the sequence over time intervals and makes the features encoded into a classified vector. Then it is fed into the logistic regression layer at the top of the prediction architecture.

3   Performance Analysis

In this section, to evaluate the effectiveness of our proposed approach, we first introduce our dataset and the experimental settings. Then we present the performances evaluated by different metrics. Finally, we show the comparative results with some baselines.

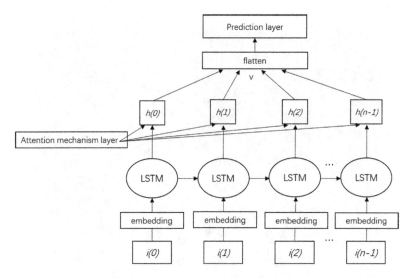

Fig. 3. The traffic prediction architecture

3.1 Datasets and Experiments Settings

(1) Dataset Description

In this study, the traffic data is collected by 855 fixed position sensors, located on the highways and roads of Gary-Chicago-Milwaukee (GCM) (consisting of 16 urbanized counties and covering 2500 miles). Each sensor collects the real-time traffic stream every 5 min, which contains attributes like longitude, latitude, length, direction, speed, volume, occupancy, congestion level, etc.

GCM highway system provides congestion levels on the different roads, which is shown in Fig. 4.

By analyzing the correlation matrix of attributes shown in Fig. 5, dark colors represent high correlation between two attributes. We select attributes (speed, travel time, volume) that are more correlated to the congestion level.

(2) Experimental settings

Our method is implemented in Keras framework. The embedding dimension is 10. We take the traffic congestion values of the first 20 days as the training set, and the next 5 days as the validation set for the purpose of tuning parameters. The number of hidden units of LSTM is 64. We then use the stochastic gradient descent (SGD) method with the RMSprop [18] is set at 0.001 to minimize the square errors between our predictions and the actual congestion levels. Moreover, the mini-batch size is set at 64.

To improve the generalization capability of our model and alleviate the overfitting problem [19], we adopted the dropout method proposed in [20, 21], which randomly drops units (along with their connections) from the network. The dropout rate of the output layer is set at 0.7.

Fig. 4. Gary-Chicago-Milwaukee (GCM) corridor transportation system

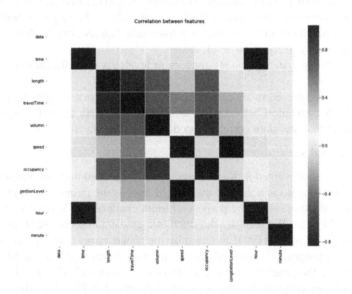

Fig. 5. Correlation between attributes

3.2   Measures

To evaluate the effectiveness of the congestion prediction, we use two performance metrics, Mean Absolute Percentage Error (MAPE) and the Root Mean Square Error (RMSE), which are defined as:

$$MAPE\left(f,\widehat{f}\right) = \frac{1}{n}\sum_{i=1}^{n}\frac{\left|f_i - \widehat{f_i}\right|}{f_i} \tag{11}$$

$$RMSE\left(f,\widehat{f}\right) = \left[\frac{1}{n}\sum_{i=1}^{n}\left(\left|f_i - \widehat{f_i}\right|\right)^2\right]^{\frac{1}{2}} \tag{12}$$

Where f is the real value of traffic congestion, and $\widehat{f}$ is the predicted value.

3.3   Experimental Results and Discussion

We compare our proposed Attention-based LSTM with several methods in predicting the short-term traffic congestion levels. We use the same dataset and measures to ensure a fair comparison.

XGBOOST: extreme gradient boosting [22]
ARIMA: autoregressive integrated moving average
KNN: K-nearest neighbors [23]
LSTM: long short-term memory network

The congestion prediction performance of the five models is listed in Table 2. Both MAPE and RMSE of attention-based LSTM are lowest among the prediction models. Therefore, our proposed method is superior to the baselines.

Table 2. Performance of different methods

Method	MAPE(%)	RMSE
XGBOOST	10.34	67.25
ARIMA	9.13	61.86
KNN	8.96	59.32
LSTM	6.21	50.32
Attention-based LSTM	6.01	48.12

Figure 6 presents the traffic congestion prediction vs. the observed congestion values collected from the data of No. IL-54 in one day. It is evident that the prediction results are satisfactory, and most of the fluctuations are captured by our Attention-based LSTM. Table 3 shows some results of congestion prediction of some sensors at 12:00 am when we set the size of time slot at 30 min. We can see the congestion trends are almost the same (0 means normal, 1 means light, 2 means medium, 3 means heavy).

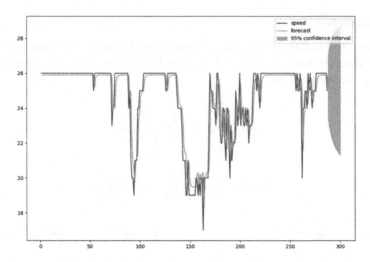

Fig. 6. Traffic congestion prediction vs. observation

Table 3. Results of congestion prediction of some sensors

Number of sensors	12:30 observation	12:30 prediction
No. IL-239	(1,1,1,1,1,1)	(1,1,1,1,1,1)
No. IL-161	(3,3,3,3,3,3)	(3,3,3,3,3,3)
No. WI-7022	(1,1,1,2,2,2)	(1,1,1,1,2,2)
No. WI-9019	(2,3,3,3,3,3)	(2,3,3,3,3,3)
No. WI-33018	(2,2,2,2,2,2)	(2,2,2,2,2,2)

4   Conclusion

In this paper, we propose an Attention-based LSTM model to predict short-term traffic congestion, which is able to capture more features at different moments and take full advantage of the time-aware traffic data. In the experimental results, both the MAPE and RMSE of our model are the lowest when compared with XGBOOST, ARIMA, KNN, LSTM models in the real traffic data from Gray-Chicago-Milwaukee (GCM) Transportation Corridor in Chicagoland. It is demonstrated that the proposed method outperforms baselines significantly.

5   Acknowledgements

This project was partially supported by Guangdong Provincial Science and Technology Project 2016B010127004 and Grants from Natural Science Foundation of China #71671178/#91546201/#61202321, and the open project of the Key Lab of Big Data Mining and Knowledge Management.

References

1. Vlahogianni, E.I., Karlaftis, M.G., Golias, J.C.: Optimized and meta-optimized neural networks for short-term traffic flow prediction: a genetic approach. Transp. Res. Part C Emerg. Technol. **13**(3), 211–234 (2005)
2. Barros, J., Araujo, M., Rossetti, R.J.F.: Short-term real-time traffic prediction methods: a survey. In: 2015 International Conference on Models and Technologies for Intelligent Transportation Systems (MT-ITS). IEEE (2015)
3. Tian, Y., Pan, L.: Predicting short-term traffic flow by long short-term memory recurrent neural network. In: 2015 IEEE International Conference on Smart City/ SocialCom/SustainCom (SmartCity). IEEE (2016)
4. Levin, M., Tsao, Y.D.: On forecasting freeway occupancies and volumes (abridgment). Transp. Res. Rec. 773, 47–49 (1980)
5. Castro-Neto, M., Jeong, Y.S., Jeong, M.K., et al.: Online-SVR for short-term traffic flow prediction under typical and atypical traffic conditions. Expert Syst. Appl. **36**(3), 6164–6173 (2009)
6. Xia, D., Wang, B., Li, H., et al.: A distributed spatial–temporal weighted model on MapReduce for short-term traffic flow forecasting. Neurocomputing **179**, 246–263 (2016)
7. Huang, W., Song, G., Hong, H., et al.: Deep architecture for traffic flow prediction: deep belief networks with multitask learning. IEEE Trans. Intell. Transp. Syst. **15**(5), 2191–2201 (2014)
8. Lv, Y., Duan, Y., Kang, W., et al.: Traffic flow prediction with big data: a deep learning approach. IEEE Trans. Intell. Transp. Syst. **16**(2), 865–873 (2015)
9. Chen, Q., Song, X., Yamada, H., et al.: Learning deep representation from big and heterogeneous data for traffic accident inference. In: AAAI, pp. 338–344 (2016)
10. Tian, Y., Pan, L.: Predicting short-term traffic flow by long short-term memory recurrent neural network. In: 2015 IEEE International Conference on Smart City/SocialCom/SustainCom (SmartCity), pp. 153–158. IEEE (2015)
11. Cui, Z., Ke, R., Wang, Y.: Deep stacked bidirectional and unidirectional LSTM recurrent neural network for network-wide traffic speed prediction. In: 6th International Workshop on Urban Computing (UrbComp 2017) (2016)
12. Wang, J., Hu, F., Li, L.: Deep bi-directional long short-term memory model for short-term traffic flow prediction. In: Liu, D., Xie, S., Li, Y., Zhao, D., El-Alfy, E.S. (eds.) ICONIP 2017. LNCS, vol. 10638, pp. 306–316. Springer, Cham (2017). https://doi.org/10.1007/978-3-319-70139-4_31
13. Kawakami, K.: Supervised sequence labelling with recurrent neural networks. Ph.D. thesis, Technical University of Munich (2008)
14. Mnih, V., Heess, N., Graves, A.: Recurrent models of visual attention. In: Advances in Neural Information Processing Systems, pp. 2204–2212 (2014)
15. Bahdanau, D., Cho, K., Bengio, Y.: Neural machine translation by jointly learning to align and translate. arXiv preprint arXiv:1409.0473 (2014)
16. Zhou, P., Shi, W., Tian, J., et al.: Attention-based bidirectional long short-term memory networks for relation classification. In: Proceedings of the 54th Annual Meeting of the Association for Computational Linguistics (Volume 2: Short Papers), vol. 2, pp. 207–212 (2016)
17. Yang, Z., Yang, D., Dyer, C., et al.: Hierarchical attention networks for document classification. In: Proceedings of the 2016 Conference of the North American Chapter of the Association for Computational Linguistics: Human Language Technologies, pp. 1480–1489 (2016)

18. Tieleman, T., Hinton, G.: Lecture 6.5-rmsprop: divide the gradient by a running average of its recent magnitude. COURSERA: Neural Netw. Mach. Learn. **4**(2), 26–31 (2012)
19. Hawkins, D.M.: The problem of overfitting. J. Chem. Inf. Comput. Sci. **44**(1), 1–12 (2004)
20. Hinton, G.E., Srivastava, N., Krizhevsky, A., et al.: Improving neural networks by preventing co-adaptation of feature detectors. arXiv preprint arXiv:1207.0580 (2012)
21. Srivastava, N., Hinton, G., Krizhevsky, A., et al.: Dropout: a simple way to prevent neural networks from overfitting. J. Mach. Learn. Res. **15**(1), 1929–1958 (2014)
22. Chen, T., Guestrin, C.: Xgboost: a scalable tree boosting system. In: Proceedings of the 22nd ACM SIGKDD International Conference on Knowledge Discovery and Data Mining, pp. 785–794. ACM (2016)
23. Habtemichael, F.G., Cetin, M.: Short-term traffic flow rate forecasting based on identifying similar traffic patterns. Transp. Res. Part C: Emer. Technol. **66**, 61–78 (2016)

Portfolio Selection Based on Hierarchical Clustering and Inverse-Variance Weighting

Andrés Arévalo[1]([✉]), Diego León[2], and German Hernandez[1]

[1] Universidad Nacional de Colombia, Bogota, Colombia
{ararevalom,gjhernandezp}@unal.edu.co
[2] Universidad Externado de Colombia, Bogota, Colombia
diego.leon@uexternado.edu.co

Abstract. This paper presents a remarkable model for portfolio selection using inverse-variance weighting and machine learning techniques such as hierarchical clustering algorithms. This method allows building diversified portfolios that have a good balance sector exposure and style exposure, respect to momentum, size, value, short-term reversal, and volatility. Furthermore, we compare performance for seven hierarchical algorithms: Single, Complete, Average, Weighted, Centroid, Median and Ward Linkages. Results show that the Average Linkage algorithm has the best Cophenetic Correlation Coefficient. The proposed method using the best linkage criteria is tested against real data over a two-year dataset of one-minute American stocks returns. The portfolio selection model achieves a good financial return and an outstanding result in the annual volatility of 3.2%. The results suggest good behavior in performance indicators with a Sharpe ratio of 0.89, an Omega ratio of 1.16, a Sortino ratio of 1.29 and a beta to S&P of 0.26.

Keywords: Portfolio construction · Portfolio selection ·
Hierarchical clustering algorithms · Inverse-variance weighting ·
Algorithmic trading

1 Introduction

Portfolio selections is an active topic on finance, and maybe, the most common problem for practitioners. On 1952, Markowitz introduced the modern portfolio theory [4] which proposed a mathematical framework, called mean-variance analysis, for assembling a portfolio of assets by solving one of the two optimization problems: To minimize the portfolio variance at a given level of expected or minimum required return. Or to maximize the portfolio expected return at a given level of expected or maximum required variance. The expected return is defined as:

$$E(R_p) = \sum_i w_i \, E(R_i) \tag{1}$$

© Springer Nature Switzerland AG 2019
J. M. F. Rodrigues et al. (Eds.): ICCS 2019, LNCS 11538, pp. 315–325, 2019.
https://doi.org/10.1007/978-3-030-22744-9_25

Where R_p is the return on the portfolio, R_i is the return on asset i and w_i is the proportion of asset i in the portfolio. Meanwhile, the variance is defined as:

$$\sigma_p^2 = \sum_i \sum_j w_i w_j \sigma_{ij} \tag{2}$$

Where σ_p^2 is the portfolio variance and σ_{ij} is the covariance of assets i and j. Figure 1 shows 500 combinations of portfolios of four assets, whose x-axis is the portfolio standard deviation and the y-axis is the portfolio return. The optimal portfolios are given by the Pareto frontier: The upper edge of the hyperbola.

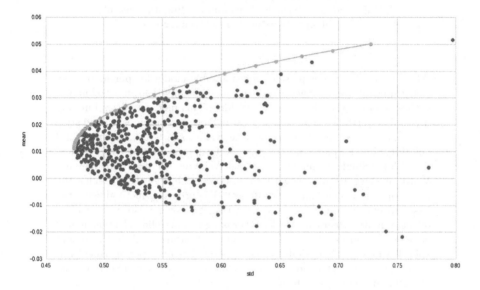

Fig. 1. Optimized markowitz portfolios

However, Markowitz' framework has issues related to instability, concentration, and under-performance given that the invertibility of the covariance matrix is required and not easy to satisfy. Therefore, [6] introduced an approach for building a diversified portfolio based on graph theory and machine-learning techniques like hierarchical clustering techniques. He presented evidence his approach produces less risky portfolios out of sample compared to traditional risk parity methods.

On [3], seven clustering techniques were tested for assembling portfolios using one-minute return data of 175 financial assets of the Russell 1000®index. The techniques were K-Means, Mini Batch K-Means, Spectral clustering, Birch and three hierarchical clustering methods (Average Linkage, Complete Linkage, and Ward's Method). Results showed that the hierarchical clustering methods had a better trade-off between risk and return.

In this work, we will extend our analysis over the hierarchical clustering techniques, expand the testing dataset to approximately 2000 assets of the U.S.

Stocks Market, and finally, propose an asset allocation tool based on inverse-variance weighting and a hierarchical clustering algorithm as an asset selection method.

This paper continues as follows: Sect. 2 presents a brief summary of hierarchical clustering methods, Sect. 3 explains the proposed method, Sect. 4 describes the experiment with real data and shows its results, and finally, Sect. 5 gives final remarks, conclusions, and further work opportunities.

2   Hierarchical Clustering Methods

Hierarchical Clustering Methods model data like a hierarchy of clusters [9]. There are two strategies for building the hierarchy: Agglomerative strategy (bottom-up approach) is that all observations start in its own cluster, and then, pairs of clusters are merged recursively. Whereas, divisive strategy (top-down approach) is that all observations start in a single cluster, and then, they are split into new clusters recursively. Divisive clustering is uncommon given that it requires an exhaustive search $\mathcal{O}(2^n)$ and not scales for large datasets [2].

On both strategies, merges and splits are determined in greedy manner by minimizing the distance (similarity) $d(u, v)$ between clusters u and v, which are determined by the linkage criterion. It is a function of the pairwise distances of observations in the clusters. The most common linkage criterion are:

- Single Linkage (Nearest Point Algorithm):

$$d(u, v) = \min(\text{dist}(u_i, v_j)) \tag{3}$$

   Where u_i is the i-th observation in the cluster u, v_j is the j-th observation in the cluster v, and $dist(a, b)$ is the euclidean, Manhattan, Mahalanobis or Maximum distance between observations a and b.
- Complete Linkage (Farthest Point Algorithm or Voor Hees Algorithm):

$$d(u, v) = \max(\text{dist}(u_i, v_j)) \tag{4}$$

- Average Linkage (UPGMA algorithm):

$$d(u, v) = \sum_{ij} \frac{\text{dist}(u_i, v_j)}{|u||v|} \tag{5}$$

   Where $|u|$ and $|v|$ are the cardinals of clusters u and v, respectively.
- Weighted Linkage (WPGMA algorithm):

$$d(u, v) = \frac{\text{dist}(s, v) + \text{dist}(t, v)}{2} \tag{6}$$

   Where u is formed by the merge between s and t.
- Centroid Linkage (UPGMC algorithm):

$$d(u, v) = ||c_u - c_v||_2 \tag{7}$$

   Where c_u and c_v are the centroids of clusters u and v, respectively.

– Median Linkage (WPGMC algorithm):

$$d(u, v) = ||c_u - c_v||_2 \tag{8}$$

$$c_u = \frac{c_s + c_t}{2} \tag{9}$$

Where u is formed by the merge between s and t, and c_s, c_t and c_u are the centroids of clusters s, t and u, respectively.
– Ward Linkage (Ward variance minimization algorithm):

$$d(u, v) = \sqrt{\frac{|v| + |s|}{T}d(v, s)^2 + \frac{|v| + |t|}{T}d(v, t)^2 - \frac{|v|}{T}d(s, t)^2} \tag{10}$$

Where u is formed by the merge between s and t, and $T = |v| + |s| + |t|$.

3   Proposed Method for Portfolio Selection

The US Stock Market lists approximately 8000 stocks which worth above 30 trillion USD [8]. However, many stocks are unsuitable for algorithmic trading or portfolio managing given its liquidity restrictions or high-risk behavior. One of the most important requirements of a portfolio is to have low-risk exposure, therefore, the universe of stocks is filtered using the following rules:

– The stock must be a common (for example, not preferred) stock, nor a depository receipt, nor a limited partnership, nor traded over the counter (OTC).
– If a company has more than one share class, the most liquid share class is chosen and the others are discarded.
– The stock must be liquid; it must have a 200-day median daily dollar volume that exceeds $2.5 Million USD.
– The stock must not be an active M&A target (Mergers and Acquisitions).
– The stock must have a market capitalization above $350 Million USD over a 20-day simple moving average.
– ETFs are excluded.

The reduced universe size ranges from 1900 to 2100 stocks. Once the universe is filtered, the distance matrix is built using the correlation matrix of the one-minute returns over the last 10 trading days. The distance matrix is defined as follows [6]:

$$D_{ij} = \sqrt{\frac{1}{2}(1 - \rho_{ij})} \tag{11}$$

Where ρ_{ij} is the Pearson correlation coefficient between the stocks i and j which ranges from -1 to 1. If this coefficient is close to 0, 1 or -1, it means uncorrelated, correlated, anti-correlated behavior, respectively. Given the fact that ρ_{ij} is bounded, D_{ij} ranges from 0 to 1. It is 0, $\sqrt{\frac{1}{2}}$ or 1 when the pair stocks are perfectly correlated, uncorrelated, and anti-correlated, respectively.

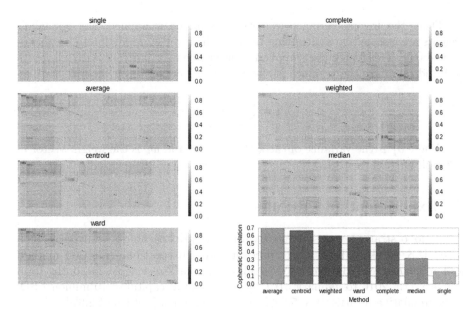

Fig. 2. Comparison of several hierarchical clustering methods

After, the distance matrix's clusters are formed using a hierarchical clustering method. The approach is to group stocks that are most similar within clusters. Figure 2 shows the comparison of seven hierarchical clustering methods: Single, Complete, Average, Weighted, Centroid, Median and Ward Linkages.

The Cophenetic Correlation Coefficient (CCC) evaluates how well the dendrogram preserved the pairwise distances between the original modelled data points [10]. It is given by [1]:

$$CCC = \frac{\sum_{i<j}(x(i,j) - \bar{x})(t(i,j) - \bar{t})}{\sqrt{[\sum_{i<j}(x(i,j) - \bar{x})^2][\sum_{i<j}(t(i,j) - \bar{t})^2]}}. \tag{12}$$

Where $x(i,j)$ is the Euclidean distance between the i-th and j-th observations. $t(i,j)$ is the dendrogrammatic distance, which is the height of the node at which these two points are first joined together, between the model points T_i and T_j. $\bar{x}$ and $\bar{t}$ is the average of all $x(i,j)$ and $t(i,j)$, respectively. Furthermore, the magnitude of CCC should be very close to 1 for a high-quality solution. Figure 2 also shows the CCC for each algorithm: The method Average has the highest CCC, meanwhile the method Single has the lowest CCC.

For each cluster, the optimal portfolio with the highest Sharpe-ratio is calculated using Markowitz theory and the critical line algorithm. Although, another more powerful method to generate the optimal portfolio within each cluster can be chosen like a multi-objective optimization, searching to optimize with liquidity and volume constraints or including aversion risk preferences or transaction costs.

Then, the Inverse-variance weighting technique is applied; the portfolio's weights are rescaled by multiplying them by the inverse proportion to its portfolio variance. This technique is applied in order to have a portfolio with a leverage of 1 and minimize the variance of the weighted average.

$$\hat{w}_k = \frac{1/\sigma_k^2}{\sum_k 1/\sigma_k^2} w_k \tag{13}$$

Where σ_k^2 is the variance of the k-th portfolio and w_k is the weight vector of the k-th portfolio's stocks.

4   Experiment and Results

A portfolio strategy was simulated with real data reaching a sample of 2,000 listed U.S. stocks. The strategy uses the previous portfolio selection method and rebalances weekly every Wednesday. The back-test took 25 months from January 6th, 2016 to January 31th, 2018 and initial capital of 10 million USD. The cumulative returns were 5.89%, namely, an annual return of 2.9%.

Fig. 3. Cumulative returns

Figure 3 shows the total percentage return of the portfolio from the start to the end of the back-test. Also, it compares the evolution against the Standard

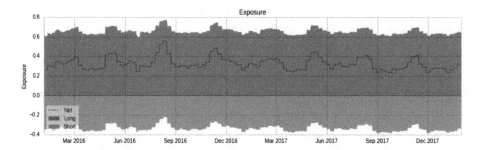

Fig. 4. Exposure

& Poor's 500 Index (S&P 500) which is the most representative index of the American stock market. It is based on the market capitalizations of 500 large companies listed on the New York Stock Exchange (NYSE) or Nasdaq Stock Market (NASDAQ). The maximal draw-down was −3.4%. Figure 4 shows strategy exposure over the back-test period. The strategy traded with an average leverage of 1 and used short and long positions.

Figure 5 shows the six-month rolling standard deviation of the portfolio's returns. The portfolio had annual volatility of 3.2% which is lower to the benchmark volatility and is a desired quality for low-risk portfolios. Meanwhile, Fig. 6 presents the six-month rolling Sharpe ratio which measure of risk-adjusted performance, which divides the portfolio's excess return over the risk-free rate by the portfolio's standard deviation. The portfolio had an average Sharpe ratio of 0.89 and a Calmar ratio of 0.83, an Omega ratio of 1.16, and a Sortino ratio of 1.29.

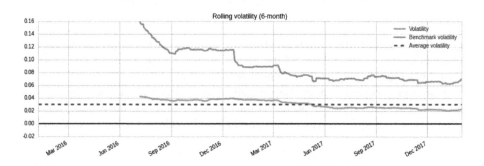

Fig. 5. Rolling volatility

Another desired quality is that portfolios must be diversified over different economic sectors. Traditionally, the portfolio selection satisfies this need manually splitting the market into sectors using subjective experts' criteria. But the clustering techniques allows removing this human parametrization because those techniques are able to learn and identify the economy sectors from data for

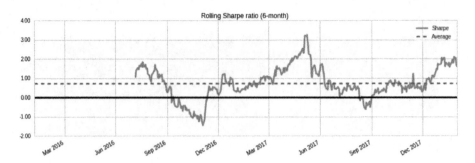

Fig. 6. Rolling sharpe

● Basic Materials	● Consumer Cyclical	● Financial Services	● Real Estate
1.41 %	4.02 %	3.89 %	2.40 %
● Consumer Defensive	● Health Care	● Utilities	● Energy
2.47 %	1.44 %	2.19 %	2.31 %
● Industrials	● Technology	● Communication Services	
3.41 %	3.96 %	0.22 %	

Fig. 7. Rolling 63-day mean of sector exposures

themselves without human intervention. Figure 7 shows the exposure to various economic sectors. The rolling 63-day mean of sector exposures for all standard economy sectors is below of 7%. This behavior is stable over time.

Moreover, portfolios must be diversified over different styles of exposures in order to ensure that all positions on any kind of stock have homogeneous behaviors with respect to the entire portfolio. The relevant Quantopian's styles are [7]:

- **Momentum:** The difference in return between assets on an upswing and a down-swing over 11 months.
- **Size:** The difference in returns between large capitalization and small capitalization assets.
- **Value:** The difference in returns between expensive and inexpensive assets (as measured by Price/Book ratio).
- **Short Term Reversal:** The difference in returns between assets with strong losses to reverse, and strong gains to reverse, over a short time period.

Fig. 8. Rolling 63-day mean of Style exposures

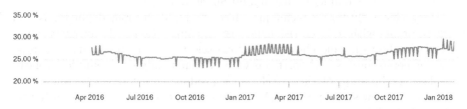

Fig. 9. Rolling 63-day mean turnover

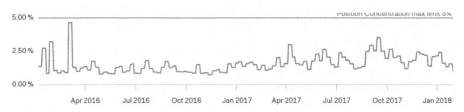

Fig. 10. Position concentration

– **Volatility:** The difference in return between high-volatility and low-volatility assets.

Figure 8 shows the portfolio style exposures. All style exposures are between −40% and 40% which is excellent for a low-risk portfolio. Figure 9 presents the rate at which assets are being bought and sold within the portfolio. The portfolio's turnover ranges from 22% to 30% with an average of 26.8%. A low turnover reduces transaction costs. Moreover, Fig. 10 shows the percentage of the portfolio invested in its most-concentrated asset. A portfolio must not have a heavy concentration because it makes high-correlated with that asset.

Finally, a portfolio must be as less as possible correlated with the market. Figure 11 shows the beta statistic. The average portfolio Beta was 0.26.

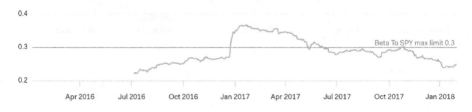

Fig. 11. Six-month rolling beta

5   Conclusions

We have tested seven hierarchical clustering techniques using actual data (sorted from best to worst performance according to CCC): Average, Centroid, Weighted, Ward, Complete, Median and Single Linkages.

Hierarchical clustering techniques allow to build diversified portfolios and achieve profits with reduced risk exposure. In conjunction with inverse-variance weighting, the technique allows a portfolio selection with the ability to consistently generate profits and portfolios with systematically stable and low volatility. The combination of these techniques produces portfolios with low sector exposure and low style exposure (Momentum, Sizes, Values, Short Term Reversal and Volatility).

Moreover, the Markowitz algorithm has issues related to instability, concentration, and under-performance given that the invertibility of the covariance matrix is required and not easy to satisfy. However, hierarchical clustering techniques do not have those issues. They are able to handle a lot of quantity of data with stable behavior.

Finally, another research opportunity would be to explore other machine learning techniques like hierarchical fuzzy clustering, to go beyond the work of [5]. Also is important to explore other methods for choosing the weight inside clusters that be more powerful than Markowitz algorithm, and other optimization objectives like Omega ratio.

References

1. Farris, J.S.: On the cophenetic correlation coefficient. Syst. Zool. **18**(3), 279–285 (1969). http://www.jstor.org/stable/2412324
2. Kaufman, L., Rousseeuw, P.J.: Finding Groups in Data: An Introduction to Cluster Analysis. Wiley, London (2009)
3. León, D., Aragón, A., Sandoval, J., Hernández, G., Arévalo, A., Niño, J.: Clustering algorithms for risk-adjusted portfolio construction. Procedia Comput. Sci. **108**, 1334–1343 (2017). https://doi.org/10.1016/j.procs.2017.05.185. International Conference on Computational Science, ICCS 2017, 12–14 June 2017, Zurich, Switzerland. http://www.sciencedirect.com/science/article/pii/S187705091730772X
4. Markowitz, H.: Portfolio selection. J. Finance **7**(1), 77–91 (1952). https://doi.org/10.1111/j.1540-6261.1952.tb01525.x

5. Nanda, S., Mahanty, B., Tiwari, M.: Clustering indian stock market data for portfolio management. Expert Syst. Appl.: An Int. J. **37**(12), 8793–8798 (2010)
6. López de Prado, M.: Building diversified portfolios that outperform out of sample. J. Portfolio Manage. **42**(4), 59–69 (2016). https://doi.org/10.3905/jpm.2016.42.4.059. http://jpm.iijournals.com/content/42/4/59
7. Quantopian Inc.: Quantopian risk model (2018). https://www.quantopian.com/papers/risk. Accessed 07 July 2018
8. Racanelli, V.J.: The U.S. stock market is now worth $30 trillion (2018). https://www.nasdaq.com/article/the-us-stock-market-is-now-worth-30-trillion-cm906996. Accessed 07 July 2018
9. Rokach, L., Maimon, O.: Clustering methods. Data Mining and Knowledge Discovery Handbook, pp. 321–352. Springer, Boston (2005). https://doi.org/10.1007/0-387-25465-X_15
10. Sokal, R.R., Rohlf, F.J.: The comparison of dendrograms by objective methods. Taxon **11**(2), 33–40 (1962). http://www.jstor.org/stable/1217208

A Computational Technique for Asian Option Pricing Model

Manisha and S. Chandra Sekhara Rao$^{(\boxtimes)}$

Department of Mathematics, Indian Institute of Technology Delhi,
Hauz Khas 110 016, New Delhi, India
srivastava.manisha99@gmail.com, scsr@maths.iitd.ac.in

Abstract. In the present work, the European style fixed strike Asian call option with arithmetic and continuous averaging is numerically evaluated where the volatility, the risk free interest rate and the dividend yield are functions of the time. A finite difference scheme consisting of second order HODIE scheme for spatial discretization and two-step backward differentiation formula for temporal discretization is applied. The scheme is proved to be second order accurate in space and time both. The numerical results are in accordance with analytical results.

Keywords: Asian option · Black-Scholes model ·
High-Order Difference approximation with Identity Expansion
(HODIE) · Two-step backward differentiation formula (BDF)

1 Introduction

Asian options are path-dependent exotic options. The payoffs of Asian options depend on some sort of average of the underlying asset price over a specified time interval. The averaging can be arithmetic (mean of the asset prices) or geometric (exponential of the mean of the logarithm of the asset prices). The price of geometric average Asian option are analytically evaluated as the geometric average of the asset prices follows the lognormal distribution [9, 16] whereas the distribution of arithmetic average of the asset prices is not explicitly known. Hence the analytical solution of arithmetic average Asian option is not known. The averaging can be weighted or unweighted. These options are among the most popular derivative securities as they reduce the risk of market manipulation by big and influential traders of the market near the maturity time. Also the average-value options are always economical than the standard European options [16].

Asian option in a complete and arbitrage free financial market consisting of a risky asset, with asset price S_τ, as the underlying security and a risk free asset with interest rate r, follows the following stochastic differential equation

$$dS_\tau = (r - D)S_\tau d\tau + \sigma dW_\tau, \tag{1}$$

© Springer Nature Switzerland AG 2019
J. M. F. Rodrigues et al. (Eds.): ICCS 2019, LNCS 11538, pp. 326–339, 2019.
https://doi.org/10.1007/978-3-030-22744-9_26

where S_τ is the asset price at time τ, r is the risk free interest rate, D is the dividend yield, σ is the market volatility and W_τ is the standard Wiener process under the risk-neutral measure.

The payoff of European call option is

$$\max(S - K, 0).$$ (2)

The payoff of the "fixed strike" Asian call option is obtained by replacing S_τ in the payoff of European call option by the average of asset prices, keeping the strike price K fixed and is given as

$$\max\left(\frac{1}{T}\int_0^T S_z dz - K, 0\right).$$ (3)

Whereas the payoff of the "floating strike" Asian call option is obtained by replacing the strike price K in the payoff of European call option by the average of asset prices and given as

$$\max\left(S - \frac{1}{T}\int_0^T S_z dz, 0\right).$$ (4)

The closed form solution of Asian option pricing model is not known.

Asian options were first explained in [13]. We consider the arithmetic average fixed strike Asian option. The existence, uniqueness and regularity of solution of arithmetic and geometric average Asian options was discussed in [2]. The partial differential equation for pricing Asian options consists of Black-Scholes equation along with an advection term, which can not be reduced to heat equation and its closed form solution is not known. Different approaches were used to solve various models for pricing Asian options such as applied statistics [19,20,26], applied probability [24,25] and mathematical analysis [14,32]. Various numerical schemes are also applied to approximate the option price. The work [3] described reducing the dimension of the partial differential equation governing the option price and then solving it using finite difference scheme. An explicit and an implicit finite difference methods were described in [21] for pricing Asian options. A hybrid finite difference along with Crank-Nicolson method was used in [6] to get a second order convergent scheme for evaluating the option price. A radial basis function based finite difference approximation for spacial operator along with θ-method for temporal approximation was applied in [17] to obtain a second order convergent scheme for pricing the Asian option. In all these works where a numerical approximation was given for pricing option value, only non-dividend paying contract was considered. Hence after the dimension reducing transformation, the reaction term was not there in the partial differential equation to be solved numerically. Also the parameters σ and r were taken as constants. In the works [22,23], these parameters along with dividend yield were taken to be functions of asset price and time variables and numerical solution of such generalized Black-Scholes model in dividend paying market was computed for different European style options. The generalizing of Black-Scholes model for pricing European call option by making the parameters r, σ and D variables was also done in [4,8,15,27,30] and then the option price was numerically approximated. The other related works were [11,29].

In the present work we have developed a numerical scheme for a generalized Asian option pricing model, transformed from two dimensional partial differential equation to one dimensional partial differential equation. Two step backward differentiation formula for temporal discretization and the High Order Difference approximation with Identity Expansion (HODIE) scheme for spacial discretization is applied simultaneously yielding second order accuracy in both time and space.

Rest of the paper is organized as follows: Sect. 2 introduces the Asian Option pricing model and the applied modifications for solving it numerically. In Sect. 3, the discrete scheme is presented. In Sect. 4, the convergence of the approximate solution is proved. Section 5 gives the numerical experiments based on our scheme and Sect. 6 concludes the paper.

2   Asian Option Pricing Model

The European style Asian arithmetic average fixed strike call option price $V(S, A, \tau)$ is determined by [13,31]

$$
V(S, A, \tau) = \exp\left(-\int_{\tau}^{T} r(z)dz\right) v(S, A, \tau),
\tag{5}
$$

where the risk free interest rate $r(\tau)$ is taken as function of time variable, T is the maturity time of the contract, A is the continuous arithmetic running sum defined as

$$
A(\tau) = \int_{0}^{\tau} S(z)dz
\tag{6}
$$

and $v(S, A, \tau)$ is governed by the following partial differential equation

$$
\frac{\partial v}{\partial \tau} + \frac{1}{2}\sigma^2(\tau)S^2\frac{\partial^2 v}{\partial S^2} + (r(\tau) - D(\tau))S\frac{\partial v}{\partial S} + S\frac{\partial v}{\partial A} - r(\tau)v = 0,
\tag{7}
$$

along with the terminal condition

$$
v(S, A, T) = \max\left(\frac{A}{T} - K, 0\right),
\tag{8}
$$

where $\sigma(\tau)$ and $D(\tau)$ are taken as functions of time, and K is the strike price. The current asset price S and the history of asset prices are independent. Hence the variables S, A and τ are independent variables. The Eq. (7) is a two dimensional ultra-parabolic partial differential equation which is numerically expensive to approximate. Inspired from Rogers and Shi [24] and Alziary, Décamps and Koehl [1], we use the following change of variables to transform the problem (7)–(8)

from final value problem to initial value problem and to reduce the dimension of the problem

$$\tilde{x} = \frac{K - A/T}{S}, \quad \tau = T - t \text{ and } v(S, A, \tau) = S\tilde{u}(\tilde{x}, t), \tag{9}$$

where $\tilde{x}$ and t are the new space and time variables respectively. Introducing the new variables

$$\hat{\sigma}(t) = \sigma(\tau),$$
$$\hat{r}(t) = r(\tau),$$
$$\hat{D}(t) = D(\tau).$$

The transformed problem is as follows

$$\frac{\partial \tilde{u}}{\partial t} = \frac{1}{2}\hat{\sigma}^2(t)\tilde{x}^2 \frac{\partial^2 \tilde{u}}{\partial \tilde{x}^2} - \left(\frac{1}{T} + (\hat{r}(t) - \hat{D}(t))\tilde{x}\right)\frac{\partial \tilde{u}}{\partial \tilde{x}}$$
$$-\hat{D}(t)\tilde{u}, \quad \tilde{x} \in (0, \infty), \ t \in (0, T), \tag{10}$$

along with the terminal condition

$$\tilde{u}(\tilde{x}, 0) = \max(-\tilde{x}, 0). \tag{11}$$

The boundary conditions for the problem (7)–(8) are well established in [12]. The left boundary condition is

$$v(0, A, \tau) = \exp\left(-\int_\tau^T r(z)dz\right)\max\left(\frac{A}{T} - K, 0\right). \tag{12}$$

The value of Asian call and put options are interrelated by the put-call parity. A closed form solution for the Asian option price for the case $A > KT$ is derived using the put-call parity [1, 10, 16], which is as follows

$$v(S, A, \tau) = \left(\frac{A}{T} - K\right)\exp\left(-\int_\tau^T r(z)dz\right)$$
$$+ \frac{S}{T}\int_\tau^T \exp\left(-\left(\int_\tau^y D(z)dz + \int_y^T r(z)dz\right)\right)dy, \quad \tau \in [0, T]. \tag{13}$$

Hence we will solve the problem (7)–(8) only for the case $A \leq KT$. Applying the transformations (9) to the formula (13), we get

$$\tilde{u}(\tilde{x}, t) = -\tilde{x}\exp\left(-\int_0^t \hat{r}(z)dz\right)$$
$$+ \frac{1}{T}\int_0^t \exp\left(-\left(\int_0^y \hat{r}(z)dz + \int_y^t \hat{D}(z)dz\right)\right)dy,$$
$$\text{for } A \leq KT, \ \tilde{x} \in [0, \infty), \ t \in [0, T]. \tag{14}$$

The left boundary condition of the problem (10)–(11) is obtained from (14)

$$\tilde{u}(0,t) = \frac{1}{T} \int_0^t \exp\left(-\left(\int_0^y \hat{r}(z)dz + \int_y^t \hat{D}(z)dz\right)\right) dy,$$

$$\text{for } A \le KT, \ t \in [0,T]. \quad (15)$$

The right boundary condition of the problem (10)–(11) for large $\tilde{x}$ is obtained from (12)

$$\tilde{u}(\infty,t) = 0, \text{ for } A \le KT, \ t \in [0,T]. \quad (16)$$

Let us introduce another change of variable

$$x = \exp(-\tilde{x}), \quad (17)$$

which converts the semi-infinite space domain $(0,\infty)$ into a finite space domain $(0,1)$ [1]. Accordingly the partial differential equation (10) is transformed as

$$\frac{\partial u}{\partial t} = \frac{1}{2}\hat{\sigma}^2(t) \left(\ln(x)\right)^2 x^2 \frac{\partial^2 u}{\partial x^2}$$
$$+ \left(\frac{1}{2}\hat{\sigma}^2(t) \left(\ln(x)\right)^2 + \frac{1}{T} - \left(\hat{r}(t) - \hat{D}(t)\right)\ln(x)\right) x \frac{\partial u}{\partial x} - \hat{D}(t)u, \quad (18)$$

and the initial and boundary conditions (11), (15) and (16) are transformed as

$$u(x,0) = 0, \ x \in [0,1], \quad (19)$$

$$u(1,t) = \frac{1}{T} \int_0^t \exp\left(-\left(\int_0^y \hat{r}(z)dz + \int_y^t \hat{D}(z)dz\right)\right) dy,$$

$$\text{for } A \le KT, \ t \in [0,T] \quad (20)$$

and

$$u(0,t) = 0, \text{ for } A \le KT, \ t \in [0,T] \quad (21)$$

respectively. Let $\Omega_x = (0,1)$ be the space domain, $\Omega_t = (0,T)$ be the time domain and $\Omega = \Omega_x \times \Omega_t$. The final problem to be solved numerically is

$$Lu(x,t) \equiv \frac{\partial u}{\partial t} - a_2(x,t)\frac{\partial^2 u}{\partial x^2} - a_1(x,t)\frac{\partial u}{\partial x} - a_0(t)u = f(x,t), \ (x,t) \in \Omega, \quad (22a)$$

where

$$a_2(x,t) = \frac{1}{2}\hat{\sigma}^2(t) \left(\ln(x)\right)^2 x^2,$$
$$a_1(x,t) = \left(\frac{1}{2}\hat{\sigma}^2(t) \left(\ln(x)\right)^2 + \frac{1}{T} - \left(\hat{r}(t) - \hat{D}(t)\right)\ln(x)\right) x,$$
$$a_0(t) = -\hat{D}(t),$$
$$f(x,t) = 0,$$

with the initial condition

$$u(x,0) = 0, \ x \in \bar{\Omega}_x, \tag{22b}$$

and the boundary conditions

$$u(0,t) = 0, \ t \in \bar{\Omega}_t \tag{22c}$$

and

$$u(1,t) = \frac{1}{T} \int_0^t \exp\left(-\left(\int_0^y \hat{r}(z)dz + \int_y^t \hat{D}(z)dz\right)\right) dy, \ t \in \bar{\Omega}_t. \tag{22d}$$

In a financial market, it is reasonable to assume that $\hat{r} \geq \hat{D}$ [28]. Also assume that

$$\left|\frac{\partial^{m+n} u}{\partial x^m \partial t^n}(x,t)\right| \leq C \text{ on } \bar{\Omega}; \ 0 \leq n \leq 3 \text{ and } 0 \leq m+n \leq 4. \tag{23}$$

Note that C is a generic constant.

3   Discretization

Discretize the domain $\bar{\Omega}$ uniformly in both the space and time directions. Let the discrete space direction be $\bar{\Omega}_h = \{x_m | m = 0, 1, ...M\}$ where M is the total number of grid intervals in space direction with uniform spacing $h = x_m - x_{m-1}$, $m = 1, 2, ...M$. Let the discrete time direction be $\bar{\Omega}^k = \{t_n | n = 0, 1, ...N\}$ where N is the total number of grid intervals in time direction with uniform spacing $k = t_n - t_{n-1}$, $n = 1, 2, ...N$. Now the problem (22a–22d) is discretized simultaneously in both the space and time directions using two-step backward differentiation formula for temporal discretization and the High Order Difference approximation with Identity Expansion (HODIE) scheme with the stencil points $\{x_{m-1}, x_m, x_{m+1}, \ m = 1, 2, ...M - 1\}$ and two nodal auxiliary points $\{x_m, x_{m+1}, \ m = 1, 2, ...M - 1\}$, for the spacial discretization. The discretization of the partial differential equation (22a) is as follows

$$\beta^n_{m,c}(\delta_t U^n_m) + \beta^n_{m,+}(\delta_t U^n_{m+1}) + [\alpha^n_{m,-} U^n_{m-1} + \alpha^n_{m,c} U^n_m + \alpha^n_{m,+} U^n_{m+1}] =$$
$$\beta^n_{m,c} f^n_m + \beta^n_{m,+} f^n_{m+1}, \ m = 1, 2, ..., M - 1, \ n = 1, 2, ..., N, \tag{24}$$

where $\alpha's$ and $\beta's$ are the HODIE coefficients to be computed, U^n_m, $m = 0, 1, ...M$, $n = 0, 1, ...N$ is the numerical approximation of the solution $u(x,t)$ of the problem (22a–22d) at the grid point (x_m, t_n), $m = 0, 1, ...M$, $n = 0, 1, ...N$ and the operator δ_t is defined as follows

$$\delta_t U^n_m = (U^n_m - U^{n-1}_m)/k, \ n = 1,$$

$$\delta_t U^n_m = \left(\frac{3}{2} U^n_m - 2U^{n-1}_m + \frac{1}{2} U^{n-2}_m\right)/k, \ n = 2, 3, ..., N,$$

We make the Eq. (24) exact on P_3, the space of polynomials with degree less than or equal to 3 and use the following normalization condition

$$\beta_{m,c}^n + \beta_{m,+}^n = 1, \ m = 1, 2, ... M - 1, \ n = 1, 2, ..., N, \tag{25}$$

to compute the HODIE coefficients uniquely, which are as follows

$$\alpha_{m,-}^n = \frac{\beta_{m,c}^n(-2a_{2,m}^n + ha_{1,m}^n) + \beta_{m,+}^n(-2a_{2,m+1}^n - ha_{1,m+1}^n)}{2h^2}, \tag{26}$$

$$\alpha_{m,+}^n = \frac{\beta_{m,c}^n(-2a_{2,m}^n - ha_{1,m}^n) + \beta_{m,+}^n(-2a_{2,m+1}^n - 3ha_{1,m+1}^n - 2h^2a_0^n)}{2h^2}, \tag{27}$$

$$\alpha_{m,c}^n = \frac{\beta_{m,c}^n(4a_{2,m}^n - 2h^2a_0^n) + \beta_{m,+}^n(4a_{2,m+1}^n + 4ha_{1,m+1}^n)}{2h^2}, \tag{28}$$

$$\beta_{m,c}^n = \frac{6ha_{2,m+1}^n + 2h^2a_{1,m+1}^n}{6ha_{2,m+1}^n + 2h^2a_{1,m+1}^n + h^2a_{1,m}^n} \tag{29}$$

and

$$\beta_{m,+}^n = \frac{h^2a_{1,m}^n}{6ha_{2,m+1}^n + 2h^2a_{1,m+1}^n + h^2a_{1,m}^n}. \tag{30}$$

Now using these coefficients, the final scheme to compute the numerical approximation to the solution of the problem (22a–22d) is as follows

$$L_h^k U_m^n \equiv \alpha_{m,-}^n U_{m-1}^n + \left(\alpha_{m,c}^n + \frac{1}{k}\beta_{m,c}^n\right) U_m^n + (\alpha_{m,+}^n + \frac{1}{k}\beta_{m,+}^n)U_{m+1}^n$$

$$= \beta_{m,c}^n \left(f_m^n + \frac{1}{k}U_m^{n-1}\right) + \beta_{m,+}^n \left(f_{m+1}^n + \frac{1}{k}U_{m+1}^{n-1}\right)$$

$$= F_m^n \text{ (say), } m = 1, 2, ... M - 1, \ n = 1, \tag{31a}$$

$$L_h^k U_m^n \equiv \alpha_{m,-}^n U_{m-1}^n + \left(\alpha_{m,c}^n + \frac{3}{2k}\beta_{m,c}^n\right) U_m^n + (\alpha_{m,+}^n + \frac{3}{2k}\beta_{m,+}^n)U_{m+1}^n$$

$$= \beta_{m,c}^n \left(f_m^n + \frac{2}{k}U_m^{n-1} - \frac{1}{2k}U_m^{n-2}\right) + \beta_{m,+}^n \left(f_{m+1}^n + \frac{2}{k}U_{m+1}^{n-1} - \frac{1}{2k}U_{m+1}^{n-2}\right)$$

$$= F_m^n \text{ (say), } m = 1, 2, ..., M - 1, \ n = 2, 3, ...N, \tag{31b}$$

$$U_m^0 = 0, \ m = 0, 1, ..., M, \tag{31c}$$

$$U_0^n = 0, \ n = 0, 1, ...N. \tag{31d}$$

$$U_M^n = \frac{1}{T} \int_0^{t_n} \exp\left(-\left(\int_0^y \hat{r}(z)dz + \int_y^{t_n} \hat{D}(z)dz\right)\right) dy, \ n = 0, 1, ...N, \tag{31e}$$

Now putting $m = 1, 2, ... M$ in (31a) and (31b) and shifting the boundary components to the right side of the equation, we can write the scheme in the matrix form as

$$A^n U^n = b^n, \quad n = 1, 2, ..., N, \tag{32}$$

where $U^n = [U_1^n, U_2^n, ..., U_{M-1}^n]^T$.

4   Error Analysis

Lemma 1. *Assume that*

$$h\beta_{m,c}^n a_{1,m}^n \le 2(\beta_{m,c}^n a_{2,m}^n + \beta_{m,+}^n a_{2,m+1}^n) + h\beta_{m,+}^n a_{1,m+1}^n,$$
$$m = 1, 2, ..., M - 1, \ n = 1, 2, ..., N, \tag{33}$$

and

$$\beta_{m,+}^n \left(-a_0^n + \frac{3}{2k} \right) \le \frac{2(\beta_{m,c}^n a_{2,m}^n + \beta_{m,+}^n a_{2,m+1}^n) + h(\beta_{m,c}^n a_{1,m}^n + 3\beta_{m,+}^n a_{1,m+1}^n)}{2h^2},$$
$$m = 1, 2, ..., M - 1, \ n = 2, 3, ..., N. \tag{34}$$

then

$$\alpha_{m,-}^n \le 0, \ m = 1, 2, ..., M - 1, \ n = 1, 2, ..., N, \tag{35}$$

$$\alpha_{m,+}^n + \frac{3}{2k}\beta_{m,+}^n \le 0, \ m = 1, 2, ..., M - 1, \ n = 2, 3, ..., N. \tag{36}$$

Proof. The inequalities (35) and (36) can be easily obtained from (26), (27), (33) and (34).

Remark 1. Similarly assume that

$$\beta_{m,+}^n \left(-a_0^n + \frac{1}{k} \right) \le \frac{2(\beta_{m,c}^n a_{2,m}^n + \beta_{m,+}^n a_{2,m+1}^n) + h(\beta_{m,c}^n a_{1,m}^n + 3\beta_{m,+}^n a_{1,m+1}^n)}{2h^2},$$
$$m = 1, 2, ..., M - 1, \ n = 1, \tag{37}$$

then

$$\alpha_{m,+}^n + \frac{1}{k}\beta_{m,+}^n \le 0, \ m = 1, 2, ..., M - 1, \ n = 1. \tag{38}$$

Lemma 2. *(Discrete Maximum Principle).* Under the assumptions of the Lemma 1, the operator L_h^k defined by (31a)–(31b) satisfies discrete maximum principle, that is, if v_m^n and w_m^n are mesh functions that satisfy $v_0^n \le w_0^n$, $v_M^n \le w_M^n$ $(n = 0, 1, ..., N)$, $v_m^0 \le w_m^0$, $(m = 0, 1, ..., M)$ and $L_h^k v_m^n \le L_h^k w_m^n$ $(m = 1, 2, ..., M - 1, \ n = 1, 2, ..., N)$, then $v_m^n \le w_m^n$ for all m, n.

Proof. The row sum of the matrices $A^n, n = 1, 2, ... N$, are

$$\alpha_{m,-}^n + \left(\alpha_{m,c}^n + \frac{3}{2k}\beta_{m,c}^n \right) + \left(\alpha_{m,+}^n + \frac{3}{2k}\beta_{m,+}^n \right) = -a_0^n + \frac{3}{2k} \tag{39}$$

$$> 0, \tag{40}$$

$m = 1, 2, ..., M - 1$, $n = 1, 2, ..., N$. $\alpha_{m,-}^n$, $m = 1, 2, ..., M - 1$, $n = 1, 2, ..., N$ are the sub-diagonal elements and $\alpha_{m,+}^n + \frac{3}{2k}\beta_{m,+}^n$, $m = 1, 2, ..., M - 1$, $n = 1, 2, ..., N$ are the super-diagonal elements of the tridiagonal matrices A^n, $n = 1, 2, ...N$. Under the hypothesis of the Lemma 1, the inequalities (35)–(36) and (39) show that the coefficient matrices corresponding to the discrete operator L_h^k are irreducible M-matrices which preserve the positivity. Hence the solution to the linear system of equations (32) exists and if v_m^n and w_m^n are mesh functions as defined in Lemma, then $v_m^n \leq w_m^n$ for all m, n.

The operator L_h^k, satisfying the discrete maximum principle, establishes the stability of the scheme.

Theorem 1. *Let u be the solution of the continuous problem (22a–22d) and U_m^n be the solution of the discrete problem (31a–31e). Then under the assumptions (23),*

$$|u_m^n - U_m^n| \leq C(k^2 + h^2), \quad m = 0, 1, ..., M, \ n = 0, 1, ..., N, \tag{41}$$

where C is a positive constant independent of k and h.

Proof. From (24) we have

$$L_h^k u_m^n - (Lu)_m^n = \frac{\beta_{m,c}^n}{k}\left(\frac{3}{2}u_m^n - 2u_m^{n-1} + \frac{1}{2}u_m^{n-2}\right)$$

$$+ \frac{\beta_{m,+}^n}{k}\left(\frac{3}{2}u_{m+1}^n - 2u_{m+1}^{n-1} + \frac{1}{2}u_{m+1}^{n-2}\right) + [\alpha_{m,-}^n u_{m-1}^n + \alpha_{m,c}^n u_m^n + \alpha_{m,+}^n u_{m+1}^n]$$

$$- \beta_{m,c}^n\left(\frac{\partial u_m^n}{\partial t} - a_{2,m}^n\frac{\partial^2 u_m^n}{\partial x^2} - a_{1,m}^n\frac{\partial u_m^n}{\partial x} - a_0^n u_m^n\right)$$

$$- \beta_{m,+}^n\left(\frac{\partial u_{m+1}^n}{\partial t} - a_{2,m+1}^n\frac{\partial^2 u_{m+1}^n}{\partial x^2} - a_{1,m+1}^n\frac{\partial u_{m+1}^n}{\partial x} - a_0^n u_{m+1}^n\right). \tag{42}$$

Applying the Taylor's expansion in two variables and taking modulus, we have

$$\left|L_h^k(u_m^n - U_m^n)\right| = \left|L_h^k u_m^n - (Lu)_m^n\right|$$

$$\leq C_1 k^2 \int_0^1 \left[\left|\frac{\partial^3 u}{\partial t^3}(x_m, t_n - ky)\right| + \left|\frac{\partial^3 u}{\partial t^3}(x_m, t_n - 2ky)\right|\right.$$

$$+ \left|\frac{\partial^3 u}{\partial t^3}(x_m + hy, t_n - ky)\right| + \left.\left|\frac{\partial^3 u}{\partial t^3}(x_m + hy, t_n - 2ky)\right|\right] dy$$

$$+ C_2 h^2 \int_0^1 \left[\left|\frac{\partial^3 u}{\partial x^2 \partial t}(x_m + hy, t_n - ky)\right| + \left|\frac{\partial^3 u}{\partial x^2 \partial t}(x_m + hy, t_n - 2ky)\right|\right.$$

$$+ \left|\frac{\partial^3 u}{\partial x^2 \partial t}(x_m + hy, t_n)\right| + \left.x_{m+1}^2\left|\frac{\partial^4 u}{\partial x^4}(x_m + hy, t_n)\right|\right] dy$$

$$\leq C_3(k^2 + h^2), \ m = 1, 2, ..., M - 1, \ n = 2, 3, ..., N. \tag{43}$$

For $n = 1$, backward Euler method is used for the computation of the discrete solution, and the one-step error of the backward Euler method is $O(k^2)$ [18]. Hence we have

$$\left|L_h^k(u_m^n - U_m^n)\right| \le C_4(k^2 + h^2), \quad m = 1, 2, ..., M - 1, \quad n = 0, 1, ..., N. \quad (44)$$

We construct the following barrier function

$$\varphi_m^n = C(k^2 + h^2)(1 + t_n) \pm (u_m^n - U_m^n),$$

and apply the Lemma 2 on φ_m^n, to show

$$|u_m^n - U_m^n| \le C(k^2 + h^2) \text{ for every } m = 1, 2, ..., M - 1, \quad n = 1, 2, ...N. \quad (45)$$

5   Numerical Experiments

Since the closed form solution of the considered Asian option pricing model is not known, we use double mesh principle to find the maximum error (E_{max}), the root mean square error (E_{rms}) and the corresponding order of convergence p_{max} and p_{rms} as follows

$$E_{max}^{M,N} = \max_{0 \le m \le M} \left|U^{M,N}(x_m, t_N) - U^{2M,2N}(x_{2m}, t_{2N})\right|,$$

$$E_{rms}^{M,N} = \sqrt{\frac{\sum_{m=0}^{M}\left[\left(U^{M,N}(x_m, t_N) - U^{2M,2N}(x_{2m}, t_{2N})\right)^2\right]}{M + 1}},$$

$$p_{max}^{M,N} = \log_2\left(\frac{E_{max}^{M,N}}{E_{max}^{2M,2N}}\right) \text{ and } p_{rms}^{M,N} = \log_2\left(\frac{E_{rms}^{M,N}}{E_{rms}^{2M,2N}}\right).$$

Example 1. Consider the initial boundary value problem (22a–22d) with $\hat{\sigma} = 0.5$, $\hat{r} = 0.09$, $\hat{D} = 0$. Take $T = 3$ and $K = 40$. The computed solution is shown in Fig. 1 and the convergence results are given in Table 1.

Example 2. Consider the initial boundary value problem (22a–22d) with $\hat{\sigma}(t) = 0.4(2 + (T - t))$, $\hat{r}(t) = 0.06(1 + t)$, $\hat{D}(t) = 0.02 \exp(-t)$. Take $T = 1$ and $K = 40$. The computed solution is shown in Fig. 2 and the convergence results are given in Table 2.

Example 3. Consider the initial boundary value problem (22a–22d) with $\hat{\sigma}(t) = 0.4(2 + \sin(T - t))$, $\hat{r}(t) = 0.06 \exp(t)$, $\hat{D}(t) = 0.02 \sin(t)$. Take $T = 1$ and $K = 40$. The computed solution is shown in Fig. 3 and the convergence results are given in Table 3.

Table 1. Maximum absolute error (E_{max}), root mean square error (E_{rms}) and corresponding orders of convergence p_{max} and p_{rms} for Example 1

M	10	10×2	10×2^2	10×2^3	10×2^4	10×2^5
N	6	6×2	6×2^2	6×2^3	6×2^4	6×2^5
E_{max}		1.4876e−02	4.3894e−03	1.1038e−03	2.6960e−04	6.6220e−05
p_{max}			1.7610	1.9915	2.0337	2.0255
E_{rms}		7.8685e−03	2.1324e−03	4.7253e−04	1.1057e−04	2.6846e−05
p_{rms}			1.8835	2.1740	2.0953	2.0422

Table 2. Maximum absolute error (E_{max}), root mean square error (E_{rms}) and corresponding orders of convergence p_{max} and p_{rms} for Example 2

M	8	8×2	8×2^2	8×2^3	8×2^4	8×2^5
N	2	2×2	2×2^2	2×2^3	2×2^4	2×2^5
E_{max}		6.2322e−02	2.6239e−02	8.4130e−03	2.1647e−03	5.1478e−04
p_{max}			1.2480	1.6411	1.9584	2.0722
E_{rms}		3.7783e−02	1.4681e−02	4.4301e−03	9.9501e−04	2.2237e−04
p_{rms}			1.3638	1.7285	2.1546	2.1617

Table 3. Maximum absolute error (E_{max}), root mean square error (E_{rms}) and corresponding orders of convergence p_{max} and p_{rms} for Example 3

M	8	8×2	8×2^2	8×2^3	8×2^4	8×2^5
N	2	2×2	2×2^2	2×2^3	2×2^4	2×2^5
E_{max}		6.2810e−02	2.6458e−02	8.5204e−03	2.1947e−03	5.2182e−04
p_{max}			1.2473	1.6347	1.9569	2.0724
E_{rms}		3.7829e−02	1.4825e−02	4.4948e−03	1.0102e−03	2.2545e−04
p_{rms}			1.3515	1.7217	2.1535	2.1639

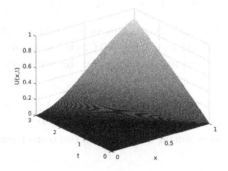

Fig. 1. Computed solution of Asian option price for Example 1

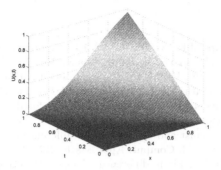

Fig. 2. Computed solution of Asian option price for Example 2

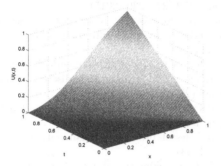

Fig. 3. Computed solution of Asian option price for Example 3

6 Conclusions

In the present work, HODIE scheme with two auxiliary points is used to achieve second order accuracy in space and two-step backward differentiation formula is applied for temporal approximation yielding second order accuracy in time also. The main features of our scheme is that it easily deals with variable coefficients occurring in the partial differential equation as we have taken volatility (σ), risk-free interest rate (r) and dividend yield (D) not as constants but as function of the time variable, which is more likely to happen in real financial market. Also this scheme can be applied simultaneously in space and time directions which leads to simpler analysis of the convergence of the solution [7]. It deals with the degeneracy issue of this problem easily without any extra efforts unlike in other works done in this field [5,6,33]. The numerical results are in accordance with the theoretical results.

Acknowledgment. The first author would like to thank National Board for Higher Mathematics, India for financial support.

References

1. Alziary, B., Décamps, J.P., Koehl, P.F.: A PDE approach to Asian options: analytical and numerical evidence. J. Bank. Financ. **21**, 613–640 (1997)
2. Barucci, E., Polidoro, S., Vespri, V.: Some results on partial differential equations and Asian options. Math. Models Methods Appl. Sci. **11**, 475–497 (2001)
3. Benhamou, E., Duguet, A.: Small dimension PDE for discrete Asian options. J. Econom. Dynam. Control **27**, 2095–2114 (2003)
4. Cen, Z., Le, A.: A robust and accurate finite difference method for a generalized Black-Scholes equation. J. Comput. Appl. Math. **235**, 3728–3733 (2011)
5. Cen, Z., Le, A., Xu, A.: Finite difference scheme with a moving mesh for pricing Asian options. Appl. Math. Comput. **219**, 8667–8675 (2013)
6. Cen, Z., Xu, A., Le, A.: A hybrid finite difference scheme for pricing Asian options. Appl. Math. Comput. **252**, 229–239 (2015)
7. Clavero, C., Gracia, J.L., Stynes, M.: A simpler analysis of a hybrid numerical method for time-dependent convection-diffusion problems. J. Comput. Appl. Math. **235**, 5240–5248 (2011)
8. Company, R., González, A.L., Jódar, L.: Numerical solution of modified Black-Scholes equation pricing stock options with discrete dividend. Math. Comput. Model. **44**, 1058–1068 (2006)
9. Conze, A., Viswanathan, R.: European path dependant options: the case of geometric averages. Finance **12**, 7–22 (1991)
10. Geman, H., Yor, M.: Bessel processes, Asian options, and perpetuities. Math. Finance **3**, 349–375 (1993)
11. Goard, J.: Exact and approximate solutions for options with time-dependent stochastic volatility. Appl. Math. Model. **38**, 2771–2780 (2014)
12. Hugger, J.: Wellposedness of the boundary value formulation of a fixed strike Asian option. J. Comput. Appl. Math. **185**, 460–481 (2006)
13. Ingersoll, J.E.: Theory of Financial Decision Making. Roman & Littlefield, Totowa (1987)
14. Ju, N.: Pricing Asian and basket options via Taylor expansion. J. Comput. Finance **5**, 79–103 (2002)
15. Kadalbajoo, M.K., Tripathi, L.P., Kumar, A.: A cubic B-spline collocation method for a numerical solution of the generalized Black-Scholes equation. Math. Comput. Model. **55**, 1483–1505 (2012)
16. Kemna, A.G.Z., Vorst, A.C.F.: A pricing method for options based on average asset values. J. Bank. Financ. **14**, 113–129 (1990)
17. Kumar, A., Tripathi, L.P., Kadalbajoo, M.K.: A numerical study of Asian option with radial basis functions based finite differences method. Eng. Anal. Bound. Elem. **50**, 1–7 (2015)
18. LeVeque, R.J.: Finite Difference Methods for Ordinary and Partial Differential Equations. Society for Industrial and Applied Mathematics (SIAM), Philadelphia, PA (2007)
19. Levy, E.: Pricing European average rate currency options. J. Int. Money Financ. **11**, 474–491 (1992)
20. Milevsky, M.A., Posner, S.E.: Asian options, the sum of lognormals, and the reciprocal gamma distribution. J. Financ. Quant. Anal. **33**, 409–422 (1998)
21. Mudzimbabwe, W., Patidar, K.C., Witbooi, P.J.: A reliable numerical method to price arithmetic Asian options. Appl. Math. Comput. **218**, 10934–10942 (2012)

22. Rao, S.C.S., Manisha: High-order numerical method for generalized Black-Scholes model. In: Proceedings of International Conference on Computational Science, San Diego, California, USA, pp. 1765–1776 (2016). Procedia Computer Science
23. Rao, S.C.S., Manisha: Numerical solution of generalized Black-Scholes model. Appl. Math. Comput. **321**, 401–421 (2018)
24. Rogers, L.C.G., Shi, Z.: The value of an Asian option. J. Appl. Probab. **32**, 1077–1088 (1995)
25. Thompson, G.W.P.: Fast narrow bounds on the value of Asian options. Technical report, Center for Financial Research, Judge Institute of Management Science, University of Cambridge (1999)
26. Turnbull, S.M., Wakeman, L.M.: A quick algorithm for pricing European average options. J. Financ. Quant. Anal. **26**, 377–389 (1991)
27. Valkov, R.: Fitted finite volume method for a generalized Black-Scholes equation transformed on finite interval. Numer. Algorithms **65**, 195–220 (2014)
28. Vázquez, C.: An upwind numerical approach for an American and European option pricing model. Appl. Math. Comput. **97**, 273–286 (1998)
29. Verrall, D.P., Read, W.W.: A quasi-analytical approach to the advection-diffusion-reaction problem, using operator splitting. Appl. Math. Model. **40**, 1588–1598 (2016)
30. Wang, S.: A novel fitted finite volume method for the Black-Scholes equation governing option pricing. IMA J. Numer. Anal. **24**, 699–720 (2004)
31. Wilmott, P., Dewynne, J., Howison, S.: Option Pricing: Mathematical Models and computation. Oxford Financial, Oxford (1998)
32. Zhang, J.E.: Pricing continuously sampled Asian options with perturbation method. J. Fut. Mark. **23**, 535–560 (2003)
33. Zvan, R., Forsyth, P.A., Vetzal, K.: Robust numerical methods for PDE models of Asian options. J. Comput. Finance **1**, 39–78 (1998)

Improving Portfolio Optimization Using Weighted Link Prediction in Dynamic Stock Networks

Douglas Castilho[1,2,3]($\boxtimes$), João Gama[2], Leandro R. Mundim[1], and André C. P. L. F. de Carvalho[1]

[1] Institute of Mathematical and Computer Sciences (ICMC),
University of São Paulo (USP), São Carlos, Brazil
douglas.castilho@usp.br, {mundim,andre}@icmc.usp.br
[2] Institute for Systems and Computer Engineering, Technology and Science,
University of Porto (UP), Porto, Portugal
jgama@fep.up.pt
[3] Laboratory of Technology and Innovation (LATIN),
Federal Institute of South of Minas Gerais (IFSULDEMINAS),
Poços de Caldas, Brazil

Abstract. Portfolio optimization in stock markets has been investigated by many researchers. It looks for a subset of assets able to maintain a good trade-off control between risk and return. Several algorithms have been proposed to portfolio management. These algorithms use known return and correlation data to build subset of recommended assets. Dynamic stock correlation networks, whose vertices represent stocks and edges represent the correlation between them, can also be used as input by these algorithms. This study proposes the definition of constants of the classical mean-variance analysis using machine learning and weighted link prediction in stock networks (method named as MLink). To assess the performance of MLink, experiments were performed using real data from the Brazilian Stock Exchange. In these experiments, MLink was compared with mean-variance analysis (MVA), a popular method to portfolio optimization. According to the experimental results, using weighted link prediction in stock networks as input considerably increases the performance in portfolio optimization task, resulting in a gross capital increase of 41% in 84 days.

Keywords: Stock market · Dynamic stock networks ·
Machine learning · Portfolio optimization

1 Introduction

Portfolio optimization in stock markets is the process of selecting a subset of assets that maintain an expected trade-off control between risk and return [16]. The portfolio selection process consists of finding, in a large collection of stocks,

© Springer Nature Switzerland AG 2019
J. M. F. Rodrigues et al. (Eds.): ICCS 2019, LNCS 11538, pp. 340–353, 2019.
https://doi.org/10.1007/978-3-030-22744-9_27

the participation (i.e. individual proportion) of each stock that minimizes the portfolio's risk at a given portfolio return, or maximizes the portfolio's return at a given risk [8]. This topic has been investigated by many researchers from many areas, such as optimization, machine learning (ML) and economics. Usually, portfolio selection algorithms use return and risk measures from a set of assets to make decision. This trade-off between risk and return is used to suggest a subset of assets that will be in the portfolio. Commonly used measures include price mean-return and price variance.

The correlation between asset prices is also important for portfolio management. It has been used in several works to create a network perspective that characterizes the complex structures in the stock market, also known as stock networks or financial networks [5,15,18]. The correlation is so important that some authors suggest the use of topological information derived from financial networks for portfolio management [13,24,32]. As an example, we can create portfolios by applying clustering algorithms or centrality measures in stock networks. However, despite its importance, we did not find works exploring the prediction of weighted links in financial networks to improve the performance of portfolio optimization algorithms. However, we found works using price and return forecasting to improve portfolio management results [19].

This study proposes a new approach to define the constants of the classic mean-variance analysis (MVA) from [16]. The new measures for expected return and asset correlation are calculated using by a new method. The proposed method, ML Weighted Link Prediction Analysis (MLink), provides information to the mathematical model of Markowitz, which is often used as basis for portfolio optimization. MLink was applied to dynamic temporal stock networks to induce a predictive model for return price forecast and weighted link prediction. We executed several experiments to assess the performance of our method. When it was compared with MVA, experimental results shows that MLink has 56% over MVA. In addition, MLink has better results than other variant methods. These variants use ARIMA, Median and Mean to return forecast combined with Weighted Link Prediction, named as ARIMA-MLink, Median-MLink and Mean-MLink, respectively.

2   Problem Definition and Related Works

The analysis of the behavior and interaction between assets in stock markets has been widely studied in the literature [3,4,30]. To formally describe it, consider i and j two distinct assets belonging to a specific set A. Let S_i and S_j be time series related to i and j, respectively. A weighted stock network can be represented by a graph $G = (V, E)$, where V is the set of assets belonging to A, $V \subset A$, and E contains all possible pairs of assets $i, j \mid i, j \in V$. The relationship between i and j is measured using correlation metrics, assigning a weight w to each edge in E. A set of time ordered sequence of weighted stock networks graphs represents Dynamic Temporal Stock Networks [10].

In this work, we want to to answer the following question: given a set of graphs $G_1, G_2, ..., G_N$ related to a temporal sequence of weighted stock networks in time

N and G_{N+1} a graph whose link weights were predicted using ML algorithms, can the combination of weighted link prediction by ML with mathematical models of portfolio optimization improve the trade-off between risk and return?

2.1 Portfolio Optimization Using Stock Network and Forecast

Several works apply ML or optimization algorithms for portfolio management [16,17]. Despite its relevance, few papers investigate the use of stock network structure to portfolio management. In [24], the authors explore centrality in complex networks to improve portfolio selection process via targeting a group of stocks belonging to certain region of the stock market network. The work [32], like this paper, analyze time evolving stock markets by using temporal network representation. The authors also propose a portfolio selection tool using temporal centrality in stock networks. A portfolio optimization based on network topology, using cross-correlation of the daily price returns for the American and Chinese stock markets to create networks, is proposed in [13]. Taking into account the importance of correlation matrices, and the possible presence of noise values in these matrices, [22] introduces an approach that allows a systematic investigation of the effect of the different sources of noise in financial correlations in the portfolio and risk management context. We also analyzed different aspects related to noise presence, like the size of time series.

Other works use price and return forecasting to improve portfolio management [19]. In [8], a neural network is used to predict future stock returns. The prediction errors are used as a risk measure. [1] also shows that, for asset allocation decisions, the use of models able to predict return is better than using historical averages. This motivated the proposal of the new method MLink that uses prediction of weighted link formation to improve results of portfolio optimization models. The problem of predicting links in weighted networks is an extension of the problem of link prediction, whose main objective is the detection of hidden links or links that will be formed [28]. In weighted link prediction problem, it is necessary to predict both link and edge weight [14].

3 Methodology

The MLink framework has two steps: (i) ML to predict stock returns and weighted links in dynamic stock networks; (ii) mathematical portfolio optimization model using stock returns and predicted weighted stock networks as input. Figure 1 illustrates this framework.

This section introduces the data set (Sect. 3.1), the weighted link prediction method (Sect. 3.2), return forecast (Sect. 3.3) and the mathematical model (Sect. 3.4).

3.1 Market Data Set

The real data set used in this work was collected from the Brazilian Stock Exchange (BM&F Bovespa) between January 2018 e October 2018. We

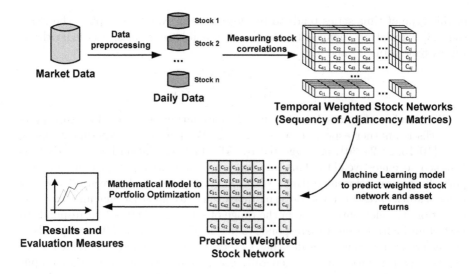

Fig. 1. Framework of the MLink

performed experiments using data regarding Bovespa Index assets (*Ibovespa*)[1]. According to its dynamical structure, this theoretical portfolio has 65 assets currently, but we used 56 assets that remained in Ibovespa list during the entire period. The list of assets used in the experiments can be seen at link below[2]. In the first step, our objective is to identifying link weight between all pairs of assets in a complete weighted network. As traditional portfolio management algorithms, we are interested in reorganize the portfolio every day. Thus, the data were processed to obtain daily price time series. Note that our approach can be easily adapted for weekly, monthly or intraday strategies, according to the main purpose of the investor. It is important to emphasize that these data include an election period with high stock price variations.

3.2 Weighted Link Prediction in Dynamic Stock Networks

In this work, we use a modified version of the method proposed in [15] to create weighted dynamic financial networks. In this method, nodes of the graph represent assets and edges represent the relationship between them. This relationship is based on price time series correlation. Let S_i and S_j both price time series with length L regarding two distinct assets i and j. We can transform these non stationary price time series into a stationary return time series using the following equation:

$$Y_i = ln(P_i(t)) - ln(P_i(t-1))$$

where, $P_i(t) \in S_i$ is the closing price of asset i at day t. In terms of market definitions, this transformation represents the logarithmic return price. The average

---

[1] www.bmfbovespa.com.br.
[2] iccs2019.douglascastilho.com.

of this type of time series tends to be close to zero. Next, we applied the Pearson [23] correlation coefficient between all possible pairs of stocks logarithmic return time series Y_i and Y_j:

$$\rho_{ij} = \frac{cov(Y_i, Y_j)}{\sqrt{var(Y_i) \cdot var(Y_j)}}$$

The number of points L to measure the correlation between all possible pairs of stocks is another aspect to be considered. We performed experiments using $L = \{10, 15, 20, 25, 30\}$, as suggested by [31]. The correlation between assets is assigned to edges weights. This information is used as input to mathematical portfolio optimization algorithm. We used correlation ρ_{ij} between all possible pairs of stocks present in the Bovespa Index to create an adjacency matrix $\mathbf{C}$. By definition, the elements ρ_{ij} are in the range of -1 to 1, where -1 corresponds to perfect anti-correlation, 1 corresponds to perfect correlation and 0 corresponds to absence of correlation. Consider that $\mathbf{C}$ represents a complete undirected stock network. Optimization models generally use similar correlation matrix as input to create a subset of assets with reasonable trade-off between return and risk. In this work, instead using the known correlation matrix at day t, we propose to use a step forward correlation matrix to improve the results of optimization portfolio. Our first main problem is to create machine learning (ML) algorithms able to induce models that can predict the weight of all edges in a future dynamic stock network Δ_{t+1}. For such, we created a method using ML to predict these values. This method uses three different sources of features to build models able to predict correlation values, represented by edge weights. In addition, since our networks are undirected, the number of weighted edges for each graph $G = (V, E)$ is given by $|E| = |V| * (|V| - 1)/2$. Thus, we have $|E| = 56 * (55)/2 = 1540$ number of values to predict for each day.

In this work, we address the weighted link prediction problem as a regression task. For such, we propose the use of supervised ML algorithms. We use three sources of features to predict weighted link formation between assets in dynamic stock networks: *(i)* complex network derived features; *(ii)* domain derived features; *(iii)* return time series forecasts. Each example in the data set is labeled with correlation value between a pair of stocks i and j for a future given period. Next, we present the set of input features used to train the ML algorithms.

Network Derived Features are computed at each iteration using complex network statistical measures. The measures can be divided according to the level of analysis to be performed: at the node-level, where nodes represent assets, and at the link-level [21]. To create each example for the supervised learning data set, metrics related with node i and j are inserted for both nodes. Metrics related with edges are inserted calculating the measures between nodes i and j [20]. Consider $|i|$ as node degree or number of edges.

- **Node-Level Derived Features:** related to the position of the node within the overall structure of the complex network [21]. The following node metrics presented in Table 1 were used as stock i derived features.

- **Link-Level Derived Features:** related to both contents and patterns of edges in complex networks [21]. The following link metrics presented in Table 2 were calculated between i and j stocks.

Table 1. Node-Level Derived Features

Name	Description
Weighted degree	$deg_w(i) = \sum_{j \in |i|} w_{<i,j>}$
Average neighbor weighted degree	$avg_w(i) = \frac{\sum_{j \in |i|} |j| * w_{<i,j>}}{|i|}$
Propensity of i to increase its degree	$\gamma(i) = \frac{|i|}{deg_w(i)}$
Node betweenness	$b_v = \sum_{i,j \in V \setminus v} \frac{\sigma_{ij}(v)}{\sigma_{ij}}$, where $\sigma_{ij}(v)$ is the number of shortest weighted paths between i and j passing through v
Node eigenvector	$x_i \frac{1}{\lambda} \sum_{j=1}^{n} d_{ij} x_j$, where d_{ij} represents an entry of the adjacency matrix C and λ denotes the largest eigenvalue

Table 2. Link-Level Derived Features

Name	Description
Correlation value	value of C_ij
Edge betweenness	$b_e = \sum_{i,j \in V} \frac{\sigma_{ij}(e)}{\sigma_{ij}}$, where $\sigma_{ij}(e)$ is the number of shortest weighted paths between i and j passing through edge e
Same Louvain community [2]	value 1 if i and j belongs to the same Louvain community, 0 otherwise
Same Girvan-Newman community [9]	value 1 if i and j belongs to the same Girvan-Newman community, 0 otherwise
Preferential attachment weighted	$PA(i,j) = deg_w(i) * deg_w(j)$

Domain Features are computed at each day using a set of Technical Analysis Indicators (TAI). An indicator can be defined as a series of data points derived from assets price information applying a mathematical formula [26]. These metrics were calculated using the same set of daily price time series used to create the networks. The relationship between these domain features and the graph networks can be seen as characteristics to describe nodes in a complex network. The domain features used are: Relative Strength Index (RSI), Simple Moving Average (SMA), Exponential Moving Average (EMA), Moving Average Convergence/Divergence (MACD), Average Directional Movement Index (ADX), Aroon Indicator (Aroon), Bollinger Bands (BB), Commodity Channel Index (CCI), Chande Momentum Oscillator (CMO), Rate of Change (ROC) and Average True Range (ATR). More information regarding how to calculate these features can be found in [27] and [26].

To create each example for the ML algorithm training data set, the subset of network derived features is concatenated to the subset of domain derived features. Thus, TAI related to assets i and j are inserted for both nodes.

3.3   Logarithmic Return Forecast Using Machine Learning

Another issue to be considered in our MLink framework is the stock logarithmic return. According to the mathematical optimization model, which will be presented in the following section, it is necessary a measure of return and risk of each asset. Usually, the mean of the returns and the standard deviation are used in optimization algorithms to estimating return and future risk, respectively. For this, we used a ML algorithm (MLP) to forecast the logarithmic return of all assets. For comparative analysis, we also used three statistical methods to estimate assets return. These methods are:

- MLP - Multilayer Perceptron neural network - has powerful approximation capabilities and its self-adaptive data driven modelling approach allow them great flexibility in modelling time series data [12]. The MLP network used has one hidden layer with 5 neurons and is trained using the resilient backpropagation algorithm, a fast weighted update mechanism to feedforward artificial neural networks [25].
- ARIMA - Autoregressive Integrated Moving Average - models are fitted to the time series data to predict future points in the series [6].
- Mean - Simple mean price return using time series data.
- Median - Simple median price return using time series data.

Based on each method result, we then calculated the risk measure, which is given by standard deviation of logarithmic return time series. These return forecasts are used as input to mathematical portfolio optimization model and also as input feature to Mlink to predict weighted link in the stock network. To create each example for the supervised learning data set regarding weighted link prediction in MLink, these forecasted return values of each pair of assets i and j are concatenated to input set of features.

3.4   Mathematical Model to Portfolio Optimization

In [16] Markowitz proposed the first mean-variance model that served as the basis for the Modern Portfolio Theory in financial management. In this theory, an investor wishes to distribute an initial wealth in a set of investments in order to minimize the risk and maximize the return. Naturally, these two objectives are conflicting because if there is a minimum risk investment and maximum return the decision is trivial. Usually the higher the risk the higher the expected return.

For this, [16] proposed a bi-objective quadratic programming model for find the Markowitz efficient front. In mathematical terms: given n assets with return vector $\mu \in \mathbb{R}^n$, estimated covariance matrix $\sigma \in \mathbb{R}^{n \times n}$, and the invested fraction of each asset in optimal portfolio is $x \in \mathbb{R}^n$. To computing the Markowitz efficient front is given by maximizing expected return for a given level of the risk (mean-variance model 1) or minimizing the risk for a given level of the expected return (mean-variance model 2).

$$\text{Maximize} \quad \mathcal{E} = \sum_{i=1}^{n} x_i \mu_i$$

$$\text{Subject to:} \quad \sum_{i=1}^{n} \sum_{j=1}^{N} x_i x_j \sigma_{ij} \leq v^2, \tag{1}$$

$$\sum_{i=1}^{n} x_i = 1,$$

$$x_i \geq 0, \ \forall \, i = 1, ..., n.$$

In mean-variance model 1 the expected value of the portfolio $(\mathcal{E})$ is maximized, subject to a minimum variation (v^2), the sum from fraction of the portfolio is equal 1, and no investment can be negative.

$$\text{Minimize} \quad v^2 = \sum_{i=1}^{n} \sum_{j=1}^{n} x_i x_j \sigma_{ij}$$

$$\text{Subject to:} \quad \sum_{i=1}^{n} x_i \mu_i \geq \mathcal{E}, \tag{2}$$

$$\sum_{i=1}^{n} x_i = 1,$$

$$x_i \geq 0, \ \forall \, i = 1, ..., n.$$

In mean-variance model 2 the objective is minimize the standard deviation (v^2) subject to a level of return $(\mathcal{E})$. Note that, if we vary the of standard deviation (v^2) in mean-variance model 1 or the desired level return $(\mathcal{E})$ in mean-variance model 2 we can build the Markowitz efficient front, i.e. the trade-off between risk and expected return.

[11] examines an alternative formulation for the problem using a measures of absolute and relative risk aversion. Consider u is a von Neumann-Morgenstern utility function. The absolute risk aversion is defined by $R_a = \frac{u''(w)}{u'(w)}$, where w is the valuation of the portfolio. The formulation result is presented below:

$$\text{Maximize} \quad F = \sum_{i=1}^{n} x_i \mu_i - \frac{R_a}{2} \sum_{i=1}^{n} \sum_{j=1}^{N} x_i x_j \sigma_{ij}$$

$$\text{Subject to:} \quad \sum_{i=1}^{n} x_i = 1 \tag{3}$$

$$x_i \geq 0, \ \forall \, i = 1, ..., n.$$

For the model 3 the objective is maximized the expected return of portfolio less $\frac{R_a}{2}$ time standard deviation of portfolio. According to [11] the utility functions (R_a) of negative exponential energy generate very risk-averse portfolios. Thus, the efficient boundary can be obtained for values of $R_a > 0$. The empirical results indicate that: risky portfolios have values of $R_a \leq 2$; moderate risk portfolios have $2 \leq R_a \leq 4$; risk-averse portfolio have $R_a \geq 4$.

4   Experiments

In this section we present results separately. First, we present results regarding to price return forecast using different lengths of L. Second, we present weighted link prediction results related to stock networks. We used the same time series length L to perform both return forecast and weighted link prediction experiments. Finally, we present financial results comparing MLink with Ibovespa, MVA and the three proposed variants ARIMA-MLink, Median-MLink and Mean-MLink.

4.1   Machine Learning for Return Forecast

In this section we present a set of experimental results using different methods to asset return forecast. We used different time series lengths $L = 10, 15, 20, 25, 30$ as input to time series forecast algorithms. Figure 2 shows a comparative result of Mean Absolute Error (MAE) evaluation metric [29]. For each length L, we plotted the Cumulative Distribution Function (CDF) related to each return forecast model. Thus, we can see the behavior of each model when different time series lengths are used.

An important result presented in Fig. 2 is that Mean and Median have great MAE values compared with ML model (MLP). ARIMA has better results in small L values, but in $L = 30$ present the worst result. Besides that, MLP has good results for large values of L.

4.2   Machine Learning for Weighted Link Prediction

This section presents results related to weighted link prediction in dynamic stock networks. To perform these experiments, the training and test set was built using sliding window [26]. For such, we used 30 daily graph snapshots in the training set. The test set corresponds to the next trading daily data. The sliding window moves one day ahead to create new training and test set.

We used XGboost [7] as main ML model to weighted link prediction. It is a fast, a highly effective and widely used machine learning method. We did not perform an exhaustive search for model parameters because this is not our main objective. Our intention is to show how predictive a machine learning model can be using the set of features that we proposed. The set of model parameters are:

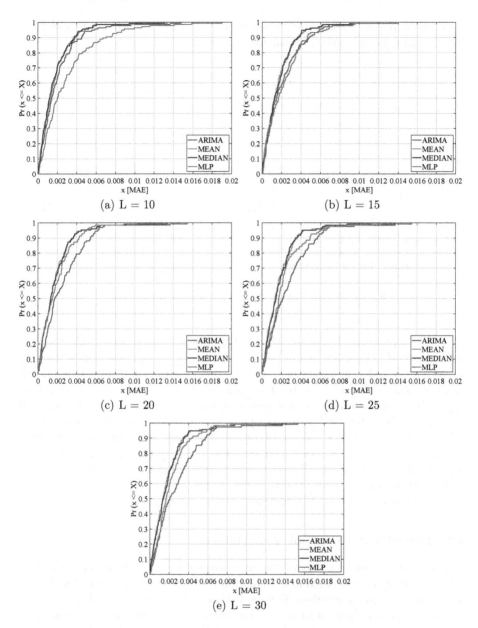

Fig. 2. Return forecast results for each size of L

- booster = "gbtree";
- objective = "reg:linear",
- eta = 0.05,
- max_depth = 2,
- min_child_weight = 100,

Figure 3 presents comparative results using CDF for both MAE and Root Mean Square Error (RMSE) evaluation metrics.

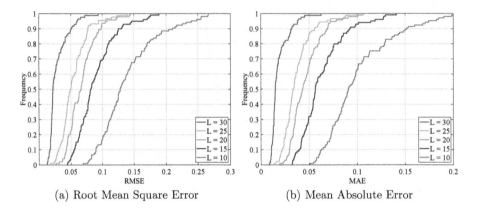

(a) Root Mean Square Error   (b) Mean Absolute Error

Fig. 3. Weight link prediction using different sizes of L

Figure 3 presents significant weighted link prediction results using the proposed method. This comparison between the behaviors of the machine learning model for the different values of L allows us to visualize that, for greater L values, the model can better predict the values of the edges weights (correlations). A possible explanation is that: the greater the size of L, more stable the financial network tends to be, facilitating the prediction of edges weights. With smaller values of L, the network tends to be more unstable. Note that the value of L influences both return forecast and weighted link prediction. Considering that edge weights are between -1 and 1, MAE results for $L = 30$ are almost 95% under 0.05, which is very expressive in terms of weighted link prediction. At each day, the model predicts 1540 edge weights (correlations).

4.3   Portfolio Optimization Experiments

This section presents a comparative experiment using $L = 30$, which is the best return forecast and weighted link prediction results. We executed the Markowitz model using results of weighted link prediction and all methods applied to return forecast. Figure 4 shows a financial simulated gross return for each approach. Each execution uses 84 trading days. For a utility functions ($R_a = 2$) a threshold value between a moderate risk and risky.

Figure 4 shows that our MLink method outperforms MVA in over 56% comparing accumulated gross return. ARIMA-MLink, Median-MLink and Mean-MLink also has better results than MVA and Bovespa Index. This is an impressive result in terms of financial return.

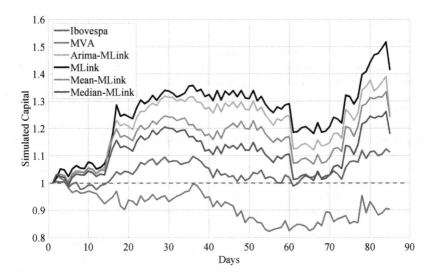

Fig. 4. Comparing accumulated return for each approach

5   Conclusions and Future Works

This study proposed determining the constants of the MVA from [16] using machine learning and a new weighted link prediction in stock networks defined in this paper as MLink. Portfolio optimization models use data from the past to create portfolios with a good trade-off between return and risk. Using the correlation between asset price series in the stock networks, when vertices represent assets and edges represent the correlation between them, the proposed method can predict the weights of all edges in dynamic stock networks. Like the MVA, the proposed method also use return forecasts. The experimental results show that using both return forecast and weighted link prediction data, the proposed method performance is superior to the performance obtained by the MVA. The experiments show that the MLink capital increases almost 41.34% in 84 days, a difference of 56.68% for the MVA (which reduced the capital in 9.79%).

These findings open a range of possibilities for future works. Several analyzes will be carried out, such as the use of different optimization models and run experiments using known market data sets for portfolio optimization. In addition, we can use other ways of predicting edge weights, such as tensor or deep learning, trying to improve weighted link prediction results. In addition, we can make available the created data set for other researchers evaluate their optimization models.

Acknowledgments. The authors would like thank CAPES, FAPESP (2013/07375-0), Intel - FAFQ, National Council for Scientific and Technological Development - CNPq - Brazil (202006/2018-2) and Federal Institute of Education, Science and Technology of South of Minas Gerais - IFSULDEMINAS - for their support.

References

1. Bessler, W., Wolff, D.: Portfolio optimization with return prediction models evidence for industry portfolios. In: World Finance and Banking Symposium (2015)
2. Blondel, V.D., Guillaume, J.L., Lambiotte, R., Lefebvre, E.: Fast unfolding of communities in large networks. J. Stat. Mech: Theory Exp. **2008**(10), P10008 (2008)
3. Bonanno, G., Caldarelli, G., Lillo, F., Mantegna, R.N.: Topology of correlation-based minimal spanning trees in real and model markets. Phys. Rev. E **68**(4), 046130 (2003)
4. Bonanno, G., Caldarelli, G., Lillo, F., Micciche, S., Vandewalle, N., Mantegna, R.N.: Networks of equities in financial markets. Eur. Phys. J. B-Condens. Matter Complex Syst. **38**(2), 363–371 (2004)
5. Bonanno, G., Lillo, F., Mantegna, R.N.: High-frequency cross-correlation in a set of stocks. Quant. Financ. **1**(1), 96–104 (2001)
6. Brockwell, P.J., Davis, R.A.: Introduction to Time Series and Forecasting. STS. Springer, Cham (2016). https://doi.org/10.1007/978-3-319-29854-2
7. Chen, T., Guestrin, C.: Xgboost: a scalable tree boosting system. In: Proceedings of the 22nd ACM SIGKDD International Conference on Knowledge Discovery and Data Mining, pp. 785–794. ACM (2016)
8. Freitas, F.D., De Souza, A.F., de Almeida, A.R.: Prediction-based portfolio optimization model using neural networks. Neurocomputing **72**(10–12), 2155–2170 (2009)
9. Girvan, M., Newman, M.E.: Community structure in social and biological networks. Proc. Nat. Acad. Sci. **99**(12), 7821–7826 (2002)
10. Holme, P., Saramäki, J.: Temporal networks. Phys. Rep. **519**(3), 97–125 (2012)
11. Kallberg, J.G., Ziemba, W.T.: Comparison of alternate utility functions in portfolio selection problems. Manage. Sci. **29**, 1257–1276 (1983)
12. Kourentzes, N., Barrow, D., Crone, S.: Neural network ensemble operators for time series forecasting. Expert Syst. Appl. **41**(9), 4235–4244 (2014)
13. Li, Y., Jiang, X.F., Tian, Y., Li, S.P., Zheng, B.: Portfolio optimization based on network topology. Phys. A: Stat. Mech. Appl. **515**, 671–681 (2019)
14. Lü, L., Zhou, T.: Link prediction in weighted networks: the role of weak ties. EPL (Europhys. Lett.) **89**(1), 18001 (2010)
15. Mantegna, R.N.: Hierarchical structure in financial markets. Eur. Phys. J. B-Condens. Matter Complex Syst. **11**(1), 193–197 (1999)
16. Markowitz, H.: Portfolio selection. J. Financ. **7**(1), 77–91 (1952)
17. Markowitz, H.: Portfolio Selection: Efficient Diversfication of Investments, vol. 7. Wiley, New York (1959)
18. Marti, G., Nielsen, F., Bińkowski, M., Donnat, P.: A review of two decades of correlations, hierarchies, networks and clustering in financial markets. arXiv preprint arXiv:1703.00485 (2017)
19. Mishra, S.K., Panda, G., Majhi, B.: Prediction based mean-variance model for constrained portfolio assets selection using multiobjective evolutionary algorithms. Swarm Evol. Comput. **28**, 117–130 (2016)
20. Narasimhan, S.: Link prediction in dynamic networks. Ph.D. thesis (2015)
21. Oliveira, M., Gama, J.: An overview of social network analysis. Wiley Interdisc. Rev.: Data Min. Knowl. Discov. **2**(2), 99–115 (2012)
22. Pafka, S., Kondor, I.: Estimated correlation matrices and portfolio optimization. Phys. A: Stat. Mech. Appl. **343**, 623–634 (2004)

23. Pearson, K.: Note on regression and inheritance in the case of two parents. Proc. Royal Soc. London **58**, 240–242 (1895)
24. Peralta, G., Zareei, A.: A network approach to portfolio selection. J. Empirical Financ. **38**, 157–180 (2016)
25. Riedmiller, M., Braun, H.: A direct adaptive method for faster backpropagation learning: the RPROP algorithm. In: IEEE International Conference on Neural Networks, pp. 586–591. IEEE (1993)
26. Silva, E., Brandao, H., Castilho, D., Pereira, A.C.: A binary ensemble classifier for high-frequency trading. In: 2015 International Joint Conference on Neural Networks (IJCNN). IEEE (2015)
27. Silva, E., Castilho, D., Pereira, A., Brandao, H.: A neural network based approach to support the market making strategies in high-frequency trading. In: 2014 International Joint Conference on Neural Networks (IJCNN). IEEE (2014)
28. Wang, P., Xu, B., Wu, Y., Zhou, X.: Link prediction in social networks: the state-of-the-art. Sci. China Inf. Sci. **58**(1), 1–38 (2015)
29. Willmott, C.J., Matsuura, K.: Advantages of the mean absolute error (MAE) over the root mean square error (RMSE) in assessing average model performance. Climate Res. **30**(1), 79–82 (2005)
30. Yang, C., Chen, Y., Niu, L., Li, Q.: Cointegration analysis and influence rank— a network approach to global stock markets. Phys. A: Stat. Mech. Appl. **400**, 168–185 (2014)
31. Yang, Y., Yang, H.: Complex network-based time series analysis. Phys. A: Stat. Mech. Appl. **387**(5–6), 1381–1386 (2008)
32. Zhao, L., Wang, G.J., Wang, M., Bao, W., Li, W., Stanley, H.E.: Stock market as temporal network. Phys. A: Stat. Mech. Appl. **506**, 1104–1112 (2018)

Track of Computational Optimization, Modelling and Simulation

Comparison of Constraint-Handling Techniques for Metaheuristic Optimization

Xing-Shi He[1], Qin-Wei Fan[1], Mehmet Karamanoglu[2], and Xin-She Yang[2(✉)]

[1] College of Science, Xi'an Polytechnic University,
Xi'an 710048, People's Republic of China
[2] School of Science and Technology, Middlesex University, London, UK
x.yang@mdx.ac.uk

Abstract. Many design problems in engineering have highly nonlinear constraints and the proper handling of such constraints can be important to ensure solution quality. There are many different ways of handling constraints and different algorithms for optimization problems, which makes it difficult to choose for users. This paper compares six different constraint-handling techniques such as penalty methods, barrier functions, ϵ-constrained method, feasibility criteria and stochastic ranking. The pressure vessel design problem is solved by the flower pollination algorithm, and results show that stochastic ranking and ϵ-constrained method are most effective for this type of design optimization.

Keywords: Constraint-handling techniques · Feasibility ·
Flower pollination algorithm · Nature-inspired computation ·
Optimization

1 Introduction

A vast majority of problems in engineering and science can be formulated as optimization problems with a set of inequality and equality constraints. However, such problems can be challenging to solve, not only because of the high nonlinearity of problem functions, but also because of the complex search domain shapes enclosed by various constraints. Consequently, both the choice of optimization algorithms and the ways of handling complex constraints are crucially important. Efficient algorithms may not exist for a given type of problem. Even with an efficient algorithm for a given problem, different ways of handling constraints may lead to varied accuracy. Thus, in addition to the comparison of different algorithms, a systematic comparison of constraint-handling techniques is also needed [3,9,18,19,29].

There are many different algorithms for solving optimization problems [6,11]. One of the current trends is to use nature-inspired optimization algorithms to solve global optimization problems [28], and the algorithms such as genetic

© Springer Nature Switzerland AG 2019
J. M. F. Rodrigues et al. (Eds.): ICCS 2019, LNCS 11538, pp. 357–366, 2019.
https://doi.org/10.1007/978-3-030-22744-9_28

algorithm, differential evolution [24], particle swarm optimization [15], firefly algorithm and cuckoo search [28], and flower pollination algorithm [26] have demonstrated their flexibility and effectiveness. Thus, we will mainly use nature-inspired metaheuristic algorithms in this paper, and, more specifically, we will use the recent flower pollination algorithm (FPA) because its effectiveness and proved convergence [13,26,27]. In addition, even with any efficient algorithm, the constraints of optimization problems must be handled properly so that feasible solutions can be easily obtained. Otherwise, many solution attempts may be wasted and constraints may be violated [11,29]. There are many different constraint-handling techniques in the literature [8,12,16,17,23,32], and our focus will be on the comparison of different constraint-handling techniques for solving global optimization problems using metaheuristic algorithms.

Therefore, this paper is organized as follows. Section 2 provides a general formulation of optimization problems with a brief introduction to the flower pollination algorithm (FPA). Section 3 outlines different constraint-handling techniques. Section 4 uses FPA to solve a pressure vessel design problem with different ways of handling constraints where comparison of results will be presented. Finally, Sect. 5 concludes with some discussions.

2   Optimization

2.1   General Formulation

Though optimization problems can take many different forms in different applications, however, it can be formulated as a mathematical optimization problem in a D-dimensional design space as follows:

$$\text{minimize } f(\boldsymbol{x}), \quad \boldsymbol{x} = (x_1, x_2, ..., x_D) \in \mathbb{R}^D, \tag{1}$$

subject to

$$\phi_i(\boldsymbol{x}) = 0, \quad (i = 1, 2, ..., M), \tag{2}$$

$$\psi_j(\boldsymbol{x}) \leq 0, \quad (j = 1, 2, ..., N), \tag{3}$$

where $\boldsymbol{x}$ is the vector of D design variables, and $\phi_i(\boldsymbol{x})$ and $\psi_j(\boldsymbol{x})$ are the equality constraints and inequality constraints, respectively. Classification of different optimization problems can be based on the problem functions. If these functions ($f(\boldsymbol{x})$, $\phi_i(\boldsymbol{x})$ and $\psi_j(\boldsymbol{x})$) are all linear, there are some efficient algorithms such as simplex methods. If problem functions are nonlinear, they may be more difficult to solve, though there are a wide range of techniques that can be reasonably effective [6,15,29]. However, global optimality may not be guaranteed in general.

2.2   Flower Pollination Algorithm

The flower pollination algorithm (FPA) as a population-based algorithm has been inspired by the characteristics of the pollination processes of flowering plants [26,27]. The main steps of the FPA have been designed to mimic some key characteristics of the pollination process, including biotic and abiotic pollination, flower constancy co-evolved between certain flower species and pollinators such as insects and animals, and the movement ranges of flower pollen of different flower species.

Briefly speaking, if x is the position vector that represents a solution in the design space to an optimization problem, this vector can be updated by

$$x_i^{t+1} = x_i^t + \gamma L(\nu)(g_* - x_i^t), \tag{4}$$

which mimics the global step in the FPA. Here g_* is the best solution found so far in the whole population of n different candidate solutions, while γ is a scaling parameter, and $L(\nu)$ is a vector of random numbers, drawn from a Lévy distribution characterized by an exponent of ν.

Though the Lévy distribution is defined as

$$L(s) = \begin{cases} \sqrt{\frac{\gamma}{2\pi}} e^{-\frac{\gamma}{2(s-\mu)}} \frac{1}{(s-\mu)^{3/2}}, & (0 < \mu < s < +\infty), \\ 0, & \text{otherwise}, \end{cases} \tag{5}$$

which has an exponent of $3/2$, it can be generalised with an exponent of $1 \leq \nu \leq 2$ in the following form:

$$L(s, \nu) \sim \frac{A\nu\Gamma(\nu)\sin(\pi\nu/2)}{\pi|s|^{1+\nu}}, \tag{6}$$

where $s > 0$ is the step size, and A is a normalization constant. The Γ-function is given by

$$\Gamma(z) = \int_0^\infty u^{z-1}e^{-u}du. \tag{7}$$

In the special case when $z = k$ is an integer, it becomes $\Gamma(k) = (k-1)!$. The average distance d_L or search radius covered by Lévy flights takes the form

$$d_L^2 \sim t^{3-\nu}, \tag{8}$$

which increases typically faster than simple isotropic random walks such as Brownian motion because Lévy flights can have a few percent of moves with large steps in addition to many small steps [21].

The current solution x_i^t as a position vector can be modified locally by varying step sizes

$$x_i^{t+1} = x_i^t + U(x_j^t - x_k^t), \tag{9}$$

where U is a vector with each of its components being drawn from a uniform distribution. Loosely speaking, x_j^t and x_k^t can be considered as solutions representing pollen from different flower patches in different regions.

Due to the combination of local search and long-distance Lévy flights, FPA can usually have a higher capability for exploration. A recent theoretical analysis using Markov chain theory has confirmed that FPA can have guaranteed global convergence under the right conditions [13]. There are many variants of the flower pollination algorithm and a comprehensive review can be found in [2]. Due to its effectiveness, FPA has been applied to solve a wide range of optimization problems in real-world applications such as economic dispatch, EEG identification, and multiobjective optimization [1,22,27].

3 Constraint-Handling Techniques

There are many different constraint-handling techniques in the literature, ranging from traditional penalty methods and Lagrangian multipliers to more sophisticated adaptive methods and stochastic ranking [8,18,23]. In essence, penalty methods neatly transform a constrained optimization problem into a corresponding, unconstrained one by transforming its constraints in the revised objective, in terms of some additional penalty terms, and these penalty terms are usually functions of constraints. The advantage of this is that the optimization problem becomes unconstrained and thus the search domain has a regular shape without changing the locations of the optimality, but this modifies its original objective landscape, which may become less smooth. In addition, more parameters such as the penalty constants are introduced into the problem, and their values need to be set or tuned properly. In many cases, they can work surprisingly well if proper values are used, and the transformed unconstrained problem can be solved effectively by various optimization methods very accurately [11,29].

In this study, we aim to compare a few methods of handling constraints, and they are barrier functions, static penalty method, dynamic penalty method, feasibility method, ϵ-constrained method, and stochastic ranking.

3.1 Static Penalty and Dynamic Penalty Methods

Among various forms of the penalty method, the Powell-Skolnick approach [20] incorporates all the constraints with feasibility

$$\rho(\boldsymbol{x}) = \begin{cases} 1 + \mu\big[\sum_{j=1}^{N} \max\{0, \psi_j(\boldsymbol{x})\} + \sum_{i=1}^{M} |\phi_i(\boldsymbol{x})|\big], & \text{if not feasible,} \\ f(\boldsymbol{x}), & \text{if feasible,} \end{cases} \tag{10}$$

where the constant $\mu > 0$ is fixed, and thus this method is a static penalty method. This approach ranks the infeasible solution with a rank in the range from 1 to ∞, assuming the lower ranks correspond to better fitness for minimization problems.

In general, the penalty-based method transform the objective $f(\boldsymbol{x})$ into a modified objective Θ in the following form:

$$\Theta(\boldsymbol{x}) = f(\boldsymbol{x})[\text{objective}] + P(\boldsymbol{x})[\text{penalty}], \tag{11}$$

where the penalty term $P(\boldsymbol{x})$ can take different forms, depending on the actual ways or variants of constraint-handling methods. For example, a static penalty method uses

$$P(\boldsymbol{x}) = \sum_{i=1}^{M} \mu_i \phi_i^2(\boldsymbol{x}) + \sum_{j=1}^{N} \nu_j \max\{0, \psi_j(\boldsymbol{x})\}^2, \tag{12}$$

where $\mu_i > 0, \nu_j > 0$ are penalty constants or parameters. In order to avoid too many penalty parameters, a single penalty constant $\lambda > 0$ can be used, so that we have

$$P(\boldsymbol{x}) = \lambda \Big[\sum_{i=1}^{M} \phi_i^2(\boldsymbol{x}) + \sum_{j=1}^{N} \max\{0, \psi_j(\boldsymbol{x})\}^2 \Big]. \tag{13}$$

Since λ is fixed, independent of the iteration t, this basic form of penalty is the well-known static penalty method.

Studies show that it may be advantageous to vary λ during iterations [14,17], and the dynamic penalty method uses a gradually increasing λ in the following form [14]:

$$\lambda = (\alpha t)^{\beta}, \tag{14}$$

where $\alpha = 0.5$ and $\beta = 1, 2$ are used.

There are other forms of penalty functions. Recent studies suggested that adaptive penalty can be effective with varying penalty strength by considering the fitness of the solutions obtained during iterations [4,5,10].

3.2   Barrier Function Method

Though the equality constraints can be handled using Lagrangian multipliers, the inequalities need to be handled differently. One way is to use the barrier function [6], and the logarithmic barrier functions can be written as

$$L(\boldsymbol{x}) = -\mu \sum_{j=1}^{N} \log \Big[-\psi_j(\boldsymbol{x}) \Big], \tag{15}$$

where $\mu > 0$ can be varied during iterations (t). Here, we will use $\mu = 1/t$ in our implementations.

3.3   Feasibility Criteria

A feasibility-based constraint-handling technique, proposed by Deb [12], uses three feasible criteria as selection mechanisms: (1) the feasible solution is chosen first among one feasible solution and one infeasible solution; (2) the solution with a better (lower for minimization) objective value is preferred if two feasible solutions are compared; and (3) among two infeasible solutions, the one with the lower degree of constraint violation is preferred.

The degree of the violation of constraints can be approximately measured by the penalty term

$$P(\boldsymbol{x}) = \sum_{i=1}^{M} |\phi_i(\boldsymbol{x})| + \sum_{j=1}^{N} \max\{0, \psi_j(x)\}^2. \tag{16}$$

Such feasibility rules can loosely be considered as fitness ranking and preference of low constraint violation. Obviously, such feasibility rules can be absolute or relative, and thus can be extended to other forms [17].

3.4 Stochastic Ranking

Stochastic ranking (SR), developed by Runarsson and Yao in 2000 [23], is another constraint-handling technique, which becomes promising. In stochastic ranking, a control parameter $0 < p_f < 0$ is pre-defined by the user to balance feasibility and infeasibility, while no penalty parameter is used. The choice and preference between two solutions are mainly based on their relative objective values and the sum of constraint violations. Ranking of solutions can be done by any sorting algorithms such as the bubble sort.

The main step involves first to draw a uniformly-distributed u and compare with the pre-defined p_f. If $u < p_f$ or both solutions are feasible, then swap them if $f(\boldsymbol{x}_j) > f(\boldsymbol{x}_i)$. If both solutions are infeasible, swap if $P(\boldsymbol{x}_j) > P(\boldsymbol{x}_i)$. The aim is to select the minimum of the objective values and the lower degree of sum of the constraint violations.

The ranking is carried out according to the probability p_s

$$p_s = p_o p_f + p_v (1 - p_f), \tag{17}$$

where p_o is the probability of individual winning, based on its objective value, while p_v is the probability of winning of that individual solution, based on the violation of the constraints [23]. The probability of selection or winning among k comparison pairs among n solutions is based on a binomial distribution

$$p_w(k) = \frac{n!}{k!(n-k)!} p_s^k (1 - p_s)^{n-k}. \tag{18}$$

According to the value suggested by Runarsson and Yao [23], $p_f = 0.425$ will be used in this study.

3.5 The ϵ-Constrained Approach

Another technique for handling constraints, called the ϵ-constrained method, was developed by Takahama and Sakai [25], which consists of two steps: the relaxation limits for feasibility consideration and lexicographical ordering. Basically, two solutions $\boldsymbol{x}_i$ and $\boldsymbol{x}_j$ can be compared and ranked by their objective values $f(\boldsymbol{x}_i)$ and $f(\boldsymbol{x}_j)$ and constraint violation ($P(\boldsymbol{x}_i)$ and $P(\boldsymbol{x}_j)$). That is

$$\{f(\boldsymbol{x}_i), P(\boldsymbol{x}_i)\} \le \epsilon \{f(\boldsymbol{x}_j), P(\boldsymbol{x}_j)\}, \tag{19}$$

which is equivalent to the following conditions:

$$\begin{cases} f(\boldsymbol{x}_i) \le f(\boldsymbol{x}_j), & \text{if both } P(\boldsymbol{x}_i), P(\boldsymbol{x}_j) \le \epsilon \\ f(\boldsymbol{x}_i) \le f(\boldsymbol{x}_j), & \text{if } P(\boldsymbol{x}_i) = P(\boldsymbol{x}_j), \\ P(\boldsymbol{x}_i) \le P(\boldsymbol{x}_j), & \text{otherwise.} \end{cases} \tag{20}$$

Loosely speaking, the parameter $\epsilon \ge 0$ controls the level of comparison. In case of ϵ is very large, the comparison is mainly about objective values, while $\epsilon = 0$ corresponds to an ordering rule so that the objective minimization is preceded by lower or minimal degrees of the constraint violation [25, 31].

4   Numerical Experiments and Comparison

In order to compare how these constraint-handling methods perform, we should use different case studies and different algorithms. However, due to the limit of space, here we only present the results for a design case study solved by the flower pollination algorithm. The optimal design of pressure vessels is a mixed integer programming, and it is a well-known benchmark in metaheuristic optimization for validating evolutionary algorithms.

4.1   Pressure Vessel Design

The pressure vessel design problem is a well-known benchmark that has been used by many researchers, and this problem is a mixed-type with four design variables. The overall design objective is to minimize the total cost of a cylindrical vessel, subject to some pre-defined volume and stress constraints. The four design variables are the thickness d_1 and d_2 for the head and body of the vessel, respectively, the inner radius r of the cylindrical section, and the length W of the cylindrical part [7,8]. The objective is to minimize the cost:

$$\text{minimize } f(\boldsymbol{x}) = 06224rWd_1 + 1.7781r^2d_2 + 19.64rd_1^2 + 3.1661Wd_1^2, \quad (21)$$

subject to four constraints:

$$g_1(\boldsymbol{x}) = -d_1 + 0.0193r \leq 0, \quad g_2(\boldsymbol{x}) = -d_2 + 0.00954r \leq 0, \quad (22)$$

$$g_3(\boldsymbol{x}) = -\frac{4\pi r^3}{3} - \pi r^2 W - 1296000 \leq 0, \quad g_4(\boldsymbol{x}) = W - 240 \leq 0. \quad (23)$$

The simple limits for the inner radius and length are: $10.0 \leq r, W \leq 200.0$.

However, due to some manufacturability requirements, it is necessary to set the thickness (d_1 and d_2) to be the integer multiples of a basic thickness of 0.0625 in.. That is

$$1 \times 0.0625 \leq d_1, d_2 \leq 99 \times 0.0625. \quad (24)$$

With four variables and four constraints, it seems not so hard to solve the problem. However, the first two variables are discrete, which makes the problem become a mixed integer programming problem. This benchmark has been studied extensively by many researchers [7,28]. For many years, the true optimal solutions were not known due to the nonlinearity in its objective and constraints.

Now the true global optimal solution [30], based on the analytical analysis, is $f_{\min} = 6059.714335$ with $d_1 = 0.8125$, $d_2 = 0.4375$, $r = 40.098446$ and $W = 176.636596$. This allows us to compare the obtained solutions with the true solution in this study.

4.2   Comparison

Most penalty methods used in the literature require a high number of iterations, typically from 10 000 or 50 000 up to even 250 000 or 500 000 so as to get sufficient accurate results [8,12]. However, in order to see how these methods evolve throughout iterations, a much lower number of iterations will be used here. In fact, we will use $t_{max} = 10000$, this allows us to see how errors will evolve over time for different methods. Other parameters are: $\lambda = 10^5$ for static penalty, $\alpha = 0.5$ and $\beta = 2$ for dynamic penalty. $\mu = 1/t$ is used for the barrier function, and $p_f = 0.425$ is used for stochastic ranking. In addition, $\epsilon = 1$ is used for the ϵ-constrained method. For the FPA, parameters are: population size $n = 40$, $p_a = 0.25$, $\gamma = 0.1$, and $\nu = 1.5$.

There are many different ways to compare simulation results, and the ranking results can largely depend on the performance measures used for comparison. Here, we will use the modified offline error E, similar to the error defined by Ameca-Alducin et al. [3]. We have $E = \frac{1}{N_{max}} \sum_{t=1}^{N_{max}} |f_{min} - f_*^{(t)}|$ where N_{max} is the maximum number of iterations and we use $N_{max} = 10000$. Here, $f_*^{(t)}$ is the best solution found by an algorithm during iteration t, and f_{min} is the known best solution from the literature, and it is the global minimum, based on analytical results for the pressure vessel design problem [30].

Table 1. Mean errors of the pressure vessel objective with 20 independent runs.

Method	Iteration ($t = 5000$)	Iteration $t = 10000$
Static penalty	416.1	322.6
Dynamic penalty	368.7	317.8
Barrier function	497.9	421.3
Feasibility approach	402.7	310.2
ϵ-constrained	341.5	171.9
Stochastic ranking	332.4	169.3

Six different constraint-handling methods are implemented in this study, and all methods can find the optimal solution $f_{min} = 6059.714$ for $t_{max} = 10000$. The results of 20 independent runs and the mean errors of the pressure vessel design objective values from the true optimal value are summarized in Table 1. As we can see, the errors are decreasing as iteration t becomes larger. Both stochastic ranking and ϵ-constrained method obtained the best results, while the feasibility approach is very competitive. Barrier function approach seems to give the worse results. Both static penalty and dynamic penalty can work well, though dynamic penalty is better than static penalty.

5   Conclusions

This paper has compared six different constraint-handling techniques in the context of bio-inspired algorithms and nonlinear pressure vessel designs. The pressure vessel design problem is a nonlinear, mixed-integer programming problem and has been solved by using the FPA. The emphasis has been on the comparison of different ways of handling constraints. Our results have shown that both stochastic ranking and ϵ-constrained method obtained the best results.

Further studies will focus on the more extensive tests of different constraint-handling techniques and different algorithms over a wide range of benchmarks and design problems. More detailed parametric studies will also be carried out so as to gain insight into advantages and disadvantages as well as robustness of different constraint-handling techniques.

References

1. Abdelaziz, A.Y., Ali, E.S., Abd Elazim, S.M.: Combined economic and emission dispatch solution using flower pollination algorithm. Int. J. Electr. Power Energy Syst. **80**(2), 264–274 (2016)
2. Alyasseri, Z.A.A., Khader, A.T., Al-Betar, M.A., Awadallah, M.A., Yang, X.-S.: Variants of the flower pollination algorithm: a review. In: Yang, X.-S. (ed.) Nature-Inspired Algorithms and Applied Optimization. SCI, vol. 744, pp. 91–118. Springer, Cham (2018). https://doi.org/10.1007/978-3-319-67669-2_5
3. Ameca-Alducin, M.-Y., Hasani-Shoreh, M., Blaidie, W., Neumann, F., Mezura-Montes, E.: A comparison of constraint handling techniques for dynamica constrained optimization problems. arXiv:1802.05825 (2018). Accessed 12 Feb 2019
4. Barbosa, H.J.C., Lemonge, A.C.C.: A new adaptive penalty scheme for genetic algorithms. Inf. Sci. **156**(5), 215–251 (2003)
5. Bean, J.C., Hadj-Alouane, A.B.: A dual genetic algorithm for bounded integer programs. Technical report RT 92–52. Department of Industrial and Operations Engineering, University of Michigan, Ann Arbor, USA (1992)
6. Boyd, S., Vandenberge, L.: Convex Optimization. Cambridge University Press, Cambridge (2004)
7. Cagnina, L.C., Esquivel, S.C., Coello, C.A.: Solving engineering optimization problems with the simple constrained particle swarm optimizer. Informatica **32**(3), 319–326 (2008)
8. Coello, C.A.C.: Use of a self-adaptive penalty approach for engineering optimization problems. Comput. Ind. **41**(2), 113–127 (2000)
9. Coello, C.A.C.: Theoretical and numerical constraint-handling techniques used with evolutionary algorithms: a survey of the state of the art. Comput. Methods Appl. Mech. Eng. **191**(11–12), 1245–1287 (2002)
10. Coit, D.W., Smith, A.E., Tate, D.M.: Adaptive penalty methods for genetic optimization of constrained combinatorial problems. INFORMS J. Comput. **6**(2), 173–182 (1996)
11. Deb, K.: Optimization for Engineering Design: Algorithms and Examples. Prentice-Hall, New Delhi (1995)
12. Deb, K.: An efficient constraint handling method for genetic algorithms. Comput. Methods Appl. Mech. Eng. **186**(2–4), 311–338 (2000)
13. He, X.S., Yang, X.S., Karamanoglu, M., Zhao, Y.X.: Global convergence analysis of the flower pollination algorithm: a discrete-time Markov chain approach. Procedia Comput. Sci. **108**(1), 1354–1363 (2017)
14. Joines, J., Houck, C.: On the use of non-stationary penalty functions to solve nonlinear constrained optimization problems with GA's. In: Fogel, D. (ed.) Proceedings of the First IEEE Conference on Evolutionary Computation, pp. 579–584. IEEE Press, Orlando (1994)

15. Kennedy, J., Eberhart, E.C., Shi, Y.: Swarm Intelligence. Academic Press, London (2001)
16. Koziel, S., Michalewicz, Z.: Evolutionary algorithms, homomorphous mappings, and constrained parameter optimization. Evol. Comput. **7**(1), 19–44 (1999)
17. Mezura-Montes, E.: Constraint-Handling in Evolutionary Optimization. SCI, vol. 198. Springer, Berlin (2009). https://doi.org/10.1007/978-3-642-00619-7
18. Mezura-Montes, E., Coello, C.A.C.: Constraint-handling in nature-inspired numerical optimization: past, present and future. Swarm Evol. Comput. **1**(4), 173–194 (2011)
19. Michalewicz, Z., Schoenauer, M.: Evolutionary algorithms for constrained parameter optimization problems. Evol. Comput. **4**(1), 1–32 (1996)
20. Powell, D., Skolnick, M.M.: Using genetic algorithms in engineering design optimization with non-linear constraints. In: Forrest, S. (ed.) Proceedings of the Fifth International Conference on Genetic Algorithms, pp. 424–431. Morgan Kaufmann, San Mateo (1993)
21. Pavlyukevich, I.: Lévy flights, non-local search and simulated annealing. J Comput. Phys. **226**(2), 1830–1844 (2007)
22. Rodrigues, D., Silva, G.F.A., Papa, J.P., Marana, A.N., Yang, X.S.: EEG-based person identificaiton through binary flower pollination algorithm. Expert Syst. Appl. **62**(1), 81–90 (2016)
23. Runarsson, T.P., Yao, X.: Stochastic ranking for constrained evolutionary optimization. IEEE Trans. Evol. Comput. **4**(3), 284–294 (2000)
24. Storn, R., Price, K.: Differential evolution: a simple and efficient heuristic for global optimization over continuous spaces. J. Global Optim. **11**(4), 341–359 (1997)
25. Takahama, T., Sakai, S.: Solving constrained optimization problems by the ϵ-constrained particle swarm optimizer with adaptive velocity limit control. In: Proceedings of the IEEE Conference on Cybernetics and Intelligent Systems (CIS 2006), 7–9 June 2006, pp. 683–689. IEEE Publication, Bangkok (2006)
26. Yang, X.-S.: Flower pollination algorithm for global optimization. In: Durand-Lose, J., Jonoska, N. (eds.) UCNC 2012. LNCS, vol. 7445, pp. 240–249. Springer, Heidelberg (2012). https://doi.org/10.1007/978-3-642-32894-7_27
27. Yang, X.S., Karamanoglu, M., He, X.S.: Flower pollination algorithm: a novel approach for multiobjective optimization. Eng. Optim. **46**(9), 1222–1237 (2014)
28. Yang, X.-S. (ed.): Cuckoo Search and Firefly Algorithm: Theory and Applications. SCI, vol. 516. Springer, Cham (2014). https://doi.org/10.1007/978-3-319-02141-6
29. Yang, X.S.: Nature-Inspired Optimization Algorithms. Elsevier, London (2014)
30. Yang, X.S., Huyck, C., Karamanoglu, M., Khan, N.: True global optimality of the pressure vessel design problem: a benchmark for bio-inspired optimisation algorithms. Int. J. Bio-Inspired Comput. **5**(6), 329–335 (2013)
31. Yang, X.S.: Optimization Technqiues and Applications with Examples. Wiley, Hoboken (2018)
32. Yeniay, O.: Penalty function methods for constrained optimization with genetic algorithms. Math. Computat. Appl. **10**(1), 45–56 (2005)

Dynamic Partitioning of Evolving Graph Streams Using Nature-Inspired Heuristics

Eneko Osaba[1(✉)], Miren Nekane Bilbao[2], Andres Iglesias[3,4], Javier Del Ser[1,2], Akemi Galvez[3,4], Iztok Fister Jr.[5], and Iztok Fister[5]

[1] TECNALIA, 48160 Derio, Spain
eneko.osaba@tecnalia.com
[2] University of the Basque Country (UPV/EHU), 48013 Bilbao, Spain
[3] Universidad de Cantabria, 39005 Santander, Spain
[4] Toho University, Funabashi, Japan
[5] University of Maribor, Maribor, Slovenia

Abstract. Detecting communities of interconnected nodes is a frequently addressed problem in situation that be modeled as a graph. A common practical example is this arising from Social Networks. Anyway, detecting an optimal partition in a network is an extremely complex and highly time-consuming task. This way, the development and application of meta-heuristic solvers emerges as a promising alternative for dealing with these problems. The research presented in this paper deals with the optimal partitioning of graph instances, in the special cases in which connections among nodes change dynamically along the time horizon. This specific case of networks is less addressed in the literature than its counterparts. For efficiently solving such problem, we have modeled and implements a set of meta-heuristic solvers, all of them inspired by different processes and phenomena observed in Nature. Concretely, considered approaches are Water Cycle Algorithm, Bat Algorithm, Firefly Algorithm and Particle Swarm Optimization. All these methods have been adapted for properly dealing with this discrete and dynamic problem, using a reformulated expression for the well-known modularity formula as fitness function. A thorough experimentation has been carried out over a set of 12 synthetically generated dynamic graph instances, with the main goal of concluding which of the aforementioned solvers is the most appropriate one to deal with this challenging problem. Statistical tests have been conducted with the obtained results for rigorously concluding the Bat Algorithm and Firefly Algorithm outperform the rest of methods in terms of Normalized Mutual Information with respect to the true partition of the graph.

Keywords: Bio-inspired computation · Nature-inspired heuristics · Evolving graphic streams · Community detection

© Springer Nature Switzerland AG 2019
J. M. F. Rodrigues et al. (Eds.): ICCS 2019, LNCS 11538, pp. 367–380, 2019.
https://doi.org/10.1007/978-3-030-22744-9_29

1   Introduction

With the impactful arrival of Social Networks, a remarkable number of tools and methods have been developed for excerpting information and insights from the multiple interrelations between their users [1]. In this regard, the knowledge that can be drawn using these methods range from evaluating the influence of a specific node in the whole graph (*centrality*), to enriched ways of network visualizing or the finding of shortest paths amidst a pair or groups of nodes. As has been seen in late years, all this information can be used for interesting practical goals, such as the identification of radicalization risk [2,3], or child abuse detections [4,5].

Among all the valuable knowledge that can be inferred from these Social Networks, the detection of different communities within the nodes is probably one of the most recurrent tasks, being the main focus of lots of recently developed scientific studies. Specifically, a *community* refers to a group of nodes which meet the general principles of strong intra-connectivity (strong links between members of the same community) and weak inter-connectivity (weak connectivity with nodes belonging to other partitions). Furthermore, the redefinition of these measured parameters leads to the characterization of different networks (weighted, multiple edges, directed, self loops), quantifying the cohesiveness of any candidate community.

Additionally, diverse efficient metrics have been projected in the literature for evaluating the quality of proposed partitions. Each of these metrics take different assumptions for measuring the connectivity, yielding to a single quality value for the community. Permanence [6], Surprise [7] and Newman and Girvan's Modularity [8] are some frequently used examples. In this specific study, the last-mentioned Modularity is considered.

In terms of computational solvers, many different approaches have been proposed in recent years for finding communities towards (explicitly or implicitly) optimizing one of the aforementioned metrics. Related with the research presented in this manuscript, a growing community is currently emerging devoted to adapting well-known (or even develop new) heuristic optimization algorithms, directly adopting one modularity metric as objective function. Many of these works can be found in the recent literature, focused on assorting combinations of algorithmic approaches, network instances and quality measurement metric functions. In this context, Genetic Algorithms (GA) crop up as one of the most often employed methods for discovering partitions in networks of different characteristics [9,10]. Besides GAs, many techniques that fall inside the umbrella of Evolutionary Computation and Swarm Intelligence have been proposed in last years, with the main goal of solving same or similar problem. Some of these solvers are the Ant Colony Optimization [11] or Particle Swarm Optimization [12]. Furthermore, interestingly for the scope of this work, a growing strand of the related literature is currently committed to the adaptation of modern nature-inspired algorithms for community detection in networks. Some examples are the Firefly Algorithm [13], Bat Algorithm [14] or Artificial Bee Colony [15].

At this point, it is worth to mention a specific case of networks, which are characterized by their dynamism. Since time immemorial, relationships between human beings tend to be different over time. In this sense, people are used to strengthen existing (or build new) relations over their whole lives, meanwhile some others are weakened (or even broken up). For this reason, if we analyze the relationship history of a single person in a long enough time lapse, we will surely find some dynamism. Of course, this evolution is also reflected in Social Networks. This way, Dynamic Networks are special cases of graphs in which the number of links among nodes, the strength of these links or even the number of nodes can suffer changes along the time. Thus, a Dynamic Network can be seen as an Evolving Graph Stream, in which the evolution of the network is described step by step.

Dynamic Networks have also been the subject matter of a recently published interesting works, focused on dynamic community finding [16]. In [17], for example, a multi-objective Bat Algorithm is presented for dealing with this problem. GAs have also been occasionally adapted to this kind of graphs, as can be seen in [18] or [19]. In any case, the amount of scientific material related to this topic is much fewer than the one associated to stationary networks.

With all this, the research presented in this paper aims at taking a step further over this scarce state of the art by elaborating on several directions: (1) we face the problem of finding communities in dynamic networks, far less used than stationary ones; (2) we adopt the Hamming Distance as a metric to assess the similarity between different solutions and partitions, and (3) we evaluate these algorithmic features with a group of ad-hoc adopted nature-inspired solvers: Water Cycle Algorithm (WCA, [20]), Bat Algorithm (BA, [21]), Firefly Algorithm (FA, [22]) and Particle Swarm Optimization (PSO, [23]). Along the paper, details on how these methods have been adapted to the proposed problem are exposed, as well as a justification of their expected benefits. In order to measure the performance of each method, results reached over 12 synthetically generated datasets are compared and discussed, based on their efficiency on discovering their ground-of-truth partition. Furthermore, with the intention of drawing fair and rigorous conclusions, two different statistical tests (Friedman's and Holm's) are employed with the obtained outcomes.

The rest of the paper is structured as follows: in the next Sect. 2, the problem of detecting communities in dynamic networks is formulated. After that, the heuristic solvers are described in Sect. 3, while the experimentation is displayed in Sect. 4. This manuscript ends with conclusions and further work in Sect. 5.

2   Problem Statement

In order to properly deal with the aforementioned community detection problem, we start by modeling the dynamic network as a graph $\mathcal{G} \doteq \{\mathcal{V}, \mathcal{E}\}$, where $\mathcal{V}$ represent the group of $|\mathcal{V}| = V$ vertex or nodes of the network, whilst $\mathcal{E}$ corresponds to a set that describes the dynamic situation of the links (or edges connecting every pair of nodes) along the whole time horizon. At this point, it is interesting to clarify that the time horizon is comprised by a set of graph snapshots, each one describing the specific situation of the network at one exact moment. This way, $\mathcal{E} \doteq \{e_1, \ldots, e_N\}$, corresponding e_n to the set of edges at timestamp n. In our study, we have established N in 30 and 40.

Another noteworthy point is that, while the relation between the nodes vary along the time, the number and situation of all vertex remains constant along the time horizon. Furthermore, the weight of each link connecting every pair of nodes v and v' is $w_{v,v'} = 1$. We also assume that $w_{v,v} = 0$ (i.e. no self-loops) and that $w_{v,v'} = 0$ if nodes v and v' are not connected. For notational convenience, we define an adjacency matrix $\mathbf{W}$ given by $\mathbf{W} \doteq \{w_{v,v'} : v, v' \in \mathcal{V}\}$, and fulfilling $\mathrm{Tr}(\mathbf{W}) = 0$. Finally, symmetry is always assumed in $\mathcal{G}$, meaning that $w_{v,v'} = w_{v',v}$. From now on, and in order to properly contemplate the dynamism inherent to the problem, weights are represented as $w_{v,v'}^n$, depicting the weight at timestamp n.

With all this, the problem of finding communities in the graph $\mathcal{G}$ is conceived in this study as the partition of the vertex set $\mathcal{V}$ into a number of non-empty and disjoint groups, each with a non-fixed size. Assuming that M is the number of groups of partition $\widetilde{\mathcal{V}} \doteq \{\mathcal{V}_1, \ldots, \mathcal{V}_M\}$, such that $\cup_{m=1}^{M} \mathcal{V}_m = \mathcal{V}$ and $\mathcal{V}_m \cap \mathcal{V}_{m'} = \emptyset \; \forall m' \neq m$ (i.e. no overlapping communities). We can thus denote the community to which node v belongs as $\mathcal{V}^v \in \widetilde{\mathcal{V}}$. Analogously, from now on, the set of partitions will be represented as $\mathcal{V}_n$, depicting $\widetilde{\mathcal{V}_n} \doteq \{\mathcal{V}_1^n, \ldots, \mathcal{V}_M^n\}$ the group communities found at a specific timestamp n. This way, dynamism can be contemplated in this formulation.

Furthermore, as has been advanced in the introduction, the Newman and Girvan's Modularity formula has been adopted for measuring the quality of a given partition. This well-known function has been employed in myriad of works before, and its adequacy has been proven extensively [24–26]. This way, the measure of modularity for the considered community can be computed by:

$$Q(\widetilde{\mathcal{V}_n}) \doteq \frac{1}{2|E_n|} \sum_{ij} \left[w_{v,v'}^n - \frac{k_i^n k_j^n}{2|E_n|} \right] \delta(v, v')_n, \qquad (1)$$

where k_i^n is the degree of node i, $|E_n|$ is the total number of edges in the network, and $\delta(v, v')_n$ represents the Kronecker delta symbol, all of them contextualized in timestamp n. Clearly explained, $\delta(v, v')_n$ is a binary function $\delta : \mathcal{V}_n \times \mathcal{V}_n \mapsto \{0, 1\}$, such that $\delta(v, v_n') = 1$ if $\mathcal{V}_n^v = \mathcal{V}_n^{v'}$ as per the partition set by $\mathcal{V}_n$ (and 0 otherwise). All of them also contextualized by the timestamp n.

Therefore, detecting a *good* partition $\widetilde{\mathcal{V}_n^*}$ of the considered network $\mathcal{G}$ can be casted as:

$$\widetilde{\mathcal{V}_n^*} = \arg \max_{\widetilde{\mathcal{V}_n} \in \mathcal{B}_V} Q(\widetilde{\mathcal{V}_n}), \qquad (2)$$

denoting $\mathcal{B}_V$ the group of possible partitions of $\mathcal{V}_n$ elements into nonempty subgroups (i.e. the solution space of the above combinatorial problem). The specific cardinality of this set is huge, which is given by the V-th Bell number [27]. This means that if we consider a network composed by $V = 20$ nodes, it can be partitioned in approximately $517.24 \cdot 10^{12}$ different manners. Thus, considering that a separated evaluation can be computed in 1 microsecond on average, we would need more than one and a half years to check all possible combinations. This situation confirms the necessity of using a heuristic method for the efficient exploration of the solution space.

Finally, it is worth-mentioning at this point that the main goal of this paper is to solve the described problem in n consecutive timestamps, trying to reach the highest precision possible. To do that, it is important to consider different degrees of similarity between adjacent graph snapshots n and $n + 1$.

3   Proposed Nature-Inspired Solvers

With the aim of properly deal with the above described problem, several nature-inspired methods have been proposed. Prior to the description of each considered solver, some common design aspects are described in what follows, related to the encoding strategy, solution repair mechanism and the method used for comparing different solutions.

Being one of the most important aspects while heuristic developing, it is interesting to mention that *label-based representation* [28] has been adopted for encoding purposes. This way, each solutions is codified as a permutation $\mathbf{x} = [c_1, c_2, \ldots, c_V]$ of V integers from the range $[1, \ldots, V]$, where V denotes the number of nodes in the network. Furthermore, c_v denotes the cluster label to which node v belongs to. For instance, considering a $V = 12$ network, a possible solution could be $\mathbf{x} = [1, 2, 2, 1, 1, 2, 2, 3, 2, 3, 3, 3]$, which means that the partition depicted is $\widetilde{\mathcal{V}} = \{\mathcal{V}_1, \mathcal{V}_2, \mathcal{V}_3\}$, where $\mathcal{V}_1 = \{1, 4, 5\}$, $\mathcal{V}_2 = \{2, 3, 6, 7, 9\}$ and $\mathcal{V}_3 = \{8, 10, 11, 12\}$.

With the intention of avoiding ambiguities in the representation, a repairing mechanism has been built, which is partly inspired from the one presented in [29]. Thanks to this procedure, which is applied to every newly created solution, ambiguities generated by solutions such as $\mathbf{x} = [4, 2, 2, 4, 4, 2, 2, 3, 2, 3, 3, 3]$ and $\mathbf{x} = [7, 1, 1, 7, 7, 1, 1, 4, 1, 4, 4, 4]$ (which represent the same partition) are solved, transforming both of them to the above shown $\mathbf{x} = [1, 2, 2, 1, 1, 2, 2, 3, 2, 3, 3, 3]$.

An additional important aspect of the developed methods is how the similarity between different solutions is measured. This similarity is the basis of the movement strategies inherent to each of the proposed techniques. Thus, the well-known Hamming Distance has been chosen for this purpose. This function has previously used in several studies with the same objective, as can be seen in [30], verifying its adequacy in this context. Concretely, Hamming Distance is calculated as the number of non-corresponding elements between two individuals. This way, and considering two partitions $\mathbf{x} = [1, 2, \mathbf{2}, 1, \mathbf{2}, 2, 2, 3, 2, 3, \mathbf{1}, \mathbf{1}]$ and $\mathbf{x} = [1, 2, \mathbf{1}, 1, \mathbf{1}, 2, 2, 3, 2, 3, \mathbf{3}, \mathbf{3}]$ their Hamming Distance $D_H(\mathbf{x}, \mathbf{x}')$ would equal 4.

Finally, four movement functions have been implemented for evolving individuals along the search process. The use of these operators is based on the distance between two different partitions. Specifically, these functions are named CE_1,

CE_3, CC_1 and CC_3. On the one hand, the subscript represents the number of randomly selected nodes, which are extracted from its corresponding community. On the other hand, in CE_* operators the taken nodes are re-inserted in already existing clusters, while in CC_* nodes can be inserted also in newly generated ones.

Now, the details of the considered metaheuristics are introduced:

WCA: The Water Cycle Algorithm was firstly proposed in [31] for solving continuous optimization problems. As has been made in other scientific works [32], a discrete adaptation has been conducted for properly facing the problem addressed in this research. Regarding this approach, and laying aside the aspects mentioned in the beginning of this section, the most crucial mechanism to implement is the way in which streams and rivers flow to their corresponding leading individual. Thus, and following the same philosophy of the basic WCA, the movement of each stream $p_{str} \in \mathcal{P}_{str}$ towards its river $\lambda(p_{str})$ at each generation $t \in \{1, \dots, T\}$ is set to:

$$\mathbf{x}^{p_{str}}(t+1) = \Psi\left(\mathbf{x}^{p_{str}}(t), \min\left\{V, \lfloor rand \cdot \theta \cdot D_H(\mathbf{x}^{p_{str}}(t), \mathbf{x}^{\lambda(p_{str})}(t)) \rfloor\right\}\right), \quad (3)$$

where $rand$ is a continuous random variable uniformly distributed in $\mathbb{R}[0, 1]$, θ is a heuristic parameter. Furthermore, $\Psi(\mathbf{x}, Z) \in \{CE_1, CE_3, CC_1, CC_3\}$, each one parametrized by the number of times Z this function is applied to $\mathbf{x}$. This way, the best movement resulting from all the Z movements carried out on $\mathbf{x}$ is selected as output. The same logic is adopted for the movements of a river or a stream towards the sea, just replacing $\mathbf{x}^{\lambda(p_{str})}(t)$ by $\mathbf{x}^{p_{sea}}(t)$.

Additionally, the *inclination* mechanism recently proposed in [32] is also used in the developed discrete WCA, with the main goal of boosting the exploration ability of the method. This simple mechanism endows the algorithm with the intelligence of properly selection the movement operator to use at each iteration for every individual. This election depends on the specific situation of each raindrop. Concretely, each time an individual is about to perform a movement, the aforementioned *inclination* $\xi(\mathbf{x}, \mathbf{x}')$ is computed using as reference the $D_H(\mathbf{x}, \mathbf{x}')$ to its designated river/sea $\mathbf{x}'$. Particularly, $\xi(\cdot, \cdot)$ is equal to $V/D_H(\cdot, \cdot)$. Accordingly, the bigger $D_H(\cdot, \cdot)$ is, the higher $\xi(\cdot, \cdot)$ should be, forcing the method to perform a *fast move* with a higher probability. On the contrary, if $D_H(\cdot, \cdot)$ is small the inclination decreases, suggesting that the search is in a promising area of the solution space, and performing a *slow move* with higher probability. In the present study, four different movement functions have been considered, deeming CC_* as *fast moves*, and CE_* as *slow moves*. Finally, the evaporation and raining procedures remain in the same way as in the original WCA. Specifically, the raining process comprises a number R of consecutive CC_3 movements.

BA: As for the WCA, the classic BA was firstly introduced for solving continuous optimization problems. For this reason, a discrete adaptation has been conducted in order to correctly face the problem addressed in this paper. As in many other adaptations [33], each bat represents a feasible solution of the problem. Both concepts of loudness A_i and pulse emission r_i have been considered in exactly the same form as in the basic version of the method. Furthermore, in order to simplify the complexity of the approach, frequency f_i parameter has not been deemed. Finally, the velocity v_i has been adapted, considering the Hamming Distance as measure function to evaluate the similarity between two bats. This way, $v_i^t = \text{Random}[1, D_H(\mathbf{x_i}, \mathbf{x_*})]$. In other words, the v_i of a bat i at time step t is a random number, which follows a discrete uniform distribution between 1 and the difference between this i and the best bat of the swarm. Finally, the way in which a bat moves is determined similarly to Expression (3), using v_i as the number of movements considered. Furthermore, the inclination concept is also implemented in this discrete version of the BA, also following the same philosophy as for the WCA, and using the best bat as reference.

FA: Again, the classic FA cannot be applied directly to address discrete problems. For this reason, some modifications have been performed over the original version of the FA. As for the BA, each firefly in the swarm represents a solution for the problem. Additionally, the concept of light absorption is considered, which is essential for the adjustment of fireflies' attractiveness. As has been mentioned, the distance between two different individuals is calculated by the Hamming Distance. Finally, the movement of a firefly attracted to another brighter one is determined following the same logic depicted in Expression (3). Besides that, when a firefly is prepared to perform a movement to another firefly, it examines its distance. If it is higher than $V/2$, it can be assumed that it is far from its counterpart. Therefore, it carries out a *wide move*, using a CC_* operator. Otherwise, a *short move* is performed by a CE_* function. This mechanism has been added aiming the adapt the *inclination* functionality above described.

PSO: The last considered approach is the well-known Particle Swarm Optimization, which has been already applied to discrete problems in multiple times [34,35]. Taking as inspiration previous discrete adaptations of the PSO, each particle also deems a feasible solution for the dealt problem. Velocity parameter v_i has been considered analogously to what has been done for the BA. Additionally, both movement operators and inclination mechanism have also been contemplated for the PSO in the same way as for the FA, WCA and BA. Finally, Hamming Distance has been taken as similarity measurement function.

4 Experimentation and Results

With the aim of properly evaluating the performance of the four developed solvers, computer experiments have been run using a heterogeneous set of synthetically generated network instances. All these instances have been created using the well accepted DANCer platform [36,37], and with the aim of covering a diverse set of common situations in dynamic environments. Specifically, the benchmark is composed by 12 different 100-nodeddatasets. The name of each instance is built joining the values of five different parameters:

- *Size of the problem*: In all cases this value is 100.
- *Communities*: The number of communities that compose the ground of truth solution.
- *Generations*: Number of generations run for each Graph Snapshot e_i.
- *Variability*: The difference between adjacent graph snapshot e_i and e_{i+1}. This parameter can adopt three different values: Slight (variety of 5% between adjacent snapshots), Medium (variety of 10%), and Dramatic (variety of 20%).
- *Transition*: Each instance is composed by 30 canonical snapshots, divided into two different families of 15 timestamps. If *Transition* takes *Abrupt* value, the transition between both families is directly made after the 15th timestamp. These instances are comprised just by these 30 snapshots. On the hand, if *Transition* takes *Gradual* value, this transition is made gradually, introducing 10 additional snapshots between the last timestamp of the first family, and the first timestamps of the second one. These datasets are finally composed by 40 snapshots.

This network construction approach allows us to measure the performance of the developed solvers over *noisy* versions of a graph characterized by a controlled underlying community distribution. This method opposes to the common practice by which the comparison is done based on the fitness value obtained by each technique. Finally, 10 independent runs have been executed for each solver and dataset, aiming at reaching statistically reliable insights. Regarding the ending criterion, it depends on both *Generations* and *Transition* parameters of the instance. Thus, depending on the values taken by these parameters, solvers end after *600, 800, 1500* or *2000* generations. The population size has been established in 50 for each method. In the concrete case of WCA, the number of rivers has been established in 9 (approximately 20% of the whole population), leading to a number of 40 streams. On the other hand, the maximum distance for evaporation) and R have been respectively set to 5% and a uniform random value from $\mathbb{N}[0, \lfloor 0.5V \rfloor]$. On the other hand, for FA $\gamma = 0.95$. Finally, for BA $\alpha = \beta = 0.98$, $A_i^0 = 1.0$ and $r_i^0 = 0.1$. For the development and parameterization of these methods, the guidelines given in [32, 33, 38] have been followed.

In Table 1, outcomes (average/best/standard deviation) obtained by the four solvers are shown. Each of these values are divided into two different sub-values, each one depicting separately the performance for the first and the second family. As has been mentioned, each dataset is composed by 30 canonical Graph Snapshots (plus 10 transitional ones for *Gradual* instances), which belong to two different families. All results are shown in terms of the Normalized Mutual Information (NMI) with respect to the *ground of truth* partition of the specific timestamp. This means that the average depicted represents the mean NMI value for all the 15 timestamps belonging to the same family. Analogously, best values depict the maximum value reached at any timestamp of the whole family. The NMI score measures the level of agreement between two community partitions:

Table 1. Obtained NMI results (average/best/standard deviation) using WCA, BA, FA and PSO. Best average results have been highlighted in bold.

Instance	WCA Avg	Best	Std	BA Avg	Best	Std	FA Avg	Best	Std	PSO Avg	Best	Std
100_7_20_Sli_Abr	0.617-0.439	0.724-0.632	0.029-0.036	**0.668**-0.484	0.811-0.676	0.035-0.062	0.646-0.571	0.739-0.682	0.038-0.034	0.573-0.460	0.665-0.608	0.024-0.0334
100_7_20_Sli_Grad	0.619-0.482	0.772-0.632	0.030-0.034	**0.676**-0.461	0.757-0.588	0.027-0.053	0.658-0.571	0.760-0.681	0.034-0.036	0.580-0.463	0.657-0.590	0.021-0.042
100_7_50_Sli_Abr	0.671-0.529	0.784-0.683	0.034-0.035	**0.699**-0.513	0.755-0.670	0.026-0.041	0.655-0.577	0.760-0.681	0.034-0.036	0.640-0.492	0.759-0.664	0.031-0.045
100_7_50_Sli_Grad	0.662-0.540	0.784-0.708	0.029-0.034	**0.689**-0.500	0.777-0.635	0.032-0.054	0.650-0.580	0.751-0.689	0.033-0.030	0.638-0.486	0.758-0.700	0.031-0.039
100_8_20_Med_Abr	0.619-0.445	0.860-0.648	0.029-0.040	**0.660**-0.500	0.869-0.653	0.040-0.042	0.633-0.519	0.811-0.618	0.030-0.033	0.561-0.474	0.698-0.597	0.027-0.030
100_8_20_Med_Grad	0.620-0.463	0.840-0.620	0.027-0.040	**0.654**-0.460	0.861-0.652	0.041-0.043	0.643-0.526	0.799-0.642	0.031-0.040	0.571-0.464	0.701-0.571	0.026-0.031
100_8_50_Med_Abr	0.681-0.495	0.861-0.663	0.029-0.036	**0.710**-0.483	0.870-0.635	0.036-0.035	0.640-0.522	0.802-0.657	0.031-0.036	0.601-0.498	0.835-0.633	0.033-0.034
100_8_50_Med_Grad	0.674-0.492	0.861-0.684	0.030-0.039	**0.683**-0.482	0.861-0.691	0.032-0.049	0.650-0.524	0.812-0.644	0.031-0.032	0.612-0.466	0.848-0.610	0.030-0.034
100_9_20_Dram_Abr	0.593-0.474	0.726-0.631	0.029-0.038	**0.639**-0.501	0.820-0.684	0.034-0.039	0.607-0.569	0.741-0.657	0.030-0.027	0.535-0.539	0.643-0.657	0.029-0.029
100_9_20_Dram_Grad	0.589-0.515	0.696-0.641	0.029-0.033	**0.641**-0.543	0.793-0.697	0.025-0.063	0.606-0.570	0.753-0.706	0.031-0.029	0.538-0.538	0.643-0.644	0.024-0.030
100_9_50_Dram_Abr	0.642-0.538	0.791-0.684	0.028-0.036	**0.677**-0.550	0.811-0.675	0.026-0.026	0.614-0.559	0.765-0.641	0.031-0.028	0.586-0.529	0.727-0.646	0.032-0.039
100_9_50_Dram_Grad	0.638-0.544	0.820-0.694	0.024-0.030	**0.651**-0.548	0.826-0.664	0.042-0.038	0.620-0.553	0.764-0.643	0.035-0.026	0.585-0.514	0.701-0.641	0.028-0.031
Friedman's non-parametric test (mean ranking)												
Rank	2.5-3.0			1.0-2.83			2.5-1.0			4.0-3.16		

if $NMI(\widetilde{\mathcal{V}}, \widetilde{\mathcal{V}}') = 1$ both distributions $\widetilde{\mathcal{V}}$ and $\widetilde{\mathcal{V}}'$ are equal to each other. This also means that lower values denote that there are differences between them.

A first analysis reveals that BA is the best alternative in the 100% of the cases for the first family, while the FA emerges as the best alternative in all the datasets for the second families. On the other hand, the performance of PSO is much lower than the other alternatives, while WCA is one step behind BA and FA. In this sense, we can see how WCA is much worse than the BA for the first families but similar to FA in this context. On the contrary, it is much worse than FA for second families while it obtains similar outcomes than BA. Although it may seem unintuitive, this switch in the quality of results has a logical explanation, which is based in the fact that BA is a better for the exploitation of the solution space, while FA shows a better adapting capacity thank to its enhanced exploratory ability.

Following the guidelines in [39] and [40], two different tests have been carried out to resolve the statistical relevance of the reported performance gaps. To begin with, the Friedman's non-parametric test for multiple comparison allows proving if there are significant differences in the results obtained by all reported methods. Last row of Table 1 displays the mean ranking returned by this non-parametric test for each of the compared algorithms and families (the lower the rank, the better the performance). Thus, for the first family the Friedman statistic (distributed according to χ^2 with 4 degrees of freedom) was equal to 32.4. Furthermore, the confidence interval has been set to 99%, being 13.27 the critical point in a χ^2 distribution with 4 degrees of freedom. Since $32.4 > 13.27$, it can be concluded that there are significant differences among the results, thus BA can be regarded as the method having the lowest rank. Regarding the second family, and taking into account also the results shown in Table 1, the Friedman statistic test is 22.0. Again, since $22.0 > 13.27$, the same conclusion is also applicable in this case.

The second statistical test is the Holm's post-hoc test. For correctly conducting this test, BA has been set as the control algorithm for the first family, whilst for the second FA has been established. Table 2 gathers the unadjusted and adjusted p-values obtained through the application of Holm's post-hoc procedure. From these p-values it can be concluded that BA, for the first case, and

FA, for the second one, are significantly better than their counterparts at a 95% confidence level, since all p values are lower than 0.05.

Finally, and seeking for the completeness of the present research, we show in Figs. 1 and 2 the evolution of the NMI and modularity along the whole execution for the instance 100_7_50_*Sli_Abr*. In Fig. 1 the performance of the BA is depicted, while Fig. 2 is devoted to FA. Both graphs represent the performance presented over the 10 runs.

Table 2. Unadjusted and adjusted p-values obtained as a result of the application of Holm's post-hoc procedure using BA and FA as control algorithms.

Fam1 (BA as control)			Fam2 (FA as control)		
Algorithm	Unadjusted p	Adjusted p	Algorithm	Unadjusted p	Adjusted p
WCA	0.004427	0.008853	WCA	0.000148	0.000296
FA	0.004427	0.008853	BA	0.000504	0.000504
PSO	0.0	0.0	PSO	0.000039	0.000118

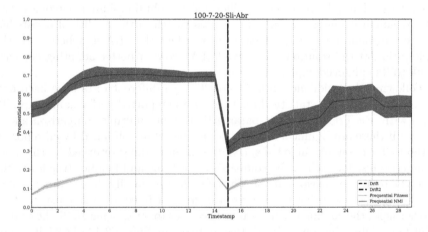

Fig. 1. Evolution of the NMI and fitness obtained by BA for 100_7_50_*Sli_Abr*. Blue line denotes the NMI. Red line depicts the modularity. Vertical grey line represents a timestamp alteration. Vertical black line points the family change. (Color figure online)

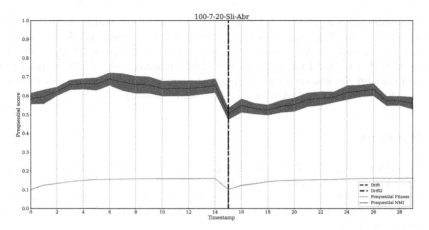

Fig. 2. Evolution of the NMI and fitness obtained by FA for 100_7_50_*Sli_Abr*. Blue line denotes the NMI. Red line depicts the modularity. Vertical grey line represents a timestamp alteration. Vertical black line points the family change. (Color figure online)

5 Conclusions and Future Research Lines

In this study, community finding in dynamic graphs has been dealt by using four different nature-inspired meta-heuristics: Water Cycle Algorithm, Bat Algorithm, Firefly Algorithm and Particle Swarm Optimization. For this purpose, the detection of optimal partitions has been modeled as a discrete optimization problem, using an adapted Newman and Girvan's Modularity as evaluation function. All the four deemed methods have been adapted to face the particularities of the solution space, such as the potential representational ambiguity of label encoding and the definition of distance between solutions to the problem. The performance of each approach has been evaluated using a benchmark composed by 12 dynamic networks, all of them comprised by 100 nodes, using as comparison criterion the Normalized Mutual Information (or NMI) regarding their *ground of truth* partition. Obtained outcomes demonstrated that BA and FA dominate over their counterparts with statistical significance.

As future work, we plan to conduct further efforts in different directions. The most imminent one is the adaption of additional nature-inspired, evolutionary and swarm intelligent methods, such as the Cuckoo Search [41] or Grey Wolf Optimizer [42]. The performance shown by these techniques applied to other optimization problems [43–46] lead us to consider them as potentially promising methods. Moreover, we will also consider larger network instances that the ones employed in this paper. Finally, we have the firm intention of exploring the hybridization of these heuristic with local search techniques, trying to mimic the operation of other heuristics found in the related state of the art, such as recently contributed message passing procedures [47] and other techniques renowned for their good scalability [48].

Acknowledgements. E. Osaba and J. Del Ser would like to thank the Basque Government for its funding support through the EMAITEK program. A. Iglesias and A. Galvez acknowledge the financial support from the projects TIN2017-89275-R (AEI/FEDER, UE) and PDE-GIR (H2020, MSCA program, ref. 778035). Iztok Fister and Iztok Fister Jr. acknowledge the financial support from the Slovenian Research Agency (Research Core Founding No. P2-0041 and P2-0057).

References

1. Bello-Orgaz, G., Jung, J.J., Camacho, D.: Social big data: recent achievements and new challenges. Inf. Fusion **28**, 45–59 (2016)
2. Lara-Cabrera, R., Pardo, A.G., Benouaret, K., Faci, N., Benslimane, D., Camacho, D.: Measuring the radicalisation risk in social networks. IEEE Access **5**, 10892–10900 (2017)
3. Torregrosa, J., Panizo, Á.: RiskTrack: assessing the risk of Jihadi radicalization on Twitter using linguistic factors. In: Yin, H., Camacho, D., Novais, P., Tallón-Ballesteros, A.J. (eds.) IDEAL 2018. LNCS, vol. 11315, pp. 15–20. Springer, Cham (2018). https://doi.org/10.1007/978-3-030-03496-2_3
4. Westlake, B.G., Bouchard, M.: Liking and hyperlinking: community detection in online child sexual exploitation networks. Soc. Sci. Res. **59**, 23–36 (2016)
5. Villar-Rodríguez, E., Del Ser, J., Torre-Bastida, A.I., Bilbao, M.N., Salcedo-Sanz, S.: A novel machine learning approach to the detection of identity theft in social networks based on emulated attack instances and support vector machines. Concurr. Comput. Pract. Exp. **28**(4), 1385–1395 (2016)
6. Chakraborty, T., Srinivasan, S., Ganguly, N., Mukherjee, A., Bhowmick, S.: On the permanence of vertices in network communities. In: ACM SIGKDD International Conference on Knowledge Discovery and Data Mining, pp. 1396–1405. ACM (2014)
7. Aldecoa, R., Marín, I.: Deciphering network community structure by surprise. PloS One **6**(9), e24195 (2011)
8. Newman, M.E., Girvan, M.: Finding and evaluating community structure in networks. Phys. Rev. E **69**(2), 026113 (2004)
9. Rizman Žalik, K.: Evolution algorithm for community detection in social networks using node centrality. In: Bembenik, R., Skonieczny, Ł., Protaziuk, G., Kryszkiewicz, M., Rybinski, H. (eds.) Intelligent Methods and Big Data in Industrial Applications. SBD, vol. 40, pp. 73–87. Springer, Cham (2019). https://doi.org/10.1007/978-3-319-77604-0_6
10. Pizzuti, C., Socievole, A.: A genetic algorithm for community detection in attributed graphs. In: Sim, K., Kaufmann, P. (eds.) EvoApplications 2018. LNCS, vol. 10784, pp. 159–170. Springer, Cham (2018). https://doi.org/10.1007/978-3-319-77538-8_12
11. Pizzuti, C.: Evolutionary computation for community detection in networks: a review. IEEE Trans. Evol. Comput. **22**(3), 464–483 (2018)
12. Rahimi, S., Abdollahpouri, A., Moradi, P.: A multi-objective particle swarm optimization algorithm for community detection in complex networks. Swarm Evol. Comput. **39**, 297–309 (2018)
13. Del Ser, J., Lobo, J.L., Villar-Rodriguez, E., Bilbao, M.N., Perfecto, C.: Community detection in graphs based on surprise maximization using firefly heuristics. In: IEEE Congress on Evolutionary Computation (CEC), pp. 2233–2239. IEEE (2016)

14. Hassan, E.A., Hafez, A.I., Hassanien, A.E., Fahmy, A.A.: A discrete bat algorithm for the community detection problem. In: Onieva, E., Santos, I., Osaba, E., Quintián, H., Corchado, E. (eds.) HAIS 2015. LNCS (LNAI), vol. 9121, pp. 188–199. Springer, Cham (2015). https://doi.org/10.1007/978-3-319-19644-2_16

15. Saoud, B.: Networks clustering with bee colony. Artif. Intell. Rev. 1–13 (2018)

16. Rossetti, G., Cazabet, R.: Community discovery in dynamic networks: a survey. ACM Comput. Surv. (CSUR) **51**(2) (2018). Article No. 35

17. Messaoudi, I., Kamel, N.: A multi-objective bat algorithm for community detection on dynamic social networks. Appl. Intell. **49**(6), 2119–2136 (2019)

18. Li, Z., Liu, J.: A multi-agent genetic algorithm for community detection in complex networks. Phys. A Stat. Mech. Appl. **449**, 336–347 (2016)

19. Folino, F., Pizzuti, C.: An evolutionary multiobjective approach for community discovery in dynamic networks. IEEE Trans. Knowl. Data Eng. **26**(8), 1838–1852 (2014)

20. Eskandar, H., Sadollah, A., Bahreininejad, A., Hamdi, M.: Water cycle algorithm - a novel metaheuristic optimization method for solving constrained engineering optimization problems. Appl. Soft Comput. **110**(111), 151–166 (2012)

21. Yang, X.S.: A new metaheuristic bat-inspired algorithm. In: González, J.R., Pelta, D.A., Cruz, C., Terrazas, G., Krasnogor, N. (eds.) NICSO 2010. SCI, vol. 284, pp. 65–74. Springer, Heidelberg (2010). https://doi.org/10.1007/978-3-642-12538-6_6

22. Yang, X.S.: Firefly algorithm, stochastic test functions and design optimisation. Int. J. Bio-Inspired Comput. **2**(2), 78–84 (2010)

23. Kennedy, J.: Particle swarm optimization. In: Sammut, C., Webb, G.I. (eds.) Encyclopedia of Machine Learning, pp. 760–766. Springer, Boston (2011). https://doi.org/10.1007/978-0-387-30164-8_630

24. Leicht, E.A., Newman, M.E.: Community structure in directed networks. Phy. Rev. Lett. **100**(11), 118703 (2008)

25. Chakraborty, T., Dalmia, A., Mukherjee, A., Ganguly, N.: Metrics for community analysis: a survey. ACM Comput. Surv. (CSUR) **50**(4) (2017). Article No. 54

26. Chen, M., Kuzmin, K., Szymanski, B.K.: Community detection via maximization of modularity and its variants. IEEE Trans. Comput. Soc. Syst. **1**(1), 46–65 (2014)

27. Harris, J.M., Hirst, J.L., Mossinghoff, M.J.: Combinatorics and Graph Theory, vol. 2. Springer, New York (2008). https://doi.org/10.1007/978-0-387-79711-3

28. Hruschka, E.R., Campello, R.J., Freitas, A.A., et al.: A survey of evolutionary algorithms for clustering. IEEE Trans. Syst. Man Cybern. Part C (Appl. Rev.) **39**(2), 133–155 (2009)

29. Falkenauer, E.: Genetic Algorithms and Grouping Problems. Wiley, New York (1998)

30. Osaba, E., Del Ser, J., Camacho, D., Galvez, A., Iglesias, A., Fister, I., Fister, I.: Community detection in weighted directed networks using nature-inspired heuristics. In: Yin, H., Camacho, D., Novais, P., Tallón-Ballesteros, A.J. (eds.) IDEAL 2018. LNCS, vol. 11315, pp. 325–335. Springer, Cham (2018). https://doi.org/10.1007/978-3-030-03496-2_36

31. Eskandar, H., Sadollah, A., Bahreininejad, A., Hamdi, M.: Water cycle algorithm-a novel metaheuristic optimization method for solving constrained engineering optimization problems. Comput. Struct. **110**, 151–166 (2012)

32. Osaba, E., Del Ser, J., Sadollah, A., Bilbao, M.N., Camacho, D.: A discrete water cycle algorithm for solving the symmetric and asymmetric traveling salesman problem. Appl. Soft Comput. **71**, 277–290 (2018)

33. Osaba, E., Yang, X.S., Fister Jr., I., Del Ser, J., Lopez-Garcia, P., Vazquez-Pardavila, A.J.: A discrete and improved bat algorithm for solving a medical goods distribution problem with pharmacological waste collection. Swarm Evol. Comput. **44**, 273–286 (2019)

34. Chen, A.L., Yang, G.K., Wu, Z.M.: Hybrid discrete particle swarm optimization algorithm for capacitated vehicle routing problem. J. Zhejiang Univ.-Sci. A **7**(4), 607–614 (2006)

35. Zhong, Y., Lin, J., Wang, L., Zhang, H.: Discrete comprehensive learning particle swarm optimization algorithm with metropolis acceptance criterion for traveling salesman problem. Swarm Evol. Comput. **42**, 77–88 (2018)

36. Largeron, C., Mougel, P.N., Benyahia, O., Zaïane, O.R.: Dancer: dynamic attributed networks with community structure generation. Knowl. Inf. Syst. **53**(1), 109–151 (2017)

37. Benyahia, O., Largeron, C., Jeudy, B., Zaïane, O.R.: DANCer: dynamic attributed network with community structure generator. In: Berendt, B., et al. (eds.) ECML PKDD 2016. LNCS (LNAI), vol. 9853, pp. 41–44. Springer, Cham (2016). https://doi.org/10.1007/978-3-319-46131-1_9

38. Osaba, E., Yang, X.S., Diaz, F., Onieva, E., Masegosa, A.D., Perallos, A.: A discrete firefly algorithm to solve a rich vehicle routing problem modelling a newspaper distribution system with recycling policy. Soft Comput. **21**(18), 5295–5308 (2017)

39. Derrac, J., García, S., Molina, D., Herrera, F.: A practical tutorial on the use of nonparametric statistical tests as a methodology for comparing evolutionary and swarm intelligence algorithms. Swarm Evol. Comput. **1**(1), 3–18 (2011)

40. Osaba, E., Carballedo, R., Diaz, F., Onieva, E., Masegosa, A., Perallos, A.: Good practice proposal for the implementation, presentation, and comparison of meta-heuristics for solving routing problems. Neurocomputing **271**, 2–8 (2018)

41. Yang, X.S., Deb, S.: Cuckoo search via lévy flights. In: 2009 World Congress on Nature & Biologically Inspired Computing, NaBIC 2009, pp. 210–214. IEEE (2009)

42. Mirjalili, S., Mirjalili, S.M., Lewis, A.: Grey wolf optimizer. Adv. Eng. Softw. **69**, 46–61 (2014)

43. Precup, R.E., David, R.C., Petriu, E.M., Szedlak-Stinean, A.I., Bojan-Dragos, C.A.: Grey wolf optimizer-based approach to the tuning of PI-fuzzy controllers with a reduced process parametric sensitivity. IFAC-PapersOnLine **49**(5), 55–60 (2016)

44. Precup, R.E., David, R.C., Petriu, E.M.: Grey wolf optimizer algorithm-based tuning of fuzzy control systems with reduced parametric sensitivity. IEEE Trans. Ind. Electron. **64**(1), 527–534 (2017)

45. Yang, X.S., Deb, S., Mishra, S.K.: Multi-species cuckoo search algorithm for global optimization. Cogn. Comput. **10**(6), 1085–1095 (2018)

46. He, X.-S., Wang, F., Wang, Y., Yang, X.-S.: Global convergence analysis of Cuckoo search using Markov theory. In: Yang, X.-S. (ed.) Nature-Inspired Algorithms and Applied Optimization. SCI, vol. 744, pp. 53–67. Springer, Cham (2018). https://doi.org/10.1007/978-3-319-67669-2_3

47. Shi, C., Liu, Y., Zhang, P.: Weighted community detection and data clustering using message passing. J. Stat. Mech. Theory Exp. **2018**(3), 033405 (2018)

48. Lu, H., Halappanavar, M., Kalyanaraman, A.: Parallel heuristics for scalable community detection. Parallel Comput. **47**, 19–37 (2015)

Bat Algorithm for Kernel Computation in Fractal Image Reconstruction

Akemi Gálvez[1,2], Eneko Osaba[3], Javier Del Ser[3,4], and Andrés Iglesias[1,2(✉)]

[1] Department of Information Science, Faculty of Sciences,
Narashino Campus Toho University, 2-2-1 Miyama, Funabashi 274-8510, Japan
[2] Department of Applied Mathematics and Computer Science,
E.T.S.I. Caminos, Canales y Puertos Universidad de Cantabria,
Avda. de los Castros, s/n, 39005 Santander, Spain
iglesias@unican.es
[3] TECNALIA Research & Innovation, 48160 Derio, Spain
[4] University of the Basque Country (UPV/EHU), Bilbao, Spain

Abstract. Computer reconstruction of digital images is an important problem in many areas such as image processing, computer vision, medical imaging, sensor systems, robotics, and many others. A very popular approach in that regard is the use of different kernels for various morphological image processing operations such as dilation, erosion, blurring, sharpening, and so on. In this paper, we extend this idea to the reconstruction of digital fractal images. Our proposal is based on a new affine kernel particularly tailored for fractal images. The kernel computes the difference between the source and the reconstructed fractal images, leading to a difficult nonlinear constrained continuous optimization problem, solved by using a powerful nature-inspired metaheuristics for global optimization called the bat algorithm. An illustrative example is used to analyze the performance of this approach. Our experiments show that the method performs quite well but there is also room for further improvement. We conclude that this approach is promising and that it could be a very useful technique for efficient fractal image reconstruction.

Keywords: Image processing · Image reconstruction · Affine kernel · Fractal image · Bat algorithm

1 Introduction

Computer reconstruction of digital images is a classical problem in fields such as image processing and computer vision. The topic has gained strong relevance during the last few decades owing to its important applications in several areas, including medical imaging (computer tomography, magnetic resonance), sensor systems, robotics, smart cities, internet of things, and many others. Roughly speaking, the problem consists of reproducing a given image described in terms of digital data (typically, raster or bitmapped images) by following procedures

© Springer Nature Switzerland AG 2019
J. M. F. Rodrigues et al. (Eds.): ICCS 2019, LNCS 11538, pp. 381–394, 2019.
https://doi.org/10.1007/978-3-030-22744-9_30

involving either a set of equations and operators, or a set of rules, or some kind of heuristics (even sometimes combinations of them). In this paper, we are interested in this problem for the case of fractal images, which exhibit a property called *self-similarity*, meaning that the images follow (at least, approximately) a self-similar pattern across different scales [2,4].

Several methods have been traditionally applied to the image reconstruction problem. When dealing with fractal images, some popular methods include the Brownian motion, escape-time fractals, finite subdivision rules, L-systems, strange attractors of dynamical systems [11], and many others [2,8,13]. For a general (non-fractal) image, other methods based on image processing techniques are more commonly applied [10]. Among them, a popular approach in image processing is the use of different kernels for various morphological image processing operations such as dilation, erosion, blurring, sharpening, and so on. In this work, we are interested to follow this approach regarding its potential application to the case of fractal images.

In this paper, we introduce a new method for digital fractal image reconstruction. Our proposal is based on a new affine kernel particularly tailored for fractal images. The kernel computes the difference between the source and the reconstructed fractal images, according to a given metrics. This leads to a difficult nonlinear constrained continuous optimization problem that has been proved to be not well suited for classical mathematical optimization techniques. To tackle this issue, we make use of a powerful nature-inspired metaheuristics for global optimization called bat algorithm (see Sect. 3 for details).

The structure of this paper is as follows: Sect. 2 summarizes the mathematical background required to follow the paper. Section 3 describes the main features of the bat algorithm, the global optimization metaheuristics used in this paper. Our proposed method is described in detail in Sect. 4 and then applied to an illustrative example in Sect. 5. The paper closes with the conclusions and some ideas for future work.

2   Basic Concepts and Definitions

2.1   Digital Images

In this work, we consider a *digital image* $\mathcal{I}$ to be numerically represented as a two-dimensional raster or bitmapped image. We exclude in our study other possible computer representations such as vector images. The *convolution operator* of two functions ϕ and ψ, denoted by $\phi \otimes \psi$, is a mathematical operation describing how the shape of one function is modified by the other. Analytically, it is given by an integral transform of both functions defined as:

$$(\phi \otimes \psi)(\rho) = \int_{-\infty}^{\infty} \phi(\tau)\psi(\rho - \tau)d\tau \tag{1}$$

In the context of image processing, the convolution operator is carried out in a discrete fashion by using a kernel applied on a given image $\mathcal{I}$ via matrix convolution. Let $\mathcal{K}$ be such a kernel. The convolution is given by:

$$\mathcal{I}'_{x,y} = (\mathcal{K} \otimes \mathcal{I})_{x,y} = \sum_{\alpha=-\mu}^{\mu} \sum_{\beta=-\nu}^{\nu} \mathcal{K}_{\alpha,\beta} \mathcal{I}_{x-\alpha,y-\beta} \tag{2}$$

where $\mathcal{I}'$ is the transformed image, the subscripts indicate the image pixels, and $\mathcal{K} = \{\mathcal{K}_{\alpha,\beta}\}_{\alpha,\beta}$, with $-\mu \leqslant \alpha \leqslant \mu, \nu \leqslant \beta \leqslant \nu$. Depending on the particular purposes, different kernels can be considered: for instance, classical operations in morphological image processing such as dilation and erosion are expressed by specific filtering kernels operating on a input binary image. Other operations such as opening, closing, and boundary detection can be obtained as a combination of such kernels (see [10] for details).

2.2 Fractal Images

In this paper, a *digital fractal image* is defined as a digital image with the property of self-similarity and whose fractal dimension is larger than its topological dimension [4,12]. Suppose a set of affine mappings $\Lambda = \{\Lambda_1, \ldots, \Lambda_\eta\}$ defined on a complete metric space $\mathcal{M} = (\Omega, \Psi)$, where $\Omega \subset \mathbb{R}^n$ and Ψ is a distance on Ω. Such affine mappings Λ_κ can be represented by a 3×3 augmented matrix $\Theta_\kappa = \{\theta^\kappa_{i,j}\}_{i,j=1,2,3}$ in homogeneous coordinates, with: $\theta^\kappa_{3,j} = \delta_{3,j}$, where δ represents the Kronecker delta. In that case, $\Lambda_\kappa(A) = \Theta_\kappa.A^*$, $\forall A \subset \mathbb{R}^2$, where the superscript $*$ denotes the augmented matrix. We assume that all mappings Λ_κ are contractive, with contractivity factor $\lambda_\kappa > 0$.

Consider now the set of all compact subsets of the plane, $\mathcal{H}$. We can define the Hutchinson operator, Ξ as:

$$\Xi(S) = \bigcup_{\kappa=1}^{\eta} \Lambda_\kappa(S) \tag{3}$$

for each $S \in \mathcal{H}$. Since all Λ_κ are contractions, this operator Ξ is also a contraction in $\mathcal{H}$ with the induced Hausdorff metric [3,14]. Then, according to the fixed point theorem, Ξ has a unique fixed point, called the *attractor of* Λ.

The reconstruction of digital fractal images is driven by a famous result by Barnsley called the Collage Theorem [2]. Roughly speaking, it states that every digital image can be represented as the attractor of a system Λ. In particular, given a non-empty $B \in \mathcal{H}$, the induced Hausdorff metric $H(.,.)$ on $\mathcal{H}$, a non-negative real threshold value $\epsilon \geqslant 0$, and a system of affine contractive mappings Λ with contractivity factor $0 < \lambda < 1$, given by: $\lambda = \max\{\lambda_\kappa\}_{\kappa=1,\ldots,\eta}$, if $H(B, \Xi(B)) \leqslant \epsilon$ then $H(B, \mathcal{A}) \leqslant \dfrac{\epsilon}{1-\lambda}$, where $\mathcal{A}$ is the attractor of Λ, or equivalently: $H(B, \mathcal{A}) \leqslant \dfrac{1}{1-\lambda} H\left(B, \bigcup_{\kappa=1}^{\eta} \phi_\kappa(B)\right)$.

3   The Bat Algorithm

The *bat algorithm* is a bio-inspired swarm intelligence algorithm originally proposed by Yang in 2010 to solve continuous optimization problems [21–23]. The algorithm is based on the echolocation behavior of microbats, which use a type of sonar called *echolocation*, with varying pulse rates of emission and loudness, to detect prey, avoid obstacles, and locate their roosting crevices in the dark. The idealization of the echolocation of microbats is as follows:

1. Bats use echolocation to sense distance and distinguish between food, prey and background barriers.
2. Each virtual bat flies randomly with a velocity $\mathbf{v}_i$ at position (solution) $\mathbf{x}_i$ with a fixed frequency f_{min}, varying wavelength λ and loudness A_0 to search for prey. As it searches and finds its prey, it changes wavelength (or frequency) of their emitted pulses and adjust the rate of pulse emission r, depending on the proximity of the target.
3. It is assumed that the loudness will vary from an (initially large and positive) value A_0 to a minimum constant value A_{min}.

Some additional assumptions are advisable for further efficiency. For instance, we assume that the frequency f evolves on a bounded interval $[f_{min}, f_{max}]$. This means that the wavelength λ is also bounded, because f and λ are related to each other by the fact that the product $\lambda.f$ is constant. For practical reasons, it is also convenient that the largest wavelength is chosen such that it is comparable to the size of the domain of interest (the search space for optimization problems). For simplicity, we can assume that $f_{min} = 0$, so $f \in [0, f_{max}]$. The rate of pulse can simply be in the range $r \in [0, 1]$, where 0 means no pulses at all, and 1 means the maximum rate of pulse emission.

With these idealized rules indicated above, the basic pseudo-code of the bat algorithm is shown in Algorithm 1. Basically, the algorithm considers an initial population of $\mathcal{P}$ individuals (bats). Each bat, representing a potential solution of the optimization problem, has a location $\mathbf{x}_i$ and velocity $\mathbf{v}_i$. The algorithm initializes these variables with random values within the search space. Then, the pulse frequency, pulse rate, and loudness are computed for each individual bat. Then, the swarm evolves in a discrete way over iterations, like time instances until the maximum number of iterations, $\mathcal{G}_{max}$, is reached. For each generation g and each bat, new frequency, location and velocity are computed according to the following evolution equations:

$$f_i^g = f_{min}^g + \beta(f_{max}^g - f_{min}^g) \tag{4}$$

$$\mathbf{v}_i^g = \mathbf{v}_i^{g-1} + [\mathbf{x}_i^{g-1} - \mathbf{x}^*] f_i^g \tag{5}$$

$$\mathbf{x}_i^g = \mathbf{x}_i^{g-1} + \mathbf{v}_i^g \tag{6}$$

where $\beta \in [0, 1]$ follows the random uniform distribution, and $\mathbf{x}^*$ represents the current global best location (solution), which is obtained through evaluation of the objective function at all bats and ranking of their fitness values. The

---

Require: (Initial Parameters)
    Population size: $\mathcal{P}$; Maximum number of iterations: $\mathcal{G}_{max}$; Loudness: $\mathcal{A}$
    Pulse rate: r ; Maximum frequency: f_{max} ; Dimension of the problem: d
    Objective function: $\phi(\mathbf{x})$, with $\mathbf{x} = (x_1, \ldots, x_d)^T$; Random number: $\theta \in U(0, 1)$
 1: $g \leftarrow 0$
 2: Initialize the bat population $\mathbf{x}_i$ and $\mathbf{v}_i$, $(i = 1, \ldots, n)$
 3: Define pulse frequency f_i at $\mathbf{x}_i$
 4: Initialize pulse rates r_i and loudness $\mathcal{A}_i$
 5: **while** $g < \mathcal{G}_{max}$ **do**
 6:     **for** $i = 1$ **to** $\mathcal{P}$ **do**
 7:         Generate new solutions by using eqns. (4)–(6)
 8:         **if** $\theta > r_i$ **then**
 9:             $\mathbf{s}^{best} \leftarrow \mathbf{s}^g$ //select the best current solution
10:             $\mathbf{ls}^{best} \leftarrow \mathbf{ls}^g$ //generate a local solution around $\mathbf{s}^{best}$
11:         **end if**
12:         Generate a new solution by local random walk
13:         **if** $\theta < \mathcal{A}_i$ **and** $\phi(\mathbf{x_i}) < \phi(\mathbf{x}^*)$ **then**
14:             Accept new solutions, increase r_i and decrease $\mathcal{A}_i$
15:         **end if**
16:     **end for**
17:     $g \leftarrow g + 1$
18: **end while**
19: Rank the bats and find current best $\mathbf{x}^*$
20: **return** $\mathbf{x}^*$

Algorithm 1: Bat algorithm pseudocode

superscript $(.)^g$ is used to denote the current generation g. The best current solution and a local solution around it are probabilistically selected according to some given criteria. Then, search is intensified by a local random walk. For this local search, once a solution is selected among the current best solutions, it is perturbed locally through a random walk of the form: $\mathbf{x}_{new} = \mathbf{x}_{old} + \epsilon \mathcal{A}^g$, where ϵ is a uniform random number on $[-1, 1]$ and $\mathcal{A}^g = <\mathcal{A}_i^g>$, is the average loudness of all the bats at generation g. If the new solution achieved is better than the previous best one, it is probabilistically accepted depending on the value of the loudness. In that case, the algorithm increases the pulse rate and decreases the loudness. This process is repeated for the given number of iterations. In general, the loudness decreases once a new best solution is found, while the rate of pulse emission decreases. For simplicity, the following values are commonly used: $\mathcal{A}_0 = 1$ and $\mathcal{A}_{min} = 0$, assuming that this latter value means that a bat has found the prey and temporarily stop emitting any sound. The evolution rules for loudness and pulse rate are as: $\mathcal{A}_i^{g+1} = \alpha \mathcal{A}_i^g$ and $r_i^{g+1} = r_i^0[1 - exp(-\gamma g)]$ where α and γ are constants. Note that for any $0 < \alpha < 1$ and any $\gamma > 0$ we have: $\mathcal{A}_i^g \to 0$, $r_i^g \to r_i^0$ as $g \to \infty$. Generally, each bat should have different values for loudness and pulse emission rate, which can be achieved by randomization. To this aim, we can take an initial loudness $\mathcal{A}_i^0 \in (0, 2)$ while the initial emission rate r_i^0 can be any value in the interval $[0, 1]$. Loudness and emission rates will be updated only if the new solutions are improved, an indication that the bats are moving towards the optimal solution.

Bat algorithm is a very promising method that has already been successfully applied to several problems, such as multilevel image thresholding [1], economic dispatch [18], B-spline curve reconstruction [15], optimal design of structures in civil engineering [17], robotics [20], fuel arrangement optimization [16], planning of sport training sessions [5], transport [19], and many others. The interested reader is also referred to the general paper in [24] for a comprehensive review of the bat algorithm, its variants and other interesting applications.

4   The Method

4.1   Optimization Problem

Suppose that we are given a digital fractal image, $\mathcal{I}$. The Collage Theorem states that $\mathcal{I}$ can be closely approximated by an iterative process driven by a set of contractive affine mappings, $\mathbf{\Lambda} = \{A_1, \dots, A_\eta\}$, on the two-dimensional real plane. In particular, for any arbitrary $S_0 \in \mathcal{H}$, consider $S_j = \Lambda_\kappa(S_{j-1}) = \mathbf{\Theta}_\kappa.S_{j-1}$, where κ is randomly chosen from the set of indices $\{1, \dots, \eta\}$ according to a set of probabilities $\mathcal{W} = \{\omega_1, \dots, \omega_\eta\}$, with $\sum_{\kappa=1}^{\eta} \omega_\kappa = 1$, for each iteration step j. Then, the sequence $\{S_j\}_j$ converges to $\mathcal{I}$ as $j \to \infty$.

In other words, any given digital fractal image $\mathcal{I}$ can be accurately approximated by the action of a finite collection of affine kernels $\{\mathbf{\Theta}_\kappa\}_{\kappa=1,\dots,\eta}$ according to a similarity function $\mathcal{S}$, which measures the graphical distance between $\mathcal{I}$ and the reconstructed image $\mathcal{I}' = \bigcup_{\kappa=1}^{\eta} \Lambda_\kappa(\mathcal{I})$. In line with this, the problem consists of computing the kernels $\mathbf{\Theta}_\kappa$ and can be formulated as the following optimization problem:

$$\underset{\{\Theta_{i,j}^\kappa\}, \{\omega_\kappa\}}{minimize} \, \mathcal{S}\left(\mathcal{I}, \bigcup_{\kappa=1}^{\eta} \Lambda_\kappa(\mathcal{I})\right) \tag{7}$$

The minimization in Eq. (7) is a continuous nonlinear constrained optimization problem, because all free variables $\{\Theta_{i,j}^\kappa\}_{i,j,\kappa}, \{\omega_\kappa\}_\kappa$ are real-valued and must satisfy the condition that the corresponding functions Λ_κ have to be contractive. It is also a multimodal problem, as there can be several global or local minima of the similarity function. The problem is so difficult that only partial solutions have been reported so far in the literature. However, the general problem still remains unsolved. In this paper we address this problem by applying the bat algorithm described in previous section.

4.2   The Procedure

In our method, we consider an initial population of χ individuals called bats, $\{\mathcal{B}_i^0\}_{i=1,\dots,\chi}$, where each bat is a real-valued vector comprised of all free variables in Eq. (7) and the superscrit denotes the iteration number. These individuals are initialized with uniform random values in $[-1, 1]$ for the variables in $\{\Theta_{i,j}^\kappa\}_{i,j,\kappa}$,

Fig. 1. Six different individuals (bats) from the initial random population.

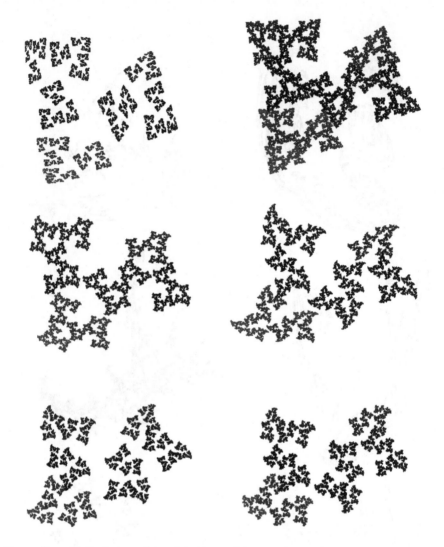

Fig. 2. (l-r, t-b) Evolution of the global best of the population from 100 to 600 iterations with step size 100, respectively.

and in $[0, 1]$ for the $\{\omega_\kappa\}_\kappa$, such that $\sum_{\kappa=1}^{\eta} \omega_\kappa^i = 1$. After this initialization step, we compute the contractive factors λ_κ and reinitialize all functions Λ_κ with $\lambda_\kappa \geqslant 1$ to ensure that only contractive functions are included in the initial population. Regarding the fitness function, it is given by the Hamming distance: the fractal images are stored as binary bitmap images for a given resolution defined by a mesh size parameter, m_s. Then, we divide the number of mismatches between the original and the reconstructed matrices by the total number of boxes in the image. This yields the *normalized similarity error rate index* (NSERI) between both images, denoted by $|\mathcal{S}(\mathcal{I}, \mathcal{I}')|$. This is the fitness function used in this work.

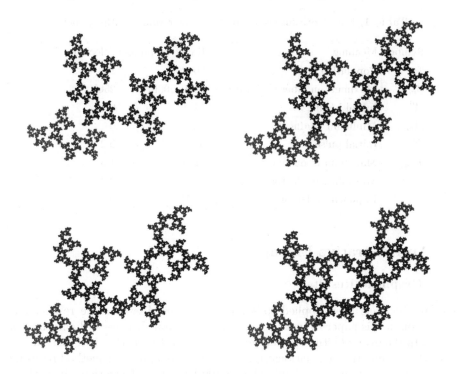

Fig. 3. *(cont'd)* (l-r, t-b) Evolution of the global best of the population from 800 to 1400 iterations with step size 200, respectively.

4.3   Parameter Tuning

A critical issue when working with swarm intelligence techniques is the parameter tuning, which is well-known to be problem-dependent. Our choice has been fully empirical, based on computer simulations for different parameter values. The different parameters used in this work are arranged in rows in Table 1. For each parameter, the table shows (in columns) its symbol, meaning, range of values, and the parameter value chosen in this paper. Regarding the stopping criterion, our method is run for a fixed number of iterations, $\mathcal{G}_{max}$. From our experiments, we found that $\mathcal{G}_{max} = 2500$ iterations is enough to reach convergence in all our simulations, so this is the value used in this work. Finally, our method requires to define the mesh size, m_s, set to $m_s = 100$ in this work.

With this choice of parameter values, we run the bat algorithm iteratively. Positions and velocities of the bats are computed according to the evolution Eqs. (4)–(6) and then ranked according to the fitness function explained above. This iterative process stops when the maximum number of iterations $\mathcal{G}_{max}$ is reached. The best solution achieved at the final iteration is taken as the solution of the optimization problem.

Table 1. Bat algorithm parameters and their values in this paper.

Symbol	Meaning	Range of values	Selected value
$\mathcal{P}$	Population size	50–200	100
$\mathcal{G}_{max}$	Maximum number of iterations	1000–5000	2500
$\mathcal{A}^0$	Initial loudness	$(0, 2)$	0.5
$\mathcal{A}_{min}$	Minimum loudness	$[0, 1]$	0
r^0	Initial pulse rate	$[0, 1]$	0.2
f_{max}	Maximum frequency	$[0, 10]$	1.5
α	Multiplicative factor	$(0, 1)$	0.3
γ	Exponential factor	$[0, 1]$	0.2

5   An Illustrative Example

5.1   Graphical Results

Our method has been applied to several examples. However, we restrict our discussion in this paper to just one illustrative example because of limitations of space. In the example, the original image, shown in Fig. 4(top), is reconstructed with three affine transformations Λ_κ, $\kappa = 1, 2, 3$. We apply our method by using an initial population of randomly chosen 100 bats. For illustration, six of them are displayed in Fig. 1. As the reader can see, they are visually very different to each other, and all them are very far from the original source image. Then, our method is applied for $\mathcal{G}_{max} = 2500$ iterations as described above.

Figures 2 and 3 show the evolution of the global best of the population at specific iteration values, ranging from 100 to 600 with step size 100, and then from 800 to 1400 with step size 200. From the picture, we can see that the global best is very far from the source image at initial stages of the method, leading to images that do not really resemble the goal image. However, as the number of iterations increases, the global best image is getting visually closer to the intended image. Also, note that the variation of the global shape of the image over the iterations is more dramatic at initial stages, corresponding to a higher explorative phase, while it varies slightly at later iterations, where the image approaches to the target image by small incremental improvements of some local features, corresponding to the exploitative phase of the method. Figure 4 (bottom) shows the reconstructed image after the convergence is reached. From Fig. 4 we can see that the final reconstructed image is very similar visually to the source image, capturing faithfully all major features of a very complicated and irregular shape. This means that our method is able to reconstruct the general shape of the given image with a high visual accuracy. The corresponding convergence diagram is shown in Fig. 5.

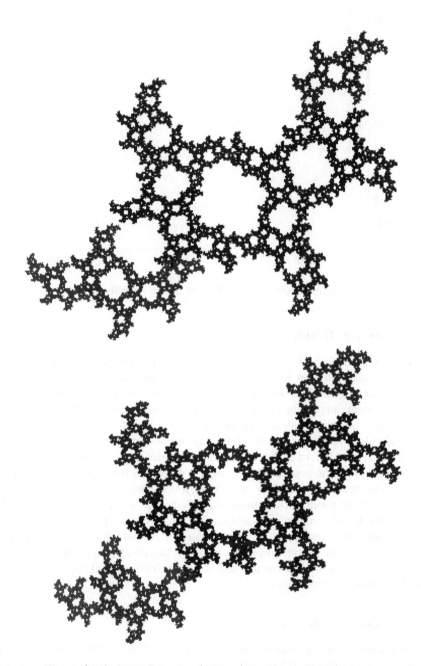

Fig. 4. (top) Original image; (bottom) best reconstructed image.

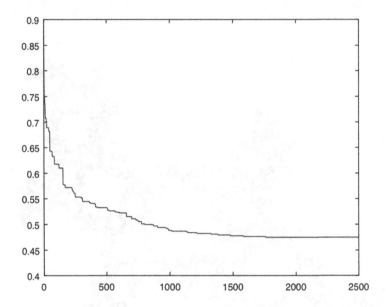

Fig. 5. Convergence diagram of the NSERI fitting error for 2500 iterations.

5.2   Numerical Results

Regarding the numerical results, the similarity error between the original and the reconstructed images is 0.4748 according to our metric, meaning that we got a 47% of mismatches between both images for the given resolution. This result may seem surprising in the light of the good visual results, but it must be taken into account that our metrics computes the differences based on the numerical values on the grid. Therefore, any minor distortion of the image (e.g., displacement, rotation, or scaling) can yield substantial increases in the similarity error, even though the general shape might still be well replicated. Furthermore, even if these variations happen at a local level, they have a dramatic effect on the numerical results. Of course, this effect can be partially alleviated by considering a less demanding fitness function. However, we preferred to preserve this more stringent metric in order to push our method further looking for a higher accuracy. As a conclusion, in spite of the good graphical results, the numerical results show that the method is not optimal yet and there is probably room for further improvement.

5.3   Computational Issues

All computations in this paper have been performed on a 2.6 GHz Intel Core i7 processor with 16 GB of RAM. The source code has been implemented by the authors in the native programming language of the popular scientific program *Matlab version 2015a* using the numerical libraries for fractals in [6,7,9]. Regarding the CPU times, they depend on the complexity of the image, the resolution of the mesh, and other factors. For illustration, each single execution takes about 25–30 min. In general, we noticed that the method is time-consuming for very high resolution images. This is the case for the image in our example, which is drawn with 5×10^5 points.

6 Conclusions and Future Work

This paper introduces a new approach for digital fractal image reconstruction. The method is based on a new affine kernel inspired by those in morphological image processing but specifically designed for fractal images. This approach leads to a difficult multimodal nonlinear continuous optimization, solved by using a powerful nature-inspired metaheuristics: the bat algorithm. An illustrative example is used to analyze the performance of this approach. Our experiments show that the method obtains very good visual results. However, the numerical results are not optimal yet, suggesting that there is also room for further improvement. We conclude that this approach is promising and it could potentially become (after further improvement to reduce the computing times and enhance its numerical accuracy) a very useful technique in the context of fractal image reconstruction.

Regarding our future work, we want to modify our method to improve our numerical and graphical results. In addition to a more optimized fitness function, we are interested to hybridize the bat algorithm with local search procedures to enhance the exploitation abilities of the method in the neighborhood of the local optima for higher accuracy. We also wish to extend our results to the cases of non-binary and colored images, with the possible addition of an extra color channel. Reducing our CPU times for better performance is also part of our plans for future work in the field.

Acknowledgments. Akemi Gálvez and Andrés Iglesias acknowledge the financial support from the project PDE-GIR of the European Union's Horizon 2020 research and innovation programme under the Marie Sklodowska-Curie grant agreement No. 778035, and from the Spanish Ministry of Science, Innovation and Universities (Computer Science National Program) under grant #TIN2017-89275-R of the Agencia Estatal de Investigación and European Funds FEDER (AEI/FEDER, UE). Eneko Osaba and Javier Del Ser would like to thank the Basque Government for its funding support through the EMAITEK program.

References

1. Alihodzic, A., Tuba, M.: Improved bat algorithm applied to multilevel image thresholding. Sci. World J. **2014**, 16 (2014). Article ID 176718
2. Barnsley, M.F.: Fractals Everywhere, 2nd edn. Academic Press, San Diego (1993)
3. Elton, J.H.: An ergodic theorem for iterated maps. Ergodic Theory Dynam. Syst. **7**, 481–488 (1987)
4. Falconer, K.: Fractal Geometry: Mathematical Foundations and Applications, 2nd edn. Wiley, Chichester (2003)

5. Fister, I., Rauter, S., Yang, X.-S., Ljubic, K., Fister Jr., I.: Planning the sports training sessions with the bat algorithm. Neurocomputing **149**(Part B), 993–1002 (2015)

6. Gálvez, A.: IFS Matlab generator: a computer tool for displaying IFS fractals. In: Proceedings of the ICCSA 2009, pp. 132–142. IEEE CS Press, Los Alamitos (2009)

7. Gálvez, A., Iglesias, A., Takato, S.: Matlab-based $K_ET$pic add-on for generating and rendering IFS fractals. In: Ślęzak, D., Kim, T., Chang, A.C.-C., Vasilakos, T., Li, M.C., Sakurai, K. (eds.) FGCN 2009. CCIS, vol. 56, pp. 334–341. Springer, Heidelberg (2009). https://doi.org/10.1007/978-3-642-10844-0_40

8. Gálvez, A., Iglesias, A., Takato, S.: $K_ET$pic Matlab binding for efficient handling of fractal images. Int. J. Future Gener. Commun. Netw. **3**(2), 1–14 (2010)

9. Gálvez, A., Kitahara, K., Kaneko, M.: *IFSGen4*^{LATEX}: interactive graphical user interface for generation and visualization of iterated function systems in LATEX. In: Hong, H., Yap, C. (eds.) ICMS 2014. LNCS, vol. 8592, pp. 554–561. Springer, Heidelberg (2014). https://doi.org/10.1007/978-3-662-44199-2_84

10. Gonzalez, R.C., Woods, R.E.: Digital Image Processing, 4th edn. Pearson Prentice Hall, Upper Saddle River (2017)

11. Gutiérrez, J.M., Iglesias, A.: A mathematica package for the analysis and control of chaos in nonlinear systems. Comput. Phys. **12**(6), 608–619 (1998)

12. Gutiérrez, J.M., Iglesias, A., Rodríguez, M.A.: A multifractal analysis of IFSP invariant measures with application to fractal image generation. Fractals **4**(1), 17–27 (1996)

13. Gutiérrez, J.M., Iglesias, A., Rodríguez, M.A., Rodríguez, V.J.: Generating and rendering fractal images. Math. J. **7**(1), 6–13 (1997)

14. Hutchinson, J.E.: Fractals and self similarity. Indiana Univ. Math. J. **30**(5), 713–747 (1981)

15. Iglesias, A., Gálvez, A., Collantes, M.: Multilayer embedded bat algorithm for B-spline curve reconstruction. Integr. Comput.-Aided Eng. **24**(4), 385–399 (2017)

16. Kashi, S., Minuchehr, A., Poursalehi, N., Zolfaghari, A.: Bat algorithm for the fuel arrangement optimization of reactor core. Ann. Nucl. Energy **64**, 144–151 (2014)

17. Kaveh, A., Zakian, P.: Enhanced bat algorithm for optimal design of skeletal structures. Asian J. Civ. Eng. **15**(2), 179–212 (2014)

18. Latif, A., Palensky, P.: Economic dispatch using modified bat algorithm. Algorithms **7**(3), 328–338 (2014)

19. Osaba, E., Yang, X.S., Diaz, F., Lopez-Garcia, P., Carballedo, R.: An improved discrete bat algorithm for symmetric and asymmetric traveling salesman problems. Eng. Appl. Artif. Intell. **48**, 59–71 (2016)

20. Suárez, P., Iglesias, A., Gálvez, A.: Make robots be bats: specializing robotic swarms to the bat algorithm. Swarm Evol. Comput. **44**(1), 113–129 (2019)

21. Yang, X.S.: A new metaheuristic bat-inspired algorithm. Stud. Comput. Intell. **284**, 65–74 (2010)

22. Yang, X.S.: Bat algorithm for multiobjective optimization. Int. J. Bio-Inspired Comput. **3**(5), 267–274 (2011)

23. Yang, X.S., Gandomi, A.H.: Bat algorithm: a novel approach for global engineering optimization. Eng. Comput. **29**(5), 464–483 (2012)

24. Yang, X.S.: Bat algorithm: literature review and applications. Int. J. Bio-Inspired Comput. **5**(3), 141–149 (2013)

Heuristic Rules for Coordinated Resources Allocation and Optimization in Distributed Computing

Victor Toporkov$^{(\boxtimes)}$ and Dmitry Yemelyanov

National Research University "Moscow Power Engineering Institute", Moscow, Russia
{ToporkovVV,YemelyanovDM}@mpei.ru

Abstract. In this paper, we consider heuristic rules for resources utilization optimization in distributed computing environments. Existing modern job-flow execution mechanics impose many restrictions for the resources allocation procedures. Grid, cloud and hybrid computing services operate in heterogeneous and usually geographically distributed computing environments. Emerging virtual organizations and incorporated economic models allow users and resource owners to compete for suitable allocations based on market principles and fair scheduling policies. Subject to these features a set of heuristic rules for coordinated compact scheduling are proposed to select resources depending on how they fit a particular job execution and requirements. Dedicated simulation experiment studies integral job flow characteristics optimization when these rules are applied to conservative backfilling scheduling procedure.

Keywords: Distributed computing · Resource allocation · Scheduling · Slot · Backfilling · Economic model · Optimization

1 Introduction and Related Works

Modern high-performance distributed computing systems (HPCS), including Grid, cloud and hybrid infrastructures provide access to large amounts of resources [1,2]. These resources are typically required to execute parallel jobs submitted by HPCS users and include computing nodes, data storages, network channels, software, etc.

There are two important classes of users' parallel jobs. Bags of tasks (BoT) represent parallel applications incorporating a large number of independent or weakly connected tasks. Typical examples of BoT are parameter sweeps, Monte Carlo simulations or exhaustive search. Workflows consist of multiple tasks with control or data dependencies. Such applications may be presented as directed graphs and represent complex computational or data processing problems in many domains of science [2–4].

Most BoT and workflow applications require some assurances of quality of services (QoS) from the computing system. In order to ensure QoS requirements

© Springer Nature Switzerland AG 2019
J. M. F. Rodrigues et al. (Eds.): ICCS 2019, LNCS 11538, pp. 395–408, 2019.
https://doi.org/10.1007/978-3-030-22744-9_31

and constraints, a coordinated allocation of suitable resources should be performed [5–7]. Most QoS requirements are based on either time or cost constraints such as total job execution cost, deadline, response time, etc. [8–11].

Some of the most important efficiency indicators of a distributed computational environment include both system resources utilization level and users' jobs time and cost execution criteria [4, 8, 9, 12]. In distributed environments with non-dedicated resources, such as utility Grids, the computational nodes are usually partly utilized and reserved in advance by jobs of higher priority [10]. Thus, the resources available for use are represented with a set of slots - time intervals during which the individual computational nodes are capable to execute parts of independent users' parallel jobs. These slots generally have different start and finish times and a performance difference. The presence of a set of slots impedes the problem of coordinated selection of the resources necessary to execute the job-flow from computational environment users. Resource fragmentation also results in a decrease of the total computing environment utilization level [12, 13].

High-performance distributed computing systems organization and support bring certain economical expenses: purchase and installation of machinery equipment, power supplies, user support, etc. As a rule, HPCS users and service providers interact in economic terms and the resources are provided for a certain payment. In such conditions, resource management and job scheduling based on the economic models is considered as an efficient way to take into account contradictory preferences of computing participants [3, 14–16].

A metascheduler or a metabroker is considered as an intermediate link between the users, local resource management and job batch processing systems [8, 17]. It defines uniform rules of a resource sharing and consumption to improve the overall scheduling efficiency [12, 13, 16].

The main contribution of this paper is a set of heuristic rules for a coordinated resources allocation for parallel jobs execution. The algorithm takes into account the system slots configuration as well as individual jobs features: size, runtime, cost, etc. When used in HPCS metaschedulers during the resources allocation step, it may improve overall system utilization level by matching jobs with resources and providing better jobs placement.

The rest of the paper is organized as follows. Section 2 presents resources allocation problem in relation to job-flow scheduling algorithms and backfilling (Subsect. 2.2). Different approaches for a coordinated resources allocation are described and proposed in Sect. 3. Section 4 contains algorithms implementation details along with simulation results and analysis. Finally, Sect. 5 summarizes the paper and describes further research topics.

2   Resources Allocation for Job-Flow Scheduling

2.1   Computing Model

In order to cover a wide range of computing systems we consider the following model for a heterogeneous resource domain.

Constituent computing nodes of a domain have different usage costs and performance levels. A space-shared resources allocation policy simulates a local queuing system (like in CloudSim [14,15] or SimGrid [4,18]) and, thus, each node can process only one task at any given time. Economic scheduling model [14,15] assumes that users and resource owners operate with some currency to coordinate resources allocation transactions. This model allows to regulate interaction between different organizations and to settle on fair equilibrium prices for resources usage.

Thus we consider a set R of heterogeneous computing nodes with different performance p_i and price c_i characteristics.

A node may be turned off or on by the provider, transferred to a maintenance state, reserved to perform computational jobs. Thus each node has a local utilization schedule known in advance for a considered scheduling horizon time L.

The execution cost of a single task depends on the allocated node's price and execution time, which is proportional to the node's performance level. In order to execute a parallel job one needs to allocate the specified number of simultaneously idle nodes ensuring user requirements from the resource request. The resource request specifies number n of nodes required simultaneously, their minimum applicable performance p, job's computational volume V and a maximum available resources allocation budget C. These parameters constitute a formal generalization for resource requests common among distributed computing systems and simulators [12,14,18].

In heterogeneous environment the required window length is defined based on a slot with the minimum performance. For example, if a window consists of slots with performances $p \in \{p_i, p_j\}$ and $p_i < p_j$, then we need to allocate all the slots for a time $T = \frac{V}{p_i}$. In this way V really defines a computational volume for each single job subtask. Common start and finish times ensure the possibility of internode communications during the whole job execution. The total cost of a window allocation is then calculated as $C = \sum_{i=1}^{n} T * c_i$.

2.2 Job-Flow Scheduling and Backfilling

The simplest way to schedule a job-flow execution is to use the First-Come-First-Served (FCFS) policy. However this approach is inefficient in terms of resources utilization and Backfilling [19] was proposed to improve system utilization.

Backfilling procedure makes use of advanced resources reservations which is an important mechanism preventing starvation of jobs requiring large number of computing nodes. Resources reservations in FCFS may create idle slots in the nodes' local schedules thus decreasing system performance. So the main idea behind backfilling is to backfill jobs into those idle slots to improve the overall system utilization. And the backfilling procedure implements this by placing smaller jobs from the back of the queue to these idle slots ahead of the priority order.

There are two common variations to backfilling - conservative and aggressive (EASY). Conservative Backfilling enforces jobs' priority fairness by making sure that jobs submitted later can't delay the start of jobs arrived earlier. EASY Backfilling aggressively backfills jobs as long as they do not delay the start of the single currently reserved jobs. Conservative Backfilling considers jobs in the order of their arrival and either immediately starts a job or makes an appropriate reservation upon the arrival. The jobs priority in the queue may be additionally modified in order to improve system-wide job-flow execution efficiency metrics. Under default FCFS policy the jobs are arranged by their arrival time. Other priority reordering-based policies like Shortest job First or eXpansion Factor may be used to improve overall resources utilization level [9,10,13].

Multiple Queues backfilling separates jobs into different queues based on metadata, such as jobs resource requirements: small, medium, large, etc. The idea behind this metaheuristic is that earlier arriving jobs and smaller-sized jobs should have higher execution priority. The number of queues and the strategy for dividing tasks among them can be set by the system administrators. Sometimes different queues may be assigned to a dedicated resource domain segments and function independently. In a single domain the metaheuristic cycles through the different queues in a round-robin fashion and may consider more jobs from the queues with smaller-sized tasks [13].

The look-ahead optimizing scheduler [10] implements dynamic programming scheme to examine all the jobs in the queue in order to maximize the current system utilization. So, instead of scanning queue for single jobs suitable for the backfilling, look-ahead scheduler attempts to find a combination of jobs that together will maximize the resources utilization.

2.3   Resources Selection Algorithms

Backfilling as well as many other job-flow scheduling algorithms in fact describe a general procedure determining high level policies for jobs prioritization and advanced resources reservations. However, the resources selection and allocation step remains sidelined since its more system specific nature. Consequently resource selection algorithms specifications usually either too hardware specific or lack certain restrictions or model features in order to cover a broader class of computing systems.

On the other hand, applying different resources allocation policies based on system or user preferences may affect scheduling results not only for individual jobs but for a whole job-flow.

In [6,7] we presented a Slot Subset Allocation (SSA) dynamic programming scheme for resources selection in heterogeneous computing environments based on economic principles. In a general case system nodes may be shared and reserved in advance by different users and organizations (including resource owners). So it's convenient to represent all available resources as a set of time-slots (see Sect. 2.1). Each slot corresponds to one computing node on which it is allocated. SSA algorithm takes these time slots as input and performs resources selection for a specified job in accordance with the computing model and constraints described in Sect. 2.1. The resulting window satisfies user QoS requirements from the resource request and may be reserved for the job execution.

Additionally SSA may perform window search optimization by a general additive criterion $Z = \sum_{i=1}^{n} z(s_i)$, where $z(s_i) = z_i$ is a target optimization characteristic value provided by a single slot s_i of window W. For this purpose SSA implements the following dynamic programming recurrent scheme to allocate n-size window with a maximum total cost C from m simultaneously available slots:

$$f_i(C_j, n_k) = \max\{f_{i-1}(C_j, n_k), f_{i-1}(C_j - c_i, n_k - 1) + z_i\}, \qquad (1)$$
$$k = 1, \ldots, n, i = 1, \ldots, m, j = 1, \ldots, C,$$

where $f_i(C_j, n_k)$ defines the maximum Z criterion value for n_k - size window allocated out of first i available slots for a budget C_j. After the forward induction procedure (1) is finished, the maximum criterion value can be found as $Z_{\max} = f_m(C, n)$. Corresponding resources are then obtained by a backward induction procedure.

These criterion values z_i may represent different slot characteristics: time, cost, power, hardware and software features, etc. Thus SSA-based resources allocation is proved to be a flexible tool for a preference-based job-slow execution [6].

3 Coordinated Resources Allocation Heuristics

3.1 Dependable Job Placement Problem

One important aspect for a resources allocation efficiency is the resources placement in regard to an actual slots configuration. So as a practical implementation for a general z_i parameter maximization we propose to study a resources allocation placement problem. Figure 1 shows Gantt chart of 4 slots co-allocation (hollow rectangles) in a computing environment with resources pre-utilized with local and high-priority jobs (filled rectangles).

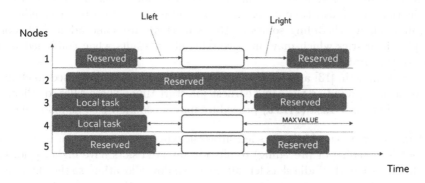

Fig. 1. Dependable window co-allocation metrics

As can be seen from Fig. 1, even using the same computing nodes (1, 3, 4, 5) there are usually multiple window placement options with respect to the slots start time. The slots' actual placement generally may affect such job execution properties as cost, finish time, computing energy efficiency, etc. Besides that, slots proximity to neighboring tasks reserved on the same computing nodes may affect the efficiency of the resources utilization. For example, reserving a slot close to a neighboring task may increase resources load by minimizing the corresponding node's idle time. On the other hand, leaving larger idle distances to the occupied or reserved slots sometimes may prove practical for the subsequent queue jobs scheduling.

For a quantitative placement criterion for each window slot we can estimate times to the previous task finish time: L_{left} and to the next task start time: L_{right} (Fig. 1). Using these values we consider the following criteria for the whole window allocation optimization:

- $L_\Sigma = \frac{1}{n} \sum_{i=1}^{n} (L_{lefti} + L_{righti})$ represents average time distance between window and the neighboring tasks reserved on the same nodes;
- $L_{\min\Sigma} = \frac{1}{n} \sum_{i=1}^{n} \min(L_{lefti}, L_{righti})$ displays average time distance to the nearest neighboring tasks.

Based on L_Σ or $L_{\min\Sigma}$ criteria idle space minimization or maximization strategies can be implemented using SSA algorithm.

3.2   Job Placement Heuristics

However these L_Σ or $L_{\min\Sigma}$ criteria alone can't improve the whole job-flow scheduling solution according to the conventional makespan or average finish time criteria. Preliminary experiments showed 30%–100% longer makespan for $L_{\min\Sigma} \rightarrow$ min resources allocation strategy compared to a traditional backfilling procedure [9,13] with jobs finish time minimization.

The reason for this result is that finish time minimization criterion in some degree incorporates and combines L_{left} minimization for early start time with a suitable (by performance) resources types selection. Consequently a greedy application of a finish time criterion in backfilling procedure provides efficient overall job-flow scheduling solution. But still there are some additional more complex heuristics which may improve scheduling results when combined with a finish time criterion.

For example in [13] a special set of *breaking a tie* rules is proposed to choose between slots providing the same earliest job start time. These rules for Picking Earliest Slot for a Task (PAST) procedure may be summarized as following.

1. Minimize number of idle slots left after the window allocation; i.e. slots adjacent (succeeding or preceding) to already reserved slots have higher priority.
2. Maximize length of idle slots left after the window allocation; so the algorithm tends to left longer slots for the subsequent jobs in the queue.

With similar intentions we propose the following Coordinated Placement (CoP) heuristic rules slightly different from PAST [13].

0. Prioritize slots allocated on nodes with lower performance. The main idea is that when deciding between two slots providing the same window finish time it makes sense to leave higher performance slot vacant for the subsequent jobs. This breaking a tie principle is applicable for heterogeneous resources environments and do not consider slots placement configuration. However, during the preliminary simulations this heuristic alone was able to noticeably improve scheduling results so we will use it as an addition to the placement rules.

1. Prioritize slots with relatively small distances to the neighbor tasks: $L_{lefti} \ll T$ or $L_{righti} \ll T$. The general idea is similar to the first rule in PAST, but CoP don't expect perfect match and defines threshold values for a satisfactory window fit.

2. Penalize slots leaving significant, but insufficient to execute a full job distances L_{lefti} or L_{righti}. For example when $\frac{T}{3} > L_{righti} > \frac{T}{5}$, the resulting slot may be to short to execute any of the subsequent jobs and will remain idle, thus, reducing the resources overall utilization.

3. On the other hand equally prioritize slots leaving sufficient compared to the job's runtime distances L_{lefti} or L_{righti}. For example with $L_{lefti} > T$.

So the main idea behind CoP is to fill in the gaps in the resources reservation schedule by providing quite an accurate resources allocation matching jobs runtime. Unlike PAST, CoP do not expect perfect matches but makes realistic heuristic decisions to minimize general resources fragmentation.

A simple resources allocation example on Fig. 2 demonstrates differences between these approaches. Figure 2 represents a computing environment segment consisting of eight nodes with some resources already allocated or reserved for six jobs A–F. Each job is represented as a filled rectangle (or a set of rectangles) spanning in both resource and time axes.

Consider a scenario when a next job requires three simultaneously available slots and the earliest finish time is achievable by using slots 1, 2, 3 or 4 from Fig. 2. In this case backfilling without any heuristic *breaking a tie* rules will choose slots 1, 2 and 3 just according to a simple slots order. PAST would choose slots 1, 3 and 4, minimizing resources fragmentation and leaving longer slot for the subsequent jobs. CoP would allocate slots 2, 3 and 4 as they provide better fit for the job, while slot 1 if allocated will leave two short slots likely unprofitable for future use.

4 Simulation Study

4.1 Implementation and Simulation Details

Based on heuristic rules described in Sect. 3.2 we implemented the following scheduling algorithms and criteria for SSA-based resources allocation.

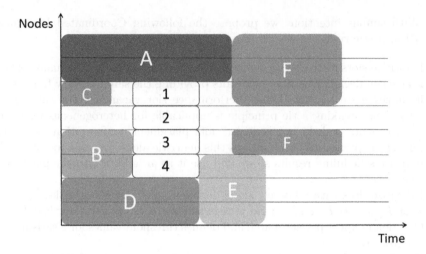

Fig. 2. Job placement heuristic rules example

1. Firstly we consider two conservative backfilling variations. *BFstart* successively implements start time minimization for each job during the resources selection step. As SSA performs criterion maximization, *BFstart* criterion for i-th slot has the following form: $z_i = -s_i.startTime$

   By analogy *BFfinish* implements a more solid strategy of a finish time minimization which is different from *BFstart* in computing environments with heterogeneous resources. *BFfinish* criterion for SSA algorithm is the following: $z_i = -s_i.finishTime$

2. PAST-like backfilling approach has a more complex criterion function which may be described with the following set of rules:

   (a) $z_i = -s_i.finishTime$; finish time is the main criterion value
   (b) $z_i = z_i - \alpha_1 * s_i.nodePerformance$; node performance amendment
   (c) $if(L_{righti} == 0) : z_i = z_i + \delta_1$; PAST rule 1
   (d) $if(L_{lefti} == 0) : z_i = z_i + \delta_1$; PAST rule 1
   (e) $z_i = z_i - \alpha_2 * L_{righti}$; PAST rule 2

3. CoP resources allocation algorithm for backfilling may be represented with the following criterion calculation:

   (a) $z_i = -s_i.finishTime$; finish time is the main criterion value
   (b) $z_i = z_i - \alpha_1 * s_i.nodePerformance$; node performance amendment
   (c) $if(L_{righti} < \epsilon_1 * T) : z_i = z_i + \delta_1$; CoP rule 1
   (d) $if(L_{lefti} < \epsilon_1 * T) : z_i = z_i + \delta_1$; CoP rule 1
   (e) $if(L_{righti} > \epsilon_2 * T \& L_{righti} < \epsilon_3 * T) : z_i = z_i - \delta_1$; CoP rule 2
   (f) $if(L_{lefti} > \epsilon_2 * T \& L_{lefti} < \epsilon_3 * T) : z_i = z_i - \delta_1$; CoP rule 2
   (g) $if(L_{righti} > T) : z_i = z_i + \delta_2$; CoP rule 3
   (h) $if(L_{lefti} > T) : z_i = z_i + \delta_2$; CoP rule 3

4. Finally as an additional reference solution we simulate another abstract back-filling variation *BFshort* which is able to reduce each job runtime for 1% during the resources allocation step. In this way each job will benefit not only from its own earlier completion time, but from earlier completion of all the preceding jobs.

The criteria for PAST and CoP contain multiple constant values defining rules behavior, namely $\alpha_1, \alpha_2, \delta_1, \delta_2, \epsilon_1, \epsilon_2, \epsilon_3$. ϵ_i coefficients define threshold values for a satisfactory job fit in CoP approach. α_i and δ_i define each rule's effect on the criteria and are supposed to be much less compared to z_i in order to break a tie between otherwise suitable slots. However their mutual relationship implicitly determine rules' priority which can greatly affect allocation results. Therefore there are a great number of possible α_i, δ_i and ϵ_i values combinations providing different PAST and CoP implementations. Based on heuristic considerations and some preliminary experiment results the values we used during the present experiment are presented in Table 1.

Table 1. PAST and CoP parameters values

Constant	α_1	α_2	δ_1	δ_2	ϵ_1	ϵ_2	ϵ_3
Value	0.1	0.0001	1	0.1	0.03	0.2	0.35

Because of heuristic nature of considered algorithms and their speculative parametrization (see Table 1) hereinafter by PAST [13] we will mean PAST-like approach customly implemented as an alternative to CoP.

4.2 Simulation Results

The experiment was prepared as follows using a custom distributed environment simulator [12, 16]. For our purpose, it implements a heterogeneous resource domain model: nodes have different usage costs and performance levels. A space-shared resources allocation policy simulates a local queuing system (like in CloudSim or SimGrid [14,18]) and, thus, each node can process only one task at any given simulation time. The execution cost of each task depends on its execution time, which is proportional to the dedicated node's performance level. The execution of a single job requires parallel execution of all its tasks. More details regarding the simulation computing model were provided in Sect. 2.1.

Besides that, the simulator implements a graphical interface representing Gantt diagram for the resulting job-flow scheduling outcome.

During each simulation experiment a new instance for the computing environment segment consisting of 32 heterogeneous nodes was automatically generated. Each node performance level is given as a uniformly distributed random value in the interval [2, 16]. This configuration provides a sufficient resources diversity level while the difference between the highest and the lowest resource performance levels will not exceed one order.

In this environment we considered job queue with 50, 100, 150 and 200 jobs accumulated at the start of the simulation. The jobs are arranged in queues by priority and no new jobs are submitted during the queue execution. Such scheduling problem statement allows to statically evaluate algorithms' efficiency in conditions with different resources utilization level. The jobs were generated with the following resources request requirements: number of simultaneously required nodes is uniformly distributed in interval $n \in [1; 8]$, computational volume $V \in [60; 1200]$ also contribute to a wide diversity in user jobs.

The results of 2000 independent simulation experiments are presented in Tables 2 and 3. Each simulation experiment includes computing environment and job queue generation, followed by a scheduling simulation independently performed using considered algorithms. The main scheduling results are then collected and contribute to the average values over all experiments.

Table 2 contain average finish time provided by algorithms *BFstart*, *BFfinish*, *BFshort*, PAST and CoP for different number of jobs pre-accumulated in the queue.

Table 2. Simulation results: average job finish time

Jobs N_Q	BFstart	BFfinish	BFshort	PAST	CoP
50	318,8	302,1	298,8	300,1	298
100	579,2	555	549,2	556,1	550,7
150	836,8	805,6	796,8	809	800,6
200	1112	1072,7	1060,3	1083,3	1072,2

As it can be seen, with a relatively small number N_Q of jobs in the queue, both CoP and PAST provide noticeable advantage by nearly 1% over a strong *BFfinish* variation and CoP even surpasses *BFshort* results. At the same time less successful *BFstart* approach provides almost 6% later average completion time highlighting difference between a good (*BFfinish*) and a regular (*BFstart*) possible scheduling solutions. So *BFshort*, CoP and PAST advantage should be evaluated against this 6% interval.

However with increasing the jobs number CoP advantage over *BFfinish* decreases and tends to zero when $N_Q = 200$. This trend may be observed on Fig. 3 presenting relative finish time advantage over *BFfinish* for all considered algorithms. *BFshort* graph is represented as an almost straight line 1% above reference *BFfinish* solution, which is expected by design.

CoP graph starts above *BFshort* and gradually decreases to the *BFfinish* 0% line. However as PAST average performance decreases contemporaneously with CoP, latter maintains 0.5%–1% advantage over PAST for all considered simulation experiments.

The performance decrease trend for PAST and CoP heuristics may be explained by increasing accuracy requirements for jobs placement caused with increasing N_Q number. Indeed, when considering for some intermediate job resource selection the more jobs are waiting in the queue the higher the probability that some future job will have a better fit for current resource during the backfilling procedure. In order to adapt to higher resources utilization levels, threshold parameters ϵ_i may be changed to encourage even better job placement fits during the resources allocation. In a general case all the algorithms' parameters $\alpha_i, \delta_i, \epsilon_i$ (more details we provided in Sect. 3.2) should be refined to correspond to the actual computing environment utilization level.

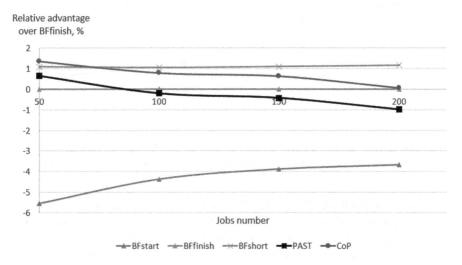

Fig. 3. Simulation results: relative advantage over BFfinish by jobs finish time criterion depending on the jobs queue size

Table 3. Simulation results: average job finish time for jobs distributed over half of a makespan interval

Jobs N_Q	BFstart	BFfinish	BFshort	PAST	CoP
50	381,7	375,4	371,8	371,8	369,5
100	672,5	662,6	656,9	657,9	653,4
150	942,4	922,6	915,8	921	914,9
200	1208,2	1184,2	1173,2	1184,1	1173,8

However if we distribute jobs arrival time over some interval we may derive similar results as average number of jobs waiting in the queue for backfilling will be less. In the following experiment we performed job queue scheduling with the same settings except that jobs had random arrival times in the range up to half of the makespan obtained during the first experiment. Corresponding average job finish times from another 2000 simulations is presented in Table 3.

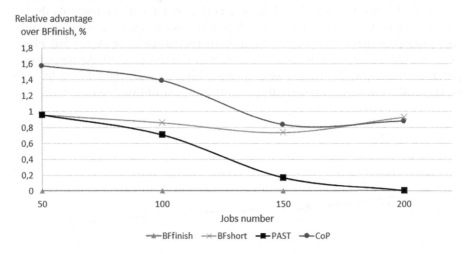

Fig. 4. Simulation results: relative advantage over *BFfinish* by jobs finish time criterion in scenario with jobs distributed over half of a makespan interval

Based on data from Table 3 Fig. 4 shows relative advantage over *BFfinish* against finish time criterion for *BFshort*, PAST and CoP aprroaches in scenario with jobs' arrival times dynamically distributed over a period of time. The trend is different from the static scenario (Fig. 3) as CoP maintains 1% advantage even for 200 jobs in the queue.

One important property of the proposed heuristic approach is that it generally preserves integral job-flow execution parameters of the base scheduling algorithm: jobs' priority and processing order, average execution cost. For example, maximum average difference in jobs-flow execution cost between CoP and *BFfinish* over all 4000 simulations reaches 0.25%.

5   Conclusion

In this work, we address the problem of a coordinated resources allocation for parallel jobs scheduling optimization in heterogeneous computing environments. Modern job-flow scheduling algorithms optimize integral job-flow scheduling characteristics mainly by determining jobs prioritization and execution order, leaving resources selection step aside as too system specific. Based on a *Slots Subset Allocation* resources selection algorithm, we propose and implement a set of heuristic job placement rules for jobs' Coordinated Placement (CoP). The main idea behind CoP approach is to fill in the gaps in the resources utilization schedule by allocating resources tailored to particular jobs runtimes.

Simulation study shows overall job-flow scheduling efficiency improvement just by using CoP rules during the resources allocation step in conservative backfilling. The advantage over a basic resources allocation strategy reaches 1.5% against average job-flow finish time criteria. At the same time CoP preserves job-flow execution parameters, jobs priorities and processing order.

In our further work, we will refine job placement heuristics to include and balance user preferences with global scheduling criteria during the resources allocation step.

Acknowledgments. This work was partially supported by the Council on Grants of the President of the Russian Federation for State Support of Young Scientists (YPhD-2979.2019.9), RFBR (grants 18-07-00456 and 18-07-00534) and by the Ministry on Education and Science of the Russian Federation (project no. 2.9606.2017/8.9).

References

1. Lee, Y.C., Wang, C., Zomaya, A.Y., Zhou, B.: Profit-driven scheduling for cloud services with data access awareness. J. Parallel Distrib. Comput. **72**(4), 591–602 (2012)
2. Bharathi, S., Chervenak, A.L., Deelman, E., Mehta, G., Su, M., Vahi, K.: Characterization of scientific workflows. In: 2008 Third Workshop on Workflows in Support of Large-Scale Science, pp. 1–10 (2008)
3. Rodriguez, M.A., Buyya, R.: Scheduling dynamic workloads in multi-tenant scientific workflow as a service platforms. Futur. Gener. Comput. Syst. **79**(P2), 739–750 (2018)
4. Nazarenko, A., Sukhoroslov, O.: An experimental study of workflow scheduling algorithms for heterogeneous systems. In: Malyshkin, V. (ed.) PaCT 2017. LNCS, vol. 10421, pp. 327–341. Springer, Cham (2017). https://doi.org/10.1007/978-3-319-62932-2_32
5. Netto, M.A.S., Buyya, R.: A flexible resource co-allocation model based on advance reservations with rescheduling support. Technical report, GRIDSTR-2007-17, Grid Computing and Distributed Systems Laboratory, The University of Melbourne, Australia, 9 October 2007
6. Toporkov, V., Yemelyanov, D.: Dependable slot selection algorithms for distributed computing. In: Zamojski, W., Mazurkiewicz, J., Sugier, J., Walkowiak, T., Kacprzyk, J. (eds.) DepCoS-RELCOMEX 2018. AISC, vol. 761, pp. 482–491. Springer, Cham (2019). https://doi.org/10.1007/978-3-319-91446-6_45
7. Toporkov, V., Yemelyanov, D.: Optimization of resources selection for jobs scheduling in heterogeneous distributed computing environments. In: Shi, Y., et al. (eds.) ICCS 2018. LNCS, vol. 10861, pp. 574–583. Springer, Cham (2018). https://doi.org/10.1007/978-3-319-93701-4_45
8. Kurowski, K., Nabrzyski, J., Oleksiak, A., Weglarz, J.: Multicriteria aspects of grid resource management. In: Nabrzyski, J., Schopf, J.M., Weglarz J. (eds.) Grid Resource Management. State of the Art and Future Trends, pp. 271–293. Kluwer Academic Publishers (2003)
9. Srinivasan, S., Kettimuthu, R., Subramani, V., Sadayappan, P.: Characterization of backfilling strategies for parallel job scheduling. In: Proceedings of the International Conference on Parallel Processing, ICPP 2002 Workshops, pp. 514–519 (2002)

10. Shmueli, E., Feitelson, D.G.: Backfilling with lookahead to optimize the packing of parallel jobs. J. Parallel Distrib. Comput. **65**(9), 1090–1107 (2005)
11. Menasc'e, D.A., Casalicchio, E.: A framework for resource allocation in grid computing. In: 12th Annual International Symposium on Modeling, Analysis, and Simulation of Computer and Telecommunications Systems, MASCOTS 2004, Volendam, The Netherlands, pp. 259–267 (2004)
12. Toporkov, V., Toporkova, A., Tselishchev, A., Yemelyanov, D., Potekhin, P.: Heuristic strategies for preference-based scheduling in virtual organizations of utility grids. J. Ambient. Intell. Hum.Ized Comput. **6**(6), 733–740 (2015)
13. Khemka, B., et al.: Resource management in heterogeneous parallel computing environments with soft and hard deadlines. In: Proceedings of 11th Metaheuristics International Conference, MIC 2015 (2015)
14. Calheiros, R.N., Ranjan, R., Beloglazov, A., De Rose, C.A.F., Buyya, R.: CloudSim: a toolkit for modeling and simulation of cloud computing environments and evaluation of resource provisioning algorithms. J. Softw. Pract. Exp. **41**(1), 23–50 (2011)
15. Samimi, P., Teimouri, Y., Mukhtar, M.: A combinatorial double auction resource allocation model in cloud computing. J. Inf. Sci. **357**(C), 201–216 (2016)
16. Toporkov, V., Yemelyanov, D., Toporkova, A.: Fair scheduling in grid VOs with anticipation heuristic. In: Wyrzykowski, R., Dongarra, J., Deelman, E., Karczewski, K. (eds.) PPAM 2017. LNCS, vol. 10778, pp. 145–155. Springer, Cham (2018). https://doi.org/10.1007/978-3-319-78054-2_14
17. Rodero, I., Villegas, D., Bobroff, N., Liu, Y., Fong, L., Sadjadi, S.: Enabling interoperability among grid meta-schedulers. J. Grid Comput. **11**(2), 311–336 (2013)
18. Casanova, H., Giersch, A., Legrand, A., Quinson, M., Suter, F.: Versatile, scalable, and accurate simulation of distributed applications and platforms. J. Parallel Distrib. Comput. **74**(10), 2899–2917 (2014)
19. Jackson, D., Snell, Q., Clement, M.: Core algorithms of the maui scheduler. In: Feitelson, D.G., Rudolph, L. (eds.) JSSPP 2001. LNCS, vol. 2221, pp. 87–102. Springer, Heidelberg (2001). https://doi.org/10.1007/3-540-45540-X_6

Nonsmooth Newton's Method: Some Structure Exploitation

Alberto De Marchi[(✉)] and Matthias Gerdts

Department of Aerospace Engineering, Bundeswehr University Munich,
Werner-Heisenberg-Weg 39, 85577 Neubiberg, Germany
{alberto.demarchi,matthias.gerdts}@unibw.de

Abstract. We investigate real asymmetric linear systems arising in the search direction generation in a nonsmooth Newton's method. This applies to constrained optimisation problems via reformulation of the necessary conditions into an equivalent nonlinear and nonsmooth system of equations. We propose a strategy to exploit the problem structure. First, based on the sub-blocks of the original matrix, some variables are selected and ruled out for a posteriori recovering; then, a smaller and symmetric linear system is generated; eventually, from the solution of the latter, the remaining variables are obtained. We prove the method is applicable if the original linear system is well-posed. We propose and discuss different selection strategies. Finally, numerical examples are presented to compare this method with the direct approach without exploitation, for full and sparse matrices, in a wide range of problem size.

Keywords: Structure exploitation · Linear algebra ·
Nonsmooth Newton's method · Nonlinear optimization

1 Introduction

In this paper, we consider the real square nonsymmetric possibly large sparse linear system

$$
\begin{bmatrix} Q & A^\top & C^\top \\ A & & \\ -SC & & T \end{bmatrix} \begin{pmatrix} x \\ y \\ z \end{pmatrix} = \begin{pmatrix} f \\ g \\ h \end{pmatrix} \tag{1}
$$

where $Q \in \mathbb{R}^{n_x \times n_x}$, $A \in \mathbb{R}^{n_a \times n_x}$, $C \in \mathbb{R}^{n_c \times n_x}$ and S, $T \in \mathbb{R}^{n_c \times n_c}$ are given matrices and $f \in \mathbb{R}^{n_x}$, $g \in \mathbb{R}^{n_a}$, $h \in \mathbb{R}^{n_c}$ are given vectors (n_x, n_a, n_c being some positive integers); S and T are non-zero diagonal matrices. The contribution of this paper is the exploitation of the structure in problem (1) and its transformation into a smaller symmetric linear system, with a saddle-point structure, from which the solution to (1) can be easily recovered. In particular, two stages are discussed. First, a reduction step generates a smaller linear system and a way to recover eliminated variables from the solution of this reduced system. This

© Springer Nature Switzerland AG 2019
J. M. F. Rodrigues et al. (Eds.): ICCS 2019, LNCS 11538, pp. 409–420, 2019.
https://doi.org/10.1007/978-3-030-22744-9_32

step exploits the fact that matrix T is diagonal, and then symbolically solve for (some of) the variables in z. Several different reduction strategies are discussed and compared. The second step aims at rewriting the linear system in a symmetric form, allowing to adopt solvers for symmetric systems, which are usually more time and memory efficient. Despite these advantages, some computational overhead is needed, especially in the reduction step, which might introduce a break-even point, that is, this exploitation may pay off, e.g., in terms of computational time, only under certain conditions, e.g., large instances. In fact, an optimal reduction strategy might exists, depending on the specific properties of the problem; indeed, it may even depend on the specific values of the entries. Throughout the paper, we investigate the influence of the reduction strategy on the performance of the aforementioned two-steps exploitation; however, a detailed optimization of the reduction policy is beyond the scope of this paper. We point out that the proposed method could be combined with constraint-reduction approaches as, e.g., those presented in [5,12].

Once the original problem (1), say $Vd = r$ for brevity, has been transformed, a reduced symmetric linear system, say $\hat{V}\hat{d} = \hat{r}$, is to be solved. To this end, any method can be adopted. The choice may depend on the problem, in particular on its size, fill-in, sparsity pattern, accuracy requirements and memory constraints, availability of good preconditioners, and so on. Within this work, we compare the effectiveness and the limitations of the proposed method for structure exploitation when a direct solver is adopted to tackle the linear system.

1.1 Motivation

Linear systems with the form (1) arise, e.g., from nonlinear complementarity problems [6], nonlinear optimization problems with inequality constraints [7] and discretized optimal control problems with state and control constraints [8, 9]. Usually, these are reformulated through the Karush-Kuhn-Tucker (KKT) necessary optimality conditions, then equivalently transformed into a nonlinear system of equations with the so called NCP-functions [16] and finally solved with a nonsmooth version of Newton's method [14]. Some globalization strategies [9,11] and results in functions spaces [18] have been reported. It has been shown

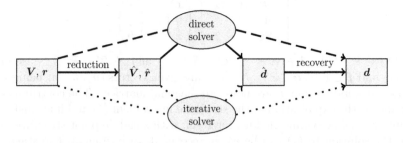

Fig. 1. Solution diagram: direct methods with (solid) and without (dashed) structure exploitation are compared; iterative methods (dotted) are not considered here.

that, for some NCP-functions, this approach is equivalent to a primal-dual active set strategy [10]. Indeed, different NCP-functions exhibit different properties and might affect the convergence behaviour [1,16]. With reference to the problem (1), matrices Q, A and C can be considered as iterate-dependent linear-quadratic approximations of an underlying nonlinear problem, while diagonal matrices S and T originate from the NCP-function adopted and vectors f, g and h are the residuals of the aforementioned nonlinear system of equations.

Example. Let us consider a quadratic program (QP) with linear equality and inequality constraints. Hence, we seek an $x \in \mathbb{R}^{n_x}$ minimizing $\frac{1}{2}x^\top Q x + q^\top x$, subject to constraints $Ax = a$, $Cx \leq c$, where Q, A, C and q, a, c are given matrices and vectors, respectively. Here the inequalities are understood componentwise. Linear constraints ensure that regularity conditions are met, then the KKT conditions are necessary for optimality; these read:

$$Qx + q + A^\top \lambda + C^\top \mu = 0 \tag{2a}$$

$$Ax = a \tag{2b}$$

$$0 \leq \mu \quad \perp \quad Cx \leq c \tag{2c}$$

where λ and μ denote the Lagrange multipliers. In (2c), inequality and complementarity constraints hold componentwise. Let us consider an NCP-function $\varphi : \mathbb{R}^2 \to \mathbb{R}$, e.g., the original or the penalized Fischer-Burmeister function [1,7], which by definition satisfies

$$\varphi(a,b) = 0 \quad \Leftrightarrow \quad 0 \leq a \perp b \geq 0 \tag{3}$$

for any pair (a,b). Thanks to this property, the KKT system (2) is equivalently rewritten as a nonlinear system $\psi(z) = 0$, collecting vector $z = (x, \lambda, \mu) \in \mathbb{R}^{n_z}$, $n_z := n_x + n_a + n_c$, and with vector-valued function $\psi : \mathbb{R}^{n_z} \to \mathbb{R}^{n_z}$ defined by

$$\psi(z) = \begin{pmatrix} Qx + q + A^\top \lambda + C^\top \mu \\ Ax - a \\ \varphi(c - Cx, \mu) \end{pmatrix} \tag{4}$$

Here the NCP-function φ applies componentwise. A (globalized) possibly nonsmooth Newton's-type method generates a sequence $\{z^k\}$ through the recurrence $z^{k+1} = z^k + \alpha^k d^k$, $k = 0, 1, 2, \ldots$, where the step length $\alpha^k > 0$ is determined, e.g., by a line-search procedure of Armijo's type and the search direction d^k is the solution of the linear equation $V^k d = -\psi(z^k)$ [7–9,11,14]. The matrix V^k is an element of the Clarke's generalized Jacobian (that is the convex hull of the Bouligand differential [2]) of ψ at z^k, namely $V^k \in \partial \psi(z^k)$ [8,11]. The NCP-function φ is the only element in (4) which can possibly make function ψ nonsmooth. Hence, from (4), one obtains

$$V^k = \begin{bmatrix} Q & A^\top & C^\top \\ A & & \\ -S^k C & & T^k \end{bmatrix} \tag{5}$$

with diagonal matrices $S^k = \text{diag}\left(s_1^k, s_2^k, \ldots, s_{n_c}^k\right)$ and $T^k = \text{diag}\left(t_1^k, t_2^k, \ldots, t_{n_c}^k\right)$, whose entries are pairwise coupled via the (possibly generalized) differential of the NCP-function φ. Defining $v^k := c - Cx^k$ the inequality constraint violation at the k-th iterate, for $i = 1, 2, \ldots, n_c$, the coupling for the i-th inequality constraint reads [1,7,8]:

$$\left(s_i^k, t_i^k\right) \in \partial\varphi\left(v_i^k, \mu_i^k\right) \tag{6}$$

We remark that matrices S^k and T^k are diagonal because the NCP-function φ applies componentwise in (4). Then, the linear system $V^k d = -\psi(z^k)$ to compute the search direction d^k corresponds exactly to problem (1).

1.2   Outline

This work is organized as follows. Section 2 outlines the structure exploitation strategy and introduces an underlying assumption. In Sects. 2.1 and 2.2 the two main steps are developed and discussed. Section 3 validates the proposed approach numerically, for both full and sparse matrices, with different reduction strategies, showing effectiveness and limitations of the proposed algorithm. Section 4 concludes the paper and presents ideas for future research.

2   Structure Exploitation

Problem (1), also denoted $Vd = r$ for brevity, can be directly solved via any linear algebra package, e.g., MA48 [4], PARDISO [15], SUPERLU [13], `mldivide` in MATLAB [17]. However, we aim at exploiting our knowledge about the structure of matrix V; computational effort and achieved accuracy might benefit from this, especially for large-scale and sparse linear systems. Firstly, we notice that matrix V and vector r are often computed blockwise and then assembled. Hence, matrices Q, A, C and vectors s, t, f, g and h are here considered as the starting ingredients for solving (1). Overall, two directions are explored, mainly exploiting the diagonal structure of S and T. In Sect. 2.1 a reduction step is discussed, eliminating some variables and introducing a smaller asymmetric linear system. Then, in Sect. 2.2, this is transformed into a symmetric one which is equivalent. Nonetheless, these operations for reorganizing the linear system and successive recovering of variables constitutes an overhead of computation. This means that these procedures might be worthy only for certain problems, likely large instances with lots of inequality constraints. In Sect. 3, numerical tests show that this break-even point corresponds to relatively small problem instances (for the tested implementation).

We point out that the proposed method relies on the following assumption on the diagonals of S and T; it reads:

Assumption 1. For $i = 1, 2, \ldots, n_c$, it holds $(s_i, t_i) \neq (0, 0)$.

This is a mild requirement, in that it corresponds to problem (1) to be well-posed. One can show the following result:

Lemma 1. *If problem* (1) *admits a unique solution, then Assumption 1 holds.*

Proof (by contradiction). Let us assume there exist d unique solution to (1) and i such that $(s_i, t_i) = (0,0)$. Hence, the row of V corresponding to the i-th inequality constraint consists of zeros only. Then, the matrix V is rank deficient and the problem is undetermined. Two cases are possible, depending on the value of h_i on the right-hand side. If $h_i = 0$, then (1) admits infinitely many solutions, hence solution d is not unique. If $h_i \neq 0$, then the linear system is unsolvable (impossible) and d cannot be a solution.   □

Remark 1. Assumption 1 requires a mild condition to be satisfied by the NCP-function φ. For instance, a sufficient condition is that for any given pair $(a, b) \in \mathbb{R}^2$ there exists a pair $(s, t) \in \partial\varphi(a, b)$ such that $(s, t) \neq (0,0)$; this allows to choose always suitable entries for S and T. The Fischer-Burmeister function and the max function, among other NCP-functions, have this property.

Let us denote $\mathcal{I} := \{1, 2, \ldots,, n_c\}$ the index set for the inequality constraints, $\mathcal{I}_{re}^0 := \{i \in \mathcal{I} \mid t_i \neq 0\}$ and $\mathcal{I}_{sy}^0 := \{i \in \mathcal{I} \mid s_i \neq 0\}$ the largest index sets that allow respectively the reduction step and the symmetrization step, discussed below. Thanks to Assumption 1, these satisfy $\mathcal{I}_{re}^0 \cup \mathcal{I}_{sy}^0 = \mathcal{I}$. Let us consider an index subset $\mathcal{I}_{re} \subseteq \mathcal{I}_{re}^0$, sufficiently large to satisfy $\mathcal{I}_{re} \cup \mathcal{I}_{sy}^0 = \mathcal{I}$. Then, for the associated complement $\mathcal{I}_{\overline{re}} := \mathcal{I}\backslash\mathcal{I}_{re}$, it holds $\mathcal{I}_{\overline{re}} \subseteq \mathcal{I}_{sy}^0$. With this construction, it is possible to apply the reduction step, ruling out a given set $\mathcal{I}_{re}$ of variables, and subsequently the symmetrization step on the linear system with the remaining variables, namely those in $\mathcal{I}_{\overline{re}}$.

Remark 2. We stress that in general it is $\mathcal{I}_{re}^0 \cap \mathcal{I}_{sy}^0 \neq \emptyset$ and hence the choice of $\mathcal{I}_{re}$ is not unique. This suggests there could be an optimal reduction strategy, possibly dependent on V and with some degree of computation awareness. However, this issue is beyond the scope of this paper.

In Sect. 3, we compare the following definitions of $\mathcal{I}_{re}$ through numerical investigations:

$$\mathcal{I}_{re}^t := \left\{i \in \mathcal{I} \,\middle|\, |t_i| \geq \epsilon\right\} \tag{7a}$$

$$\mathcal{I}_{re}^s := \left\{i \in \mathcal{I} \,\middle|\, |s_i| \leq \epsilon\right\} \tag{7b}$$

$$\mathcal{I}_{re}^{ts} := \left\{i \in \mathcal{I} \,\middle|\, |t_i| \geq |s_i|\right\} \tag{7c}$$

where $\epsilon > 0$ is a given, sufficiently small value, introduced as a numerical tolerance in (7a)–(7b). For $\epsilon \to 0^+$, these sets approach the largest and the smallest possible reduction sets, respectively, namely reducing the most and the least of the variables. Instead, the set defined in (7c) represents an arbitrary trade-off, introduced for the sake of comparison; see Fig. 2.

Remark 3. One could think about performing either the reduction or symmetrization step. However, (i) under Assumption 1, once the system is reduced,

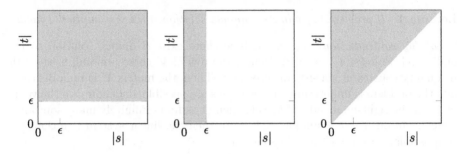

Fig. 2. Reduction sets (7) in the $|s|$-$|t|$ plane: $\mathcal{I}_{re}^t$ (left), $\mathcal{I}_{re}^s$ and $\mathcal{I}_{re}^{ts}$ (right).

the symmetrization step is straightforward, inexpensive and likely effective; (ii) the symmetrization step might be impossible without preliminary reduction, depending on the invertibility of S.

2.1   Reduction

The idea behind the reduction step stems from the observation that problem (1) may be separable, i.e. that it may be possible to compute the value of some variables *a posteriori*, namely once the others are given. In fact, a solution to problem (1) must satisfy

$$Tz = SCx + h \tag{8}$$

where matrix T is diagonal. Given an index set $\mathcal{I}_{re} \subseteq \mathcal{I}_{re}^0$, it is possible to compute z_i, for every $i \in \mathcal{I}_{re}$, from (8) once the solution vector x is known. To be sure, let us build matrices $T_{re} := \operatorname{diag}\left(t_i \mid i \in \mathcal{I}_{re}\right)$ and $T_{\overline{re}} := \operatorname{diag}\left(t_i \mid i \notin \mathcal{I}_{re}\right)$ and define z_{re} and $z_{\overline{re}}$ the corresponding vectors of unknown variables which can and cannot be reduced, respectively. Then, partitioning the linear system (1) accordingly with these definitions yields:

$$\begin{bmatrix} Q & A^\top & C_{re}^\top & C_{\overline{re}}^\top \\ A & & & \\ -S_{re}C_{re} & & T_{re} & \\ -S_{\overline{re}}C_{\overline{re}} & & & T_{\overline{re}} \end{bmatrix} \begin{pmatrix} x \\ y \\ z_{re} \\ z_{\overline{re}} \end{pmatrix} = \begin{pmatrix} f \\ g \\ h_{re} \\ h_{\overline{re}} \end{pmatrix} \tag{9}$$

where matrices C_{re}, $C_{\overline{re}}$, S_{re}, $S_{\overline{re}}$ and vectors h_{re} and $h_{\overline{re}}$ are constructed analogously, based on $\mathcal{I}_{re}$. The matrix T_{re} is nonsingular, by definition, and then, from (9), one can formally solve for z_{re}, obtaining

$$z_{re} = T_{re}^{-1}\left(S_{re}C_{re}x + h_{re}\right), \tag{10}$$

whose evaluation is straightforward because T_{re} is diagonal. Plugging (10) back into (9) leads to a smaller linear system, after rearrangements, without reduced variables z_{re}, namely:

$$\begin{bmatrix} \hat{Q} & A^\top & C_{\overline{re}}^\top \\ A & & \\ -S_{\overline{re}}C_{\overline{re}} & & T_{\overline{re}} \end{bmatrix} \begin{pmatrix} x \\ y \\ z_{\overline{re}} \end{pmatrix} = \begin{pmatrix} \hat{f} \\ g \\ h_{\overline{re}} \end{pmatrix} \tag{11}$$

where matrix $\hat{Q}$ and vector $\hat{f}$ are defined by:

$$\hat{Q} := Q + C_{\text{re}}^\top T_{\text{re}}^{-1} S_{\text{re}} C_{\text{re}} \tag{12a}$$

$$\hat{f} := f - C_{\text{re}}^\top T_{\text{re}}^{-1} h_{\text{re}} \tag{12b}$$

The larger the set $\mathcal{I}_{\text{re}}$, the more reduced variables, the smaller the obtained linear system (11). In turn, the computation of $\hat{Q}$ may be costly, involving a matrix-matrix multiplication, Eq. 12a. Also, for sparse problems, the fill-up of matrix $\hat{Q}$ may become significant. These drawbacks suggest there might be a trade-off in the reduction step, and hence an optimal reduction strategy, as pointed out in Remark 2.

2.2   Symmetrization

Linear systems with a symmetric matrix can be solved more efficiently, in terms of time and memory. In order to get a symmetric matrix out of (11), it would suffice to left-multiply the rows associated with inequality constraints, namely with $z_{\overline{\text{re}}}$, by the inverse of $-S_{\overline{\text{re}}}$. As discussed above, it is $\mathcal{I}_{\overline{\text{re}}} \subseteq \mathcal{I}_{\text{sy}}^0$, hence the matrix $S_{\overline{\text{re}}}$ is nonsingular, by construction; moreover, its inversion is straightforward, since it is diagonal. Then, the reduced symmetric linear system $\hat{V}\hat{d} = \hat{r}$ reads:

$$\begin{bmatrix} \hat{Q} & A^\top & C_{\overline{\text{re}}}^\top \\ A & & \\ C_{\overline{\text{re}}} & & -S_{\overline{\text{re}}}^{-1} T_{\overline{\text{re}}} \end{bmatrix} \begin{pmatrix} x \\ y \\ z_{\overline{\text{re}}} \end{pmatrix} = \begin{pmatrix} \hat{f} \\ g \\ -S_{\overline{\text{re}}}^{-1} h_{\overline{\text{re}}} \end{pmatrix} \tag{13}$$

The matrix $\hat{V}$ is symmetric and smaller than V; the vector $\hat{d}$ collects the unknowns corresponding to optimization variables (x), equality constraints' multipliers (y) and not-reduced inequality constraints' multipliers ($z_{\overline{\text{re}}}$).

Remark 4. In (12)–(13), the matrix-matrix products $T_{\text{re}}^{-1} S_{\text{re}}$, $S_{\overline{\text{re}}}^{-1} T_{\overline{\text{re}}}$ and the matrix-vector products $T_{\text{re}}^{-1} h_{\text{re}}$, $S_{\overline{\text{re}}}^{-1} h_{\overline{\text{re}}}$ can be evaluated as entry-wise vector-vector products. In fact, this is possible because matrices S_{re}, $S_{\overline{\text{re}}}$, T_{re} and $T_{\overline{\text{re}}}$ are diagonal. Furthermore, one can exploit this feature by choosing a specific multiplication ordering, aiming at the lowest possible computational complexity.

3   Numerical Results

This Section reports and discusses the results obtained from a MATLAB [17] implementation of Algorithm 1, considering Remark 4. The plain code (as well as a Julia 1.0 and a Python 3.6 implementation) are publicly available [3].

We are interested in comparing the computation time for solving problem (1), through direct methods, with and without the proposed structure exploitation method, see Fig. 1 above. Also, we investigate how it is affected by the problem size $N := n_x + n_a + n_c$ and the relative number of equality and inequality constraints, $\alpha := n_a/n_x$ and $\gamma := n_c/n_x$, respectively.

Algorithm 1. Abstract structure-exploiting linear solver.

Input: $Q, A, C, s, t, f, g, h; \epsilon$
Output: x, y, z
$\mathcal{I}_{\mathrm{re}} \leftarrow s, t, \epsilon;$ // reduction strategy, Eq. 7
$C_{\mathrm{re}}, s_{\mathrm{re}}, t_{\mathrm{re}}, h_{\mathrm{re}}, C_{\overline{\mathrm{re}}}, s_{\overline{\mathrm{re}}}, t_{\overline{\mathrm{re}}}, h_{\overline{\mathrm{re}}} \leftarrow C, s, t, h, \mathcal{I}_{\mathrm{re}};$ // partitioning
$\hat{Q} \leftarrow Q, C_{\mathrm{re}}, t_{\mathrm{re}}, s_{\mathrm{re}} ;$ // Eq. 12a
$\hat{f} \leftarrow f, C_{\mathrm{re}}, t_{\mathrm{re}}, h_{\mathrm{re}} ;$ // Eq. 12b
$\hat{V} \leftarrow \hat{Q}, A, C_{\overline{\mathrm{re}}}, s_{\overline{\mathrm{re}}}, t_{\overline{\mathrm{re}}} ;$ // Eq. 13
$\hat{r} \leftarrow \hat{f}, g, s_{\overline{\mathrm{re}}}, h_{\overline{\mathrm{re}}} ;$ // Eq. 13
$x, y, z_{\overline{\mathrm{re}}} \leftarrow \hat{V}, \hat{r} ;$ // linear system
$z_{\mathrm{re}} \leftarrow C_{\mathrm{re}}, s_{\mathrm{re}}, t_{\mathrm{re}}, h_{\mathrm{re}}, x ;$ // recovering, Eq. 10
$z \leftarrow z_{\mathrm{re}}, z_{\overline{\mathrm{re}}}, \mathcal{I}_{\mathrm{re}};$ // assembling

---

A problem instance consists of matrices Q, A, C and vectors s, t (the diagonal of S and T, respectively), f, g and h. In the case of full matrices, starting from given values of N, α and γ, an instance is generated as follows:

$$n_x = \left[\frac{N}{1 + \alpha + \gamma} \right]$$

$$n_a = [\alpha n_x]$$

$$n_c = N - n_x - n_a$$

$$\bar{Q}_{ij} \sim \mathcal{N}(0,1) \qquad i = 1, \ldots, n_x, \; j = 1, \ldots, n_x$$

$$Q = \frac{1}{2} \left(\bar{Q} + \bar{Q}^\top \right)$$

$$A_{ij} \sim \mathcal{N}(0,1) \qquad i = 1, \ldots, n_a, \; j = 1, \ldots, n_x$$

$$C_{ij} \sim \mathcal{N}(0,1) \qquad i = 1, \ldots, n_c, \; j = 1, \ldots, n_x$$

$$\rho_i \sim \sqrt{\mathcal{U}(0,1)} \qquad i = 1, \ldots, n_c$$

$$\theta_i \sim \mathcal{U}(0, 2\pi) \qquad i = 1, \ldots, n_c$$

$$s_i = 1 + \rho_i \cos \theta_i \qquad i = 1, \ldots, n_c$$

$$t_i = 1 + \rho_i \sin \theta_i \qquad i = 1, \ldots, n_c$$

$$f_i \sim \mathcal{N}(0,1) \qquad i = 1, \ldots, n_x$$

$$g_i \sim \mathcal{N}(0,1) \qquad i = 1, \ldots, n_a$$

$$h_i \sim \mathcal{N}(0,1) \qquad i = 1, \ldots, n_c$$

where $\mathcal{N}(\mu, \sigma)$ denotes the normal continuous probability distribution with mean value μ and standard deviation σ, and $\mathcal{U}(a, b)$ the uniform distribution with support in $[a, b]$. Entries of S and T are pairwise coupled in that they are sampled from a disk in the s-t plane, centered in $(1, 1)$ with unitary radius, with uniform probability distribution. This setting is motivated by and mimics the generalized differential of the Fischer-Burmeister function [8]. Both, the direct and the structure-exploiting methods setup the linear system starting from these inputs. Notice that the reduced approach does not build V nor r, but their reduced

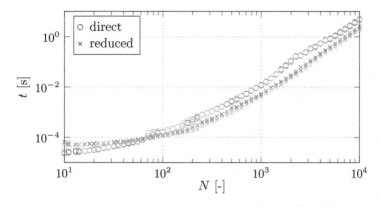

Fig. 3. Execution time: direct approach (t_{dir}, dot) and reduced approach (t_{red}, cross) with reduction set $\mathcal{I}_{\mathrm{re}}^{\mathrm{t}}$; full matrices. Median value. (Color figure online)

and symmetric counterpart $\hat{V}$ and $\hat{r}$. In our implementation, the direct method builds V and r and then adopts the `mldivide` routine to solve $Vd = r$. Instead, for the reduced approach (with full matrices), the `linsolve` routine is adopted and explicitly informed that matrix $\hat{V}$ is symmetric. The problem size N varies between 10 and 10^4 for full and between 10^3 and $2 \cdot 10^4$ for sparse matrices. For each problem size, a set of 100 problem instances are generated (only 10 if $N > 10^4$), checked for ill-conditioning and eventually solved. The composition of constraints is chosen to be $(\alpha, \gamma) \in \{(0, 1), (0, 1.5), (0.5, 1), (0.5, 1.5)\}$ (colored in blue, red, green and violet, respectively). The index sets defined in (7) are adopted and compared, with the tolerance $\epsilon = 10^{-3}$. Sparse matrices are generated in such a way that they approximately have 10 entries for each row; this makes the number of nonzero entries to increase linearly and not quadratically with the problem size N.

The computation time for the direct and reduced case are depicted in Fig. 3, considering full matrices and the (large) index set $\mathcal{I}_{\mathrm{re}}^{\mathrm{t}}$. This gives an idea about the adopted implementation and computing hardware; also, one can guess the computational complexity of the underlying algorithm for solving a linear system. As expected in Sect. 2, the overhead due to partitioning, reducing and recovering, introduces a break-even point, at around $N = 60$ (for $\mathcal{I}_{\mathrm{re}}^{\mathrm{t}}$ and $\mathcal{I}_{\mathrm{re}}^{\mathrm{ts}}$, but not for $\mathcal{I}_{\mathrm{re}}^{\mathrm{s}}$); hence, the reduced approach is not beneficial only for small-sized problems. This and other considerations can be drawn based on Fig. 4, where it is depicted the (median value of the) ratio of the execution time with the reduced approach, t_{red}, with the different reduction strategies, over the direct one, t_{dir}. Therein, the break-even point corresponds to the unitary ratio; also, the additional computational burden is significant for low values of N. For large N, instead, the ratio decreases to approximately one-half for $\mathcal{I}_{\mathrm{re}}^{\mathrm{t}}$ and $\mathcal{I}_{\mathrm{re}}^{\mathrm{ts}}$, while for $\mathcal{I}_{\mathrm{re}}^{\mathrm{s}}$ it stays around the unit. As one could expect, the reduction set $\mathcal{I}_{\mathrm{re}}^{\mathrm{s}}$ is not as effective as $\mathcal{I}_{\mathrm{re}}^{\mathrm{t}}$ and $\mathcal{I}_{\mathrm{re}}^{\mathrm{ts}}$ because it does not benefit very much from the reduction step, in that it eliminates only few variables. The relative number

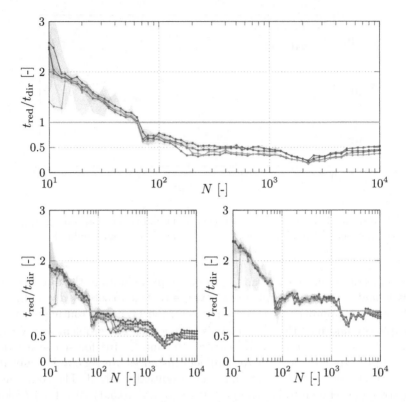

Fig. 4. Ratio of execution time: $\mathcal{I}_{re}^{t}$ (top), $\mathcal{I}_{re}^{ts}$ (bottom left), $\mathcal{I}_{re}^{s}$ (bottom right); full matrices. Median value (line) and 9%–91% quantiles (filled area). (Color figure online)

of constraints has an impact on the execution time but it does not drastically affect the overall behaviour (for both, full and sparse case). In fact, all else being equal, either decreasing the number of equalities and increasing the number of inequalities reduces the execution time ratio, meaning that the reduced approach is more effective and worthy for (large) problems with many inequality constraints. For what concerns the case of sparse matrices, similar observations are valid, see Fig. 5. In order to show the distribution of the results obtained from the executed tests, along with the median value, the 9% and 91% quantiles are also reported. For full matrices, Fig. 4, the distribution is relatively narrow, while for sparse matrices, Fig. 5, the results are relatively scattered. Thus, we argue the sparsity pattern greatly affects the computation time. Nevertheless, for relatively large sparse matrices, the ratio t_{red}/t_{dir} approaches one-half and promisingly decreases.

These numerical results suggest the set $\mathcal{I}_{re}^{t}$ defined in (7a) to be the most effective reduction strategy among those tested. In fact, it generates the smallest linear system and then post-solves the most variables. However, as argued in

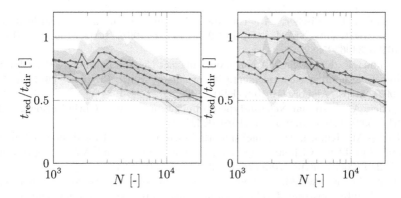

Fig. 5. Ratio of execution time: $\mathcal{I}_{re}^t$ (top) and $\mathcal{I}_{re}^{ts}$ (bottom); sparse matrices. Median value (line) and 9%–91% quantiles (filled area). (Color figure online)

Sect. 2.1, we claim it might be not the case for (much) larger problem instances, because of the required overhead for the reduction step.

4   Conclusions

This paper proposed and studied a structure-exploiting approach for solving linear systems arising in the context of nonsmooth Newton's method. The applicability of this method was established under well-posedness of the original problem. Numerical examples showed that the developed approach reduces the computational time, with both, full and sparse linear systems. Some of the tested reduction strategies resulted in halving the execution time.

Analogous ideas apply when an iterative linear solver is of choice, e.g., for very large systems; tailored preconditioners are subject of future research. It remains to assess the effectiveness and the drawbacks of the method when embedded into larger routines for numerical optimization. Moreover, it would be interesting to investigate an optimal reduction strategy, possibly computationally aware.

Acknowledgements. A.D.M. heartily thanks his *marafiki* for the unforgettable memories and wishes them a happy marriage.

References

1. Chen, B., Chen, X., Kanzow, C.: A penalized Fischer-Burmeister NCP-function. Math. Program. **88**(1), 211–216 (2000). https://doi.org/10.1007/PL00011375
2. Clarke, F.H.: Optimization and Nonsmooth Analysis. Wiley, New York (1983)
3. De Marchi, A.: Code supporting "Nonsmooth Newton's method: some structure exploitation", November 2018. https://doi.org/10.5281/zenodo.1486064
4. Duff, I.S., Reid, J.K.: MA48 – A Fortran code for direct solution of sparse unsymmetric linear systems of equations. Technical report, RAL-93-072, Rutherford Appleton Laboratory, October 1993

5. Facchinei, F., Fischer, A., Kanzow, C.: On the accurate identification of active constraints. SIAM J. Optim. **9**(1), 14–32 (1998). https://doi.org/10.1137/S1052623496305882

6. Facchinei, F., Kanzow, C.: A nonsmooth inexact Newton method for the solution of large-scale nonlinear complementarity problems. Math. Program. **76**(3), 493–512 (1997). https://doi.org/10.1007/BF02614395

7. Fischer, A.: A special Newton-type optimization method. Optimization **24**, 269–284 (1992). https://doi.org/10.1080/02331939208843795

8. Gerdts, M., Kunkel, M.: A nonsmooth Newton's method for discretized optimal control problems with state and control constraints. J. Ind. Manag. Optim. **4**(2), 247–270 (2008). https://doi.org/10.3934/jimo.2008.4.247

9. Gerdts, M., Kunkel, M.: A globally convergent semi-smooth Newton method for control-state constrained DAE optimal control problems. Comput. Optim. Appl. **48**(3), 601–633 (2011). https://doi.org/10.1007/s10589-009-9275-0

10. Hintermüller, M., Ito, K., Kunisch, K.: The primal-dual active set strategy as a semismooth Newton method. SIAM J. Optim. **13**(3), 865–888 (2002). https://doi.org/10.1137/S1052623401383558

11. Jiang, H.: Global convergence analysis of the generalized Newton and Gauß-Newton methods of the Fischer-Burmeister equation for the complementarity problem. Math. Oper. Res. **24**(3), 529–543 (1999). http://www.jstor.org/stable/3690647

12. Laiu, M.P., Tits, A.L.: A constraint-reduced MPC algorithm for convex quadratic programming, with a modified active set identification scheme. Comput. Optim. Appl. (2019). https://doi.org/10.1007/s10589-019-00058-0

13. Li, X.S.: An overview of SuperLU: algorithms, implementation, and user interface. ACM Trans. Math. Softw. **31**(3), 302–325 (2005)

14. Qi, L., Sun, J.: A nonsmooth version of Newton's method. Math. Program. **58**(1), 353–367 (1993). https://doi.org/10.1007/BF01581275

15. Schenk, O., Gärtner, K.: Solving unsymmetric sparse systems of linear equations with PARDISO. Future Gener. Comput. Syst. **20**(3), 475–487 (2004). https://doi.org/10.1016/j.future.2003.07.011

16. Sun, D., Qi, L.: On NCP-functions. Comput. Optim. Appl. **13**(1), 201–220 (1999). https://doi.org/10.1023/A:1008669226453

17. The MathWorks, Inc.: MATLAB 9.3 Release 2017b, Natick, Massachusetts, United States, September 2017

18. Ulbrich, M.: Nonsmooth Newton-like methods for variational inequalities and constrained optimization problems in function spaces. Ph.D. thesis, Technische Universität München, February 2002

Fully-Asynchronous Cache-Efficient Simulation of Detailed Neural Networks

Bruno R. C. Magalhães[1], Thomas Sterling[2], Michael Hines[3],
and Felix Schürmann[1(✉)]

[1] Blue Brain Project, École polytechnique fédérale de Lausanne Biotech Campus,
1202 Geneva, Switzerland
[2] CREST - Center for Research in Extreme Scale Technologies, Indiana University,
Bloomington, IN 47404, USA
[3] Department of Neuroscience, Yale University, New Haven, CT 06510, USA

Abstract. Modern asynchronous runtime systems allow the re-thinking of large-scale scientific applications. With the example of a simulator of morphologically detailed neural networks, we show how detaching from the commonly used bulk-synchronous parallel (BSP) execution allows for the increase of prefetching capabilities, better cache locality, and a overlap of computation and communication, consequently leading to a lower time to solution. Our strategy removes the operation of collective synchronization of ODEs' coupling information, and takes advantage of the pairwise time dependency between equations, leading to a fully-asynchronous exhaustive yet not speculative stepping model. Combined with fully linear data structures, communication reduce at compute node level, and an earliest equation steps first scheduler, we perform an acceleration at the cache level that reduces communication and time to solution by maximizing the number of timesteps taken per neuron at each iteration.

Our methods were implemented on the core kernel of the NEURON scientific application. Asynchronicity and distributed memory space are provided by the HPX runtime system for the ParalleX execution model. Benchmark results demonstrate a superlinear speed-up that leads to a reduced runtime compared to the bulk synchronous execution, yielding a speed-up between 25% to 65% across different compute architectures, and in the order of 15% to 40% for distributed executions.

1 Introduction

Asynchronous runtime systems built on a global memory address space (GAS) opens up new possibilities for numerical resolutions without synchronization barriers at the core and compute node level, and allow for a substantial reduction of runtime by better utilizing the CPU's prefetching capabilities and cache-level acceleration. Our use case is the simulation of morphologically detailed neural networks, categorized with the following properties: (1) neurons are branched representations of spatially discretized capacitors with ionic current channels; (2) neurons are represented by Ordinary Differential Equations (ODEs) that

© Springer Nature Switzerland AG 2019
J. M. F. Rodrigues et al. (Eds.): ICCS 2019, LNCS 11538, pp. 421–434, 2019.
https://doi.org/10.1007/978-3-030-22744-9_33

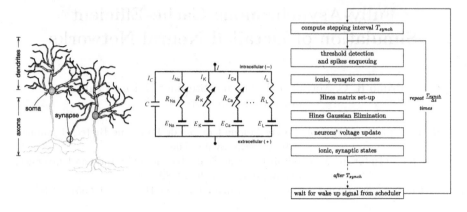

Fig. 1. Left: Model representation of two neurons and a synapse. Each neuron includes an axonic branch (south of soma, pictured in light) and a spatially discretized representation of a tree of dendrite compartments (in dark). A synapse is a connection between an axon and a dendrite of different neurons. Middle: the RC circuit representing the electrical activity on the membrane of a single compartment, between the intra and extracellular spaces. Right: The workflow of the algorithm. A neuron computes the stepping interval T_{synch} from the synaptic dependencies time instants, and performs $T_{synch}/\Delta t$ steps of length Δt.

define the current on the capacitor and the voltage-dependent opening of each ion channel; and (3) ODEs are coupled with a time dependency based on the synaptic connectivity between neurons. For clarity, refer to Fig. 1 (left) for a schematic representation of the underlying model.

Due to the high complexity of the data representation—including topological structure, biological mechanisms, synaptic connectivity and external currents—simulations are computationally very costly. State of the art approaches for the acceleration of large neural simulations rely on common parallel and distributed computing techniques. Multi-core and multi-compute node acceleration can be found in NEURON [1]. Complementary efforts rely on Single Instruction Multiple Data (SIMD or vectorization) optimization of state variables replicated across ODEs [2]. Acceleration of small datasets of detailed neuron models have been explored with branch-parallelism [3] (single-core, Single Instruction Single Data, multiple compute nodes), and improved by Magalhaes et al. [4] (with added multi-core, SIMD, and distributed computation). Volumetric decomposition and tessellation with parallel processing of spatial regions has been presented by Kozloski et al. [5].

Similar to most large-scale scientific simulation approaches, synchronization of neurons in existing methods follows the Bulk Synchronous Parallel (BSP) model of computation: execution is split in time grids of equidistant intervals, a period of time with duration equivalent to the minimum synaptic delay across all pairs of neurons in the system. Synaptic communication is typically performed with Message Passing Interface (MPI). It has been shown that, for extremely

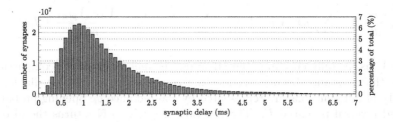

Fig. 2. Distribution of synaptic delays in terms of count (left y-axis) and percentage (right y-axis) of all synapses on a network of 219.247 neurons, extracted from a biologically inspired digital reconstructed model of the rodent neocortex from Markram et al. [6]. Histogram contains one bin per interval of 0.1 ms. The leftmost bar ($x = 0.1$ ms) represents the communication step size of state of the art implementations following the Bulk Synchronous Parallel model.

large networks of compute nodes, the synchronous collective communication can account for over 10% of the overall runtime [2]. This limitation is difficult to overcome in current approaches, as acceleration of the computation of complex models above one-tenth of real time is difficult, due to latency of inter-process communication [7].

In that line of thought, this work presents an **exhaustive yet not speculative** execution model that improves cache locality and provides cache-level acceleration by removing synchronous communication steps, and introducing a fully-asynchronous execution model that advances ODEs timestepping beyond synchronization barriers, based on the time couplings between equations. Our strategy includes five components. At first, (1) a fully-asynchronous stepping protocol that allows elements to perform several timesteps without collective synchronisation. Cache locality is improved by (2) a fully linear memory representation of the data structure, including vector, map and priority queue containers, and is further increased by (3) a computation scheduler that tracks the time progress of ODEs in time and advances the earliest element to its furthest instant in time. Network communication on distributed executions is minimized by (4) a point-to-point fully-asynchronous protocol that signals elements' time advancement to its dependees laid out in a Global Memory Address Space, and by (5) a local communication reduce operation at every compute node—herewith also referred to as locality.

We implemented our methods on the core computation of the NEURON simulator, available as open source [8], with communication, synchronization, and threading enabled by the HPX-5 runtime library [9], demonstrating a shorter time to solution on a wide range of architectures.

1.1 Mathematical Formulation

The main function that describes the currents passing through the membrane of a capacitor n (also referred to as compartment) is described by:

$$C_n \frac{dV_n}{dt} = -\sum_i g_i x_i (V_n - E_i) - \sum_{c:p(c)=n} \frac{V_c - V_n}{r_c} - \frac{V_n - V_{p(n)}}{r_{p(n)}} + I_n(t) \qquad (1)$$

where V_n is the difference in potential across the membrane of the compartment, and r the resistance between connecting compartments, when available. The activity of different ions are represented by conductance g_i, opening probability x_i, and reversal potential E_i. The function $p(c) : \mathbb{N} \to \mathbb{N}$ returns the id of the parent compartment of a given compartment c. Refer to Fig. 1 (middle) for the electrical model of the mathematical equation. The first right-hand side term refers to the ionic currents passing through the membrane, described by the Hodgkin-Huxley (HH) model [10]. The voltage-dependent variables x_i describe the opening of the ion channels as a voltage-gated first-order ODE and for brevity were omitted. The fixed step size of the numerical resolution is defined as the time interval *small enough* to capture the dynamics of the biological mechanism with the fastest kinetics—typically the fast Potassium channels—and is set in our model to 0.025 ms. The second term extends the representation of a neuron to a branched morphology, by adding the neighbouring compartments' contributions according to the neuronal cable theory for multiple compartments [11]. To allow the removal of the spatial interpolation of state along each compartment, long compartments are divided into a sequence of smaller ones, and—as a result of their small length—assume that the average state of a compartment along its length is accurately represented by the state of a compartment at its center, and needs only interpolation at consecutive discrete time intervals. The final right hand side term $I(t)$ refers to external currents from time driven events such as injected current stimuli and synaptic activity. The synaptic delay for a given synapse connecting a pre- to a post-synaptic neuron is determined by the time required for the information following an Action Potential (spike) from the pre-synaptic neuron axon to reach its target post-synaptic neuron dendrite.

We apply the simplification that the spike propagation along the axon is stereotypic and that it can be approximated by converting the path from the soma to the synapse to a delay interval after which a simple event is delivered to the synapse. In the model of Markram et al. [6], the minimum synaptic delay in our model is set to 0.1 ms or equivalently 4 compute steps—refer to Fig. 2 for details—and accounts for circa 0.13% of all the synaptic delays. The communication of spikes at the end of every minimum synaptic delay time frame, allows the update of neuron states in the subsequent period without loss of information.

2   Methods

Significant cache acceleration is difficult to achieve for scientific problems defined by complex data representations. Typically, the main principles to improve cache-efficiency are based on the following rules: using smaller data types and organizing the data so that memory alignment holes are reduced; avoiding the use of algorithms and data structures that exhibit irregular memory access patterns; using linear data structures, i.e. serial memory representations that improve access patterns; and improving spatial locality, by using each cache line to the maximum extent once it has been mapped to the cache. Following this reasoning, the next section details the implementation of our cache-efficiency methods. For completion, the workflow of the scheduled stepping and the kernels of individual compute steps discussed hereafter are presented in Fig. 1 (right).

2.1 Linear Data Structures

To avoid fragmentation of data layouts in memory due to dynamic allocations and optimize cache memory reutilization, we implemented a fully linear neuron representation, including class variables and containers. Because the number of elements in the containers are either fixed or defined by a predictable worst case scenario, the size of the container data structures can be computed beforehand. The description of the containers follows in the following paragraphs.

Linear Vector: implemented as a serialization of the `std::vector` class with the meta data, address of array, and elements of the array placed on a sequential memory space. An illustration of the linear vector data structure is displayed in Fig. 3 (a).

Linear Map: an unordered map structure storing the mapping of a key to a value or to an array of values. A search for a given key is performed with a binary search across all (ordered) keys, thus yielding similar computational complexity as the `std::map` implementation with a red-black tree, at $O(log\ n)$. The index of a key refers to the count and the pointer to the elements for that key. The memory layout is presented in Fig. 3 (b). Moreover, the linear data representation of the map values allows for operations such as minimum value, maximal value and value query to be performed with the same efficiency as a vector.

Linear Priority Queue storing time-driven events as pairs of delivery time and destination. Capable of handling dynamic insertion and removal of events throughout the simulation on a queue of time ordered events. Our implementation relies on a map of circular arrays of ordered time events per pre-synaptic id (the key field). Circular arrays are dimensioned by a pre-computed maximum size, defined by the maximum number of events that can occur during the time window that two given neurons can be set apart at any time throughout the execution. As an example, for a given synaptic connectivity $A \rightarrow B$ with minimum synaptic delay of 1 ms and the converse $B \rightarrow A$ of 5 ms, the maximum stepping time window between both is 6 ms long. To retrieve all subsequent events to be delivered in the following step, the algorithm loops through all keys, collects all events in the interval, and returns the time-sorted list of events. This replaces the iterative peak/top and pop operations underlying regular queue implementations. The memory layout is presented in Fig. 3 (c). At the level of each key, given a pre-synaptic neuron id, the list of future events is retrieved in the pop-push interval of elements in the respective circular array. Push (pop) operations will increment the push (pop) offset variable and insert (retrieve) the element

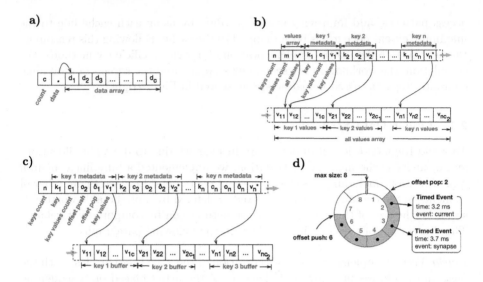

Fig. 3. Memory representation of linear data structures. Gray arrows represent connections between contiguous memory regions. (a) linear vector; (b) linear map; (c) linear priority queue; (d) a circular array representing a sample entry in the priority queue.

in that position. For completion, Fig. 3 (d) displays an example of the circular array memory structure for a given key.

As a side note, cache-optimized implementations of priority queues such as funnel heap, calendar queue or other cache-oblivious queues [12] improve memory access pattern yet do not guarantee fully-linear memory allocation. For the sake of comparison, the computational complexity of both ours and the standard library `std::priority_queue` implementations are similar, requiring the retrieval of all events within the next timestep ($O(k)$ for a loop through the all k queues and extraction of the first element on the circular arrays), plus a sorting operation (with worst-case scenario $O(n \log n)$) for a solution of size n, compared to the standard library implementation requiring a complexity in the order of $O(n \log n)$ for n retrievals.

2.2   Time-Based Elements Synchronization and Stepping

To allow for a flexible progress of neurons in time that detach from the constraints of the minimum synaptic delay across all pairs of neurons in the system (0.1 ms or 0.13% of total delays, shown previously in Fig. 2), we introduce a graph of time dependencies between neurons that allows for a given post-synaptic neuron to advance in time based on their pre-synaptic dependencies' progress. The result is an exhaustive stepping mechanism, that maximises the number of steps per neuron and the simulation time held on CPU cache. The pre- to post-synaptic neuron time updates are provided by an active asynchronous pairwise neuron notification messaging framework. Stepping notifications from a pre- to a post-

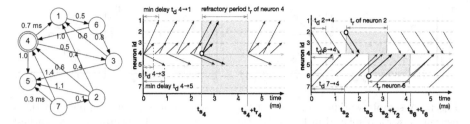

Fig. 4. A representative schema of the algorithm for dependency based synchronization of stepping. Left: a sample network of neurons (vertices 1–7). Arrow heads (tails) connect to post- (pre-) synaptic neurons. Labels on edges describe the minimum synaptic delay from a pre- to a post-synaptic neuron. Center: outgoing communication for neuron 4. Arrow tail (head) represents a message to the source (destination) neuron. A neuron transmits the time step allowed by the post-syn. neuron, given by his present time plus the minimum transmission delay the a post-synaptic neuron—represented by t_d $pre \rightarrow post$ and conforming to the graph on the left. Spike notifications (t_s, circles) allow post-synaptic neuron to freely proceed to a time equivalent to the spike time plus the refractory period (t_r) of the pre-synaptic neuron. Right: incoming communication for neuron 4. A post-synaptic neuron actively receives progress notifications and keeps track of the maximum step allowed based on pre-synaptic neuron status.

synaptic neuron are sent at a period defined by their minimum synaptic delay. At the onset of every computation step, a neuron notifies its post-synaptic neuron ids of its stepping if necessary, and stores in a queue the next stepping when notification is required. To reduce communication, the transmission of a spike is also handled as a stepping notification by the post-synaptic size. As a problem-specific optimization, communication is further reduced by taking into account the refractory period, i.e. an interval after a spike during which a neuron is unable to spike again. A schematic workflow of the time-dependency algorithm is presented in Fig. 4. The fully-asynchronous stepping yields a more flexible threading by completely removing collective synchronization barriers, less often communication as the pairwise communication delays are generally two orders of magnitude longer than the global minimum transmission delay and a full overlap of computation and communication. To maximise the number of steps taken on any run, a neuron scheduler allows for an optimal decision of the next neuron to step, by keeping track of the progress of neurons. This topic is covered next.

2.3   Neuron Scheduler

To maximise cache efficiency, a scheduler was implemented to control and trigger the advancement of neurons in time based on their simulation time. At every iteration, the scheduler (one per locality) actively picks the earliest neuron in time and triggers its stepping. On multi-core architectures, a multi-threaded version of the scheduler allows for several neurons to be launched in parallel. A mutual exclusion control object (mutex) initiated with a counter equal to the number of threads serves as progress control gate. When all threads have

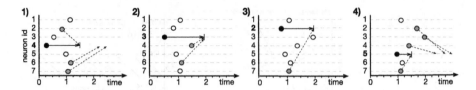

Fig. 5. A sample workflow of 4 iterations of the neuron scheduler applied to the 7 neuron network displayed in Fig. 4. On the top-left (frame 1), neuron 4 is the earliest in time (coloured black) and is allowed to proceed to time 1.5 ms, dictated by the transmission delay of the pre-synaptic neurons 2, 6 and 7 (coloured gray). The same logic follows in the following iterations, with neurons 3, 2 and 5 being the next ones to advance, as pictured in frame (2) and (3), respectively.

been assigned a neuron, the scheduler waits on the mutex. Upon the end of the stepping from a neuron, its thread goes dormant and atomically decrements the mutex counter, waking up the scheduler, and updating its progress in the scheduler's progress map. At the onset of stepping, a neuron queries the time allowed by its pre-synaptic dependencies and performs all necessary steps. An example of scheduled stepping is illustrated in Fig. 5.

2.4   Communication Reduce

Global memory address space (GAS) on the Parallax execution model allows for remote thread execution across multiple objects (neurons) distributed across several localities. On a single locality, each message incurs the overhead of a lightweight thread, as GAS addresses are an abstraction to local memory. However, on a distributed execution, each call is an instantiation of a procedure in an object held possibly in a different locality. Therefore, large amount of object-to-object communication may become a bottleneck by saturating the network bandwidth. This issue is trivial to overcome on MPI-based implementations, as the sender is responsible for buffering, packing and initiating the communication, while the converse operations must follow from the receiver. On the Parallax runtime system, its resolution is not as simple, as data representation in GAS arrays remove the locality-awareness of each object in a distributed array. To reduce the overhead of the high amount of point-to-point (inter-neuron) messaging, an extra layer of communication was introduced. Notifications of stepping and spikes for several post-synaptic neurons are packed at the onset of communication as single packets to remote localities. At the recipient side, a mapping of pre-synaptic id to the list of local GAS addresses, allows message to be unfolded and locally spawn to the recipient GAS addresses in the locality. This method replaces n remote communications by a single remote communication with n local lightweight threads spawn. For completion, Fig. 6 provides an illustration of the communication reduce methods.

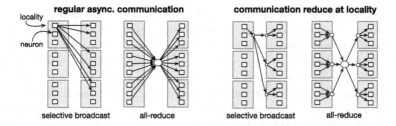

Fig. 6. A sample diagram of the communication required for a selective broadcast and an all-reduce operation with regular (left) versus locality-reduced (right) communication.

3 Results

Our strategy was implemented in the core computation of the NEURON scientific application, available as open source [8]. Communication, synchronization and memory allocations performed with MPI, OpenMP and malloc, were replaced by the equivalent HPX counterparts. Both our and reference implementations follow the same numerical resolution. The benchmark use case is the simulation of 100 ms of electrical activity of a morphologically detailed neural network of layer 4 and 5 cells of the rodent brain, extracted from the model of Markram et al. [6], with the distribution of synaptic connectivity previously presented in Fig. 2. To demonstrate general applicability of our methods to a wide range of compute architectures, we utilised four different compute architectures with high variability in processor architecture, CPU frequency, memory bandwidth and cache: an Intel Sandy Bridge E5-2670 with 16 cores at 2.6 GHz, a Cray XE6 compute node with an AMD Opteron 6380 with 16 cores at 2.5 GHz each, an Intel Knights Landing (KNL) Xeon Phi with 64 cores at 1.3 GHz, and an Intel Xeon Gold 6140 with 18 cores at 2.3 GHz. The L1, L2 and L3 cache sizes for the architectures are: 448 KB, 3.5 MB and 35MB for the Intel E5; 768 KB, 16 MB and 16 MB for the Opteron; 16 KB, 1 MB and 32 MB for the Intel KNL; and 576 KB, 18 MB and 24.75 MB for the Xeon 6140. Each representation of a neuron requires a total memory of 4 to 12 MB. Distributed execution were executed on 32 compute nodes of Cray XE6 compute nodes, with specialized Infiniband network hardware for efficient point-to-point communication. We benchmarked the efficiency of each feature individually. The performance analysis of individual components follows in the following paragraphs.

Linear Containers: Cache efficiency of linear containers was measured with the *likwid* suite for performance monitoring and benchmarking [13] on the Xeon 6140 processor. The performance counters account for the containers performance only, isolating linear structures performance from other features. The benchmark test bench compares cache efficiency of linear versus standard library's containers. The estimated amounts of read/write workload are: a spike or event notification (loop through a map of post-synaptic neuron information) at approx.

every 15 ms; a delivery of an event—spike information, external currents, time notification—at circa every 0.05 ms, requiring a query to the priority queue; a computation of max time time step allowed by querying the map of time instant per pre-synaptic neuron at every timestep (0.025 ms); and an insertion of future events to be delivered at almost every time step (a push of a time event yo the priority queue). The results of cache efficiency on the BSP-based stepping protocol, with 4 continuous steps per neuron, and a communication interval at every 0.1 ms, is provided in Table 1 (top). Results demonstrate lower time to solution of circa 4× on the linear implementations versus standard library's, caused by: (1) less instructions, suggesting a more efficient implementation; (2) less data volume across different cache levels and system, suggesting higher reutilisation of data structures across all memory layers; and (3) less memory data volume, suggesting a more compact representation of data leading to more information loaded per cache line. As a relevant remark, Layer 3 cache in the Xeon 6140 architecture is a *victim cache*, or a refill path of CPU cache. Thus, the L2/L3 data volume is higher in our implementation due to demotions of L2 data to L3 instead of main RAM, representing an advantageous behaviour compared to the reference implementation.

Neuron Scheduler and Asynchronous Stepping: Our analysis was extended with asynchronous stepping. Neuron step scheduling for *earliest neuron steps first* was enabled and the distribution of steps size for different input datasets is presented in Fig. 7 (c). The step sizes vary depending on the circuit size due to increased inter-neuron connectivity for larger circuits. In practice, increased number of neurons leads to a possibly increased amount of pre-synaptic connectivity, and a higher probability of having a smaller minimum synaptic delay for a given pair of neurons, leading to smaller stepping intervals. We performed a similar cache efficiency benchmark for the asynchronous execution model, and the details are provided in the bottom of Table 1. Results of linear vs std implementations follow in line with the BSP use case, displaying better memory access and lower time to solution when comparing linear vs std container implementations. Asynchronous scheduled stepping yields circa 5–10% lower runtime and a much more efficient memory access compared with the previous BSP benchmark, on both linear and std implementations.

Communication Reduce: The reduce of communications at locality level was measured in terms or runtime and number of point-to-point (p2p) and reduce operations on a similar test bench, and executed on 32 nodes of the Cray XE6 architecture. A benchmark compares the reduced vs non-reduced (simple) communication implementations, measured on the BSP execution model—with a point-to-point communication of synapses and a reduce operation for control gate of neurons time advancement—and the asynchronous model presented, where p2p communication guides synaptic activity and neurons stepping notifications. The results are provided in Table 2, and suggest a significant reduction of communication workload and runtime, on both the BSP and asynchronous execution models. The communication workload gap between reduced and

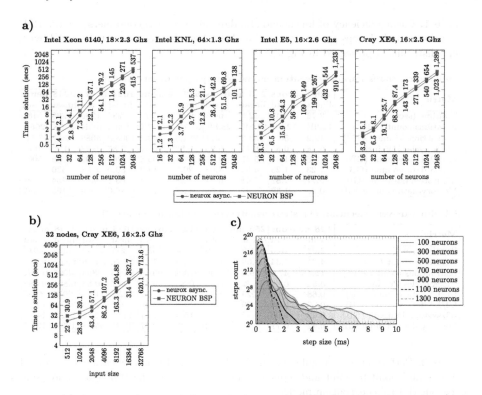

Fig. 7. (a) Time to solution of the methods presented (neurox async.) and the Bulk Synchronous Parallel equivalent (NEURON BSP) on the simulation of 100 ms of the electrical activity of differently sized neural networks, on four different hardware specifications. (b) Benchmark results for the simulation of 100 ms of electrical activity of an increasing number of neurons extracted, on a network of 32 Cray XE6 compute nodes. (c) Distribution of maximum step size allowed when following the earliest neuron steps first scheduler in the network with synaptic delays represented in Fig. 2.

non-reduced implementations increases with the circuit size, as more neurons incur more synaptic activity and communication. An acceleration of circa 5%–10% is visible when moving from a BSP to an asynchronous execution model.

Acceleration on Single Compute Nodes: The benchmark for a single compute node of the four aforementioned architectures is displayed in Fig. 7 (top) and compares our methods (neurox async.) with the reference solution (NEURON BSP), for an increasing number of interconnected neurons. The results demonstrate that the speed-up achieved decreases as we increase the number of neurons in the dataset. This property is due to the reduction of maximal step allowed by the neuron scheduler as we increase the number of neurons, as presented in Fig. 7 (c). On the Intel Xeon 6140, the methods yield a speed-up between 31%—for the largest network of 2048 neurons—and 51% for the network of 16 neurons. The

Table 1. Cache efficiency of linear and standard library (std) containers, for the BSP execution model (4 steps per neuron, top) and the Asynchronous execution model (with steps distribution presented in Fig. 7).

Bulk Synchronous Parallel execution model (4 steps per iteration)

Metric	128 neurons		256 neurons		512 neurons		1024 neur.		2048 neur.	
	linear	std	linear	std	linear	std	linear	std	linear	std
Runtime (secs)	2.13	14.42	12.5	64.3	63.7	278	294	1206	1298	5182
Iterations count ($\times 10^3$)	12.9K	12.9K	25.8K	25.8K	51.7K	51.7K	103K	103K	206K	206K
Instructions count ($\times 10^9$)	12.2	50.9	53.2	221	231.5	953.4	1003.2	4089	4327	17.5K
Clock cycles Per Instr.	0.54	0.85	0.71	0.90	0.82	0.87	0.87	0.88	0.90	0.89
L1/L2 data volume (GB)	1.16	1.52	5.47	9.53	32.1	90.6	266	902	2138	5065
L2/L3 data volume (GB)	1.23	1.23	4.64	4.08	20.3	14.7	80.9	56.8	330	233
L3/system data vol. (GB)	0.77	1.81	3.49	6.94	15.8	27.3	63.8	95.4	254	346
Memory data volume (GB)	0.90	1.39	2.87	4.50	11.5	16.1	46.0	58.0	163	222

Scheduler-driven execution (4+ steps per iteration, following Figure 7)

Metric	128 neurons		256 neurons		512 neurons		1024 neur.		2048 neur.	
	linear	std	linear	std	linear	std	linear	std	linear	std
Runtime (secs)	2.03	13.6	11.9	60.9	60.4	263.6	277	1143	1222	4913
Iterations count ($\times 10^3$)	4.34	4.34	8.69	8.69	17.39	17.39	34.76	34.76	69.45	69.45
Instructions count ($\times 10^9$)	11.4	47.9	49.9	209	218.3	901.5	948.3	3868	4096	16.4K
Clock cycles Per Instr.	0.54	0.85	0.72	0.87	0.83	0.87	0.87	0.88	0.891	0.888
L1/L2 data volume (GB)	0.68	0.96	4.29	8.34	29.2	78.2	252.9	818.5	2036	4655
L2/L3 data volume (GB)	0.63	0.48	2.60	1.67	13.9	6.10	59.3	24.8	249.3	109.6
L3/system data vol. (GB)	0.43	0.96	2.10	3.95	10.6	13.7	43.03	42.06	172.3	148.1
Memory data volume (GB)	0.42	0.77	1.54	2.42	7.33	9.32	32.48	35.18	123.2	121.2

Table 2. Performance of regular versus locality-reduced communication in terms of runtime and point-to-point and reduce communications, on the BSP (top) and asynchronous (bottom) execution models.

BSP execution; 32 compute nodes; p2p comm. for spiking, reduce at every 0.1ms

Metric	512 neurons		1024 neurons		2048 neurons		4096 neurons		8192 neurons	
	reduce	simple	reduce	simple	reduce	simple	reduce	simple	reduce	simple
Runtime (secs)	3.90	4.07	4.93	5.51	7.48	8.70	12.96	15.66	28.38	31.61
point-to-point count	2168	2327	7543	8855	24.3K	33.4K	70.1K	124K	188K	480K
reduce comm. count	100	1600	100	3200	100	6400	100	12.8K	100	25.6K

Asynchronous Execution; 32 compute nodes; p2p for spiking and stepping notification

Metric	512 neur.		1024 neur.		2048 neurons		4096 neurons		8192 neurns	
	reduce	simple	reduce	simple	reduce	simple	reduce	simple	reduce	simple
Runtime (secs)	3.60	3.80	4.07	4.42	6.66	6.53	12.14	13.27	26.75	28.31
point-to-point count	623K	665K	2.34M	2.72M	8.25M	11.09M	44.77M	25.79M	71.75M	181.46M

speed-ups for the remaining architectures are 36%–65% for the KNL, 35%–54% on the Intel E5, and 26%–31% on the Cray XE6.

Acceleration on Distributed Executions: In order to understand whether the single node advantages of the asynchronous execution hold in a distributed setting with multiple nodes, we extended our benchmark to a network of 32 nodes of the Cray XE6 architectures. Similarly to the single compute node use case, the test bench provides the runtime for an increasing number of neurons, in this case for a fixed network of 32 compute nodes. The results are presented in Fig. 7 (b), and display a speed-up of 16% for the largest dataset of 32768 neurons, up to 40% for 256 neurons i.e. one neuron per core per locality.

4 Conclusions

In this article, we explore the capabilities of new runtime systems for the numerical simulation of large systems of ODEs. We present an asynchronous model of execution with the capability of removal of global synchronization barriers, leading to better cache-efficiency and lower time to solution, due to long timestepping of individual equations based on their time coupling information. We detail the implementation of a fully-asynchronous, cache-accelerated, parallel and distributed simulation strategy supported by the HPX runtime system for the ParalleX execution model, providing a Global Address Memory space, remote procedure calls and asynchrony capabilities. Five components are introduced and detailed: (1) a linear data representation of a vector, map and priority queue containers that allow fully sequential instantiation of data structures in memory; (2) an exhaustive yet not speculative stepping of individual equations based on its time dependencies, supported by (3) a point-to-point communication protocol that actively notifies time dependencies of time advancements of their dependees and allows for the full overlap of computation and communication; (4) an object scheduler that further improves cache locality by maximising the number of steps per run by tracking equations progress throughout the execution; and (5) a local communication reduce operation that translates point-to-point to point-to-locality communication in a global address memory space.

Our methods were implemented on the core computation of the NEURON scientific application and tested on a biologically-inspired branched neural network. We analyse and demonstrate the efficiency of the features introduced in terms of communication, cache efficiency, patterns of data loading, and time to solution. Benchmark results demonstrate a significant speed-up in runtime in the order of 25% to 65% across different compute architectures and up to 40% on distributed executions. To finalize, most techniques presented follow from first principles in Computer Science, and can therefore be applied to a wide range of scientific problem domains.

Acknowledgements. The work was supported by funding from the ETH Domain for the Blue Brain Project (BBP). The super-computing infrastructures were provided by the Blue Brain Project at EPFL and Indiana University. A portion of Michael Hines efforts was supported by NINDS grant R01NS11613.

References

1. Hines, M.L., Carnevale, N.T.: The neuron simulation environment. Neural Comput. **9**(6), 1179–1209 (1997)
2. Ovcharenko, A., et al.: Simulating morphologically detailed neuronal networks at extreme scale. Advances in Parallel Computing (2015)

3. Hines, M.L., Markram, H., Schürmann, F.: Fully implicit parallel simulation of single neurons. J. Comput. Neurosci. **25**(3), 439–448 (2008)
4. Magalhaes, B., Hines, M., Sterling, T., Schuermann, F.: Asynchronous SIMD-enabled branch-parallelism of morphologically-detailed neuron models (2019, unpublished)
5. Kozloski, J., Wagner, J.: An ultrascalable solution to large-scale neural tissue simulation. Front. Neuroinform. **5**, 15 (2011). https://doi.org/10.3389/fninf.2011.00015
6. Markram, H., et al.: Reconstruction and simulation of neocortical microcircuitry. Cell **163**(2), 456–492 (2015)
7. Zenke, F., Gerstner, W.: Limits to high-speed simulations of spiking neural networks using general-purpose computers. Front. Neuroinform. **8**, 76 (2014). http://www.frontiersin.org/neuroinformatics/10.3389/fninf.2014.00076/abstract
8. Blue Brain Project: Coreneuron - simulator optimized for large scale neural network simulations. https://github.com/bluebrain/CoreNeuron
9. Sterling, T., Anderson, M., Bohan, P.K., Brodowicz, M., Kulkarni, A., Zhang, B.: Towards exascale co-design in a runtime system. In: Exascale Applications and Software Conference, Stockholm, Sweden (2014)
10. Hodgkin, A.L., Huxley, A.F.: A quantitative description of membrane current and its application to conduction and excitation in nerve. J. Physiol. **117**(4), 500–544 (1952)
11. Niebur, E.: Neuronal cable theory. Scholarpedia **3**(5), 2674 (2008). Revision 121893
12. Arge, L., Bender, M.A., Demaine, E.D., Holland-Minkley, B., Munro, J.I.: Cache-oblivious priority queue and graph algorithm applications. In: Proceedings of the Thirty-Fourth Annual ACM Symposium on Theory of Computing, pp. 268–276. ACM (2002)
13. Treibig, J., Hager, G., Wellein, G.: LIKWID: a lightweight performance-oriented tool suite for x86 multicore environments. In: 2010 39th International Conference on Parallel Processing Workshops (ICPPW), pp. 207–216 . IEEE (2010)

Application of the Model
with a Non-Gaussian Linear Scalar Filters
to Determine Life Expectancy, Taking
into Account the Cause of Death

Piotr Sliwka[(✉)] [iD]

Faculty of Mathematics and Natural Sciences, School of Exact Sciences,
Cardinal S. Wyszynski University, Warsaw, Poland
p.sliwka@uksw.edu.pl

Abstract. It is well-known that civilization diseases shorten life expectancy. The most common causes of death in Poland, both for women and men, are cancer and cardiovascular disease. The aim of the article is to use the non-Gaussian scalar filter model to determine life expectancy based on death rates after eliminating one of the above causes of death. Based on the obtained results, it can be stated that depending on the sex and type of the cause of death, the life expectancy may extend to several years.

Keywords: Life tables · Life expectancy · A cause of death ·
Forecasting of mortality rates · Ito stochastic differential equations ·
Hybrid mortality models

JEL Classifications: C32 · C53 · J11

1   Introduction

The creation of the first tables of life goes back to the 17th century. E. Halley used the death records available in the years 1687–91 of the inhabitants of Wrocław, on the basis of which he built the first life tables. The basis of the life table is the set of deaths at the age of x completed years (usually for one-year age ranges x from 0 to at least 100 years). Due to different length of life and gender diversity, separate life tables are built for women and for men. Existing expectancy life tables give the expected number of complete years remaining to live e_x for a person at age x without considering the cause of death. The following assumption can, therefore, be made: elimination of the cause of death extends e_x (mortality occurs as a result of natural death).

The purpose of this article is to try to estimate how many years life can last longer if mortality does not occur due to a specific disease, but because of natural death. The appointment of a precise e_x requires accurate data on the number of people in the cohort who died at the age of x due to the cause of

© Springer Nature Switzerland AG 2019
J. M. F. Rodrigues et al. (Eds.): ICCS 2019, LNCS 11538, pp. 435–449, 2019.
https://doi.org/10.1007/978-3-030-22744-9_34

y. Obtaining such data with the current restrictions of law in relation to The General Data Protection Regulation (GDPR-RODO) is very difficult. On the other hand, on the website of the Statistics Poland (GUS) there are data on the number of deaths in particular age groups (0, 1, 2, 3, 4, 5–9, 10–14, ..., 90–95 and 95 and more years) and at the level defined by the International Statistical Classification of Diseases and Related Health Problems in Poland after revision since 1997. These data allow the percentage of deaths to be determined in the case of a selected cause in each calendar year for a fixed age group (for these data, linear interpolation was used for each age of age group x). On this basis, it is possible to correct the number of people in the cohort whose death will not occur due to a given cause, but due to natural death, and then set the corrected death rates. There are not many articles dedicated to modelling mortality rates and determining e_x with the assumption given above. According to the best knowledge of the author, the methods proposed in the literature for determining e_x including the cause of death are most often based on the Lee-Carter model [13] and its mutations ([1,3–6,14,15,17–19]). In some articles, modelling and forecasting changes in mortality due to the established cause of death, time series techniques (e.g. ARIMA(1, 1, 1), [7]) are more often used than stochastic processes (e.g. birth and death process [2]). In others, instead of e_x, the rate of mortality of the number of people susceptible to a given disease ending in death is indicated (e.g. [12]), which also allows life expectancy to be determined for the studied cohort. However, no one has used the scalar model where a stochastic process is a colored noise modeled by a scalar linear filter with white noise input described by a scalar linear stochastic differential equations with constant coefficients ([9,16,20,21]). The usefulness and advantages of this proposition in relation to the Lee-Carter model were shown, among others, in [22].

The paper is organized as follows. In Sect. 2 basic notations and definitions of stochastic hybrid systems are introduced. In Sect. 3 materials and methods are presented: data set, the continuous non-Gaussian excitation model, the procedure of parameters estimation and the determination of submodels based on switching points to obtain hybrid model. In the case without restrictions on parameters, the standard estimation methods (such as: maximum likelihood or least squares) are used. Section 4 compares the empirical model with the theoretical model described in Sect. 3 and discusses the obtained results. Section 5 with general conclusions ends the article.

2 Mathematical Preliminaries

Throughout this paper we use the following notation. Let $|\cdot|$ and $<\cdot>$ be the Euclidean norm and the inner product in $\mathbb{R}^n$, respectively. We mark $\mathbb{R}_+ = [0,\infty)$, $\mathbb{T} = [t_0,\infty)$, $t_0 \geq 0$. Let $\Xi = (\Omega, \mathcal{F}, \{\mathcal{F}_t\}_{t\geq 0}, \mathbb{P})$ be a complete probability space with a filtration $\{\mathcal{F}_t\}_{t\geq 0}$ satisfying usual conditions. Let $\sigma(t) : \mathbb{R}_+ \to \mathbb{S}$ be the switching rule, where $\mathbb{S} = \{1,\ldots,N\}$ is the set of states. We denote switching times as $\tau_1, \tau_2, \ldots$ and assume that there is a finite number of switches on every finite time interval. Let $W_k(t)$ $(k = 1,\ldots,M)$ be the independent Brownian motions. We assume that processes $W_k(t)$ and $\sigma(t)$ are both $\{\mathcal{F}_t\}_{t\geq 0}$ adapted.

By the stochastic hybrid system we call the vector Itô stochastic differential equations with a switching rule described by

$$dx(t) = \mathbf{f}(\mathbf{x}(t), t, \sigma)dt + \sum_{k=1}^{M} \mathbf{g}_k(\mathbf{x}(t), t, \sigma)dW_k(t), \quad (\sigma(t_0), \mathbf{x}(t_0)) = (\sigma_0, \mathbf{x}_0), \quad (1)$$

where $\mathbf{x} \in \mathbb{R}^n$ is the state vector, $(\sigma_0, \mathbf{x}_0)$ is an initial condition, $t \in \mathbb{T}$ and M is a number of Brownian motions. $\mathbf{f}(\mathbf{x}(t), t, \sigma(t))$ and $\mathbf{g}_k(\mathbf{x}(t), t, \sigma(t))$ are defined by sets of $\{f(\mathbf{x}(t), t, l)\}$ and $\{\mathbf{g}_k(\mathbf{x}(t), t, l)\}$, respectively i.e. $\mathbf{f}(\mathbf{x}(t), t, \sigma(t)) = \mathbf{f}(\mathbf{x}(t), t, l)$, $\mathbf{g}_k(\mathbf{x}(t), t, \sigma(t)) = \mathbf{g}_k(\mathbf{x}(t), t, l)$ for $\sigma(t) = l$. Functions $\mathbf{f} : \mathbb{R}^n \times \mathbb{T} \times \mathbb{S} \to \mathbb{R}^n$ and $\mathbf{g}_k : \mathbb{R}^n \times \mathbb{T} \times \mathbb{S} \to \mathbb{R}^n$ are locally Lipschitz and such that $\forall l \in \mathbb{S}, t \in \mathbb{T} \mathbf{f}(\mathbf{0}, t, l) = \mathbf{g}_k(\mathbf{0}, t, l) = \mathbf{0}, k = 1, \dots, M$. These conditions together with these enforced on the switching rule $\sigma(t)$ ensure that there exists a unique solution to the hybrid system (1).

Hence it follows that Eq. (1) can be treated as a family (set) of subsystems defined by

$$dx(t, l) = \mathbf{f}(\mathbf{x}(t), t, l)dt + \sum_{k=1}^{M} \mathbf{g}_k(\mathbf{x}(t), t, l)dW_k(t), \quad l \in \mathbb{S} \qquad (2)$$

where $\mathbf{x}(t, l) \in \mathbb{R}^n$ is the state vector of l-subsystem.

We assume additionally that the trajectories of the hybrid system are continuous. It means, when the stochastic system is switched from l_1 subsystem to l_2 subsystem in the moment τ_j, then

$$\mathbf{x}(\tau_j, l_1) = \mathbf{x}(\tau_j, l_2), \quad l_1, l_2 \in \mathbb{S}. \qquad (3)$$

3   Materials and Methods

3.1   Data

From the HMD database, both data describing death rates of men and women from 2002–2016 were taken, as well as for each year 1958–2016 and for each age X (X = 0, ..., 110) the number of people (l_x) surviving age X (eg for men from 2016: $l_0 = 100000, l_1 = 99541, \dots, l_1 00 = 731, \dots$). Based on l_x, q_x as the probability of death in the period up to 1 year was determined, and next μ_x as death rates computed. Using the data of Statistics Poland regarding the number of deaths due to cardiovascular disease (cause C) and cancer (cause I), the percentages of these deaths in the number of all deaths for each calendar year were separately determined. These percentages were used to correct (usually increase) l_x - the number of survivors aged x, and consequently, q_x and μ_x. The modelling of adjusted mortality ratios μ_x was based on the non-Gaussian scalar filter model, whose analysis and purposefulness of the application in the present study was included in the works of, among others, [22–24], while the general form is contained in Subsect. 3.2.

3.2   Model with a Non-Gaussian Linear Scalar Filters (Non-Gaussian LSF)

We consider a family of mortality model with a continuous non-Gaussian scalar linear filter described by

$$\mu_x(t,l) = \mu_{x0}^l \exp\{\alpha_x^l t + \sum_{i=1}^{3} q_{x_i}^l y^i(t,l)\}, \tag{4}$$

$$dy(t,l) = -\beta_{x_1}^l y(t,l)dt + \gamma_{x_1}^l dW(t), \tag{5}$$

Introducing new variables $y_1(t,l) = y(t,l), y_2(t,l) = y^2(t,l), y_3(t,l) = y^3(t,l)$ and applying Ito formula we obtain

$$dy_2(t,l) = [-2\beta_{x_1}^l y_2(t,l) + (\gamma_{x_1}^l)^2]dt + 2\gamma_{x_1}^l y_1(t,l)dW(t), \tag{6}$$

$$dy_3(t,l) = [-3\beta_{x_1}^l y_3(t,l) + 3(\gamma_{x_1}^l)^2 y_1(t,l)]dt + 3\gamma_{x_1}^l y_2(t,l)dW(t), \tag{7}$$

where $\mu_x(t,l)$ is a stochastic process representing a mortality rate for a person aged x at time t, α_x^l, $\beta_{x_1}^l$, $q_{x_1}^l$, $q_{x_2}^l$, $q_{x_3}^l$, μ_{x0}^l, $\gamma_{x_1}^l$ are constant parameters, $l \in \mathbb{S}$; $W(t)$ is a standard Wiener process.

Taking natural logarithm of both sides of Eq. (4) and applying Ito formula we find

$$d\ln\mu_x(t,l) = [\alpha_x^l - (\beta_{x_1}^l q_{x_1}^l - 3(\gamma_{x_2}^l)^2)y_1(t,l)$$

$$- (2\beta_{x_1}^l q_{x_1}^l - 6(\gamma_{x_2}^l)^2)y_2(t,l) - (\gamma_{x_2}^l)^2 - 3\beta_{x_1}^l q_{x_3}^l y_3(t,l)]\, dt \tag{8}$$

$$+ [\gamma_{x_1}^l q_{x_1}^l + 2\gamma_{x_1}^l q_{x_2}^l y_2(t,l) + 3\gamma_{x_1}^l q_{x_3}^l y_3(t,l)]\, dW(t)$$

Introducing a new vector state

$$\mathbf{z}_x(t,l) = [z_{x_1}(t,l), z_{x_2}(t,l), z_{x_3}(t,l), z_{x_4}(t,l)]^T$$
$$= [\ln\mu_x(t,l), y_1(t,l), y_2(t,l), y_3(t,l)]^T, \tag{9}$$

Equations (8) and (5)–(7) one can rewrite in a vector form

$$
d\mathbf{z}_x(t,l) =
\begin{bmatrix}
0 & -\beta_{x_1}^l q_{x_1}^l + 3(\gamma_{x_1}^l)^2 & -2\beta_{x_1}^l q_{x_2}^l + 6(\gamma_{x_1}^l)^2 & -3\beta_{x_1}^l q_{x_3}^l \\
0 & -\beta_{x_1}^l & 0 & 0 \\
0 & 0 & -2\beta_{x_1}^l & 0 \\
0 & 3(\gamma_{x_1}^l)^2 & 0 & -3\beta_{x_1}^l
\end{bmatrix}
\mathbf{z}_x(t,l)dt
$$

$$
+
\begin{bmatrix}
\alpha_x^l + q_{x_2}^l(\gamma_{x_1}^l)^2 \\
0 \\
(\gamma_{x_1}^l)^2 \\
0
\end{bmatrix}
dt
$$

$$
+
\begin{bmatrix}
\gamma_{x_1}^l q_{x_1}^l + 2\gamma_{x_1}^l q_{x_2}^l y_1(t,l) + 3\gamma_{x_1}^l q_{x_3}^l y_2(t,l) \\
\gamma_{x_1}^l \\
+2\gamma_{x_1}^l y_1(t,l) \\
+3\gamma_{x_1}^l y_2(t,l)
\end{bmatrix}
dW(t)
\tag{10}
$$

The unknown parameters are

$$
\ln\mu_0^l, \alpha_x^l, \beta_{x_1}^l, q_{x_1}^l, q_{x_2}^l, q_{x_3}^l, \gamma_{x_1}^l.
$$

Using the method of the moment equations (see Appendix 3 in [22]) we find the nonstationary solutions of the first and second moment of the process $z_{x_1}(t,l), l \in \mathbb{S}$ (see Appendix 4 in [22])

$$
E[z_{x_1}(t,l)] = \alpha_x^l t + \alpha_{0_x}^l,
\tag{11}
$$

$$
E[z_{x_1}^2(t,l)] = (\alpha_x^l)^2 t^2 + 2\alpha_x^l \alpha_{0_x}^l t - 2\alpha_x^l \frac{(\gamma_{x_1}^l)^2}{2\beta_{x_1}^l} t + c_{0_x}^l
\tag{12}
$$

where $\alpha_{0_x}^l$ and $c_{0_x}^l$ are constants of integration.

3.3   The Procedure of Parameters Estimation and the Determination of Submodels (Based on Switching Points)

To find the parameters estimation and the determination of the switching points for non-Gaussian linear scalar filters (non-Gaussian LSF) we use similar procedure to the one described in [22].

Due to the limited number of observations (from 2002 to 2016) the parameter estimation procedure was performed for two types of models, namely:

- for the moment model with non-Gaussian LSF without switchings (for only one l)
- for the moment model with non-Gaussian LSFs with switchings, i.e. for $l \in \mathbb{S}$ using the estimation methods for each subsystem (next subsection).

Parameters Estimation. We note that the first and second moments of $z_{x_1}(t, l) = \ln \mu_x(t, l)$ depend on only six parameters α_x^l, $\alpha_{0_x}^l = \ln \mu_{x0}^l(t)$, $c_{0_x}^l$, $q_{x_2}^l$, $\beta_{x_1}^l$, $\frac{(\gamma_{x_1}^l)^2}{2\beta_{x_1}^l}$ and does not depend on the others, namely $q_{x_1}^l$, $q_{x_2}^l$, $q_{x_3}^l$. As it was shown in [22], only two parameters: α_x^l and $\alpha_{0_x}^l = \ln \mu_{x_0}^l(t), l \in \mathbb{S}$ are used and are found separately from minimization of the following square criterion

$$I_1 = \left(E[z_{x_1}(t, l)] - \alpha_x^l t - \alpha_{0_x}^l \right)^2. \tag{13}$$

Next, we assume for simplicity that $q_{x_1}^l = q_{x_2}^l = q_{x_3}^l = 1$. Then from the second moments of $z_{x_1}^2(t, l)$, i.e. $E[z_{x_1}^2(t, l)]$ we find the two parameters p_1^l and p_2^l, where $p_1^l = \frac{(\gamma_{x_1}^l)^2}{2\beta_{x_1}^l}$ and $p_2^l = c_{0_x}^l$, the relationship of which is nonlinear, namely

$$E[z_{x_1}^2(t, l)] = (\alpha_x^l)^2 t^2 + 2\alpha_x^l \alpha_{0_x}^l t - 2\alpha_x^l p_1^l t + p_2^l \tag{14}$$

Hence, the square criterion has the form

$$I_2 = \left(E[z_{x_1}^2(t, l)] - (\alpha_x^l)^2 t^2 - 2\alpha_x^l \alpha_{0_x}^l t + 2\alpha_x^l p_1^l t - p_2^l \right)^2 \tag{15}$$

In this case, all parameters $(\alpha_x^l, \alpha_{0_x}^l, p_1^l$ and $p_2^l)$ in the formula (14)–(15) based on the numerical algorithm of nonlinear minimization with additional conditions of $\alpha_{0_x}^l$ parameters $(\forall x\ \alpha_{0_x}^l < 0)$ were assessed. The algorithm works by generating a population of random starting points and next uses a local optimization method from each of the starting points to converge to a local minimum. As the solution, the best local minimum was chosen.

The Procedure of the Determination of Switching Time Points. To identify the switching time points s_t the procedure based on the Chow test [8] (which allows to assess whether the respective regression coefficients are different for split data sets) due to limited series of time observations only on three- and six-years intervals was used.

Step 1. Split the 2002–2016 mortality data (source: [11]) into two groups of intervals. The first group consists of six-years intervals e.g.:

$$\tilde{\tau}_6(1) = \{2002, 2003, \ldots, 2007\}, \quad \ldots, \quad \tilde{\tau}_6(10) = \{2011, 2012, \ldots, 2016\}$$

The second group consists of three-years intervals e.g.:

$$\tilde{\tau}_3(1) = \{2002, 2003, 2004\}, \ldots, \quad \tilde{\tau}_3(13) = \{2010, 2011, \ldots, 2015\}$$

Note: $\tilde{\tau}_3(1) \cup \tilde{\tau}_3(4) = \tilde{\tau}_6(1), \ldots$ and so on. In the next steps of the algorithm, the following sets of indices will be considered: $\tilde{\tau}_3(i), \tilde{\tau}_3(i+3)$ and $\tilde{\tau}_6(i)$ for i = 1, $\ldots$, 10.

Step 2. For $i = 1, l = 1$:

Using estimated parameters $\widehat{\alpha_0^l}$ and $\widehat{\alpha_1^l}$ of the regression model: $\mu_{x,t} = \alpha_0^l + \alpha_1^l t + \varepsilon_t$ for the years 2002–2016 ($t \in 1, \ldots, 15$) we determine three types of sums of residual squares (based on the above regression):

- the first one for the 3-element subinterval $S_{\tilde{\tau}_3(i)} = \sum_{k \in \tilde{\tau}_3(i)} e_k^2$,
- the second one for the 3-element subinterval $S_{\tilde{\tau}_3(i+3)} = \sum_{k \in \tilde{\tau}_3(i+3)} e_k^2$,
- the third one for $S_{\tilde{\tau}_6(i)} = \sum_{k \in \tilde{\tau}_6(i)} e_k^2$ for $i = 1, \ldots, 10$.

Step 3. To test the existence of a switching point, we propose to apply the Chow test of statistic F_{emp} [8] based on the Fisher-Snedecor distribution F:

$$F_{emp,i} = \frac{\frac{S_{\tilde{\tau}_6(i)} - S_{\tilde{\tau}_3(i)} - S_{\tilde{\tau}_3(i+3)}}{m}}{\frac{S_{\tilde{\tau}_3(i)} + S_{\tilde{\tau}_3(i+3)}}{n_1 + n_2 - 2m}} \tag{16}$$

where: m is the number of the estimated parameters (with intercept), $n_1 = n_2 = 3$ are numbers of observations in two neighbor rolling subintervals.

If $F_{emp,i} > F_{r_1, r_2, \alpha}$ (alternatively: $p - value \leq \alpha$, where α is the level of significance; usually $\alpha = 0.05$) then reject null hypothesis H_0 with the set of statistical hypotheses as follows:

$$H_0 : \alpha_{0_x}^l = \alpha_{0_x}^{l+1} \wedge \alpha_x^l = \alpha_x^{l+1} \wedge c_{0_x}^l = c_{0_x}^{l+1} \quad \text{against the alternative} \quad H_1 : \neg H_0$$

and accept as the switching point the first element of the set $\tilde{\tau}_3(i+3)$, where: $F_{r_1, r_2, \alpha}$ is a value of theoretical Fisher-Snedecor distribution F with $r_1 = m$ and $r_2 = n_1 + n_2 - 2m$ at significance level α.

If we have rejected H_0 then we have found a switching point s_l between subsystem l and subsystem $l + 1$ and we add it to the set of switching points, $l = l + 1$.

Step 4. Go back to Step 2, $i = i + 1$, repeat Step 2 and Step 3 until i = 13.

Step 5. Finally, we have created the set of switching points s_j, $j = 1, \ldots, N - 1$ and the corresponding N intervals of the mortality data.

4   Results

Based on empirical central death rates $\mu_{x,t}$ for all ages x ($x = 0, \ldots, 100$) the parameters of the models non-Gaussian SLF without and with switchings (in the case that at least one switching point has appeared) were evaluated and two sets of theoretical mortality rates $\widehat{\mu_{x,t}}$ were determined.

Selected results for a 40, 65, 67 and 70 year old woman and man are shown in Tables 1 and 2.

Table 1. Chow test values, 3- and 6-year intervals 2002–2016 (woman-W, man-M).

Sex	Age	02–07	03–08	04–09	05–10	06–11	07–12	08–13	09–14	10–15	11–16
W	40	0.65	1.42	**141.32**	0.30	1.25	4.35	2.37	11.49	8.88	3.31
	65	2.86	10.92	3.05	0.14	6.26	1.96	0.03	2.90	1.81	0.09
	67	4.57	5.10	3.32	6.38	5.41	1.11	1.04	0.23	1.72	1.93
	70	3.69	0.70	0.03	2.70	1.42	0.03	2.09	7.47	3.97	0.30
M	40	1.36	1.05	**127.50**	9.97	2.54	**28.52**	4.31	5.37	17.02	2.64
	65	**93.61**	12.62	0.62	0.11	3.61	0.98	0.06	1.04	0.14	1.57
	67	0.23	0.27	1.15	4.38	0.59	1.32	1.10	0.69	0.07	6.41
	70	2.20	0.12	0.96	0.23	1.43	0.13	0.49	2.36	0.27	0.46

Table 2. Life expectancy e_x: all causes of death, after removing cause C ($e_{x,C}$) and I ($e_{x,I}$) separately - selected years of life for women (e_{xW}) and men (e_{xM}).

age x	e_{xW}	$e_{xW,C}$	$e_{xW,I}$	e_{xM}	$e_{xM,C}$	$e_{xM,I}$
40	42.56	48.31	50.27	35.57	42.88	44.80
65	20.13	24.91	27.91	15.86	23.10	25.10
67	18.56	23.17	26.34	14.65	21.79	23.87
70	16.25	20.60	24.03	12.89	19.87	22.08

Due to the small number of observations (only 15 years), there was no switching point in every age group. Based on the results of the Chow test in Table 1 for selected age groups, it can be seen that there is only one switching point for women aged 40 and men aged 65 years, while for men aged 40 years, there are two switching points. Removal of the cause of death as expected generally extends the life expectancy. If the cause of death C is removed, the average life expectancy will increase, depending on the age group, from 4.35 to 5.75 years for women and from 6.98 to 7.31 for men. If the cause of I is removed, the average life expectancy will be extended by approx. 7.8 years for women and approx. 9.2 for men (see Table 2). In addition, it can be seen that in all cases empirical mortality rates without taking into account the cause of death decrease over time, which means an increase in life expectancy.

Value of empirical, theoretical death rates $\widehat{\mu_x}$, $\widehat{\mu_{x,C}}$ (without the cause of death C) and $\widehat{\mu_{x,I}}$ (without the cause of death I) determined by the nGLSF model (nGLSFC - without the cause of death C, nGLSFI - without the cause of death I) and forecasts from 2017 to 2025 (denoted by an additional letter f, i.e. nGLSFf) for women (W) and men (M) are included in Figs. 1, 2, 3, 4, 5, 6, 7 and 8.

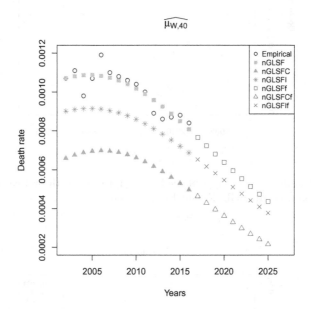

Fig. 1. Values of death rates for women aged 40 - empirical and theoretical values as well as forecasts based on the nGLSF model respectively

The following conclusions can be drawn from the Figs. 1, 2, 3, 4, 5, 6, 7 and 8:

1. For a 40-year-old woman, the difference between $\widehat{\mu_{40}}$ and $\widehat{\mu_{40,C}}$, as well as between $\widehat{\mu_{40}}$ and $\widehat{\mu_{40,I}}$ is more or less stable, while for men it is decreasing, which means that C and I definitely increase their share in the total number of deaths and they are definitely the dominant causes of deaths in this age group.
2. For a 65-year-old woman, the difference between $\widehat{\mu_{65}}$ and $\widehat{\mu_{65,C}}$ is more or less stable, whereas between $\widehat{\mu_{65}}$ and $\widehat{\mu_{65,I}}$ it decreases, thus becoming the dominant cause of death in time, while in the case of men, the trend is slightly decreasing.
3. For a 67-year-old woman and man, the situation is similar to the case of a 65-year-old.

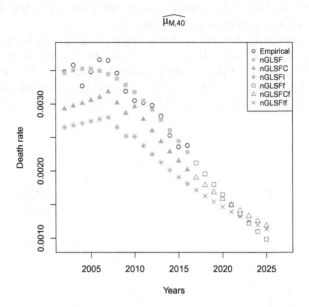

Fig. 2. Values of death rates for men aged 40 - empirical and theoretical values as well as forecasts based on the nGLSF model respectively

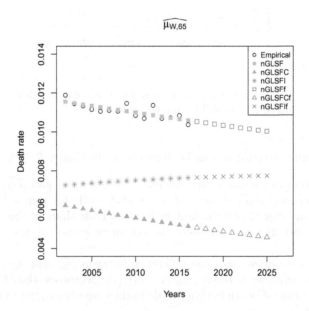

Fig. 3. Values of death rates for women aged 65 - empirical and theoretical values as well as forecasts based on the nGLSF model respectively

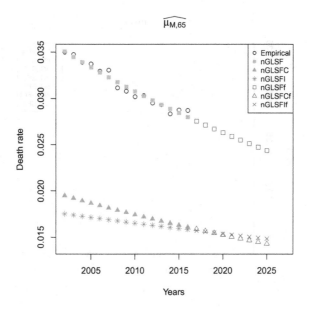

Fig. 4. Values of death rates for men aged 65 - empirical and theoretical values as well as forecasts based on the nGLSF model respectively

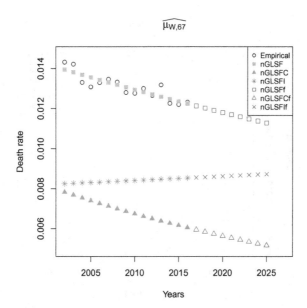

Fig. 5. Values of death rates for women aged 67 - empirical and theoretical values as well as forecasts based on the nGLSF model respectively

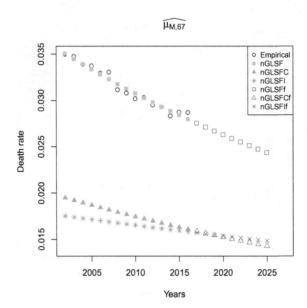

Fig. 6. Values of death rates for men aged 67 - empirical and theoretical values as well as forecasts based on the nGLSF model respectively

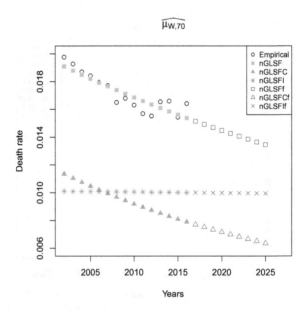

Fig. 7. Values of death rates for women aged 70 - empirical and theoretical values as well as forecasts based on the nGLSF model respectively

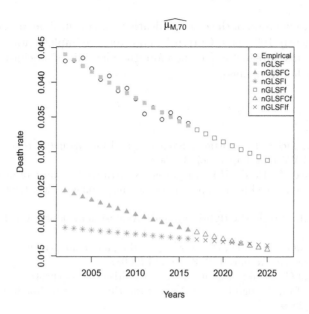

Fig. 8. Values of death rates for men aged 70 - empirical and theoretical values as well as forecasts based on the nGLSF model respectively

4. In the case of a 70-year-old woman the difference between $\widehat{\mu_{70}}$ and $\widehat{\mu_{70,C}}$ to 2008 is slightly decreasing, and since 2008 it is slightly growing, while the difference between $\widehat{\mu_{70}}$ and $\widehat{\mu_{70,I}}$ is decreasing until 2012, then more or less stable; in the case of men, the trend is slightly diminishing with a decreasing disparity between $\widehat{\mu_{70,C}}$ and $\widehat{\mu_{70,I}}$

where

$\widehat{\mu_{x,C}}$ - death rate without a cardiological cause,
$\widehat{\mu_{x,I}}$ - death rate without a cancer cause.

5   Conclusions

The purpose of this article was to try to estimate how many years life can last longer if death does not occur because of a specific disease, but because of natural death using a non-Gaussian linear scalar filter model. Determining the exact life expectancy e_x of individual people requires accurate data on the number of people in the cohort who died in the age of x due to the cause of C or I. Obtaining such data (e.g. from a hospital) is very difficult due to the current restrictions of law and in connection with the Act on protection of personal data GDPR. Nevertheless, the attempt to determine life expectancy with the exclusion of death due to C or I illness using the proposed model and the method of estimation seems realistic. Thus, the results obtained in the article should be treated as an approximation of the real life expectancy e_x. Determining the "real" e_x after exclusion of the cause of death would occur if, in fact, one observes a cohort for nearly 100 years. Currently, however, the implementation of such an experience seems unrealistic and therefore the methods of stochastic simulation should be further developed.

References

1. Booth, H., Tickle, L.: Mortality modelling and forecasting: a review of methods. In: ADSRI Working Paper, vol. 3 (2008)
2. Boumezoued, A., Hardy, H.L., El Karoui, N., Arnold, S.: Cause-of-death mortality: what can be learned from population dynamics? Insur. Math. Econ. **78**, 301–315 (2018)
3. Boukas, E.K.: Stochastic Hybrid Systems: Analysis and Design. Birkhauser, Boston (2005)
4. Cairns, A.J.G., Blake, D., Dowd, K.: Modelling and management of mortality risk: a review. Scand. Actuar. J. **2–3**, 79–113 (2008)
5. Cairns, A.J.G., et al.: A quantitative comparison of stochastic mortality models using data from England and Wales and the United States. North Am. Actuar. J. **13**, 1–35 (2009)
6. Cairns, A.J.G., Blake, D., Dowd, K., Coughlan, G.D., Epstein, D.: Mortality density forecasts: an analysis of six stochastic mortality models. Insur. Math. Econ. **48**, 355–367 (2011)
7. Cao, H., Wang, J., et al.: Trend analysis of mortality rates and causes of death in children under 5 years old in Beijing, China from 1992 to 2015 and forecast of mortality into the future: an entire population-based epidemiological study. BMJ Open **7** (2017). https://doi.org/10.1136/bmjopen-2017-015941
8. Chow, G.C.: Tests of equality between sets of coefficinets in two linear regressions. Econometrica **28**, 591–605 (1960)
9. Giacometti, R., Ortobelli, S., Bertocchi, M.: A stochastic model for mortality rate on Italian data. J. Optim. Theory App. **149**, 216–228 (2011)
10. Gompertz, B.: On the nature of the function expressive of the law of human mortality and on a new mode of determining life contingencies. R. Soc. Lond. Philos. Trans. Ser. A Math. Phys. Eng. Sci. Ser. A **CXV**, 513–580 (1825)
11. GUS: Central Statistical Office of Poland (2015). http://demografia.stat.gov.pl/bazademografia/TrwanieZycia.aspx
12. Jahangiri, K., Aghamohamadi, S., Khosravi, A., Kazemi, E.: Trend forecasting of main groups of causes-of-death in Iran using the Lee-Carter model. Med. J. Islam. Repub. Iran **32**, 124 (2018)
13. Lee, R.D., Carter, L.: Modeling and forecasting the time series of U.S. mortality. J. Am. Stat. Assoc. **87**, 659–671 (1992)
14. Lee, R.D., Miller, T.: Evaluating the performance of the Lee-Carter method for forecasting mortality. Demography **38**, 537–549 (2001)
15. Maccheroni, C., Nocito, S.: Backtesting the Lee-Carter and the Cairns-Blake-Dowd stochastic mortality models on Italian death rates. In: CeRP Working Papers, vol. 166 (2017)
16. Milevsky, M.A., Promislov, S.D.: Mortality derivatives and the option to annuitise. Insur. Math. Econ. **29**, 299–318 (2001)
17. Pollard, J.H.: Projection of age-specific mortality rates. Popul. Bull. U. N. **21–22**, 55–69 (1987)

18. Renshaw, A., Haberman, S.: Lee-Carter mortality forecasting with age-specific enhancement. Insur. Math. Econ. **33**, 255–272 (2003)
19. Renshaw, A., Haberman, S.: A cohort-based extension to the Lee-Carter model for mortality reduction factor. Insur. Math. Econ. **38**, 556–570 (2006)
20. Rossa, A., Socha, L.: Proposition of a hybrid stochastic Lee-Carter mortality model. Metodoloski zvezki (Adv. Methodol. Stat.) **10**, 1–17 (2013)
21. Rossa, A., Socha, L., Szymanski, A.: Hybrid Dynamic and Fuzzy Models of Mortality, 1st edn. WUL, Lodz (2018)
22. Sliwka, P., Socha, L.: A proposition of generalized stochastic Milevsky-Promislov mortality models. Scand. Actuar. J. **8**, 706–726 (2018)
23. Sliwka, P.: Proposed methods for modeling the mortgage and reverse mortgage installment. In: Recent Trends in the Real Estate Market and Its Analysis, pp. 189–206. Oficyna Wydawnicza SGH, Warszawa (2018)
24. Sliwka, P.: Application of the Markov chains in the prediction of the mortality rates in the generalized stochastic Milevsky-Promislov model. In: Trends in Biomathematics: Mathematical Modeling for Health, Harvesting, and Population Dynamics. Springer (2019, in press). ISBN 978-3-030-23432-4

Improving ODE Integration on Graphics Processing Units by Reducing Thread Divergence

Thomas Kovac[1,2(✉)], Tom Haber[1], Frank Van Reeth[1], and Niel Hens[2,3]

[1] Expertise Centre for Digital Media, Hasselt University, Hasselt, Belgium
{thomas.kovac,tom.haber,frank.vanreeth}@uhasselt.be
[2] Center for Statistics, Hasselt University, Hasselt, Belgium
[3] Chermid, Vaccine and Infectious Disease Institute, University of Antwerp, Antwerp, Belgium
niel.hens@uantwerpen.be

Abstract. Ordinary differential equations are widely used for the mathematical modeling of complex systems in biology and statistics. Since the analysis of such models needs to be performed using numerical integration, many applications can be gravely limited by the computational cost. This paper present a general-purpose integrator that runs massively parallel on graphics processing units. By minimizing thread divergence and bundling similar tasks using linear regression, execution time can be reduced by 40–80% when compared to a naive GPU implementation. Compared to a 36-core CPU implementation, a 150 fold runtime improvement is measured.

Keywords: Pharmacometrics · Epidemiology · Parallelism · High-performance computing · Graphics processing units

1 Introduction

Systems of coupled ordinary differential equations (ODEs) are widely used for mathematical modeling of complex systems in epidemiology [6,8], biology [9] and pharmacology [5]. The analysis of such models is usually performed by numerical integration, since analytical solutions can not be derived in general. Additionally, many applications need to carry out a massive amount of simulations for which the computational cost can be a serious limitation.

Performing such analyses on a central processing unit (CPU) can be quite time-consuming, even on today's multi-core processors. Graphics processing units (GPU) have moved from being common computer graphics and image processing instruments to powerful general-purpose devices [12]. GPUs are single-instruction multiple-data (SIMD) devices, consisting out of hundreds of cores, that give access to tera-scale performance on common workstations. The architecture is ideally suited for executing identical and independent operations on different data, such as image processing.

© Springer Nature Switzerland AG 2019
J. M. F. Rodrigues et al. (Eds.): ICCS 2019, LNCS 11538, pp. 450–456, 2019.
https://doi.org/10.1007/978-3-030-22744-9_35

This paper mainly focuses on ODE simulations that can be performed independently and embarrassingly parallel. Many applications already exhibit this behavior or the underlying algorithm can often be changed to be more favorable to this situation. For example: sensitivity and parameter sweep analysis inherently require the evaluation of many parameters. Optimization and Bayesian sampling algorithms are typically sequential in nature, but parallel alternatives exist for both: evolutionary/genetic algorithms [4,7,8] and sequential Monte-Carlo methods [3,11]. Alternatively, computing expectations [15] over differential equation based models is again massively parallel.

While the embarrassingly parallel nature matches perfectly with the GPUs capabilities, branching in all but the simplest integration methods can cause divergent program paths and a significant drop in performance [1]. The proposed method rearranges tasks among available threads such that they are less likely to diverge. A linear regression model is constructed to predict which tasks are similar in behavior and grouped together accordingly.

2   Background

2.1   GPU

NVIDIA GPUs are comprised out of different layers of parallel processing; the top level consists of Streaming Multiprocessors (SMs). Blocks of a kernel are mapped onto an SM, which in turn executes it using user-allocated threads. The multiprocessor creates, manages, schedules, and executes threads in groups of 32 parallel threads called warps. For optimal performance, threads within a warp are required to execute the same instruction at any given time. As a result, whenever two or more threads diverge, operations of both branches need to be executed for all warp threads. This can lead to a serious drop in performance [1].

2.2   Integration Methods

Among numerical integration algorithms, Runge-Kutta methods are a family of explicit iterative methods, with a wide variety of orders and schemes [13]. The Dormand-Prince method [2], also known *DOPRI*, is a fifth-order method where the step-size is adjusted by the truncation error, which is approximated by the difference between the fourth and fifth-order estimates. MATLAB's *ode45* is also an implementation of the DOPRI method.

While these methods cannot cope well with *stiff* ODEs, many statistical or biological models only exhibit stiffness at extreme parameter values.

2.3   Epidemiological Models

Epidemiologists use mathematical modeling of infectious diseases to improve insight into disease dynamics, resulting in the creation of more effective vaccines and antiviral drugs, better intervention/vaccination programs [6,8,15].

One epidemiological model is the Susceptible-Exposed-Infected-Recovered (SEIR) model which describes the flow of individuals through these mutually exclusive disease states. This model is extensible to more complex diseases by adding compartments. Santermans et al. [15] studied the Ebola outbreak of 2015 in West Africa. Equation 1 models SEIR dynamics over time, where $S(t)$, $E(t)$, $I(t)$, and $R(t)$ are the number of susceptibles, exposed, infected and recovered, respectively and $N(t) = S(t) + E(t) + I(t) + R(t)$ denotes population size.

$$
\begin{cases}
\frac{dS(t)}{dt} = -\beta(t)S(t)\frac{I(t)}{N(t)}, \\
\frac{dE(t)}{dt} = \beta(t)S(t)\frac{I(t)}{N(t)} - \gamma E(t), \\
\frac{dI(t)}{dt} = \gamma E(t) - \alpha I(t) - \sigma I(t), \\
\frac{dR(t)}{dt} = \sigma I(t).
\end{cases}
\tag{1}
$$

3   Related Work

Seen et al. [16] show that a Runge-Kutta-Fehlberg method ($RK45$) with adaptive step-size on GPU outperforms a CPU implementation, given that the problem dimensions are large enough, as in 200 equations, or more. Having a Runge-Kutta implementation with adaptive time steps, however, suffers from a phenomenon called variable task-length [10]. The step-size modification can vary between individual GPU threads, resulting in warp divergence and loss in performance. Murray et al. [10] suggest bundling multiple data items into each thread, allowing a thread to immediately advance onto the next task once an item is completed.

Stone et al. [18] implemented two parallel strategies; a so-called "one-thread" method and "one-block" method. The former method employs one thread to solve an ODE, the latter uses an entire block of threads. Both $RK45$ and CVODE integration methods were ported to GPU using aforementioned parallel strategies. Significant speedups were reported of all GPU adaptations, but without thread divergence even greater speedups are possible.

Stone et al. [17] emphasize that an efficient and effective ODE integrator must employ the available instruction-level parallelism of the underlying hardware as well as the numerical efficiency. Having implemented a non-stiff Runge-Kutta ODE solver on both GPU and Xeon Phi, the authors report the GPU version being slower than the Xeon Phi version as thread divergence caused by the adaptive step-sizes negatively impacts performance.

4   Methods

Related work shows multiple successful Runge-Kutta method ports to GPU. Solutions where each thread solves an ODE suffer from performance loss due to varying length of tasks caused by thread divergence. Results that employ multiple threads to solve an ODE overcome this problem [8,17], however not nearly as many ODEs can be solved as compared to one-threaded solutions. In

Algorithm 1. Dormand-Prince

1: **function** INTEGRATE(t, tOut)
2:     **while** t <= tOut **do**   ▷ try to make a step with size dt
3:         ynew, error ← TryStep(dt)
4:         **if** error ≤ rtol **then**   ▷ step was successful
5:             t ← t + dt
6:             y ← ynew
7:             dt ← GrowStepsize(dt, error)   ▷ potentially grow stepsize
8:         **else**
9:             dt ← ShrinkStepsize(dt, error)   ▷ step failed, shrink stepsize
        return y

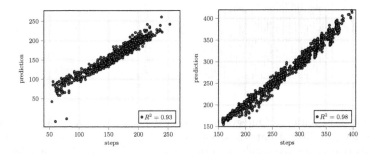

Fig. 1. Number of actual steps plotted against the prediction for the SEIR model (left) and the Nimotuzumab model (right). For both models that the number of integration steps required, based on the parameters, can be accurately predicted.

this paper, the DOPRI integration method, an NVIDIA Tesla P100 GPU, a one-threaded solution are employed and thread divergence is strongly reduced.

As can be seen in Algorithm 1, there are two causes of thread divergence in one-threaded solutions: the test whether or not a step was successful and the number of steps taken during integration. The former is less of a problem since the branches are very small and care was taken to move all common operations out of them. The latter is the main cause of performance loss since it results in threads that idle for a long time. The number of steps required can wildly differ from parameter to parameter. Ensuring that threads performing similar tasks by bundling them by the required number of steps, accomplished by sorting followed by partitioning, automatically reduces thread divergence as all threads within a warp execute the same instructions.

A linear regression model is used to predict the number of integration steps a task will require. This model is trained a-priori on a small set of parameters (1000) and is used at runtime to group tasks that are similar in number of steps. Transforming the rate parameters to log-space results in a higher predictive performance. To demonstrate general applicability, the number of steps is also predicted for the Nimotuzumab model [14]. Figure 1 shows that the model accurately predicts the number of integration steps required, given the parameters, for both the SEIR and Nimotuzumab model.

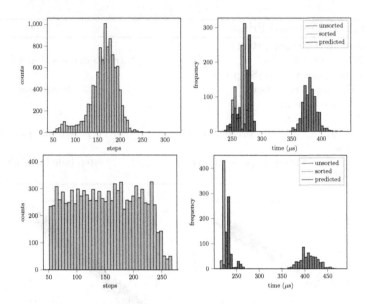

Fig. 2. Left: distributions of integrator steps of the original dataset sampled from the prior [15] and an altered dataset where the parameters were re-sampled such that the distribution is approximately uniform. Right: distribution of runtime for random subsets of the original dataset, bundled using a-priori knowledge of the number of steps and bundled using the predictor for both datasets.

5   Results

All results are based on the SEIR model from Santermans et al. [15]. 10K parameters were sampled from their prior to create a realistic dataset. Since runtime depends heavily on the parameters, random subsets of 7168 parameters are created and performance distributions are shown. The predictor is always trained on random subsets of 1000 parameters.

The top left plot of Fig. 2 shows the distribution of the number of steps attempted by the integrator. While the average number of steps is 160, the spread is quite significant. The top right plot compares distributions of runtime for random subset of parameters, bundled using the predictor as well as bundled employing a-priori knowledge of the number of steps. Bundling tasks according to similarity clearly has a positive effect on performance. On average, a 40% increase is observed and the predictor performs slightly worse compared to the case with a-priori knowledge. Compared to a CPU implementation (36-core Intel Skylake Xeon 6140; 192 GB RAM), a 150-fold improvement in runtime is measured.

To indicate the impact of the distribution of steps, the original dataset is re-sampled such that the distribution of steps is approximately uniform. Bottom plots of Fig. 2 show the performance distribution for the uniform dataset. On average, the performance is increased by 80%.

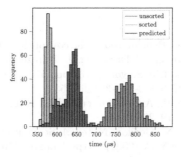

Fig. 3. Runtime distribution for an implementation forcing fixed step-sizes for all threads within the same warp and a comparison with bundling the tasks.

Clearly, when the distribution of steps is extremely peaked, the performance difference will be quite small. In the worst case, every parameter will require the same number of steps with no change in performance due to bundling.

Figure 3 shows performance for a implementation which forces the integrator to use the same step-size for all threads in a warp similar to Murray et al. [10]. After attempting a step and computing the error, a local shuffle operation is performed to compute the largest error and all threads continue with this error instead. As a result, all threads will take the same branch and make the same decision in terms of step-size. However, due to the additional cost of the local shuffle as well as the extra steps taken by the integrator, this method is roughly 2× slower. Even for this implementation bundling the parameters helps.

6   Conclusion and Future Work

GPUs are massively parallel single-instruction multiple-data devices capable of tera-scale performance. However, the small differential equation based models typically used in statistics and biology are not ideally suited for these devices due to the branching nature of numerical integration algorithms. This branching causes SIMD units to diverge, resulting in a significant performance drop. By rearranging the order of tasks, bundling similar tasks together, this problem can be alleviated to some extent. A linear regression model, trained on a small set of parameters, is used for predicting the similarity of tasks. The runtime improvement depends on the distribution of the number of steps in the tasks: the more spread out the distribution, the bigger the expected improvement. The experiments show improvements of 40% to 80%.

The global relation between parameters and number of steps can sometimes be hard to capture with a predictive model. However, optimization algorithms and Bayesian samplers typically explore the parameter space only locally. Therefore, future work will look into an online learning method which captures only this local behavior and requires no a-priori training.

References

1. Bialas, P., Strzelecki, A.: Benchmarking the cost of thread divergence in CUDA. In: Wyrzykowski, R., Deelman, E., Dongarra, J., Karczewski, K., Kitowski, J., Wiatr, K. (eds.) PPAM 2015. LNCS, vol. 9573, pp. 570–579. Springer, Cham (2016). https://doi.org/10.1007/978-3-319-32149-3_53
2. Dormand, J.R., Prince, P.: A family of embedded Runge-Kutta formulae. J. Comput. Appl. Math. **6**(1), 19–26 (1980)
3. Doucet, A., de Freitas, N., Gordon, N. (eds.): Sequential Monte Carlo Methods in Practice. Springer, New York (2001). https://doi.org/10.1007/978-1-4757-3437-9
4. Eiben, A.E., Smith, J.E.: Introduction to Evolutionary Computing. Springer, Heidelberg (2003). https://doi.org/10.1007/978-3-662-05094-1
5. van der Graaf, P.H., Benson, N.: Systems pharmacology: bridging systems biology and pharmacokinetics-pharmacodynamics (PKPD) in drug discovery and development. Pharm. Res. **28**(7), 1460–1464 (2011)
6. Hens, N., Shkedy, Z., Aerts, M., Faes, C., Van Damme, P., Beutels, P.: Modeling Infectious Disease Parameters Based on Serological and Social Contact Data: A Modern Statistical Perspective, vol. 63. Springer Science & Business Media, New York (2013)
7. Kennedy, J., Eberhart, R.: Particle swarm optimization (1995)
8. Kovac, T., Haber, T., Van Reeth, F., Hens, N.: Heterogeneous computing for epidemiological model fitting and simulation. BMC Bioinform. **19**(1), 101 (2018)
9. Murray, J.D.: Mathematical Biology I. An Introduction. Interdisciplinary Applied Mathematics, vol. 17, 3rd edn. Springer, New Yor (2002). https://doi.org/10.1007/b98868
10. Murray, L.: GPU acceleration of Runge-Kutta integrators. IEEE Trans. Parallel Distrib. Syst. **23**(1), 94–101 (2012)
11. Nemeth, B., Haber, T., Liesenborgs, J., Lamotte, W.: Relaxing scalability limits with speculative parallelism in sequential Monte Carlo. In: 2018 IEEE International Conference on Cluster Computing (CLUSTER). IEEE, September 2018
12. Owens, J.D., et al.: A survey of general-purpose computation on graphics hardware (2007)
13. Press, W.H., Teukolsky, S.A., Vetterling, W.T., Flannery, B.P.: Numerical Recipes: The Art of Scientific Computing, 3rd edn. Cambridge University Press, New York (2007)
14. Rodríguez-Vera, L., et al.: Semimechanistic model to characterize nonlinear pharmacokinetics of nimotuzumab in patients with advanced breast cancer. J. Clin. Pharmacol. **55**(8), 888–898 (2015)
15. Santermans, E., et al.: Spatiotemporal evolution of Ebola virus disease at subnational level during the 2014 West Africa epidemic: model scrutiny and data meagreness. PLoS One **11**(1), e0147172 (2016)
16. Seen, W.M., Gobithaasan, R.U., Miura, K.T., Ismail, M.T., Ahmad, S., Rahman, R.A.: GPU acceleration of Runge Kutta-Fehlberg and its comparison with Dormand-Prince method. In: AIP Conference Proceedings, vol. 1605, no. 1, pp. 16–21 (2014)
17. Stone, C.P., Alferman, A.T., Niemeyer, K.E.: Accelerating finite-rate chemical kinetics with coprocessors: comparing vectorization methods on GPUs, MICs, and CPUs. Comput. Phys. Commun. **226**, 18–29 (2018)
18. Stone, C.P., Davis, R.L.: Techniques for solving stiff chemical kinetics on graphical processing units. J. Propuls. Power **29**(4), 764–773 (2013)

Data Compression for Optimization of a Molecular Dynamics System: Preserving Basins of Attraction

Michael Retzlaff[1]([✉]), Todd Munson[2], and Zichao (Wendy) Di[2]

[1] Department of Mathematics and Statistics,
University of Maryland Baltimore County, Baltimore, MD, USA
`mretzla1@umbc.edu`
[2] Mathematics and Computer Science Division, Argonne National Laboratory,
Lemont, IL, USA

Abstract. Understanding the evolution of atomistic systems is essential in various fields such as materials science, biology, and chemistry. The gold standard for these calculations is molecular dynamics, which simulates the dynamical interaction between pairs of molecules. The main challenge of such simulation is the numerical complexity, given a vast number of atoms over a long time scale. Furthermore, such systems often contain exponentially many optimal states, and the simulation tends to get trapped in local configurations. Recent developments leverage the existing temporal evolution of the system to improve the stability and scalability of the method; however, they suffer from large data storage requirements. To efficiently compress the data while retaining the basins of attraction, we have developed a framework to determine the acceptable level of compression for an optimization method by application of a Kantorovich-type theorem, using binary digit rounding as our compression technique. Choosing the Lennard-Jones potential function as a model problem, we present a method for determining the local Lipschitz constant of the Hessian with low computational cost, thus allowing the use of our technique in real-time computation.

Keywords: Lossy compression · Basins of attraction ·
Nonlinear optimization · Lennard-Jones potential

1 Introduction

Simulating atomistic evolution is essential for predicting materials properties for use in materials science, chemistry, and biology. Molecular dynamics (MD), the gold standard for atomistic simulations, simulates the interactions of atoms and molecules for a fixed period of time [1,14]. The limited temporal scale is inherent from the system's sequential nature but poses great challenges for observing various transitions on MD time scales. Recent advances leverage statistics and massively parallel computers to accelerate the simulations, such as the accelerated

© Springer Nature Switzerland AG 2019
J. M. F. Rodrigues et al. (Eds.): ICCS 2019, LNCS 11538, pp. 457–470, 2019.
https://doi.org/10.1007/978-3-030-22744-9_36

molecular dynamics method [15,18] and the parallel replica dynamics method (ParRep) [19]. In particular, the parallel trajectory splicing method (ParSplice) [12], as a generalization of ParRep, has drawn much attention in recent years because of its superior computational performance by utilizing large-scale high-performance computers to parallelize the state-to-state simulations in the time domain. One major burden, however, is the need to store all the local minimizers computed in the long MD simulation, since they are used as the initial guess for the next state simulation. Since the number of local minimizers is exponential in the number of atoms, storage space must be conserved, even though each individual minimizer is a small set of numbers (e.g., 3D coordinates of the atom positions). Lossy compression of these local minimizers can potentially result in significant savings. By compressing the values, we can cache more values locally on the compute node and send/receive less information from the global database. However, a critical step is to identify the basins of local convergence for a particular numerical method. As long as the compressed data stays in the same basin, we can recover the local minimizer by applying the numerical method to the stored iterate.

In this work, for an atomic system and Newton's method as the underlying solver, we employ a Kantorovich-type theorem for studying the basins of attraction of potential energy minima. We propose a mathematical framework to estimate bounds for the compression level of an optimizer when using binary digit rounding as the compression routine. One critical step is to calculate the required information, namely, the local Lipschitz-type constant of the function's derivative. As the system gets large, an analytic expression for this information is likely impossible. Therefore, a robust method to approximate this information is required. By exploring the topology of the potential energy landscape, we propose an efficient way to estimate the required constant. We prototype our method on the simple, yet common, Lennard-Jones potential function and validate our approach on various numerical examples.

2   Minimization of Lennard-Jones Potential

Given a configuration of n atoms in a cluster $\mathbf{X} = \{\mathbf{x}_1, \ldots, \mathbf{x}_n\}$, we consider the Lennard-Jones (LJ) potential function [8], a simplified model that simulates the potential energy of the cluster based on inter-atom distances:

$$\hat{V}_n(\mathbf{X}) = 4\varepsilon \sum_{i<j} \left(\left(\frac{\sigma}{\hat{d}(\mathbf{x}_i, \mathbf{x}_j)} \right)^{12} - \left(\frac{\sigma}{\hat{d}(\mathbf{x}_i, \mathbf{x}_j)} \right)^{6} \right), \tag{2.1}$$

where $\hat{d}(\mathbf{x}_i, \mathbf{x}_j)$ is the Euclidean distance between atoms $\mathbf{x}_i, \mathbf{x}_j \in \mathbb{R}^3$. The physical constants ε and σ are the depth of the potential well and inter-atom reaction limit, respectively, both depending on the type of atom. Because of its importance in computational chemistry, finding optimal configurations that locally minimize the potential energy (2.1) remains an active research area. For example, Maranas et al. [10] proposed an exotic optimization algorithm to find many

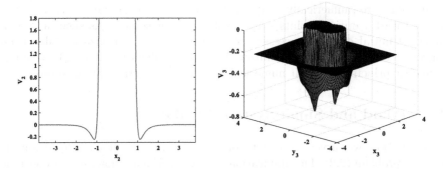

Fig. 1. Left: landscape of function V_2 given x_2 as its only free variable. Right: landscape of function V_3 with reduced dimension by setting $x_2 = 2^{1/6}$ (its equilibrium distance) with DOF x_3, y_3.

stationary points, and Asenjo et al. [3] studied the mapping of the basins of attraction for various optimization algorithms.

In this work, we follow the longstanding convention in [7] and consider the reduced-unit optimization problem[1]

$$\min_{\mathbf{X}} V_n(\mathbf{X}) = \sum_{i<j} \left(\frac{1}{d(\mathbf{x}_i, \mathbf{x}_j)^{12}} - \frac{1}{d(\mathbf{x}_i, \mathbf{x}_j)^6} \right), \tag{2.2}$$

whereby we set $d(\mathbf{x}_i, \mathbf{x}_j) = \dfrac{\hat{d}(\mathbf{x}_i, \mathbf{x}_j)}{\sigma}$ and $V_n = \dfrac{\hat{V}_n}{4\varepsilon}$ to eliminate the parameters in (2.1). The coordinates of the atoms $\mathbf{x}_i = (x_i, y_i, z_i)$, $i = 1, \ldots, n$ are unknown variables; and we fix atom 1 at $(0, 0, 0)$, atom 2 at $(x_2, 0, 0)$, and atom 3 at $(x_3, y_3, 0)$. In addition to eliminating the need to postprocess a collection of distances in order to visualize the system, using coordinates in this manner offers some mathematical advantages. First, for $n \geq 3$ the function V_n has $3n - 6$ degrees of freedom (DOF), while the pairwise distance formulation has $O(n^2)$ DOF. Second, using coordinates eliminates the need for both configuration feasibility and non-negative constraints on the variables, resulting in an unconstrained optimization problem. Third, fixing the coordinates of atoms 1–3 in the described manner ensures distinguishable minimizers with reduced quantity. However, the reduced LJ potential function V_n still has $O(\exp(n^2))$ unique minimizers [7], and even enumerating the quantity of local minima of a cluster has been shown to be an NP-hard problem [21]. Furthermore, each unique minimizer has at least $O(n^3)$ equivalent permutations (for example, by interchange of atom numbering and rotation).

Figure 1 illustrates the functions V_2 and V_3, respectively. We can see that minimizing even such small systems can be challenging. First, when any pairwise atom distance approaches zero, there exists a singularity in V_n. Second, when any pairwise distance $d(\mathbf{x}_i, \mathbf{x}_j)$ is large, the norm of the corresponding gradient

---

[1] Simplifying the units in this manner does not alter the topology of the critical points.

components are (numerically) very small. That is, there can be an infinite number of (nearly) critical points when $\|\mathbf{x}_i\| \to \infty$ for some i. As such, a careful choice of initial guess is necessary to ensure that the optimization algorithm converges to the desirable minimizer. Many researchers (e.g., [17,23]) leverage geometric principles in their numerical methods to generate initial iterates.

3   Method and Small System Validation

In this section, we survey existing theorems that can guarantee the local convergence of problem (2.2). The application of these theorems allows us to determine a maximal amount that a local minimizer can be perturbed under a lossy compression technique, while preserving its basin of attraction. That is, the given optimization algorithm applied to the compressed data point converges to the same optimal point. To quantify the amount of compression in terms of the acceptable perturbation, we use binary digit rounding as a simple compression technique.

For the optimization algorithm applied to recover the minimizer, we use Newton's method to solve the first-order optimality conditions of (2.2), since convergence can be guaranteed under certain conditions when the iterates are *near* a root. Among the most well-known convergence results for Newton's method, the Kantorovich Theorem (KT) [9] continues to draw attention by many researchers (e.g., [2,13,24]). The crux of the KT result guarantees that a solution $\mathbf{X}^*$ of $U_n(\mathbf{X}) = V_n'(\mathbf{X}) = 0$ exists in a neighborhood of a point $\mathbf{X}^{(0)}$ under some assumptions, and that Newton's method converges to $\mathbf{X}^*$ when $\mathbf{X}^{(0)}$ is the initial iterate. Note that KT does not require knowledge of the optimal point, only information based on the initial iterate, a workflow opposite to our compression framework where $\mathbf{X}^*$ is given. Instead, we exploit a variant of KT called the *ball of convergence about* $\mathbf{X}^*$ that assumes its prior knowledge.

First, we define $B(\mathbf{X}^*, r)$ as an open ball in a Banach space $\mathcal{B}$ with center $\mathbf{X}^*$ and radius $r > 0$, that is, $B(\mathbf{X}^*, r) = \{\mathbf{X} \in \mathcal{B} : \|\mathbf{X} - \mathbf{X}^*\| < r\}$. In our case, for any U_n, $n \geq 3$, we have $\mathcal{B} = \mathbb{R}^{3n-6}$, and we choose the Euclidean norm on $\mathcal{B}$, abbreviating $\|\mathbf{X}\| := \|\mathbf{X}\|_2$. We use the same convention for the compatible matrix operator norm, namely, $\|U_n'(\mathbf{X})\| := \|U_n'(\mathbf{X})\|_2 = \max_{\mathbf{Y} \in \mathcal{B}} \dfrac{\|U_n'(\mathbf{X})\mathbf{Y}\|_2}{\|\mathbf{Y}\|_2}$.

Theorem 1 *(Rheinboldt [16]).* For $U_n : N \subseteq \mathbb{R}^m \to \mathbb{R}^m$, *suppose* $U_n(\mathbf{X}^*) = 0$ *and that* $(U_n'(\mathbf{X}^*))^{-1}$ *exists. Furthermore, for some* $r_1 > 0$, *there exists a constant* $L_* > 0$ *such that*

$$\|(U_n'(\mathbf{X}^*))^{-1}(U_n'(\mathbf{X}) - U_n'(\mathbf{Y}))\| \leq L_* \|\mathbf{X} - \mathbf{Y}\| \tag{3.3}$$

for all $\mathbf{X}, \mathbf{Y} \in B(\mathbf{X}^*, r_1)$. *Then, if* $r_1 < \frac{2}{3L_*}$, *Newton's method converges uniquely to* $\mathbf{X}^*$ *for any* $\mathbf{X}^{(0)} \in B(\mathbf{X}^*, r_1)$.

Alternatively, we have the following.

Theorem 2 *(Wang [20]).* *For* $U_n : N \subseteq \mathbb{R}^m \to \mathbb{R}^m$, *suppose* $U_n(\mathbf{X}^*) = 0$ *and that* $(U_n'(\mathbf{X}^*))^{-1}$ *exists. Furthermore, for some* $r_2 > 0$, *there exists a constant* $L_{**} > 0$ *such that*

$$\|(U_n'(\mathbf{X}^*))^{-1}U_n'(\mathbf{X}) - I\| \le L_{**}\|\mathbf{X} - \mathbf{X}^*\| \tag{3.4}$$

for all $\mathbf{X} \in B(\mathbf{X}^*, r_2)$. *Then, if* $r_2 \le \frac{2}{L_{**}}$, *Newton's method converges uniquely to* $\mathbf{X}^*$ *for any* $\mathbf{X}^{(0)} \in B(\mathbf{X}^*, r_2)$.

A ball $B(\mathbf{X}^*, r)$ satisfying either Theorems 1 or 2 is a *ball of convergence* for $\mathbf{X}^*$, and any compressed minimizer that lies in $B(\mathbf{X}^*, r)$ *preserves the basin of attraction* of $\mathbf{X}^*$ under Newton's method.

Notice that the given theorems require that $U_n(\mathbf{X}^*)$ be exactly zero. For computer/numerical calculations this assumption is almost never true. To adapt this condition, one can perform a comprehensive convergence test for terminating the iterates to ensure that the sequences $\{U_n(\mathbf{X}^{(k)})\}$ and $\{\mathbf{X}^{(k)}\}$ are both converging. This numerical issue is negligible when accurate minimizers are computed. All computed $\mathbf{X}^*$ for our numerical tests are accurate to at least 10 decimal places.

While different lossy compression techniques can be implemented depending on their intended use, our work focuses on providing a theoretical bound on the maximum amount of perturbation that any lossy compression scheme can afford, in order to preserve a basin of attraction. We use significant binary digit truncation/rounding as the underlying compression technique since it is computationally cheap to implement and easy to quantify. Furthermore, the resulting relative perturbations are small and easy to control, since rounding a real number to c significant binary digits (bits) produces a maximum relative perturbation of $\frac{2^{-(c+1)}}{1 + 2^{-(c+1)}}$. For example, rounding to 4 significant bits has a maximum relative perturbation of 3.03%, and to 8 bits with more than 0.0195 %, a potentially significant reduction in memory from the standard 52 significant bits of IEEE double precision. We use $\mathbf{X}^{(0),c}$ to denote $\mathbf{X}^*$ compressed to c significant bits, which is used as the initial guess for Newton's method.

To understand the usage of the stated theory and how it may be applied at large scales, we offer a brief analysis of the 2-atom (1 free variable) LJ potential function V_2. We apply Newton's method to the first-order conditions of (2.2) by solving $0 = U_2(\mathbf{X}) = U_2(x_2) = -12x_2^{-13} + 6x_2^{-7}$, for $x_2^* = 2^{1/6}$. To apply Theorems 1 and 2, we must choose radii r_1 and r_2 small enough to meet their corresponding Lipschitz inequality. This step may not be straightforward, however, because both L_* and L_{**} depend on r. In this one-dimensional case, we can explicitly solve for these radii because we know that the maximum values of $|U_2'|$ and $|U_2''|$ are attained at $2^{1/6} - r$ for any $0 < r < 2^{1/6}$. We have the following:

$$\left| \frac{-2184(2^{1/6} - r_1)^{-15} + 336(2^{1/6} - r_1)^{-9}}{156(2^{1/6})^{-14} - 42(2^{1/6})^{-8}} \right| < \frac{2}{3r_1} \implies r_1 < 0.02432,$$

$$\left| \frac{156(2^{1/6} - r_2)^{-14} - 42(2^{1/6} - r_2)^{-8}}{r_2(156(2^{1/6})^{-14} - 42(2^{1/6})^{-8})} - \frac{1}{r_2} \right| \le \frac{2}{r_2} \implies r_2 \le 0.06276.$$

Therefore, given the radii needed for different truncations $r_c = |x_2^{(0),c} - x_2^*|$, either the Rheinboldt or the Wang criterion can guarantee that Newton's method converges to $x_2^* = 2^{1/6}$ for compression as far as 4 bits ($r_c = 0.0025$). Note that Newton's method succeeds beyond the theoretical guarantee for all compressed $x_2^{(0),c}$ down to 1 bit, but it does fail for $c = 0$, where $x_2^{(0),c} = 2.0$ and the Newton iterates coincide with large x_2 of nearly critical configurations.

In this case, choosing Theorems 1 or 2 does not make a difference in the amount of binary digit compression. However, the ball of convergence provided by Theorem 2 can be three times more generous, given the different prefactors in the radius conditions. More specifically, if the derivative U_n' is linear, for any configurations $\mathbf{X}, \mathbf{Y}$, we have

$$\frac{\|U_n'(\mathbf{X}) - U_n'(\mathbf{X}^*)\|}{\|\mathbf{X} - \mathbf{X}^*\|} = \frac{\|U_n'(\mathbf{X}) - U_n'(\mathbf{Y})\|}{\|\mathbf{X} - \mathbf{Y}\|}.$$

Provided that $U_n'(\mathbf{X})$ is approximately linear near $\mathbf{X}^*$, then

$$\sup_{\mathbf{X} \in B(\mathbf{X}^*, r)} \frac{\|(U_n'(\mathbf{X}^*))^{-1} U_n'(\mathbf{X}) - I\|}{\|\mathbf{X} - \mathbf{X}^*\|} \approx \sup_{\mathbf{X}, \mathbf{Y} \in B(\mathbf{X}^*, r)} \frac{\|(U_n'(\mathbf{X}^*))^{-1} (U_n'(\mathbf{X}) - U_n'(\mathbf{Y}))\|}{\|\mathbf{X} - \mathbf{Y}\|}.$$

Thus $r_2 \approx 3r_1$, suggesting that Wang's theorem can offer a ball of convergence up to 3 times the size of Rheinboldt's. Furthermore, the univariate calculation of L_{**} in (3.4) is more easily *tractable* than the bivariate calculation required in (3.3), in the sense that $\mathbf{Y} = \underset{\mathbf{X} \in B(\mathbf{X}^*, r)}{\arg\sup} \left\{ \dfrac{\|(U_n'(\mathbf{X}^*))^{-1} U_n'(\mathbf{X}) - I\|}{\|\mathbf{X} - \mathbf{X}^*\|} \right\}$ can be described as a single configuration(s), leading to better estimation of $L_{**}(n, r)$. For these reasons, we focus on applying Wang's theorem through the rest of the paper.

Given the goal to provide a radius r_2 guaranteed to satisfy Theorem 2, so that Newton's method converges to $\mathbf{X}^*$ for any initial guess $\mathbf{X}^{(0),c} \in B(\mathbf{X}^*, r_2)$, we propose Algorithm 1 as the first contribution.

Algorithm 1. Derivation of allowable radius for compression of $\mathbf{X}^*$

1: Given a local minimizer $\mathbf{X}^*$ and radius $r > 0$ of perturbation in lossy compression, calculate

$$L_{**}(n, r) = \sup_{\mathbf{X} \in B(\mathbf{X}^*, r)} \frac{\|(U_n'(\mathbf{X}^*))^{-1} U_n'(\mathbf{X}) - I\|}{\|\mathbf{X} - \mathbf{X}^*\|}. \tag{3.5}$$

2: Define the radius of the convergence ball $r_2 = \min \left\{ r, \dfrac{2}{L_{**}(n, r)} \right\}$.

3: *(optional)* If either $r \ll \dfrac{2}{L_{**}(n, r)}$ or $\dfrac{2}{L_{**}(n, r)} \ll r$, choose $\hat{r}$ between r and $\dfrac{2}{L_{**}(r)}$, and return to step 1 using $\hat{r}$.

4: Any compressed value $\mathbf{X}^{(0),c} \in B(\mathbf{X}^*, r_2)$ converges to $\mathbf{X}^*$ using Newton's method.

Figure 2 illustrates the results of applying Algorithm 1 to U_6 for various radii to find an acceptable convergence ball radius. The curve for $2/L_{**}$ is due to the approximation described in the forthcoming sections. Note that an acceptable radius can always be derived after one guess of r; but by following step 3 and refining to $\hat{r}$ based on the results of step 1, continual improvement can be realized until the user decides to stop.

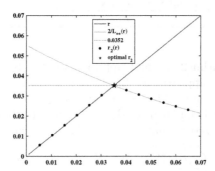

Fig. 2. Behavior of Algorithm 1 for U_6. For any r, the result is the minimum of r and $2/L_{**}(r)$, which for incremental r is shown by the black markers. The maximal r based on L_{**}^{est} is 0.0352.

4 Approximation of L_{**}

The 2-atom case allows us to explicitly solve for the radius of convergence because we know the location of $\mathrm{argsup}\{L_{**}(2,r)\}$. As the system becomes large $(n \geq 3)$, however, finding an analytic solution for this information is exceedingly difficult. The critical step is to find $L_{**}(n,r)$ for a given r satisfying (3.4). For any r less than the distance to the nearest critical configuration, we assume that the supremum for (3.5) is located at the boundary of the ball.[2] Then, problem (3.5) is equivalent to finding the configuration $\mathbf{X}$ which maximizes

$$L_{**}(n,r) = \max_{\mathbf{X} \in \partial B(\mathbf{X}^*,r)} \left\{ \frac{\|(U_n'(\mathbf{X}^*))^{-1} U_n'(\mathbf{X}) - I\|}{r} \right\}, \tag{4.6}$$

where ∂B denotes the boundary of the ball B.

It is straightforward that certain entries of $U_n'(\mathbf{X})$ approach infinity if $\mathbf{X}$ approaches a *singular configuration* $\mathbf{X}^{(s)}$, where two or more atoms coincide and V_n returns infinite energy. Therefore, any singular configuration maximizes $\|U_n'(\mathbf{X}) - U_n'(\mathbf{X}^*)\|$. Furthermore, the shortest route to perturb a configuration to be singular is to move the two closest atoms toward each other. These observations inspire us to propose an efficient approach to approximate L_{**} in Algorithm 2, which is denoted as L_{**}^{est}.

---

[2] We observe this assumption to be true for all U_n through numerical experimentation.

Algorithm 2. Approximation of L_{**} via the shortest path

1: Choose a radius of perturbation r.

2: Choose the closest pair of atoms in the $\mathbf{X}^*$ configuration (with respect to the perturbation required to move the pair to singularity).

3: Find $\mathbf{X}^{(p)}$ by perturbing $\mathbf{X}^*$ with only the chosen pair of atoms from step 2, toward each other along their shortest path, in the amount of r (when $n \geq 6$ the path is linear and atom movement is equal at distance of $r/2$).

4: Compute $L_{**}^{est}(n, r) = \dfrac{\|(U_n'(\mathbf{X}^*))^{-1} U_n'(\mathbf{X}^{(p)}) - I\|}{r}$.

To validate Algorithm 2, we first approximate L_{**} by random sampling, a practical approach to approximate Lipschitz-type constants [22]. Specifically, given a minimizer $\mathbf{X}^*$ and radius r, we generate $K = 10^6$ vectors $\{\mathbf{d}_k, k = 1, \ldots, K\}$, where each component of $\mathbf{d}_k$ is a uniformly chosen random number in $[-1, 1]$, and normalized so that $\|\mathbf{d}_k\| = 1$. Then we compute

$$
L_{**}^{rand}(n, r) = \max_{k=1\ldots K} \left\{ \frac{\|(U_n'(\mathbf{X}^*))^{-1} U_n'(\mathbf{X}_k) - I\|}{r} \quad : \quad \begin{matrix} \mathbf{X}_k = \mathbf{X}^* + r\mathbf{d}_k, \\ \|\mathbf{d}_k\| = 1 \end{matrix} \right\}. \quad (4.7)
$$

Figure 3 tracks the configurations $\mathbf{X}^{(r)} = \operatorname{argmax}\left\{L_{**}^{rand}(3, r)\right\}$ for increasing radii, ranging from 0.01 to 1.05, incremented by 0.02. Again, we enforce the semi-fixed positions for atoms 1–3 as described earlier. $\mathbf{X}^*$ is the equilateral triangle labeled by the stars. Each $\mathbf{X}^{(r)}$ configuration consists of a cyan marker (position of atom 3), a magenta marker (position of atom 2), and the fixed atom 1. As r increases, atom 3 of $\mathbf{X}^{(r)}$ moves along the cyan path and atom 2 of $\mathbf{X}^{(r)}$ moves along the magenta path, where for both atoms a darker shade corresponds to a configuration with larger r. Eventually, the two atoms meet at $(0.84, 0, 0)$ when $r = 1.05$, which coincides with the nearest singular configuration of $\mathbf{X}^*$. As hypothesized, the observed path of the $\mathbf{X}^{(r)}$ follows the shortest path as described in Algorithm 2.

The tendency for the configurations $\mathbf{X}^{(r)}$ to track along the shortest path continues for larger systems, illustrated in Fig. 4 for U_4 (left) and U_6 (right). In each plot, 40 configurations of $\mathbf{X}^{(r)} = \operatorname{argmax}\left\{L_{**}^{rand}(n, r)\right\}$ are displayed, corresponding to $0 \leq r \leq 0.8$. As before, any $\mathbf{X}^{(r)}$ configuration consists of an atom of each color, as well as atom 1, while the star markers indicate the equilibrium positions of the atoms, which in these cases form a regular tetrahedron and a regular octahedron, respectively. With U_4, the position of atom 2 remains near its $\mathbf{X}^*$ position in all $\mathbf{X}^{(r)}$, while atoms 3 and 4 perturb toward each other as r increases. For U_6, atoms 2 and 3 remain near their respective $\mathbf{X}^*$ coordinates in all $\mathbf{X}^{(r)}$, while either the pair of atoms 4 and 5 or the pair of atoms 5 and 6 perturb toward each other as r increases. Specifically, for any r, the $\mathbf{X}^{(r)}$ configuration has atoms 5 and 6 perturb toward each other at an approximate distance of $r/2$ each, while atoms 1 to 4 remain nearly stationary, or, atoms 4 and 5 move together about $r/2$ and the remaining atoms perturb very little. This alternating behavior occurs as the nearest singular configuration to $\mathbf{X}^*$ is

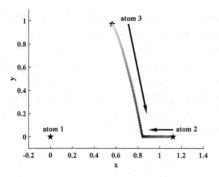

Fig. 3. For $0.01 \leq r \leq 1.05$, 53 configurations of $\mathbf{X}^{(r)}$ which maximize $L_{**}^{rand}(3, r)$. Each $\mathbf{X}^{(r)}$ configuration consists of one cyan marker (atom 3), one magenta marker (atom 2), and atom 1. Darker markers indicate an $\mathbf{X}^{(r)}$ from a greater r, while the star markers are the position of the atoms in the optimal configuration. As r increases, the $\mathbf{X}^{(r)}$ follow the shortest path toward the nearest singular configuration of $\mathbf{X}^*$.

not unique, since both the 4–5 and 5–6 pair singularities are equidistant from $\mathbf{X}^*$. Note that once the system is large enough ($n \geq 6$), the path followed by $\mathbf{X}^{(r)}$ toward the nearest singular configuration is linear, since the nearest pairs of atoms each have three free variables.

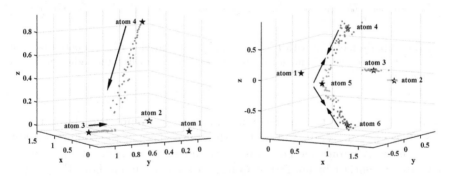

Fig. 4. For $n = 4$ (left) and $n = 6$ (right), 40 configurations of $\mathbf{X}^{(r)}$ which maximize $L_{**}^{rand}(n, r)$, for $0 \leq r \leq 0.8$. Atoms of the $\mathbf{X}^{(r)}$ are coordinated by color tone. The perturbed configurations follow the shortest path toward the nearest singular configuration $\mathbf{X}^{(s)}$, as indicated by the arrows. For U_6, there are two $\mathbf{X}^{(s)}$, so the $\mathbf{X}^{(r)}$ follows both paths, alternating by chance.

Now, we compare the two estimations of L_{**}, from the proposed Algorithm 2 and from random sampling due to (4.7). We focus on the cases $n \geq 6$ whose shortest paths are strictly linear. Taking the global minimum configurations from [5] for each $\mathbf{X}^*$, in Fig. 5, we compare the behavior of L_{**}^{rand} and L_{**}^{est} for

U_6, U_8, and U_{10}, separately in each column.[3] The top row shows for the cases when $r \leq 0.8$, which is the approximate distance to the nearest singularity for these $\mathbf{X}^*$. The bottom row zooms in to the cases when $r \leq 0.05$. The curve of $L = 2/r$ is drawn so that an estimate for the radius of convergence can be inferred from the plots by its intersection with the L_{**} curves. It is clear that $L_{**}^{est}(n, r) \geq L_{**}^{rand}(n, r)$ for all test cases, and therefore $L_{**}^{est}(n, r)$ is a more accurate estimate. For brevity, the cases $n = 7, 9$ are not plotted here, but consistent behavior is observed at all radii.

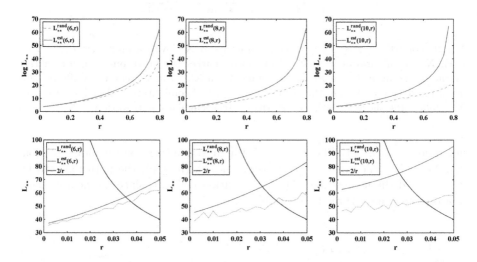

Fig. 5. Comparison of L_{**}^{est} with L_{**}^{rand} for U_6, U_8, U_{10}, showing L_{**}^{est} is a better estimate for L_{**}. The curve of $2/r$ is included since $L_{**} \leq 2/r$ determines the radius of the ball of convergence.

Next, we compute the corresponding binary digit compression levels for systems of size $n = 6, \ldots, 10$, and we test the convergence of the compressed minimizers using Newton's method. Table 1 catalogs these results. Its first column shows the system size by number of atoms; the second column shows the returned radius of the ball of convergence (to the nearest 0.002) after applying the proposed method, and the third column shows the corresponding compression levels in terms of number of bits. The last column shows that, in practice, Newton's method still converges to the same $\mathbf{X}^*$ with even higher compression levels than the theoretical guarantee.

5   Numerical Results in Higher Dimensions

In this section, we apply the proposed method to larger atomic systems to show its scalability. For the optimal configuration chosen, we use the global minimums

---

[3] Similar results hold for other local minimizers.

Table 1. For $n = 6, \ldots, 10$, the radii of convergence, maximal compression ability based on L_{**}^{est}, and convergence results for the compressed minimizers $\mathbf{X}^{(0),c}$ using Newton's method. The theoretical c_t obtained by our method provides a guarantee for compression so that $c \geq c_t$ implies that $\mathbf{X}^{(0),c}$ converges to $\mathbf{X}^*$ with Newton's method.

n	max $r : r \leq 2/L_{**}^{est}$	theoretical c_t (in bits): $\|\mathbf{X}^{(0),c} - \mathbf{X}^*\| \leq r$	practical c_p (in bits): $\mathbf{X}^{(0),c} \to \mathbf{X}^*$
6	0.034	5	2:24
7	0.034	6	2:24
8	0.030	6	4:24
9	0.026	7	4:24
10	0.026	6	5:24

tabulated in [5] and first published in [7,11]. Because of the partially fixed positions of atoms 1,2, and 3, we observe that some permutations of a minimizer $\mathbf{X}^*$ (in terms of interchanging atom numbering) can cause the spectral radius of $(U_n'(\mathbf{X}^*))^{-1}$ to be multiple orders of magnitude larger than others, resulting in significantly larger $L_{**}(r)$ than other permutations. This observation is particularly true when atom 3 is nearly on the x-axis. Therefore, we choose permutations of these minimizers so that the matrix $(U_n'(\mathbf{X}^*))^{-1}$ has lower spectral radius.

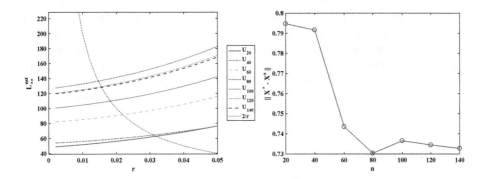

Fig. 6. Left: $L_{**}^{est}(r)$ values for the U_{20} to U_{140} functions. The curve of $2/r$ is included because $L_{**} \leq 2/r$ determines the radius of the ball of convergence. Right: distances to nearest singularity $\|\mathbf{X}^* - \mathbf{X}^{(s)}\|$ for each $\mathbf{X}^*$. An increase in $L_{**}(n,r)$ corresponds to a decrease in $\|\mathbf{X}^* - \mathbf{X}^{(s)}\|$.

For systems with $n = 20, 40, 60, 80, 100, 120$, and 140 atoms, Fig. 6 (left) shows the shortest-path estimates $L_{**}^{est}(n,r)$, respectively. For a fixed radius r, the corresponding $L_{**}^{est}(n,r)$ are increasing in n for $n \leq 80$, but the values plateau for $80 \leq n \leq 140$. These changes in $L_{**}^{est}(n,r)$ correspond to a decrease and leveling off of the distances to nearest singularity $\|\mathbf{X}^* - \mathbf{X}^{(s)}\|$ for each $\mathbf{X}^*$, which are plotted in Fig. 6 (right). It follows from Hoare [6] that the smallest inter-atom distances in any n-atom structure cannot fall far below the equilibrium

Table 2. Compression ability based on L_{**}^{est} and corresponding convergence results of Newton's method for U_{20} through U_{140} functions.

n	max $r : r \leq 2/L_{**}^{est}$	$L_{**}^{est}(r)$	$L_{**}^{rand}(r)$	theoretical c_t (in bits): $\|\mathbf{X}^{(0),c} - \mathbf{X}^*\| \leq r$	practical c_p (in bits): $\mathbf{X}^{(0),c} \to \mathbf{X}^*$
20	0.032	62.89	30.22	6	5:24
40	0.030	63.96	25.28	7	5:24
60	0.020	90.39	25.08	8	5:24
80	0.014	136.0	30.01	10	5:24
100	0.018	109.7	23.42	9	5:24
120	0.014	128.1	23.87	10	6:24
140	0.014	127.0	23.18	10	6:24

distance of $2^{1/6}$, so that these observed distances to singularity have approached a lower bound. Indeed, for the 1000 atom minimizer recorded in [23], the minimal distance to singularity is 0.733. Further, we have $L_{**}^{est}(1000, 0.014) = 135.5$, a corresponding maximum compression level $c_t = 13$, and Newton's method succeeds for all $\mathbf{X}^{(0),c}$ for $c \geq 9$. As both this minimal distance to singularity and L_{**}^{est} are consistent with the 80–140 atom cases, this is an indication that our proposed method scales to much larger atom systems.

Now, we further emphasize the accuracy of L_{**}^{est} by comparing it to the random sampling approximation from (4.7), again with $K = 10^6$ guesses. The comparison is done at the radius that we deduced is optimal using Algorithm 1 with L_{**}^{est}. As shown in Table 2, the shortest-path estimates are significantly higher than the random sampling estimates in all test cases. The corresponding compression levels deduced from L_{**}^{est} are reported in the fifth column. We apply Newton's method on $\mathbf{X}^{(0),c}$ from all compression levels, where the last column shows that Newton's method converges to the same $\mathbf{X}^*$ for compression levels exceeding the theoretical guarantee.

6   Conclusions

In this work, we propose a framework to guide the amount of compression that can be applied to an optimal atom configuration with the Lennard-Jones potential function, so that the compressed value converges to the same optimal configuration under Newton's method. Furthermore, we develop a shortest-path estimate by exploring the topology of the Lennard-Jones function to provide the necessary information required by the framework. We show that the shortest-path estimate outperforms the estimate from random sampling. We also demonstrate the reliability and stability of the proposed framework for systems of varying size, so that the provided compression level guarantees the preservation of the basins of attraction.

Our proposed framework also can be applied in a wider range of applications. One potential application is the compression of the large neural nets in deep learning. As the trained neural net is moved from a resource-rich setting to different model architectures with restrictive computation constraints such as memory and speed, the large-scale model with billions of trained weights and hyperparameters needs to be compressed. In this case, one can study the compression accuracy given the proposed framework so that certain accuracy can be preserved from the perspective of optimization.

Some gaps between the theory and numerical approximation need to be studied that are beyond the scope of this work. For example, a rigorous proof for the optimality of the proposed shortest-path estimate is desirable, to offer a truly definitive convergence guarantee. If such a result is unavailable, an error bound could be developed that measures the gap between the true L_{**} and estimate L_{**}^{est}, which is guaranteed to be only a lower bound for L_{**}. A similar error analysis should investigate the effect of using approximations for the inverse Hessian $(U'(\mathbf{X}^*))^{-1}$ and operator norm $\| \cdot \|_2$ in the computation of L_{**}^{est}, as in practice either are likely to be precisely computed. Furthermore, one can extend the proposed framework to local convergence theory for quasi-Newton methods (e.g., [4]) to preserve the corresponding basins of attraction.

Acknowledgments. We thank Florian Potra and Stefan Wild for their discussions and insights on this work. This work was supported by the Exascale Computing Project (17-SC-20-SC), a collaborative effort of two U.S. Department of Energy organizations (Office of Science and the National Nuclear Security Administration) responsible for the planning and preparation of a capable exascale ecosystem, including software, applications, hardware, advanced system engineering, and early testbed platforms, in support of the nation's exascale computing imperative. The material was also based in part on work supported by the U.S. Department of Energy, Office of Science, under contract DE-AC02-06CH11357.

References

1. Alder, B.J., Wainwright, T.E.: Studies in molecular dynamics, I: general method. J. Chem. Phys. **31**(2), 459–466 (1959)
2. Argyros, I.K., George, S.: Ball convergence comparison between three iterative methods in banach space under hypothese only on the first derivative. Appl. Math. Comput. **266**, 1031–1037 (2015)
3. Asenjo, D., Stevenson, J.D., Wales, D.J., Frenkel, D.: Visualizing basins of attraction for different minimization algorithms. J. Phys. Chem. B **117**(42), 12717–12723 (2013)
4. Dembo, R.S., Eisenstat, S.C., Steihaug, T.: Inexact Newton methods. SIAM J. Numer. Anal. **19**(2), 400–408 (1982)
5. Doye, J.: Table of Lennard-Jones cluster global minima. http://www-wales.ch.cam. ac.uk/~jon/structures/LJ/tables.150.html
6. Hoare, M.R.: Structure and dynamics of simple microclusters, pp. 49–135. Wiley (2007)
7. Hoare, M., Pal, P.: Physical cluster mechanics: statics and energy surfaces for monatomic systems. Adv. Phys. **20**(84), 161–196 (1971)
8. Jones, J.E.: On the determination of molecular fields, II: from the equation of state of a gas. Proc. R. Soc. Lond. A **106**(738), 463–477 (1924)

9. Kantorovich, L.V.: Functional analysis and applied mathematics. Uspekhi Mat. Nauk **3**(6), 89–185 (1948)
10. Maranas, C.D., Floudas, C.A.: A global optimization approach for Lennard-Jones microclusters. J. Chem. Phys. **97**(10), 7667–7678 (1992)
11. Northby, J.: Structure and binding of Lennard-Jones clusters: $13 \leq N \leq 147$. J. Chem. Phys. **87**(10), 6166–6177 (1987)
12. Perez, D., Cubuk, E.D., Waterland, A., Kaxiras, E., Voter, A.F.: Long-time dynamics through parallel trajectory splicing. J. Chem. Theory Comput. **12**(1), 18–28 (2015)
13. Potra, F.: A superquadratic variant of Newton's method. SIAM J. Numer. Anal. **55**(6), 2863–2884 (2017)
14. Rahman, A.: Correlations in the motion of atoms in liquid argon. Phys. Rev. **136**(2A), A405 (1964)
15. Sørensen, M.R., Voter, A.F.: Temperature-accelerated dynamics for simulation of infrequent events. J. Chem. Phys. **112**(21), 9599–9606 (2000)
16. Traub, J.F., Woźniakowski, H.: Convergence and complexity of Newton iteration for operator equations. Technical report, Carnegie Mellon University, Department of Computer Science, Pittsburgh, PA (1977)
17. Uppenbrink, J., Wales, D.J.: Packing schemes for Lennard-Jones clusters of 13 to 150 atoms: minima, transition states and rearrangement mechanisms. J. Chem. Soc. Faraday Trans. **87**, 215–222 (1991)
18. Voter, A.F.: Hyperdynamics: accelerated molecular dynamics of infrequent events. Phys. Rev. Lett. **78**(20), 3908 (1997)
19. Voter, A.F.: Parallel replica method for dynamics of infrequent events. Phys. Rev. B **57**(22), R13985 (1998)
20. Wang, X.: Convergence of Newton's method and uniqueness of the solution of equations in Banach space. IMA J. Numer. Anal. **20**(1), 123–134 (2000)
21. Wille, L.T., Vennik, J.: Computational complexity of the ground-state determination of atomic clusters. J. Phys. A: Math. Gen. **18**(8), L419 (1985)
22. Wood, G., Zhang, B.: Estimation of the Lipschitz constant of a function. J. Glob. Optim. **8**(1), 91–103 (1996)
23. Xiang, Y., Cheng, L., Cai, W., Shao, X.: Structural distribution of Lennard-Jones clusters containing 562 to 1000 atoms. J. Phys. Chem. A **108**(44), 9516–9520 (2004)
24. Yamamoto, T.: Historical developments in convergence analysis for Newton's and Newton-like methods. In: Numerical Analysis: Historical Developments in the 20th Century, pp. 241–263. Elsevier (2001)

An Algorithm for Hydraulic Tomography Based on a Mixture Model

Carlos Minutti[1]([✉]), Walter A. Illman[2], and Susana Gomez[1]

[1] Institute of Research in Applied Mathematics and Systems,
National University of Mexico, Mexico City, Mexico
`carlos.minutti@iimas.unam.mx`
[2] Department of Earth and Environmental Sciences, University of Waterloo,
Waterloo, ON, Canada

Abstract. Hydraulic Tomography (HT) has become one of the most robust methods to characterize the heterogeneity in hydraulic parameters such as hydraulic conductivity and specific storage. However, in order to obtain high resolution hydraulic parameter estimates, several pumping/injection tests with sufficient monitoring data are necessary. In highly heterogeneous media, even with large numbers of measurements, the resolution may not be enough for predicting contaminant transport behavior. In addition, during inverse modeling, the groundwater flow equation is solved numerous times, thus the computational burden could be large, especially for a large, three-dimensional, transient model.

In this work we present a new approach to model aquifer heterogeneity, based on a Gaussian Mixture Model (GMM) to parameterize the K field, which significantly reduces the number of parameters to be estimated during the inversion process. In addition, a new objective function based on the spatial derivatives of hydraulic heads is introduced.

The developed approach is tested with synthetic data and data from a previously conducted sandbox experiments. Results indicate that the new approach improves the accuracy of the K heterogeneity map produced through HT and reduces the computational effort. For two dimensional synthetic experiments, this approach was able to achieve a significant reduction in the error for K field estimation as well as computational time compared to a geostatistical inversion approach. Similar results were also achieved when the approach was tested using pumping test data conducted in a synthetic aquifer constructed in the laboratory.

Keywords: Hydraulic tomography · Gaussian mixture ·
Aquifer heterogeneity

1 Introduction

During a hydraulic tomography experiment, water is sequentially pumped from or injected into an aquifer at different intervals of the aquifer. During each pumping/injection event, hydraulic head responses of the aquifer at different intervals

© Springer Nature Switzerland AG 2019
J. M. F. Rodrigues et al. (Eds.): ICCS 2019, LNCS 11538, pp. 471–486, 2019.
https://doi.org/10.1007/978-3-030-22744-9_37

are monitored, yielding a set of head/discharge (or recharge) data. By sequentially pumping/injecting water at one interval and monitoring the steady state head responses at others, many head/discharge (recharge) data sets are obtained (Yeh and Liu [11]).

These head values can be compared with results given by the groundwater flow equation (the forward model) that describes the head response taking into account the hydraulic conductivity (K) values and the pumping/injection rate. The goal of hydraulic tomography (HT) is to find the K values throughout the aquifer that minimizes the difference between observed and simulated head values and this process is also known as the inverse problem.

In order to estimate the K values, many approaches have been developed. For example, Hoeksema and Kitanidis [4] used cokriging as an approach to model the K heterogeneity of the aquifer, but this approach relies on the availability of enough head and K measurements. Also, because cokriging is a linear estimator, its application and accuracy is limited due to the non-linear nature of the problem.

Gottlieb and Dietrich [3] minimized the quadratic error between the measured and the simulated head values using Tikhonov regularization to penalize how much the solution deviates from a geological *a priori* guess. The K value of each element in a finite element methods is optimized. The smoothness and accuracy of the solution can be affected by the amount of available data and how good the prior guess is. Also, the optimization of each K value is computationally expensive.

Yeh *et al.* [10] developed an iterative stochastic estimation approach called the Successive Linear Estimator (SLE) where the cokriged log-K field is updated at each iteration with new covariances and cross covariance estimates, in order to try to capture the non-linear relationship between the head values and K. Yeh and Liu [11] then extended the SLE approach by sequentially incorporating data sets from multiple pumping/injection tests in the inverse model, which resulted in a Sequential Successive Linear Estimator (SSLE).

Illman *et al.* [5] reported that the order in which data is included into the SSLE algorithm has a significant effect in the K estimates. Later, based on SSLE, Xiang *et al.* [9] presented the Simultaneous Successive Linear Estimator (SimSLE) which uses all data sets simultaneously in the inversion algorithm.

A number of studies on HT have been published which demonstrate the robust performance of SSLE and SimSLE. In particular, Illman *et al.* [6,7], and Berg and Illman [1] validated steady state hydraulic tomography (SSHT) based on SSLE with data collected from laboratory sandbox experiments. Xiang *et al.* [9] used SimSLE to reproduce the K field of synthetic data as well as data from an actual sandbox experiment. Zhao *et al.* [14] improved the results of Illman *et al.* [6,7], and Berg and Illman [1] by including geological data as a prior distribution of K in SimSLE.

Berg and Illman [2] compared different methods of modeling aquifer heterogeneity at a highly heterogeneous field site through data collected at the North Campus Research Site (NCRS) on the University of Waterloo campus, and showed that transient hydraulic tomography (THT) analysis with SSLE performed the best. Subsequently, Zhao and Illman [12,13] performed SSHT and THT analyses of pumping tests at the NCRS and improved the results of Berg and Illman [2] by including additional pumping tests and geological data as the initial distribution of K in the inversion process. They compared these results with the effective parameter and geological zonation approaches, showing that HT with SimSLE performs the best in terms of model calibration and validation. The inclusion of geological data was useful in producing less smooth hydraulic parameter distributions and also showed how the initial distribution of K produces different solutions of the K field, suggesting the necessity of taking into account the problem of local optimal solutions.

Regardless of the good results in HT that have been achieved with geostatistical inversion approaches such as SSLE and SimSLE, computational time could be an issue in the case of large, three-dimensional, transient models. In addition, geostatistical inverse approaches can depend significantly on the prior distribution of K, and the algorithms may suffer from slow convergence, especially for highly heterogeneous aquifers, because the hydraulic parameters are continuously updated to capture the non-linear dependence of hydraulic parameters and head.

In order to model the spatial distribution of K, in this work we propose a new algorithm, the GM inversion algorithm, to solve the inverse problem for HT applications. This algorithm is based on the statistical concept of a mixture model, specifically a Gaussian Mixture Model (GMM), which will be used to parameterize the K field.

In the proposed inversion algorithm, we introduce a new objective function, based on the spatial derivatives of head values. This function allows the inversion algorithm to increase the accuracy of K estimates and it makes use of a regularization term to address the ill-posed problem to estimate more parameters than available observations.

The results of the new inversion algorithm are compared with the results of SSLE using synthetic experiments as well as a laboratory sandbox aquifer. With this new approach, we were able to achieve good estimates of the K field, reducing the computational time as well as the amount of data needed to perform HT, even if the spatial distribution of K is complex or very heterogeneous.

2 The GM Inversion Algorithm

2.1 Parameterization of the Conductivity Field

The parameterization of the K field refers to the process in which a set of values, i.e. the parameters, models the K field of the aquifer, with each K field represented with a different set of parameter values.

Although the value of these parameters are not known a priori, it is possible to estimate them by comparing the different values of pressure (head) generated using a forward model (the groundwater flow equation), against the observed data measured during the pumping/injection tests, *i.e.*, through the minimization of the error defined in an objective function.

The usual and the simplest selection for the parameterization of the K field is $G(\boldsymbol{\theta}) = \boldsymbol{K}$, *i.e.*, the parameterization is exactly $\boldsymbol{K}$, consisting of a vector of local K estimates. However, to optimize each K value for a K field can be computationally expensive, and possibly an ill-posed problem, thus we propose to use a parameterization function G, based on a Gaussian Mixture Model.

Gaussian Mixture Model. In statistics, a mixture model is a probabilistic model used to represent different sub-populations within a larger population, through the combination of many probability density functions where each component of the function could be interpreted as a "probability" or the likelihood of the existence of each sub-population in each spatial coordinate.

In this case, the main population would be the aquifer material in the porous media that has a larger volume or the one that is more spread out and the sub-populations would be any other materials found among the main population.

Although a mixture describes a probability distribution, its value can be rescaled to describe different attributes, as a density or K values. In this approach, it describes the logarithm of K (log-K).

If Gaussian distributions are used for the mixture model (GMM), this is known as the Gaussian Mixture Model, which is represented by:

$$G(\boldsymbol{\theta}) = G(X, \mu_i, \Sigma_i) = \sum_{i=1}^{N_k} \omega_i N(X, \mu_i, \Sigma_i) \tag{1}$$

where ω_i is a weight for the i-th Gaussian distribution with mean μ_i and covariance matrix Σ_i. N_k is the number of components in the mixture, and X is the vector of spatial coordinates.

If the K field has a complex spatial distribution K that does not resemble Gaussian kernels, it can be approximated by increasing the number of components in the mixture: as higher the number of components, the higher the complexity of the K field that can be modeled.

In order to model the spatial distribution of K with a GMM, we propose to approximate the true K field, $\boldsymbol{K}$, in the GM algorithm by:

$$\log(\boldsymbol{K}) \approx G(\boldsymbol{\theta}^*) - \log(\bar{K}) \tag{2}$$

where $\bar{K}$ is the effective or mean K value of the aquifer and $G(\boldsymbol{\theta})$ is the GMM that minimizes the proposed objective function.

If Eq. (2) is used to approximate the spatial distribution of a two-dimensional K field, a set of bi-variate Gaussian kernels are needed to generate the GMM. For the GMM, the parameter N_k is the most important, as it is the number of components/materials in the mixture and the higher its value the greater the level of complexity that the GMM can model. Under certain assumptions, each component in the mixture could be considered as a layer of the material.

The rest of parameters in the GMM are estimated by the optimization of an objective function, and these parameters control the spatial correlation of the heterogeneity in the estimated K field. In particular, Σ_i, the covariance matrix, contains the variance in the spatial (horizontal and vertical) directions (σ_x^2, σ_z^2), and the correlation between them (ρ_{xz}). In the case of layered materials, one can expect to have a larger horizontal variance (this constraint can also be included in the optimization process), because the layer is typically longer in the horizontal direction rather than the vertical. μ_i is the spatial location or *centroid* for each component in the horizontal and vertical directions, while ω_i is the weight of each component in the mixture. This value is directly related with the K value of each component/material.

2.2 Objective Function

Even though the proposed approach is closely related to cluster analysis and statistical classification, the main difference is that the K field to be approximated with a GMM is not known *a priori*, so it is necessary to estimate it, by using hydraulic head as a proxy variable of K.

Typically the error to be minimized in the inverse problem of HT would be:

$$F_{LS}(\boldsymbol{\theta}) = \sum_{i=1}^{n}(h_i(\boldsymbol{\theta}) - h_i^*)^2 \tag{3}$$

where h_i^* are the observed head values for all the pumping/injection tests and $h_i(\boldsymbol{\theta}) = H_i(G(\boldsymbol{\theta}))$ are the head values simulated using the forward model, where $G(\boldsymbol{\theta})$ is the estimated log-K field.

In addition, here, we include the spatial derivatives of head values to the objective function:

$$F_d(\boldsymbol{\theta}) = \phi_f F_{LS}(\boldsymbol{\theta}) + \phi_x \sum_{i=1}^{n}\left(\frac{d}{dx}h_i(\boldsymbol{\theta}) - \frac{d}{dx}h_i^*\right)^2 + \phi_z \sum_{i=1}^{n}\left(\frac{d}{dz}h_i(\boldsymbol{\theta}) - \frac{d}{dz}h_i^*\right)^2 \tag{4}$$

where ϕ_i are weights such that $\sum_i \phi_i = 1$, and x, z are the horizontal and vertical spatial directions.

The spatial derivatives are especially useful, because they are more sensitive to the locations of the Gaussian components, being easier for the optimization algorithm to place each component depending on the head change between two observed head values.

A downside for the use of the derivatives is that the noise level in the observed head values could be increased, but this problem is addressed by choosing a small value of N_k, or by using a noise reduction method such as splines or the one proposed in Minutti *et al.* [8], which is designed to address noisy data in inverse problem when the derivative is required.

Regularization. A measure of smoothness of the K field, is added as a regularization term in the objective function. This term can lead to a faster optimization process avoiding to "visit" many local optima that only fit parts of the calibration data. The smoothness of the solution is measured by:

$$S(\boldsymbol{\theta}) = \sum_{i=1}^{m} \left(\frac{d}{dx} G(\boldsymbol{\theta}) \right)_i^2 + \sum_{i=1}^{m} \left(\frac{d}{dy} G(\boldsymbol{\theta}) \right)_i^2 \tag{5}$$

where m is the number of K values used for the solution of the forward model.

Finally, the objective function of the optimization problem can be written as:

$$F(\boldsymbol{\theta}) = (1 - \lambda)F_d(\boldsymbol{\theta}) + \lambda S(\boldsymbol{\theta}) \tag{6}$$

where λ ($0 \leq \lambda < 1$), the regularization parameter is especially useful if there are few data measurements and/or there is noisy data.

Even though λ and N_k can be used as smoothing parameters, these two values represent different types of smoothness. High values of λ enforces K fields to have a low variance, *i.e.* more homogeneous aquifers, while small N_k values tend to group different layers of materials with similar K, but it allows high variance between groups. In other words, λ can be used to describe intralayer heterogeneity and N_k for interlayer heterogeneity.

3   Experiments

In order to test the capabilities of the GM algorithm, data from a sandbox and some synthetic experiments are used. These experiments intend to measure (1) the effect of the amount of data used; (2) the effect of noisy data; (3) the speed and accuracy in the estimation of the K field and (4) the performance when the spatial distribution of the K fields are Non-Gaussian. We utilize VSAFT (Xiang et al. [9]), a previously utilized geostatistical inversion model, to solve the same problems for the purpose of comparing the results obtained through the GM algorithm.

For all experiments we use a two-dimensional heterogeneous aquifer that is based on the structure of the sandbox experiment presented by Illman et al. [7] (Fig. 1). This sandbox consists of 18 layers of sandy material, that measures 193.0 cm in length, 82.6 cm in height, and is 10.2 cm thick.

In order to simulate a synthetic scenario of this sandbox experiment, the aquifer was discretized into 741 elements (a grid of 19 by 39) with element dimensions of 4.1 cm by 4.1 cm. In terms of boundary conditions, the top, left and right boundaries of the aquifer are set as constant head boundaries, while the bottom is treated to be a no-flow boundary.

For each element the K values originates from the measured values presented in Table 1 of Berg and Illman [1], where the log-mean K value of its corresponding layer, is used. For the cases of an element that corresponds to more than one layer of material, the log-mean K value of those layers was used.

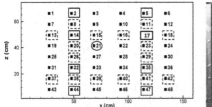 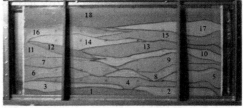

Fig. 1. Schematic diagram of the synthetic heterogeneous aquifer and Photograph of the synthetic heterogeneous aquifer from Illman *et al.* [7].

The above process results in a K field with a variance of $\sigma^2_{\log(K)} = 0.1361$ (low-heterogeneity case). A high-heterogeneity case was simulated by re-scaling each K value (keeping the same mean log-K value), resulting in a variance of $\sigma^2_{\log(K)} = 0.8508$.

For the forward simulation and inverse modeling, a saturated and isotropic steady state forward model is used. Synthetic data are generated by simulating, nine pumping tests on the computer, with 48 observed head values for each one (8 rows and 6 columns), using a constant pumping rate of $Q = 1.25$ cm^3/s. The data from the first 4 pumping tests are used as input for inverse modeling (this data is called the HT data set) and the remaining data from tests 5 to 9 (validation data set) were used for validation of the estimated field.

Finally, for the sandbox experiment, eight pumping tests were used as input to the inversion algorithms and the results are compared with the ones reported by Illman *et al.* [5].

4   Results

For each experiment, the Mean Squared Error (MSE) is calculated for the HT data set $(\mathrm{E}(H_{HT}))$ and for the validation data set $(\mathrm{E}(H_V))$. For the synthetic experiments, the MSE of the log-K is also estimated $(\mathrm{E}(K))$ as well as a Relative MSE measure $(\mathrm{E}_\sigma(K))$, which takes into account the variance of the log-K values, being $\mathrm{E}_\sigma(K) = \mathrm{E}(K)/\sigma^2_{\log(K)}$. This allows one to compare the errors between the low and the high heterogeneity cases.

4.1   Synthetic Experiments

Low-Heterogeneity Case: Table 1 presents the results of the estimated K fields for the case $\sigma^2_{\log(K)} = 0.1361$, with VSAFT2 and GM as inversion methods, using head values from 1 to 4 pumping tests as input data.

If only one pumping test is used, it can be seen that the error measurements for K and for the head values in the HT data set, are very similar. However the K field estimated with GM, has a higher variance, more similar to the true variance, and a smaller error of the head values in the validation data set.

Table 1. CPU-Time and different error measurements for the HT problem, in a synthetic aquifer with a low-heterogeneity K field ($\sigma^2_{\log(K)} = 0.1361$).

N_{tests}	Method	$\hat{\sigma}^2_{\log(K)}$	E(H_{HT})	E(H_V)	E(K)	E$_\sigma$(K)	CPU-Time
1	VSAFT	0.0254	0.0018	0.0471	0.1012	0.7436	0.62
	GM	0.0632	0.0012	0.0178	0.0926	0.6804	0.17
2	VSAFT	0.0201	0.0058	0.0534	0.0950	0.6980	1.06
	GM	0.0799	0.0015	0.0085	0.0359	0.2638	0.23
3	VSAFT	0.0305	0.0032	0.0263	0.0711	0.5224	1.84
	GM	0.0997	0.0012	0.0018	0.0156	0.1146	0.25
4	VSAFT	0.0456	0.0017	0.0094	0.0481	0.3534	2.84
	GM	0.1165	0.0012	0.0011	0.0096	0.0705	0.53

N_{tests} = Number of pumping tests used, Method = Inversion method, $\hat{\sigma}^2_K$ = Variance of the estimated K field, E(H_{HT}) = MSE of the Head values used in the inversion process, E(H_V) = MSE of the Head values used for validation purposes, E(K) = MSE of the log-K values, E$_\sigma$(K) = E(K)/$\sigma^2_{\log(K)}$, CPU-Time = Minutes used to estimate the K field.

As the number of pumping tests is increased, the errors of K and of the head values, tend to be more significant between both methods. When 4 pumping tests are used, GM only has one fifth of the error of K and uses one fifth for the CPU-Time, in contrast with VSAFT.

Figure 2 presents the K tomograms generated with each inversion method, the scatter plot of the observed vs estimated K values and the drawdowns when 4 pumping test were used. Although both methods detect the main features of heterogeneity in K, the level of resolution obtained with GM is higher than the one obtained with VSAFT, suggesting that the K tomogram generated with GM better approximates the true K field. It can be seen that GM reproduces better the low and high K areas.

High-Heterogeneity Case: Table 2 presents the results of the estimation of the K field when $\sigma^2_{\log(K)} = 0.8508$ with VSAFT and GM as inversion methods, using 1 to 4 pumping tests as input data.

Similar to the low-heterogeneity case, when only one pumping test was used for inverse modeling, the MSE of both inversion approaches are similar, but when the number of pumping tests increases, the differences between the two methods become more significant.

Examination of Table 2 reveals that when 4 pumping tests are used for inverse modeling, the GM algorithm has less than one fifth of the error of K than VSAFT, using less than one third of the computational time. The K tomograms for this case are presented in Fig. 3.

Mean Performance: Using the results of Tables 1 and 2, we compare the performance of both inversion algorithms for the low and high heterogeneity

Table 2. CPU-Time and error measurements for the HT problem in a synthetic aquifer with a high-heterogeneity K field ($\sigma^2_{\log(K)} = 0.8508$).

N_{tests}	Method	$\hat{\sigma}^2_{\log(K)}$	E(H_{HT})	E(H_V)	E(K)	E$_\sigma$(K)	CPU-Time
1	VSAFT	0.1053	0.0243	0.1621	0.6848	0.8049	0.62
	GM	0.5253	0.0047	0.1853	0.5138	0.6039	0.37
2	VSAFT	0.0827	0.0758	0.2145	0.6350	0.7464	1.06
	GM	0.3857	0.0479	0.2544	0.3452	0.4057	0.28
3	VSAFT	0.1358	0.0391	0.0975	0.5331	0.6266	1.80
	GM	0.6654	0.0170	0.0155	0.1096	0.1288	0.87
4	VSAFT	0.2043	0.0215	0.0414	0.4079	0.4794	2.88
	GM	0.6647	0.0080	0.0023	0.0702	0.0825	0.82

N_{tests} = Number of pumping tests used, Method = Inversion method, $\hat{\sigma}^2_K$ = Variance of the estimated K field, E(H_{HT}) = MSE of the Head values used in the inversion process, E(H_V) = MSE of the Head values used for validation purposes, E(K) = MSE of the log-K values, E$_\sigma$(K) = E(K)/$\sigma^2_{\log(K)}$, CPU-Time = Minutes used to estimate the K field.

cases, where E$_\sigma$(K) is the main measure for the error of K. Figure 4 presents the average E$_\sigma$(K) value and the average CPU-Time by the number of pumping tests used for each inversion algorithm.

Based on this synthetic study, the results for the inversion of 4 pumping tests reveal that GM has on average one fifth the estimation error of VSAFT, using one fourth of the computation time. It is also important to note that the error achieved with GM using 2 pumping tests, is lower than the one with VSAFT using 4 pumping tests, *i.e.*, GM was able to achieve a similar error than VSAFT using only 2 pumping tests instead of 4 tests, with less than one tenth of the computational time.

4.2 Sandbox Case

Table 3 presents the results of the estimation of the K field for the sandbox experiment conducted by Illman *et al.* [7] using VSAFT and GM as inversion methods.

It can be seen that with 8 pumping tests the amount of CPU-Time required with GM is almost one fifth of the time used with VSAFT and the variance of the estimated K field is similar in both cases, as well as the error of the head values.

Even though it is not possible to calculate the estimation error of the K field, there are some indications of how good the estimation is. For example, the mean value of log-K is reported in Table 3 as log(K_G), and in both cases, the estimations are similar to the value of -2.56, reported in Illman *et al.* [7] using 48 core samples. Although the variances of log-K are slightly different to the value estimated using core samples ($\sigma^2_{\log(K)} = 0.868$), the variance obtained with the GM algorithm is closer to that value.

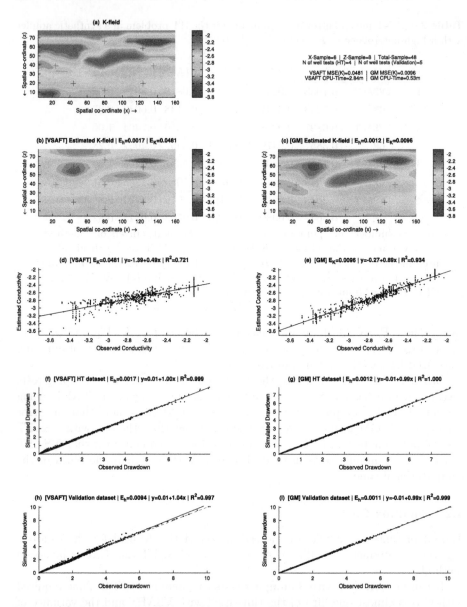

Fig. 2. (a) Synthetic K field of **low-heterogeneity** and the log-K tomograms obtained with (b) VSAFT and (c) GM as inversion methods, **using 4 injection tests** with 48 head values per test. The blue crosses mark the location of the injection wells and the red crosses the location of the injection tests used for validation (h)-(i). The dashed line is the 45-degree line, representing a perfect match, the solid line is a linear model fit with the slope, intercept, and the coefficient of determination (R^2) provided.

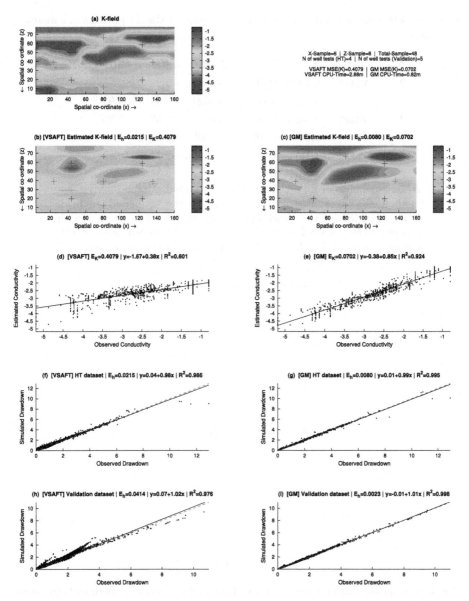

Fig. 3. (a) Synthetic K field of **high-heterogeneity** and the log-K tomograms obtained with (b) VSAFT and (c) GM as inversion methods, **using 4 injection tests** with 48 head values per test. The blue crosses mark the location of the injection wells and the red crosses the location of the injection tests used for validation (h)-(i). The dashed line is the 45-degree line, representing a perfect match, the solid line is a linear model fit with the slope, intercept, and the coefficient of determination (R^2) provided.

Table 3. CPU-Time and different error measurements for the HT problem in a sandbox experiment.

N_{tests}	Method	$\hat{\sigma}^2_{\log(K)}$	$E(H_{HT})$	$E(H_V)$	$\log(K_G)$	CPU-Time
2	VSAFT	0.7053	0.0102	0.0552	−2.28	2.58
	GM	0.7498	0.0088	0.0521	−2.34	0.92
4	VSAFT	0.8424	0.0066	0.0720	−2.32	5.70
	GM	0.7611	0.0105	0.0509	−2.28	1.18
8	VSAFT	1.1272	0.0081	0.0464	−2.31	10.83
	GM	1.0370	0.0054	0.0417	−2.35	2.40

N_{tests} = Number of different pumping tests used, Method = Inversion method that was used, $\hat{\sigma}^2_K$ = Variance of the estimated K field, $E(H_{HT})$ = MSE of the Head values that were used in the inversion process, $E(H_V)$ = MSE of the Head values that were left apart to validation purposes, $\log(K_G)$ is the mean value of the log-K field and CPU-Time = Minutes used to estimate the K field.

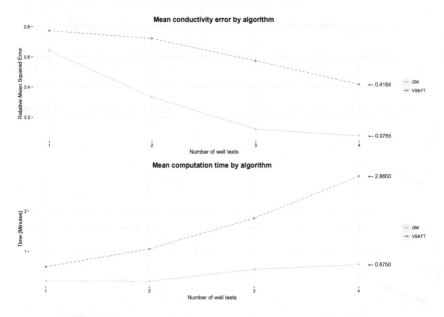

Fig. 4. Average estimation error and CPU-Time of the synthetic experiments by inversion algorithm

In addition, Fig. 5 shows the K tomograms generated with each inversion method and a photograph of the synthetic aquifer is overlain. The high K areas are indicated in red, where it is possible to see that each of these areas correspond to a different layer of sand in the sandbox. An estimated value of K of each sand type is presented in Illman *et al.* [7] (Table 1) with the four highest K sand types include (from highest to lower average K), the 16/30, 20/30, 20/40 and

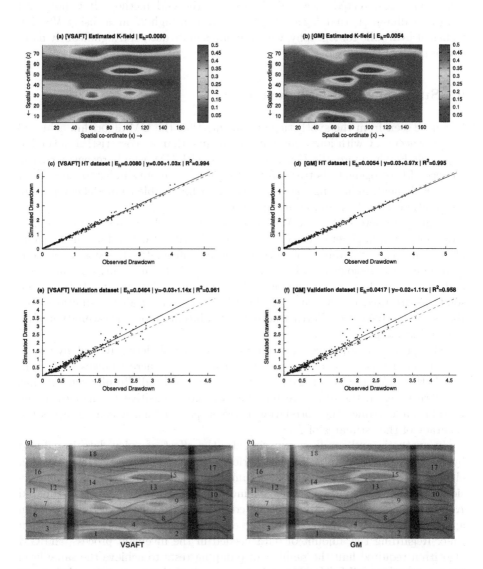

Fig. 5. K tomograms of the sandbox experiment obtained with (a) VSAFT and (b) GM as inversion methods, using 8 injection tests with 47 head values per test. Overlaid picture of the tomograms and the sandbox photograph (g)–(h). The dashed line is the 45-degree line, representing a perfect match.

#12 sands. Examination of Fig. 5 reveals that the 16/30 sand is the only type that is not shown in red in the K tomograms. This could be due to the fact this is the sand type with the second smallest volume in the sandbox.

VSAFT detected as high K areas, all layers with sand types 20/30, 20/40 and #12 with the only exception of layer 13, which is a sand 20/30. Nevertheless, this layer was detected as a high K area by the GM method. It is not until Berg and Illman [1] that layer 13 is detected as a high K area, using VSAFT transient hydraulic tomography, which requires more data and significantly more computational time.

5 Summary and Conclusions

It can be seen from the synthetic experiments that the GM algorithm achieved a higher accuracy with lower computational time than a geostatistical inversion approach. Furthermore, the K fields estimated with GM usually reconstructs the pattern of heterogeneity better, with values of σ_{logK}^2 more similar to variance of the true K field, indicating that the GM algorithm is able to model complex K distributions with less computational effort.

Some parameters of the GMM describes the interlayer and intralayer heterogeneity and, in general, the GM inversion algorithm, depending on the value of N_k (and the availability of prior geological information), keeps a relationship between a geological and a geostatistical model. The multistart approach address the problem of multiple local/global optimal solutions, in addition these solutions are used to estimate a K field by consensus of all the solutions, the Bayes conductivity field. Furthermore, these solutions allow the estimation of an uncertainty map.

The proposed objective function uses the spatial derivatives of hydraulic head to detect more precisely, the spatial locations where changes in K lead to a better reproduction of the observed changes in head. Results suggest that the derivatives in the objective function are more sensitive to changes in the K values and reduce the correlation between parameters, which increases the accuracy of the estimates of K.

In synthetic and sandbox experiments, the effect of noisy data is reduced by changing the N_k parameter. The performance gap between the two inversion algorithms increases as the number of pumping test also increases. The GM inversion algorithm has shown a significant increase in the accuracy of the estimated K field using less computational time, compared with a geostatistical inversion approach, having up to one fifth of the error in one fourth of the computation time. Regarding the amount of data needed in the inversion process, the GM algorithm required half the number of pumping tests to achieve the same level of accuracy than SSLE/VSAFT, using one tenth of the computational time.

Due to the nature of the GM algorithm to group areas of similar K value, if the aquifer have very spread out areas of different materials, the algorithm could need a large number of components in order to capture the aquifer heterogeneity, resulting in a large number of parameters to be estimated. In addition, a mixture model which includes different probability distributions (*e.g.* Gamma or Beta distributions), could increase the performance of the algorithm, making it able to reproduce even more complex spatial distributions of K, with a lower number parameters.

These results are promising but further testing of the GM algorithm is required. In particular, more extensive testing is necessary on non-Gaussian K fields, and under three-dimensional and transient flow conditions.

References

1. Berg, S.J., Illman, W.A.: Capturing aquifer heterogeneity: comparison of approaches through controlled sandbox experiments. Water Resour. Res. **47**(9) (2011). https://doi.org/10.1029/2011WR010429
2. Berg, S.J., Illman, W.A.: Comparison of hydraulic tomography with traditional methods at a highly heterogeneous site. Groundwater **53**(1), 71–89 (2015). https://doi.org/10.1111/gwat.12159
3. Gottlieb, J., Dietrich, P.: Identification of the permeability distribution in soil by hydraulic tomography. Inverse Probl. **11**(2), 353 (1995). http://stacks.iop.org/0266-5611/11/i=2/a=005
4. Hoeksema, R.J., Kitanidis, P.K.: An application of the geostatistical approachto the inverse problem in two-dimensional groundwater modeling. Water Resour. Res. **20**(7), 1003–1020 (1984). https://doi.org/10.1029/WR020i007p01003
5. Illman, W.A., Craig, A.J., Liu, X.: Practical issues in imaging hydraulic conductivity through hydraulic tomography. Groundwater **46**(1), 120–132 (2008). https://doi.org/10.1111/j.1745-6584.2007.00374.x
6. Illman, W.A., Liu, X., Craig, A.: Steady-state hydraulic tomography in a laboratory aquifer with deterministic heterogeneity: multi-method and multiscale validation of hydraulic conductivity tomograms. J. Hydrol. **341**(3), 222–234 (2007). https://doi.org/10.1016/j.jhydrol.2007.05.011.   http://www.sciencedirect.com/science/article/pii/S0022169407002818
7. Illman, W.A., Zhu, J., Craig, A.J., Yin, D.: Comparison of aquifer characterization approaches through steady state groundwater model validation: a controlled laboratory sandbox study. Water Resour. Res. **46**(4) (2010). https://doi.org/10.1029/2009WR007745
8. Minutti, C., Gomez, S., Ramos, G.: A machine-learning approach for noise reduction in parameter estimation inverse problems, applied to characterization of oil reservoirs. J. Phys.: Conf. Ser. **1047**, 012010 (2018). https://doi.org/10.1088/1742-6596/1047/1/012010
9. Xiang, J., Yeh, T.C.J., Lee, C.H., Hsu, K.C., Wen, J.C.: A simultaneous successive linear estimator and a guide for hydraulic tomography analysis. Water Resour. Res. **45**(2) (2009). https://doi.org/10.1029/2008WR007180
10. Yeh, T.C.J., Jin, M., Hanna, S.: An iterative stochastic inverse method: conditional effective transmissivity and hydraulic head fields. Water Resour. Res. **32**(1), 85–92 (1996). https://doi.org/10.1029/95WR02869
11. Yeh, T.C.J., Liu, S.: Hydraulic tomography: development of a new aquifer test method. Water Resour. Res. **36**(8), 2095–2105 (2000). https://doi.org/10.1029/2000WR900114

12. Zhao, Z., Illman, W.A.: On the importance of geological data for three-dimensional steady-state hydraulic tomography analysis at a highly heterogeneous aquifer-aquitard system. J. Hydrol. **544**, 640–657 (2017). https://doi.org/10.1016/j.jhydrol.2016.12.004.   http://www.sciencedirect.com/science/science/article/pii/S002216941630796X

13. Zhao, Z., Illman, W.A.: Three-dimensional imaging of aquifer and aquitard heterogeneity via transient hydraulic tomography at a highly heterogeneous field site. J. Hydrol. **559**, 392–410 (2018). https://doi.org/10.1016/j.jhydrol.2018.02.024. http://www.sciencedirect.com/science/article/pii/S0022169418301008

14. Zhao, Z., Illman, W.A., Berg, S.J.: On the importance of geological data for hydraulic tomography analysis: Laboratory sandbox study. J. Hydrol. **542**, 156–171 (2016). https://doi.org/10.1016/j.jhydrol.2016.08.061. http://www.sciencedirect.com/science/article/pii/S0022169416305510

Rapid Multi-band Patch Antenna Yield Estimation Using Polynomial Chaos-Kriging

Xiaosong Du[1], Leifur Leifsson[1(✉)], and Slawomir Koziel[2]

[1] Iowa State University, Ames, IA 50011, USA
leifur@iastate.edu
[2] Reykjavik University, Menntavegur 1, 101 Reykjavik, Iceland

Abstract. Yield estimation of antenna systems is important to check their robustness with respect to the uncertain sources. Since the Monte Carlo sampling-based real physics simulation model evaluations are computationally intensive, this work proposes the polynomial chaos-Kriging (PC-Kriging) metamodeling technique for fast yield estimation. PC-Kriging integrates the polynomial chaos expansion (PCE) as the trend function of Kriging metamodel since the PCE is good at capturing the function tendency and Kriging is good at matching the observations at training points. The PC-Kriging is demonstrated with an analytical case and a multi-band patch antenna case and compared with direct PCE and Kriging metamodels. In the analytical case, PC-Kriging reduces the computational cost by around 42% compared with PCE and over 94% compared with Kriging. In the antenna case, PC-Kriging reduces the computational cost by over 60% compared with Kriging and over 90% compared with PCE. In both cases, the savings are obtained without compromising the accuracy.

Keywords: Yield estimation · Microstrip multi-band patch antenna · Monte Carlo sampling · PCE · Kriging · PC-Kriging

1 Introduction

Yield is the metric for checking the reliability of antenna system with respect to the uncertainties due to the manufacturing process [1, 2]. In particular, yield is the percentage of designs satisfying the design specifications. The process of yield estimation can be completed by running arbitrary number of high-fidelity simulation models [1], such as full-wave electromagnetic (EM) model [3], using Monte Carlo sampling (MCS) [4]. The high-fidelity physics model evaluations are typically time-consuming, making the MCS-based yield estimation computationally prohibitive.

Metamodeling techniques [5, 6] are widely used to alleviate the computational burden. There are generally two types of metamodels, data-fit metamodels [7] and multi-fidelity metamodels [8]. Data-fit metamodels utilize the high-fidelity physics-based simulation model evaluations as training points, while the multi-fidelity metamodels can make use of physics-based simulation models of varying degree of accuracy. Multi-fidelity metamodels can be efficient when fast and good low-fidelity models

© Springer Nature Switzerland AG 2019
J. M. F. Rodrigues et al. (Eds.): ICCS 2019, LNCS 11538, pp. 487–494, 2019.
https://doi.org/10.1007/978-3-030-22744-9_38

are available. Data-fit metamodeling is more versatile because only one level of simulation model is needed.

Advanced data-fit metamodels have been successfully used for antenna system modeling and design at reduced computational costs. Rama Sanjeeva Reddy et al. [9] introduced the radial basis function neural network into design of multiple function antenna arrays and obtained a success rate as high as 98%. Koziel et al. [10] constructed the fast data-fit Kriging metamodel as part of multi-objective design optimization of antennas handling arbitrary number of objective functions. Du et al. [11] introduced the PCE method for statistical metamodeling of the far field radiated by antennas undergoing random disturbances and validated the PCE model with a deformable canonical antenna.

This work introduces the PC-Kriging metamodel [12] for the yield estimation of multi-band patch antenna systems. PCE [13] is well-known for capturing the tendency of the objective function, whereas Kriging [14] handles the observation values at training points well. The PC-Kriging technique aims at integrating the advantages of both metamodeling methods expecting fewer training points required for constructing a reliable and fast model in lieu of the computationally expensive high-fidelity simulation model. This work demonstrates the PC-Kriging technique for the yield estimation of a multi-band patch antenna case.

The remainder part of this paper is organized as follows. Section 2 provides the details formulating the yield estimation of antenna. Section 3 describes the metamodeling methodologies, including Kriging, PCE and PC-Kriging, utilized in this work. Then all metamodeling techniques are demonstrated and compared on numerical examples in Sect. 4. This papers ends with conclusion in Sect. 5.

2   Yield Estimation of Antennas

Let the antenna response of interest, evaluated using EM simulation models, be denoted by $\mathbf{R}(\mathbf{x})$, and $\mathbf{x}$ is the vector containing deterministic/uncertain design parameters. Let $\mathbf{x}^0$ represent the nominal design under ideal conditions. Let $d\mathbf{x}$ be the disturbance due to the manufacturing tolerances or uncertainties existing in the antenna system, and can be sample using pre-define empirical probabilistic distributions. Therefore, the actual designs taking the tolerances and uncertainties under consideration can be represented as $\mathbf{x}^0 + d\mathbf{x}$.

Now a counting function $H(\mathbf{x})$ can be set up as [2]

$$H(\mathbf{x}) = \begin{cases} 1, & \text{if } \mathbf{R}(\mathbf{x}) \text{ satisfied the design specifications} \\ 0, & \text{otherwise} \end{cases} \tag{1}$$

Then the yield at the nominal design introduced above, i.e., the percentage of satifying designs out of the total designs, can be given as

$$Y(\mathbf{x}^0) = [\sum_{j=1}^{N} H(\mathbf{x}^0 + d\mathbf{x}^j)]/N, \tag{2}$$

where $dx^j, j = 1, 2, \ldots, N$, are the disturbances with pre-assigned empirical probabilistic distributions as introduced above.

3 Methods

This section describes the mathematical details of formulating the metamodeling methods, including Kriging, PCE and PC-Kriging. This work considers the response feature approach proposed by Koziel et al. [2], which can reduce the complexity of problem constructing metamodel for response of interest at specific frequencies rather than modeling the whole signal.

3.1 Kriging

Kriging metamodeling technique is a type of Gaussian process regression, which takes the training points as the realization of the unknown random process. The generalized Kriging formulation [14] is the sum of a trend function $\mathbf{f}^T(\mathbf{x})\boldsymbol{\beta}$ and a Gaussian deviation term $Z(\mathbf{x})$ as follows

$$M^{Kr}(\mathbf{x}) = \mathbf{f}^T(\mathbf{x})\boldsymbol{\beta} + Z(\mathbf{x}), \tag{3}$$

where $\mathbf{f}(\mathbf{x}) = \left[f_0(\mathbf{x}), \ldots, f_{p-1}(\mathbf{x}) \right]^T \in \mathbb{R}^p$ is defined with a set of the regression basis functions, $\boldsymbol{\beta} = \left[\beta_0(\mathbf{x}), \ldots, \beta_{p-1}(\mathbf{x}) \right]^T \in \mathbb{R}^p$ denotes the vector of the corresponding coefficients, and $Z(\mathbf{x})$ denotes a stationary random process with zero mean, variance and nonzero covariance. In this work, Gaussian exponential correlation function is adopted with the form

$$R[\mathbf{x}, \mathbf{x}'] = \sigma^2 \exp\left[-\sum_{k=1}^{m} \theta_k \left| x_k - x_k' \right|^2 \right], \tag{4}$$

where $\boldsymbol{\theta} = [\theta_1, \theta_2, \ldots, \theta_m]^T$ denotes the vectors of unknown hyperparameters to be tuned. The Kriging predictor for any untried $\mathbf{x}$ can be written as

$$M^{Kr}(\mathbf{x}) = \mathbf{f}^T(\mathbf{x})\widehat{\boldsymbol{\beta}} + \mathbf{r}^T(\mathbf{x})\mathbf{R}^{-1}(\mathbf{M}_S - \mathbf{F}\widehat{\boldsymbol{\beta}}), \tag{5}$$

where a linear trend function $\mathbf{f} = [1, x_1, x_2, \ldots, x_m]^T$ is used in this work, $F_{ij} = f_j(x_i)$ where $i = 1, 2, \ldots, N, j = 1, 2, \ldots, N+1$, N is the total number of training points, $\widehat{\boldsymbol{\beta}}$ comes from generalized least squares estimation, $\mathbf{r}$ is the correlation vector between the point to be predicted ($\mathbf{x}_{\text{pred}}$) and training set points, here $r_i = R(\mathbf{x}_{\text{pred}}, \mathbf{x}_i; \boldsymbol{\theta})$, $\mathbf{R}$ is the correlation matrix among training points with $R_{ik} = R(\mathbf{x}_i, \mathbf{x}_k; \boldsymbol{\theta})$ where $i, k = 1, 2, \ldots, N$, $\mathbf{M}_S$ is the model response of the training points. $\boldsymbol{\beta}$ and σ^2 are given by

$$\boldsymbol{\beta} = (\mathbf{F}^T\mathbf{R}^{-1}\mathbf{F})^{-1}\mathbf{G}^T\mathbf{R}^{-1}\mathbf{M}_S, \tag{6}$$

and

$$\sigma^2 = 1/N(\mathbf{M}_S - \mathbf{G}\boldsymbol{\beta})^T \mathbf{R}^{-1}(\mathbf{M}_S - \mathbf{G}\boldsymbol{\beta}). \tag{7}$$

The maximum likelihood estimation on $\boldsymbol{\theta}$ is found by solving

$$\boldsymbol{\theta} = \arg\min(\frac{1}{2}\log(\det(R)) + \frac{N}{2}\log(2\pi\sigma^2) + N/2). \tag{8}$$

3.2   Polynomial Chaos Expansion

PCE has the generalized formulation as follows [13]

$$M^{PC}(\mathbf{x}) = \sum_{i=1}^{\infty} \alpha_i \boldsymbol{\Psi}_i(\mathbf{x}), \tag{9}$$

where $\mathbf{x} \in \mathbb{R}^m$ is a vector with random independent components described by a probability density function $f_{\mathbf{X}}$, $M^{PC}(\mathbf{x})$ is a map of $\mathbf{x}$, i is the index of ith polynomial term, $\boldsymbol{\Psi}_i$ are multivariate polynomial basis functions, whereas α_i are their corresponding expansion coefficient. In practice, a truncated form of the PCE is used

$$M^{PC}(\mathbf{x}) = \sum_{i=1}^{P} \alpha_i \boldsymbol{\Psi}_i(\mathbf{x}), \tag{10}$$

where $M^{PC}(\mathbf{x})$ is the approximate truncated PCE model, and P is the total number of sample points, which can be calculated as

$$P = \frac{(p+n)!}{p!n!}, \tag{11}$$

where p is the order of the PCE, and n is the total number of random input variables.

The coefficient vector $\boldsymbol{\alpha}$ is found by solving a least-squares minimization problem

$$\widehat{\boldsymbol{\alpha}} = \arg\min E[\boldsymbol{\alpha}^T \boldsymbol{\Psi}(\mathbf{x}) - M(\mathbf{x})]. \tag{12}$$

In this work, the least-angle regression (LARS) method is used to solve (12) by adding an L_1 penalty term

$$\widehat{\boldsymbol{\alpha}} = \arg\min E[\boldsymbol{\alpha}^T \boldsymbol{\Psi}(\mathbf{x}) - M(\mathbf{x})] + \lambda\|\boldsymbol{\alpha}\|_1, \tag{13}$$

where λ is a penalty factor, $\|\boldsymbol{\alpha}\|_1$ is the L_1 norm of the coefficients of PCE.

3.3 Polynomial Chaos-Kriging

PC-Kriging [12] is a recently developed class of metamodels that integrates the PCE and Kriging metamodels. In particular, PCE is utilized as the trend function for the Kriging metamodel. The modeling flow is as follows:

1. Obtain observations (training points) from the physics-based simulation model.
2. Generate a PCE model following Sect. 3.2.
3. In Step 2, LARS technique selects the "important" basis terms, meaning those most correlated with the model response.
4. Plug those "important" basis terms into (5), then construct the Kriging model.

4 Numerical Examples

The proposed PC-Kriging metamodeling technique is demonstrated on two numerical examples in this section. The first example is the modeling of a short column function which was first utilized by Eldred et al. [15] for demonstrating uncertainty quantification. The second example is a multi-band patch antenna system which has normal distributions of zero mean and standard deviation of 0.08 mm modeling the disturbances [2].

4.1 Short Column Function

The short column function [15] models a structural column with uncertainties due to the material properties. The function is given as

$$f(\mathbf{x}) = 1 - \frac{4M}{bh^2 Y} - \frac{P^2}{b^2 h^2 Y^2}, \tag{14}$$

where b is the width of the cross section and equals 5 mm, h is the depth of the cross section and equals 15 mm, Y, M and P are the uncertain parameters in this case and $Y \sim$ Lognormal (5, 0.5) MPa is the yield stress, $M \sim$ Normal (2,000, 400) MNm is the bending moment, and $P \sim$ Normal (500, 100) MPa is the axial force.

In this case, we set up the 1% of standard deviation (σ) of the testing points as the accepted root mean squared error (RMSE). Figure 1 shows the plot of the RMSE of all three metamodeling techniques versus the number of training points. The plot shows that all metamodeling approaches can reduce the RMSE when increasing the total number of training points. The Kriging, PCE and PC-Kriging metamodels, however, need different number of samples to reach the 1% $\sigma_{testing}$ accuracy. In particular, Kriging needs around 1,200 training points and PCE around 120 training points, whereas PC-Kriging requires only around 70 training points. Thus, PC-Kriging needs around 42% fewer samples than PCE and around 94% fewer than Kriging. In this case, the PC-Kriging metamodel at each number of training points utilizes a 14th degree of the PCE as the trend function.

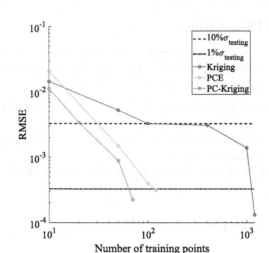

Fig. 1. Metamodeling accuracy versus the computational cost.

4.2 Multi-band Patch Antenna

The geometry of the microstrip dual-band patch antenna utilized in this work is given in Fig. 2. The antenna is implemented on a 0.762 mm thick Taconic RF-35 dielectric substrate ($\varepsilon_r = 3.5$). The independent geometry parameters are $\mathbf{x} = [L\, l_1\, l_2\, l_3\, l_4\, W\, w_1\, w_2\, g]^T$. The EM model $\mathbf{R}$ is implemented in CST [1, 2]. The nominal design, corresponding to the antenna resonances allocated at the frequencies 2.4 GHz and 5.8 GHz, is $\mathbf{x}^0 = [14.18\, 3.47\, 12.44\, 5.06\, 15.56\, 0.65\, 8.29\, 5.60]^T$ (all dimensions in mm).

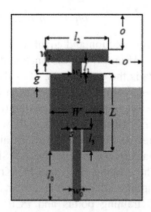

Fig. 2. Geometry of the dual-band patch antenna.

The antenna yield is estimated for the following specs: $|S_{11}| \leq -10\,\text{dB}$ for both 2.4 GHz and 5.8 GHz. It is assumed that Gaussian distribution of the geometry deviation vector $d\mathbf{x}$ has a zero mean and a standard variance of 0.08 mm. The

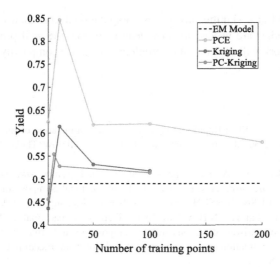

Fig. 3. Convergence of yield estimation as a function of the number of training points for the considered metamodeling techniques as well as direct EM-based Monte Carlo simulation.

parametric study on the convergence of the yield value versus the number of training points is shown in Fig. 3. The PCE, Kriging and PC-Kriging metamodeling approaches are compared with the direct Monte Carlo sampling technique involving 500 EM evaluations of **R**.

Table 1. Computational cost for satisfactory yield estimation of the multi-band patch antenna.

Geometry	Methodology	Yield estimation	Number of samples
Gaussian $\sigma = 0.08$ mm	EM Model	0.490	500
	PCE	0.580	200
	Kriging	0.532	50
	PC-Kriging	0.528	20

As shown in Table 1, to reach satisfactory yield estimations, the PCE and Kriging require around 200 and 50 training points, respectively. The proposed PC-Kriging requires only 20 training points. Thus, in this case the PC-Kriging needs over 90% fewer samples than PCE and more than 60% fewer samples than Kriging.

5 Conclusion

The PC-Kriging metamodeling technique has been proposed for rapid multi-band patch antenna yield estimation. PC-Kriging aims at combining the advantages of both PCE and Kriging metamodels for a further reduction on the computational cost. The results of multi-band patch antenna yield estimation show that PC-Kriging can be used to estimate the yield at a significantly lower computational cost than using Kriging or PCE. Further studies are needed to fully determine how the well proposed approach works. Future work will also consider problems of higher complexity.

References

1. Koziel, S., Bandler, J.W.: Rapid yield estimation and optimization of microwave structures exploiting feature-based statistical analysis. IEEE Trans. Microw. Theory Tech. **63**(1), 107–114 (2015)
2. Koziel, S., Bekasiewicz, A.: Rapid statistical analysis and tolerance aware design of antennas by response feature surrogates. In: 2017 IEEE International Symposium on Antennas and Propagation and USNC/URSI National Radio Science Meeting, pp. 2199–2200 (2017)
3. Nordin, M.S.A., Rahman, N.H.A., Ali, M.T., Sharatol Ahmad Shal, A.A., Ahmad, M.R.: Performance comparison of various textile composition and structure through full-wave electromagnetic simulation and measurement. J. Telecommun. Electron. Comput. Eng. **10**(1-6), 55–58 (2018)
4. Datta, T., Kumar, A.N., Chockalingam, A., Rajan, S.B.: A novel monte-carlo-sampling-based receiver for large-scale uplink multiuser MIMO systems. IEEE Trans. Veh. Technol. **62**(7), 3019–3038 (2013)
5. Koziel, S., Leifsson, L.: Surrogate-Based Modeling and Optimization, Applications in Engineering. Springer, New York (2013). https://doi.org/10.1007/978-1-4614-7551-4
6. Koziel, S., Leifsson, L., Yang, X.: Simulation-Driven Modeling and Optimization. Springer, New York (2014). https://doi.org/10.1007/978-1-4614-7551-4
7. Ulaganathan, S., Koziel, S., Bekasiewicz, A., Couckuyt, I., Laermans, E., Dhaene, T.: Data-driven model based design and analysis of antenna structures. IET Microw. Antennas Propag. **10**(13), 1428–1434 (2016)
8. Liu, B., Koziel, S., Zhang, Q.: A multi-fidelity surrogate-model-assisted evolutionary algorithm for computationally expensive optimization problems. J. Comput. Sci. **12**, 28–37 (2016)
9. Rama Sanjeeva Reddy, B., Vakula, D., Sarma, N.V.S.N.: Design of multiple function antenna array using radial basis function neural network. J. Microw. Optoelectron. Electromagn. Appl. **12**(1), 210–216 (2013)
10. Koziel, S., Bekasiewicz, A., Szczepanski, S.: Multi-objective design optimization of antennas for reflection, size, and gain variability using kriging surrogates and generalized domain segmentation. Int. J. RF Microw. Comput.-Aided Eng. **28**(5), 1–11 (2018)
11. Du, J., Roblin, C.: Statistical modeling of disturbed antennas based on the polynomial chaos expansion. IEEE Antennas Wirel. Propag. Lett. **16**, 1843–1846 (2017)
12. Schobi, R., Sudret, B., Wiart, J.: Polynomial-Chaos-based kriging. Int. J. Uncertain. Quantif. **5**, 171–193 (2015)
13. Blatman, G.: Adaptive sparse polynomial chaos expansions for uncertainty propagation and sensitivity analysis. Ph.D. thesis, Blaise Pascal University (2009)
14. Sacks, J., Welch, W.J., Mitchell, T.J., Wynn, H.P.: Design and analysis of computer experiments. Stat. Sci. **4**(4), 409–423 (1989)
15. Eldred, M.S., Agarwal, H., Perez, V.M., Wojtkiewicz Jr., S.F., Renaud, J.E.: Investigation of reliability method formulations in DAKOTA/UQ. Struct. Infrastruct. Eng. **3**(3), 199–213 (2007)

Accelerating Limited-Memory Quasi-Newton Convergence for Large-Scale Optimization

Alp Dener[✉] and Todd Munson

Argonne National Laboratory, Argonne, IL 60439, USA
adener@anl.gov

Abstract. Quasi-Newton methods are popular gradient-based optimization methods that can achieve rapid convergence using only first-order derivatives. However, the choice of the initial Hessian matrix upon which quasi-Newton updates are applied is an important factor that can significantly affect the performance of the method. This fact is especially true for limited-memory variants, which are widely used for large-scale problems where only a small number of updates are applied in order to minimize the memory footprint. In this paper, we introduce both a scalar and a sparse diagonal Hessian initialization framework, and we investigate its effect on the restricted Broyden-class of quasi-Newton methods. Our implementation in PETSc/TAO allows us to switch between different Broyden class methods and Hessian initializations at runtime, enabling us to quickly perform parameter studies and identify the best choices. The results indicate that a sparse Hessian initialization based on the diagonalization of the BFGS formula significantly improves the base BFGS methods and that other parameter combinations in the Broyden class may offer competitive performance.

1 Introduction

Quasi-Newton methods are a variation of Newton's method where the Jacobian or the Hessian is approximated using the secant condition. Since their inception in the late-1950s by Davidon [10,11] and Fletcher and Powell [19], quasi-Newton methods have been widely used in solving nonlinear systems of equations, especially in optimization applications. For a comprehensive review of these methods, see Dennis and Moré [15] and Nocedal and Wright [32].

Our interest in quasi-Newton methods is motivated by the computational cost and difficulty in calculating exact Hessians for large-scale or partial differential equation (PDE)-constrained optimization problems. In particular, for reduced-space methods for PDE-constrained problems where the PDE constraint is eliminated via the implicit function theorem, constructing exact second-order information at each iteration requires as many adjoint solutions as the number of optimization variables [33]. Computing Hessian-vector products without

The U.S. government retains certain licensing rights. This is a U.S. government work and certain licensing rights apply 2019

J. M. F. Rodrigues et al. (Eds.): ICCS 2019, LNCS 11538, pp. 495–507, 2019.
https://doi.org/10.1007/978-3-030-22744-9_39

computing the Hessian itself can be computationally cheaper [3,24,28], but the matrix-free nature of this approach poses additional difficulties in preconditioning the systems [13,14].

Limited-memory quasi-Newton methods circumvent these issues by directly constructing approximations to the inverse Hessian using only first-order information; however, they also typically exhibit slower convergence than truncated-Newton methods [27]. Our goal is to investigate the so-called restricted Broyden class of quasi-Newton methods and develop new strategies to accelerate their convergence in order to minimize the number of function and gradient evaluations.

For a given bound-constrained optimization problem,

$$
\begin{aligned}
\underset{x}{\text{minimize}} \quad & f(x), \\
\text{s.t.} \quad & x_l \leq x \leq x_u,
\end{aligned}
\tag{1}
$$

with $f : \mathbb{R}^n \rightarrow \mathbb{R}$ as the objective function and $g_k = \nabla f(x_k)$ as its gradient at the kth iteration, Broyden's method [4] constructs the approximate Hessian with the update formula,

$$
B_{k+1} = B_k - \frac{B_k s_k s_k^T B_k}{s_k^T B_k s_k} + \frac{y_k y_k^T}{y_k^T s_k} + \phi(s_k^T B_k s_k) v_k v_k^T,
\tag{2}
$$

where $s_k = x_{k+1} - x_k$, $y_k = g_{k+1} - g_k$,

$$
v_k = \frac{y_k}{y_k^T s_k} - \frac{B_k s_k}{s_k^T B_k s_k},
$$

and ϕ is a scalar parameter. In the active-set approach, the inverse of this approximate Hessian is applied to the negative gradient and a projected line search is performed along the resulting step direction. The process repeats until the projected gradient norm is reduced below a prescribed tolerance or an iteration limit is reached. Many methods are available for estimating the index set of active variables [2,7,26,31]; however, our focus in the present work is the quasi-Newton approximation.

Methods in the Broyden class are defined by the different scalar values of the parameter ϕ. The restricted Broyden class, in particular, limits the choice of ϕ to the range $[0,1]$, which guarantees that the updates are symmetric positive-definite provided that $s_k^T y_k > 0$. The most well-known methods of this type are the Broyden-Fletcher-Goldfarb-Shanno (BFGS) [5,18,22,34] and Davidon-Fletcher-Powell (DFP) [11] methods, which are recovered with $\phi = 0$ and $\phi = 1$, respectively. The restricted Broyden-class formulation in (2) can also be rewritten as a convex combination of the BFGS and DFP methods, such that

$$
B_{k+1} = (1 - \phi)B_{k+1}^{BFGS} + \phi B_{k+1}^{DFP}.
\tag{3}
$$

The limited-memory variant of the restricted Broyden update takes the form

$$
B_{k+1} = B_0 + \sum_m \left[\frac{y_m y_m^T}{y_m^T s_m} - \frac{B_m s_m s_m^T B_m}{s_m^T B_m s_m} + \phi(s_m^T B_m s_m) v_m v_m^T \right],
\tag{4}
$$

where B_0 is an initial Hessian, $m = \{\max(0, k - M + 1) \ldots k\}$ is the index set for the sequence of quasi-Newton updates, and M is the maximum number of updates to be applied to the initial Hessian (i.e.: the "memory" size). Since Newton-type optimization algorithms seek to apply the inverse of the Hessian matrix to the gradient, the Sherman-Morrison-Woodbury formula [25] is utilized to update an approximation to the inverse of the Hessian matrix directly, such that

$$H_{k+1} = H_0 + \sum_m \left[\frac{s_m s_m^T}{s_m^T y_m} - \frac{H_m y_m y_m^T H_m}{y_m^T H_m y_m} + \psi_m (y_m^T H_m y_m) w_m w_m^T \right], \quad (5)$$

where

$$w_m = \frac{s_m}{y_m^T s_m} - \frac{H_m y_m}{y_m^T H_m y_m}$$

and

$$\psi_m = \frac{(1 - \phi)(y_m^T s_m)^2}{(1 - \phi)(y_m^T s_m)^2 + \phi(y_m^T H_m y_m)(s_m^T B_m s_m)}.$$

In practice, the limited-memory formula is often implemented in a matrix-free fashion where only M of the (s_k, y_k) vector pairs are stored and the action of the approximate Hessian – the product between the approximate inverse of the Hessian and a given vector – is defined by multiplying (5) with a vector. With this approach, the H_m terms inside the summation recurse into their own quasi-Newton formulas and are implemented with nested loops. Some special cases, such as the BFGS formula, can be unrolled into two independent loops that minimize the number of operations [8]. For a more comprehensive look at this approach, we refer the reader to [17].

In this paper, we investigate a framework for constructing scalar or sparse diagonal choices for the initial Hessian B_0 in (4). Our software implementation of (5) including the sparse Hessian initialization is available as part of PETSc/TAO Version 3.10 [1,12]. We leverage PETSc extensibility to explore values of ϕ and other parameters associated with the initial Hessian at runtime to study the convergence and performance of our approach on the complete set of 119 bound-constrained CUTEst test problems [23].

2   Hessian Initialization

The choice of a good initial Hessian H_0 in limited-memory quasi-Newton methods is critically important to the quality of the Hessian approximation. It has been well documented that the scaling of the approximate Hessian depends on this choice and dramatically affects convergence [21,29]. Our goal is to develop a modular framework that can generate effective initializations that preserve the symmetric positive-definite property of the restricted Broyden class methods and are easily invertible as part of efficient matrix-free applications of limited-memory quasi-Newton formulas (e.g., two-loop L-BFGS inversion [8]). To that end, we construct our initial Hessians in scalar and sparse diagonal forms, the latter of which is based on the restricted Broyden-class formula.

The first step is to address the starting point x_0 when there is no accumulated information with which to construct either scalar or diagonal initializations. It is common for the matrix at iteration 0 to be set to a multiple $B_0 = \rho_0 I$ of the identity that promotes acceptance of the unit-step length by the line search; however, no good general strategy exists for choosing a suitable value for ρ_0. Gilbert and Lemaréchal [20] proposed $\rho_0 = 2\Delta/\|g_0\|_2^2$, where Δ is a user-supplied parameter that represents the expected decrease in $f(x)$ at the first iteration. We use

$$\rho_0 = \begin{cases} 2/\|g_0\|_2^2 & \text{for } f(x_0) = 0 \\ 2|f(x_0)|/\|g_0\|_2^2 & \text{otherwise,} \end{cases} \tag{6}$$

which has proven to be an effective choice across our numerical experiments and eliminates a user-defined parameter from the algorithm. This choice also appears to be related to more recent investigations into scaled gradient descent steps with an a priori estimation of the local minimum [9], with $f(x^*) = 0$ where x^* is the minimizer. Both the scalar and sparse diagonal B_0 constructions we introduce below leverage this initial scalar choice.

2.1   Scalar Formulation

Scalar Hessian initializations restrict the estimate to a positive scalar multiple of the identity matrix, such that $B_0 = \rho_k I$ during iteration k. For BFGS matrices, a common and well-understood choice has been

$$\rho_k = \frac{y_k^T y_k}{y_k^T s_k}, \tag{7}$$

which is an approximation to an eigenvalue of $\nabla^2 f(x_k)$ [32].

Our scalar construction begins with the recognition that (7) is the positive solution to the scalar minimization problem

$$\rho_k = \operatorname*{argmin}_{\rho > 0} \|\frac{1}{\rho} y_k - s_k\|_2^2, \tag{8}$$

which is also a least-squares solution to the secant equation $B_0^{-1} y_k = s_k$ with $B_0 = \rho I$ [6]. We then introduce a new parameter $\alpha \in [0, 1]$ such that $B_0 = \rho^{2\alpha - 1} I$ and we solve the modified least-squares problem,

$$\rho_k = \operatorname*{argmin}_{\rho > 0} \|\rho^{-\alpha} y_k - \rho^{-(1-\alpha)} s_k\|_2^2. \tag{9}$$

After constructing the optimality conditions and solving for ρ, we arrive at the following values:

1. If $\alpha = 0$, then

$$\rho_k = \frac{y_k^T s_k}{s_k^T s_k}$$

2. If $\alpha = 1/2$, then

$$\rho_k = \sqrt{\frac{y_k^T y_k}{s_k^T s_k}}$$

3. If $\alpha = 1$, then

$$\rho_k = \frac{y_k^T y_k}{y_k^T s_k}$$

Note: this value corresponds to the commonly used eigenvalue estimate in (7).

4. Otherwise, ρ_k is the positive root of the quadratic equation,

$$\alpha(y_k^T y_k)\rho^2 - (2\alpha - 1)(y_k^T s_k)\rho + (\alpha - 1)(s_k^T s_k) = 0.$$

Since $s_k^T s_k$ and $y_k^T y_k$ cannot be negative and are zero only for a zero step length, the scalar Hessian approximation preserves symmetric positive-definiteness for any (s_k, y_k) update that satisfies the Wolfe conditions.

2.2 Sparse Diagonal Formulation

The sparse diagonal formulation constructs an initial Hessian as a diagonal matrix, $B_0 = \text{diag}(b_k)$, at iteration k defined by and stored as the vector of diagonal entries b_k. Specifically, we construct this diagonal vector using the full-memory restricted Broyden formula in (2), such that

$$
\begin{aligned}
b_{k+1} = b_k + (1 - \theta)&\left[\frac{y_k \circ y_k}{y_k^T s_k} - \frac{(b_k \circ s_k)^2}{s_k^T (b_k \circ s_k)}\right] \\
+ \theta&\left[\left(\frac{1}{y_k^T s_k} + \frac{s_k^T(b_k \circ s_k)}{(y_k^T s_k)^2}\right)(y_k \circ y_k) - \frac{2(s_k \circ b_k \circ y_k)}{y_k^T s_k}\right].
\end{aligned}
\tag{10}
$$

This expression is the expanded version of the convex combination notation in (3), where $(1 - \theta)$ and θ correspond to the BFGS and DFP components, respectively. Since we compute only diagonal entries, all matrix-vector products have been replaced by Hadamard products with the previous diagonal. As in Broyden's method, $\theta = 0$ corresponds to a pure BFGS formulation, while $\theta = 1$ recovers DFP.

This initialization is a full-memory approach; the diagonal entries of B_0 are explicitly stored in b_k and updated with every accepted new iterate. Consequently, B_0 contains information from all iterates traversed in the optimization instead of only the last M iterates stored for the limited-memory formula, but without the large memory cost of storing dense matrices.

Gilbert and Lemaréchal have explored a similar initial Hessian using diagonalizations of the BFGS formula only [20] and reported the need to rescale the diagonal to account for the inability to rapidly modify it in large steps. To that end, we redefine the initial Hessian as $B_0 = \sigma_k^{2\alpha-1}\mathrm{diag}(b_k)$ and compute the rescaling factor σ_k by seeking the least-squares solution to the secant equation, $B_0^{-1}y_k = s_k$, such that

$$\sigma_k = \arg\min_{\sigma}||\sigma^{-\alpha}(b_k^{-0.5} \circ y_k) - \sigma^{-(1-\alpha)}(b_k^{0.5} \circ s_k)||_2^2. \tag{11}$$

Note that the expression inside the l_2-norm is equivalent to the secant equation in residual form, restructured so that the solution can be more easily expressed in the form of quadratic roots. The solution yields the following values:

1. If $\alpha = 0$, then

$$\sigma_k = \frac{y_k^T s_k}{s_k^T(b_k \circ s_k)}.$$

2. If $\alpha = 1/2$, then

$$\sigma_k = \sqrt{\frac{y_k^T(b_k^{-1} \circ y_k)}{s_k^T(b_k \circ s_k)}}.$$

3. If $\alpha = 1$, then

$$\sigma_k = \frac{y_k^T(b_k^{-1} \circ y_k)}{y_k^T s_k}.$$

4. Otherwise, σ_k is the positive root of the quadratic equation,

$$\alpha\left[y_k^T(b_k^{-1} \circ y_k)\right]\sigma^2 - (2\alpha - 1)(y_k^T s_k)\sigma + (\alpha - 1)\left[s_k^T(b_k \circ s_k)\right] = 0.$$

As with the scalar initialization, the sparse diagonal B_0 remains positive definite for any (s_k, y_k) pair that satisfies the Wolfe conditions. Nonetheless, we have encountered cases where numerical problems surface in finite precision arithmetic. Therefore, we safeguard all the methods by checking whether the denominators are equal to zero and setting their value to a small constant, 10^{-8}, if so.

3   Numerical Studies

We now investigate the numerical performance of our proposed scalar and sparse Hessian initializations and study the parameter space of the user-controlled scalar factors to determine useful recommendations. Our quasi-Newton implementation in PETSc/TAO utilizes an active-set estimation based on the work of Bertsekas [2] and is discussed in further detail in the TAO manual [12]. The step direction is globalized via a projected Moré-Thuente line search [30], which is capable of taking step lengths greater than 1.

Our numerical experiments are based on 119 bound-constrained problems from the CUTEst test set [23], covering a diverse range of problems from 2 to 10^5 variables. In all cases presented in this section, we set the quasi-Newton memory size to $M = 5$ updates, limit the maximum number of iterations to 1,000, and require convergence to an absolute tolerance of $||g^*||_2 \leq 10^{-6}$.

Performance profiles are constructed by using the methodology proposed by Dolan and Moré [16]. For a given CUTEst problem $p \in \mathcal{P}$ and solver configuration $c \in \mathcal{C}$, we define a cost measure

$t_{p,c} =$ function evaluations required to solve problem p with configuration c

and normalize it by the best configuration for each problem, such that

$$r_{p,c} = \frac{t_{p,c}}{\min\{t_{p,\hat{c}} : \hat{c} \in \mathcal{C}\}}.$$

Performance of each configuration is then given by

$$P_c(\pi) = \frac{1}{n_p}\text{size}\{p \in \mathcal{P} : r_{p,c} \leq \pi\},$$

which describes the probability for configuration $c \in \mathcal{C}$ to have a cost ratio $r_{p,c}$ that is within a factor of $\pi \in \mathbb{R}$ of the best configuration.

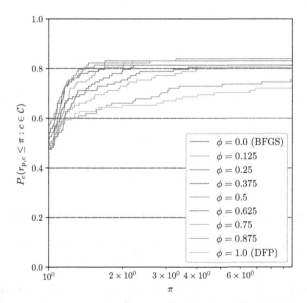

Fig. 1. Parameter study for restricted Broyden convex combination factor.

We begin our analysis with a sweep through the ϕ parameter space in the restricted Broyden updates in Fig. 1. For this study we turn off all Hessian initialization (i.e., $H_0 = I$) and investigate only the relative performances of the raw Broyden-class methods. Note that $\phi = 0$ and $\phi = 1$ are included, which correspond to the BFGS and DFP methods, respectively. The results indicate that ϕ values in range $(0, 0.5]$ produce Hessian approximations that outperform BFGS, with $\phi = 0.5$ yielding the best performance.

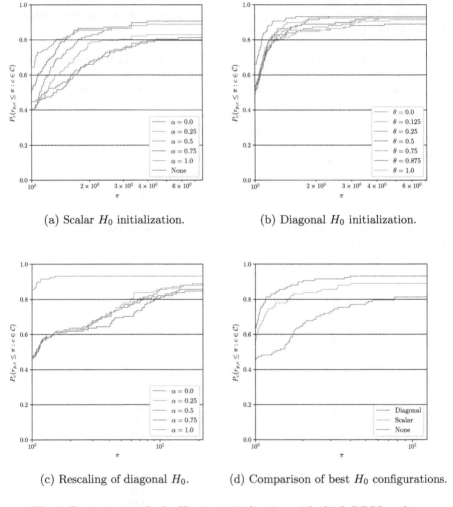

(a) Scalar H_0 initialization.   (b) Diagonal H_0 initialization.

(c) Rescaling of diagonal H_0.   (d) Comparison of best H_0 configurations.

Fig. 2. Parameter study for Hessian initialization with the L-BFGS update.

For the next study, we used the L-BFGS method as the basis for analyzing the effects of different H_0 initialization methods. We start with the scalar H_0 in Fig. 2a and investigate the effect of the α parameter. Here, setting $\alpha = 1.0$

recovers the widely used $y_k^T s_k / y_k^T y_k$ factor that approximates an eigenvalue of $\nabla^2 f(x_k)^{-1}$. As expected, this choice yields the best scaled Hessian approximation and the lowest number of function evaluations on most problems. Surprisingly, however, the results indicate that $\alpha = 0.75$ remains competitive in cost, while converging on two additional test problems.

Figure 2b shows the relative performances of the sparse diagonal H_0 initialization. For this study, we fix the rescaling parameter to $\alpha = 1.0$. We investigate the effect of the rescaling parameter below. Not surprisingly, the results indicate that $\theta = 0.0$, which corresponds to a "full-memory" BFGS diagonal, produce the best convergence improvement for L-BFGS updates.

Fixing $\theta = 0.0$ as the best case for the sparse diagonal H_0, we now perform a parameter sweep through the rescaling term in Fig. 2c. The results indicate that $\alpha = 1.0$ produces the most well-scaled initialization for the sparse diagonal case. Additional experiments not shown have failed to recover a better diagonal initialization at different α and θ combinations.

Table 1. Select problems for comparison of H_0 methods in L-BFGS.

	# of vars.			# of iterations		
	Total	Free	Active	$H_0 = I$	$H_0 = \rho_k I$	$H_0 = \sigma_k \mathrm{diag}(h_k)$
EXPLIN	1200	52	1148	213	170	109
EXPQUAD	1200	1119	81	N/A	314	113
BDEXP	5000	5000	0	36	18	16
TORSIONB	5625	3624	1852	165	151	132
JNLBRNGA	10000	6359	3641	N/A	327	299
OBSTCLBL	10000	7057	2943	131	130	113

In Fig. 2d, we compare the best-case configurations for both H_0 initialization types to the raw L-BFGS results. The sparse diagonal initialization enables L-BFGS to solve over 90% of the bound-constrained CUTEst test set in under 1,000 iterations and accelerates convergence on all problems, offering a significant improvement over both the scalar and identity initialization methods. Statistics for a subset of the problems from this plot are available in Table 1.

We also explore additional H_0 parameters to accelerate convergence of DFP and the best Broyden method at $\phi = 0.5$. Figure 3a and b show the parameter study for the scalar initialization in the Broyden and DFP methods, respectively. The best case for DFP at $\alpha = 0$ yields a scalar H_0 that is the dual of the best scalar term for BFGS (i.e.: interchanging roles for s_k and y_k). This mimics the duality between the DFP and BFGS formulas themselves. Additionally, the best case for Broyden's method matches the α parameter to the convex combination term of $\phi = 0.5$. This observation suggests that the best scalar initialization parameter, α, for any member of the Broyden-class method may be the same as the convex combination term ϕ that defines the method.

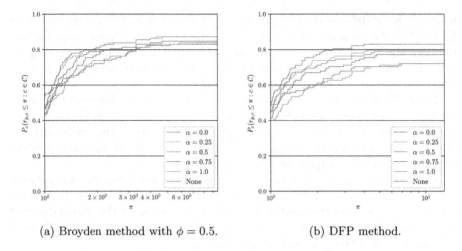

(a) Broyden method with $\phi = 0.5$.   (b) DFP method.

Fig. 3. Scalar Hessian initialization for Broyden ($\phi = 0.5$) and DFP methods.

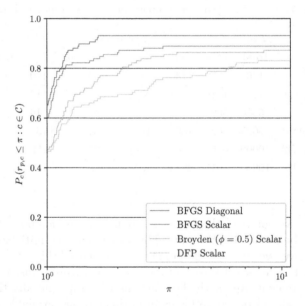

Fig. 4. Comparison of best H_0 definitions for selected quasi-Newton methods.

In Fig. 4, we compare these scalar H_0 terms with one another and against the best BFGS initializations above. Results show that selecting the correct scalar H_0 for each method significantly improves all quasi-Newton methods tested and makes the $\phi = 0.5$ Broyden method competitive with BFGS in the number of problems solved. Our observations indicate that a more comprehensive parameter study may reveal other Broyden-class methods with different convex combination and Hessian initialization terms that offer competitive or better performance than BFGS on some problems. Preliminary numerical experiments we have conducted suggest that the DFP method does not benefit from a diagonal Hessian initialization; however, we aim to utilize our flexible quasi-Newton framework to explore the diagonal H_0 with other Broyden-class methods in the near future.

4 Conclusions

We have introduced a flexible framework for constructing both scalar and sparse diagonal, positive-definite Hessian initializations for limited-memory quasi-Newton methods based on the restricted Broyden-class updates. Our implementation in PETSc/TAO allows us to rapidly change parameters and shift between different members of the Broyden-class methods and select the form for the Hessian initializations at runtime.

Our numerical experiments indicate that intermediate values of ϕ in the Broyden-class outperform the base BFGS formula for a significant subset of the bound-constrained CUTEst problems. We also compare different scalar initializations for different quasi-Newton methods; the results suggest that the best possible α parameter in our H_0 formulation tracks with the convex combination parameter ϕ that defines members of the Broyden-class.

We demonstrate that the diagonal Hessian initialization successfully accelerates BFGS convergence at minimal additional memory and algebra cost compared with scalar initializations. Our preliminary experience testing similar initializations with DFP and other Broyden-class methods suggests that other parameter values may reveal alternative quasi-Newton methods that are competitive with BFGS on large-scale optimization problems. We hope to leverage our flexible Broyden-class quasi-Newton algorithm to further investigate these possibilities in the future.

Acknowledgments. This work was supported by the Exascale Computing Project (17-SC-20-SC), a collaborative effort of two U.S. Department of Energy organizations (Office of Science and the National Nuclear Security Administration) responsible for the planning and preparation of a capable exascale ecosystem, including software, applications, hardware, advanced system engineering, and early testbed platforms, in support of the nation's exascale computing imperative.

References

1. Balay, S., et al.: PETSc users manual. Technical report ANL-95/11 - Revision 3.10, Argonne National Laboratory (2018). http://www.mcs.anl.gov/petsc
2. Bertsekas, D.P.: Projected Newton methods for optimization problems with simple constraints. SIAM J. Control Optim. **20**(2), 221–246 (1982)
3. Biros, G., Ghattas, O.: Parallel Lagrange-Newton-Krylov-Schur methods for PDE-constrained optimization. Part i: The Krylov-Schur solver. SIAM J. Sci. Comput. **27**(2), 687–713 (2005)

4. Broyden, C.G.: A class of methods for solving nonlinear simultaneous equations. Math. Comput. **19**(92), 577–593 (1965)
5. Broyden, C.G.: The convergence of a class of double-rank minimization algorithms 1. General considerations. IMA J. Appl. Math. **6**(1), 76–90 (1970)
6. Burke, J.V., Wiegmann, A., Xu, L.: Limited memory BFGS updating in a trust-region framework. Technical report, Department of Mathematics, University of Washington (2008)
7. Byrd, R.H., Lu, P., Nocedal, J., Zhu, C.: A limited memory algorithm for bound constrained optimization. SIAM J. Sci. Comput. **16**(5), 1190–1208 (1995)
8. Byrd, R.H., Nocedal, J., Schnabel, R.B.: Representations of Quasi-Newton matrices and their use in limited memory methods. Math. Program. **63**(1–3), 129–156 (1994)
9. D'Alves, C.: A Scaled Gradient Descent Method for Unconstrained Optimization Problems With A Priori Estimation of the Minimum Value. Ph.D. thesis (2017)
10. Davidon, W.C.: Variable metric method for minimization, argonne natl. Technical report No. ANL-5990, Argonne National Laboratory (1959)
11. Davidon, W.C.: Variable metric method for minimization. SIAM J. Optim. **1**(1), 1–17 (1991)
12. Dener, A., et al.: Tao users manual. Technical report ANL/MCS-TM-322 - Revision 3.10, Argonne National Laboratory (2018). http://www.mcs.anl.gov/petsc
13. Dener, A., Hicken, J.E.: Matrix-free algorithm for the optimization of multidisciplinary systems. Struct. Multidiscip. Optim. **56**(6), 1429–1446 (2017)
14. Dener, A., Hicken, J.E., Kenway, G.K., Martins, J.: Enabling modular aerostructural optimization: individual discipline feasible without the Jacobians. In: 2018 Multidisciplinary Analysis and Optimization Conference, p. 3570 (2018)
15. Dennis Jr., J.E., Moré, J.J.: Quasi-Newton methods, motivation and theory. SIAM Rev. **19**(1), 46–89 (1977)
16. Dolan, E.D., Moré, J.J.: Benchmarking optimization software with performance profiles. Math. Program. **91**(2), 201–213 (2002)
17. Erway, J.B., Marcia, R.F.: On solving large-scale limited-memory quasi-Newton equations. Linear Algebr. Appl. **515**, 196–225 (2017)
18. Fletcher, R.: A new approach to variable metric algorithms. Comput. J. **13**(3), 317–322 (1970)
19. Fletcher, R., Powell, M.J.: A rapidly convergent descent method for minimization. Comput. J. **6**(2), 163–168 (1963)
20. Gilbert, J.C., Lemaréchal, C.: Some numerical experiments with variable-storage quasi-Newton algorithms. Math. Program. **45**(1–3), 407–435 (1989)
21. Gill, P.E., Leonard, M.W.: Reduced-hessian quasi-Newton methods for unconstrained optimization. SIAM J. Optim. **12**(1), 209–237 (2001)
22. Goldfarb, D.: A family of variable-metric methods derived by variational means. Math. Comput. **24**(109), 23–26 (1970)
23. Gould, N.I., Orban, D., Toint, P.L.: CUTEst: a constrained and unconstrained testing environment with safe threads for mathematical optimization. Comput. Optim. Appl. **60**(3), 545–557 (2015)
24. Haber, E., Ascher, U.M.: Preconditioned all-at-once methods for large, sparse parameter estimation problems. Inverse Probl. **17**(6), 1847 (2001)
25. Hager, W.W.: Updating the inverse of a matrix. SIAM Rev. **31**(2), 221–239 (1989)
26. Hager, W.W., Zhang, H.: A new active set algorithm for box constrained optimization. SIAM J. Optim. **17**(2), 526–557 (2006)
27. Hicken, J., Alonso, J.: Comparison of reduced-and full-space algorithms for PDE-constrained optimization. In: 51st AIAA Aerospace Sciences Meeting including the New Horizons Forum and Aerospace Exposition, p. 1043 (2013)

28. Hinze, M., Pinnau, R.: Second-order approach to optimal semiconductor design. J. Optim. Theory Appl. **133**(2), 179–199 (2007)
29. Liu, D.C., Nocedal, J.: On the limited memory BFGS method for large scale optimization. Math. Program. **45**(1–3), 503–528 (1989)
30. Moré, J.J., Thuente, D.J.: Line search algorithms with guaranteed sufficient decrease. ACM Trans. Math. Softw. (TOMS) **20**(3), 286–307 (1994)
31. Moré, J.J., Toraldo, G.: On the solution of large quadratic programming problems with bound constraints. SIAM J. Optim. **1**(1), 93–113 (1991)
32. Nocedal, J., Wright, S.J.: Numerical Optimization, 2nd edn. Springer, New York (2006). https://doi.org/10.1007/978-0-387-40065-5
33. Papadimitriou, D., Giannakoglou, K.: Direct, adjoint and mixed approaches for the computation of Hessian in airfoil design problems. Int. J. Numer. Methods Fluids **56**(10), 1929–1943 (2008)
34. Shanno, D.F.: Conditioning of quasi-Newton methods for function minimization. Math. Comput. **24**(111), 647–656 (1970)

Reduced-Cost Design Optimization of High-Frequency Structures Using Adaptive Jacobian Updates

Slawomir Koziel[1,2(✉)] ⓘ, Anna Pietrenko-Dabrowska[2] ⓘ,
and Leifur Leifsson[3] ⓘ

[1] Engineering Optimization and Modeling Center,
School of Science and Engineering, Reykjavík University,
Menntavegur 1, 101 Reykjavík, Iceland
koziel@ru.is
[2] Faculty of Electronics Telecommunications and Informatics,
Gdansk University of Technology, Narutowicza 11/12, 80-233 Gdansk, Poland
annadabrl@pg.edu.pl
[3] Department of Aerospace Engineering, Iowa State University,
Ames, IA 50011, USA
leifur@iastate.edu

Abstract. Electromagnetic (EM) analysis is the primary tool utilized in the design of high-frequency structures. In vast majority of cases, simpler models (e.g., equivalent networks or analytical ones) are either not available or lack accuracy: they can only be used to yield initial designs that need to be further tuned. Consequently, EM-driven adjustment of geometry and/or material parameters of microwave and antenna components is a necessary design stage. This, however, is a computationally expensive process, not only because of a considerable computational cost of high-fidelity EM analysis but also due to a typically large number of parameters that need to be adjusted. In particular, conventional numerical optimization routines (both local and global) may be prohibitively expensive. In this paper, a reduced-cost trust-region-based gradient search algorithm is proposed for the optimization of high-frequency components. Our methodology is based on a smart management of the system Jacobian enhancement which combines: (i) omission of (finite-differentiation-based) sensitivity updates for variables that exhibit small (relative) relocation in the directions of the corresponding coordinate system axes and (ii) selective utilization of a rank-one Broyden updating formula. Parameter selection for Broyden-based updating depends on the alignment between the direction of the latest design relocation and respective search space basis vectors. The proposed technique is demonstrated using a miniaturized coupler and an ultra-wideband antenna. In both cases, significant reduction of the number of EM simulations involved in the optimization process is achieved as compared to the benchmark algorithm (computational speedup of 60% on average). At the same time, degradation of the design quality is minor.

Keywords: Electromagnetic simulation · Microwave engineering ·
Antenna design · Design optimization · Gradient-based search ·
Jacobian update · Trust region framework

© Springer Nature Switzerland AG 2019
J. M. F. Rodrigues et al. (Eds.): ICCS 2019, LNCS 11538, pp. 508–522, 2019.
https://doi.org/10.1007/978-3-030-22744-9_40

1 Introduction

Similarly as in many other engineering disciplines, design of high-frequency systems (e.g., microwave [1], antenna [2], photonic [3], etc.) is heavily based on computer simulation tools. While in some cases, analytical representations (e.g., array factor models for antenna array radiation patterns [4], or coupling-matrix-based models for microwave filters [5]) or equivalent network models (e.g., for transmission-line-based circuits [6]) are available, the ultimate level of accuracy and generality can be achieved by full-wave electromagnetic (EM) simulation [7]. As a matter of fact, EM analysis is mandatory for design verification of vast majority of contemporary high-frequency components and devices, which is due to their geometrical complexity. The latter is a consequence of more and more stringent performance requirements imposed on the components, demands for multi-band operation [8], additional functionalities (e.g., band notches in wideband antennas [9]), and, most importantly, compact size [10–13]. It is especially the miniaturization requirement that results in the development of topologically complex layouts exhibiting considerable EM cross-coupling effects, impossible to be adequately accounted for using analytical or equivalent circuit models. Examples include microwave components exploiting a slow-wave phenomenon [14] or defected ground structures [15], as well as compact antennas incorporating stepped-impedance feeds [16] or modified radiators [17].

Whenever electromagnetic simulation is involved, some variation of parametric optimization is necessary in order to obtain the best possible performance of the component at hand. Application of conventional numerical optimization methods, either local [18] or global [19], faces considerable challenges which come from several sources: (i) a high cost of individual EM analysis, (ii) a necessity of simultaneous handling of multiple performance figures and constraints, and (iii) a large number of simulations required to converge to the optimized design. The latter is partially due to the fact that geometries of modern high frequency structures are typically described by many parameters [20, 21]. All of these factors result in a considerable computational cost of the optimization process, often prohibitive. It should be mentioned here that a commonly used workaround these issues is an interactive design that combines engineering experience and parameter sweeping (one, maximum two parameters at a time). This is a laborious process yet incorporation of the expert knowledge permitted, in many cases, finding a satisfactory design relatively quickly. However, the aforementioned increase of component complexity, has made this approach questionable, especially in the context of controlling multiple objectives, handling constraints, and operating in highly-dimensional parameter spaces.

Meanwhile, several methods have been developed and applied to reduce the cost of numerical optimization. These include utilization of adjoint sensitivities (in the context of local design [22]), machine learning methods (for global optimization [23]), as well as various surrogate-assisted techniques [24–26]. The last group of approaches comes in many variations and involves both data-driven surrogates (e.g., kriging [27], Gaussian process regression [28], or polynomial response surfaces [29]), and physics-based ones. In the latter case, the surrogate is constructed from the underlying low-fidelity model (space mapping [30], response correction techniques [31], feature-based optimization [32]). For the sake of computational efficiency, the low-fidelity models should be cheap (e.g., analytical or equivalent network ones). However, in many situations, e.g., antennas or various classes of miniaturized microwave components, the only reliable option for low-fidelity models are those obtained from coarse-discretization EM simulations [33]. Here, the overhead related to optimizing the surrogate model (often carried out using conventional algorithms) determines the overall design expense.

As discussed above, reducing the number of EM simulations required by an optimization routine is important for both direct and surrogate-assisted design procedures. In this paper, a low-cost trust-region-based gradient search algorithm is proposed for the optimization of high-frequency structures. The foundation of the method is an appropriate management of the system response sensitivity (here, Jacobian) updates. This management scheme involves a few mechanisms, including an omission of (finite-differentiation-based) sensitivity updates for variables that exhibit small (relative) relocation in the directions of the corresponding coordinate system axes as well as a selective application of a rank-one Broyden updating formula. The Broyden-based updates are performed for parameters whose corresponding basis vectors are sufficiently well aligned with the direction of the latest design relocation. For the sake of demonstration, two high-frequency structures are considered: a miniaturized coupler implemented using compact microstrip resonant cells (CMRCs), and an ultra-wideband antenna operating in 3.1 GHz to 10.6 GHz frequency range. In both cases, application of the proposed algorithm leads to a significant reduction of the number of EM simulations necessary for identifying the optimized design. The average computational speedup is as high as 60% as compared to the reference algorithm. The improvement of computational efficiency is achieved with only minor degradation of the design quality.

2   Reduced-Cost Design Optimization Through Jacobian Update Management

In this section, the proposed procedure for expedited design optimization of high-frequency structures is formulated and explained. Here, local optimization is considered with the trust-region gradient search utilizing numerical derivatives used as a reference algorithm. The section starts by formulating the high-frequency design optimization problem, followed by a brief description of the reference algorithm, the proposed sensitivity updating scheme, as well as the complete optimization framework.

2.1   High-Frequency Design Problem Formulation

The computational model of the high-frequency structure of interest will be denoted as $R(x)$, where x stands for the adjustable variables of the problem (typically, geometry parameters). R represents the model responses which are typically frequency characteristics such as scattering parameters (e.g., reflection S_{11}, transmission S_{21}, etc.), gain, efficiency, radiation pattern, power split ration, and so on. The computational model is assumed to be evaluated using a full-wave electromagnetic (EM) analysis. Normally, computational cost of EM simulation is considerable, ranging from a few dozens of seconds to a few hours, depending on the structure complexity, its electrical size (i.e., physical dimensions as compared to the guided wavelength at the operating frequency), as well as other components that need to be included in the model (e.g., SMA connectors, housing, installation fixtures, etc.).

The design problem can be formulated as

$$\boldsymbol{x}^* = \arg\min_{\boldsymbol{x}} U(\boldsymbol{R}(\boldsymbol{x})) \tag{1}$$

where $\boldsymbol{x}^*$ is the optimum design to be found and U is a scalar objective function encoding the performance specifications imposed on the structure at hand.

Given a large variety of high-frequency structures and the figures of interest involved, it is obvious that the objective function is very much problem dependent. For the sake of illustration we will discuss two types of objective functions, pertinent to particular illustration cases considered in Sect. 3.

In many situations, especially related to the design of antenna structures but also certain other components such as impedance transformers or microwave filters, it is important to improve the in-band matching (which is equivalent to minimization of the reflection response S_{11} or reducing the return loss). In this case, the objective function may be defined as

$$U(\boldsymbol{R}(\boldsymbol{x})) = \max_{f \in F} |S_{11}(\boldsymbol{x},f)| \tag{2}$$

where $|S_{11}(\boldsymbol{x},f)|$ stands for the reflection as a function of optimization variables $\boldsymbol{x}$ and frequency f, with F being the frequency range of interest (e.g., 3.1 GHz to 10.6 GHz in case of UWB antennas). The problem (1), (2) is thus formulated in a minimax sense.

The objective function (2) addresses a single performance figure. However, in majority of practical cases, there is a need to control several figures. Representative examples are microwave couplers, where a typical design problem requires maximization of the bandwidth BW (usually, symmetric with respect to the operating frequency f_0), obtaining the required (e.g., equal) power split $d_S = |S_{21}| - |S_{31}|$ at f_0, as well as the allocation of the matching and isolation characteristics ($|S_{11}|$ and $|S_{41}|$) minima close to f_0. In this case, assuming implicit constraint handling, we may define U as

$$U(\boldsymbol{R}(\boldsymbol{x})) = -BW(\boldsymbol{x}) + \sigma_1 d_S(\boldsymbol{x})^2 + \sigma_2 (f_{min.S_{11}}(\boldsymbol{x}) - f_0)^2 + \sigma_3 (f_{min.S_{41}}(\boldsymbol{x}) - f_0)^2 \tag{3}$$

where $f_{min.S11}$ and $f_{min.S41}$ are the frequencies corresponding to $|S_{11}|$ and $|S_{41}|$ minima, respectively, and σ_k are penalty coefficients. A penalty function approach as in (3) is a convenient way of handling expensive constraints, especially when dealing with EM simulation models [34, 35].

2.2   Trust-Region Gradient Search

The reference optimization algorithm for this paper is a conventional trust-region (TR)-based gradient-search procedure [18]. The TR algorithm yields a series of approximations $x^{(i)}$, $i = 0, 1, \ldots$, to the optimum design x^*, by solving sub-problems

$$x^{(i+1)} = \arg \min_{x;\ -d^{(i)} \le x - x^{(i)} \le d^{(i)}} U(L^{(i)}(x)) \qquad (4)$$

In (4), $L^{(i)}(x) = R(x^{(i)}) + J_R(x^{(i)}) \cdot (x - x^{(i)})$ is a first-order Taylor expansion of the computational model R at the current iteration point $x^{(i)}$. The Jacobian J_R can be evaluated using adjoint sensitivities if available; however, in vast majority of practical cases of high-frequency structure design it is estimated through finite differentiation (FD). The reason is that the adjoint technology is not supported (with rare exceptions) by commercial EM simulation packages. FD incurs the cost of additional n EM analyses (n being dimensionality of the design space) per algorithm iteration. Obviously, reducing this overhead would lead to computational savings, which is the main subject of this paper.

It should also be noticed that the trust region size in (4) does not take a traditional form of $||x - x^{(i)}|| \le \delta^{(i)}$ (Euclidean norm with scalar TR radius), but it is an interval defined through the size vector $d^{(i)}$ (adjusted using the standard rules [18]). The inequalities $-d^{(i)} \le x - x^{(i)} \le d^{(i)}$ are understood component-wise. This type of the TR region allows —by making the initial size vector $d^{(0)}$ proportional to the design space sizes—for ensuring a similar treatment of variables with significantly different ranges, a situation common in the antenna or microwave design. For example, the lengths of the transmission line components are typically in the range of millimeters or tens of millimeters, whereas the line widths or spacings between them may be small fractions of millimeters.

2.3   Reduced-Cost TR Algorithm Using Adaptive Jacobian Updates

The computational cost of the design optimization involving the conventional TR algorithm is primarily determined by the cost of Jacobian estimation performed using the finite differentiation. Here, a reduced-cost TR algorithm using adaptive Jacobian updates is introduced which allows us to notably decrease the number of EM simulations necessary to obtain the optimal design. The computational savings result from the reduction of the overall number of FD calculations. The algorithm combines two independent procedures for suppressing Jacobian updates: an accelerated update procedure (AUP) and Broyden update procedure (BUP). The former is based on a relative relocation of the design variable vector between iterations, and the optimization run history. The latter adopts the Broyden updating formula for the selected design variables, depending on the alignment between the most recent design relocation and the coordinate system axes. When used separately, each procedure allows for achieving considerable computational savings, however, at the expense of a slight degradation of the design quality, as shown in Sect. 3. Furthermore, a combination of both procedures further expedites the optimization process while ensuring satisfactory design quality.

The essential stages of the proposed reduced-cost TR algorithm are shown in the form of a flow diagram in Fig. 1.

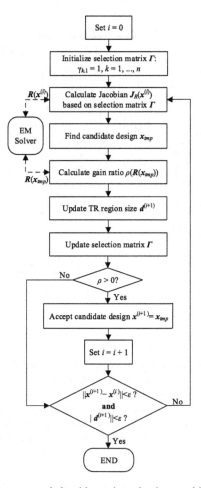

Fig. 1. Flow diagram of the proposed algorithm using adaptive sensitivity updating scheme. The following notation is used: Γ – selection matrix containing information about the Jacobian updates, $d^{(i+1)}$ – TR region size in the $(i + 1)$th iteration, ρ – gain ratio (deciding about iteration acceptance and $d^{(i+1)}$ update), ε – algorithm termination threshold.

The information about either performing (1) or omitting (0) FD calculation of the system response Jacobian J_R is stored in a selection matrix Γ. The complete description of the procedure of creating the matrix Γ is provided in Sect. 2.4, along with the account of the two component procedures: accelerated update procedure and Broyden update procedure. Here, the major outline of the modified TR algorithm is given. The selection matrix Γ is initialized as a column vector, $n \times 1$, with all its entries set to ones: $\gamma_{k.1} = 1, k = 1, \ldots, n$. This implies that in the first iteration, the initial estimate of the entire Jacobian $\boldsymbol{J_R}$ is obtained with FD. Upon each successful iteration, the matrix is extended by an additional column, utilized to govern the Jacobian update in the upcoming iteration (as depicted in Fig. 2).

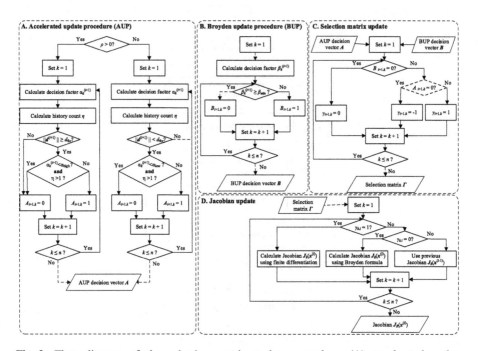

Fig. 2. Flow diagram of the selection matrix update procedure: (A) accelerated update procedure, (B) Broyden update procedure, (C) selection matrix $\boldsymbol{\Gamma}$ update based on AUP and BUP, D) Jacobian update based on $\boldsymbol{\Gamma}$. The following notation is used: (i) AUP parameters: $\alpha_k^{(i+1)}$ – decision factor for kth geometry parameter and $(i + 1)$th iteration; η – history count, α_{high}, α_{low} – threshold values, A – AUP decision vector; (ii) BUP parameters: $\beta_k^{(i+1)}$ – decision factor for kth geometry parameter and $(i + 1)$th iteration, β_{min} – alignment threshold value, $\boldsymbol{B}$ – BUP decision vector.

After the Jacobian update is completed, a candidate design $\boldsymbol{x}_{tmp}$ is found by solving (4), with the objective function U described by (2) or (3). Then, a gain ratio $\rho = (U(\boldsymbol{R}(\boldsymbol{x}_{tmp}) - U(\boldsymbol{R}(\boldsymbol{x}^{(i)})) / (\boldsymbol{L}^{(i)}(\boldsymbol{x}_{tmp}) - \boldsymbol{L}^{(i)}(\boldsymbol{x}^{(i)}))$ is calculated, and the TR region size $\boldsymbol{d}^{(i+1)}$ for the next $(i+1)$th iteration is adjusted. If the gain ratio is positive, the candidate step is accepted and the next iteration is executed, unless the termination criterion is satisfied, i.e., $\left\| \boldsymbol{d}^{(i+1)} \right\| < \varepsilon$ and $\left\| \boldsymbol{x}^{(i+1)} - \boldsymbol{x}^{(i)} \right\| < \varepsilon$, where ε is the algorithm termination threshold.

2.4   Selection Matrix Updating Procedure

In this section, a procedure for updating the selection matrix $\boldsymbol{\Gamma}$ is described in detail. It combines two separate routines that control the Jacobian update: the accelerated (AUP) and the Broyden (BUP) update procedure. The former pinpoints the parameters that exhibit small relative design changes between iterations, therefore implying that the calculation of the respective part of the Jacobian can be omitted.

Whereas, the latter identifies the parameters for which the corresponding basis vectors are sufficiently well aligned with the most current design relocation vector. For these parameters, calculation of the respective Jacobian columns will be executed using the Broyden formula instead of FD. The outcomes of both procedures are then combined to create the selection matrix. The two procedures, along with the resulting Jacobian update procedure are presented in the form of a flow diagram in Fig. 2.

Accelerated Update Procedure. The accelerated procedure, depicted in Panel A of Fig. 2, accommodates the changes of geometry parameters observed throughout the optimization course. These are quantified by calculating a relative design change of the kth parameter, $k = 1, \ldots, n$, w.r.t. the TR region size in the ith iteration

$$\alpha_k^{(i+1)} = \left| x_k^{(i+1)} - x_k^{(i)} \right| / d_k^{(i)}, \qquad k = 1, \ldots, n, \tag{5}$$

where $x_k^{(i)}$, $x_k^{(i+1)}$ and $d_k^{(i)}$ refer to the kth components of vectors $\boldsymbol{x}^{(i)}$, $\boldsymbol{x}^{(i+1)}$, and $\boldsymbol{d}^{(i)}$, respectively. Let us also denote by $\boldsymbol{J}_k$ the kth column of the Jacobian $\boldsymbol{J_R}$. The decision factors $\alpha_k^{(i+1)}$ are utilized to determine whether $\boldsymbol{J}_k$, pertaining to the kth parameter of the structure at hand, is to be calculated using FD in the next, $(i + 1)$th, iteration. Additionally, the optimization run history is inspected, in order to ensure that the part of the Jacobian $\boldsymbol{J}_k$ is computed at least once every few iterations.

The update history is examined in the course of the last N iterations (N being the algorithm control parameter), and a history count η is defined as a total number of iterations, among the last N iterations, in which the Jacobian update using FD was carried out. Both the decision factors $\alpha_k^{(i+1)}$ and the history count η are then translated into a binary AUP decision vector $\boldsymbol{A}$ that stores the information about the update for the upcoming iteration. Let us denote by $A_{i+1,k}$ the kth entry of the vector $\boldsymbol{A}$ referring to the $(i + 1)$th iteration. If the entry of the vector $\boldsymbol{A}$ equals zero, i.e., $A_{i+1,k} = 0$, AUP indicates that, for the kth parameter of the structure at hand, the respective Jacobian column $\boldsymbol{J}_k$ does not have to be calculated with FD in the next iteration; otherwise, i.e., if $A_{i+1,k} = 1$, FD is suggested.

As shown in Fig. 2, at the beginning of AUP, a decision factor $\alpha_k^{(i+1)}$ (see (5)) and history count η are calculated for all parameters. Then, the entries of the decision vector $\boldsymbol{A}$ are established depending on whether the iteration was successful ($\rho > 0$) or not ($\rho \leq 0$). If the iteration was accepted, $A_{i+1,k}$ is set to 0 in the two following cases:

1. For all parameters, if the TR size in the next iteration $\left\| \boldsymbol{d}^{(i+1)} \right\|$ is small, below a user-specified threshold d_{thr},

2. For the selected parameters: if the TR region size in the next iteration $\left\|\boldsymbol{d}^{(i+1)}\right\|$ exceeds the threshold d_{thr}, and if, for a given parameter k, the Jacobian column $\boldsymbol{J}_k$ was: (i) calculated with FD at least once in the last N iterations and (ii) the factor $\alpha_k^{(i+1)}$ is below the user-specified threshold α_{high}.

In the case of the rejected iteration, (i.e., $\rho \leq 0$), as a rule, the decision vector established in the previous iteration, and stored in the selection matrix $\boldsymbol{\Gamma}$, is not altered. However, if $\left\|\boldsymbol{d}^{(i+1)}\right\|$ is below the threshold d_{thr}, $A_{i+1,k}$ is set to 0, when simultaneously the two following conditions are met: (i) $\alpha_k^i < \alpha_{\text{low}}$ and (ii) $\boldsymbol{J}_k$ was calculated with FD at least once in the last N iterations. The user defined thresholds fulfill the condition: $0 < \alpha_{\text{low}} < \alpha_{\text{high}} < 1$, in order to ensure more frequent updates, if the iteration was unsuccessful.

Broyden Update Procedure. Here, the Broyden update procedure, shown in Panel B of Fig. 2, is described in detail. During BUP, the parameters for which the alignment between the most recent design relocation and the coordinate system axes is satisfactory, are selected. Then, the corresponding columns of the circuit response Jacobian $\boldsymbol{J}_R$, are calculated using the Broyden formula (BF) instead of FD. Here, we adopt a rank-one Broyden update [36]

$$\boldsymbol{J}_R^{(i+1)} = \boldsymbol{J}_R^{(i)} + \frac{\left(\boldsymbol{f}^{(i+1)} - \boldsymbol{J}_R^{(i)} \cdot \boldsymbol{h}^{(i+1)}\right) \cdot \boldsymbol{h}^{(i+1)T}}{\boldsymbol{h}^{(i+1)T}\boldsymbol{h}^{(i+1)}}, \qquad i = 0, 1, \ldots \qquad (6)$$

where the following notation is used: $\boldsymbol{f}^{(i+1)} = \boldsymbol{R}\left(\boldsymbol{x}^{(i+1)}\right) - \boldsymbol{R}\left(\boldsymbol{x}^{(i)}\right)$, and $\boldsymbol{h}^{(i+1)} = \boldsymbol{x}^{(i+1)} - \boldsymbol{x}^{(i)}$. Note that (6) only updates the Jacobian in the one-dimensional subspace spanned by $\boldsymbol{h}^{(i)}$ and at least n executions of (6) are necessary to obtain $\boldsymbol{J}_R$ information within the entire space. Consequently, in higher-dimensional spaces, unsatisfactory results are usually obtained when the Jacobian is calculated solely using (6). Here, $\boldsymbol{J}_R$ updating involves both FD and BF.

Let us denote by $\boldsymbol{e}^{(k)}$ the standard basis vectors, i.e., $\boldsymbol{e}^{(k)} = [0 \ \ldots \ 0 \ 1 \ 0 \ \ldots \ 0]^T$ with 1 on the kth position. In the BUP procedure, for each parameter k, the values of the alignment factors $\beta_k^{(i+1)} = \left|\boldsymbol{h}^{(i+1)T}\boldsymbol{e}^{(k)}\right| / \left\|\boldsymbol{h}^{(i+1)}\right\|$ are calculated. The factors $\beta_k^{(i+1)}$ act as a quantification measure of the alignment between the current design relocation $\boldsymbol{h}^{(i+1)}$ and the respective basis vectors $\boldsymbol{e}^{(k)}$. Note that $0 \leq \beta_k^{(i+1)} \leq 1$ ($\beta_k^{(i+1)} = 0$ and $\beta_k^{(i+1)} = 1$, if $\boldsymbol{h}^{(i+1)}$ and $\boldsymbol{e}^{(k)}$ are orthogonal and co-linear, respectively). For a given parameter k, the respective entry of the BUP decision vector $B_{i+1,k}$ is set to 0, if the alignment $\beta_k^{(i+1)}$ is better than a user-defined alignment acceptance threshold β_{min}, otherwise $B_{i+1,k} = 1$; β_{min} is a control parameter for BUP: the higher β_{min}, the more stringent the condition for using BF gets, which likely leads to lower computational savings but higher quality of the obtained solution.

Jacobian Update. Upon performing AUP and BUP, the selection matrix $\boldsymbol{\Gamma}$ is altered and the Jacobian update is performed accordingly. $\boldsymbol{\Gamma}$ is extended by adding the $(i + 1)$th column after each successful iteration. The added column is computed from both AUP and BUP in the following manner (see Panel C of Fig. 2). For a given index k, the entry $\gamma_{i+1,k}$ is set to 0, if the $B_{i+1,k}$ equals 0 (disregarding the value of $A_{i+1,k}$). In that case, the respective Jacobian column J_k is estimated using BF. Otherwise, if $B_{i+1,k} = 1$ and $A_{i+1,k} = 0$, then $\gamma_{i+1,k}$ is set to -1 which indicates that the Jacobian from the previous iteration is to be used, i.e., $J_k(x^{(i+1)}) = J_k(x^{(i)})$. Finally, if $B_{i+1,k} = A_{i+1,k} = 1$, J_k is obtained using FD. Thus, FD is performed exclusively if both routines conclude it is mandatory. Consequently, substantial computational savings are secured, as it is confirmed by the results presented in Sect. 3. At the same time, the observed degradation of the solution quality is practically acceptable. The algorithmic flow of the Jacobian update procedure is shown in Panel D of Fig. 2.

3   Numerical Results and Benchmarking

Here, two high-frequency structures are considered as a benchmark set: an equal-split rat-race coupler (RRC) composed of compact microstrip resonant cells (CMRCs) [37] shown in Fig. 3(a), and a wideband antenna [38] (Fig. 3(b)). The first structure (RRC) is implemented on Taconic RF-35 substrate ($h = 0.762$ mm, $\varepsilon_r = 3.5$, $\tan\delta = 0.0018$). The RRC geometry is described by a parameter vector $x = [\, w_1 \quad l_1 \quad w_2 \quad l_2 \quad w_3 \,]^T$. Relative dimensions are $l_3 = 19w_1 + 18w_2 + w_3 - l_1$, $l_4 = 5w_1 + 6w_2 + l_2 + w_3$, $l_5 = 3w_1 + 4w_2$ and $w_4 = 9w_1 + 8w_2$ (all in mm). The coupler is supposed to operate at $f_0 = 1$ GHz and it has been optimized for maximum bandwidth (defined at -20 dB level of matching and isolation), and symmetric around f_0.

The antenna [37] is also implemented on RF-35 substrate, and it has the following independent geometry parameters $x = [L_0 \ dR \ R \ r_{rel} \ dL \ dw \ L_g \ L_1 \ R_1 \ dr \ c_{rel}]^T$. The antenna has been optimized for minimum reflection within the UWB frequency range (3.1 GHz to 10.6 GHz). The computational model is implemented in CST Microwave Studio and evaluated using its transient solver.

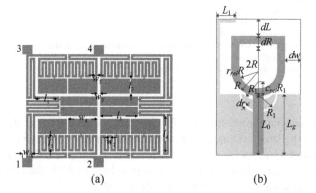

(a)   (b)

Fig. 3. High-frequency structures used for verification of the proposed algorithm: (a) CMRC-based miniaturized microstrip rat-race coupler [37], and (b) wideband antenna [38].

For both structures, ten algorithm runs have been executed with random initial designs. The results for the proposed algorithm are presented in Table 1 and compared to the results obtained with the reference TR algorithm. The following acceptance threshold values were used $\varphi_0 = 0$, 0.025, 0.05, 0.1, 0.2 and 0.3. The higher the threshold value, the lower the savings are, which is due to the more stringent condition for applying the Broyden update formula (and FD performed more frequently). The results for $\varphi_0 = 0$ (Broyden-only Jacobian updates) are enclosed in order to demonstrate that this version does not yield acceptable design quality. The frequency characteristics before and after optimization have been shown in Fig. 4. The proposed combined algorithm delivers considerable reduction of the number of EM simulations needed to find the optimized solution even for the highest value of the threshold $\varphi_0 = 0.3$ (around 48% for the coupler and around 57% for the antenna. The presented results also reveal an important advantage of the proposed algorithm, namely that the solution quality is stable, virtually independent of the acceptance threshold value apart from when $\varphi = 0$. This applies to both the objective function value (bandwidth BW in the case of the coupler and maximum in-band reflection S_{11} for the antenna) and the standard deviation of the respective objective functions. The latter is used as a measure of the results repeatability.

Table 1. Performance statistics of the proposed algorithm

Algorithm		Compact RRC				UWB antenna			
		Cost[a]	Cost savings[b] [%]	BW[c] [GHz]	$SD(B)$[d] [GHz]	Cost[a]	Cost savings[b] [%]	Max $\lvert S_{11}\rvert$[e] [dB]	SD (max$\lvert S_{11}\rvert$)[f] [dB]
Reference		43.0	–	0.27	0.01	111.2	–	−14.9	0.6
φ_0	0[$]	15.9	63.0	0.17	0.12	27.4	75.4	−13.3	1.3
	0.025	17.4	59.5	0.19	0.10	31.0	72.1	−13.4	1.2
	0.05	20.3	52.8	0.22	0.10	35.5	68.1	−13.5	1.2
	0.1	22.0	48.8	0.19	0.11	43.0	61.3	−13.6	1.2
	0.2	22.6	47.4	0.20	0.11	51.1	54.0	−13.6	1.2
	0.3	22.4	47.9	0.20	0.11	47.4	57.1	−13.4	1.0

[a]Number of EM simulations averaged over 10 algorithm runs (random initial points);
[b]Percentage-wise cost savings w.r.t. the reference algorithm;
[c]Objective function values for the compact RRC (bandwidth BW in GHz);
[d]Standard deviation of BW in dB across 10 algorithm runs;
[e]Objective function values for the UWB antenna (maximum in-band reflection S_{11} in dB);
[f]Standard deviation of S_{11} in dB across 10 algorithm runs;
[$] Broyden-only Jacobian updates meaning no FD used whatsoever.

The most suitable threshold value for the RRC is $\varphi_0 = 0.05$: it delivers the best bandwidth and the smallest value of the bandwidth standard deviation, as well as cost savings of around 53%. As far as the UWB antenna is concerned, the value $\varphi_0 = 0.025$ seems to be most appropriate: it delivers savings as high as 72%, while both the design quality and its repeatability are degraded only slightly.

In order to provide more in-depth analysis of the results, a supplementary Table 2 is provided, in which the results obtained with the sole usage of Broyden update procedure (without the accelerated procedure) are shown. The results presented in Table 2

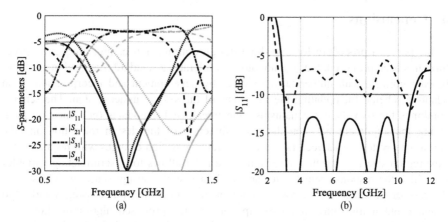

Fig. 4. Reflection characteristics of the considered high-frequency components for the representative algorithm runs: (a) compact RRC, (b) UWB antenna: initial and optimized design are marked gray and black, respectively. The vertical line in (a) indicates the required operating frequency f_0, whereas the horizontal line in (b) the design specifications.

Table 2. Performance statistics of the algorithm utilizing only BUP

Algorithm		Compact RRC				UWB antenna			
		Cost[a]	Cost savings[b] [%]	BW[c] [GHz]	SD(B)[d] [GHz]	Cost[a]	Cost savings[b] [%]	Max $|S_{11}|$[e] [dB]	SD (max$|S_{11}|$)[f] [dB]
Reference		43.0	–	0.27	0.01	111.2	–	−14.9	0.6
φ_0	0[$]	15.9	63.0	0.18	0.11	26.5	76.2	−13.3	1.7
	0.025	13.4	68.8	0.20	0.06	37.5	66.3	−13.9	1.3
	0.05	28.9	32.8	0.23	0.04	47.9	56.9	−14.0	0.9
	0.1	27.0	37.2	0.22	0.05	58.4	47.5	−13.7	1.1
	0.2	42.6	0.9	0.22	0.05	75.9	31.7	−14.3	0.9
	0.3	41.3	4.0	0.21	0.06	89.3	19.7	−14.2	0.8

[a]Number of EM simulations averaged over 10 algorithm runs (random initial points);
[b]Percentage-wise cost savings w.r.t. the reference algorithm;
[c]Objective function values for the compact RRC (bandwidth BW in GHz);
[d]Standard deviation of BW in dB across 10 algorithm runs;
[e]Objective function values for the UWB antenna (maximum in-band reflection S_{11} in dB);
[f]Standard deviation of S_{11} in dB across 10 algorithm runs;
[$] Broyden-only Jacobian updates meaning no FD used whatsoever.

exhibit more pronounced dependence of the design quality on the acceptance threshold value when compared to the combined (AUP and BUP) algorithm (see Table 1). As φ_0 increases, the better the quality of the solution is, as expressed by the objective function value and its standard deviation across ten algorithm runs. At the same time, the cost savings drop significantly (to as low as 4% for the RRC and to around 20% for the antenna). Thus, the introduction of the second, AUP procedure allowed for higher savings delivering similar design quality.

4   Conclusions

In the paper, a procedure for expedited design optimization of high-frequency structures has been proposed. Our methodology accelerates a standard trust-region gradient-based algorithm with numerical derivatives by development and incorporation of an adaptive scheme for updating the Jacobians of the system under design. The scheme combines the two main mechanisms, selective Broyden updates (governed by measuring the alignment between the latest direction of the design relocation and the coordinate system basis vectors), as well as relative design changes with respect to the trust region size in respective directions. The latter allows for discriminating variables for which the response sensitivities are stable across the design space, which is often the case of antennas and microwave components. As demonstrated using a wideband antenna and a miniaturized microstrip coupler, the proposed algorithm allows for considerable computational savings without compromising the design quality. Various design trade-offs controlled by user-defined algorithm parameters have been investigated as well. The algorithm can be used not only to expedite direct optimization of high-fidelity EM models, but also to speed up surrogate-assisted procedures involving variable-fidelity simulations. The future work will be focused on incorporating the proposed methodology into surrogate-based optimization frameworks.

Acknowledgement. The authors thank Dassault Systemes, France, for making CST Microwave Studio available. This work is partially supported by the Icelandic Centre for Research (RANNIS) Grant 174114051, and by National Science Centre of Poland Grant 2015/17/B/ST6/01857.

References

1. Koziel, S., Yang, X.S., Zhang, Q.J. (eds.): Simulation-Driven Design Optimization and Modeling for Microwave Engineering. Imperial College Press, London (2013)
2. Hao, Z.C., He, M., Fan, K., Luo, G.: A planar broadband antenna for the E-band gigabyte wireless communication. IEEE Trans. Ant. Prop. **65**(3), 1369–1373 (2017)
3. Yao, Z., et al.: Integrated silicon photonic microresonators: emerging technologies. IEEE J. Sel. Top. Quantum Electron. **24**(6), 1–24 (2018)
4. Mailloux, R.J.: Phased Array Antenna Handbook, 2nd edn. Artech House, Boston (2005)
5. Simpson, D.J., Psychogiou, D.: Coupling matrix-based design of fully reconfigurable differential/balanced RF filters. IEEE Microw. Wirel. Comp. Lett. **28**(10), 888–890 (2018)
6. Bandler, J.W., et al.: Space mapping: the state of the art. IEEE Trans. Microw. Theory Tech. **52**(1), 337–361 (2004)
7. Sevgi, L.: Electromagnetic Modeling and Simulation. IEEE Press Series on Electromagnetic Wave Theory (2014)
8. Guo, D., He, K., Zhang, Y., Song, M.: A multiband dual-polarized omnidirectional antenna for indoor wireless communication. IEEE Ant. Wirel. Prop. Lett. **16**, 290–293 (2017)
9. Wen, D., Hao, Y., Munoz, M.O., Wang, H., Zhou, H.: A compact and low-profile MIMO antenna using a miniature circular high-impedance surface for wearable applications. IEEE Trans. Ant. Prop. **66**(1), 96–104 (2018)
10. Wu, J., Sarabandi, K.: Compact omnidirectional circularly polarized antenna. IEEE Trans. Ant. Prop. **65**(4), 1550–1557 (2017)

11. Ding, K., Gao, C., Qu, D., Yin, Q.: Compact broadband MIMO antenna with parasitic strips. IEEE Ant. Wirel. Prop. Lett. **16**, 2349–2353 (2017)
12. Shao, W., He, J., Wang, B.-Z.: Compact rat-race ring coupler with capacitor loading. Microw. Opt. Technol. Lett. **52**(1), 7–9 (2010)
13. Tseng, C.-H., Chang, C.-L.: A rigorous design methodology for compact planar branch-line and Rat-Race couplers with asymmetrical T-structures. IEEE Trans. Microw. Theory Tech. **60**(7), 2085–2092 (2012)
14. Wang, J., Wang, B.-Z., Guo, Y.X., Ong, L.C., Xiao, S.: Compact slow-wave microstrip rat-race ring coupler. Electron. Lett. **43**(2), 111–113 (2007)
15. Verma, S., Rano, D., Hashmi, M.S.: A novel miniaturized band stop filter using fractal type defected ground structure (DSG). In: IEEE Asia Pacific Microwave Conference (APMC), pp. 799–802 (2017)
16. Haq, M.A., Koziel, S., Cheng, Q.S.: Miniaturization of wideband antennas by means of feed line topology alterations. IET Microw. Ant. Prop. **12**(3), 2128–2134 (2018)
17. Vendik, I.B., Rusakov, A., Kanjanasit, K., Hong, J., Filonov, D.: Ultrawideband (UWB) planar antenna with single-, dual- and triple-band notched characteristic based on electric ring resonator. IEEE Ant. Wirel. Prop. Lett. **16**, 1597–1600 (2017)
18. Conn, A.R., Gould, N.I.M., Toint, P.L.: Trust Region Methods. MPS-SIAM Series on Optimization (2000)
19. Lalbakhsh, A., Afzal, M.U., Esselle, K.P.: Multiobjective particle swarm optimization to design a time-delay equalizer metasurface for an electromagnetic band-gap resonator antenna. IEEE Ant. Wirel. Prop. Lett. **16**, 915 (2017)
20. Ullah, U., Koziel, S.: A broadband circularly polarized wide-slot antenna with a miniaturized footprint. IEEE Ant. Wirel. Prop. Lett. **17**, 2454–2458 (2018)
21. Liu, Y.Y., Tu, Z.H.: Compact differential band-notched stepped-slot UWB-MIMO antenna with common-model suppression. IEEE Ant. Wirel. Prop. Lett. **16**, 593–596 (2017)
22. Koziel, S., Bekasiewicz, A.: Rapid design optimization of antennas using variable-fidelity EM models and adjoint sensitivities. Eng. Comp. **33**(7), 2007–2018 (2016)
23. Xiao, L.Y., Shao, W., Ding, X., Wang, B.Z.: Dynamic adjustment kernel extreme learning machine for microwave component design. IEEE Trans. Microw. Theory Tech. **66**(10), 4452–4461 (2018)
24. Koziel, S., Ogurtsov, S.: Antenna Design by Simulation-Driven Optimization. Surrogate-Based Approach. Springer, New York (2014). https://doi.org/10.1007/978-3-319-04367-8
25. Koziel, S., Bekasiewicz, A.: Rapid microwave design optimization using adaptive response scaling. IEEE Trans. Microw. Theory Tech. **64**(9), 2749–2757 (2016)
26. Zhang, J., Zhang, C., Feng, F., Zhang, W., Ma, J., Zhang, Q.J.: Polynomial chaos-based approach to yield-driven EM optimization. IEEE Trans. Microw. Theory Tech. **66**(7), 3186–3199 (2018)
27. De Villiers, D.I.L., Couckuyt, I., Dhaene, T.: Multi-objective optimization of reflector antennas using kriging and probability of improvement. In: International Symposium on Antennas and Propagation, San Diego, USA, pp. 985–986 (2017)
28. Jacobs, J.P.: Characterization by Gaussian processes of finite substrate size effects on gain patterns of microstrip antennas. IET Microw. Ant. Prop. **10**(11), 1189–1195 (2016)
29. Easum, J.A., Nagar, J., Werner, D.H.: Multi-objective surrogate-assisted optimization applied to patch antenna design. In: International Symposium on Antennas and Propagation, San Diego, USA, pp. 339–340 (2017)
30. Zhu, J., Bandler, J.W., Nikolova, N.K., Koziel, S.: Antenna optimization through space mapping. IEEE Trans. Ant. Prop. **55**(3), 651–658 (2007)

31. Koziel, S., Leifsson, L.: Simulation-Driven Design by Knowledge-Based Response Correction Techniques. Springer, New York (2016). https://doi.org/10.1007/978-3-319-30115-0
32. Koziel, S.: Fast simulation-driven antenna design using response-feature surrogates. Int. J. RF Microw. CAE **25**(5), 394–402 (2015)
33. Koziel, S., Bekasiewicz, A.: Multi-objective design of antennas using surrogate models. World Scientific (2016)
34. Koziel, S., Kurgan, P.: Compact cell topology selection for size-reduction-oriented design of microstrip rat-race couplers. Int. J. RF Microw. CAE **28**(5), e21261 (2018)
35. Koziel, S., Kurgan, P.: Inverse modeling for fast design optimization of small-size rat-race couplers incorporating compact cells. Int. J. RF Microw. CAE **28**(5), e21240 (2018)
36. Nocedal, J., Wright, S.J.: Numerical Optimization, 2nd edn. Springer, New York (2006). https://doi.org/10.1007/978-0-387-40065-5
37. Koziel, S., Bekasiewicz, A., Kurgan, P.: Rapid design and size reduction of microwave couplers using variable-fidelity EM-driven optimization. Int. J. RF Microw. CAE **26**(1), 27–35 (2016)
38. Alsath, M.G.N., Kanagasabai, M.: Compact UWB monopole antenna for automotive communications. IEEE Trans. Ant. Prop. **63**(9), 4204–4208 (2015)

An Algorithm for Selecting Measurements with High Information Content Regarding Parameter Identification

Christian Potthast$^{(\boxtimes)}$ (iD)

Corporate Sector Research and Advance Engineering, Robert Bosch GmbH,
Robert-Bosch-Campus 1, 71272 Renningen, Germany
christian.potthast@de.bosch.com

Abstract. Reducing the measurement effort that is made for identification of parameters is an important task in some fields of technology. This work focuses on calibration of functions running on the electronic control unit (ECU), where measurements are the main expense factor. An algorithm for information content analysis of recorded measurement data is introduced that places the calibration engineer in the position to shorten future test runs. The analysis is based upon parameter sensitivities and utilizes the Fisher-information matrix to determine the value of certain measurement portions with respect to parameter identification. By means of a simple DC motor model the algorithm's working principle is illustrated. The first use on a real ECU function achieves a measurement time reduction of 67% while a second use case opens up new features for the calibration of connected cars.

Keywords: Parameter identification · Fisher-information matrix ·
Local sensitivity · Measurement information ·
Measurement period reduction

1 Introduction

The software of an electronic control unit (ECU) in a passenger car is an extensive program consisting of a couple of self-contained functions with strict interfaces and a vast number of parameters. These functions are designed to run in real-time and to model complex physical behaviour with a rough model structure only. But nevertheless a high flexibility regarding the output quantities is achieved at the same time since much data is stored in lookup tables. Most of the parameters for ECU functions result from these 1D- or 2D-lookup tables.

Since ECU models strongly rely on data tables, adapted product-specific parameter calibration has to be done again and again for each vehicle type making this process even more time-consuming and costly. By far the most expensive task in calibration are the measurements, i.e. driving on public roads or on test areas, and in the majority of cases more measurement time is carried out and recorded than would actually be necessary.

© Springer Nature Switzerland AG 2019
J. M. F. Rodrigues et al. (Eds.): ICCS 2019, LNCS 11538, pp. 523–536, 2019.
https://doi.org/10.1007/978-3-030-22744-9_41

In practice the tuning process of the parameter values is either done manually or by means of an optimization algorithm. This work is mostly related to cases where an optimizer is used since manual calibration requires a more rigid measurement schedule with long holding times at fixed operating points and hence leaving lesser leeway for compression. Investigations of the optimization algorithm itself are not subject of this article because prior studies have shown that a gradient-based optimizer for nonlinear least-squares curve fitting problems is doing well on the special class of ECU functions.

This paper presents an algorithm that analyses existent measurements piecewise and separates important sections providing new information from sections without content of further value. The main aim is to give advice on how to shorten similar test runs in the future. Secondly, the algorithm may indicate possibilities on how to speed up the parameter optimization. However, this only works out if measurement parts can be left out for simulation.

The basic technology is the well-known Fisher-information matrix, which has been used intensively for the design of experiments [1,2] in various scientific fields. The Fisher matrix is also used for prioritization of parameters in order to identify most sensitive parameters first [3,4] and hereby supporting a more target-oriented estimation process. However, the design of test runs from scratch by utilizing typical eigenvalue-based criteria is impractical for isolated ECU functions: a real test run produces input signals for the function that cannot be defined prior to the real experiment because of influences from the environment, the test track or road traffic.

The following section states the theoretical background before the algorithm is described in Sect. 3 by means of a simple DC motor model. The successful application is demonstrated in Sect. 4 for a new calibration feature of the Connected Car and for a real ECU function, where a drastic time reduction of 67% is achieved.

2   Theoretical Background

For this paper a nonlinear dynamic model in state space formulation is assumed

$$
\begin{aligned}
\dot{\boldsymbol{x}}(t) &= \boldsymbol{f}(t, \boldsymbol{x}(t), \boldsymbol{u}(t), \boldsymbol{\theta}), \quad t > 0, \quad \boldsymbol{x}(0) = \boldsymbol{x_0} \\
\boldsymbol{y}(t) &= \boldsymbol{h}(t, \boldsymbol{x}(t), \boldsymbol{u}(t), \boldsymbol{\theta}), \quad t \geq 0,
\end{aligned}
\tag{1}
$$

where $\boldsymbol{x} \in \mathbb{R}^n$ is the vector of states, $\boldsymbol{y} \in \mathbb{R}^m$ is the vector of model outputs, $\boldsymbol{u} \in \mathbb{R}^q$ is the vector of model inputs and $\boldsymbol{\theta} \in \mathbb{R}^L$ is the parameter vector. The initial conditions for the states are $\boldsymbol{x_0} \in \mathbb{R}^n$.

Calibration aims at finding the unknown parameters $\boldsymbol{\theta}$, which result in the best fit between model outputs $\boldsymbol{y}(t, \boldsymbol{\theta})$ and measured responses $\boldsymbol{y}^M(t)$ from the real process under consideration. The error $\boldsymbol{e}(t, \boldsymbol{\theta})$ denotes the difference between measured and simulated outputs:

$$
\boldsymbol{e}(t, \boldsymbol{\theta}) = \boldsymbol{y}^M(t) - \boldsymbol{y}(t, \boldsymbol{\theta}).
\tag{2}
$$

The root mean square error (RMSE) is typically used to quantify the deviation of a simulated signal consisting of N instants of time obtained with parameter set $\boldsymbol{\theta}$ from a measurement of the i-th quantity of interest within the output vector $\boldsymbol{y}$:

$$RMSE(\boldsymbol{\theta}) = \sqrt{\frac{\sum\limits_{k=1}^{N} e_i^2(t_k, \boldsymbol{\theta})}{N}}. \tag{3}$$

2.1 The Error Stochastics

Assuming that real measurements include all kinds of systematic errors as well as noise from the sensor, the measurement $\boldsymbol{y}^M$ is considered to have a deterministic and a stochastic part:

$$\boldsymbol{y}^M(t) = \boldsymbol{y}^{M,det}(t) + \boldsymbol{\epsilon}(t). \tag{4}$$

Putting Eq. (4) into Eq. (2) gives:

$$e(t, \boldsymbol{\theta}) = \boldsymbol{y}^{M,det}(t) - \boldsymbol{y}(t, \boldsymbol{\theta}) + \boldsymbol{\epsilon}(t) = \boldsymbol{\eta}(t, \boldsymbol{\theta}) + \boldsymbol{\epsilon}(t). \tag{5}$$

Thus, the total error consists of a deterministic part $\boldsymbol{\eta}$ and a stochastic part $\boldsymbol{\epsilon}$. The stochastic part $\boldsymbol{\epsilon}$ is assumed to be Gaussian white noise, where the samples are independent of each other and have a normal distribution with zero mean:

$$\epsilon_i(t_k) \sim \mathcal{N}\left(0, \sigma_{\epsilon_i}^2\right) \quad \text{for } i = 1, \ldots, m \text{ and } k = 1, \ldots, N. \tag{6}$$

The probability density function $p(\epsilon_i(t_k))$ for one time sample t_k from the i-th output signal therefore is:

$$p(\epsilon_i(t_k)) = \frac{1}{\sqrt{2\pi\sigma_{\epsilon_i}^2}} \exp\left(-\frac{\epsilon_i^2(t_k)}{\sigma_{\epsilon_i}^2}\right). \tag{7}$$

For one time sample t_k and multiple output signals $(m > 1)$ it holds:

$$p(\boldsymbol{\epsilon}(t_k)) = \prod_{i=1}^{m} p(\epsilon_i(t_k)) = (2\pi)^{-\frac{m}{2}} \left(\det \boldsymbol{C_\epsilon}\right)^{-\frac{1}{2}} \exp\left(-\frac{1}{2} \sum_{i=1}^{m} \frac{\epsilon_i^2(t_k)}{\sigma_{\epsilon_i}^2}\right), \tag{8}$$

where $\boldsymbol{C_\epsilon} \in \mathbb{R}^{m \times m}$ is the diagonal covariance matrix of the independent measurement error

$$\boldsymbol{C_\epsilon} = \mathrm{E}\left[(\boldsymbol{\epsilon} - \mathrm{E}(\boldsymbol{\epsilon})) \cdot (\boldsymbol{\epsilon} - \mathrm{E}(\boldsymbol{\epsilon}))^\top\right] = \mathrm{E}\left[\boldsymbol{\epsilon} \cdot \boldsymbol{\epsilon}^\top\right] = \begin{pmatrix} \sigma_{\epsilon_1}^2 & 0 & \cdots & 0 \\ 0 & \sigma_{\epsilon_2}^2 & & \\ \vdots & & \ddots & \\ 0 & & & \sigma_{\epsilon_m}^2 \end{pmatrix}. \tag{9}$$

For multiple time samples $(N > 1)$ and multiple output signals $(m > 1)$ it holds:

$$p(\boldsymbol{\epsilon}) = \prod_{k=1}^{N} p(\boldsymbol{\epsilon}(t_k)) = (2\pi)^{-\frac{m \cdot N}{2}} \prod_{k=1}^{N} (\det \boldsymbol{C_\epsilon})^{-\frac{1}{2}} \exp\left(-\frac{1}{2} \sum_{k=1}^{N} \sum_{i=1}^{m} \frac{\epsilon_i^2(t_k)}{\sigma_{\epsilon_i}^2}\right). \tag{10}$$

In parameter estimation the exact (or true) values θ^* of the parameter vector θ are unknown. If during estimation the iteratively determined parameter values are close to the true values and under the assumption that the model represents all systematic properties of the process, i.e.

$$\theta \to \theta^* : \quad \eta(t, \theta) = 0 \quad \Rightarrow \quad \epsilon(t) = y^M(t) - y(t, \theta^*) \tag{11}$$

the density function for $\theta \to \theta^*$ can be approximated in the following form:

$$p(\theta) = (2\pi)^{-\frac{m \cdot N}{2}} \prod_{k=1}^{N} (\det C_\epsilon)^{-\frac{1}{2}} \exp\left(-\frac{1}{2} \sum_{k=1}^{N} \sum_{i=1}^{m} \frac{\left(y^M(t_k) - y(t_k, \theta) \right)^2}{\sigma_{\epsilon_i}^2} \right). \tag{12}$$

2.2   The Fisher-Information Matrix

The general formula for the Fisher-information matrix at a parameter vector $\hat{\theta}$ is derived from the probability density function and is as follows:

$$\mathcal{I}(\hat{\theta}) = \mathrm{E}\left[\left. \frac{\partial \log p(\theta)}{\partial \theta} \right|_{\hat{\theta}} \cdot \left. \frac{\partial \log p(\theta)}{\partial \theta} \right|_{\hat{\theta}}^{\top} \right]. \tag{13}$$

Due to the log function the gradient from Eq. (12) has the compact form:

$$\begin{aligned} \frac{\partial \log p(\theta)}{\partial \theta} &= \sum_{k=1}^{N} \sum_{i=1}^{m} \frac{y^M(t_k) - y(t_k, \theta)}{\sigma_{\epsilon_i}^2} \cdot \frac{\partial y(t_k, \theta)}{\partial \theta} \\ &= \sum_{k=1}^{N} \left(\frac{\partial y(t_k, \theta)}{\partial \theta} \right)^{\top} \cdot C_\epsilon^{-1} \cdot \left(y^M(t_k) - y(t_k, \theta) \right). \end{aligned} \tag{14}$$

Some further calculations and simplifying assumptions stated in [3] lead to

$$\begin{aligned} \mathcal{I}(\hat{\theta}) &= \sum_{k=1}^{N} \left(\left. \left(\frac{\partial y(t_k, \theta)}{\partial \theta} \right)^{\top} \right|_{\hat{\theta}} \cdot C_\epsilon^{-1} \cdot \left. \left(\frac{\partial y(t_k, \theta)}{\partial \theta} \right) \right|_{\hat{\theta}} \right) \\ &= \sum_{k=1}^{N} \left(S_y^{\top}|_{t_k, \hat{\theta}} \cdot C_\epsilon^{-1} \cdot S_y|_{t_k, \hat{\theta}} \right). \end{aligned} \tag{15}$$

The Fisher-information matrix $\mathcal{I} \in \mathbb{R}^{L \times L}$ is a symmetric positive semidefinite matrix, which can be calculated easily as summation over all instants of time. The only necessary quantities are the constant covariance matrix of measurement noise C_ϵ and the time-variant output sensitivity matrix S_y. Both are briefly described in Sects. 2.4 and 2.5.

In order to achieve comparability between parameters often the normalized Fisher matrix $\mathcal{I}_n(\theta)$ is used. Since parameter values may differ by several orders of magnitude, $\mathcal{I}(\theta)$ is multiplied from left and right with a diagonal matrix with parameter values on its main diagonal

$$\mathcal{I}_n(\theta) = \theta^{\top} I_L \, \mathcal{I}(\theta) I_L \, \theta, \tag{16}$$

where I_L is the $L \times L$ identity matrix. It is worth mentioning that parameter values of zero pose a problem to the described normalization because the corresponding normalized sensitivity is set to zero and erroneous zero rows and columns appear in I_n. The problem may be handled by setting zero parameters to small values instead or, if reasonable, by renouncing normalization at all. Although the selection algorithm introduced in Sect. 3 is based upon the normalized Fisher matrix this is not absolutely necessary since the algorithm uses a parameter individual assessment, where the comparability between parameters is virtually unnecessary.

2.3 Parameter Variances

The Cramér-Rao inequality

$$C_\theta \geq I^{-1}(\theta^*) \tag{17}$$

says that the inverse Fisher-information matrix is a lower bound for the covariance matrix C_θ of the parameter estimation error [5–7], where C_θ is defined by:

$$C_\theta = \mathrm{E}\left[(\theta - \theta^*) \cdot (\theta - \theta^*)^\top\right]. \tag{18}$$

The algorithm for evaluation of information content of measurements introduced in Sect. 3 uses the square roots of the main diagonal elements of I (or I_n respectively), i.e. the standard deviations for individual parameters j

$$\sigma_{\theta_j} = \sqrt{I_{jj}^{-1}(\theta^*)}, \quad j = 1, \ldots, L. \tag{19}$$

2.4 Covariance of the Measurement

There are several reasonable possibilities to determine the covariance of the uncoupled measurement error C_ϵ used in Eq. (15). Three are specified below:

(a) Stationary Measurement Phase. If the measurement contains stationary periods of time the variance can simply be determined as sample variance

$$\sigma_{\epsilon_i}^2 = \frac{1}{N_{stat}} \sum_{k=1}^{N_{stat}} \left(y_{i,stat}^M(t_k) - \bar{y}_{i,stat}^M\right)^2 \tag{20}$$

from the N_{stat} measurement samples assumed to be stationary or as in [8]: with $N_{stat} - 1$ in the denominator of Eq. (20). The constant value $\bar{y}_{i,stat}^M$ is the mean value of the N_{stat} measured samples under consideration.

(b) Difference Between Measurement and Simulation. In dynamic measurements it is often impossible to find time periods, which can be regarded as stationary. In these cases a formula proposed in [9] may be used:

$$\sigma_{\epsilon_i}^2 = \frac{1}{N - L} \sum_{k=1}^{N} \left(y_i^M(t_k) - y_i(t_k, \theta)\right)^2. \tag{21}$$

(c) Data Sheet. The third possibility is to exploit given or known information about the applied sensor, e.g. from the sensor manufacturer's data sheet or just from experiential knowledge of experts.

2.5   Sensitivities

It should be pointed out that in this context the sensitivities are the local output sensitivities and should not be mixed up with the global sensitivities, whose meaning and method of calculation is completely different [10]. Basically there are three different options for determining the local output sensitivities:

(a) Sensitivity Differential Equation System. Defining shorter forms of state and output sensitivity as

$$S_x = \frac{\partial x}{\partial \theta} \ , \quad S_y = \frac{\partial y}{\partial \theta} \tag{22}$$

both can be calculated using the sensitivity differential equation system (SDES)

$$\begin{aligned} \dot{S}_x &= \frac{\partial f}{\partial x} S_x + \frac{\partial f}{\partial \theta} \ , \quad t > 0 \ , \quad S_x(0) = \frac{\partial x_0}{\partial \theta} \\ S_y &= \frac{\partial h}{\partial x} S_x + \frac{\partial h}{\partial \theta} \ , \quad t \geq 0. \end{aligned} \tag{23}$$

Equation system (23) is usually created with a computer algebra system using symbolic differentiation and solved numerically afterwards.

The advantage of solving the SDES over other methods like the simple difference quotient (see following paragraph) is that the resulting sensitivities are very precise, if the numerical integration is handled well. The main drawback of this kind of calculation is that the model must be able to be formulated as continuous ODE system as stated in Eq. (1). For many practical models this is already a criterion of exclusion, because models of real processes in industry often contain switching parts like lookup tables or other types of switching operations that require case-by-case analysis. Furthermore solving of the SDES together with the model equations can be computationally expensive due to a large number of state variables.

(b) Difference Quotient. If model equations are not accessible directly or switching parts are included, often the simple difference quotient, which is basically equivalent to an external numerical differentiation, is used. The sensitivity vector s_j, i.e. the derivative of all outputs to a single parameter θ_j, arises from:

$$s_j(t) = \frac{\partial y(t, \theta)}{\partial \theta_j} \approx \frac{y(t, \theta + \Delta\theta_j e_j) - y(t, \theta)}{\Delta\theta_j} \ , \quad j = 1, \ldots, L, \tag{24}$$

where $\Delta\theta_j$ is the deflection of the j-th parameter that can be positive (forward) or negative (backward). $e_j \in \mathbb{R}^L$ is the j-th unit vector. The major drawback of this procedure is the accuracy that decreases with larger absolute deflections $|\Delta\theta_j|$. On the other hand very small deflections lead to catastrophic cancellation [11]. In addition to simple forward and backward differences also methods of higher order (two-sided and fourth-order differences) are common. A guiding value for choosing $\Delta\theta_j$ depending on machine precision, reference point and method order is given in [12].

(c) Automatic Differentiation. Another important variant for calculating sensitivities is automatic differentiation (AD). Assuming the model is available as computer code (in the first instance C-Code) the idea is to create a second code that calculates the necessary derivatives by exploiting the chain rule additionally to the model equations itself, see [13,14] and many others.

3   The Selection Algorithm

The purpose of the selection algorithm is to find measurements or just measurement parts with high information content with respect to a subsequent parameter optimization. A reverse interpretation, i.e. finding the least informative parts, is also reasonable and intended. Since the Fisher matrix is summed up over time and can be interpreted as measure for information content, every time sample increases information. Information increase in turn is reflected in parameter variance decrease, which is used as measure for importance. For practical reasons measurements are analysed section-wise instead of sample-wise from start to end. Based on a threshold value the proposed algorithm decides from section to section whether it is important or not. Algorithm 1 gives the selection process as pseudocode.

Since the decision about section importance is made immediately and together with past sections the algorithm is suitable and explicitly designed for online usage. However, for the sake of clarity Algorithm 1 demonstrates the offline use case. The first step (line 1) is partitioning the measurement in N_s sections. It is worth mentioning that these sections may, but need not be of equal length. The first section acts as initialization for a global Fisher matrix representing the collected information of all important sections (line 2). The outer for-loop (line 5) runs over all sections and combines the new information with the already selected one (line 7). Inside the inner for-loop, that runs over all parameters, a section is chosen as important if at least one parameter related standard deviation decreases by at least the predefined percentage threshold value δ_{thr} (line 10). A positive decision for any parameter updates the global information (line 11), while a negative decision dismisses the combined information without global update.

Before a continued and summarizing discussion of assumptions and limits of the proposed algorithm is given in Sect. 3.2, the following Sect. 3.1 demonstrates algorithm as well as results with the help of a simple DC motor model.

Algorithm 1. Calculate set of important sections $\mathcal{S}$

1: Split measurement(s) in N_s sections
2: $\mathcal{I}^{global}$ ← calculate Fisher-information matrix for section 1 acc. to Eq. (15)
3: σ_θ^{global} ← main diagonal from inverted $\mathcal{I}^{global}$ (Eq. 19)
4: $\mathcal{S}$ ← {1}   // set section 1 as first element of set of important sections
5: **for** $i = 2$ **to** N_s **do**
6:    $\mathcal{I}^{local}$ ← calculate Fisher-information matrix for section i acc. to Eq. (15)
7:    $\mathcal{I}^{comb.}$ ← $\mathcal{I}^{global} + \mathcal{I}^{local}$
8:    $\sigma_\theta^{comb.}$ ← main diagonal from inverted $\mathcal{I}^{comb.}$ (Eq. 19)
9:    **for** $j = 1$ **to** L **do**
10:       **if** $\sigma_{\theta_j}^{comb.} \leq \sigma_{\theta_j}^{global} \cdot (1 - \delta_{thr}/100)$ **then**
11:          $\mathcal{I}^{global}$ ← $\mathcal{I}^{comb.}$; σ_θ^{global} ← $\sigma_\theta^{comb.}$
12:          $\mathcal{S}$ ← $\mathcal{S} \cup \{i\}$ // add section i to set of important sections
13:          **break**
14:       **end if**
15:    **end for**
16: **end for**
17: **return** $\mathcal{S}$

3.1 Exemplary DC Motor Model

The exemplary motor model is an adapted version from [15] with external load torque and mechanical transmission. The model equations are as follows:

$$\dot{\boldsymbol{x}}(t) = \begin{bmatrix} \dot{I}(t) \\ \dot{\omega}(t) \end{bmatrix} = \begin{bmatrix} -R/L & -c/L \\ c/J & -D/J \end{bmatrix} \cdot \begin{bmatrix} I(t) \\ \omega(t) \end{bmatrix} + \begin{bmatrix} 1/L & 0 \\ 0 & -1/(i \cdot J) \end{bmatrix} \cdot \begin{bmatrix} U(t) \\ T_L(t) \end{bmatrix}$$

$$\boldsymbol{y}(t) = \begin{bmatrix} I(t) \\ \omega_m(t) \end{bmatrix} = \begin{bmatrix} 1 & 0 \\ 0 & 30/\pi \end{bmatrix} \cdot \begin{bmatrix} I(t) \\ \omega(t) \end{bmatrix}. \tag{25}$$

The equivalent circuit is shown in Fig. 1(a). Model inputs are voltage U and load torque T_L, outputs are armature current I and angular velocity ω_m in the non-SI unit rpm. All parameters (R, L, D, J, i, c) are assumed to be unknown. In order to design the model being more realistic in terms of the target applications (ECU functions) the motor constant is modelled as 1D-lookup table $c = c(U)$, i.e. the actual value of c depends on the input voltage and has to be interpolated between supporting points as shown in Fig. 1(b).

Assuming that there is no need for the 1D-lookup table to fulfil a particular shape requirement (e.g. monotony), it is transformed into six independent parameters $c_1, \ldots, c_6$ in place of c and hence extending the length of parameter vector $\boldsymbol{\theta}$ from 6 to 11.

Figure 2 shows the preconceived inputs as well as the output measurements generated synthetically by a simulation with true parameters $\boldsymbol{\theta}^*$ (black line) with superimposed white noise in accordance with Eq. (4). The green lines show simulation results obtained with parameter start vector $\boldsymbol{\theta}_0$. Amongst others Table 1 lists the values of $\boldsymbol{\theta}_0$ and $\boldsymbol{\theta}^*$.

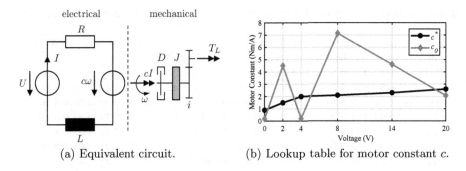

(a) Equivalent circuit.

(b) Lookup table for motor constant c.

Fig. 1. Input and output signals for the DC motor model.

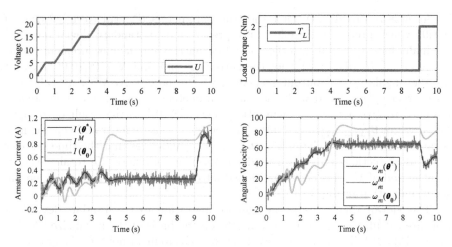

Fig. 2. Input and output signals for the DC motor model. (Color figure online)

Since the model outputs have different ranges of values, the normalized root mean square error (NRMSE), see Eq. (26), is used as scalar dimensionless error measure for both outputs together.

$$NRMSE(\boldsymbol{\theta}) = \sqrt{\frac{\sum_{k=1}^{N} \frac{\left(I^M(t_k) - I(t_k, \boldsymbol{\theta})\right)^2}{\Delta I^2} + \sum_{k=1}^{N} \frac{\left(\omega_m^M(t_k) - \omega_m(t_k, \boldsymbol{\theta})\right)^2}{\Delta \omega_m^2}}{2N}} \tag{26}$$

In contrast to Eq. (3) the residuals are normalized with the corresponding ranges of values ΔI and $\Delta \omega_m$ that are obtained from the deterministic measurement parts $I(t, \boldsymbol{\theta}^*)$ and $\omega_m(t, \boldsymbol{\theta}^*)$:

$$\begin{aligned} \Delta I = \Delta I(\boldsymbol{\theta}^*) = \max\left(I(t, \boldsymbol{\theta}^*)\right) - \min\left(I(t, \boldsymbol{\theta}^*)\right) \\ \Delta \omega_m = \Delta \omega_m(\boldsymbol{\theta}^*) = \max\left(\omega_m(t, \boldsymbol{\theta}^*)\right) - \min\left(\omega_m(t, \boldsymbol{\theta}^*)\right). \end{aligned} \tag{27}$$

Table 1. Different predefined/optimized parameter sets with resulting NRMSE values

Symbol	R	L	D	J	i	c_1	c_2	c_3	c_4	c_5	c_6	$NRMSE$
unit	Ω	H	kg/(m²s)	kg m²	–			Nm/A				–
θ^*	9.0	2.0	0.1	0.1	1.2	0.9	1.5	2.0	2.1	2.3	2.6	0.0499
θ_0	1.8	7.0	0.2	0.05	4.2	0.18	4.5	0.2	7.14	4.6	2.08	0.5760
$\bar{\theta}_{\{1,...,37\}}$	8.67	2.05	0.10	0.10	1.18	0.00	2.20	1.99	2.13	2.30	2.62	0.0497
$\bar{\theta}_{\mathcal{S}}$	8.13	2.01	0.10	0.10	1.32	0.00	2.35	1.96	2.15	2.33	2.63	0.0504
$\bar{\theta}_{\mathcal{R}1}$	7.60	2.22	0.10	0.09	4.20	7.93	1.70	2.01	2.18	2.35	2.67	0.1155
$\bar{\theta}_{\mathcal{R}2}$	8.39	2.04	0.10	0.11	1.19	0.29	2.42	2.04	2.35	2.04	2.63	0.0550
$\bar{\theta}_{\mathcal{R}3}$	9.84	1.99	0.10	0.09	4.20	5.54	26.91	2.07	2.05	2.21	2.59	0.1194

As it is recorded in Table 1 the true parameters θ^* lead to a NRMSE of 0.0499 which pretty much reflects the predetermined standard deviations that originally have been put into the noise distributions for the measurement generation, i.e. $\mathcal{N}_I\left(0,(0.05 \cdot \Delta I)^2\right)$ for the current output and $\mathcal{N}_{\omega_m}\left(0,(0.05 \cdot \Delta \omega_m)^2\right)$ for the angular speed output respectively. The largest NRMSE in Table 1 is unsurprisingly produced with the start parameters θ_0, which are intentionally designed to give a worse fit, while the lowest NRMSE is the result of an optimized vector $\bar{\theta}_{\{1,\,...,\,37\}}$ based upon the complete measurement and hence can be regarded as reference. The index set $\{1,...,37\}$ indicates that the time samples from all 37 sections have been used for optimization, since the total measurement period of 10 s has been partitioned into 37 sections of nearly equal length, see visualization in Fig. 3 (left).

The relevant quantities for the selection algorithm are the standard deviations that are calculated according to Eq. (19) on the basis of the normalized Fisher matrix Eq. (16). The required covariance of measurement noise C_ϵ is a diagonal matrix with the variances used in the already mentioned noise distributions $\mathcal{N}_I$ and $\mathcal{N}_{\omega_m}$ on the main diagonal. The required sensitivities are calculated using difference quotient formula Eq. (24) since the tabled motor constant c complicates the creation of the SDES, see Eq. system (23), drastically. The threshold value δ_{thr} is set to 60 %.

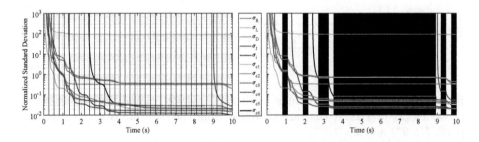

Fig. 3. Normalized standard deviations if the complete measurement is considered (left) and if only the most important sections (set $\mathcal{S}$) are considered (right).

Figure 3 depicts the resulting parameter variations: on the left every instant of time is used for calculation while on the right it is assumed that the selection algorithm has been applied and each not selected section is skipped and highlighted black. As can be seen in Fig. 3 (right) the deviations are monotonically decreasing within each of the 11 sections in set S while remaining constant in the unpicked ones. It should be mentioned that all standard deviations start at ∞ and only accept real numbers when the corresponding parameter becomes sensitive for the first time.

The optimization result with all sections in S is given in Table 1 followed by three results for performance comparison purposes. The latter use section sets $\mathcal{R}_1$, $\mathcal{R}_2$ and $\mathcal{R}_3$ containing as many randomly selected sections as contained in S. Of course, the given NRMSE values are calculated on the basis of the whole measurement and not only for the underlying measurement subset, i.e. the parameter set after optimization is used for a subsequent simulation of the whole 10 s and the result is put into Eq. (26). It turns out that the measurement parts in S produce a fit only slightly worse than the reference, while a random selection is less reliable performing sometimes comparably well ($\mathcal{R}_2$) and sometimes significantly worse ($\mathcal{R}_1$, $\mathcal{R}_3$).

3.2   Assumptions and Limitations

Some characteristics of the selection algorithm deserve further discussion:

- In view of measurement period reduction an unambitious partitioning generally leads to a couple of measurement snippets that are hard or impossible to combine into a realizable test run. Therefore a more anticipatory segmentation is essential for practical usage.
- In view of computation time reduction during optimization an adequate segmentation is also important since stopping the simulation and restarting at arbitrary points without having simulated the parts in between changes the characteristics of a dynamic model.
- The threshold value δ_{thr} is the main tuning parameter of the algorithm. It may vary between 0 % (select everything) and 100 % (select nothing except for Sect. 1). The choice depends on the desired share of important segments. Values between 40 % and (more ambitious) 80 % have turned out to be sensible.
- The measurement section order is crucial for the resulting set S. For online usage this is logical and unavoidable if an immediate decision is necessary. For offline usage it always has to be taken into account that the algorithm's decision is based upon the preceding measurements.
- Since the theory only holds for models without systematic errors, see Eq. (11), and true parameters, the used start parameter values should not be too far away from the true ones. Real applications in the context of ECU function calibration allow the use of start parameters from previous projects and therefore satisfy this precondition.

– It would be unreasonable to expect that the algorithm always finds either the best sections or only the necessary ones. It must be kept in mind that the basis is just a stochastic analysis not necessarily valid for every single spot check. Also the fitting quality depends strongly on the last link in the chain: the optimization algorithm.

4   Real Industry Applications

Two applications show how the algorithm can be applied beneficially in industry.

4.1   Shortening of a Standard Measurement Program

The goal of the first application is to shorten an established measurement program used for calibration of an ECU function that calculates temperatures in the exhaust gas system of a passenger car. The task for the calibration engineer is to carry out the measurements and tune 59 parameters so that the model predicts measured temperatures as good as possible. The original program requires a car ride of several hours in total. For reasons of confidentiality more details about the model's interior structure and the measuring process cannot be revealed.

Although the function has two output signals (temperatures of exhaust gas and exhaust pipe), Fig. 4 only shows gas temperatures for the benefit of a more compact presentation. The upper digram in Fig. 4 shows the standard test run consisting of 11 independent real measurements separated by vertical lines. The measurement order does not matter, they are just plotted one after another in the same diagram in order to save space and to show the total length.

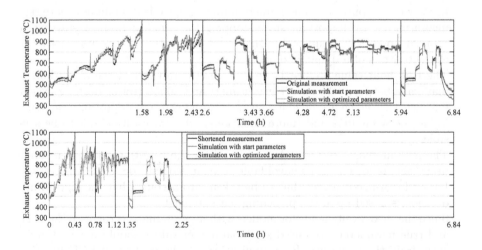

Fig. 4. Temperature characteristics of original and shortened measurement program.

The segmentation for Algorithm 1 is done in such a way that some sections, which are necessary for preserving the conventional test run structure, are pre-selected. Using a threshold value $\delta_{thr} = 40\,\%$ the result from Algorithm 1 after some manual tuning leads to a shorter test run that was put into practice, see lower graph in Fig. 4.

Two optimizations of original and shortened test run result in different parameter sets reducing the initial error from approx. $60\,°C$ to $11\,°C$ in both cases. Applied to a third comparison measurement the two parameter sets produce similar good fits. The RMSE for parameters from the standard test is $11.78\,°C$ and the parameters from the short test run lead to $12.67\,°C$. The slight worsening of almost $1\,°C$ is very acceptable in view of the achieved time reduction of 67% and the related cost saving.

4.2 Online Parametrization of Connected Cars

The second application deals with calibration data selection in the context of the Connected Car [16]. The use case is to decide online which parts of the car's recorded data is of value for calibration. The benefit from connecting cars in this setting is that parametrization can be carried out not only for a single car but for a fleet as well. Measurements can be transferred to a central cloud space or server and updated parameters can be sent back and flashed into the ECU software. The need for an importance decision of measurement data right in the car is caused by limitations of the data transmission via mobile net: the limiting factor is either bandwith (3G network) or cost (4G network).

The Connected Car application of this measurement selection algorithm is also subject of a related published patent application [17].

5   Conclusion

In this paper an algorithm was presented that distinguishes important measurement parts from unimportant ones. The algorithm may either be used online during the test run for immediate decisions as used e.g. for reducing measurement data being transferred via mobile network in the Connected Car application, see Sect. 4.2. Or it may be used offline for analysing and shortening of pre-built measurement plans as shown for the exhaust temperature ECU function, see Sect. 4.1.

Currently the simplest method for sensitivity calculation is used: the difference quotient. Especially in the context of ECU software it should be switched to automatic differentiation. Future work will concentrate on a modified version of the algorithm focussing on offline uses cases where the order in which the algorithm reads the measurement sections should no longer affect its selection result. The future calibration trend towards Big Data (i.e. measure and record everything) will increase the need for thinning out measurements and will open up further beneficial applications for the selection of important measurements parts.

References

1. Atkinson, A.C.: The usefulness of optimum experimental designs. J. Roy. Stat. Soc. Ser. B (Methodol.) **58**(1), 59–76 (1996)
2. Goodwin, G. C.: Identification: experiment design. In: System and Control Encyclopedia, vol. 4, pp. 2257–2264. Pergamon Press, Oxford (1987)
3. Majer, C.P.: Parameterschätzung, Versuchsplanung und Trajektorienoptimierung für verfahrenstechnische Prozesse. Fortschritt-Berichte Reihe 3, No. 538. VDI-Verlag Düsseldorf (1998)
4. Schmidt, A.P., Bitzer, M., Imre, Á.W., Guzzella, L.: Experiment-driven electrochemical modeling and systematic parameterization for a lithium-ion battery cell. J. Power Sources **195**, 5071–5080 (2010). https://doi.org/10.1016/j.jpowsour.2010.02.029
5. Ljung, L.: System Identification: Theory for the User, 2nd edn. Prentice Hall, Upper Saddle River (2006)
6. Norton, J.P.: An Introduction to Identification. Academic Press, London (1986)
7. Goodwin, G.C., Payne, R.L.: Dynamic System Identification: Experiment Design and Data Analysis, vol. 136. Academic Press, New York (1977)
8. Zwillinger, D.: CRC Standard Mathematical Tables and Formulae. CRC Press, Boca Raton (1995)
9. Schittkowski, K.: Experimental design tools for ordinary and algebraic differential equations. Ind. Eng. Chem. Res. **46**(26), 9137–9147 (2007). https://doi.org/10.1021/ie0703742
10. Saltelli, A., et al.: Global Sensitivity Analysis. The Primer. Wiley, Hoboken (2008)
11. Forsythe, G.E.: Pitfalls in computation, or why a math book isn't enough. Am. Math. Mon. **77**(9), 931–956 (1970). https://doi.org/10.1080/00029890.1970.11992636
12. Schittkowski, K.: NLPQLP A Fortran Implementation of a Sequential Quadratic Programming Algorithm with Distributed and Non-Monotone Line Search - UserâĂŹs Guide, Version 4.2 (2014)
13. Griewank, A., Walther, A.: Evaluating Derivatives: Principles and Techniques of Algorithmic Differentiation, 2nd edn. Society for Industrial and Applied Mathematics, Philadelphia (2008)
14. Rall, L.B., Corliss, G.F.: An introduction to automatic differentiation. In: Computational Differentiation: Techniques Applications and Tools, pp. 1–18. SIAM, Philadelphia (1996)
15. Toliyat, H.A., Kliman, G.B.: Handbook of Electric Motors, 2nd edn. CRC Press, Boca Raton (2004)
16. Coppola, R., Morisio, M.: Connected car: technologies, issues, future trends. In: ACM Computing Surveys (CSUR), vol. 49, no. 3, pp. 46:1–46:36. ACM, New York (2016). https://doi.org/10.1145/2971482
17. Robert Bosch GmbH: Method and device for influencing a vehicle behavior. Inventors: Potthast, C., Bleile, T., Hagen, L., Petridis, K., Bitzer, M., Jarmolowitz, F. Filing date: 28 November 2016. Patent application no. WO 2017/093156 A1. Publication date: 8 June 2017

Optimizing Parallel Performance of the Cell Based Blood Flow Simulation Software HemoCell

Victor Azizi Tarksalooyeh[1(✉)], Gábor Závodszky[1], and Alfons G. Hoekstra[1,2]

[1] Computational Science Lab, Institute of Informatics,
University of Amsterdam, Amsterdam, The Netherlands
`v.w.azizitarksalooyeh@uva.nl`
[2] ITMO University, Saint-Petersburg, Russian Federation

Abstract. Large scale cell based blood flow simulations are expensive, both in time and resource requirements. HemoCell can perform such simulations on high performance computing resources by dividing the simulation domain into multiple blocks. This division has a performance impact caused by the necessary communication between the blocks. In this paper we implement an efficient algorithm for computing the mechanical model for HemoCell together with an improved communication structure. The result is an up to 4 times performance increase for blood flow simulations performed with HemoCell.

Keywords: Blood flow simulation · High performance computing · Computational optimization

1 Introduction

Blood flow simulation remains an area of active research. Many interesting properties have been identified with the help of simulations [4,6,9–11,13]. There is an increasing interest in blood flow simulations in which the blood cells (red blood cells, platelets, white blood cells) are fully resolved [3,8,15,16]. These simulations can be used to understand and find underlying mechanics of complex behaviour of blood flows including but not limited to platelet margination [9], the formation of the cell free layer [6], the Fåhræus–Lindqvist effect [2], the behaviour in microfluidic devices or the behaviour around micromedical implants [1,5]. Simulations that model blood as a pure fluid flow are not able to recover these intricate properties of blood flow.

One of the challenges of suspension simulation codes is to parallelize them such that interesting systems with sufficient number of cells (>1000 cells) can be simulated for an extended duration (>0.1 s) in a reasonable time span (<5 days). Only a few open source solutions exist for suspension simulations that need to implement a complex mechanical model for the simulated cells, HemoCell [16] and Palabos-LAMMPS [14] are examples of available open-source codes that can

© Springer Nature Switzerland AG 2019
J. M. F. Rodrigues et al. (Eds.): ICCS 2019, LNCS 11538, pp. 537–547, 2019.
https://doi.org/10.1007/978-3-030-22744-9_42

be used to simulate blood flow. Other codes exists but are not (yet) available as open-source.

HemoCell is a software package that is developed at the University of Amsterdam that is able to simulate blood flow at high shear rates (>1000 s^{-1}) and with a high number of cells (>1000 cells). In this paper we present HemoCell an highly efficient parallel code for blood flow suspension simulations.

HemoCell is built on top of Palabos [7] and offers support for complex suspension simulations. Palabos is a general purpose lattice Boltzmann solver with high performance computing capabilities. We will shortly introduce HemoCell and its underlying models, followed by a discussion of challenges and solutions for efficient parallel simulations. These include boundary communication of processors for the suspension part, efficiently storing relevant information while avoiding global communication, and efficiently computing the complex material model associated with the cells within HemoCell. Next, we discuss the theoretical and practical implications of the methods we used to implement the suspension simulation software within HemoCell and provide performance measurements.

1.1   HemoCell

HemoCell [16] is an open source parallel code for simulating blood flows with fully resolved cells that is built as a library on top of Palabos [7]. Palabos is a versatile library which can be used to solve pure fluid flow problems with the lattice Boltzmann method (LBM). Palabos offers relevant multi-processing abilities. HemoCell implements the cell mechanics simulations and their coupling to the fluid using the immersed boundary method (IBM), see also Fig. 1.

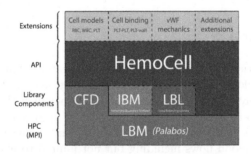

Fig. 1. Overview of the Palabos and HemoCell libraries

HemoCell uses data parallelism to distribute the workload over many cores. With the help of PalaBos HemoCell can divide the flow domain into multiple rectangular blocks of each which represents a processor These domains are called atomic blocks (AB). ABs are abstracted away from the user through the use of functionals, which can be used to perform operations on a domain without knowing about the underlying distributed structure. Furthermore, each simulation can have multiple fields, which span the whole domain and represent a specific part of the simulation. In Hemocell two fields are used, a fluid field and a cell field. Palabos takes care of the boundary communication between processors for the fluid field. HemoCell takes care of the cell field and of the communication between the two fields as required by the immersed boundary method [12].

The cells consist of vertices which are connected through links that make up a boundary to represent a cell in the fluid. A RBC in HemoCell has 1280 vertices. A complex mechanical model is used to calculate forces [16]. This mechanical model requires that a cell is present on both processors whenever it is crossing a boundary. This results in the two main bottlenecks and thus challenges for HemoCell.

1. The material model of the cells needs to be calculated efficiently.
2. Dividing the cell field into multiple processors is complex because the material model requires duplication of cells over boundaries.

2   Calculating the Mechanical Model of a Cell

The cells within HemoCell are implemented as vertices and connections that form a triangulated mesh. These cells compute the forces acting on its vertices through a mechanical model [16]. Figure 2 shows a mesh used to represent a red blood cell.

Fig. 2. Mesh representing red blood cell in HemoCell

Závodszky et al. [16] model the forces acting on the vertices of a cell as follows:

$$F_{\text{total}} = F_{\text{link}} + F_{\text{bend}} + F_{\text{volume}} + F_{\text{area}} + F_{\text{visc}} \tag{1}$$

Below we list all five forces and explain in detail what information is needed to compute them.

1. The link force F_{link} acts along the edges and between neighbouring points. The force on a single vertex i (F_{link}^i) can be described as follows:

$$F_{link}^i = \sum_{n=1}^{m} C_{link} \frac{E_{i,i_n} - |i_n^x - i^x|}{E_{i,i_n}} \qquad (2)$$

Where E_{i,i_n} is the equilibrium length between two vertices, i^x is the location of vertex i, m is the number of direct neighbours of vertex i and i_n is the n'th direct neighbour of vertex i. C_{link} consists of all the constant terms that do not change during a simulation as explained by Závodszky et al. [16].

2. The bending force F_{bend} uses patches which are defined as a plane that goes through the average location of all direct neighbours of vertex i. The normal direction of this plane is defined as the average normal of all neighbouring triangles that include vertex i. The distance along the normal direction of this plane towards vertex i is used to calculate the bending force on vertex i, a negative term is added to the neighbours of i to make the force zero-sum.

$$F_{bend}^i = C_{bend} \left(E_i^{patch} - \left(\frac{\sum_{n=1}^{m} i_n^x}{m} - i^x \right) \cdot \left(\frac{\sum_{n=1}^{m} \mathbf{normal}(t_i^n)}{L} \right) \right)$$
$$- \sum_{n=1}^{m} \frac{1}{N_i^m} F_{bend}^{i_n} \qquad (3)$$

Where E_i^{patch} is the equilibrium distance between the patch and the vertex i along the patch normal. t_i^n is the n'th triangle that is a direct neighbour of vertex i. $\mathbf{normal}()$ returns the normal pointing outward from a triangle. L is the length of the summation of the normal vectors of all the triangles that are part of the patch, thus this division results in a unit vector along the average normal direction. The dot product results in a length term along the patch normal. N_i^m is the number of direct vertex neighbours of i_n. C again of all the constant terms that do not change during a simulation.

3. The area force F_{area} acts on all the triangles that are part of the mesh. Therefore the force on a single vertex is a sum over all neighbouring triangles:

$$F_{area}^i = \sum_{n=1}^{m} \mathbf{C}_{area} \left(\frac{E_{i_t^n}^{area} - \mathbf{area}(i_t^n)}{E_{i_t^n}^{area}} \right) (i^x - \mathbf{middle}(i_t^n)) \qquad (4)$$

Where $\mathbf{area}()$ calculates the area of a triangle, $E_{i_t^n}^{area}$ is the equilibrium value for the area of triangle i_t^n. $\mathbf{middle}()$ calculates the average of the three triangle vertices of triangle i_t^n. $\mathbf{C}_{area}()$ is a function that takes the area ratio as input and outputs a force coefficient.

4. The volume force F_{volume} results from the total volume of the cell, thus information about all vertices is needed. The force is distributed over the vertices proportional to the area of the direct neighbouring triangles of that vertex.

$$F_{volume}^i = \frac{\mathbf{volume}(cell^i) - E_{cell^i}^{volume}}{E_{cell^i}^{volume}} \sum_{n=1}^{m} C_{volume} \frac{\mathbf{area}(t_i^n)}{E_{t_i^n}^{area}} \mathbf{normal}(t_i^n) \qquad (5)$$

Where **volume** () calculates the volume of a complete cell, this function needs every vertex of the cell as input. $E_{\text{cell}^i}^{\text{volume}}$ is the equilibrium volume of cell^i and **normal**() is the normal direction of triangle t_i^n.

5. The viscous force F_{visc} limits the relative velocity of neighbouring vertices connected with an edge.

$$F_{\text{visc}}^i = \sum_{n=1}^{m} C_{\text{visc}} \cdot \left((v^i - v_n^i) \cdot \left(\frac{i_n^x - i^x}{|i_n^x - i^x|} \right) \right) \cdot \left(\frac{i_n^x - i^x}{|i_n^x - i^x|} \right) \tag{6}$$

Where v^v and v_n^i are the velocity of vertex i and i_n respectively. $\sum_{n=1}^{m}$ sums over all direct vertex neighbours of i.

2.1   Implementation of the Mechanical Model

The formulas for calculating force on each independent vertex are explained above. Between the calculation of the separate forces there are some overlaps, for example the calculation of the area of a triangle is used for both the volume and area forces Eqs. 4 and 5. This leaves room for optimization within implementing the calculations. In Fig. 3 a pseudo code of the implementation is shown. In this implementation we have tried to calculate each necessary value only once. Furthermore, we try to minimize the number of loops. Most notably in the first loop which calculates F_{area} all the necessary calculations for F_{volume} are stored for the second loop. In addition F_{link} and F_{visc} are calculated in the same loop as well.

```
for triangle in cell.triangles:
  volume += volume_from_triangle(triangle);
  normal,area,center = triangle_properties(triangle)
  area_force = ((area - eq(area))/eq(area)) * C_area
  for vertex in triangle:
    vertex.force += (center-vertex)*area_force
volume_force = ((volume - eq(volume))/eq(volume)) * C_volume
for triangle in cell.triangles:
  triangle_volume_force = triangle_volume_force_formula()
  for vertex in triangle:
    vertex.force += triangle_volume_force
for vertex in cell:
  for neighbour in vertex:
    middle += neighbour
  vertex.force +=bending_force_formula(middle, vertex)
for edge in cell:
  vertex.force += link_force_formula(edge)
  vertex.force += visc_force_formula(edge)
```

Fig. 3. Pseudocode explaining how we optimized the calculation of the mechanical model within HemoCell.

3 Implementation of the Cell Field Communication Structure

When the cell field is divided up into multiple atomic blocks it becomes necessary to implement a communication structure. For a regular fluid field this simply constitutes to communicating the values of the fluid cells in the boundary layer to their corresponding neighbours. However it is not so simple for the cell field. The number of vertices in a communication boundary can change over time and therefore the communication size is not static but dynamic. Furthermore at every communication step it has to be determined which vertices are present within a communication boundary and which vertices are not.

Cells need information from all their vertices to calculate the mechanical forces. Almost all forces (F_{area}, F_{link}, F_{bend}, F_{visc}) that act on the vertices only need information from their direct neighbours to be calculated. However the volume force F_{volume} needs information of all the vertices of the cell to be calculated. Therefore whenever a single vertex of a cell is present in an atomic block, the boundaries must include every other vertex of the corresponding cell as well. This means that the size of the boundary must be larger than the largest possible diameter of a cell. Figure 4 shows that a larger boundary size means that the number of neighbours and thus the communication will increase if the atomic blocks get too small.

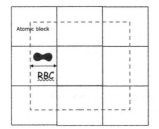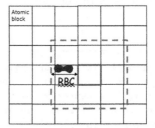

Fig. 4. Visualization of the boundary size needed for the cell field.

There is a simple way to implement this boundary, namely by communication of vertices in the boundary. We will use this communication pattern as the base upon which we propose improvements, see Fig. 5. In the naïve implementation firstly all neighbours are determined that overlap with the boundary of the atomic block. Within HemoCell a RBCs (the largest cell) can stretch up to 12 μm. Thus all neighbours within a 12 μm range send the vertices corresponding to the overlap they have with the boundary. This method has two drawbacks: First a lot of unnecessary data is communicated and second when the boundary size is larger than an atomic block the number of neighbours with which communication is necessary grows, usually in the form of $(2N+1)^3 - 1$ Where N is the number of neighbours in a single direction. So going from $N = 1$ to $N = 2$ creates $124 - 26 = 98$ extra neighbours.

```
neighbours = block.neighbours(12) for neighbour in neighbours:
  send_particles = block.findparticles(intersect(block, neighbour)
  MPI_Isend(neighbour,send_particles.size())
  MPI_Isend(neighbour,send_particles)
While (MPI_WaitAny(neighbours)):
  MPI_Recv(neighbour, size)
  MPI_Recv(neighbour, recv_particles, size)
  block.add_particles(recv_particles)
```

```
neighbours = block.neighbours(1)
  #Same communication pattern as top code block
  #But with a boundary of size one
Neighbours = block.neighbours(12) requested_cells =
block.findlocalcellIds() for neighbour in neighbours:
  MPI_Isend(neighbour,requested_cells)
for neighbour in neighbours:
  MPI_Probe(neighbour) #Get any neighbour
  MPI_Recv(neighbour, requested_cells)
  send_particles =
block.findParticlesFromCells(requested_cells)
  MPI_Isend(neighbour, send_particles)
for (neighbour in neighbours):
  MPI_Probe()              #Get any neighbour
  MPI_Irecv(neighbour,recv_buffer)
for (neighbour in neighbours):
  MPI_WaitAny(receive) #Wait for any receive
  block.addParticles(recv_buffer)
MPI_WaitAll(sends)
```

Fig. 5. The top block shows in pseudocode a naïve implementation of the boundary communication. The bottom block shows our optimized implementation of the boundary communication algorithm.

We implemented an improved and consequently faster method to communicate vertices in boundaries. The main idea is to only communicate vertices of cells that are needed. For this an extra communication step needs to be implemented. In this extra communication step an atomic block sends a list with all the IDs of the cells that need to be communicated to its neighbours. In the next communication step only these vertices are communicated. It is not possible to get rid of the inefficient boundary communication entirely as vertices very close to the domain are needed for non-local force calculations (e.g. inter cellular forces). However, this is much more efficient if only a very small boundary needs to be communicated.

Fig. 6. The domain with which the simulations are performed with a differing number of processors

3.1   Comparison Between Naïve and Optimized Implementation of the Boundary Communication Algorithm

To test the performance gain of our optimized boundary communication algorithm we have set up a simulation which is executed both with the naïve and the optimized implementation. The simulation consists of a cubic $128\ \mu m^3$ volume that is periodic in all directions. Within this volume 7736 red blood cells are present. Figure 6 shows the simulated domain. An external body force is applied to drive the cell suspension inside the volume. The volume is simulated for $0.1\,s$ and statistics are collected over the whole duration. The results are plotted in Fig. 7.

The results show a significant improvement of HemoCell in two ways. Firstly, the base performance has improved by $\approx36\%$, this can be deducted from the difference in wall clock time per iteration in Fig. 7 for 8 cores. Secondly, the strong scaling (dividing the same domain into more smaller atomic blocks) properties are better. In the worst case ($512\ \mu m^3$ per atomic block) the edges of an atomic block are only $8\ \mu m$ long. This means that the boundary of each block overlaps with 124 neighbours. In this case we see a performance improvement of ≈4 times over the naïve version. Over the whole range we see that our improved communication performs better.

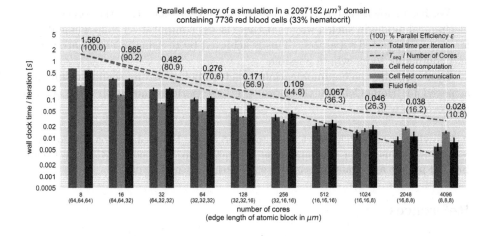

a)

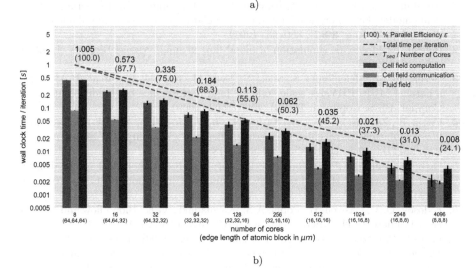

b)

Fig. 7. Statistics for each of the simulations. The fluid part is handled by Palabos. The dotted line shows perfect linear scaling. (a) shows the statistics for the naïve implementation of the communication. (b) shows the statistics for our improved implementation.

4   Conclusions

Improving the performance of fully resolved blood flow simulations allows us to perform simulations up to 4 times faster. For a simulation of 1 s a total number of 10 million timesteps is required. This means that the improved version of HemoCell only needs one day to complete this simulation with ABs of 512 μm³, whereas the naïve version would need four days.

We have shown that it is possible to merge the calculation the forces of the mechanical model in such a way that there is less computation than when all the forces are computed separately. This is achieved by re-using intermediate values and combining loops where possible.

By improving the communication structure better strong scaling results are achieved for HemoCell. Furthermore, the base performance with large ABs is improved as well.

Acknowledgments. This work was supported by the European Union Horizon 2020 research and innovation programme under grant agreement no. 675451, the Comp-BioMed project and grant agreement no. 671564, the ComPat project.

References

1. Augsburger, L., Reymond, P., Rufenacht, D., Stergiopulos, N.: Intracranial stents being modeled as a porous medium: flow simulation in stented cerebral aneurysms. Ann. Biomed. Eng. **39**(2), 850–863 (2011)
2. Bagchi, P.: Mesoscale simulation of blood flow in small vessels. Biophys. J. **92**(6), 1858–1877 (2007)
3. Bernaschi, M., Melchionna, S., Succi, S., Fyta, M., Kaxiras, E., Sircar, J.: MUPHY: a parallel MUlti PHYsics/scale code for high performance bio-fluidic simulations. Comput. Phys. Commun. **180**(9), 1495–1502 (2009)
4. Czaja, B., Závodszky, G., Azizi Tarksalooyeh, V., Hoekstra, A.: Cell-resolved blood flow simulations of saccular aneurysms: effects of pulsatility and aspect ratio. J. Roy. Soc. Interface **15**(146), 20180485 (2018)
5. Farb, A., Burke, A.P., Kolodgie, F.D., Virmani, R.: Pathological mechanisms of fatal late coronary stent thrombosis in humans. Circulation **108**(14), 1701–1706 (2003)
6. Fedosov, D.A., Caswell, B., Popel, A.S., Karniadakis, G.E.: Blood flow and cell-free layer in microvessels. Microcirculation **17**(8), 615–628 (2010)
7. Latt, J.: Palabos, parallel lattice Boltzmann solver (2009). https://palabos.org
8. Moeendarbary, E., Ng, T.Y., Zangeneh, M.: Dissipative particle dynamics: introduction, methodology and complex fluid applications - a review. Int. J. Appl. Mech. **01**(04), 737–763 (2009)
9. Mountrakis, L., Lorenz, E., Hoekstra, A.G.: Where do the platelets go? A simulation study of fully resolved blood flow through aneurysmal vessels. Interface Focus **3**(2), 20120089 (2013)
10. Mountrakis, L., Lorenz, E., Hoekstra, A.G.: Scaling of shear-induced diffusion and clustering in a blood-like suspension. EPL (Europhys. Lett.) **114**(1), 14002 (2016)
11. Ouared, R., Chopard, B.: Lattice Boltzmann simulations of blood flow: non-Newtonian rheology and clotting processes. J. Stat. Phys. **121**(1–2), 209–221 (2005)
12. Peskin, C.S.: The immersed boundary method. Acta Numerica **11**, 479–517 (2002)
13. Skorczewski, T., Erickson, L., Fogelson, A.L.: Platelet motion near a vessel wall or thrombus surface in two-dimensional whole blood simulations. Biophys. J. **104**(8), 1764–1772 (2013)
14. Tan, J., Sinno, T.R., Diamond, S.L.: A parallel fluid-solid coupling model using LAMMPS and Palabos based on the immersed boundary method. J. Comput. Sci. **25**, 89–100 (2018)

15. Ye, T., Phan-Thien, N., Lim, C.T.: Particle-based simulations of red blood cells-a review. J. Biomech. **49**(11), 2255–2266 (2016)
16. Závodszky, G., van Rooij, B., Azizi, V., Hoekstra, A.: Cellular level in-silico modeling of blood rheology with an improved material model for red blood cells. Front. Physiol. **8**, 563 (2017)

Surrogate-Based Optimization of Tidal Turbine Arrays: A Case Study for the Faro-Olhão Inlet

Eduardo González-Gorbeña[1]([⊠]) [iD], André Pacheco[1] [iD],
Theocharis A. Plomaritis[2] [iD], Óscar Ferreira[1], Cláudia Sequeira[1],
and Theo Moura[1] [iD]

[1] Centre for Marine and Environmental Research, Universidade do Algarve,
Ed. 7, Campus de Gambelas, 8005-139 Faro, Portugal
{egeisenmann, ampacheco, oferreir, cdsequeira,
tgmoura}@ualg.pt
[2] Faculty of Marine and Environmental Science, Department of Earth Science,
Universidad de Cádiz, Campus Rio San Pedro (CASEM),
Puerto Real, 11510 Cádiz, Spain
haris.plomaritis@uca.es,
http://www.cima.ualg.pt/

Abstract. This paper presents a study for estimating the size of a tidal turbine array for the Faro-Olhão Inlet (Potugal) using a surrogate optimization approach. The method compromises problem formulation, hydro-morphodynamic modelling, surrogate construction and validation, and constraint optimization. A total of 26 surrogates were built using linear RBFs as a function of two design variables: number of rows in the array and Tidal Energy Converters (TECs) per row. Surrogates describe array performance and environmental effects associated with hydrodynamic and morphological aspects of the multi inlet lagoon. After validation, surrogate models were used to formulate a constraint optimization model. Results evidence that the largest array size that satisfies performance and environmental constraints is made of 3 rows and 10 TECs per row.

Keywords: Hydro-morphodynamic modelling · Marine renewable energy · Ria Formosa

1 Introduction

The sustainable development of island or isolated communities is getting in the agenda of governments all around the world. In Europe, the European Commission is promoting initiatives to help islands generate their own sustainable, low-cost energy [1]. Tidal current energy is a form of marine renewable energy, which converts the kinetic energy of the tides into electricity using Tidal Energy Converters (TECs). As tides are perfectly predictable and have a significant energy density, tidal current energy becomes a reliable source of clean energy if compared with other sources of renewable energy, such as waves or wind. The main drawback is that it is very site specific, i.e. there are not too many places around the world with feasible conditions for commercial

© Springer Nature Switzerland AG 2019
J. M. F. Rodrigues et al. (Eds.): ICCS 2019, LNCS 11538, pp. 548–561, 2019.
https://doi.org/10.1007/978-3-030-22744-9_43

exploration. As with wind energy, in order to decrease the levelized cost of electricity (LCOE) tidal turbines need to be grouped in arrays. Reasons that served to conduct research in TEC arrays optimization.

Initially, optimization strategies that were developed to answer the TEC array problem have been inspired by the wind energy industry, whose primary objective is to maximize power output by reducing wake interferences between turbines within the array [2]. For this purpose, analytical wake models were developed to model these effects, being the Jensen wake model [3] one of the most popular methods. The main difference between the effects on flow due to wind and tidal turbines is that the former occupies a little portion of the vertical profile of the flow, while the latter, usually, occupies more than 1/3 of the flow depth. This causes that the flow is not only affected downstream the turbines but also upstream and around the turbines. Flow effects are more significant when increasing the number of turbines, which can be felt far away from the array deployment and can result in environmental impacts. It is for this reason that simple analytical models like those used in [2] are not suitable for tidal array arrays, especially when turbines are placed in complex environments. Therefore, any proposed tidal energy array design method has to be fully coupled with the flow to ensure a proper optimization, to take the most of the resource while trying to reduce as much as possible detrimental environmental impacts.

In the literature, there are two approaches that tackle this problem. Both approaches are fully coupled by using numerical models that solve the shallow water equations. In the first approach, developed by Funke et al. [4], the gradient of the array power output is computed using the adjoint technique of variational calculus. The advantage of this approach is that the computational cost is independent of the number of turbines that compose the array and allows the free position of each device within the domain defined. The main drawbacks of this approach are that it requires the development of an adjoint solver, which might be difficult depending on the software being used to solve the hydrodynamics, and that the iterative process yields an optimum solution not giving the possibility to explore the design variable domain by changing the constraint values without repeating the calculation. The inclusion of environmental constraints is still under development. For the moment, environmental impacts are only considered in terms of the effects on flow velocities at specific regions [5].

On the other hand, in the approach of González-Gorbeña et al. [6], the optimum tidal array is searched by means of surrogates based on a set of expensive computer experiments, a method known as surrogate-based optimization [7]. The SBO method consists in fitting a mathematical function to approximate a more time-consuming function. The main advantages are that once a surrogate is validated, the whole design variable space can be explored instantaneously, thus giving the possibility to assess efficiently the effect of changing the constraints limits on the objective function. Moreover, it can be implemented in any set of data, independently of the software in which the responses are generated, and surrogates can represent any environmental impact that the simulations can generate, e.g. shear velocities [8], tide discharges and morphological changes [9]. The main disadvantage of the SBO approach is that the number of simulations that are necessary to build a surrogate is a function of the number of design variable defined in the problem. If the free positioning of turbines is considered, this approach is impractical for the optimization of large arrays, as the position of each device is defined by at least a pair of design variables (i.e. the x and y horizontal coordinates).

In this paper, it is presented a case study applying the SBO approach where it is estimated the maximum size of a tidal array for the Faro-Olhão Inlet considering performance and environmental constraints. The paper has the following organization: Sect. 2 describes the case study region; Sect. 3 details the SBO approach adopted; Sect. 4 presents the results of the design space exploration and optimization models; finally, Sect. 5 provides a discussion of the results and the conclusion of the study.

2 Site Description

The Ria Formosa is coastal lagoon system with multiple inlets placed in Southern Portugal (Fig. 1). The system has two peninsulas and five islands enclosing an area with, sand flats, salt marshes and a complex system of tidal channels. Small communities of fishermen live on these islands. There are six inlets that connect the lagoon with the ocean; two of them are stabilized with jetties at both sides (Faro-Olhão and Tavira inlets) and the rest are free to migrate (Ancão, Armona, Fuseta and Lacém). The Faro-Olhão Inlet and the Armona Inlet together represent almost 90% of the total tidal prism of the lagoon. During spring and neap tides the Faro-Olhão Inlet provides 61% and 45%, respectively, of the overall tidal prism, while the Armona Inlet accounts 23% and 40%, correspondingly [10]. The dynamics that force these discharges are due to the semi-diurnal tides that the region experiences, with average elevation ranges of 2.8 m for spring tides and 1.3 m for neap tides. Both inlets have an ebb dominated behavior (i.e. shorter ebb duration generate higher mean ebb velocity). However, at the Faro-Olhão Inlet, the flood prism is considerably greater than the ebb prism, therefore sediment deposition occurs inside the lagoon, while for the Armona Inlet sediment flushes seaward. As a result, due to the importance of the Faro-Olhão Inlet, any alteration of the inlet could disrupt the dynamic equilibrium of the whole system which can result in adverse environmental impacts.

3 Surrogate-Based Optimisation Approach

Surrogate-based optimization has been applied in various branches of knowledge, including aeronautical [11], automotive [12], and telecommunications [13], among others. As previously said, it is a very popular method when time consuming computer simulations are involved in the design process. The SBO approach compromises problem formulation, design of experiments, computer simulations, surrogate construction and validation, and mathematical optimization. Figure 2 summarizes the SBO approach flowchart and the following subsections detail each of the steps adopted to estimate the maximum capacity of the Faro-Olhão Inlet for tidal energy extraction.

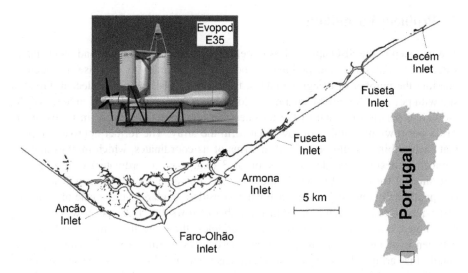

Fig. 1. Location map of the region of study and 3D model of Evopod[TM] 35 kW (E35) tidal energy converter.

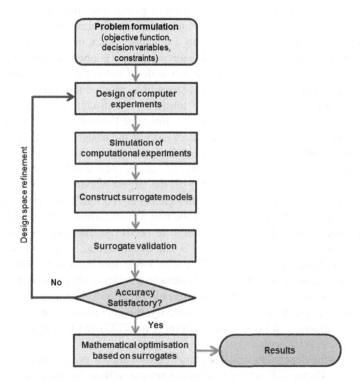

Fig. 2. Surrogate based optimization methodology flowchart.

4 Problem Formulation

The first step in the SBO approach is to define what are the dependent and independent variables that describe the problem to be solved. In order to decrease as much as possible the number of design variables, these need to be carefully selected. Given a site with potential for a tidal stream development, arrays can be defined in terms of the individual position of each of the turbines that compose the farm or in terms of the TECs per row and number of rows that form the array. The former approach implies that each turbine should be defined in terms of its coordinates, which implies that the number of design variable by a factor of 3 (e.g. x-y-z coordinates) or 2 (e.g. x-y coordinates) for each TEC within the array. As mentioned above, this will lead to a computationally unfordable approach when considering a large array made of hundreds of devices. Instead, given a predefined number of rows and TECs per row, arrays can be denoted as a function of two design variables, these are the longitudinal spacing between rows and the lateral spacing among devices within a row [5]. This approach entails that arrays should have a uniform distribution inline (i.e. downstream TECs have 0° phase difference respect upstream devices) or staggered (i.e. downstream TECs have 90° phase difference respect upstream devices). In order to overcome this problem, TEC arrays can be defined in terms of four design variables, these are: the longitudinal, lateral, vertical and staggered spacings [8]. The approaches of [5] and [8] imply the use of continuous variables, which entails the re-meshing of the domain for each computer simulation. This represents a drawback when modelling environmental hydraulics of complex lagoons where domain meshing is a laborious task and TECs are modelled in a sub-grid scale (i.e. the length scale of the computational grid is larger than the length scale of the TEC). In this study, it is adopted the strategy presented in [9], where a tidal array was defined in terms of two discrete design variables, the number of rows and the number of TECs per row. The Evopod^{TM} 35 kW TEC (Fig. 1) was selected to form the array. The limits of each of the design variables were defined considering model grid discretization, hydrodynamic conditions, turbine operation specifications and geometric constraints i.e. depth and width of the Faro-Olhão Channel. As it can be observed from Fig. 3, the largest resource is at the inlet throat, where occurrences of tidal current velocities greater than 0.7 m/s are larger than 60% of the time. It was here that was positioned the first row of turbines with a minimum and maximum number of 6 TECs and 11 TECs, respectively. This implies that the minimum and maximum lateral spacing is 3 (to avoid turbine collision) and 6 rotor diameters (blockage effects are negligible with larger spacings), correspondingly. On the other hand, considering a uniform longitudinal spacing of 20 rotor diameters (to ensure wake recovery) and a minimum threshold of 25% of the time with occurrences of flow velocities greater than the cut-in velocity, the maximum number of array rows was set to 13. Figure 3 displays the deposition of the TEC rows for a hypothetical array of 13 rows.

Regarding the dependent variables, array performance is usually measured in terms of its Capacity Factor, CF, which denotes the percent of time that the array is operating at rated power. Therefore, the performance of the array can be described by the overall array efficiency as well as by the average of the efficiencies of array rows. Assess and quantify the effects of TEC arrays on the environment is a more difficult task, principally in what regards to identify what are the adverse impacts and, consequently, to define the thresholds of what is considered a negative impact or not. Therefore, environmental impacts will be project-specific. In this particular study, a set of environmental impacts related with hydrodynamic and morphological effects was defined as function of the above mentioned design variables.

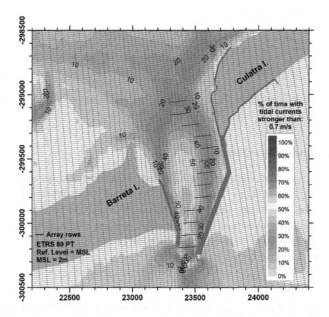

Fig. 3. Contour map showing percent of time with tidal currents above 0.7 m/s.

4.1 Design of Computer Experiments

A design of experiments (DoE) consists in generating a set of observation points relating the independent variables to adequately describe the design space. The objective of the DoE is to minimize as much as possible the error between the real-function (i.e. the computer simulation) and the fitted function the (i.e. the surrogate) using the least number of observations. In the literature, there is a lot of research to attain this goal, where new methods are proposed frequently. For a classical review on this matter see Ref. [14]. DoEs can have a pre-defined number of sample plans or can follow an adaptive-sequential approach. There are no rules to define the minimum number of sample points but, generally, is not less10 than times the number of design variable [15]. Following the conclusions of [16], in this study, a fixed sample plan size with 15 points per variable was chosen to characterize the design space plus 3 additional points to validate the surrogates. Sample points were selected manually and Fig. 4 shows the initial design plan.

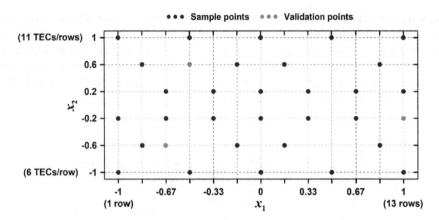

Fig. 4. Initial design of experiments sample plan (blue points) and validation points (red points). (Color figure online)

4.2   Computational Simulations

After the sample plan was defined, the computer experiments were executed using the modelling package Delft3D [17] in its two-dimensional depth averaged version (2DH). Full details about the set-up, TEC parameterization, calibration and validation of the hydro-morphodynamic model of the entire Ria Formosa are available in Ref. [9].

A total of 26 responses were obtained for each of the 33 simulations compromising performance and environmental indicators. The performance indicators are: the overall array capacity factor, CF_{Array}; and average capacity factor of each array row, CF_i. The environmental indicators are: (i) percent variations of cumulative flood and ebb instantaneous discharges ($\Delta\Sigma Q_i$) throughout a spring tide cycle at Faro-Olhão and Armona tidal inlets and for the whole lagoon system; (ii) percent changes of the cumulative flood and ebb instantaneous discharges ratio between Armona and the Faro-Olhão inlets ($\Delta\Sigma Q_{Ar}/\Delta\Sigma Q_{FO}$); and (iii) net sediment volume differences (ΔV) and variations in average depth changes (Δh_{avg}) for the Armona Inlet and the Faro-Olhão Inlet flood delta.

In order to assess the array and its rows performance as well as the array impacts on hydrodynamics and morphodynamics of Ria Formosa natural park, a set of threshold limits have been defined and summarised in Tables 1, 2 and 3.

Table 1. Constraints values for capacity factors of the entire array, CF_{Array}, and for each row of the array, CF_i.

Constr. N°	1	2	3	4	5	6	7	8	9	10	11	12	13	14
Constraint	CF_{Array} [%]	CF_1 [%]	CF_2 [%]	CF_3 [%]	CF_4 [%]	CF_5 [%]	CF_6 [%]	CF_7 [%]	CF_8 [%]	CF_9 [%]	CF_{10} [%]	CF_{11} [%]	CF_{12} [%]	CF_{13} [%]
Value	12.5	15	12	10	10	10	10	10	10	10	10	10	10	10

Table 2. Environmental constraints values for flood and ebb spring discharges ($\Delta\Sigma Q$) of Armona ($_{Ar}$) and Faro-Olhão ($_{FO}$) inlets, together with their discharges fraction ($\Sigma Q_{Ar}/\Sigma Q_{FO}$). Subscripts e and f depict for ebb and flood tides, respectively.

Constr. N°	15	16	17	18	19	20	21	22
Constr.	$\Delta\Sigma Q_{f,FO}[\%]$	$\Delta\Sigma Q_{f,Ar}[\%]$	$\Sigma Q_{f,Ar}/\Sigma Q_{f,FO}[\%]$	$\Delta\Sigma Q_{f,all}[\%]$	$\Delta\Sigma Q_{e,FO}[\%]$	$\Delta\Sigma Q_{e,Ar}[\%]$	$\Sigma Q_{e,Ar}/\Sigma Q_{e,FO}[\%]$	$\Delta\Sigma Q_{e,all}[\%]$
Value	−10	10	7.5	−5	−5	2.5	2.5	−2.5

Table 3. Environmental constraints values for morphological aspects. Δh_{avg} and ΔV_{net} depict the average depth and net sediment volume variations, respectively, for Armona ($_{Ar}$) and Faro-Olhão ($_{FO}$) inlets.

Constr. N°	23	24	25	26
Constraint	$\Delta V_{net,FO}[\text{m}^3/\text{yr}]$	$\Delta V_{net,Ar}[\text{m}^3/\text{yr}]$	$\Delta h_{avg,FO}[\text{cm}/\text{yr}]$	$\Delta h_{avg,Ar}[\text{cm}/\text{yr}]$
Value	−10,000	40,000	−5	15

4.3 Surrogate Construction and Validation

The data generated though the computational simulations was used to train a set of surrogates that approximate the entire design variable space. In the literature, there are many candidates to use as surrogates For a review on surrogates, readers can refer to [18]. In the present study, linear Radial Basis Functions (RBFs) were used to build the surrogates.

For exact data interpolation in a multi-dimensional space, RBF is a popular technique [19]. A response, y, is related to a vector of input variables, $\mathbf{x}$, through a linear combination of the basis functions. As a real valued function, RBF data points, $\mathbf{x}$, affect their distance, r, from another data point, $\mathbf{x}_i$, named a center. The Euclidian norm, $r = \|\mathbf{x} - \mathbf{x}_i\|$ norm represents the distance between the two points. Therefore, a data point in a data set will affect to a greater extent the nearer points than the faraway points, in such a way that $\phi(\mathbf{x}, \mathbf{x}_i) = \phi(r) = \phi(\mathbf{x} - \mathbf{x}_i)$. In this manner, the way data points are related depends on the basis function selected. Some of the most common basis functions are: linear, cubic, Gaussian, thin-plate-spline, inverse multiquadric, and multiquadric.

Consequently, it is possible to construct the approximation response function, $\hat{y}_i$, using the following expression:

$$\hat{y}_j = \mathbf{w}^\mathrm{T}\, \phi = \sum_{i,j=1}^{n} w_i\, \phi\left(\left\|\mathbf{x}_j - \mathbf{x}_i\right\|\right), \quad i,j = 1,2,...,n \tag{1}$$

here, $\mathbf{w}$ is a vector comprising the weights, w_i, for the linear arrangement of basis vectors that are in the vector ϕ; which provides the expression,

$$\mathbf{y} = \mathbf{w}^\mathrm{T}\, \Phi \tag{2}$$

where $\mathbf{y}$ represents the vector with the results from the computer model and Φ is the matrix containing the combination of linear basis vectors.

Finally, weights values can be obtained employing the least squares estimator given by Eq. (3), i.e.

$$\hat{\mathbf{w}} = \left(\mathbf{\Phi}^T \mathbf{\Phi}\right)^{-1} \mathbf{\Phi}^T \mathbf{y} \tag{3}$$

Once the surrogates were built, the prediction capabilities of each of the surrogates were assessed using a leave-k-out ($k = 3$) cross validation technique [20].

The Normalised Root Mean Square Error (NRMSE) and the Normalised Maximum Error (NMAXE) were used to assess the predictive capabilities of the surrogates. Forrester et al. [20] suggests that values of a NRMSE < 0.1 and NRMSE < 0.02 imply surrogates with reasonable and exceptional predictive abilities, respectively. In this study, NRMSE and NMAXE values for all surrogates were below 0.01.

4.4   Constraint Optimisation

After validation, surrogate models were used to formulate a multi-objective constrained optimisation model that maximises the number of tidal turbines and associated power output for the Faro-Olhão Inlet subject to performance and environmental restrictions. The mathematical model is given by:

$$\text{Maximise } \left\{x_1 x_2, P_{Array}\right\} \tag{4}$$

Subject to:

$$P_{Array} = CF_{Array}\left(\mathbf{x}\right) x_1 x_2 P_r t \tag{5}$$

$$\hat{g}_i\left(\mathbf{x}\right) \geq b_i, \quad \forall i \in \left\{1,\ldots,x_2 + 1\right\} \tag{6}$$

$$\hat{g}_i\left(\mathbf{x}\right) \geq b_i, \quad \forall i \in \left\{15, 18, 19, 22, 23, 25\right\} \tag{7}$$

$$\hat{g}_i\left(\mathbf{x}\right) \geq b_i, \quad \forall i \in \left\{16, 17, 20, 21, 24, 26\right\} \tag{8}$$

$$x_1 \in \left\{1, .., 13\right\}, x_2 \in \left\{6, \ldots, 11\right\} \tag{9}$$

Equation (4) defines the objective functions to be maximized. Equation (5) defines how to calculate the overall power output of the array, where P_r depicts the turbine's rated (35 kW), and t, the time interval to calculate power production (364 days.yr-1). Equation (6) defines the set of constraints related with array and row efficiency (Table 1). Equations (7), and (8) represent the set of environmental constraints, i.e. the values of the constraints in Tables 2 and 3. Finally, Eq. (9) declares the values of the design variables, i.e. the number of array rows and TECs per row. Notice that the integer value of the design variable representing the number of array row defines the rows that are active. The first row of the array is placed at the inlet throat, then subsequent rows are placed consecutively toward the interior of the lagoon.

5   Design Space Exploration and Optimisation Results

In order to understand how sensitive the responses are to the design variables, the domain space of each of the surrogates were explored for all the possible combinations

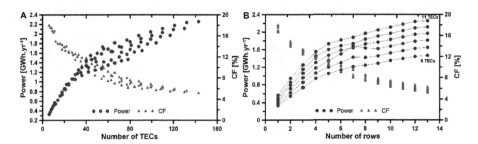

Fig. 5. Array capacity factor and annual power output related with (A) quantity of TECs, and (B) quantity of rows.

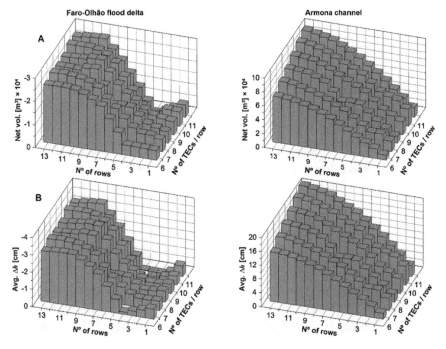

Fig. 6. Bar charts illustrating: (A) sedimentation and erosion net volume changes and (B) average depth variations for the Faro-Olhão flood delta (left) and Armona inlet (right). Each of the bars represent one of the possible solutions (i.e. 78 array layouts. Negative and positive values indicate decrease and increase of scalar quantities, respectively. Notice that the Armona Inlet experiences erosion and the Faro-Olhão flood delta suffers sedimentation.

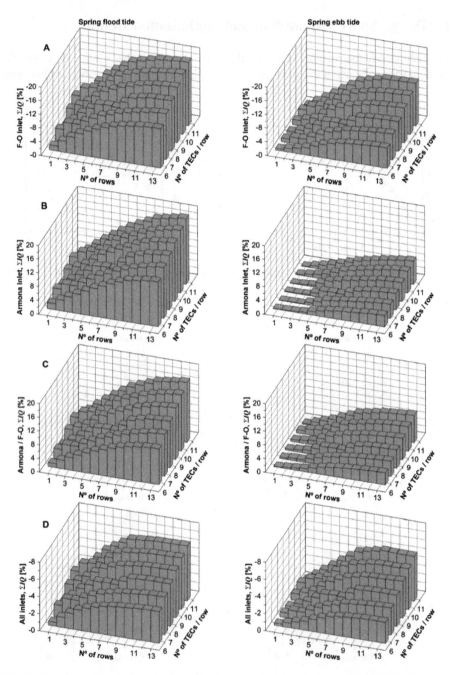

Fig. 7. Bar charts illustrating: (A and B) changes of immediate cumulative discharges and (C) discharge fractions for the Armona and Faro-Olhão, and (D) variations of immediate cumulative discharges for all inlets during flood (left) and ebb (right). Each of the bars represent one of the possible solutions (i.e. 78 array layouts. The baseline fraction between Armona and Faro-Olhão inlets for ebb and flood discharges are 35.8% and 37.8%, respectively.

Table 4. Values of the design variables, x_1 and x_2, array capacity factor, CF_{Array}, and power output, P_{Array}, for the optimal solution, x^*.

x_1 [rows]	x_2 [TECs/row]	x^* [TECs]	CF_{Array} [%]	P_{Array} [GWh.yr^{-1}]
3	10	30	12.6	1.20

Table 5. Constraints values model results for capacity factors of the entire array, CF_{Array}, and for each row of the array, CF_i.

Constr. N°	1	2	3	4	5	6	7	8	9	10	11	12	13	14
Constr.	CF_{Array} [%]	CF_1 [%]	CF_2 [%]	CF_3 [%]	CF_4 [%]	CF_5 [%]	CF_6 [%]	CF_7 [%]	CF_8 [%]	CF_9 [%]	CF_{10} [%]	CF_{11} [%]	CF_{12} [%]	CF_{13} [%]
Value	12.6	15.3	12.2	10.1	–	–	–	–	–	–	–	–	–	–

Table 6. Models results of the environmental constraints for flood and ebb spring discharges ($\Delta\Sigma Q$) of Armona ($_{Ar}$) and Faro-Olhão ($_{FO}$) inlets, together with their discharges fraction ($\Sigma Q_{Ar}/\Sigma Q_{FO}$). Subscripts e and f depict for ebb and flood tides, respectively.

Constr. N°	15	16	17	18	19	20	21	22
Constraint	$\Delta\Sigma Q_{f,FO}$ [%]	$\Delta\Sigma Q_{f,Ar}$ [%]	$\Sigma Q_{f,Ar}/\Sigma Q_{f,FO}$ [%]	$\Delta\Sigma Q_{f,all}$ [%]	$\Delta\Sigma Q_{e,FO}$ [%]	$\Delta\Sigma Q_{e,Ar}$ [%]	$\Sigma Q_{e,Ar}/\Sigma Q_{e,FO}$ [%]	$\Delta\Sigma Q_{e,all}$ [%]
Value	−9.4	7.8	7.1	−3.7	−4.0	1.0	1.9	−1.9

Table 7. Model results of the environmental constraints for morphological aspects. Δh_{avg} and ΔV_{net} depict the average depth and net sediment volume variations, respectively, for Armona ($_{Ar}$) and Faro-Olhão ($_{FO}$) inlets.

Constraint N°	23	24	25	26
Constraint	$\Delta V_{net,FO}$ [m^3/yr]	$\Delta V_{net,Ar}$ [m^3/yr]	$\Delta h_{avg,FO}$ [cm/yr]	$\Delta h_{avg,Ar}$ [cm/yr]
Value	−3013	37019	0.1	8.4

of the discrete values of the design variables, $x_1 = [1, \ldots, 13]$ and $x_2 = [6, \ldots, 11]$. Figures 5, 6 and 7 summarize graphically the results obtained.

By exploring the design variable domain, it was possible to conclude that: (1) the most productive arrays were those with smaller number of rows; (2) the number of TECs per row does not affect the capacity factor as much as the number of array rows does; (3) once a certain quantity of TECs in an array is achieved, the placement of additional turbines does not contribute significantly to increase the overall power production; (4) the magnitude of the environmental responses were proportional to the quantity of TECs within the array, except for the morphological changes of the Faro-Olhão flood delta, which were more influenced by the number of rows than by the number of TECs (Table 4).

Under the constraints imposed, the results from the optimisation models revealed that arrays with 30 TECs distributed in 3 rows with 10 devices each were the maximum allowable array size for the Faro-Olhão Inlet satisfying the performance constraints (Table 5) and without adversely affecting the environment (Tables 6 and 7).

6 Conclusions

In this study, the maximum permissible capacity of the Faro-Olhão Inlet for tidal current energy exploration was estimated by means of using a surrogate-based optimisation. Tidal arrays were defined as a function of a pair of design variables, these are: (1) number of TEC rows and (2) number of TECs per array row. A 2DH hydro-morphodynamic model was used to compute a set of scenarios to assess performance and environmental effects of tidal arrays in Ria Formosa. Linear RBF were selected to build surrogates from the responses of the numerical simulations. Finally, employing validated surrogates, a multi-objective optimisation model was formulated to maximize array quantity of TECs and power output, while minimizing detrimental environmental impacts. Results suggest that the optimum solution consist of an array with 10 TECs per row and a total of 3 rows.

The main advantages of using the surrogate-based approach for optimizing tidal energy arrays can be summarized as follows:

- The SBO approach results very useful when the use expensive computational simulations is involved;
- High complex response that are challenging to model with simple analytical functions can be approximated by using surrogates based on high definition numerical models;
- Both the objective function and the constraints can be represented by surrogates;
- Allows to formulate constraint optimisation models;
- The optimisation model can incorporate environmental and performance constraints; and
- In multidimensional problems, the SBO approach allows to search efficiently (i.e. using less computational resources) the whole design space.

Acknowledgements. Eduardo González-Gorbeña has received funding for the OpTiCA project (http://msca-optica.eu/) from the Marie Skłodowska-Curie Actions of the European Union's H2020-MSCA-IF-EF-RI-2016/GA#: 748747. The paper is a contribution to the SCORE project, funded by the Portuguese Foundation for Science and Technology (FCT–PTDC/AAG-TEC/1710/2014). André Pacheco was supported by the Portuguese Foundation for Science and Technology under the Portuguese Researchers' Programme 2014 entitled "Exploring new concepts for extracting energy from tides" (IF/00286/2014/CP1234).

References

1. European Commission: Factsheet: Clean Energy for Islands Initiative. https://ec.europa.eu/commission/sites/beta-political/files/clean-energy-islands-initiative_en.pdf
2. Brutto, O.A.L., Thiébot, J., Guillou, S.S., Gualous, H.: A semi-analytic method to optimize tidal farm layouts–application to the Alderney Race (Raz Blanchard), France. Appl. Energy **183**, 1168–1180 (2016)
3. Shakoor, R., Hassan, M.Y., Raheem, A., Wu, Y.K.: Wake effect modeling: a review of wind farm layout optimisation using Jensen's model. Renew. Sustain. Energy Rev. **58**, 1048–1059 (2016)

4. Funke, S.W., Farrell, P.E., Piggott, M.D.: Tidal turbine array optimisation using the adjoint approach. Renew. Energy **63**, 658–673 (2014)
5. Du Feu, R.J., et al.: The trade-off between tidal-turbine array yield and impact on flow: a multi-objective optimisation problem. Renew. Energy **114**, 1247–1257 (2017)
6. González-Gorbeña, E., Qassim, R.Y., Rosman, P.C.: Optimisation of hydrokinetic turbine array layouts via surrogate modelling. Renew. Energy **93**, 45–57 (2016)
7. Queipo, N.V., Haftka, R.T., Shyy, W., Goel, T., Vaidyanathan, R., Tucker, P.K.: Surrogate-based analysis and optimisation. Prog. Aerosp. Sci. **41**(1), 1–28 (2005)
8. González-Gorbeña, E., Qassim, R.Y., Rosman, P.C.: Multi-dimensional optimisation of tidal energy converters array layouts considering geometric, economic and environmental constraints. Renew. Energy **116**, 647–658 (2018)
9. González-Gorbeña, E., Pacheco, A., Plomaritis, T.A., Ferreira, Ó., Sequeira, C.: Estimating the optimum size of a tidal array at a multi-inlet system considering environmental and performance constraints. Appl. Energy **232**, 292–311 (2018)
10. Pacheco, A., Vila-Concejo, A., Ferreira, Ó., Dias, J.A.: Assessment of tidal inlet evolution and stability using sediment budget computations and hydraulic parameter analysis. Mar. Geol. **247**, 104–127 (2008)
11. Koziel, S., Tesfahunegn, Y., Leifsson, L.: Variable-fidelity CFD models and co-Kriging for expedited multi-objective aerodynamic design optimisation. Eng. Comput. **33**(8), 2320–2338 (2016)
12. Acar, E., Tüten, N., Güler, M.A.: Design optimisation of an automobile torque arm using global and successive surrogate modeling approaches. Proc. Inst. Mech. Eng. Part D: J. Automob. Eng. **233**(6), 1453–1465 (2018). https://doi.org/10.1177/0954407018772739
13. Koziel, S., Ogurtsov, S.: Surrogate-based optimization. In: Koziel, S., Ogurtsov, S. (eds.) Antenna Design by Simulation-Driven Optimization. SO, pp. 13–24. Springer, Cham (2014). https://doi.org/10.1007/978-3-319-04367-8_3
14. Sacks, J., Welch, W.J., Mitchell, T.J., Wynn, H.P.: Design and analysis of computer experiments. Stat. Sci. **4**, 409–423 (1989)
15. Loeppky, J.L., Sacks, J., Welch, W.J.: Choosing the sample size of a computer experiment: a practical guide. Technometrics **51**(4), 366–376 (2009)
16. Eisenmann, E.G.G., Qassim, R.Y., Rosman, P.C.C.: A metamodel simulation based optimisation approach for the tidal turbine location problem. Aquat. Sci. Technol. **3**(1), 33–58 (2015)
17. Delft3D Open Source Community. https://oss.deltares.nl/web/delft3d/home
18. Fang, K.-T., Li, R., Sudjianto, A.: Design and Modeling for Computer Experiments. Computer Science & Data Analysis Series. Chapman & Hall/CRC, Boca Raton (2006)
19. Buhmann, M.D.: Radial Basis Functions: Theory and Implementations. Cambridge University Press, Cambridge (2003)
20. Forrester, A., Sobester, A., Keane, A.: Engineering Design via Surrogate Modelling: A Practical Guide. Wiley, Chichester (2008)

Time-Dependent Link Travel Time Approximation for Large-Scale Dynamic Traffic Simulations

Genaro Peque Jr.[(⊠)], Hiro Harada, and Takamasa Iryo

Department of Civil Engineering, Kobe University, Kobe, Japan
gpequejr@panda.kobe-u.ac.jp,
171t139t@stu.kobe-u.ac.jp, iryo@kobe-u.ac.jp

Abstract. Large-scale dynamic traffic simulations generate a sizeable amount of raw data that needs to be managed for analysis. Typically, big data reduction techniques are used to decrease redundant, inconsistent and noisy data as these are perceived to be more useful than the raw data itself. However, these methods are normally performed independently so it wouldn't compete with the simulation's computational and memory resources.

In this paper, we propose a data reduction technique that will be integrated into a simulation process and executed numerous times. Our interest is in reducing the size of each link's time-dependent travel time data in a large-scale dynamic traffic simulation. The objective is to approximate the time-dependent link travel times along the y - axis to reduce memory consumption while insignificantly affecting the simulation results. An important aspect of the algorithm is its capability to restrict the maximum absolute error bound which avoids theoretically inconsistent results which may not have been accounted for by the dynamic traffic simulation model. One major advantage of the algorithm is its efficiency's independence from the input data complexity such as the number of sampled data points, sampled data's shape and irregularity of sampling intervals. Using a 10×10 grid network with variable time-dependent link travel time data complexities and absolute error bounds, the dynamic traffic simulation results show that the algorithm achieves around 80%–90% of link travel time data reduction using a small amount of computational resource.

Keywords: Large-scale dynamic traffic simulation ·
Piecewise linear approximation · Route planning · Parallel computing

1 Introduction

1.1 Background

Large-scale dynamic traffic simulations are becoming widespread partly due to the exponential growth of the computer's processing power, memory capacity and parallelization capability. Along with it is the increasing need to manage the sizeable amount of raw data that it generates. Typically, big data reduction techniques are used to decrease redundant, inconsistent and noisy data as it is perceived to be more useful than the raw data itself. However, these methods are normally performed independently

© Springer Nature Switzerland AG 2019
J. M. F. Rodrigues et al. (Eds.): ICCS 2019, LNCS 11538, pp. 562–576, 2019.
https://doi.org/10.1007/978-3-030-22744-9_44

so it wouldn't compete with the simulation's computational and memory resources. A major challenge is when data reduction is integrated into the simulation process that is executed numerous times since it needs to be simple, fast and efficient.

In this paper, we are interested in reducing the size of each link's time-dependent travel time data at each iteration of a large-scale dynamic traffic simulation. The simulation is planned to be executed using a parallel computer with a distributed memory architecture. Large-scale dynamic traffic simulations are normally conducted on massive networks with substantial number of links (e.g. a road network of Western Europe has 42.6 million links) [1, 2]. Since travel time is a function of a single variable that is strictly monotone (time), the problem can be defined as a piecewise linear approximation of the time-dependent link travel time data. There are already many existing algorithms developed for this problem but most are interested in retaining the data's significant features. Moreover, the space and time complexities of these algorithms are usually dependent on the input complexity such as the number of sampled data points, sample data's shape and irregularity of sampling intervals.

The piecewise linear approximation of polygons has several applications in various fields such as pattern recognition [3], motion planning in robotics [4], vector graphics and cartographic generalization [5], among others. The goal is to reduce the size or complexity of a given data as much as possible while retaining its significant features. One of the widely known piecewise linear approximation algorithms that have been studied in various fields is the Ramer-Douglas-Peucker algorithm [6–9]. This algorithm decimates a curve composed of line segments into a similar curve with fewer points. The major advantage of this algorithm is its efficiency's independence from the input data complexity. However, the algorithm is still computationally inefficient for large-scale simulations and its error criterion isn't well-suited for the reduction of time-dependent link travel time data since it isn't possible to restrict the absolute error bound. This makes it inaccurate and inconsistent when applied to multiple links with different input complexities.

Piecewise linear approximation algorithms usually rely on error bounds or complexity minimization that further depends on the context of the application. In this paper, the application lies in the context of link travel time approximation for time-dependent route planning in large-scale dynamic traffic simulations. Increasing amounts of time-dependent travel time data are usually generated as the number of iterations or the length of the simulation time period of the traffic simulation increases. Depending on the design of the traffic simulator [10–12], some of the data may be, among other things, redundantly generated, insignificantly noisy and/or slowly changing (Fig. 1). All of which can be approximated according to an absolute error bound without significantly affecting the simulation results. Thus, an absolute error bound restriction capability is an integral part of the approximation method in order to maintain an accurate and consistent simulation result. Typically, travel time data is generated by a time-dependent piecewise linear function that is strictly monotone in time. This implies that time-dependent link travel time approximation is a special case of the piecewise linear approximation of polygons; the piecewise linear curve is an open polyline and a function of one variable that is strictly increasing.

There are several papers focusing on the approximation of piecewise linear functions that might be used for large-scale dynamic traffic simulations with absolute error

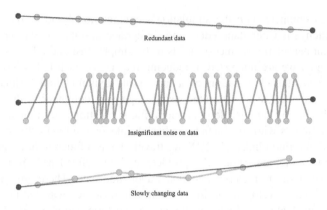

Fig. 1. Some types of data that can be approximated.

restriction capabilities [13–15]. For example, in Tomek [13] two simple heuristic algorithms were proposed for functions of one variable. Both algorithms used a limit on the absolute error values. The first algorithm was fast and gave satisfactory results for sufficiently smooth functions while the second algorithm was slower but gave better approximations for less well-behaved functions. In Imai and Iri [14], using the idea of Suri [16], they developed an algorithm for the edge-visibility problem [17] which produces an optimal solution for a piecewise linear function of one variable that is strictly monotone. The idea of the algorithm is to compute a minimum link path through a tunnel using a light source covering the entry of a tunnel. The tunnel is produced by an absolute upper and lower bound on the function given by an exogenously supplied absolute error value. Then, a light source illuminates a part of the tunnel and divides it into several invisible parts and a visible part. The intersection of boundaries of an invisible part and a visible part containing the exit of the tunnel is called a window where the first point of the approximated function lies. From the window, another light source illuminates part of the tunnel which creates another window. This process continues until the other end of the tunnel is reached. A major advantage of this algorithm is its $O(N)$ complexity where N is the number of points representing the piecewise linear curve.

In the transportation context, Neubauer [15] applied Imai and Iri's algorithm on a real-world time-dependent link travel time data. The algorithm was primarily used to preprocess the time-dependent travel time data to reduce its size and use it for route planning. However, there are three main concerns of its use in the approximation of time-dependent link travel time data. First, although Imai and Iri's [14] algorithm is simple and optimal given an absolute error value, lower bounds for link travel times need to be checked and possibly adjusted as it may produce values lesser than some links' free-flow travel times or even produce negative values. Neubauer [15] addressed this by using only relative upper bounds. When lower bounds were considered, it yielded first-in, first-out (FIFO) violations. Second, the calculation of line distances, intersections and angles to create a link's travel time estimate still requires a substantial amount of computational and memory resources. Furthermore, the algorithm's efficiency is dependent on the input complexity, specifically the sample data's shape. Third, an integral part of the algorithm is to create a polygon by offsetting the piecewise linear curve above and below by a constant value. This implies that 2 copies of the N points of the piecewise linear curve (above and below) also needs to be stored in the memory during the approximation process. This is challenging when used during a large-scale dynamic traffic simulation with limited memory capacity such as a parallel computer with distributed memory architecture.

1.2   Contribution of This Paper

In this paper, a piecewise linear approximation algorithm is developed to reduce the size of the time-dependent link travel time data generated by the large-scale dynamic traffic simulation. Our main objective is to reduce each link's memory consumption at each iteration given that the maximum error bound can be restricted to avoid theoretically inconsistent results not accounted for by the dynamic traffic simulation model.

We propose an algorithm that only requires linear interpolation calculations with respect to the given input points. The motivation is to develop an algorithm that is simple, fast and whose efficiency is independent of the input complexity such as the number of sampled data points, sampled data's shape and irregularity of the sampling intervals. Additionally, the algorithm should be capable of absolute error bound restriction to avoid inaccurate and inconsistent results. The idea is to use linear interpolation along the y - axis to calculate and reduce the Euclidean distance of the real link travel time and approximated link travel time. The approximated time-dependent link travel data will then be retrieved by each driver for route planning. Subsequently, we will show that only the Euclidean distances from the given input points to the estimated line segments need to be checked against the criterion function because these are the only points where the absolute maximum error for each ordered subset can occur. Furthermore, since linear interpolation only calculates points between any of the given input points, the estimated link travel time values are assured to be bounded by these points in all directions. One disadvantage of the algorithm is that the points of the approximated function are restricted to the subset of the input function. It doesn't allow arbitrary points [16, 18, 19]. However, this is insignificant for our application.

1.3   Outline of This Paper

This paper is structured as follows. In the next section, a brief description of a dynamic traffic simulation and the importance of link travel time data approximation is presented. In Sect. 3, we formally define the piecewise linear approximation problem and introduce some notations. Additionally, the proposed piecewise linear approximation algorithm for time-dependent link travel time data reduction is introduced including the method in which travel time is retrieved after the approximation. In Sect. 4, using a 10×10 grid network with variable time-dependent link travel time data complexities and error bounds, results show that the algorithm is very simple, fast and efficient both in time and space and is highly suitable for integration in large-scale dynamic traffic simulation. A summary of the importance of our contribution and conclusion is presented in the last section.

2 Large-Scale Dynamic Traffic Simulation

2.1 Travel Time Approximation in a Dynamic Traffic Simulation

Traffic simulators that can perform large-scale dynamic traffic simulations are becoming increasingly popular as these provide more detailed means to represent the interaction between travel choices, traffic flows, and time and cost measures in a temporally coherent manner. In most cases, these are iteratively conducted and involves an interplay of the vehicular traffic loading and travelers' route assignments [20] until a stopping criterion is met (Fig. 2).

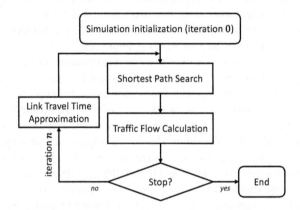

Fig. 2. Dynamic traffic simulation flowchart.

An iteration in a simulation usually represents a time period (from peak hours to an entire day). Within this time period, time-dependent travel time data is generated for each link sampled in specific time intervals. Thus, the longer the simulation time period, the larger the size of the time-dependent link travel time data that needs to be stored for travelers' time-dependent route planning. Although it might not be necessary to store the data of each iteration for route planning, these are usually necessary for post-processing analysis.

In large-scale dynamic traffic simulations, massive networks will have substantial number of links that generate time-dependent travel time data. This would require a large amount of memory for storage (depending on the simulation time period and the number of iterations) which is usually a very limited resource. However, some of these sampled data may be, among other things, redundantly generated, insignificantly noisy and/or slowly changing (Fig. 1). All of which can be approximated according to an absolute maximum error bound without significantly affecting the simulation results. Thus, other than requiring a simple, fast and efficient algorithm, a necessary feature is the capability of the algorithm to restrict the absolute error bound consistently and accurately. In some dynamic traffic simulations, travelers' route assignments are assigned based on the shortest paths to their destinations (i.e. a deterministic route choice model) [21]. This in turn depends on the time-dependent travel time data of the links along these routes. An inaccurate or inconsistent approximation would lead to a different simulation result if a deterministic route choice model is a requirement. More importantly, even if a deterministic route choice model isn't a requirement it may still lead to theoretically inconsistent results not accounted for by the dynamic traffic simulation model.

Therefore, we are motivated in reducing the size of the time-dependent link travel time data given that we are able to restrict the absolute error bounds of the approximation process.

3   The Piecewise Linear Approximation Algorithm

3.1   Piecewise Linear Approximation

The problem of retaining significant features of polygonal curves through iterative approximation was formalized by Ramer [6]. The problem is to approximate a polygon represented by points using an iterative approximation algorithm that produces another polygon with a lesser number of points. More formally, an arbitrary two-dimensional plane curve is represented by an ordered set C of N consecutive points along the curve. The points in set C can be interpreted as the vertices p_i of a polygon P with $N-1$ edges. A polygon $\tilde{P}$ with a reduced number $\tilde{N}-1$ of edges, whose vertices p_k coincide with the vertices of P, then corresponds to an ordered subset $\tilde{C}$ of points such that $\tilde{C} \subseteq C$. The problem is to find an ordered subset $\tilde{C}$ of the ordered set C such that the polygon $\tilde{P}$ with vertices $p_k \in \tilde{C}$ approximates the curve based on an application-dependent error criterion. A desirable property for the ordered subset $\tilde{C}$ is that its size be as small as possible.

The ordered subset $\tilde{C}$ of vertices p_k divides the set C into ordered subsets $S_k \in S$ of the following form,

$$S_k = \{p_s, p_{s+1}, \ldots, p_{|S_k|}\}, \tag{1}$$

where $p_s = p_{k-1}$ and $p_{|S_k|} = p_k$; $s = 1, 2, \ldots, |S_k|$ and $k = 1, 2, \ldots, \tilde{N}$. The points p_s and p_{s+1} are consecutive points in polygon P and the points p_{k-1}, p_k are consecutive points in polygon $\tilde{P}$. The ordered subset $\tilde{C}$ also partitions the polygon $\tilde{P}$ into curve segments and S_k contains the points belonging to the k - th curve segment. Then, the following constraints hold between the ordered sets $C, \tilde{C}$ and S_k,

$$\bigcup_{k=2}^{\tilde{N}} S_k = C \text{ and } \left(\bigcup_{k=2}^{\tilde{N}} (S_k \cap S_{k+1}) \right) \cup \{p_1\} \cup \{p_N\} = \tilde{C}. \tag{2}$$

Constraints for the generation of the polygon $\tilde{P}$ can now be conveniently included into the conditions for the subsets S_k and $\tilde{C}$. Conditions on the subsets S_k correspond to conditions for the approximation of a curve segment $S_k \in C$ by a straight-line segment $\overline{p_{k-1} p_k}$ and conditions on $\tilde{C}$ can be regarded as global properties.

Generally, the objective is to find a subset $\tilde{C} \subseteq C$ of vertices with a minimum number of elements satisfying a specified criterion function $g(S_k) \leq \theta$, where θ is a constant, given a polygon P with a set of vertices $C = \{p_1, \ldots, p_i, \ldots, p_N\}$. The solution to the problem with an unknown number of inequalities is complicated but methods such as dynamic programming [9] or graph searching techniques [22] can be used to get optimal solutions at the expense of additional computational resource. An example of a criterion function $g(S_k)$ used by Ramer [6] is the maximum-distance, $g(S_k) = \max(dist(p_{k-1}, p_k)) \leq \alpha$ where $dist(p_1, p_2)$ is the Euclidean distance between the points p_1 and p_2; $p_i = (x_{p_i}, y_{p_i})$. In Ramer [6], an iterative approximation method was used to divide C into subsets which were tested against the maximum-distance criterion. If the subset satisfied the criterion, the subset was retained. Otherwise, the subset was divided and the resulting subsets were tested. The iteration was terminated when all subsets S_k satisfied the criterion $g(S_k) \leq \theta$. Various application-dependent criteria have been developed to test subsets such as perimeter error [9], error of area [5], mean square error or maximum deviation. Another possible criterion not based on error is to minimize the size of $\tilde{C}$ to a desirable number of points. In this paper, we deal with the former, i.e. $g(S_k) \leq \theta$.

3.2   Vertical Linear Interpolation Algorithm

The polygon P in this study is a special case of the piecewise linear approximation problem in [6] due to its special structure. The y coordinates of the x and y coordinates of points p_i in the ordered set C are defined by a function $y = f(x)$ where x is strictly monotone, $p \equiv (x, y) \in \mathbb{R}^2_+$ and $f : \mathbb{R}_+ \to \mathbb{R}_+$. This implies that polygon P is an open polyline with no intersecting edges. From now on this will be denoted as P'. The values of the x and y coordinates for each $p_i \in C$ will be denoted as x_{p_i} and y_{p_i}, respectively.

For a time-dependent link travel time data, an estimated $\tilde{y}_p$ value is a time-dependent travel time for a particular time $\tilde{x}_p$ of a link along a certain route of a traveler's destination in a specific iteration. Hence, in the piecewise linear approximation of the time-dependent travel time data defined by the polyline P', we propose an algorithm which linearly interpolates the values $y_{\tilde{p}_k}$ of projected points $\tilde{p}_k$ to the line $\overline{p_{k-1}p_k}$ from each point p_s in the ordered subset S_k for all $S_k \in S$ as shown in Fig. 3 above. Linear interpolation is chosen because it is very simple and fast to calculate and its efficiency is independent of the input complexity. The algorithm is characterized by the projection of each $p_s \in S_k$ to the line $\overline{p_{k-1}p_k}$ orthogonal to the x-axis. The value of a projection from p_s is calculated using the point's x_{p_s} value and is determined by the equation,

$$y_{\tilde{p}_k} = \left| y_{p_{k-1}} + (x_{p_s} - x_{p_{k-1}}) \left(\frac{y_{p_k} - y_{p_{k-1}}}{x_{p_k} - x_{p_{k-1}}} \right) \right|. \tag{3}$$

From the values created by Eq. (3), we are interested in the criterion function $g(S_k)$ which represents the line with the maximum length among all of the lines projected by all the points $p_s \in S_k$ to the line $\overline{p_{k-1}p_k}$. This is given by $dist(p_s, \tilde{p}_k) = \left| y_{p_s} - y_{\tilde{p}_k} \right|$

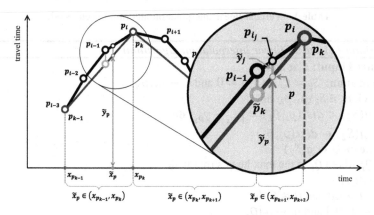

Fig. 3. Link travel time data approximated by the vertical linear interpolation algorithm.

where p_s and $\tilde{p}_k$ have both the same x_{p_s} value. More formally, the criterion function is defined by the equation,

$$g(S_k) = \max_{p_s}\{dist(p_s, \tilde{p}_k)|p_s \in S_k\} \le \alpha\beta = \theta, \qquad (4)$$

where the α is a constant and β is a scaling factor. The scaling factor β is necessary when the same constant α is used for different polylines P' with different properties. Then, each of the $g(S_k)$ value for each S_k is added into a max heap Q where its root node Q_r has a value of $\tilde{g} = \max_k g(S_k)$. The variable $\tilde{g}$ represents the maximum Euclidean distance of $\tilde{P}'$ from P' along the y - axis.

The criterion in the selection of the next ordered subset $S_{\tilde{k}}$ to divide into two smaller ordered subsets $S_{\tilde{k}_1}$ and $S_{\tilde{k}_2}$ is given by $S_{\tilde{k}} = \arg\max_{S_k} g(S_k)$ where $S_{\tilde{k}} = S_{\tilde{k}_1} + S_{\tilde{k}_2} - \left(S_{\tilde{k}_1} \cap S_{\tilde{k}_2}\right), \tilde{k}_1 - 1 = \tilde{k} - 1$ and $\tilde{k}_2 = \tilde{k}$. Additionally, the ordered subsets $S_{\tilde{k}_1}$ and $S_{\tilde{k}_2}$ replaces $S_{\tilde{k}}$ in the set S. The point $p_{\tilde{i}}$ where the ordered subset $S_{\tilde{k}}$ is divided at is denoted by,

$$p_{\tilde{i}} = \arg\max_{p_s}\{dist(p_s, \tilde{p}_k)|p_s \in S_k\}, \qquad (5)$$

where $S_{\tilde{k}_1} = \left\{p_{\tilde{k}_1-1}, \ldots, p_{\tilde{i}}\right\}$, $S_{\tilde{k}_2} = \left\{p_{\tilde{i}}, \ldots, p_{\tilde{k}_2}\right\}$ and $p_{\tilde{i}} = S_{\tilde{k}_1} \cap S_{\tilde{k}_2}; \tilde{k} - 1 < \tilde{i} < \tilde{k}$. When $S_{\tilde{k}}$ is divided into $S_{\tilde{k}_1}$ and $S_{\tilde{k}_2}$, it creates new $g\left(S_{\tilde{k}_1}\right)$ and $g\left(S_{\tilde{k}_2}\right)$. However, both $g\left(S_{\tilde{k}_1}\right)$ and $g\left(S_{\tilde{k}_2}\right)$ aren't guaranteed to be less than $\tilde{g}$. Thus, both $g\left(S_{\tilde{k}_1}\right)$ and $g\left(S_{\tilde{k}_2}\right)$ need to be evaluated and placed into the max heap Q. The process of ordered subset selection and division is repeated until the stopping criterion $\tilde{g} \le \alpha\beta$ is satisfied or the max heap is empty.

Table 1 below summarizes the vertical linear interpolation algorithm (Algorithm 1) through a pseudocode.

Table 1. Vertical linear interpolation algorithm pseudocode.

Algorithm 1 (Vertical linear interpolation algorithm)
Required input: C, α and β
Initialization: $S_2 = C$, $g(S_2) = 0$ and $\tilde{C} \leftarrow p_1, p_N$
for each $p_s \in S_2 \backslash p_1, p_N$ **do**
if $g(S_2) < dist(p_s, \tilde{p}_2) = \left\| y_{p_s} - y_{\tilde{p}_2} \right\|$ **do**
$g(S_2) \leftarrow dist(p_s, \tilde{p}_2)$
max heap, $Q \leftarrow g(S_2)$
$\tilde{g} \leftarrow Q_r$ where Q_r is the max heap's root node
$\tilde{C} \leftarrow p_{\tilde{\imath}}[Q_r]$
while $\tilde{g} > \alpha\beta$ and $Q \neq \emptyset$ **do**
$\tilde{k} \leftarrow k[Q_r]$ and $p_{\tilde{\imath}} \leftarrow p_{\tilde{\imath}}[Q_r]$
remove Q_r from Q
split $S_{\tilde{k}}$ into $S_{\tilde{k}_1}$ and $S_{\tilde{k}_2}$ along $p_{\tilde{\imath}}$
replace $S_{\tilde{k}} \in S$ with $S_{\tilde{k}_1}$ and $S_{\tilde{k}_2}$
for each $p_{s_i} \in S_{\tilde{k}_i} \backslash p_{\tilde{k}_i - 1}, p_{\tilde{k}_i}, i \in \{1, 2\}$ **do**
if $g(S_{k_i}) < dist(p_{s_i}, \tilde{p}_{k_i}) = \left\| y_{p_{s_i}} - y_{\tilde{p}_{k_i}} \right\|$ **do**
$g(S_{k_i}) \leftarrow dist(p_{s_i}, \tilde{p}_{k_i})$
for each $S_{k_i}, i \in \{1, 2\}$ **do**
max heap, $Q \leftarrow g(S_{k_i})$
$\tilde{g} \leftarrow Q_r$
$\tilde{C} \leftarrow p_{\tilde{\imath}}[Q_r]$
Return $\tilde{C}$

In Fig. 4 below, a visual demonstration of how the vertical linear interpolation algorithm iteratively approximates an open polyline P' is shown. Iteration 1 shows S_k as the entire polyline P' and its line segment $\overline{p_{k-1} p_k}$ approximation (red line). Additionally, it shows that vertical lines orthogonal to the x - axis (blue lines) are projected from the points $p_s \in S_k$ (black dots) to the line segment $\overline{p_{k-1} p_k}$. The vertical line with the maximum length $\tilde{g}$ (green line) is also shown. The set S_k is then divided into two line segments $\overline{p_{k-1} p_k}$ and $\overline{p_k p_{k+1}}$ in iteration 2 from the point $p_{\tilde{\imath}}$ at which $\tilde{g}$ occurred in the previous iteration. This process is repeated for each S_k where $\tilde{g}$ occurs until the stopping criterion $\tilde{g} \leq \alpha\beta$ is met. After the algorithm terminates, the points of the polyline $\tilde{P}'$ denoted by the ordered subset $\tilde{C}$ is returned.

A condensed algorithm flowchart for Algorithm 1 is shown in Fig. 5 below. The algorithm is performed to each link in the network at each iteration using the criterion function $\tilde{g} \leq \alpha\beta$. The algorithm flowchart is a generalized representation of a piecewise linear approximation algorithm where the type of estimation (violet box) and stopping criterion (diamond) can be replaced based on the application.

The results below show that the maximum Euclidean distance between the real and estimated time-dependent link travel time data produced by Algorithm 1 are less than

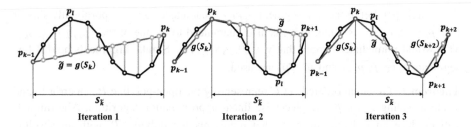

Fig. 4. Iterative piecewise linear approximation of an open polyline using Algorithm 1. (Color figure online)

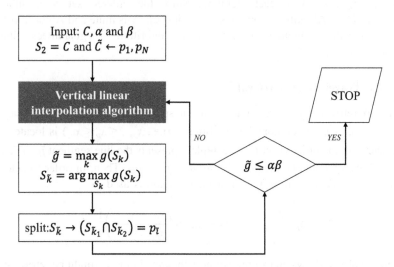

Fig. 5. Algorithm flowchart for the vertical linear interpolation algorithm.

or equal to the maximum error bound $\alpha\beta$. Moreover, it also shows that Algorithm 1 only needs to check $dist(p_i, \tilde{p}_k) = |y_{p_i} - y_{\tilde{p}_k}|$ for all p_i in order and not any interpolated values to satisfy this.

Lemma 1. Using Algorithm 1, the Euclidean distances between any real link travel time $\tilde{y}_{p_{i_j}}$ belonging to the line segments $\overline{p_s p_{s+1}}, \overline{p_{s+1} p_{s+2}}, \ldots, \overline{p_{|S_k|-1} p_{|S_k|}}$, where $(p_s, p_{s+1}) \in S_k \times S_k$, and its corresponding estimated travel time $\tilde{y}_p$ in the line segment $\overline{p_{k-1} p_k}$ along the y - axis is bounded by $\tilde{y}_j \in [0, g(S_k)]$ for any S_k.

Proof. For a calculated $S_k = \{p_s, p_{s+1}, \ldots, p_{|S_k|}\}$, we know that $p_s = p_{k-1}$ and $p_{|S_k|} = p_k$ in the line segment $\overline{p_{k-1} p_k}$. Hence, if the point p in $\tilde{x}_p$ is either $p = p_s$ or $p = p_{|S_k|}$, then $\tilde{y}_j = 0$. Then, it just needs to be shown that for any $p_{k-1} < p < p_k$, $\tilde{y}_j$ is bounded by $(0, g(S_k))$ and $g(S_k)$ will occur at one of the points $p_s \in S_k$.

Let us create triangles by connecting line segments $\overline{p_s \tilde{p}_k}$, $\overline{p_k p_k}$ and $\overline{p_{k-1} p_s}$, $\forall p_s \in S_k$. For each triangle, each $\overline{p_s \tilde{p}_k}$ is orthogonal to the x - axis with length $dist(p_s, \tilde{p}_k) = \tilde{y}_j$. By the triangle side splitter theorem and Pythagorean theorem, any line segment inside the triangle parallel to the line segment $\overline{p_s \tilde{p}_k}$ is guaranteed to be proportionally shorter than $\overline{p_s \tilde{p}_k}$. Since S_k is finite, $g(S_k)$ is also guaranteed to be one of the $\{\overline{p_s \tilde{p}_k}|p_s \in S_k\}$. This would immediately imply that $\tilde{y}_j \in [0, g(S_k)], \forall S_k$. Similar arguments hold if line segments $\overline{p_s \tilde{p}_k}$, $\overline{\tilde{p}_k p_{k-1}}$ and $\overline{p_{k-1} p_s}$ are used.

Theorem 1. Given an ordered set C representing the open polyline P', an error bound α and a scaling factor β, the piecewise linear approximation algorithm, Algorithm 1, will produce an ordered subset $\tilde{C}$ that represents the polyline $\tilde{P}'$ with an absolute maximum Euclidean distance bounded by $\alpha\beta$ from P'.

Proof. From Lemma 1, we know that $g(S_k)$ will only occur at one of the given input points $p_s \in S_k, \forall S_k \in S$ at each iteration. Since the ordered set S is finite and $\tilde{g} = \max_k g(S_k)$, if Algorithm 1 only tests $\tilde{g}$ against $\alpha\beta$, N is finite and Algorithm 1 only stops when all calculated $\tilde{g} \leq \alpha\beta$, then, Algorithm 1 guarantees that all $dist(p_i, \tilde{p}_k) \leq \alpha\beta, \forall p_i \in C$.

3.3   Link Travel Time Retrieval

In order to retrieve the travel time $\tilde{y}_p$ for a specific time $\tilde{x}_p$ in a link, the tuple $(x_{p_{k-1}}, x_{p_k}) \in \tilde{C} \times \tilde{C}$ which contains the time $\tilde{x}_p$ (i.e., $x_{p_{k-1}} \leq \tilde{x}_p \leq x_{p_k}$) is located using binary search in the ordered set $\tilde{C}$. This tuple represents the line segment $\overline{p_{k-1} p_k}$ in the polyline $\tilde{P}'$ which contains the point p, i.e. $p \in [p_{k-1}, p_k]$. Then, in order to determine the link travel time $\tilde{y}_p$ at time $\tilde{x}_p$, linear interpolation is used,

$$\tilde{y}_p = \left| y_{p_{k-1}} + (\tilde{x}_p - x_{p_{k-1}}) \left(\frac{y_{p_k} - y_{p_{k-1}}}{x_{p_k} - x_{p_{k-1}}} \right) \right|. \tag{6}$$

Figure 3 above shows an example of how the travel time $\tilde{y}_p$ might be retrieved from the ordered subset $\tilde{C}$ along the line segment $\overline{p_{k-1} p_k}$ for a specific time $\tilde{x}_p$. Additionally, it shows that the error between the estimated link travel time $\tilde{y}_p$ and the real link travel time $\tilde{y}_{p_{ij}}$ is given by $\tilde{y}_j \leq \alpha\beta$. Each link travel time retrieval has a complexity of $O(\log \tilde{N})$ which implies that the lesser the size of the ordered subset $\tilde{C}$, the better.

4   Numerical Simulation

Algorithm 1 was tested on a 10×10 grid network with 5 different absolute error bounds; α equal to 1, 30, 180 (3 min), 300 (5 min) and 600 (10 min) seconds. Additionally, travel time sampling intervals were randomly performed between 1 and 20 s.

A plot of the last iteration of the dynamic traffic simulation shown in Fig. 6 below had a one-hour simulation time period and was iteratively calculated 100 times using 500 different random demand patterns from 9 randomly selected origin-destination (OD) pairs and one CPU. The scaling factor β was set equal to 1. In the plot of the figure, if all the values of the real and estimated link travel time data are the same (*No approximation*), it will appear as a straight line with a 45° angle. The more the absolute error value is increased, the more obvious the difference between the real and estimated travel time values become (e.g. error differences based on the absolute maximum error bounds). In the simulation, there were no noticeable difference in calculation time between the *No approximation* and *10-minute error* since only a few links (around 35% of the total links in the network) of the 10 × 10 grid network were used and the simulation time period was very short. Furthermore, the time-dependent link travel time data of an unused link appeared to be approximated by two points (i.e. the start and end points) which only required a single iteration of the approximation algorithm. The average data reduction percentage of the simulation was at 98%.

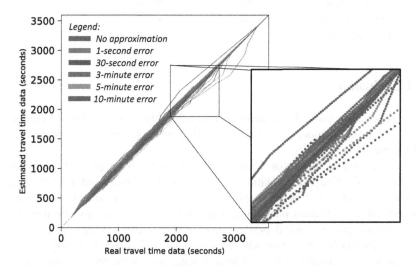

Fig. 6. Real and estimated link travel time data of all links sampled at random intervals.

In order to suitably check the effect of the piecewise linear approximation on the dynamic traffic simulation calculation time, 400 randomly selected OD pairs were fixed and a 24-h simulation time period was iteratively calculated 100 times using 16 CPUs. The dynamic traffic simulation utilized around 98% of the total links in the network. Two peak periods where simulated in the 24-h simulation time period, from 7:00 AM to 9:00 AM and 5:00 PM to 7:00 PM.

A table showing the calculation time and average data reduction percentage of the entire dynamic traffic simulation is presented below.

Table 2 above shows that the entire 24-h simulation iteratively calculated 100 times took around 90 min without the time-dependent link travel time approximation. Using a 1-second absolute error bound on the piecewise linear approximation of each link, a significant savings in memory is evident (82.18%) while only increasing the calculation time to 4 min. A 1 s error bound would normally be enough to ensure that the simulation results will be insignificantly affected by the time-dependent link travel time approximation. As the value of $\alpha\beta$ is increased, the average data reduction rate also increased and the calculation time decreased, however, the time-dependent travel time data's accuracy decreased. This is expected since the number of iterations required by the vertical linear approximation algorithm also decreases and thus, the number of points in the set $\tilde{C}$ decreases.

Table 2. Calculation time and average data reduction in a dynamic traffic simulation.

Approximation error $(\alpha\beta)$	Calculation time (s)	Average data reduction (%)
No approximation	5301	0.00
10-minute error	5305	99.91
5-minute error	5315	98.90
3-minute error	5350	95.10
30-second error	5486	89.72
1-second error	5539	82.18

5   Conclusion

In this paper a piecewise linear approximation algorithm was developed to reduce the size of the time-dependent link travel time data generated by a large-scale dynamic traffic simulation.

We were able to show that the developed algorithm is simple and fast because it only requires the linear interpolation of the given input points and a simple criterion function to conduct the approximation. This makes it very suitable for large-scale dynamic traffic simulation integration. Correspondingly, it can also be used to pre-process or post-process time-dependent link travel time data. Furthermore, we showed that it is efficient because even with its simplicity and speed, it is very effective in reducing the time-dependent travel time data regardless of the input complexity. More importantly, the algorithm is capable of producing accurate and consistent approximations at each iteration by restricting the absolute maximum error. This is very important for traveler's route planning since it avoids theoretically inconsistent results unaccounted for by the simulation model or even avoids altering simulation results.

Results from our dynamic traffic simulations showed that the algorithm was able to accurately and consistently restrict the absolute maximum error of the real and estimated time-dependent link travel time data. Moreover, it was able to reduce the link travel time's average space consumption by up to 82% using an absolute maximum error value of $\alpha\beta = 1$ s while only increasing the calculation time of the dynamic traffic simulation to 4 min. A 1 s error bound is enough to ensure that the simulation results will be insignificantly affected by the time-dependent link travel time approximation.

Acknowledgement. This work was supported by the Post K computer project (Priority Issue 3: Development of Integrated Simulation Systems for Hazard and Disaster Induced by Earthquake and Tsunami).

References

1. Delling, D., Wagner, D.: Time-dependent route planning. In: Ahuja, R.K., Möhring, R.H., Zaroliagis, C.D. (eds.) Robust and Online Large-Scale Optimization. LNCS, vol. 5868, pp. 207–230. Springer, Heidelberg (2009). https://doi.org/10.1007/978-3-642-05465-5_8
2. Delling, D., Schultes, D., Wagner, D.: Highway hierarchies star. In: 9th DIMACS Implementation Challenge (2006)
3. Prasad, D., Leung, M., Quek, C., Cho, S.: A novel framework for making dominant point detection methods non-parametric. Image Vis. Comput. **30**(11), 843–859 (2012)
4. Nguyen, V., Gachter, S., Martinelli, A., Tomatis, N., Siegwart, R.: A comparison of line extraction algorithms using 2D range data for indoor mobile robotics. Auton. Robot. **23**(2), 97 (2007)
5. Visvalingam, M. Whyatt, J.: Line generalisation by repeated elimination of the smallest area. Technical report, Discussion Paper, Cartographic Information Systems Research Group (CISRG), The University of Hull 10 (1992)
6. Ramer, U.: An iterative procedure for the polygonal approximation of plane curves. Comput. Graph. Image Process. **1**, 244–256 (1972)
7. Douglas, D., Peucker, T.: Algorithms for the reduction of the number of points required to represent a digitized line or its caricature. Can. Cartogr. **10**(2), 112–122 (1973)
8. Duda, R., Hart, P.: Pattern Classification and Scene Analysis. Wiley, New York (1973)
9. Sato, Y.: Piecewise linear approximation of plane curves by perimeter optimization. Pattern Recognit. **25**(12), 1535–1543 (1992)
10. Horni, A., Nagel, K., Axhausen, K.: The Multi-Agent Transport Simulation MATSim. Ubiquity Press, London (2016)
11. Krajzewicz, D., Hertkorn, G., Rossel, C., Wagner, P.: SUMO (Simulation of Urban MObility) - an open-source traffic simulation. In: Al-Akaidi, A. (ed.) Proceedings of the 4th Middle East Symposium on Simulation and Modelling (MESM 2002), Sharjah, United Arab Emirates, pp. 183–187 (2002)
12. Smith, L., Beckman, R., Baggerly, K., Anson, D., Williams, M..: TRANSIMS: TRansportation ANalysis and SIMulation System: Project Summary and Status (1995)
13. Tomek, I.: Two algorithms for piecewise-linear continuous approximation of functions of one variable. IEEE Trans. Comput. **C-23**(4), 445–448 (1974)
14. Imai, H., Iri, M.: An optimal algorithm for approximating a piecewise linear function. J. Inf. Process. **9**(3), 159–162 (1987)
15. Neubauer, S.: Space Efficient approximation of piecewise linear functions. Student Research Project (Studienarbeit), Universitat Karlsruhe (TH) (2009)
16. Suri, S.: A linear time algorithm with minimum link paths inside a simple polygon. Comput. Vis. Graph. Image Process. **35**(1), 99–110 (1986)
17. Guibas, L., Hershberger, J., Leven, D., Sharir, M., Tarjan, R.: Linear time algorithms for visibility and shortest path problems inside simple polygons. In SCG 1986: Proceedings of the Second Annual Symposium on Computational Geometry, New York, NY, USA, pp. 1–13 (1986)
18. Mitchell, J., Polishchuk, V., Sysikaski, M.: Minimum-link paths revisited. Comput. Geom. **47**(6), 651–667 (2014)
19. Kostitsyna, I., Loffler, M., Polishchuk, V., Frank, S.: On the complexity of minimum-link path problems. J. Comput. Geom. **8**(2), 80–108 (2016)

20. Chiu, Y.C., et al.: Dynamic traffic assignment: a primer. Transportation Research Circular E-C153, Transportation Research Board, Washington, DC (2011)
21. Cantarella, G., Cascetta, E.: Dynamic processes and equilibrium in transportation networks: towards a unifying theory. Transp. Sci. **29**(4), 305–329 (1995)
22. Nilsson, N.: Problem-Solving Methods in Artificial Intelligence. McGraw Hill, New York (1971)

Evaluation of the Suitability of Intel Xeon Phi Clusters for the Simulation of Ultrasound Wave Propagation Using Pseudospectral Methods

Filip Vaverka[1(✉)], Bradley E. Treeby[2], and Jiri Jaros[1]

[1] Faculty of Information Technology, Centre of Excellence IT4Innovations, Brno University of Technology, Bozetechova 2, 612 00 Brno, Czech Republic
{ivaverka,jarosjir}@fit.vutbr.cz
[2] Medical Physics and Biomedical Engineering, Biomedical Ultrasound Group, University College London, Malet Place Eng Bldg, London WC1E 6BT, UK
b.treeby@ucl.ac.uk

Abstract. The ability to perform large-scale ultrasound simulations using Fourier pseudospectral methods has generated significant interest in medical ultrasonics, including for treatment planning in therapeutic ultrasound and image reconstruction in photoacoustic tomography. However, the routine execution of such simulations is computationally very challenging. Nowadays, the trend in parallel computing is towards the use of accelerated clusters where computationally intensive parts are offloaded from processors to accelerators. During last five years, Intel has released two generations of Xeon Phi accelerators. The goal of this paper is to investigate the performance on both architectures with respect to current processors, and evaluate the suitability of accelerated clusters for the distributed simulation of ultrasound propagation using Fourier-based methods. The paper reveals that the former version of Xeon Phis, the Knight's Corner architecture, suffers from several flaws that reduce the performance far below the Haswell processors. On the other hand, the second generation called Knight's Landing shows very promising performance comparable with current processors.

Keywords: Ultrasound simulations · Pseudospectral methods · k-Wave toolbox · Intel Xeon Phi · KNC · KNL · MPI · OpenMP · Performance evaluation · Scaling

1 Introduction

There are many medical applications of ultrasound ranging from ultrasound and photoacoustic imaging [2] through to neurostimulation and neuromodulation [23] to direct treatment using high intensity focused ultrasound (HIFU) [5,14]. The common characteristic of all these applications is the reliance on fast, accurate

© Springer Nature Switzerland AG 2019
J. M. F. Rodrigues et al. (Eds.): ICCS 2019, LNCS 11538, pp. 577–590, 2019.
https://doi.org/10.1007/978-3-030-22744-9_45

and versatile ultrasound propagation models in biological tissue [17]. A typical scenario consists of modeling a nonlinear ultrasound wave propagating from one or more ultrasound sources through a heterogeneous medium with a power law absorption and eventually recorded by one or more ultrasound sensors or within a given region of interest.

Computational speed is still a concern even though supercomputing facilities are used. The fundamental issue is the size of the computational domain compared to the highest wavelength modeled. This challenge has raised a lot of interest across the ultrasound, mathematics and high performance computing communities. As a consequence, several ultrasound modeling packages have been released, see [8] for a recent review.

One promising approach to discretizing the acoustic governing equations is the pseudospectral time-domain (PSTD) and k-space pseudospectral time-domain (KSTD) methods [21]. The main benefit is the exponential convergence with increasing spatial resolution which can significantly reduce memory requirements for large 3D simulations. The KSTD method is considered more accurate than the PSTD method because it uses a semi-analytical time-stepping schemes [20], whereas the pseudospectral method uses a finite-difference approximation. Consequently, the KSTD method allows for a larger time step.

Unfortunately, the relaxation in the required discretization for the PSTD and KSTD schemes compared to conventional finite-difference schemes is somewhat counteracted by the introduction of a global trigonometric basis and the use of the fast Fourier transform (FFT) to compute spatial gradients. For PSTD schemes, the FFTs are one-dimensional (in the direction of the required gradient). However, for the KSTD scheme, the introduction of the k-space correction means the FFTs are performed in three-dimensions. The scaling on parallel systems is then inherently limited by the necessity of performing distributed matrix transpositions over all subdomains [11] as part of the 3D FFT. Although a lot of work on efficient distributed FFTs has been carried out (FFTW [6], P3DFFT [16], PFFT [18], AccFFT [7] or multi-GPU CUDA FFT [15]), the computation time is still often determined by the communication between subdomains, which in many cases prevents the use of accelerators such as GPUs or Intel Xeon Phis.

A promising direction in joining the advantages of FDTD and PSTD methods is the decomposition of the global Fourier basis into a set of local ones [10]. This composition inherits the simplicity of the FDTD nearest neighbor halo exchange while maintaining the spectral accuracy of PSTD and KSTD methods [22].

This paper investigates the suitability of domain decomposition, implemented as part of the k-Wave toolbox [12], for deployment on cluster of Intel Xeon Phi accelerators based on both Knight's Corner (KNC) and Knight's Landing (KNL) architectures. First, the principle of local Fourier basis domain decomposition vital for the distributed computation is explained in Sect. 2. Second, the architecture of two accelerated clusters is described in Sect. 3. After that, the main components of the benchmark implementation are outlined in Sect. 4. Section 5 presents the experimental results collected on both clusters and compares the performance scaling with a CPU cluster. Finally, the most important conclusions are drawn.

2 Efficient Local Fourier Basis Domain Decomposition

The local Fourier basis domain decomposition (LFB for short) of PSTD and KSTD methods splits the 3D domain into a number of cuboid subdomains, each of which is supported by its own local Fourier basis [10]. The required global communication is consequently reduced into local direct neighbor exchange of the overlap regions. However, the split of the Fourier basis breaks the periodicity condition on local domains. To restore it, Fourier extension methods can be used [3]. The subdomains are coupled by overlap exchanges and the local subdomain periodicity is restored by multiplying with a bell function [12], see Fig. 1.

The restriction of the Fourier basis to the local subdomain has naturally a negative impact on the accuracy of the LFB method. The amount of accuracy loss depends on the overlap size and the properties of the bell function used [4]. While the overlap size can be chosen by the user as a compromise between the accuracy and the performance for any particular problem, the shape of the bell function has be optimized in advance for the whole set of overlap sizes by means of numerical optimization.

Figure 2 shows the relationship between the accuracy, the size of the overlap, and the bell function used. Figure 2a shows the dependency of the numerical error in terms of the L_∞ norm on the size of the overlap for the domain split into two subdomains (a single cut). The figure also compares two different bell functions, the well known Error (Erf) function [1] and a numerically optimized one. Figure 2b indicates the minimum overlap size required to keep the error below 10^{-3} or 10^{-4} for a given number of subdomain interfaces the wave has to cross in a single Cartesian direction. The conclusion drawn from this figure suggests that an overlap size of 8 or 20 should be chosen to keep the overall accuracy of 10^{-3} and 10^{-4} for decompositions with the total number of subdomains below 512 (8 in every Cartesian direction), respectively.

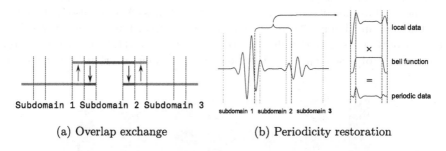

(a) Overlap exchange   (b) Periodicity restoration

Fig. 1. The principle of local Fourier basis domain decomposition shown for one spatial dimension. (a) The local subdomain is padded with an overlap from both neighboring subdomains. These overlaps are periodically exchanged. (b) After the exchange, each local subdomain is multiplied by a bell function.

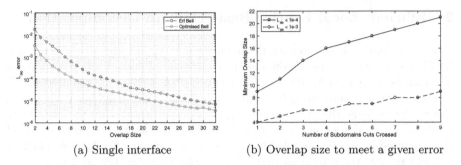

(a) Single interface   (b) Overlap size to meet a given error

Fig. 2. The accuracy of the local Fourier basis domain decomposition determined by the size of the overlap, the number of subdomain interfaces and the shape of the bell function.

The numerical model of the nonlinear wave propagation in heterogeneous absorption medium investigated in this paper is based on the governing equations derived by Treeby [21] written as three-coupled first-order partial differential equations:

$$\frac{\partial \mathbf{u}}{\partial t} = -\frac{1}{\rho_0}\nabla p + \mathbf{F}, \qquad\qquad \text{(momentum conservation)}$$

$$\frac{\partial \rho}{\partial t} = -\rho_0 \nabla \cdot \mathbf{u} - \mathbf{u} \cdot \nabla \rho_0 - 2\rho \nabla \cdot \mathbf{u} + \mathbf{M}, \qquad \text{(mass conservation)}$$

$$p = c_0^2 \left(\rho + \mathbf{d} \cdot \nabla \rho_0 + \frac{B}{2A}\frac{\rho^2}{\rho_0} - \mathrm{L}\rho \right). \qquad \text{(equation of state)} \quad (1)$$

Here $\mathbf{u}$ is the acoustic particle velocity, $\mathbf{d}$ is the acoustic particle displacement, p is the acoustic pressure, ρ is the acoustic density, ρ_0 is the ambient (or equilibrium) density, c_0 is the isentropic sound speed, and B/A is the nonlinearity parameter. Two linear source terms (force $\mathbf{F}$ and mass $\mathbf{M}$) are also included.

The computation itself consists of an iterative algorithm running over a given number of time steps (a detailed description is given in [11,12]). Each time step is composed of a sequence of element-wise operations, overlap exchanges and

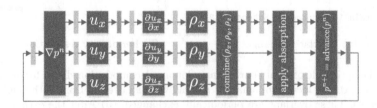

Fig. 3. Simplified computation loop governed by Eq. (1). The blue blocks denote element-wise operations, yellow 3D FFTs, and orange the overlap exchanges. (Color figure online)

local 3D FFTs, see Fig. 3 [12]. The majority of the computation time is usually spent on 3D FFTs or overlap exchanges.

3   Target Architecture

The target architectures of interest are represented by two clusters of Intel Xeon Phi accelerators. Salomon is an accelerated cluster based on the first generation of Knight's Corner architecture operated by the IT4Innovations national supercomputing center Ostrava, Czech Republic[1]. CoolMUC3 is a newer cluster based on the second generation of the Knight's Landing architecture operated by the Leibniz Rechenzentrum in Garching, Germany[2].

3.1   Architecture of Knight's Corners Cluster

Salomon consists of 1008 compute nodes, 432 of which are accelerated by Intel Xeon Phi 7120P accelerators. The architecture of Salomon's accelerated part is shown in Fig. 4. Every node consists of a dual socket motherboard populated with two Intel Xeon E5-2680v3 (Haswell) processors accompanied with 128 GB of RAM. The nodes also integrate a pair of accelerators connected to individual processor sockets via the PCI-Express 2.0 x16 interface. The communication between processor sockets and accelerators is handled by the Intel QPI interface.

The nodes are interconnected by a 7D enhanced hypercube running on the 56 Gbit/s FDR Infiniband technology. The accelerated nodes occupy a subset of the topology constituting a 6D hypercube. Every node contains a single Infiniband network interface (NIC) connected via PCI-Express 3.0 to the first socket and a service 1 Gbit/s Ethernet interface connected to the same socket. Both accelerators are capable of directly accessing the Infiniband NIC by means of Remote Direct Memory Access (RDMA).

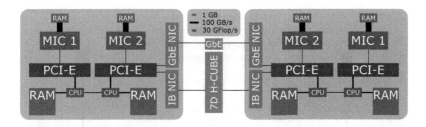

Fig. 4. The architecture of the Salomon accelerated nodes and interconnection. The size of the rectangles representing individual components is proportional to their performance, bandwidth or capacity.

---

[1] https://docs.it4i.cz/salomon/hardware-overview/.

[2] https://www.lrz.de/services/compute/linux-cluster/coolmuc3/overview_en/.

A single Intel Xeon Phi 7120P accelerator packs 61 P54C in-order cores extended by 4-wide simultaneous multithreading (SMT) and a 512-bit wide vector processing unit (VPU). The KNC cores are supported by 30.5 MB of L2 cache distributed over individual cores and interconnected via a ring bus. The memory subsystem consists of 4 memory controllers managing in total 16 GB of GDDR5 [13]. The theoretical performance and memory bandwidth of a single accelerator is over 2 TFLOP/s in single precision and 352 GB/s, respectively. A single accelerator is theoretically supposed to provide a speedup of 4× for compute bound and 5× for memory bandwidth bound applications over a single twelve core Haswell processor. The total compute power of the accelerated part of Salomon reaches one PFLOP/s.

3.2   Architecture of Knight's Landing Cluster

CoolMUC3 consists of 148 nodes equipped with Intel Xeon Phi 7210-F accelerators. The architecture of CoolMUC3 is shown in Fig. 5. The Xeon Phi generation installed in this system is the first from Intel to be stand-alone. Since a classic CPU is thus not required to control the computation node, the cluster is composed of single socket nodes populated with the KNL processors only.

The nodes are interconnected by an Intel Omnipath network forming a fat tree topology. A single node has two independent NICs connected via PCI-Express 3.0 x16 interfaces offering aggregated throughput of 25 GB/s per node.

Every KNL chip consists of 32 tiles placed in a 2D grid providing a bisection bandwidth of 700 GB/s. Each tile integrates two out-of-order 4-wide SMT cores, four 512-bit wide vector processing units and 1 MB of shared L2 cache. The theoretical performance of a single KNL chip exceeds 5 TFLOP/s in single precision. The main memory of each node is split between 16 GB of High Bandwidth Memory (HBM) and 96 GB of DDR4 providing bandwidth of 460 GB/s and 80 GB/s, respectively. Since the KSTD and PSTD solvers are proven to be memory bound, only HBM memory is used in this study. The expected speedup

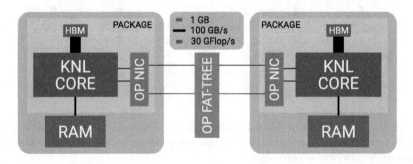

Fig. 5. The architecture of the CoolMUC3 accelerated nodes and interconnection. The size of the rectangles representing individual components is proportional to their performance, bandwidth or capacity.

over a Haswell CPU should attain a factor of 10 for compute bound and 6 for memory bound applications.

4   Implementation

The PSTD and KSTD methods (e.g., as implemented in k-Wave) are typical examples of memory bound problems with a relatively low arithmetic intensity, usually on the order of $O(\log n)$ (due to FFTs). Furthermore, the LFB domain decomposition relies on communication stages which are latency sensitive because very little communication can be overlapped. Such a combination of algorithm properties suggests the use of parallel architectures with high memory bandwidth and, ideally, a direct access to NICs. The Intel Xeon Phi accelerators look very favorable from this point of view.

The proposed implementation can be executed on both CPUs and accelerators. Although, it is possible to use any combinations of CPUs and Xeon Phis concurrently, this is not tested in this paper because no load balancing has been implemented yet (only uniform decompositions are supported). The code is logically structured into MPI processes handling single subdomains running on particular accelerators or CPUs. The work distribution within the subdomain is implemented by means of OpenMP threads and OpenMP SIMD constructs. Since realistic simulations do not require double precision, only single precision floating point operations are used in the critical path. This yields higher performance and saves valuable memory bandwidth.

Logically, the simulation code of k-Wave code boils down to a mix of element-wise operations on 3D real or complex matrices, 3D Fourier transforms and overlap exchanges. These are further explained in following subsections.

4.1   Element-Wise Operations

The element-wise operations can easily take the full advantage of the accelerator memory bandwidth and compute power because of their locality. Listing 1 shows a typical example of an element-wise computation kernel.

```
1    const float norm = 1.0f / (Nx * Ny * Nz);
2
3    #pragma omp parallel for collapse(2)
4    for(size_t z = 0; z < Nz; z++) {
5      for(size_t y = 0; y < Ny; y++) {
6        const float ePmlY = pmlY[y];
7
8        #pragma omp simd
9        for(size_t x = 0; x < Nx; x++) {
10          const size_t i = z * Ny * Nx + y * Nx + x;
11          uy[i] = ((uy[i] * ePmlY) - (norm * fftPressure[i])) * ePmlY;
12        }
13      }
14    }
```

Listing 1: Update of the acoustic velocity field in the y-axis direction with an application of the perfectly matched layer (PML) optimized for Intel Xeon Phi.

The kernel runs over three spatial dimensions starting from the most significant one to comply with the row-major array ordering. The two outermost loops are collapsed into a single one and parallelised over multiple OpenMP threads. Although the loop collapsing introduces some overhead into the calculation of z and y indices, it is vital for even distribution of the work among many parallel threads. Let us note that up to 256 threads can be executed simultaneously on the Xeon Phi while the maximum subdomain size which can fit within the accelerator memory is on the order of 400^3 grid points. The innermost loop is vectorised by means of an OpenMP SIMD pragma to ensure the full utilization of the vector units.

4.2 Fourier Transforms

The most computationally expensive part of the simulation loop consists of 14 3D fast Fourier transforms calculated over the local subdomains. Their actual implementation relies on third party libraries compatible with the FFTW interface [6], in this case the Intel MKL[3] library [9] which is believed to be well optimized for the Intel Xeon Phi architecture [24].

The algorithms to perform forward and inverse FFTs typically assume complex input and output data. However, the solutions of the wave equation require only real-valued data in the time domain. This makes the use of real-to-complex (R2C) and complex-to-real (C2R) transforms possible and reduces the temporal and spatial complexity of the FFT by a factor of two [19].

The simulation code naturally uses out-of-place transforms to preserve the input fields needed later in the simulation loop and calculates derivatives in reusable temporary matrices. Unfortunately, the implementation of the out-of-place C2R transforms in the MKL library has proved to be very inefficient on KNC, showing an almost 12× performance drop for bigger subdomains. Hence, the C2R transforms are performed in-place using a temporary matrix and the results consequently copied to the destination matrix.

4.3 Overlap Exchanges

Before every gradient calculation, it is necessary to synchronize all subdomains by exchanging the overlap regions, see the orange bars in Fig. 3. Depending on the rank of the decomposition, up to 26 mutual exchanges are performed per subdomain. The amount of data being transferred is proportional to the size of the overlap region, and also dependent on the mutual position of subdomains in the simulated space. If two subdomains only touch at the corner, only a small number of grid points is transferred ($N_d{}^3$, where d is the size of the overlap). In contrast, if two subdomains sit side by side, a large block of $N_x \times N_y \times N_d$ grid points must be transferred.

Since the overlaps have to contain the most recent data, it is difficult to hide the communication by overlapping it with useful computation. Fortunately, it is

---

[3] Intel 2017b and 2018a suite were used on Salomon and CoolMUC3 respectively.

possible to decouple the calculation of velocity gradients for each spatial dimension and overlap the data exchange with gradient calculations. This enables two out of three communications to be hidden. The same approach is applied to the medium absorption calculation where two gradients are calculated independently. In total, up to 50% of the communication time may be hidden.

Practically speaking, the communication overlapping is achieved by a combination of persistent communications and non-blocking calls provided by MPI. Listing 2 shows the principle of the communication hiding during the velocity gradient calculation. The calculation of partial derivative of the velocity along a given axis starts as soon as the overlaps arrive while the other communication can still be in flight.

```
1    // Initialization stage
2    for (auto &m: /* Velocity matrices U_x, U_y, U_z */) {
3      for(auto &n: m.getNeighbors()) {
4        MPI_Send_init(n.data, n.size, n.otherRank, /* ... */);
5        MPI_Recv_init(n.data, n.size, n.otherRank, /* ... */);
6      }
7    }
8
9    // Main simulation loop stage
10   for (auto &m: /* Velocity matrices U_x, U_y, U_z */)
11     MPI_Startall(m.getRequests().size(), m.getRequests().data());
12
13   for (auto &m: /* Velocity matrices U_x, U_y, U_z */) {
14     // Partially overlapped communication
15     MPI_Waitall(m.getRequests().size(), m.getRequests().data(),
16                 /* ... */);
17     Compute_Forward_FFT_3D(m);
18   }
```

Listing 2: The principle of communication hiding during velocity gradient calculation. Persistent communications are created in the initialization stage (lines 1–7). The exchange on multiple matrices is started at a given place in the simulation loop (lines 9–11). As soon as the communication for a given matrix finishes, the computation starts. The other transfers can be still in flight (lines 13–17).

5   Experimental Results

Numerical experiments were conduced on a various number of accelerators ranging from 1 to 16. The number of accelerators was limited by our preparatory allocations enabling the maximum use of 16 nodes on CoolMUC3 and the capacity of the express queue on Salomon (8 nodes, 2 accelerators each).

Benchmark runs of the same type were packed into single larger jobs to maintain the same MPI rank placement over the cluster between particular benchmark runs. Therefore, only a tiny variation, considered insignificant from the perspective of the overall scaling trends and even the absolute performance, may be observed between different benchmark runs. Every benchmark run consists of 100 time steps of the simulation loop summarized in Fig. 3. This number is deemed sufficient to hide any cache and communication warm-up effects.

The simulation domain was progressively expanded from $256 \times 256 \times 256$ (2^{24}) to $1024 \times 1024 \times 512$ (2^{29}) grid points by sequentially doubling the dimension sizes starting from the least significant one. The global domains were partitioned into a number of subdomains growing from 1 to 16. The numbers of subdomains for particular domain sizes were further restricted by the size of the smallest meaningful subdomain (64^3) and the largest possible subdomain ($256 \times 256 \times 512$) that can fit within the memory, excluding the overlaps. Particular subdomains were assigned either to a single accelerator or a single CPU sockets. The reason for this kind of comparison is twofold. First, the amount of communication overhead is kept the same and IT4I's allocation, and second, pricing policies take only CPU cores into account (and not the accelerator usage).

On the OpenMP level, each subdomain was processed by the optimal number of threads on a given architecture. For Haswell CPUs we used one thread per core (12 threads per CPU) whereas the optimal number of threads for a single KNC accelerator was found to be 120 (2 threads per core). Finally, KNL performed best using all 256 threads per accelerator.

5.1   Strong Scaling Evaluation

Figure 6 shows the strong scaling for investigated architectures. Although the whole range of the overlap sizes between 2 to 32 grid points was investigated, only one overlap size of 16 is presented for the sake of brevity. Scaling with small overlap sizes generally runs faster due to a higher degree of communication overlapping. For bigger overlap sizes, the absolute execution time is more influenced by the communication time and the strong scaling curves appear flatter.

Looking at Figs. 6a and b, a significant disproportion in the performance between CPUs and KNC accelerators can be observed. The execution time on KNC is between 2.2× and 4.3× longer than on CPUs. This behavior was further investigated by analyzing flat performance profiles. First, the overlap exchange among accelerators is on average 2× slower than among CPUs. This substantial overhead can be attributed to a combined effect of the additional PCI-Express communication and much slower compute cores on the accelerators responsible for packing the overlaps into MPI messages and their management. The maximum measured core-to-core bandwidth only reaches 2.65 GB/s, which is about a half of the theoretical Infiniband bandwidth. Second, the performance of the 3D FFTs very low. For the subdomain sizes examined in this section, the speedup of KNC with respect to CPU was between 0.03 for domain sizes of 64^3 and 0.4 for domain sizes of 256^3. For small domains, this is most certainly due to expected thread congestion and cache coherence effects such as false sharing. Since the Intel MKL is a closed software, it was impossible to further investigate this issue.

Apart from the poor absolute performance of KNC, the scaling factors look favorable. For the three biggest domains, the scaling factor reaches a value of 1.52 every time the number of accelerators is doubled. This yields a parallel efficiency of 76%, which is comparable to the CPU cluster.

The strong scaling achieved by the KNL cluster is significantly better, see Fig. 6c. The KNL accelerators are significantly faster than the previous generation KNC. When only a single accelerator is used, the benchmarks are completed in an order of magnitude shorter time. When communication comes into play,

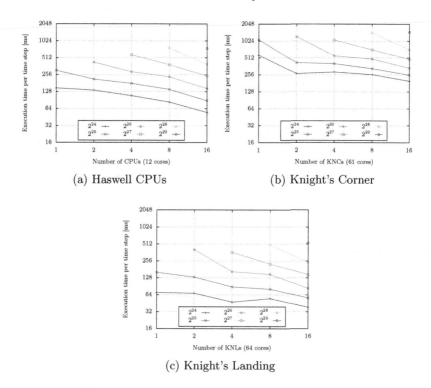

(a) Haswell CPUs

(b) Knight's Corner

(c) Knight's Landing

Fig. 6. Strong scaling with overlap size of 16 grid points.

the Omnipath interconnection shows its strengths. The average scaling factors reaches 1.62. This amounts to 4.16× speedup compared to KNCs with the Infiniband interconnect on Salomon. The comparison against the Haswell CPUs with 12 cores yields an average speedup of 1.7 in favor of the new Intel Xeon Phi accelerators.

5.2   Weak Scaling Evaluation

Figure 7 shows the weak scaling achieved on the CPUs and accelerators. Each of the plotted series corresponds to a constant subdomain size from the investigated range between $128 \times 128 \times 64$ and $256 \times 256 \times 512$ grid points. At first glance, poor weak scaling is observed for CPUs and KNLs when the simulation domain is split into fewer than 8 subdomains. This is due to the growing rank of the domain decomposition and the number of neighbors. Since the computation on KNC is much slower, there is better possibility for communication overlapping and the initial growth in the execution time is not observed.

Once a full 3D decomposition is reached, the scaling curves remain almost flat being a sight of almost perfect scaling. However, to support this statement, benchmark results using a much higher number of accelerators are needed.

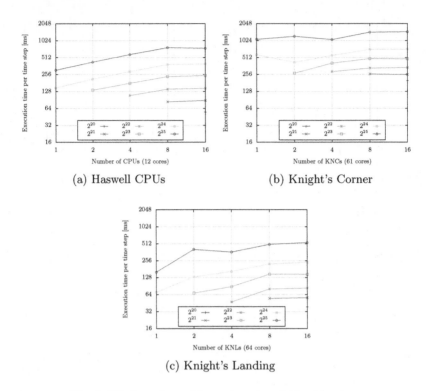

(a) Haswell CPUs   (b) Knight's Corner

(c) Knight's Landing

Fig. 7. Weak scaling with overlap size of 16 grid points.

6   Conclusion

The goal of this paper was to investigate the performance scaling and suitability of two Xeon Phi accelerated clusters for large simulations of ultrasound wave propagation using Fourier pseudospectral methods and compare the computational performance against a common CPU cluster.

Starting with the former Knight's Corner architecture of Intel Xeon Phi, we conclude that the cluster of KNCs did not come up to expectations when running the pseudospectral time domain solver of the k-Wave toolbox [12]. The biggest obstacle was the performance of the 3D fast Fourier transforms, which for the domain sizes of interest reaches only a fraction of the performance provided by CPU. This may be caused by a too many active threads when small domains are computed, and relatively small L2 caches resulting in many accesses to the main memory if the domain sizes are bigger. In future work, we would like to examine other FFT libraries such as FFTW [6] and confirm that the poor performance is caused by a bug in the Intel MKL library. Considering that this issue might be fixed in the future, the strong and weak scaling achieved on KNC promises easy deployment on all of the 432 Salomon's accelerators. Since the allocation policy in terms of core hours charged is very favorable for accelerators, the cluster of KNC can decrease the computational costs for running large scale simulations.

The performance of the Knight's landing cluster was (after the experience with its predecessor) a pleasant surprise. The performance of a single KNL accelerator is 4× higher than KNC and almost 1.7× higher than a single twelve core Haswell CPU. The strong scaling is also better with a parallel efficiency of 81%. This shows that Intel has achieved a significant improvement on the interconnection part as well. In the future work, we would like to use much higher number of the KNL accelerators to extend the scaling study and run full production simulations on CoolMUC3.

Acknowledgement. This work was supported by The Ministry of Education, Youth and Sports from the National Programme of Sustainability (NPU II) project IT4Innovations excellence in science - LQ1602 and by the IT4Innovations infrastructure which is supported from the Large Infrastructures for Research, Experimental Development and Innovations project IT4Innovations National Supercomputing Center - LM2015070. This project has received funding from the European Union's Horizon 2020 research and innovation programme H2020 ICT 2016–2017 under grant agreement No 732411 and is an initiative of the Photonics Public Private Partnership. This work was also supported by the Engineering and Physical Sciences Research Council, UK, grant numbers EP/L020262/1 and EP/P008860/1.

References

1. Andrews, L.C.: Special Functions of Mathematics for Engineers. SPIE Pub. (1997)
2. Beard, P.: Biomedical photoacoustic imaging. Interface Focus **1**(4), 602–631 (2011). https://doi.org/10.1098/rsfs.2011.0028
3. Boyd, J.P.: A comparison of numerical algorithms for fourier extension of the first, second, and third kinds. J. Comput. Phys. **178**(1), 118–160 (2002). https://doi.org/10.1006/jcph.2002.7023
4. Boyd, J.P.: Asymptotic fourier coefficients for a C∞ bell (Smoothed-"Top-Hat") & the fourier extension problem. J. Sci. Comput. **29**(1), 1–24 (2006). https://doi.org/10.1007/s10915-005-9010-7
5. Dubinsky, T.J., Cuevas, C., Dighe, M.K., Kolokythas, O., Joo, H.H.: High-intensity focused ultrasound: current potential and oncologic applications. Am. J. Roentgenol. **190**(1), 191–199 (2008)
6. Frigo, M., Johnson, S.G.: The design and implementation of FFTW3. Proc. IEEE **93**(2), 216–231 (2005)
7. Gholami, A., Hill, J., Malhotra, D., Biros, G.: AccFFT: a library for distributed-memory FFT on CPU and GPU architectures, May 2016. http://arxiv.org/abs/1506.07933v3
8. Gu, J., Jing, Y.: Modeling of wave propagation for medical ultrasound: a review. IEEE Trans. Ultrason. Ferroelectr. Freq. Control **62**(11), 1979–1992 (2015). https://doi.org/10.1109/TUFFC.2015.007034
9. Intel Corporation: Math Kernel Library 11.3 Developer Reference. Intel Corporation (2015)
10. Israeli, M., Vozovoi, L., Averbuch, A.: Spectral multidomain technique with local Fourier basis. J. Sci. Comput. **8**(2), 135–149 (1993)
11. Jaros, J., Rendell, A.P., Treeby, B.E.: Full-wave nonlinear ultrasound simulation on distributed clusters with applications in high-intensity focused ultrasound. J. High Perform. Comput. Appl. **30**(2), 137–155 (2016)

12. Jaros, J., Vaverka, F., Treeby, B.E.: Spectral domain decomposition using local fourier basis: application to ultrasound simulation on a cluster of GPUs. Supercomput. Frontiers Innov. **3**(3), 40–55 (2016)

13. Jeffers, J., Reinders, J.: Intel Xeon Phi Coprocessor High Performance Programming, vol. 1. Elsevier Inc., Waltham (2013)

14. Meairs, S., Alonso, A.: Ultrasound, microbubbles and the blood-brain barrier. Progress Biophys. Mol. Biol. **93**(1–3), 354–362 (2007)

15. Nandapalan, N., Jaros, J., Treeby, E.B., AlistairRendell, P.: Implementation of 3D FFTs across multiple GPUs in shared memory environments. In: Proceedings of the Thirteenth International Conference on Parallel and Distributed Computing, Applications and Technologies, pp. 167–172 (2012). https://doi.org/10.1109/PDCAT.2012.79. http://www.fit.vutbr.cz/research/view_pub.php?id=10171

16. Pekurovsky, D.: P3DFFT: a framework for parallel computations of Fourier transforms in three dimensions (2012). https://doi.org/10.1137/11082748X

17. Pinton, G.F., Dahl, J., Rosenzweig, S., Trahey, G.E.: A heterogeneous nonlinear attenuating full-wave model of ultrasound. IEEE Trans. Ultrason. Ferroelectr. Freq. Control **56**(3), 474–488 (2009)

18. Pippig, M.: PFFT: an extension of FFTW to massively parallel architectures. SIAM J. Sci. Comput. **35**(3), 213–236 (2013)

19. Sorensen, H., Jones, D., Heideman, M., Burrus, C.: Real-valued fast Fourier transform algorithms. IEEE Trans. Acoust. Speech Signal Process. **35**(6), 849–863 (1987). https://doi.org/10.1109/TASSP.1987.1165220

20. Tabei, M., Mast, T.D., Waag, R.C.: A k-space method for coupled first-order acoustic propagation equations. J. Acoust. Soc. Am. **111**(1 Pt 1), 53–63 (2002). https://doi.org/10.1121/1.1421344

21. Treeby, B.E., Jaros, J., Rendell, A.P., Cox, B.T.: Modeling nonlinear ultrasound propagation in heterogeneous media with power law absorption using a k-space pseudospectral method. J. Acoust. Soc. Am. **131**(6), 4324–4336 (2012). https://doi.org/10.1121/1.4712021

22. Treeby, B.E., Vaverka, F., Jaros, J.: Performance and accuracy analysis of nonlinear k-wave simulations using local domain decomposition with an 8-GPU server. Proc. Meet. Acoust. **34**(1), 022002 (2018)

23. Tufail, Y., Yoshihiro, A., Pati, S., Li, M.M., Tyler, W.J.: Ultrasonic neuromodulation by brain stimulation with transcranial ultrasound. Nat. Protoc. **6**(9), 1453–1470 (2011). https://doi.org/10.1038/nprot.2011.371

24. Wang, E., et al.: High-Performance Computing on the Intel® Xeon Phi™. Springer, Cham (2014). https://doi.org/10.1007/978-3-319-06486-4

Track of Computational Science in IoT and Smart Systems

Fog Computing Architecture Based Blockchain for Industrial IoT

Su-Hwan Jang[1], Jo Guejong[1], Jongpil Jeong[1(✉)], and Bae Sangmin[2]

[1] Department of Smart Factory Convergence, Sungkyunkwan University,
Gyeonggi-do 16419, Republic of Korea
{korjsh,jejewind,jpjeong}@skku.edu
[2] Markate Incor., Gyeonggi-do 13522, Republic of Korea
baesm@markate.net

Abstract. Industry 4.0 is also referred to as the fourth industrial revolution and is the vision of a smart factory built with CPS. The ecosystem of the manufacturing industry is expected to be activated through autonomous and intelligent systems such as self-organization, self-monitoring and self-healing. The Fourth Industrial Revolution is beginning with an attempt to combine the myriad elements of the industrial system with Internet communication technology to form a future smart factory. The related technologies derived from these attempts are creating new value. However, the existing Internet has no effective way to solve the problem of cyber security and data information protection against new technology of future industry. In a future industrial environment where a large number of IoT devices will be supplied and used, if the security problem is not resolved, it is hard to come to a true industrial revolution. Therefore, in this paper, we propose block chain based fog system architecture for Industrial IoT. In this paper, we propose a new block chain based fog system architecture for industrial IoT. In order to guarantee fast performance, And the performance is evaluated and analyzed by applying a proper fog system-based permission block chain.

Keywords: Industrial IoT · Block chain · Hyperledger Fabric

1 Introduction

This system, also known as Industrial Automation and Control Systems (IACS) or Operational Technology (OT), is well established today. The system is used in a variety of industries, including manufacturing, transportation and utilities, and is sometimes referred to as a cyber-physical system (CPS).

Internet of Things (IoT) [1] was first used in 1999 and has been applied to devices connected to the Internet in consumer, home, business and industrial environments [2]. Although IoT has a large body of literature to define its uses and typical components, little is known about how these apply in industrial settings. Therefore, IIoT, which is to be applied in industrial environment, is

© Springer Nature Switzerland AG 2019
J. M. F. Rodrigues et al. (Eds.): ICCS 2019, LNCS 11538, pp. 593–606, 2019.
https://doi.org/10.1007/978-3-030-22744-9_46

conducting researches on CPS in real time and IoS is used to provide inter-organizational and inter-organizational services by participants in the value chain. And it is said to be the core of Industry 4.0 as it is utilized [3].

Industry 4.0 is also referred to as the fourth industrial revolution and is the vision of a smart factory built with CPS. The ecosystem of the manufacturing industry is expected to be activated through autonomous and intelligent systems such as self-organization, self-monitoring and self-healing.

The Fourth Industrial Revolution is beginning with an attempt to combine the myriad elements of the industrial system with Internet communication technology to form a future smart factory. The related technologies derived from these attempts are creating new value. However, the existing Internet has no effective way to solve the problem of cyber security and data information protection against new technology of future industry. In a future industrial environment where a large number of IoT devices will be supplied and used, if the security problem is not resolved, it is hard to come to a true industrial revolution.

Currently, Internet technology is a centralized data processing process, which has the advantage of being able to be designed for each purpose in each industrial environment, but it has a disadvantage that it is very vulnerable because of centralized processing on a central server. Methods to address these drawbacks are also very limited and issues of performance load can not be avoided. So, to solve the problem, the system structure must change fundamentally. In other words, the goal of this paper is to study the need to reorganize the centralized structure into a distributed structure.

In this paper, we propose the necessary technologies and systems to build IoT - based smart factories for manufacturing companies and set the scope of the study to derive conclusions about the proposed structure through performance evaluation and analysis.

2   Related Work

2.1   Industrial IoT

There are a lot of IoT definitions in academia, but I will try to understand them through three related definitions.

(1) The definition of IoT is "an infrastructure group that interconnects connected objects and allows access to management, data mining and generated data". Linked objects are "specific functions by which sensors or actuators can communicate with other devices" [4].

(2) "The Internet of Things is broadly referred to broadly as the extension of network connectivity and computing capabilities to objects, devices, sensors, and items that are not generally considered to be computers. These smart objects. They often need to have connectivity to remote data collection, analysis and management functions." [5].

(3) "IoT represents a scenario in which all objects or 'objects' are embedded in the sensor and can automatically communicate status with other objects and automated systems in the environment. Each object represents a node of the virtual network, And continuously transmits large amounts of data to the periphery." [6].

As IoT is defined and evolved, more and more applications are being made to apply this technology in the industry. IIoT is also called the "industrial internet" in GE. How it is called, IIoT differs from other IoT applications in that it focuses on connecting machines and devices in industrial sectors such as oil and gas, power facilities, and medical care. The initial definition of IIoT has been introduced in many literature [7,8], using industry specific IoT technology in industrial environments. One of the recent studies for the IIoT environment has implemented hierarchical fog server architecture. This approach seeks to decentralize the fog server architecture to reduce communication latency and increase throughput.

IIoT in smart factories is one of the most mentioned industrial business concepts in recent years. The term IIoT refers to the Internet for industrial objects. In a broader sense, it is to connect devices in transportation, energy, and industrial sectors and sensors and other devices in vehicles to the network. The meaning of IIoT actually varies. The IIoT system is simple enough to leave home text like a connected mousetrap, but it can track and share large amounts of information, from a fully automated mass production line to maintenance, productivity, ordering and shopping, It can be as complex as it is. Currently, the level of application of IoT in industrial automation is not high overall. [9] IoT solutions for industrial automation are still evolving. The fog-based approach can meet the requirements of modern industrial systems. Most existing research, however, focuses on a centralized computing architecture that uses cloud computing to monitor data and manage control processes in industrial automation. Most of the existing approaches and solutions from the cloud computing point of view for industrial automation are focused on higher than the field level.

2.2 Fog Computing

The concept of fog computing connects these two environments as a kind of distributed network. " Fog complements the middle between data that needs to be pushed to the cloud and local analysis," said Morchiang, Dean of the College of Engineering at Purdue University and Fogg and Edge Computing researcher. The OpenFog Consortium, a group of research institutes and companies that support the standard development of fog technology, is committed to delivering fog computing to the computing community, "distributing computing, storage, control, networking services and resources from anywhere in the cloud- System-level horizontal architecture" [10].

With the fog computing framework, companies have new options for processing data wherever they are for optimal data processing. It is particularly useful in areas where data must be processed as quickly as possible. For example, there is a typical manufacturer whose connected equipment has to respond to an accident as soon as possible.

Fog computing creates a network connection with less delay between the device and the analysis endpoint. This reduces the amount of bandwidth required by sending data to the data center or to the cloud. It can also be used when the bandwidth connection to send data is difficult and needs to be handled close to the data generation site. It is also an advantage that security can be applied to the fog network from the divided network traffic to the virtual firewall to protect it.

Fog Computing is still in the early stages of commercialization, but its use cases are diverse. First, ConnectedCada. The emergence of semi-automatic and autonomous vehicles is driving up data generated by vehicles. In order for the vehicle to operate independently, local data such as the surrounding environment, driving condition, and driving direction must be analyzed locally in real time. Other data can be sent by the manufacturer to improve vehicle maintenance or track vehicle usage. The fog computing environment enables communication from both the edge (in-vehicle) as well as the endpoint (manufacturer) for all this data.

Smart cities and smart grids are also good places to apply fog computing. Like connected cars, utility systems are also increasingly leveraging real-time data for efficient operation. This data may be located in a remote area, so it is necessary to process it near the place where it was created. Data may need to be collected from multiple sensors. In both cases, the fog computing architecture presents a good solution.

Real-time analysis is used at a variety of locations, from manufacturing systems that need to respond to events as soon as they occur, to financial institutions that use real-time data to provide information necessary for transaction decisions or to monitor fraudulent transactions. Placing fog computing here makes data transfer between the data generation site and the various places where the data needs to be moved more smoothly.

The 5G mobile connection expected to be launched after 2018 is expected to spread more rapidly to fog computing. According to Andrew Anders, senior vice president of technology planning and network architecture at CenturyLink, 5G technology requires very dense antenna placement. The distance between the antennas should not exceed 20 km. In this case, by creating a fog computing architecture that includes a central control between the antenna points, it is possible to manage applications running on this 5G network and support connectivity to back-end data centers or the cloud.

The fog computing fabric has various components and functions. There are various wired and wireless granular collection endpoints, such as fog computing gateways and ruggedized routers and switching equipment that accept data collected by IoT devices. Gateway and customer premises equipment (CEP) for edge node access may also be included. At the higher end of the stack, the fog computing architecture may touch the core network and routers, and ultimately, the broadband cloud services and servers.

The Open Fog consortium, which is developing a reference architecture, introduced three objectives of fog framework development. First, the fog environment must be horizontally scalable. This means that you will support multiple vertical industry use cases. Second, it must work across the continuum from cloud to things. Third, it must be a system-level technology that extends across the network protocol from the object through the network edge to the various network protocols.

The Difference between Fog Computing and Edge Computing Helder InTunes, a senior member of Cisco's Corporate Strategy Innovation and an openfog consortium, explained that Edge Computing is a component or subset of fog computing. He thinks "fog computing is the way data is processed from where it was created to where it is stored. Edge computing only refers to processing near the point at which data is generated. Fog computing not only covers its edge processing, but it also includes the network connections needed to bring that data from the edge to the endpoint."

2.3   Blockchain

Satoshi Nakamoto has laid the foundation for block-chain technology by presenting a solution for distributed trust among unauthorized entities [11]. The first decentralized virtual currency is a bit coin. That is, the bit coin is a cipher and the block chain is a technique that supports bit coin. Cryptography is a digital currency that runs on a block chain, and many virtual currencies based on the current block chain are created, creating a huge virtual money market. Especially, they are interested in the block chain technology in the financial market, and they are investing heavily. For example, a block chain of bit coins means "not centralized", that is, not controlled by a single central authority. While the centralized system structure, the bank issues the currency, the bit coin has no central authority. Instead, the block chain of bit coins is maintained and credible by the network of people known as miners. In other words, building a block - chain - based distributed network means that a financial institution does not have to devote much effort to security and authentication for reliability.

Block Chain Technology is a revolutionary technology that will lead the next generation Internet, attracting a lot of interest from the industry and rapidly expanding its influence with the emergence of ethereum, hyperledger, ripple and R3. In addition, competition for technology and platforms is getting hotter as a start-up company that provides block chain solutions in various fields such as virtual currency, asset management, shared economy, IoT, health, and logistics.

To connect an industrial IoT to a block-chain platform, it must create a module that will integrate IoT and block-chain functionality. This module stores information in the smart space of the IIoT and replicates information from the block chain into signed transactions. Also, it will be possible to create contracts that can be automatically processed under appropriate conditions. The best platform for this kind of integration is Hyperledger Fabric/Burrow [12].

They provide a fault-tolerant consensus mechanism as well as a basic infrastructure for creating and processing smart contracts. Using these platforms, it is possible to create both types of block-chain networks, both private and public.

A public block chain is an etheric, bit coin that we commonly know. Anyone can join a block chain to produce a block, a node, or a transaction. "Anyone" has the advantage of creating a huge network, so you can have high security. But you have to use complex logic circuits to keep that "everyone" in a state of trust. And because of this, Public Block Chain has the disadvantage that Transaction Per Second (TPS) is much lower than the existing network.

A private block chain is the key to building a reliable chain of chains without having to have complex logic circuits, such as public block chains, by limiting who or what can participate in the block chain. And since only limited users are involved, it is easy to modify the software maintenance, so that the structure of smart contract can be changed and developed so that it can be used immediately in the current system.

For example, in the financial sector, the transaction record should not be open to everyone. But a third-party supervisory authority must have the authority to see it. A public block chain is a very challenging environment to implement. However, the private block chain can be easily implemented with the freedom to modify and set permissions for smart contracts. It's a technology that banks can use without having to change the existing banking laws.

2.4 Hyperledger

Hyperledger is a block-chain open source project hosted by the Linux Foundation. The goal is to create block chain technology that can be applied across industries such as finance, IoT, logistics, manufacturing, and technology industries. In addition to hyperledger, there are other block-chain platforms such as R3, Ripple and Ethereum. Hyperledger is special for the following reasons.

(1) an environment suitable for implementing a corporate business as a private block-chain platform;
(2) a technology standard that can be adopted universally for various industries, unlike other platforms specialized in finance is.

Hyperledger is committed to mass-producing enterprise block-chain technology. These include distributed ledger framework, smart contract engine, client library, a graphical interface, other utilities, and sample application. This hyperledger umbrella strategy drives the rapid development of Distributed Director technology elements while strengthening the community by reusing common infrastructure elements. Incubating projects are divided into two major divisions. One is the Hyperledger Framework and the other is the Hyperledger Tool.

Hyperledger Consensus
In the HLF, several nodes execute first, verify the result, and process the steps applied to all nodes separately. In the Execute phase, the peers execute once and compare the results to each other. Then, in the order step, the order is sorted

and transmitted to all peers. Finally, each peer applies the requested contents in order and validates them. At this last step, the update is finally made to the ledger. This Execute-Order-Validate step itself is an agreement. But to be precise, it is a bit different from the block chain algorithm we think. Since Kafka, which is often called the HLF's algorithm for solving, is actually used only in the Order stage, it is not an accurate expression.

In other words, HLF is a structure in which consensus occurs in duplicate. In the Execute and Validate phase, it performs its own Endorse process according to the endorsement policy defined in the network. It is not an algorithm, but an agreement is determined according to the 'policy' defined in the network. In the intermediate Order step, the assigned orderers in the network sort the submitted transactions in the Kafka method and the agreement between the orderers is made.

Kafka is not strictly an algorithm for block chain aggregation. Kafka is a distributed messaging system developed by Linkedin, specializing in real-time large-volume log processing. So while the other agreement algorithms are BFT - which verifies the content to be delivered - Kafka is a Crash Fault Tolerance (CFT). It ensures that only the sequence is stacked up correctly. HLF is a block chain based on genuine "consensus." The use of Kafka seems to have resulted from the architectural troubles to improve the performance of HLF. In the first 0.x HLF, PBFT was installed as a standard. Nevertheless, criticism of performance (TPS) was constantly raised, and HLF developers would have to take measures against it. It is still a powerful but effective way to reach consensus. Kafka was an attractive solution in that it could quickly deliver messages in a pull-sub-structure, minimizing the burden on the receiving peer and improving speed. Although BFT verification can not be done, it is the problem of solving the problem by using the characteristic of Permissioned Blockchain to block the risk first in the CA and further supplementing the endorsement policy without any holes.

HLF can also satisfy the requirement for BFT verification between organizations. The algorithm for HLF's consensus is modular, as developers can choose which approach to consent - note that the application of this approach is only in the Order stage. The BFT verification method that HLF basically provides is SBFT. 1.3 release criteria Although not included, it was included in the 2018 roadmap released earlier this year and was announced to be available from the 1.4 release.

Hyperledger Framework
FABRIC: Hyperledger Fabric (HLF) [13,14] is an open source project for the Hyperledger umbrella project. It is a modular, permissible block-chain platform designed to support pluggable implementations of various components such as order and membership services. With HLF, clients can manage transactions using chain codes, peers, and order services. Chaincode is the smart contract of HLF [15]. It consists of distributed code in a network of HLF run and validated by the peer to maintain the ledger, conforms to the state of the database (modeling as a key/value store) and warranty policy, and the order service generates a distributed ledger block The order of each block is also determined.

SAWTOOTH: Sawtooth is a modular platform for distributed branch construction, deployment and execution. We have successfully completed a healthcare-related POC with Intel-led projects.

IROHA: Iroha is a distributed led development project focused on mobile application development. C++ design-based, Byzantine Falut Tolerant consensus algorithm. Japanese companies such as Soramitsu, Hitachi, and NTT Data are leading.

INDY: Indy is a certification-specific project. Provides tools, libraries, and reuse components for creating and using independent digital identity records based on block chains. The best thing is that you can use it across other 'silos' such as admin domains, applications, and so on.

BURROW: A smart contract interpreter based on the Ethereum Virtual Machine (EVM) provides a built-in block-chain client.

Hyperledger Tools

COMPOSER: Hyperledger Composer is a tool that you have heard once if you are interested in developing a block chain. It is the first tool released in the Hyper Leisure project, allowing you to build a block-chain business network. It also allows the contract to flow between the smart contract and the ledger.

CELLO: Cello enables an as-a-service deployment model in a block-chain ecosystem. It is an orchestration tool that minimizes the effort of creating, managing, and deleting a block chain network.

EXPLORER: Hyperledger Explorer is a tool that allows you to quickly and easily view blocks, transactions, related data, network information, chain codes, transaction families, and other information in your ledger.

QUILT: It enables interoperate between led systems with the payment protocol ILP. That is, it allows the transfer of values between the Distributed Ledger and the General Ledger.

CALIPER: Hyperledger Caliper generates reports that contain various performance indicators such as Transactions Per Second (TPS), transaction latency, resource utilization, and so on. The results you create in Caliper are used to build the framework in other Hyperledger projects and help you choose the block chain implementation that suits your specific needs.

3   Proposed Architecture

In this paper, we design a fog computing model based on the permissible block chain for future Industrial IoT as shown in Fig. 1. Key requirements are fast exchange of data and reliable exchange of data based on low throughput and low latency. The basic flow is to create a private key on the edge device and register it with the fog node/server. Then, in the fog, the authority for the transaction of the edge device is checked by referring to the registered public key, and the initial leisure is distributed. And requests the ordering service to distribute the registered transactions to all the peers. When processed, records are added to the world state, and the peers in the same channel can request transactions such as query and resource exchange of data of the newly registered edge device.

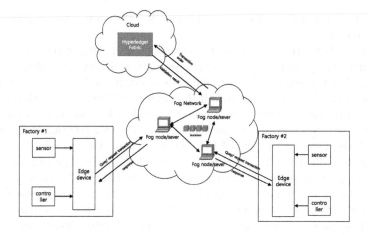

Fig. 1. Blockchain based Industrial IoT fog computing proposal model

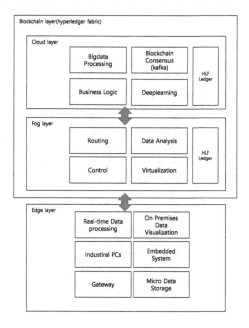

Fig. 2. Blockchain based Industrial IoT fog computing Proposal Model Feature architecture by layer

Figure 2 shows the functions to be operated in the proposed model by dividing each layer. In the edge stage, edge functions include functions such as real-time processing, small-scale data storage, and gateways. In the fog group, functions such as control, routing, data analysis, data preprocessing and virtualization are included. Algorithm of logic, block chain. Deep learning is included. By building a block-chain network with cloud and fog nodes, the nodes of each layer share a distributed ledger.

Figure 3 shows the actual configuration of Fig. 1 with hyperledger and fog computing orderer in the cloud because security and network stability must be guaranteed. It is essential that each fog node below it can be guaranteed stability and speed in a block-chain network if the orderer is distributed in the cloud and supports clustering services even if the geographical location is far away. In this paper, we have placed Endoser peers in the fog node for security of the network and deployed them as brokers of IoT device groups. The combination of these layers creates the following service and network features: The ordering service adds a block to the chain through a consensus algorithm. The more the orderer gets and the more complex the algorithm becomes, the better the stability. But the slower the speed, It handles this task with a Fog node to maintain each block chain data, and acts as a network so that the edge device can query the desired data or maintain a connection. In an industrial local network, in a local network connected by wifi and ethernet, the edge receives signals from sensors, controllers and other devices and processes them in real time.

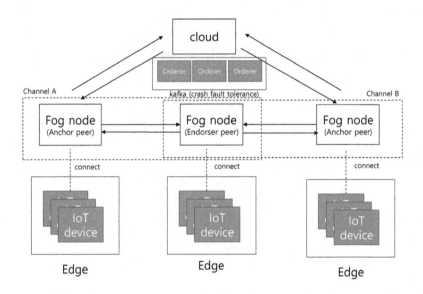

Fig. 3. Blockchain based Industrial IoT fog computing architecture

4   Evaluation

The scenarios for designing a model for evaluating performance in the Permission Block Chain based fog computing architecture are as follows.

(1) When an ordering service is established at the fog node, the evaluation of the corresponding throughput delay time and the like of the number of each authors is carried out.

(2) When an ordering service through a VM is established in the cloud, the evaluation of the corresponding throughput delay time and the like of the number of each node is performed.

For the performance evaluation, the orderer is an AWS EC2 t2.micro (Ubuntu server 16.04 LTS (HVM), SSD Volume Type Variable ECU, 1 vCPUs, 2.5 GHz, Intel Xeon Family, 1 GiB memory, EBS only), the fog server is a laptop (Core i5-7200U CPU @ 2.50 GHz 2.71 GHz Ubuntu server 16.04 LTS (VirualBox) 8 GB RAM). The performance of each virtualization instance of the Endoser peer is vCPU 2, 4 GB RAM. The block chain network is composed of Hyperledger Fabric v1.3, the sample chain code (smart contract) is installed in the network, and the chain code is operated by communicating with the REST-server through HTTP communication. In the experiment, 350 threads in the block-chain network are all executed in less than one second, and each thread repeats the next operation (HTTP Request, etc.) 20 times. TPS stands for Transactions per Second, which means the number of transactions processed per second. As a result, response time was 7209 ms, minimum 537 ms, average 4665 ms, and TPS was calculated as about 48 TPS through 7000 sample data. Figure 4 shows the TPS graph over time.

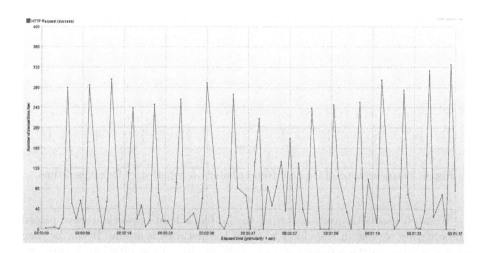

Fig. 4. TPS of fog architecture based on hyperledger for elapsed time

Figure 5 is a graph of Active Threads according to each elapsed time. We can see that there is some delay in the network when a thread group starts and ends an HTTP request. Other than that, we reliably handled threads in a block-chain network.

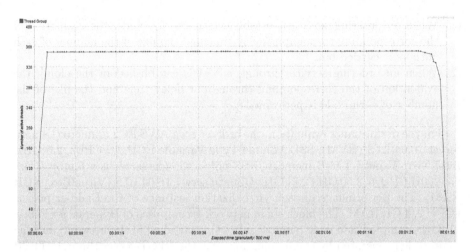

Fig. 5. Active threads of fog architecture based on hyperledger for elapsed time

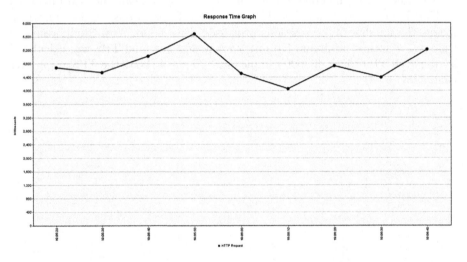

Fig. 6. Response time of fog architecture based on hyperledger for elapsed time

Figure 6 shows the response time to the blockchain network. I have placed the ordering service in the cloud to see if the network is stable. It can be seen that even though the orderer instance is the minimum specification, it supports a relatively stable network. It is analyzed that it can be used very appropriately in environments where high throughput and real-time environment are not required.

Figure 7 shows the throughput graph of a hyperledger fabric based fog network over time. Throughput is calculated as requests/unit of time. The time is calculated from the start of the first sample to the end of the last sample. This includes any intervals between samples, as it is supposed to represent the load

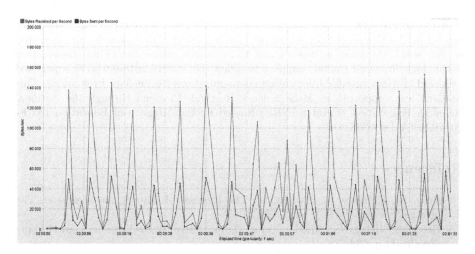

Fig. 7. Response time based on elapsed time for HTTP request

on the server. The formula is: Throughput = (number of requests)/(total time). We can see that the throughput is much lower than the bytes are received.

By placing orderer in the cloud as shown in the above figures, we can confirm that the performance is stable. Therefore, if orderers are configured with higher performance instances in the cloud, it can be seen that the block-chain network can also achieve high performance.

5   Conclusion

In this paper, we propose a system architecture that can prevent data forgery by converting existing centralized database method to distributed type based on block chain. By dividing the proposed system structure into cloud, fog, and edge, we proposed a way to organically operate the IIoT ecosystem. We also investigated whether the performance of the block - chain network agreement algorithm can be guaranteed by using public cloud resources. Resolves data validation and security based on Hypeledger block chain platform in IIoT environment (permissioned blockchain), Resolves performance load issues by not connecting IIoT devices directly to block-chained networks, Orderer is ported to the cloud to ensure stability, security and scalability. We propose that smart contract and transaction verification proceed to process in fog node to obtain network latency and throughput performance. In the future research, challenges will include research into network configuration for cost savings in the cloud, and optimization of cloud and fog node instancing performance to maximize the performance of the block-chain network hyperledger. Finally, we hope to analyze the performance and consistency of the model by applying it to actual industry environment.

Acknowledgments. This work has supported by the Gyeonggi Techno Park grant funded by the Gyeonggi-Do government (No. Y181802). This research was supported by the MSIT (Ministry of Science and ICT), Korea, under the ITRC (Information Technology Research Center) support program (IITP-2019-2018-0-01417) supervised by the IITP (Institute for Information communications Technology Promotion). This research was supported by Basic Science Research Program through the National Research Foundation of Korea (NRF) funded by the Ministry of Education (NRF-2016R1D1A1B03933828).

References

1. Rose, K., Eldridge, S., Chapin, L.: The Internet of Things: An Overview. The Internet Society (ISOC), pp. 1–50 (2015)
2. Beecham Research, M2M Sector Map. http://www.beechamresearch.com/download.aspx?id=18
3. Hermann, M., Pentek, T., Otto, B.: Design principles for industrie 4.0 scenarios. In: Proceedings of the 49th Annual Hawaii International Conference on System Sciences, HICSS 2016, pp. 3928–3937 (2016)
4. Dorsemaine, B., et al.: Internet of things: a definition and taxonomy. In: 2015 9th International Conference on Next Generation Mobile Applications, Services and Technologies, pp. 72–77. IEEE (2015)
5. Rose, K., Eldridge, S., Chapin, L.: The Internet of Things: An Overview. The Internet Society (ISOC), p. 12 (2015)
6. Satyavolu, P., Setlur, B., Thomas, P., Iyer, G.: Designing for manufacturing's "Internet of Things". Technology solutions, pp. 4–14. https://www.cognizant.com/whitepapers/Designing-for-Manufacturings-Internet-of-Things.pdf. Accessed 21 Nov 2016
7. Aberle, L.: A Comprehensive Guide to Enterprise IoT Project Success. IoT Agenda. https://internetofthingsagenda.techtarget.com/essentialguide/A-comprehensive-guide-to-enterprise-IoT-project-success
8. World Economic Forum: Industrial Internet of Things: Unleashing the Potential of Connected Products and Services. World Economic Forum Industry Agenda. http://www3.weforum.org/docs/WEFUSA_IndustrialInternet_Report2015.pdf. Accessed 11 Oct 2017
9. Da Xu, L., He, W., Li, S.: Internet of things in industries: a survey. IEEE Trans. Ind. Inform. **10**(4), 2233–2243 (2014)
10. OpenFog Consortium Architecture Working Group, et al.: OpenFog reference architecture for fog computing. OPFRA001, vol. 20817, p. 162, October 2017
11. Nakamoto, S.: Bitcoin: a peer-to-peer electronic cash system (2008)
12. Teslya, N., Ryabchikov, I.: Blockchain platforms overview for industrial IoT purposes. In: 22nd Conference of Open Innovations Association (FRUCT), pp. 250–256. IEEE (2018)
13. Vukolić, M.: Rethinking permissioned blockchains. In: Proceedings of the ACM Workshop on Blockchain, Cryptocurrencies and Contracts, pp. 3–7. ACM (2017)
14. Cachin, C.: Architecture of the hyperledger blockchain fabric. In: Workshop on Distributed Cryptocurrencies and Consensus Ledgers (2016)
15. Szabo, N.: Smart contracts: building blocks for digital markets. EX-TROPY: J. Transhumanist Thought (16) (1996)

Exploration of Data from Smart Bands in the Cloud and on the Edge – The Impact on the Data Storage Space

Mateusz Gołosz[ID] and Dariusz Mrozek[✉][ID]

Institute of Informatics, Silesian University of Technology,
ul. Akademicka 16, 44-100 Gliwice, Poland
{mateusz.golosz,dariusz.mrozek}@polsl.pl

Abstract. Wearable devices used for tracking people's health state usually transmit their data to a remote monitoring data center that can be located in the Cloud due to large storage capacities. However, the growing number of smart bands, fitness trackers, and other IoT devices used for health state monitoring pose pressure on the data centers and may raise the Big Data challenge and cause network congestion. This paper focuses on the consumption of the storage space while monitoring people's health state and detecting possibly dangerous situations in the Cloud and on the Edge. We investigate the storage space consumption in three scenarios, including (1) transmission of all data regardless of the health state and any danger, (2) data transmission after the change in person's activity, and (3) data transmission on the detection of a health-threatening situation. Results of our experiments show that the last two scenarios can bring significant savings in the consumed storage space.

Keywords: Internet of Things · IoT · Data exploration ·
Cloud computing · Edge computing

1 Introduction

In recent years we are witnessing a dynamic development of various technologies that bring a revolution in many areas of our lives. Internet of Things (IoT), as a network of electronic devices that are able to communicate, interact, and exchange data with the other, is one of the leading technologies of today that have great potential to change the image of the current world and various processes on it. The IoT enters many areas of people's life, including the development of smart buildings, home automation, designing intelligent transportation,

This work was supported by pro-quality grant for highly scored publications or issued patents of the Rector of the Silesian University of Technology, Gliwice, Poland (grant No 02/020/RGJ19/0167), and partially, by Statutory Research funds of Institute of Informatics, Silesian University of Technology, Gliwice, Poland (grant No BK/204/RAU2/2019).

J. M. F. Rodrigues et al. (Eds.): ICCS 2019, LNCS 11538, pp. 607–620, 2019.
https://doi.org/10.1007/978-3-030-22744-9_47

supporting smart manufacturing, farming, energy management, fitness tracking, and finally, monitoring our health state and detecting life-threatening situations. Smart bands, as well as fitness trackers, and smartwatches are wearable IoT devices that allow recognizing and monitoring the activities of people (e.g., walking, biking, jogging, sleeping) and some of their physiological parameters (e.g., heart rate, burned calories, and quality of sleep). Some of the devices can even measure the electrocardiography (ECG) signal. Sensor measurements are usually transmitted to another device that possesses higher compute capabilities, provides long-term storage of data, or acts as IoT gateway for sending the data to the other place (e.g., a data center). These can be smartphones, laptops, tablets, personal computers or other dedicated IoT devices. Data transmission is usually carried out with the use of a suitable, usually short-range and wireless communication protocol, like Bluetooth [3], Bluetooth Low Energy [12] (BLE), ZigBee [3,7], ANT [3,8,10], Near Field Communications (NFC) [2,9], or WiFi.

Wearable devices, such as smart bands and smartwatches, which are especially popular among young people, can be used not only for tracking the fitness of an individual person. These devices may also deliver valuable data that can be used for remote monitoring of someone's health state. They can be used to monitor older people or people after some serious health-related incidents, like a heart attack or stroke. In such scenarios, the data containing the information about the activity of the person and some parameters of the physiology are usually transmitted through the IoT gateway to the monitoring center where they are automatically analyzed in order to detect any risky situations. Detection of any danger should raise an alert and notify appropriate caregiver or relative who should react suitably to the situation. Since monitoring centers providing their services for hundreds of seniors require large storage spaces and analytical capabilities for the transmitted data, they are eagerly located in Cloud platforms. Cloud platforms provide scalable and almost unlimited resources for storing and performing various computations on the transmitted data. However, with the new applications of the IoT and the rapid growth of the number of users of IoT devices the amount of data transmitted to the Cloud increases very fast. This necessitates moving operations, like data processing (including assembling, transformation, and filtering) and data analysis (including data exploration with pre-trained machine learning models) on the Edge and partially free the Cloud from these operations. Edge computing assumes performing some of these operations on IoT devices and is an alternative to the centralized data processing and analysis performed in the Cloud. It prevents network congestion and storage space overload, and in some situations, may eliminate unnecessary latency.

In this paper, we show two alternative system architectures for monitoring the health state of older people with the Cloud-based centralized and Edge-based distributed data analysis. We investigate the impact of both architectures on (1) the speed of detection of dangers in health with the use of trained machine learning models and (2) the consumed Cloud storage space. We also propose alternative approaches for initiating Cloud-to-Device connectivity and transmitting data to the Cloud, which allow to reduce network traffic, the number of transactions, and bring savings in the consumed storage space.

2   Related Works

Transmission of data from IoT devices and storing the data in the storage space are important areas of the Internet of Things, since IoT devices may lead to Big Data challenges. This is reflected in several scientific works focused on IoT technologies. Authors of [5] have given a general proposition of an architecture for a system that would exchange data between wearable devices and computing Cloud. However, their work has been focused mostly on the concept of actively supporting health services, diagnosis of disease in particular. Moreover, no real data gathered from the implementation of such a system has been presented. In [6], authors have proposed a solution to a problem that occurs in a different area - lack of coherency in both input and output interfaces. The implemented framework standardizes data regardless of its size, source device, format, and structure. Zhu et al. in [14] propose a model of a gateway for a sensor network, but it does not provide any details on how the given data is being processed. Instead, it presents a very general hardware implementation and a general overview of network packet construction, server architecture and overall flow of the transferred data without going further into processing the data once it has been sent. Yang et al. [11] proposed a wearable ECG monitoring system that utilizes the Cloud platform. The work covers the hardware implementation and data transportation model and investigates the risk of heart disease. In [4] Doukas and Maglogiannis show the usage of the IoT and cloud computing in pervasive healthcare, but instead of an ECG examination, they propose quite a unique implementation of its own wearable sensor system. The system is integrated into a sock and consists of multiple sensors measuring values such as heartbeat, motion, and temperature. However, none of the above works go into details when it comes to storing and processing gathered data. Chen et al. [1] also describe the process of transferring data from wearable devices to computing clouds, but with consideration for an improvement of the wearable devices themselves. The main emphasis has been put onto integrating multiple sensors which are available as separate modules into versatile smart clothing that would constantly monitor various health indicator as well as environmental parameters, such as air pollution. On the other hand, except introducing an architecture of a model being able to transfer data from IoT devices to the Cloud, Zhou et al. [13] focuses on an emerging problem with the privacy of data collected by such devices. They describe an efficient way of encrypting and anonymizing data in the process.

None of the works listed above concentrates on the amounts of data produced by wearable devices and on ways of reducing them to a minimum. One of the ways includes changing the point where most of the data are being processed, moving the processing from the cloud itself to another (Edge) device which takes a part in the earlier stage on the data flow. In the next section, we present and compare the Edge and Cloud-based standard architectures for data processing and analysis.

3   Alternative Architectures for Monitoring Human Health State

The health and activity monitoring system for older adults with the data center located in the Cloud can adopt one of the two general architectures presented in Fig. 1. The main goal of the developed system was to determine whether a user of the wearable device (a monitored person) happened to be in a life-threatening situation. First of the presented architectures assumes classification of data and detection of dangerous situations in the Cloud. The second architecture assumes data classification on the field IoT devices (on the Edge). Both architectures consists of:

- a wearable device with sensors measuring various parameters,
- a smartphone with the Android operating system,
- a data center located in the Cloud.

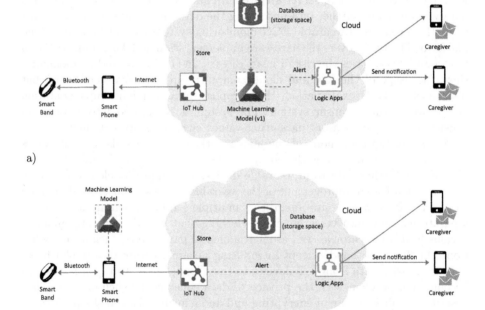

Fig. 1. Two general alternative architectures for the health state and activity monitoring system with the Machine Learning model for detection of dangerous situations implemented in the Cloud (a) and on the Edge device (b).

As the wearable device sensing various parameters of the monitored person, we decided to use the Xiaomi Mi Band 2 smart band. The smart band was selected on the basis of availability, popularity, and economic issues. One of the key points while choosing the smart band was also the possibility to access raw data from the sensors. Most smart band manufacturers do not provide the possibility to extract raw data of sensor measurements, or such a feature is limited to one extraction per 24 h. The 24-h period is too long for constant monitoring of the current status of a person. At the time of performed implementation of the monitoring system, there was no open-source wearable fitness tracker available on the market that would provide application programming interfaces (APIs) to extract raw data. We extracted the sensor measurements from the Xiaomi Mi Band 2 in a reverse engineering process because there was no officially supported method of gathering raw data from the smart band.

We were able to extract the following sensor measurements from the Xiaomi Mi Band 2 device:

- the number of steps made,
- heart rate,
- the quality of sleep,
- the activity currently performed, identified on the basis of the steps taken,
- the time of measurement.

For the extraction of data, we used the Gadgetbridge application for Android-based smartphones. The Gadgetbridge application is open-source software available on the GitHub platform. Apart from the Xiaomi Mi Band 2, it supports several different devices, however, we haven't tested them in our solution. The Gadgetbridge application was installed on the smartphone, which served as the IoT gateway mediating data transfers to the monitoring data center.

The remote monitoring center with a huge storage space was established in the Microsoft Azure cloud. Microsoft Azure provides scalable storage and computing resources. It offers a wide range of tools, programming languages and different platforms that can be used to develop the IoT solutions. The Cloud was also selected due to its high data security standards, global access to data with guaranteed bandwidth, and relatively easy and intuitive user interface. The Azure cloud was used to gather data transmitted from the IoT gateways (smartphones) and to store the data in the database storage repository. We tested two storage repositories in our system: Azure SQL Database – a relational database, and Cosmos DB – a document store. Data classification and detection of possible dangers on the basis of raw sensor readings were possible with the use of trained Machine Learning (ML) models. For this purpose, we used the Machine Learning Studio – an Azure module that allows for creation, training, testing, and manipulation of ML models.

The process of detection of dangerous situations in monitored persons involves a binary classification. The output of the process indicates that the person is safe or that there might be something wrong with the person. We tested multiple machine learning algorithms, like logistic regression, decision trees, support-vector machine to this purpose, but since all of them produce the same binary output and all of the trained models use the same input data for classification, changing the ML algorithm did not affect the taken storage space in any way. Therefore, this work will not focus on describing certain algorithms and models used. However, the place where the data classification occurs – in the Cloud or on the Edge – may significantly influence the network traffic and the number of data sent to the Cloud. The first approach (Fig. 1a) assumes that all data processing is done in the monitoring center, thus, all the data used for training and using the ML model are sent directly to the Cloud. The second approach (Fig. 1b) reduces the amount of data that needs to be sent by performing classification of the raw sensor readings on the Edge device before they are sent to the Cloud. Simplified data flow for both architectures can bee seen in Fig. 2.

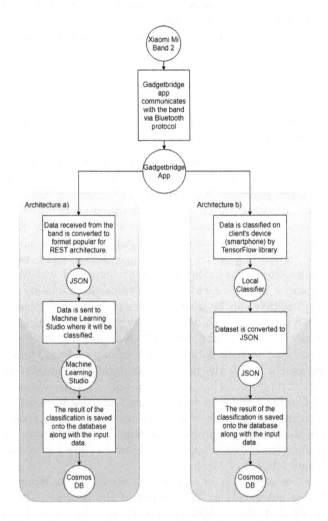

Fig. 2. Data flow from the wearable device to the database.

4   The Impact of the Architecture on the Detection Speed

Both architectures presented in Fig. 1 enable detection of possibly dangerous situations in health of monitored persons through data classification, which, depending on the adopted approach, can be done in the monitoring center located in the Cloud (Fig. 1a) or on the Edge devices (Fig. 1b). The classification model accepts sensor readings as the input data set. Each such a reading from the sensors located in the smart band posses the following attributes:

- Timestamp – a moment in time when the data reading occurred (a 10-digits integer),
- DeviceId – a unique identifier of the device connected to the Cloud;
- UserId – a unique identifier of a monitored person (user of the smart band),
- Raw intensity – an integer expressing the intensity of the performed activity, its values range between 0 and 99,
- Steps – an integer showing the number of steps per unit of time, which the monitored person made,
- Raw kind – a value describing the recognized activity performed by the user of the smart band (particular activities are represented as integers from the range of 0 and 99),
- Heart rate – a heart rate measured by the pulse sensor in the smart band.

The construction of the classifier that is used in the Cloud is presented in Fig. 3. The classifier is based on the pre-trained Decision Tree ML model and is available through Web service located in the Azure cloud. After the classification process the data set is supplemented by an additional attribute, called *healthy*. The attribute holds a binary value of 1 (reflecting the person is healthy) or 0 (reflecting possible health danger). Classification results are used to notify caregivers in case of dangers (through Web service output) and are saved in the database (through Export Data).

We prepared a simple benchmark to examine whether using local classifiers leads to significant impact on the detection speed. For this purpose, we measured the time between the moment when data acquisition began and the moment when the results of the classification were delivered to the client's IoT device. However, these results may vary depending on factors such as network bandwidth, data center load, CPU used in smartphone, therefore, presented values might distinctly differ in a scenario in which more users send data to the server at the same time or their mobile devices have more computing power. The results of the benchmark test performed for a single device connected to the Cloud (we used Motorola Moto X, 2014) are presented in Table 1.

5   The Impact of the Architecture on the Storage Space

There are various possibilities to store data from IoT devices in the Azure cloud, including BLOB storage spaces, relational databases, NoSQL databases, and file systems. Due to a partially structured nature of the produced data, we tested

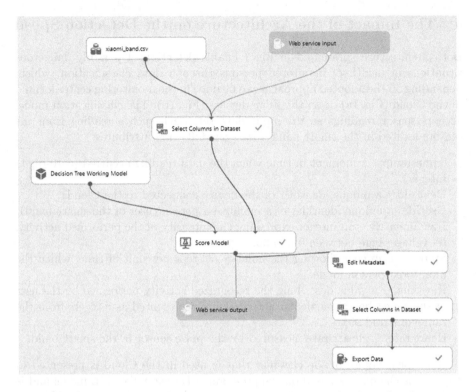

Fig. 3. Construction of the ML model providing Web service for data classification in the Cloud.

two services that offer storage space as a service, i.e., Azure SQL Database and Cosmos DB. The Azure SQL Database is a relational SQL database. It has been selected, since it provides highly available and structured data storage space among all cloud solutions. The Cosmos DB is a globally distributed, multi-model database service that supports document, key-value, graph, and columnar data models. Since data processed in our system are sent as JSON objects uniquely characterized only by the timestamp and user identifier, CosmosDB turned out to be the best NoSQL choice among available storage options in the Azure cloud. The purchase model of the Azure SQL relational database is based on the Database Transaction Units (DTUs). As a kind of currency used by the cloud customers, DTUs determine the *compute sizes* and, thus, the performance of the database, which is reflected in the compute, storage, and IO resources used by it. In Table 2 we show three different plans (also called as *tiers*) that influence capabilities, limitations, and costs for the usage of the Azure SQL relational databases.

While testing the system, we noticed that the size of data transmitted to the database and stored in it was 0.5 kB per transaction. The storage space consumed in the relational database depends on the number of transactions performed

Table 1. Comparison between performance (detection speed) of local and cloud classifiers.

ML algorithm	Average time for Azure cloud	Average time for local classifier
Linear regression	2.96 s	4.39 s
Logistic regression	3.34 s	4.89 s
Decision tree	3.28 s	5.02 s
Support vector machine	4.87 s	6.32 s
Naive Bayes	3.11 s	5.10 s

Table 2. Various tiers defining performance capabilities for a single Azure SQL database.

Tier	DTUs	Max. available storage (GB)	Min. cost per month (EUR)
Basic	up to 5	2	4.21
Standard	10–3 000	250	12.65
Premium	125–4 000	1 000	392.13

within one minute (time periods with which the Gadgetbridge application sends data to the Cloud). The time interval between successive sensor readings is one of the factors affecting the consumption of Cloud storage resources. In order to provide near real-time monitoring of a people, incoming sensor readings are processed at once and, depending on the architecture variant, whole or part of the data are immediately sent to the Cloud to be stored. For the basic variant of the Cloud-to-Device connectivity, we assumed a time interval of 1 min, which defines how often data from the sensors are gathered and processed. This is a default, assumed value for the basic scenario with constant data transmission for every further analysis presented throughout this section. Figure 4 shows the growth of data observed in the database within one hour for various time periods of sending data and the various number of connected IoT devices.

As can be expected the amount of data that must be stored grows with the number of monitored persons and frequency of sending data to the database. It significantly increases when the mobile application sends data every minute. This growth translates appropriately to the increase of the minimal number of DTUs that are needed since more data must be saved within a certain period of time. However, more frequent data transmissions allow reacting quicker in case of detected dangers.

When using the Cosmos DB, instead of Azure SQL database, we stored data as JSON files. While testing our system, we noticed that JSON objects consume only 0.22 kB per one data transaction containing sensor readings. This is less than half of the size of the storage space taken by the same data stored in the Azure SQL relational database (0.5 kB). This difference influenced the con-

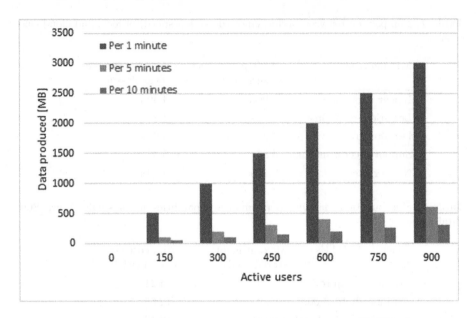

Fig. 4. The size of data produced within one hour for various time periods of successive data transmissions and the growing number of active devices and monitored persons.

sumption of the overall storage space and the cost. However, we cannot compare the costs directly (as sizes), since the pricing model of the Cosmos DB is not based on DTUs. The cost of usage of the Cosmos DB increases elastically with the number of transactions that are made to the database. In contrast to Azure SQL database, which makes the cost of storing data dependent of both the price of minimal number of DTUs needed for a database to operate and the storage space consumed, the cost of using the Cosmos DB is calculated differently on the basis the size of stored data, the number of requests, and the operational time. The following formula is used for this purpose (EUR):

$$Cost = g * 0.211 * req * h * 0.007, \tag{1}$$

where g is the consumed storage space (GB), req is the number of requests made per second, h is the number of hours when the database is active.

Assuming that both the relational database and the Cosmos DB perform the same number of transactions, we can compare costs of using both tested databases. In Fig. 5 we can be observe the minimum cost for both storage approaches for the various number of active users and different time periods of data transmission (1 and 5 min) per one hour of constant work of the particular database. As could be expected, the cost of using the Cosmos DB in the developed monitoring system is much lower than using the Azure SQL database (for the same time periods), which is even more visible for the increasing number of connected IoT devices.

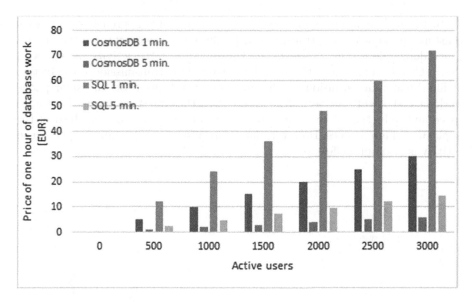

Fig. 5. Costs of using Azure SQL database and Cosmos DB (per one hour) for the various number of active users (IoT devices) and various time periods of data transmission.

The cost of storing data in one of the selected databases while constantly monitoring hundreds of persons with IoT devices remains at a reasonable level. However, it is still possible to reduce the consumption of the storage space and costs in some certain circumstances by reducing the amount of data transmitted to the Cloud. For this purpose, apart from the standard approach, when data are transmitted to the monitoring center with a given frequency, we tested two other approaches that moderate the data transmission.

The first of the implemented approaches assumes that the data from the smart band are transmitted by the IoT device only when the activity performed by a monitored person changed since the last measurement. The main idea behind this solution is that the state of the person whose life can be in danger would be very likely to change, e.g., a fainting person will probably lie down, a jogging person will probably stop and calm the heart. In this approach, the possible reduction of data transmissions and storage space consumption highly depends on the individual activity of the person during the day. Additional savings can be also expected during the night hours when most of the monitored persons should sleep for several hours. We tested this solution with the classification of the health state performed in the monitoring center, but it allowed to reduce the amount of data transmitted to the Cloud through monitoring and filtering the activity on the Edge. In the second approach, we assumed that the detection of dangers in the health state is performed on the IoT device (a smartphone). If the used classifier indicates any danger in the health state, the data are sent to the Cloud.

In Fig. 6 we can observe the comparison of all three approaches – (1) with constant, periodical data transmission to the monitoring center (with the in-Cloud danger detection), (2) with data transmission on the activity change (with the in-Cloud danger detection), and (3) with data transmission on detection of life-threatening situation (with the Edge classification on the smartphone working as the field gateway). Results show that the largest savings in the storage space and reduction of the transmitted data are achieved for the third of the implemented approaches. However, data transmission when changing performed activity is also quite effective.

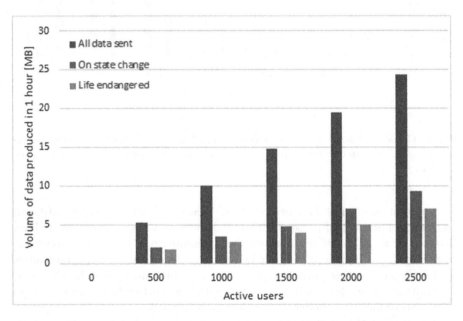

Fig. 6. Consumption of the storage space for three Device-to-Cloud connectivity approaches – constant, periodical transmission to the Cloud with detection of danger in the Cloud (All data), with data transmission only when the activity of the user changes with detection of danger in the Cloud (On state change), with data transmission only when possibly dangerous situation is detected, with detection on the Edge IoT device (Life endangered).

6   Conclusions

The growing popularity and applicability of wearable devices in monitoring people's health state leads to the increase of data that must be transmitted, stored and processed in telemonitoring data centers located in the Cloud. In consequence, this may cause network congestion and raise Big Data challenges. In this paper, we showed possible solutions of the problems by introducing event-based connectivity when the state of the person changes and by moving the burden of data processing and analysis to the Edge. Although, this may slightly decrease the speed of performed data analysis, savings in the storage space and reduction of the network traffic can be significant. Capabilities of Edge devices are also important here, since they influence, e.g., the speed of performed classification. Therefore, development of such systems may involve assembling users into a few profiled groups and apply the most cost-effective strategy to each of the group.

The results of our experiments proved that data filtering or detection of dangerous situations on the Edge device can be an effective solution not only for reducing the amount of data to be stored but also for reducing the number of transactions. Since providing a sufficient number of concurrent transactions is multiple times more expensive than storage space itself, this seems to be a proper approach. This conclusion is based on the fact that the cost of resources needed to establish a connection grows significantly faster than the cost of resources needed to save and store gathered data. This is also important for building such systems in public cloud platforms in the future.

References

1. Chen, M., Ma, Y., Li, Y., Wu, D., Zhang, Y., Youn, C.H.: Wearable 2.0: enabling human-cloud integration in next generation healthcare systems. IEEE Commun. Mag. **55**(1), 54–61 (2017). https://doi.org/10.1109/mcom.2017.1600410cm
2. Coskun, V., Ozdenizci, B., Ok, K.: A survey on Near Field Communication (NFC) technology. Wirel. Pers. Commun. **71**(3), 2259–2294 (2013). https://doi.org/10.1007/s11277-012-0935-5
3. Dementyev, A., Hodges, S., Taylor, S., Smith, J.: Power consumption analysis of Bluetooth Low Energy, ZigBee and ANT sensor nodes in a cyclic sleep scenario. In: 2013 IEEE International Wireless Symposium (IWS), pp. 1–4, April 2013. https://doi.org/10.1109/IEEE-IWS.2013.6616827
4. Doukas, C., Maglogiannis, I.: Bringing IoT and cloud computing towards pervasive healthcare. In: 2012 Sixth International Conference on Innovative Mobile and Internet Services in Ubiquitous Computing. IEEE, July 2012. https://doi.org/10.1109/imis.2012.26
5. Hassanalieragh, M., et al.: Health monitoring and management using internet-of-things (IoT) sensing with cloud-based processing: opportunities and challenges, June 2015. https://doi.org/10.1109/scc.2015.47
6. Jiang, L., Xu, L.D., Cai, H., Jiang, Z., Bu, F., Xu, B.: An IoT-oriented data storage framework in cloud computing platform. IEEE Trans. Ind. Inform. **10**(2), 1443–1451 (2014). https://doi.org/10.1109/tii.2014.2306384
7. Malhi, K., Mukhopadhyay, S.C., Schnepper, J., Haefke, M., Ewald, H.: A ZigBee-based wearable physiological parameters monitoring system. IEEE Sens. J. **12**(3), 423–430 (2012). https://doi.org/10.1109/JSEN.2010.2091719
8. Mehmood, N.Q., Culmone, R.: An ANT+ protocol based health care system. In: 2015 IEEE 29th International Conference on Advanced Information Networking and Applications Workshops, pp. 193–198, March 2015. https://doi.org/10.1109/WAINA.2015.45

9. Pang, Z., Zheng, L., Tian, J., Kao-Walter, S., Dubrova, E., Chen, Q.: Design of a terminal solution for integration of in-home health care devices and services towards the internet-of-things. Enterp. Inf. Syst. **9**(1), 86–116 (2015). https://doi.org/10.1080/17517575.2013.776118

10. Valchinov, E., Antoniou, A., Rotas, K., Pallikarakis, N.: Wearable ECG system for health and sports monitoring. In: 2014 4th International Conference on Wireless Mobile Communication and Healthcare - Transforming Healthcare Through Innovations in Mobile and Wireless Technologies (MOBIHEALTH), pp. 63–66, November 2014. https://doi.org/10.1109/MOBIHEALTH.2014.7015910

11. Yang, Z., Zhou, Q., Lei, L., Zheng, K., Xiang, W.: An IoT-cloud based wearable ECG monitoring system for smart healthcare. J. Med. Syst. **40**(12) (2016). https://doi.org/10.1007/s10916-016-0644-9

12. Zhang, T., Lu, J., Hu, F., Hao, Q.: Bluetooth low energy for wearable sensor-based healthcare systems. In: 2014 IEEE Healthcare Innovation Conference (HIC), pp. 251–254, October 2014. https://doi.org/10.1109/HIC.2014.7038922

13. Zhou, J., Cao, Z., Dong, X., Lin, X.: Security and privacy in cloud-assisted wireless wearable communications: challenges, solutions, and future directions. IEEE Wirel. Commun. **22**(2), 136–144 (2015). https://doi.org/10.1109/mwc.2015.7096296

14. Zhu, Q., Wang, R., Chen, Q., Liu, Y., Qin, W.: IOT gateway: bridgingwireless sensor networks into internet of things. In: 2010 IEEE/IFIP International Conference on Embedded and Ubiquitous Computing. IEEE, December 2010. https://doi.org/10.1109/euc.2010.58

Security of Low Level IoT Protocols

Damas P. Gruska[1] and M. Carmen Ruiz[2(✉)]

[1] Institute of Informatics, Comenius University, Bratislava, Slovakia
gruska@fmph.uniba.sk
[2] Universidad de Castilla-La Mancha, Albacete, Spain
mcarmen.ruiz@uclm.es

Abstract. Application of formal methods in security is demonstrated. Formalism for description of security properties of low level IoT protocols is proposed. It is based on security property called infinite step opacity. We prove some of its basic properties as well as we show its relation to other security notions. Finally, complexity issues of verification and security enforcement are discussed. As a working formalism timed process algebra is used.

Keywords: Formal methods · Security · Protocol · Opacity · Process algebra · Information flow

1 Introduction

The Internet of Things (IoT) is finally a reality. Smart devices, smart phones, smart cars, smart homes, smart industries, in short a smart world. There are multiple predictions which declare that there will be at least tens of billions connected devices by 2020, almost everything from personal items such as pacemakers to aircraft black boxes could all become part of this interconnected world. But each device connected increases privacy and security concerns surrounding the IoT. These concerns range from hackers stealing our data and even threatening our lives to how corporations can easily uncover private data we carelessly give them. So that, IoT transforms our whole world bringing great benefits but also new risks due to the fact that we need to worry about protecting more and more devices.

The great revolution brought about by IoT involves the emergence of new protocols, networks, sensors, devices and, of course, new security requirements. Nevertheless, all these changes are occurring at such a speed that new protocols come into operation before they are properly evaluated. Due to this lack of prior study motivated by the need for their early use, malfunctions arise when

This study was funded in part by the grant VEGA 1/0778/18, the Spanish Ministry of Science and Innovation and the European Union FEDER Funds under grant CAS18/00106 and TIN2015-65845-C3-2-R and by the JCCM regional project SBPLY/17/180501/000276.

© Springer Nature Switzerland AG 2019
J. M. F. Rodrigues et al. (Eds.): ICCS 2019, LNCS 11538, pp. 621–633, 2019.
https://doi.org/10.1007/978-3-030-22744-9_48

the protocols are already in use, causing the appearance of new versions or "patches".

A nice example how a network of IoT can be jeopardized in several ways can be found in [20]. We came to this issue even in our own works. In [13] we propose a new packet format and a new Bluetooth Low Energy (BLE) mesh topology, with two different configurations: Individual Mesh and Collaborative Mesh. To include user devices is a security challenge due to the physical accessibility to sensors, actuators and objects, and the openness of the systems, and the fact that most devices will communicate wirelessly. Other examples could be found in [4] where we deploy a collaborative mesh network based on BLE and long range wide-area network (LoRaWAN) technologies and in [22] where it is presented a BLE wearable which is aimed at enhancing working conditions and efficiency in Industry 4.0 scenarios. In both cases we feel some lack of basic security requirements. Therefore, from our experience, we can say that IoT security can not be an afterthought or an add-on and needs to be addressed from the earliest stages of development.

Simulations and tests represent the common validation techniques but we advocate the use of formal methods for the evaluations for several reasons: simulation results depend on the simulator and may vary from those obtained in real scenarios. At this point, it would be easy to argue that the results obtained by these methods will also depend on the formalism used, but in defense of formal methods, we should say that the underlying mathematical models in simulators are often not clear and are not accessible to users while formal methods are supported by rigorous mathematical basis which allow users to model all the required properties, abstracting from the details that are not relevant. Moreover, simulation and testing can only show the presence of errors, not their absence but to rule out errors we must consider all possible executions. This can be made by means of formal methods that provide correct results that cover the full behavior of the models. At present there are few works where formal methods are used in IoT and they are focused on the field of automotive.

In this paper, we focus on formal methods applied to the field of security of IoT protocols. Here communication issues are considered to be weakest point. Many new, general or proprietary protocols are often vulnerable to different types of attacks. Timing attacks represent a powerful tool for breaking "unbreakable" systems, algorithms, protocols, etc. In the literature we can find examples how strong security algorithms or its components (fixed Diffie-Hellman exponents, RSA keys, RC5 block encryption, SSH protocol or web privacy, see [1, 2, 12, 17, 23] could be compromised by timing attacks. All these attacks employ so called information flow. Information flow based security properties (see [5]) are based on an absence of information flow between private and public data or activities. Systems are considered to be secure if from observations of their public data or activities no information about private data or activities can be obtained. This approach has found many reformulations and among them opacity could be considered as the most general one (for an overview of various types of opacity of discrete event systems see [14]).

Opacity could be, in general, formulated in the following way. Let us assume a (security) property. This could be an execution of actions in a particular classified order, time stamps of actions, execution of a private action etc. Let predicate ϕ over process's traces expresses such the property. An observer cannot deduce the validity of the property ϕ just by observing system's behaviour (traces) if there are two traces w, w' such that $\phi(w)$ holds and $\phi(w')$ does not hold and the traces w, w' cannot be distinguished by the observer.

Many security properties are special cases of opacity (see, for example, [8,9]). In [10] process opacity is introduced in such a way that instead of traces produced by processes we concentrate on reachable states and their properties. We assume an intruder who wants to discover whether given process reaches a state for which some given predicate holds. In this way we could express some new security properties. On the other hand some security flaws, particularly important for IoT protocols, are not covered by this state-based security property neither by its variant called an initial state opacity, studied in [11].

In this paper we define infinite state opacity for timed process algebras, we show its properties and relation to other security notions. The property is based on a similar concept defined for discrete event structures (see [14]) but some modifications are needed. In this way we obtain the security property with some practical value, since timed process algebra, as our working formalism, can be used for description of many communication protocols. Moreover, there are well developed techniques and software tools for process algebra formal verification and they can be employed for security checking. We start with CCS (Milners's Calculus of Communicating Systems, see [19]) to which we add means to express time behaviour. Then we formalize infinite step opacity for such process algebra, called TPA.

The organization of the paper is the following. In Sect. 2 we describe the our working formalism, a special timed process algebra TPA. In next Section we present and study infinite step opacity, we show its properties, relations to other security notions and some complexity issues. In Sect. 4 we discuss obtained results and plans for future work.

2 TPA

In this section we introduce an extension of CCS (Milners's Calculus of Communicating Systems, see [19]). This extension will be called Timed Process Algebra, TPA for short, and it enriches CCS by the special time action t. This action will express elapsing of time. TPA represents a simplification of Timed Security Process Algebra (tSPA) introduced in [3]. An explicit idling operator ι used in tSPA is omitted since we allow implicit idling of processes. Processes can idle either by performing t actions or they can also idle if there is no internal communication possible. We also do not need division of actions between high and low level ones. TPA also strictly preserves *time determinancy* contrary to tCryptoSPA (see [6]).

First we assume a set of atomic actions A such that $\tau, t \notin A$. For every $a \in A$ there exists $\bar{a} \in A$ and $\bar{\bar{a}} = a$. We define $Act = A \cup \{\tau\}, At = A \cup \{t\}$,

$Actt = Act \cup \{t\}$. Let us suppose that $a, b, \ldots$ range over A, $u, v, \ldots$ range over Act, and $x, y \ldots$ range over $Actt$. We consider the signature $\Sigma = \bigcup_{n \in \{0,1,2\}} \Sigma_n$, where

$$\Sigma_0 = \{Nil\}$$
$$\Sigma_1 = \{x. \mid x \in A \cup \{t\}\}$$
$$\cup \{[S] \mid S \text{ is a function form } A \text{ to } A\}$$
$$\cup \{\backslash M \mid M \subseteq A\}$$
$$\Sigma_2 = \{\mid, +\}$$

we will write unary action operators in prefix form, the unary operators $[S], \backslash M$ in postfix form, and the rest of operators in infix form. For S, $S : A \to A$ are such that $\overline{S(a)} = S(\bar{a})$.

TPA terms over the signature Σ are defined as follows:

$$P \; ::= \; X \; \mid \; op(P_1, P_2, \ldots P_n) \; \mid \; \mu X P$$

where $X \in Var$, Var is a set of process variables, $P, P_1, \ldots P_n$ are TPA terms, $\mu X-$ is the binding construct, $op \in \Sigma$.

TPA terms without t action are CCS terms. We will use an usual definition of opened and closed terms. Closed terms which are t-guarded (each occurrence of X is within some subterm $t.A$) will be called processes.

A structural operational semantics of terms will be defined by labeled transition systems. The set of states is given by the set of terms, and actions represent labels. The transition relation $\to$ connects terms. We write $P \xrightarrow{x} P'$ instead of $(P, x, P') \in \to$ and $P \not\xrightarrow{x}$ if there is no P' such that $P \xrightarrow{x} P'$. By $P \xrightarrow{x} P'$ we indicate that the term P can perform action x to become P', $P \xrightarrow{x}$ means that P can perform the action x. The transition relation $\to$ is defined as for CCS with these new inference rules:

$$\frac{}{Nil \xrightarrow{t} Nil} \quad N$$

$$\frac{}{u.P \xrightarrow{t} u.P} \quad I$$

$$\frac{P \xrightarrow{t} P', Q \xrightarrow{t} Q', P \mid Q \not\xrightarrow{\tau}}{P \mid Q \xrightarrow{t} P' \mid Q'} \quad Par$$

$$\frac{P \xrightarrow{t} P', Q \xrightarrow{t} Q'}{P + Q \xrightarrow{t} P' + Q'} \quad ND$$

Rules N, I allow arbitrary idling for Nil process and for processes prefixed by an action from the set A. Processes cannot idle if there is a possibility of an internal communication (Par). Time determinancy is given by the last rule (ND).

In general, rules with negative premises could cause some problems (for details see [7]). But since in TPA derivations of τ actions are independent of derivations of t action, our system lacks such problems.

We write $P \xrightarrow{s}$ instead of $P \xrightarrow{x_1 x_2} \cdots \xrightarrow{x_n}$ where $s = x_1.x_2.\ldots.x_n, x_i \in Actt$. Sequence of actions s will be called a trace of P. The empty sequence of actions will be be indicated by ϵ. Let $Succ(P) = \{P' | P \xrightarrow{s} P', s \in Actt^*\}$. i.e. $Succ(P)$ is the set of all successors of P If the set of all successors of a process is finite we say that this process is finite state process.

To define transitions which hide τ action, a relation $\xRightarrow{x}$ was defined for CCS [19]. We use this idea to define transitions $(\xRightarrow{x}_M)$ which hide actions from M, $M \subseteq Actt$. We will write $P \xRightarrow{x}_M P'$ iff $P \xrightarrow{s_1} \xrightarrow{x} \xrightarrow{s_2} P'$ for $s_1, s_2 \in M^*$ and similarly for $P \xRightarrow{s}_M$. We will write $P \xRightarrow{s}_M$ iff $P \xRightarrow{s}_M P'$ for some P'. If $x \in M$ then $P \xRightarrow{\hat{x}}_M P'$ denotes the same as $P \xRightarrow{x}_M P'$. The relation $\xRightarrow{x}_M$ will be used in our definitions of security properties only for actions (or sequence of actions) not belonging to M. We will extend of $\xRightarrow{s}_M$ for sequences of actions s similarly to $\xrightarrow{s}$. Let $Tr_w(P) = \{s \in (A \cup \{t\})^* | \exists P'.P \xRightarrow{s}_{\{\tau\}} P'\}$. We say that $Tr_w(P)$ is the set of weak timed traces of process P Two processes are weakly timed trace equivalent $(P \simeq_w Q)$ iff $Tr_w(P) = Tr_w(Q)$. Now we are ready to define M-bisimulation which is an extension of bisimulation and which ignores actions from the set M.

Definition 1. *A relation $\Re \subseteq TPA \times TPA$ is called a M-bisimulation if it is symmetric and the following holds: if $(P, Q) \in \Re$ and $P \xrightarrow{x} P', x \in Actt$ then there exists a process Q' such that $Q \xRightarrow{\hat{x}}_M Q'$ and $(P', Q') \in \Re$. Two processes P, Q are called to be M-bisimilar, abbreviated $P \approx_M Q$, if there exists a M-bisimulation relating P and Q.*

Note that $\approx_{\{\tau\}}$, i.e. in the case that $M = \{\tau\}$ corresponds to weak bisimulation. Standard CCS bisimulation will be denoted by $\sim$ (see [19]).

3 Opacity

To motivate security concepts introduced later we start this section with some examples of security properties. At the beginning we mention property called Strong Nondeterministic Non-Interference (SNNI, for short, see [3]). Let us assume that actions are divided into two groups, namely low level actions (called public) L and high level actions (called private) H i.e. $L \cup H = A$. SNNI property is based on an absence of information flow between low and high level actions. Process P has SNNI property whenever an observer cannot learn whether a high level action was performed if only low level actions can be observed. This means that process has SNNI property (denoted by $P \in SNNI$) if $P \setminus H$ behaves like P for which all high level actions are hidden (i.e. replaced by action τ) for a possible intruder observing P. The hiding is expressed by binary operator $P/M, M \subseteq A$, for which it holds if $P \xrightarrow{a} P'$ then $P/M \xrightarrow{a} P'/M$ whenever $a \notin M \cup \bar{M}$ and $P/M \xrightarrow{\tau} P'/M$ whenever $a \in M \cup \bar{M}$. Formally SNNI property can be defined as follows.

Definition 2. *We say that process P has Strong Nondeterministic Non-Interference property iff $P \setminus H \simeq_w P/H$.*

Strong Nondeterministic Non-Interference property belongs to a group of so called language based properties. These properties are focused on traces of actions instead of system's states. Hence such approach is not appropriate for systems which can have sensitive, say secure, states and an intruder tries to learn whether such state is reached. In such cases state based security properties are more appropriate. They do not assume division of actions into the high and low level ones but a more general concept of observations and predicates are exploited.

Let us have a predicate over system's states. This could be capability to idle, deadlock, capability to execute only traces with a given time length, capability to perform at the same time actions form a given set, incapacity to idle (to perform t action) etc. Suppose that an intruder tries to learn whether a state for which the predicate holds has been reached. We do not assume any restrictions for predicates except that they are consistent with some behavioral equivalence. The formal definition follows.

Definition 3. *Predicate ϕ over processes is consistent with respect to relation $\cong$ if $P \cong P'$ implies that $\phi(P) \Leftrightarrow \phi(P')$.*

Behavioral equivalence $\cong$ could be bisimulation ($\sim$), weak bisimulation ($\approx_{\{\tau\}}$) etc. We can also define predicates by means $\cong$ (denoted as $\phi_{\cong}^Q$) Let us assume process Q and equivalence relation $\cong$. We define that $\phi_{\cong}^Q(P)$ holds iff $P \cong Q$.

Now we will assume intruders which can observe only some processes activities i.e. only some actions. Hence we suppose that there is a set of observable actions which can be observed and a set of hidden i.e. non-observable actions (what does not mean classified). We model such observations by relations $\overset{s}{\Rightarrow}_M$.

3.1   Infinite Step Opacity

Now we will define security property called *infinite step opacity* for TPA. It is based on a similar concept defined for discrete event structures (see [14]) but some modifications are needed. Since many protocols could be described by means of timed process algebra formalism we obtain the security property with some practical value added. Moreover, to verify this property we could exploit many software tools and techniques for process algebra formal verification.

Definition 4 (Infinite Step Opacity). *Let us assume TPA process P, a predicate ϕ over processes is infinite step opaque w.r.t. the set M if whenever $P \overset{x_1}{\Rightarrow}_M P_1 \overset{x_2}{\Rightarrow}_M P_2 \cdots \overset{x_n}{\Rightarrow}_M P_n$ for $x_i \in Actt \setminus M$, $1 \leq i \leq n$ and $\phi(P_i)$ holds then there exist also processes P_i', such that $P \overset{x_1}{\Rightarrow}_M P_1' \overset{x_2}{\Rightarrow}_M P_2' \cdots \overset{x_n}{\Rightarrow}_M P_n'$ and $\phi(P_i')$ does not hold. By ISO_M^ϕ we will denote the set of processes for which the predicate ϕ is infinite step opaque w.r.t. to the M.*

Infinite step opacity is depicted on Fig. 1. Note that contrary to Strong Nondeterministic Non-Interference property which express that an execution of secrete action can be detected by an intruder here we express that an intruder cannot detect a presence of a state satisfying ϕ during a particular step of computation.

$$P \overset{x_1}{\Longrightarrow}_M \ldots \overset{x_i}{\Longrightarrow}_M P_i \overset{x_{i+1}}{\Longrightarrow}_M \ldots P_n$$
$$\phi(P_i)$$

$$\neg\phi(P_i')$$
$$P \overset{x_1}{\Longrightarrow}_M \ldots \overset{x_i}{\Longrightarrow}_M P_i' \overset{x_{i+1}}{\Longrightarrow}_M \ldots P_n'$$

Fig. 1. Infinite step opacity

Example 1. Let us assume a system P which could be a communication protocol, description of an interface mechanism, power supply management, memory management etc. Suppose that the system could enter into a sensitive phase i.e. into a state, from which an intruder can obtain some sensitive information about this system, for example, previous communications, activities, stored private values etc. To do so the intruder usually cannot try all states because doing so she or he could be easily detected or it is just costly etc.

Many attacks use information about cash memory usage. Recently this techniques was used by attacks called Meltdown and Spectre. They exploit critical vulnerabilities in practically all today's processors which allow to obtain secrets stored in the memory of other running programs. This could be passwords stored in a password manager or browser, personal photos, emails, instant messages and even business-critical files. (see [16, 18]). A difference in time which is needed to load some value from the cash memory or from the main memory could help an attacker to obtain information whether that value was previously used or not and consequently he or she could exploit that information.

For simplicity, let us assume property ϕ of a (sub)system which says, that some private action h could be performed within two time units. Formally, $\phi(Q)$ holds iff $Q \overset{h}{\to}$ or $Q \overset{t.h}{\to}$ or $Q \overset{t.t.h}{\to}$. Attacker cannot see private actions from the set M i.e. she or he could see only actions from the set $Act \setminus M$ and tries to discover, whether the system has reached a state which satisfy property ϕ. The system is infinite step opaque, i.e. $P \in ISO_M^\phi$ if the attacker cannot learn whether system is or was in the such sensitive state.

3.2 Properties

In this subsection we present same basic properties of infinite step opacity. We start with two "inclusion" properties reflecting power of predicate ϕ as well as capability of an observer to see actions, what is expressed by a set M.

Proposition 1. *Let $\phi_1 \Rightarrow \phi_2$. Then $ISO_M^{\phi_2} \subseteq ISO_M^{\phi_1}$.*

Proof. Let $P \in ISO_M^{\phi_2}, P \overset{x_1}{\Rightarrow}_M P_1 \overset{x_2}{\Rightarrow}_M P_2 \cdots \overset{x_n}{\Rightarrow}_M P_n$ for $x_i \in (Actt \setminus M)$ and $\phi_1(P_i)$ holds. Since $\phi_1 \Rightarrow \phi_2$ and $P \in ISO_M^{\phi_2}$ then we know that there exist processes P_i', $1 \leq i \leq n$ such that $P \overset{x_1}{\Rightarrow}_M P_1' \overset{x_2}{\Rightarrow}_M P_2' \cdots \overset{x_n}{\Rightarrow}_M P_n'$ and $\neg\phi_2(P_i')$ holds. Again since $\neg\phi_2 \Rightarrow \neg\phi_1$ we have $P \in ISO_M^{\phi_1}$.

Proposition 2. *Let $M_1 \subseteq M_2$. Then $ISO_{M_1}^{\phi} \subseteq ISO_{M_2}^{\phi}$.*

Proof. Let $P \in ISO_{M_1}^{\phi}$ and $P \overset{x_1}{\Rightarrow}_{M_2} P_1 \overset{x_2}{\Rightarrow}_{M_2} P_2 \cdots \overset{x_n}{\Rightarrow}_{M_2} P_n$ for $x_i \in (Actt \setminus M_2)$ and $\phi(P_i)$ holds. Since $M_1 \subseteq M_2$ it hold also $x_i \in (Actt \setminus M_1)$ and since $P \in ISO_{M_1}^{\phi}$ then we know that there exist processes P_i', $1 \leq i \leq n$ such that $P \overset{x_1}{\Rightarrow}_M P_1' \overset{x_2}{\Rightarrow}_M P_2' \cdots \overset{x_n}{\Rightarrow}_M P_n'$ and $\neg\phi(P_i')$ holds. This means that $P \in ISO_{M_2}^{\phi}$.

An equivalence of systems is the fundamental concept in the process algebra theory. The following propositions guarantees that under some conditions infinite state opacity could be extended to all processes which are equivalent (in the sensese of bisimulation) to a process which is already infinity state opaque.

Proposition 3. *Let us assume the predicate ϕ which is consistent with respect to relation $\cong$ such that $\sim \subseteq \cong$. Let $P \in ISO_M^{\phi}$ and $P \sim Q$. Then also $Q \in ISO_M^{\phi}$.*

Proof. $Q \overset{x_1}{\Rightarrow}_M Q_1 \overset{x_2}{\Rightarrow}_M Q_2 \cdots \overset{x_n}{\Rightarrow}_M Q_n$ for $x_i \in (Actt \setminus M)$ and $\phi(Q_i)$ holds. Since $P \sim Q$ process P can perform exactly the same sequence of actions and for all processes we have $Q_i \sim P_i$ and since according the assumption $\sim \subseteq \cong$ holds and ϕ is consistent with respect to relation $\cong$ we have that $\phi(P_i)$ holds. Now since $P \in ISO_M^{\phi}$ we know that there exist P_i', $1 \leq i \leq n$ for which $P \overset{x_1}{\Rightarrow}_M P_1' \overset{x_2}{\Rightarrow}_M P_2' \cdots \overset{x_n}{\Rightarrow}_M P_n'$ and $\neg\phi(P_i')$ holds. Again, we can repeat the previous arguments to show that there exist processes Q_i', $1 \leq i \leq n$ for which $Q \overset{x_1}{\Rightarrow}_M Q_1' \overset{x_2}{\Rightarrow}_M Q_2' \cdots \overset{x_n}{\Rightarrow}_M Q_n'$ and $\neg\phi(Q_i')$ holds. Hence $Q \in ISO_M^{\phi}$.

3.3   Relation to Process Opacity

In this subsection we present a relation between infinite step opacity and process opacity. The formal definition of the later one (see [10]) follows.

Definition 5. *Let us assume process P, a predicate ϕ over processes is process opaque w.r.t. the set M if whenever $P \overset{s}{\Rightarrow}_M P'$ for $s \in (Actt \setminus M)^*$ and $\phi(P')$ holds then there exists P'' such that $P \overset{s}{\Rightarrow}_M P''$ and $\phi(P'')$ does not hold. By POp_M^{ϕ} we will denote the set of processes for which the predicate ϕ is process opaque w.r.t. to the M.*

Example 2. Let the set M contains only actions h_1 and h_2 and let the predicate ϕ says that both actions h_1 and h_2 cannot be performed. Then $P \in POp_M^{\phi}$ but $P' \notin POp_M^{\phi}$ where $P = l.Nil + l.h_1.Nil + l.h_2.Nil + l.(h_1.Nil + h_2.Nil)$ and $P' = l.Nil + l.h_1.Nil + l.h_2.Nil$. When action l is performed this process always reaches a state which satisfies ϕ and hence it cannot be considered to be safe with respect to predicate ϕ and set M. On the other side these two processes cannot be distinguished by any opacity property since the property ϕ is not a trace property (as it is property SNNI).

$$P \quad \overset{s}{\Longrightarrow}_M \quad \phi(P')$$

$$P \quad \overset{s}{\Longrightarrow}_M \quad \neg\phi(P'')$$

Fig. 2. Process opacity

Process opacity is depicted on Fig. 2.

Let ϕ is consistent with respect to $\cong$ and $\cong$ is such that it a subset of the trace equivalence (see [19]). Then we have if $P \cong P'$ then $P \in POp_M^\phi \Leftrightarrow P' \in POp_M^\phi$.

The relation between process opacity and infinite step opacity is expressed by the following proposition.

Proposition 4. *For every M and ϕ it holds $ISO_M^\phi \subseteq PO_M^\phi$. Moreover there exist M and ϕ such that $ISO_M^\phi \subset PO_M^\phi$.*

Proof. Sketch. It can be easily proved that the first part of the proposition directly follows from Definitions 4 and 5. What distinguish infinite step opacity from process opacity is a stronger requirement of the former that visible computational trace s has to be emulated even when it does not ends in a state satisfying ϕ.

On the other side, we can obtain a stronger security property than infinite step opacity by requiring both process opacity for ϕ and its negation. The corresponding proposition is the following.

Proposition 5. *For every M and total ϕ it holds $PO_M^\phi \cap PO_M^{\neg\phi} \subseteq ISO_M^\phi$. Moreover there exist M and ϕ such that $PO_M^\phi \cap PO_M^{\neg\phi} \subset ISO_M^\phi$.*

Proof. The main idea. Since ϕ is total any step reached be a process satisfy either ϕ or $\neg\phi$. The rest could be directly easily proved directly from Definitions 4 and 5.

As a consequence of Propositions 4 and 5 we obtain the following corollary.

Corollary 1. *For every M and total ϕ it holds $PO_M^\phi \cap PO_M^{\neg\phi} = ISO_M^\phi \cap ISO_M^{\neg\phi}$.*

3.4   Timing Attacks vs Non-timing Attacks

For attackers who can observe systems in real time, time attacks represent powerful tools to compromise such systems. On the other side these techniques are useless for off-line systems and hence these could be considered safe with respect timing attacks, i.e. with respect to attackers who cannot observe elapsing of (real) time. By the presented formalism we have a way how to distinguish these two cases.

Definition 6 (Immunity with respect to Timinig Attacks). *Process P is safe with respect to timing attack, ϕ and $M, t \notin M$ iff $P \notin ISO_M^\phi$ but $P \in ISO_{\{M \cup t\}}^\phi$.*

In case that an attacker can observe time behavior of systems but only with a limited accuracy we can model that behavior with a larger time granularity (bigger time units) or we could change Definition 4 in such a way that we would allow that actions x_i to be from $(Actt \setminus M) \cup \{t\}$.

3.5   Complexity

Unfortunately, infinite step opacity is undecidable in general. This fact is implied by undecidability of process opacity (see [10]) and Propositions 4 and 5.

Now we briefly sketch how to overcome this problem. First, to obtain decidability of infinite step opacity we need to restrict predicates ϕ. We will model predicates by special processes called tests. From now on, we will assume that action τ is not visible for an attacker, that is $\tau \in M$. Process called test of $'phi$ (denoted by T_ϕ) will communicates with processes P and produce a special action $\sqrt{}$ (indicating passing of the test) if the predicate ϕ holds for the processes P.

Definition 7. *We say that process T_ϕ tests predicate ϕ if $\phi(P)$ holds iff $(P|T_\phi) \setminus At \approx_t \sqrt{}.Nil$.*

We say that ϕ is the finitely definable predicate if T_ϕ is the finite state process. By means of test T_ϕ infinite step opacity could be expressed as M-bisimulation based property similarly to process opacity (see [10]). Hence due to the above sketched construction, we can obtain a decidable variant of infinite step opacity. We limit tests to be finite states processes but this limitation has no practical importance since majority of (if not all) practically important and useful properties (predicates) can be described by such tests.

Proposition 6. *Let both predicates ϕ and $\neg\phi$ are finitely definable. In this case infinite step opacity $ISOp_M^\phi$ is decidable in time $O((|A|)^3.k.l)$ for finite state processes, where k and l are numbers of states of P and the maximum of numbers of states of processes representing tests of predicates ϕ and $\neg\phi$, respectively.*

Proof. The main idea. We need to show that the relation $\approx_{M \cup \{t\}}$ can be decided in time $O((|A|)^3.k.l)$. For this we need a slight modification of the proof of complexity results for weak bisimulation (see [15]). For testing scenario see Fig. 3.

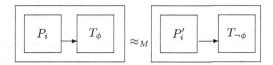

Fig. 3. Testing scenario

This methods allows us to reduce verification of security given in terms of infinite step opacity to checking bisimulation by some of existing techniques and software tools. Moreover, protocols for IoT could be defined by finite state systems as well as security properties expressed by ϕ and $\neg\phi$, hence Proposition 6 is fully applicable.

3.6 Enforcement

In case that a given process is not secure with respect to infinite step opacity for predicate ϕ and set M i.e. $P \notin ISO_M^\phi$ we can either modify its behavior. But this could be costly, difficult or even impossible, in the case that it is already part of a hardware solution, proprietary firmware etc. Or we can use supervisory control (see [21]). This method restricts system's activities in such a way that the resulting system becomes secure with respect to infinite step opacity. A supervisor can see (some) set of process's actions and can control (disable or enable) some set of process's action. In such a way supervisor restricts resulting behaviour to guarantee system's security. This represents a trade-off between functionality and security. But in many cases it does not cause a real problem.

Let us assume some communication protocol which could reach (with some very low probability) a state which cannot be considered to be secure with respect to infinite step opacity. In such case the transmission of a packet is interrupted by supervisor and it should start from the begging. What has no influence on basic functionality of the protocol but the supervisor increases its security.

Sometimes this restriction could have even smaller influence on process's behavior. Let us assume that the system can perform action a and b in an arbitrary order but only a sequence $b.a$ could leak some classified information i.e. lead to some sensitive intermediate states. If the supervisor restricts such sequence, system becomes secure but this has no significant influence on overall system's functionality. In [11] we have proposed supervisory control for process opacity. This technique could be extended also for infinite state opacity as well.

4   Discussion and Further Work

We have presented application of formal methods in security, namely we have introduced the security property called infinite step opacity. We have formalized this property in the framework of timed process algebra called TPA. This security property is very general and can cover some already defined security properties.

Moreover, by suitable choice of process's predicates we obtain property which can be effectively checked.

One of the main features that makes it so valuable in systems for IoT devices is the fact that we can model process's security with respect to a limited number of attempts to perform an attack, with respect to limited time of attack, to limited time precision of an attacker and so on.

Moreover, we can use also different process algebras than TPA. We can enrich TPA by operators expressing such "parameters" as probability, distribution, networking architecture, network capacity or throughput, power consumption and so on. In such a way also other types of attacks, which exploit information flow through various covert channels, could be described. All these is particularly challenging for systems consisting of IoT devices.

Due to complexity issues, plan to define and examine compositional properties of infinite step opacity in such a way that bottom-up design of secure processes would be easier. We also consider to work with value-passing process algebra to simplify some notations and also to express new security features. In this way we could obtain new formalism(s) capable to express other aspects of behavior of low level protocols used in the field of IoT.

For the time being, we are testing the usefulness of this algebra in our own systems deployed in our laboratory: a Wireless Sensor and Actuator Networks using BLE and TCP/IP protocols, but we intend to study real systems in a short time.

References

1. Dhem, J.-F., Koeune, F., Leroux, P.-A., Mestré, P., Quisquater, J.-J., Willems, J.-L.: A practical implementation of the timing attack. In: Quisquater, J.-J., Schneier, B. (eds.) CARDIS 1998. LNCS, vol. 1820, pp. 167–182. Springer, Heidelberg (2000). https://doi.org/10.1007/10721064_15
2. Felten, E.W., Schneider, M.A.: Timing attacks on web privacy. In: Proceedings of the 7th ACM Conference on Computer and Communications Security, CCS 2000, pp. 25–32. ACM, New York (2000). https://doi.org/10.1145/352600.352606
3. Focardi, R., Gorrieri, R., Martinelli, F.: Information flow analysis in a discrete-time process algebra. In: Proceedings 13th IEEE Computer Security Foundations Workshop, CSFW 2013, pp. 170–184, July 2000
4. Garrido-Hidalgo, C., Hortelano, D., Roda-Sanchez, L., Olivares, T., Ruiz, M.C., Lopez, V.: IoT heterogeneous mesh network deployment for human-in-the-loop challenges towards a social and sustainable industry 4.0. IEEE Access **PP**, 1 (2018)
5. Goguen, J.A., Meseguer, J.: Security policies and security models. In: 1982 IEEE Symposium on Security and Privacy, p. 11, April 1982
6. Gorrieri, R., Martinelli, F.: A simple framework for real-time cryptographic protocol analysis with compositional proof rules. Sci. Comput. Program. **50**(1), 23–49 (2004). 12th European Symposium on Programming (ESOP 2003)
7. Groote, J.F.: Transition system specifications with negative premises. Theor. Comput. Sci. **118**(2), 263–299 (1993)
8. Gruska, D.P.: Observation based system security. Fundam. Inf. **79**(3–4), 335–346 (2007)

9. Gruska, D.P.: Informational analysis of security and integrity. Fundam. Inf. **120**(3–4), 295–309 (2012)

10. Gruska, D.P.: Process opacity for timed process algebra. In: Voronkov, A., Virbitskaite, I. (eds.) PSI 2014. LNCS, vol. 8974, pp. 151–160. Springer, Heidelberg (2015). https://doi.org/10.1007/978-3-662-46823-4_13

11. Gruska, D., Ruiz, M.: Initial process security. In: Proceedings of 26th International Workshop on Concurrency, Specification and Programming, CS&P 2017 (2017)

12. Handschuh, H., Heys, H.M.: A timing attack on RC5. In: Tavares, S., Meijer, H. (eds.) SAC 1998. LNCS, vol. 1556, pp. 306–318. Springer, Heidelberg (1999). https://doi.org/10.1007/3-540-48892-8_24

13. Hortelano, D., Olivares, T., Ruiz, M.C., Garrido-Hidalgo, C., López, V.: From sensor networks to internet of things. Bluetooth low energy, a standard for this evolution. Sensors **17**, 372 (2017)

14. Jacob, R., Lesage, J.J., Faure, J.M.: Overview of discrete event systems opacity: models, validation, and quantification. Ann. Rev. Control **41**, 135–146 (2016)

15. Kanellakis, P.C., Smolka, S.A.: CCS expressions, finite state processes, and three problems of equivalence. Inf. Comput. **86**(1), 43–68 (1990)

16. Kocher, P., et al.: Spectre attacks: exploiting speculative execution. CoRR abs/1801.01203 (2018)

17. Kocher, P.C.: Timing attacks on implementations of Diffie-Hellman, RSA, DSS, and other systems. In: Koblitz, N. (ed.) CRYPTO 1996. LNCS, vol. 1109, pp. 104–113. Springer, Heidelberg (1996). https://doi.org/10.1007/3-540-68697-5_9

18. Lipp, M., et al.: Meltdown. CoRR abs/1801.01207 (2018)

19. Milner, R.: Communication and Concurrency. Prentice-Hall Inc., Upper Saddle River (1989)

20. Nassi, B., Sror, M., Lavi, I., Meidan, Y., Shabtai, A., Elovici, Y.: Piping Botnet - turning green technology into a water disaster. CoRR abs/1808.02131 (2018). http://arxiv.org/abs/1808.02131

21. Ramadge, P.J.G., Wonham, W.M.: The control of discrete event systems. Proc. IEEE **77**(1), 81–98 (1989)

22. Roda-Sanchez, L., Garrido-Hidalgo, C., Hortelano, D., Olivares, T., Ruiz, M.C.: Operable: an IoT-based wearable to improve efficiency and smart worker care services in industry 4.0. J. Sens. **2018**, 6272793:1–6272793:12 (2018)

23. Song, D.X., Wagner, D., Tian, X.: Timing analysis of keystrokes and timing attacks on SSH. In: Proceedings of the 10th Conference on USENIX Security Symposium - Volume 10, SSYM 2001. USENIX Association (2001)

FogFlow - Computation Organization for Heterogeneous Fog Computing Environments

Joanna Sendorek[✉], Tomasz Szydlo, Mateusz Windak,
and Robert Brzoza-Woch

Department of Computer Science,
AGH University of Science and Technology, Krakow, Poland
send.joanna@gmail.com

Abstract. With the arising amounts of devices and data that Internet of Things systems are processing nowadays, solutions for computational applications are in high demand. Many concepts targeting more efficient data processing are arising and among them edge and fog computing are the ones gaining significant interest since they reduce cloud load. In consequence Internet of Things systems are becoming more and more diverse in terms of architecture. In this paper we present *FogFlow* - model and execution environment allowing for organization of data-flow applications to be run on the heterogeneous environments. We propose unified interface for data-flow creation, graph model and we evaluate our concept in the use case of production line model that mimic real-world factory scenario.

Keywords: IoT · Fog computing · Stream processing ·
Data-flow graphs

1 Introduction

Internet of Things (IoT) is becoming more and more present in our reality with the development of intelligent buildings, smart cities [18] or mobile communications [2]. We are being surrounded by increasing number of electronic devices including our everyday objects, like watches or mobiles, and whole systems managing our environment, such as traffic monitoring or intelligent surveillance. Majority of those devices is able to be network-connected and this number is still growing with hardware advancements causing increasing capabilities of simplest appliances. This can be observed especially in the sensors area which gives foundations for sensor networks solutions and architectures [1]. All of those devices are potential sources of data which may be shared or gathered thanks to the network access.

Constantly growing amount of real-time data has been causing a shift towards data-orientation in the systems architectures as well as programming paradigms

© Springer Nature Switzerland AG 2019
J. M. F. Rodrigues et al. (Eds.): ICCS 2019, LNCS 11538, pp. 634–647, 2019.
https://doi.org/10.1007/978-3-030-22744-9_49

over the last few years [9]. Even with computing clouds becoming recently more developed and mature, processing data we have access to is still posing challenges. Although cloud solutions provide a way to achieve capabilities required to handle significant amounts of data, they bring in response delays, coming from communication overhead, which are often unacceptable in the real-time applications. Therefore, current research in IoT domain is often focused on finding a balance between responsiveness and robustness, resulting for instance in edge computing [8] or fog computing [3] concepts. Mentioned ideas assume that some of the computations can be moved closer to the data sources and in this scenario cloud infrastructure may be responsible only for heavy analytic and global information handling. Also, it is beneficial to move computations responsible for local decision making closer to the devices as it reduces response time. In order to fully make use of available resources, cloud, edge and fog computing should be treated as complimentary ones since each of them is most advantageous in different aspect of whole system [10].

In the area of cloud and data processing a lot have been done recently with multiple mature analytic solutions such as *Apache Spark*[1], *Google Cloud Dataflow*[2] or *Apache Flink*[3]. However, since cloud and fog should be interoperating as mentioned, it is desirable to have a generic development model for data-flow applications enabling to span fog and cloud [4]. Such a concept, but limited only to multiple cloud solutions, is realized by Apache Beam[4] - unified model which enable defining data streams and running them on different back-ends. In this paper, we take the idea of unification further and propose *FogFlow* - the model and engine enabling defining data processing pipelines able to be decomposed into the sub-pipelines that span fog or edge and the cloud. We base our concept on graph model, commonly used in data-flow applications [13,16]. Scientific contribution of this paper can be summarized as follows: (i) we define unified graph-based model for abstract definition of data processing applications; (ii) we discuss methods of decomposing and translating the graph model into the set of processing nodes appropriate for given infrastructure; (iii) we propose methodology and provide example of design and implementation data-processing application able to be run in heterogeneous IoT environments.

Organization of the paper is as follows. Section 2 describes the related work and Sect. 3 discuses FogFlow - its components, model and implementation. Section 4 describes the evaluation, while Sect. 5 concludes the paper.

2   Related Work

To our best knowledge *FogFlow*, as the model aiming at unifying data processing applications definition and enabling their execution in heterogeneous environments, is a novel solution bridging the gap of existing concepts. However,

---

[1] spark.apache.org access for 15.03.19.

[2] cloud.google.com/dataflow/ access for 15.03.19.

[3] flink.apache.org access for 15.03.19.

[4] https://beam.apache.org/get-started/beam-overview/ access for 02.02.2019.

while developing *FogFlow* we relied on current research in the area of organizing modern applications centred around IoT data processing.

Since many of nowadays systems are focused on data processing, the data-flow paradigm is adopted altogether with IoT systems in many current works and research. For example, in [7] authors describe scalable IoT framework to design logical data-flow using virtual sensors. In this solution, operators for data processing are defined and then processed in the logical data-flow. The whole modelling is based on the graph that is being executed in the proper topological order. Data-flow architecture is also commonly used for application design - in [5] author proposes data-flow architecture for smart city applications. In this solution operations can manipulate data coming from different flows.

The necessity for designing and managing strategies for IoT services placement dedicated for IoT and Fog computing has been acknowledged lately as the natural consequence of heterogeneity in the IoT environments. Since components of both IoT and Fog system infrastructures are often highly distributed and they vary in the context of resources availability and computing capabilities, efficient resource management strategies have to be implemented in order to maximize whole system response while minimizing overall solution cost and latency. *IFogSim* [6] is a simulator for IoT and fog environments which enable assessment of system efficiency due to the chosen metrics, such as energy consumption, network latency and operational cost. *IFogSim* allows for system modelling and evaluation of system focusing on resource management techniques. It is worth noting that *IFogSim* uses graph model as the system representation and makes usage of typical graph algorithm, Floyd-Warshall for all shortest path problem, in order to carry out data transmission simulation. In very recent work [11] authors present an extension to iFogSim allowing for the design of data placement strategies based on 'divide and conquer' approach. There are other methods such as [12], where authors propose a strategy of data placement in fog infrastructure and it is based on graph partitioning. They use graph as the model of infrastructure with edges standing for data-flows numbers passing between the nodes.

The concept similar to the one presented in the paper is the Distributed Dataflow (DDF) - the programming model described in [4] which aims at being used with infrastructures of both fog an cloud. DDF framework is based on the *Node-RED*[5] tool, which allows for creating high-level application definition constructed as a graph. In a similar way to *FogFlow* application parts can be placed on different elements of the infrastructure, but the same execution environment (*NodeRed*) is required on all of the devices. In contrast to that, *FogFlow* enables multiple execution environment and this is achieved, among all, via source code generation.

---

[5] https://nodered.org/.

3   *FogFlow* Structure

We propose the *FogFlow* model and execution environment, which enables the design of applications able to be decomposed onto heterogeneous IoT environments according to the chosen decomposition schema. We achieve this flexibility of running applications in the multiple environments by abstracting out the application definition from its architecture and the target implementation. In order to provide one unambiguous, well-defined model of computations, we rely on graph representation. We aim at fulfilling the following requirements for data-driven IoT applications:

1. the application definition should be infrastructure-independent and contain only logic of data processing;
2. execution of the application should be possible on different set of devices.

Those requirements are able to be fulfilled with the assumption that each data-processing step is stateless and all information required for further processing is contained in the messages passed with data. Independence between application definition and its execution is achieved by three-layered application model corresponding with three-layered module architecture. Figure 1 depicts conceptual diagram of *FogFlow* - both its architecture and application model. First and most high-level layer of *FogFlow* is *FogFlow API*, which enables creating application definition. Our API is organized in the functional style, enabling for creation of data processing pipeline reflecting successive steps. Definition created is then being **transformed** into data-flow graph representation managed by the next layer - graph module. Core function of using graphs as the intermediate application state is to provide one, unambiguous model which can be used in the further processing. In the module discussed the data-flow graphs are fetched into the form appropriate for the **modification**. This term includes both adjustments made to remove ambiguities introduced by user in the application definition and those aiming at decomposing graph as preparation for multiple

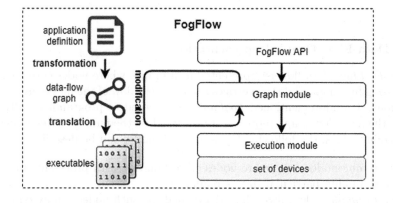

Fig. 1. Conceptual diagram of *FogFlow*

devices execution. The last phase is `translation` happening in the execution module which results in executables ready to be run on the provided devices. The same data-flow definition is able to be run in the chosen environment via appropriate `modifications` in the graph module and `translation` preparing graphs to be executed.

3.1   *FogFlow* API

FogFlow API is the part of *FogFlow* enabling for creation of data processing pipelines in the functional style. There are two main entities defined in the API:

1. `Flow` - representing the whole data processing pipeline;
2. `StreamData` - representing data stream containing data of the specified type at the given state of pipeline.

We decided upon using *Java 8* for *FogFlow API* implementation. It enables type-checking at compile time and ensuring that consecutive data processing steps are valid. Table 1 presents list of methods defined for each of the entities mentioned.

Table 1. *FogFlow API methods.*

StreamData<T>		Flow	
type	**name**	**type**	**name**
StreamData<U>	*map*	void	*setUpEnvironment*
StreamData<U>	*aggregateOnCount-SlidingWindow*	void	*executeFlow*
StreamData<U>	*aggregateOnCount-BatchWindow*	StreamData<T>	*createStreamEntry*
List<StreamData<T>>	*splitStream*	StreamData<T>	*createAsStreamsJoin*
StreamData<T>	*filter*		
void	*sink*		

3.2   Data-Flow Graph Representation

In the *FogFlow*, data-flow is represented as a graph, where nodes correspond to processing functions while the vertices represent the flow of the messages between functions. The aforementioned *FogFlow API* is used to programmatically construct the data-flow graph representation. Based on the number of incoming and out-coming edges, the following types of graph nodes can be described:

– *One-to-one* nodes - those are nodes corresponding with processing functions able to modify particular data stream;
– *One-to-many* nodes, also called `split nodes` - such nodes can execute splitting the incoming data stream into a few, based on the defined condition;

– *Many-to-one* nodes, also called `join nodes` - those are able to reverse action of split by merging few data streams into one.
– *Zero-to-one* nodes, also called `sources` - entries for data streams;
– *One-to-zero* nodes, also called `sinks` - nodes representing termination of data processing.

The described differentiation requires additional comment regarding split and join nodes. First of all, multiple out-coming edges may occur implicitly when user calls multiple one-to-one processing functions on particular stream therefore creating multiple one-to-one nodes. The difference between this situation and application of split node is that in the first case, all created nodes are assigned `same copy` of the stream as the incoming edge while the usage of split node is inevitably related with creating `different data streams` being substream of the initial one. Second issue worth noting is that `join nodes` are the only ones that enable multiple incoming edges and in consequence those are the nodes responsible for the whole graph being `directed acyclic graph` and not a `tree` in particular. From the perspective of computation results equivalence, joining streams and next creating *one-to-one* node corresponds to multiple *one-to-one* nodes being created for each of join terms. However, such a solution would be both less intuitive for user and less effective to evaluate.

Taking into account, the specific data processing that the node can carry out, we propose the following types of intermediate vertices gathered in Table 2.

Table 2. Types of graph nodes and their representation

Node	In-edges	Out-edges	Symbol
split	1	mutiple	(S)
join	multiple	1	(J)
source	0	1	◇
sink	1	0	◯

Node	In-edges	Out-edges	Symbol
window	1	1	(W)
map	1	1	(M)
filter	1	1	(F)

3.3   Graph Modifications

As discussed previously, the choice of representing consecutive data-flow transformations via graphs has been driven, among others, by the convenience of adapting the whole graph analysis and algorithms to structure computations efficiently. Treating each transformation of data as a separate graph node allows for placing them on different devices according to the acquired policy. Moreover, graph structure allow for straightforward simplifications with tree-like balancing, pruning or path reductions.

In this section, we present two examples illustrating decomposition onto two devices and simple pruning. Figure 2 depicts fragment of data-flow consisting

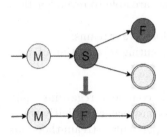 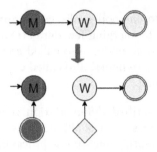

Fig. 2. Graph prunning   **Fig. 3.** Decomposition

initially of four nodes: *map* and *split* node being source for *filter* and *data sink*. The most basic pruning algorithm used in *FogFlow* is described by the pseudo-code 1 and it removes all of the branches from the graph that are not finished by a sink node. In the example, the upper-side branch of split in unused since it ends in the filter node. Therefore it is removed at this state of modification. Since the *sink node* on the bottom side of *split* has been assigned only part of the data and now this is the only successor of the *split*, in order to maintain unambiguous description, the *split node* should be now converted to *filter node*. The graph after the pruning is equivalent to the one before in the sense that it will give the same data transformation results. Nodes not participating directly in the pruning (coloured in white) remain unchanged. The question may arise when pruning will be beneficial in the practical sense. The issue with an API styled in the functional way is that even though it is intuitive to use, it lacks mechanisms imposing ambiguity. On the contrary, the graph description allows for detecting any redundancies and reducing graph to the concise form. Apart from concept illustrated by the example, other possible ways for graph data-flow simplification would be: analyses of parallel edges, avoiding duplicating vertices or path reduction. All of the mentioned modifications are valid under the assumption that nodes are stateless as mentioned earlier.

Algorithm 1. Tree pruning algorithm.

1: **function** PRUNE(G)
2:     $usedNodes \leftarrow \varnothing$
3:     **for** $graphNode$ in $G.V$ **do**
4:         **if** $graphNode$ *is sink* **then**
5:             $usedNodes \leftarrow usedNodes \cup graphNode$
6:     **for** $sinkNode$ in $usedNodes$ **do**
7:         $usedNodes \leftarrow usedNodes \cup ancestors\ of\ sinkNode$
8:     $unusedVertices \leftarrow G.V \setminus usedNodes$
9:     **return** $unusedVertices$

Figure 3 illustrates decomposition of the graph fragment onto two devices. Flow fragment consists of three nodes being structured as one pipeline. Additionally, each of the nodes has been assigned to the one the two available devices - dark nodes to one of the devices (let it be called device A) and white nodes to the other (device B). The choice of exact algorithm or policy used for node tagging and device assignment is the whole broad topic and out of scope of this paper, but there are a few approaches we consider:

- one of the algorithms described in Sect. 2 may be adopted;
- dynamic system analysis may be conducted in order to determine the devices usage and balancing may be implemented;
- heuristic aiming at moving majority of processing as close to data source as possible at each processing phase may be used - we discussed it in [14].

Regardless of the approach taken, in the example the nodes were tagged in a way resulting in the *window* and succeeding *sink* node being to be moved onto device B, while assigning all of the rest to the device A. The decomposition onto two devices in the example is crossing one edge - the one connecting *map* with the *window* node. Algorithm 2 describing the process of decomposition on the given edge. Since the effect of decomposition should be two sub-graphs - representing two independent data-flows able to be run on separated devices, this edge is being removed. In order to maintain valid graph structure and provide a way to pass data from device A to device B, two new nodes are inserted into graph: sink node on device A and corresponding source node on device B. The exact implementation of those so-called *communication nodes* depends on the devices and the established protocol, but it does not affect the decomposition itself.

Algorithm 2. Decomposition on edge algorithm.

1: **function** DECOMPOSE($G, edge, deviceA, deviceB$)
2:      $sourceVertex \leftarrow edge.source$
3:      $targetVertex \leftarrow edge.target$
4:      $newSourceVertex \leftarrow newSource(deviceB)$
5:      $newSinkVertex \leftarrow newSink(deviceA)$
6:      $G.E \leftarrow G.E \setminus edge$
7:      $G.V \leftarrow G.V \cup \{newSourceVertex, newSinkVertex\}$
8:      $G.E \leftarrow G.E \cup \{E(sourceVertex, newSinkVertex),$
          $E(targetVertex, newSourceVertex)\}$

3.4   Execution Module and Implementation

Execution module is responsible for translating data-flow graphs into executables ready for running on the given devices set. Up to this point, we have used term 'devices set' to describe multiple possible infrastructures that may be existent in the IoT system. However, in order to translate data-flow graphs into executables for heterogeneous target environments, we distinguish three types of infrastructure summarized in the Table 3. At current state of work, *FogFlow* provides execution with one runtime per type - further development regarding this area is described in Sect. 5. While implementation details are out of scope for this paper in which we focus on concept and methodology, we will briefly describe execution module structure.

Table 3. Types of infrastructure distinguished for *FogFlow*

Infrastructure	Description		Runtime
Cloud	Features	- focus on high-availability	*Apache Flink*
		- scalability	
		- possible contenerization	
	Use cases	- runnning resource-demanding algorithms	
		- correlating data from different sources	
		- data persistence	
Edge	Features	- ability to run local analytics	*Apache Edgent*
		- limited resources and small footprint	
		- having application processors	
	Use cases	- reducing data sent for further analytics	
		- local decision making and classification	
		- immediate events reaction	
MCUs	Features	- resource constrainted embedded microcontrollers	*C based*
		- impossibility of porting high-level libraries	
	Use cases	- acquiring sensor-data	
		- basic preprocessing and reformatting	

Cloud environments are the ones that are characterized by resource capabilities and global knowledge of the whole environment. Therefore, application parts best suited to be run in the cloud are the ones that correlate data from multiple sources or are highly demanding. In *FogFlow* we currently provide support for execution with *Apache Flink* framework to run such application components. We chose *Flink* because of its efficiency (in-memory speed) and interoperability with majority of commonly used cluster environments. On the contrary to cloud, edge devices lack resources required to run distributed computations, but are able to execute local analysis. We distinguish this group of devices by presence of application processors allowing for high-level programs execution. Paramount example of edge devices would be gateways passing data from lower-level devices up to the cloud. They can be used to process and filter data streams and conduct non complex local analysis leading to rapid system reaction. For edge, we decided upon using *Apache Edgent*[6] runtime to execute data-flow graphs. The lowest-level of infrastructure consitues of MCU's - resources constrained embedded microcontrollers. They lack ability to port any high-level libraries and using

---

[6] edgent.apache.org access for 15.03.2019.

two Apache technologies mentioned would be impossible. Data-flow parts placed on such devices could contain data-gathering logic and basic preprocessing such as changing format or cutting off unnecessary information. Data-flow may be able to be executed with C based runtime.

Execution module is divided into submodules (implementation providers) corresponding with the three types of runtime. Each submodule contains code that translates data-flow graph into the given target executable. When translating, we rely on the source code generation approach giving promising results when small footprint processing running on the edge is concerned [15]. Both *Flink* and *Edgent* rely on the slightly different graph models, are based on Java programming language and have application programming interfaces allowing for basic functional-style operations on data streams. With *FogFlow* it is possible to translate data-flow graphs crated with *FogFlow API* into their source code in following steps:

1. data-flow graph and chosen implementation provider are passed to the executing module;
2. data-flow graph is being translated into source code in the topological order using provider implementation of *FogFlow API*;
3. all of the external objects such as custom sources or sinks are serialized and attached to the code;
4. proper java archive file is created including only dependencies of *FogFlow API* and chosen implementation provider.

Translation targeting C code is far more complex and lower-level issue, with many aspects needed to be taken into account: inter-language typing system, objects serialization between technologies, constraints of C language or threads execution in C. We do not discuss this broad and hardware related topic in the scope of this paper and our use case does not include data-flow graph execution on MCU's.

4 Use Case and Evaluation

In this section we present an example usage of *FogFlow* to organize computations for vibrations detection of the working assembly line in our laboratory. Schema of our environment has been depicted in Fig. 4. The goal of the presented use case is to measure vibrations generated by the working assembly line. The *micromachined microelectromechanical system (MEMS)* accelerometer is gathering three dimensional data allowing for vibration measurement and it is read by connected embedded device. For each 128 samples gathered in the batch windows we calculate *Root Mean Square (RMS)* [17] defined as $\sqrt{\frac{\sum_{i=j-\gamma}^{j} x_i^2}{\gamma}}$, where γ states for the window size and in the case described equals to 128. In the local network we also have *Raspberry Pi* gateway. Another component of environment is the private cloud with database accessible from both the assembly line and gateway.

Results of the vibration analysis are then written to the database in the cloud for further analysis. Processing pipeline is defined as follows:

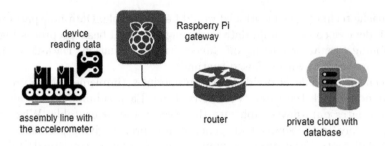

device
reading data

Raspberry Pi
gateway

assembly line with
the accelerometer

router

private cloud with
database

Fig. 4. Laboratory environment schema

1. data is received from the device with accelerometer using `MQTT` protocol;
2. data is parsed - only the component aligned with x axis is being extracted from three dimensional data and interpreted as floating-point number;
3. data is gathered in 128-elements batch windows and for each window *RMS* is calculated;
4. results are written to the database.

Listing 1.1 presents application definition corresponding with the described pipeline created with *FogFlow API*.

```
flow.setUpEnvironment();
StreamData<String> receivedMessage = flow.
    createStreamEntry(new MqttExternalReceiver());
StreamData<Double> x = receivedMessage.map(new
    MessageParser());
StreamData<Double> rms = x.aggregateOnCountBatchWindow
    (128, new RmsLambda());
rms.sink(new DatabaseWriter());
flow.executeFlow();
```

Listing 1.1. Assembly line data processing pipeline in *FogFlow API*.

Created application definition is then **transformed** into data-flow graph. At this point graph nodes are assigned to the particular devices in our laboratory and then graph is decomposed accordingly. Given the infrastructure described, we decompose data processing application with the two schemas:

1. cloud-based data processing - where all of the data from the accelerometer is received in the cloud and processed there;
2. cloud database operations with edge computing pre-processing - where all of the data is received on gateway, pre-processed there and send to cloud for further steps.

In the scenario 1, the whole pipeline is aimed at executing with *Apache Flink* and in scenario 2, the underlying data-flow graph is decomposed into two subgraphs. This process is illustrated in Fig. 5. With cloud based data processing,

Apache Flink is used for the whole execution. One thing requires discussion at this point - the impact of possible number of production lines on the processing efficiency. In such a simple scenario, we propose creating separate pipeline for each of the lines. Therefore, execution of all of them in one environment, such as *Apache Flink*, is a embarrassingly parallel problem with only potential bottleneck being database writing. Since cloud allows for high distribution of computations, significant processing capabilities may be achieved. However, major drawback of this scenario is that of all the data from accelerometer is sent to the cloud generating great amount of network traffic leading to unnecessary delays and greater costs.

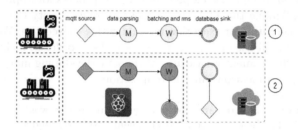

Fig. 5. Data-flow graph decomposition with two scenarios.

We measured data traffic which is being processed at different stages of dataflow. First of all, accelerometer is sending data as fast as it gets consecutive measurements, which gives approximately *950 messages per second (msg/s)* which with the MQTT protocol for communication gives a traffic of *149 596 B/s*. After batch window reduction is applied this is reduced to approximately *584 B/s*. In the second scenario, reduction of data in the batch windows is conducted by *Raspberry Pi gateway* with the *Apache Edgent*.

Table 4. Processing capabilities for *Apache Edgent*

	Simple data source	MQTT data source
Intel Core i5 2 cores	462 929 $\frac{msg}{s}$	41523 $\frac{msg}{s}$
Raspberry Pi	20 145 $\frac{msg}{s}$	9861 $\frac{msg}{s}$

We conducted artificial tests of *Apache Edgent* processing capability where we replaced MQTT data source with the data spawner. In Table 4 we gathered results obtained both with MQTT source and spawner in different setups. It is worth noting that even when running on *Raspberry Pi* it can process *9861 msg/s* with the MQTT client as the data source. That means that in this particular setup the amount of data being processed is limited only by sending speed of the device with *MEMS accelerometer*, which is determined by communication protocol. It demonstrates that even devices with relatively constrained resources are able to significantly reduce amount of data processed in the cloud. *FogFlow* enables their seamless integration with whole IoT system infrastructure.

5 Summary and Future Work

In this paper we have presented *FogFlow* which allows for defining data processing applications able to be run with different technologies with the unified interface. We have described concept of modelling such applications and decomposing them onto given set of devices. At current state of work we have focused on designing *FogFlow* and developing end-to-end process for applications design and execution. Among the goals that are our current priority there is extending existing implementation of execution module with C-based one. We plan also to create toolkit of lambdas and data providers/sinks in order to facilitate *FogFlow* *API* usage. At the moment we are also working on developing more sophisticated ways for graph modifications and decomposition.

Acknowledgment. The research presented in this paper was supported by the National Centre for Research and Development (NCBiR) under Grant No. LID-ER/15/0144 /L-7/15/NCBR/2016.

References

1. Akyildiz, I.F., Sankarasubramaniam, Y., Cayirci, E.: A survey on sensor networks. IEEE Commun. Mag. **40**(8), 102–114 (2002). https://doi.org/10.1109/MCOM. 2002.1024422
2. Aloi, G., et al.: Enabling IoT interoperability through opportunistic smartphone-based mobile gateways. J. Netw. Comput. Appl. **81**, 74–84 (2017). https://doi. org/10.1016/j.jnca.2016.10.013.   http://www.sciencedirect.com/science/article/pii/S1084804516302405
3. Bellavista, P., Berrocal, J., Corradi, A., Das, S.K., Foschini, L., Zanni, A.: A survey on fog computing for the internet of things. Pervasive Mob. Comput. **52**, 71–99 (2019). https://doi.org/10.1016/j.pmcj.2018.12.007. http://www.sciencedirect. com/science/article/pii/S1574119218301111
4. Giang, N.K., Blackstock, M., Lea, R., Leung, V.C.M.: Developing IoT applications in the fog: a distributed dataflow approach. In: 2015 5th International Conference on the Internet of Things (IOT), pp. 155–162, October 2015. https://doi.org/10. 1109/IOT.2015.7356560
5. Grégoire, J.: A data flow architecture for smart city applications. In: 2018 21st Conference on Innovation in Clouds, Internet and Networks and Workshops (ICIN), pp. 1–5, February 2018. https://doi.org/10.1109/ICIN.2018.8401639
6. Gupta, H., Vahid Dastjerdi, A., Ghosh, S.K., Buyya, R.: iFogSim: a toolkit for modeling and simulation of resource management techniques in the internet of things, edge and fog computing environments. Softw.: Pract. Exp. **47**(9), 1275–1296 (2017). https://doi.org/10.1002/spe.2509

7. Kim-Hung, L., Datta, S.K., Bonnet, C., Hamon, F., Boudonne, A.: A scalable IoT framework to design logical data flow using virtual sensor. In: 2017 IEEE 13th International Conference on Wireless and Mobile Computing, Networking and Communications (WiMob), pp. 1–7, October 2017. https://doi.org/10.1109/WiMOB.2017.8115775

8. Mao, Y., You, C., Zhang, J., Huang, K., Letaief, K.B.: A survey on mobile edge computing: the communication perspective. IEEE Commun. Surv. Tutorials **19**(4), 2322–2358 (2017, Fourthquarter). https://doi.org/10.1109/COMST.2017.2745201

9. Milutinovic, V., Kotlar, M., Stojanovic, M., Dundic, I., Trifunovic, N., Babovic, Z.: DataFlow systems from their origins to future applications in data analytics, deep learning, and the internet of things. In: Milutinovic, V., Kotlar, M., Stojanovic, M., Dundic, I., Trifunovic, N., Babovic, Z. (eds.) DataFlow Supercomputing Essentials. CCN, pp. 127–148. Springer, Cham (2017). https://doi.org/10.1007/978-3-319-66125-4_5

10. Munir, A., Kansakar, P., Khan, S.U.: IFCIoT: integrated fog cloud IoT: a novel architectural paradigm for the future internet of things. IEEE Consum. Electron. Mag. **6**(3), 74–82 (2017). https://doi.org/10.1109/MCE.2017.2684981

11. Naas, M.I., Boukhobza, J., Parvedy, P.R., Lemarchand, L.: An extension to iFogSim to enable the design of data placement strategies. In: 2018 IEEE 2nd International Conference on Fog and Edge Computing (ICFEC), pp. 1–8, May 2018. https://doi.org/10.1109/CFEC.2018.8358724

12. Naas, M.I., Lemarchand, L., Boukhobza, J., Raipin, P.: A graph partitioning-based heuristic for runtime IoT data placement strategies in a fog infrastructure. In: Proceedings of the 33rd Annual ACM Symposium on Applied Computing, SAC 2018, pp. 767–774. ACM, New York (2018). https://doi.org/10.1145/3167132.3167217

13. Sena, A.C., Vaz, E.S., França, F.M.G., Marzulo, L.A.J., Alves, T.A.O.: Graph templates for dataflow programming. In: 2015 International Symposium on Computer Architecture and High Performance Computing Workshop (SBAC-PADW), pp. 91–96, October 2015. https://doi.org/10.1109/SBAC-PADW.2015.20

14. Szydlo, T., Brzoza-Woch, R., Sendorek, J., Windak, M., Gniady, C.: Flow-based programming for IoT leveraging fog computing. In: 2017 IEEE 26th International Conference on Enabling Technologies: Infrastructure for Collaborative Enterprises (WETICE), pp. 74–79, June 2017. https://doi.org/10.1109/WETICE.2017.17

15. Szydlo, T., Sendorek, J., Brzoza-Woch, R.: Enabling machine learning on resource constrained devices by source code generation of the learned models. In: Shi, Y., et al. (eds.) ICCS 2018. LNCS, vol. 10861, pp. 682–694. Springer, Cham (2018). https://doi.org/10.1007/978-3-319-93701-4_54

16. Teranishi, Y., Kimata, T., Yamanaka, H., Kawai, E., Harai, H.: Dynamic data flow processing in edge computing environments. In: 2017 IEEE 41st Annual Computer Software and Applications Conference (COMPSAC), vol. 1, pp. 935–944, July 2017. https://doi.org/10.1109/COMPSAC.2017.113

17. Türkay, S., Akçay, H.: A study of random vibration characteristics of the quarter-car model. J. Sound Vib. **282**(1), 111–124 (2005). https://doi.org/10.1016/j.jsv.2004.02.049. http://www.sciencedirect.com/science/article/pii/S0022460X04002974

18. Zanella, A., Bui, N., Castellani, A., Vangelista, L., Zorzi, M.: Internet of things for smart cities. IEEE Internet of Things J. **1**(1), 22–32 (2014). https://doi.org/10.1109/JIOT.2014.2306328

Research and Implementation
of an Aquaculture Monitoring System Based
on Flink, MongoDB and Kafka

Yuansheng Lou[1], Lin Chen[1], Feng Ye[1,2(✉)], Yong Chen[2,3],
and Zihao Liu[4]

[1] College of Computer and Information,
Hohai University, Nanjing 211100, China
yefeng1022@hhu.edu.cn
[2] Postdoctoral Centre, Nanjing Longyuan Micro-Electronic Company,
Nanjing 211106, China
[3] Huai'an Longyuan Agricultural Technology Company,
Huai'an 223345, China
[4] Jiangsu University of Science and Technology, Zhenjiang 212000, China

Abstract. With the rapid advancement of intelligent agriculture technology, the application of IoT and sensors in aquaculture domain is becoming more and more widespread. Traditional relational database management systems cannot store the large scale and diversified sensor data flexibly and expansively. Moreover, the sensor stream data usually requires a processing operation with high throughput and low latency. Based on Flink, MongoDB and Kafka, we propose and implement an aquaculture monitoring system. Among them, Flink provides a high throughput, low latency processing platform for sensor data. Kafka, as a distributed publish-subscribe message system, acquires different sensor data and builds reliable pipelines for transmitting real-time data between application programs. MongoDB is suitable for storing diversified sensor data. As a highly reliable and high-performance column database, HBase is often used in sensor data storage schemes. Therefore, using real aquaculture dataset, the execution efficiency of some common operations between HBase and our solution are tested and compared. The experimental results show that the efficiency of our solution is much higher than that of HBase, which provided a feasible solution for the sensor data storage and processing of aquaculture.

Keywords: Aquaculture · Flink · MongoDB · Kafka · Big data

1 Introduction

With the advancement of IoT (Internet of Things) technologies, the aquaculture industry is also facing a transition from the traditional rough pattern to the refined, intelligent pattern. The IoT has been widely used in all parts of the aquaculture production such as data collection, environmental monitoring, fish fry epidemic detection [1]. Large-scale business data such as ammonia nitrogen, temperature, dissolved oxygen and pH value, is produced and needs to be processed. When storing massive

© Springer Nature Switzerland AG 2019
J. M. F. Rodrigues et al. (Eds.): ICCS 2019, LNCS 11538, pp. 648–657, 2019.
https://doi.org/10.1007/978-3-030-22744-9_50

data, the capacity of vertically increasing data nodes is limited, and the performance bottleneck will be caused in the processing of massive data [2]. Therefore, the distributed cloud storage scheme that dynamically accesses new storage nodes through horizontal expansion is an ideal solution. In addition, the heterogeneous and non-structural trend of these IoT data is obvious, and using traditional database systems for storage is no longer appropriate. NoSQL stores support dynamic data model, which can deal with the variety, complexity and later expansion of data acquisition devices in the IoT [3].

On the other hand, the real-time requirement of sensor data processing in aquaculture is very high. Taking water temperature as an example, the water temperature will directly affect the growth and health of fish. Temperature sensors frequently acquire measurement data and transmit them to data centers in the form of streams. In data centers, real-time or near-real-time applications update the display board and issue warnings about changes in water temperature to avoid losses in aquaculture [4]. Therefore, based on Flink, MongoDB and Kafka, this paper proposed and implemented a high-throughput, low-latency architecture for processing real-time streaming data of the IoT of aquaculture.

2 Background and Related Work

2.1 Apache Flink

Many systems generate continuous stream of events. If we can efficiently analyze large-scale sensor stream data, we will have a clearer and faster understanding of the system. In this context, Apache Flink [5] came into being. Flink is an open source stream processing framework that supports distributed, high performance, ready-to-use, and accurate streaming applications. Flink not only provides real-time computing that supports both high throughput and exactly-once semantics, but also provides batch processing. Flink treats batch processing (that is, processing limited static data) as a special stream processing, so that batch and stream processing can be implemented simultaneously. Its core computational construct is the Flink runtime execution engine in Fig. 1, which is a distributed system that accepts data flow programs and performs fault-tolerant execution on one or more machines. The Flink Runtime Execution Engine can run as a YARN (Yet Another Resource Negotiator) application on a cluster, on a Mesos cluster, or on a standalone machine. Figure 1 is an architectural diagram of Flink.

Existing experiments show that by avoiding flow processing bottlenecks and utilizing Flink's stateful stream processing capability, the throughput can reach about 30 times of Strom. At the same time, the exactly-once and high availability can be guaranteed [6]. Therefore, Flink is chosen as the data processing platform in this paper.

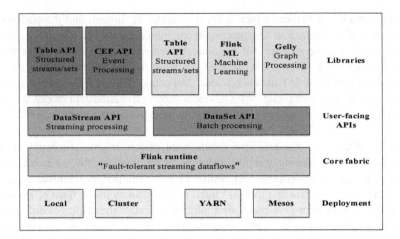

Fig. 1. The architecture diagram of Apache Flink

2.2 MongoDB

MongoDB [7] is a distributed, document-oriented, open source document database. MongoDB is the most similar to relational database among non-relational databases. It has rich functions and can even use many SQL statements for relational query, etc. Moreover, it has a variety of performance advantages of non-relational databases, such as convenient deployment, painless expansion and storage without mode. MongoDB supports a variety of common development languages, has a number of open source MongoDB frameworks, and simplifies data layer operations. Compared with the traditional MySQL database, MongoDB is much more efficient than MySQL in addition, deletion, selection and other operations [8]. In addition, MongoDB is a fast scalable database that can greatly increase the performance of Flink.

2.3 Kafka

Apache Kafka [9] is a distributed publish-subscribe messaging system. It can handle a large volume of data which enables you to send messages at end-point. Apache Kafka is developed at LinkedIn & available as an open source project with Apache Software Foundation. In this article, Kafka was chosen as the message broker. It supports Apache Flink very well. In addition, it provides precise primary semantics to ensure that each record is eventually delivered to its consumer, even in the event of a failure, and that no duplication is created in the process [10].

2.4 Related Work

At present, MapReduce programming model and Hadoop architecture are the popular processing technologies for big data in aquaculture, but Hadoop is a typical batch processing architecture for big data, and cannot meet the real-time requirements for streaming data processing [11–13]. In [14], by implementing Lambda architecture, the real-time computing platform is combined with the offline batch processing mechanism to make the agricultural big data processing framework have both the function of streaming data and the function of historical data mining. But this architecture requires two programming of the same business logic in two different APIs: one for batch computing and one for streaming computing. For the same business problem, there are two code bases, each with different vulnerabilities. Such systems are more difficult to maintain.

In terms of data storage, NoSQL is the mainstream storage scheme of IoT data. In [15], through testing and comparing the storage performance and data processing performance of MongoDB, HDFS and MySQL, experiments show that MongoDB has higher scalability and higher availability, and can store IoT data efficiently. Therefore, MongoDB is a feasible sensor data storage scheme. In [16], HBase, which is a distributed column-oriented database based on Hadoop file system, is used to store stream data. It can also be scaled horizontally like MongoDB. Therefore, this paper tested and compared the performance of HBase and MongoDB under the IoT data storage scenario.

In summary, a high throughput, low delay and easy maintenance architecture is urgently needed to deal with the stream data generated by sensors.

3 The Architecture of the Aquaculture Information Monitoring System and Implementation

3.1 Architecture

According to the characteristics of aquaculture IoT data and considering the key factors such as reliability, flexibility, scalability and load balancing, the system can be divided into four layers: data collection layer, data processing layer, business logic layer and application display layer. Figure 2 shows the whole architecture of the system.

In the data collection layer, Kafka is a pub-subscribe messaging system. We create different topics for different data stream. Data collection layer pushes JSON objects of sensor data to corresponding topics. Kafka's ZooKeeper can realize dynamic cluster expansion. Once the zookeeper changes, Kafka client can perceive and make corresponding adjustments in time. This ensures that when brokers are added or removed, they are still automatically load-balanced among themselves.

In the data processing layer, Flink receives and analyzes data in different topics, and analyzes infinite data streams by calling the DataStream API. The stream data processing program is implemented with the Java language. The processed data can flow to another message queue or be directly updated in the local database by the Flink program for historical query. To improve the fault tolerance, we adopt Flink's State management mechanism. State is the state of calculation when the stream data is saved from one event to the next event, and the calculation state can continue to accurately update the state after the failure or interruption. For storing massive amounts of sensor data, we use MongoDB for data persistence.

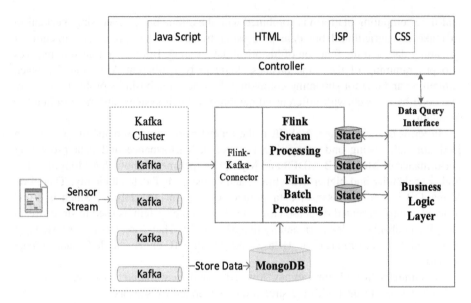

Fig. 2. The framework of aquaculture system

In the business logic layer, for the data that needs to be displayed in real time, this layer constantly obtains the stream data processed by Flink for real-time update. For historical queries, this layer provides various query interfaces for MongoDB database tables. Spring is used to manage object dependencies through dependency injection.

The application display layer uses JavaScript, HTML, CSS and JSP to implement the front-end interface. When the client makes a request, Tomcat receives the request and forwards the request to the DispatcherServlet for processing if it matches the mapping path configured by the DispatcherServlet in web.xml. The DispatcherServlet is used as the front-end controller for request distribution, and the controller is called by the browser request. Controller processes business requests and returns the corresponding view page.

3.2   System Implementation

At present, aquaculture information monitoring system adopts 30 million sensor data of an aquaculture farm to simulate stream data, and Flink is used to realize real-time update and historical query function of dashboard sensor, as shown in Fig. 3. The green part of the line chart indicates that the temperature is in normal value, and the red part indicates that the temperature is beyond normal value. By means of visualization [17], users can easily see the maximum, minimum and average temperature in different time periods.

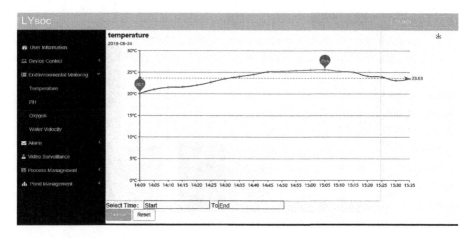

Fig. 3. Sensor real-time data display

4 Experiments and Discussion

In order to verify the performance of this storage scheme, the insertion and query performance of MongoDB and HBase were tested in different orders of magnitude. In order to verify the performance of the storage scheme in this paper, the insert and query time of MongoDB and HBase were tested under different orders of magnitude scenarios.

MongoDB and HBase clusters are deployed on four PCs of the same type. The PC's configuration is: Intel(R) Xeon(R) CPU E5645@2.40 GHz dual-core 24 CPU, Kingston DDR3 1333 MHz 8G, 500 GB SSD Flash Memory. Operating system are Ubuntu 16.04 64-bit and Linux 3.11.0 kernel. MongoDB version is 3.6.3 and HBase version is 1.2.6. The CPU utilization of each primary node is monitored by Ganglia.

4.1 Data Insertion Performance Verification

The experimental steps are as follows.

(1) 30 million pieces of sensor data are randomly selected from the monitoring system of the IoT, and then 3 groups of data are randomly selected from these data. The data volumes are 10,000 pieces, 100,000 pieces and 250,000 pieces of data respectively.

(2) The data is sent to the Topic through Kafka Producer. The MongoDB cluster with 4 nodes obtains data from Kafka Broker.

(3) Build a HBase cluster with 4 nodes under the same hardware configuration, and get data from the same Kafka Broker in (2).

(4) Record the insertion time and CPU utilization, and repeat the insertion five times to get the average.

The insert performance comparison between MongoDB and HBase is shown in Fig. 4.

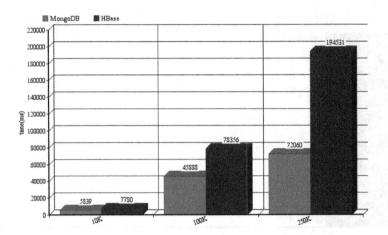

Fig. 4. Performance comparison of data insertion between MongoDB and HBase

It shows that when the data scale is small, the insertion performance of the MongoDB cluster and the HBase cluster is close; however, when the data volume is 100,000 pieces of data, MongoDB cluster data insertion performance is obviously better than that of HBase cluster; when the data volume is 250,000, the insert operation time of MongoDB cluster data is only 37.04% of that of HBase cluster. Ganglia recorded the CPU utilization in the process of data insertion, and the results are as follows.

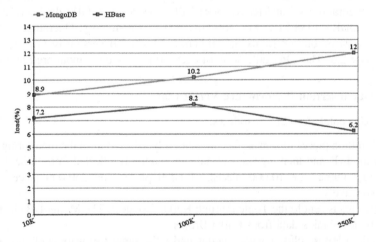

Fig. 5. CPU utilization comparison of data insertion

It can be seen that with the increase of data quantity, the CPU utilization of MongoDB slightly increase, but the change range is not large, and the change of CPU utilization of HBase is relatively stable (Fig. 5).

4.2 Data Query Performance

The test steps are as follows.

(1) Respectively, MongoDB cluster and HBase cluster are used to query 10,000, 100,000 and 250,000 pieces of data from 30 million pieces of data set in experiment 4.1.
(2) Record the insertion time and CPU utilization, and repeat the query operation five times to get the average.

The insert operation performance comparison between MongoDB and HBase is shown in Fig. 6.

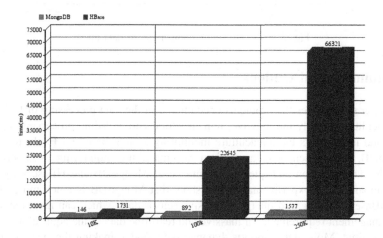

Fig. 6. Performance comparison of data query between MongoDB cluster and HBase cluster

When the data size is small, the data query performance of MongoDB cluster is close to that of HBase cluster, but always lower than that of HBase. With the increase of data volume, the query time of HBase increased greatly, while the growth of MongoDB is small and far smaller than that of HBase. When the query data volume is 250,000, the MongoDB cluster's data query time is only 2.38% of HBase cluster's. As a result, when the amount of data is large, the query speed of MongoDB is faster.

Ganglia recorded the CPU utilization in the process of data query, and the results are shown in Fig. 7.

With the increase of queries, the CPU utilization of MongoDB and HBase increase slowly, and the CPU utilization of MongoDB is significantly higher than that of HBase.

Through the above two comparative experiments, we can see that when the sensor data volume is large, the insertion and query efficiency of MongoDB cluster is much higher than that of HBase. Although the CPU utilization is slightly higher than that of HBase, it is basically between 10% and 20%. Therefore, in the case of high throughput and high concurrency, it is feasible to use MongoDB to store sensor data.

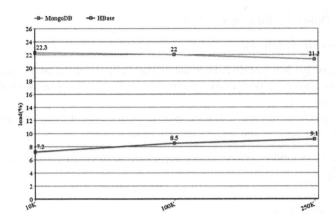

Fig. 7. CPU utilization comparison of data query

5 Summary and Outlook

This paper summarizes the problems encountered in the data processing of IoT sensor in aquaculture, namely the massive, non-structural and real-time processing, and proposes and implements the aquaculture monitoring system based on Flink, Kafka and MongoDB. Taking sensor data acquired in aquaculture monitoring as experimental data, the data insertion and query performance of MongoDB and HBase were tested and compared. The experimental results show that MongoDB is much more efficient than HBase in storing and querying massive sensor data, which can meet the storage and management requirements of massive and real-time data of the aquaculture IoT. At the same time, MongoDB supports dynamic data model, making the system easier to expand.

The subsequent research will focus on Flink, which combines distributed computing and machine learning technology to calculate the real-time data of the IoT, providing real-time disaster warning and decision-making services for aquaculture.

Acknowledgement. This work is partly supported by the 2018 Jiangsu Province Key Research and Development Program (Modern Agriculture) Project under Grant No. BE2018301, 2017 Jiangsu Province Postdoctoral Research Funding Project under Grant No. 1701020C, 2017 Six Talent Peaks Endorsement Project of Jiangsu under Grant No. XYDXX-078.

References

1. Shetty, S., Pai, R.M., Pai, M.M.M.: Energy efficient message priority based routing protocol for aquaculture applications using underwater sensor network. Wirel. Pers. Commun. **103**(2), 1871–1894 (2018)
2. Xu, X., Shi, L., He, L., Zhang, H., Ma, X.: Design and implementation of cloud storage system for farmland internet of things based on NoSQL database. Trans. CSAE **35**(1), 172–179 (2019)
3. Edward, S.G., Sabharwal, N.: Practical MongoDB. Apress, Berkeley (2015)

4. Liu, S., Xu, L., Chen, J., Li, D., Tai, H., Zeng, L.: Retracted: water temperature forecasting in sea cucumber aquaculture ponds by RBF neural network model. In: Li, D., Chen, Y. (eds.) CCTA 2012. IAICT, vol. 392, pp. 425–436. Springer, Heidelberg (2013). https://doi.org/10.1007/978-3-642-36124-1_51

5. Tanmay, D.: Learning Apache Flink. Packt Publishing, Birmingham (2017)

6. Friedman, E., Tzoumas, K.: Introduction to Apache Flink: Stream Processing for Real Time and Beyond. O'Reilly Media, Sebastopol (2016)

7. Chodorow, K.: MongoDB the Definitive Guide. O'Reilly Media, Sebastopol (2013)

8. Győrödi, C., Győrödi, R., Pecherle, G., Olah, A.: A comparative study: MongoDB vs. MySQL. In: 13th International Conference on Engineering of Modern Electric Systems (EMES), pp. 1–6. IEEE (2015)

9. Narkhede, N., Shapira, G., Palino, T.: Kafka the Definitive Guide. O'Reilly Media, Sebastopol (2013)

10. Versaci, F., Pireddu, L., Zanetti, G.: Kafka interfaces for composable streaming genomics pipelines. In: IEEE EMBS International Conference on Biomedical & Health Informatics (BHI), pp. 259–262. IEEE (2018)

11. Wang, D., Zheng, J., Wang, D., Liu, Y.: Primary research of fishery big data and application technology in China. Shangdong Agric. Sci. 48(10), 152–156 (2016)

12. Yu, Z.: Review of fishery big data. J. Anhui Agric. Sci. 45(9), 211–213 (2017)

13. Li, D., Yang, H.: State-of-the-art review for internet of things in agriculture. Trans. Chin. Soc. Agric. Mach. 49(1), 1–20 (2018)

14. Duan, Q., Liu, Y., Zhang, L., Li, D.: State-of-the-art review for application of big data technology in aquaculture. Trans. Chin. Soc. Agric. Mach. 49(06), 1–16 (2018)

15. Yang, P., Lin, J.: A scheme for massive unstructured iot data processing based on MongoDB and Hadoop. Microelectron. Comput. 35(04), 68–72 (2018)

16. Wang, Y., Chiang, Y., Wu, C., Yang, C., Chen, S., Sun, P.: The implementation of sensor data access cloud service on HBase for intelligent indoor environmental monitoring. In: 15th International Symposium on Parallel and Distributed Computing (ISPDC), pp. 234–239. IEEE (2016)

17. ECharts. https://echarts.baidu.com. Accessed 1 Feb 2019

Enhanced Hydroponic Agriculture Environmental Monitoring: An Internet of Things Approach

Gonçalo Marques[✉] [iD], Diogo Aleixo, and Rui Pitarma

Unit for Inland Development, Polytechnic Institute of Guarda,
Av. Dr. Francisco Sá Carneiro, n° 50, 6300-559 Guarda, Portugal
goncalosantosmarques@gmail.com, aleixo.da@gmail.com,
rpitarma@ipg.pt

Abstract. Hydroponic cultivation is an agricultural method where nutrients are efficiently provided as mineral nutrient solutions. This modern agriculture sector provides numerous advantages such as efficient location and space requirements, adequate climate control, water-saving and controlled nutrients usage. The Internet of things (IoT) concept assumes that various "things," which include not only communication devices but also every other physical object on the planet, are going to be connected and will be controlled across the Internet. Mobile computing technologies in general and mobile applications, in particular, can be assumed as significant methodologies to handle data analytics and data visualisation. Using IoT and mobile computing is possible to develop automatic systems for enhanced hydroponic agriculture environmental monitoring. Therefore, this paper presents an IoT monitoring system for hydroponics named *iHydroIoT*. The solution is composed of a prototype for data collection and an iOS mobile application for data consulting and real-time analytics. The collected data is stored using Plotly, a data analytics and visualisation library. The proposed system provides not only temporal changes monitoring of light, temperature, humidity, CO_2, pH and electroconductivity but also water level for enhanced hydroponic supervision solutions. The *iHydroIoT* offers real-time notifications to alert the hydroponic farm manager when the conditions are not favourable. Therefore, the system is a valuable tool for hydroponics condition analytics and to support decision making on possible intervention to increase productivity. The results reveal that the system can generate a viable hydroponics appraisal, allowing to anticipate technical interventions that improve agricultural productivity.

Keywords: Hydroponics · Internet of things · Mobile computing · Smart agriculture · Water quality monitoring

1 Introduction

Hydroponic cultivation is an agricultural method where nutrients are efficiently provided as mineral nutrient solutions with several advantages such as pest problems reduction, continuous feeding of nutrients when compared to traditional agriculture

© Springer Nature Switzerland AG 2019
J. M. F. Rodrigues et al. (Eds.): ICCS 2019, LNCS 11538, pp. 658–669, 2019.
https://doi.org/10.1007/978-3-030-22744-9_51

methods [1]. However, this technique is expensive due to the energy investment compared to conventional soil agriculture [2]. Hydroponic agriculture can produce more food at a lower cost but also have several hidden expenses [3]. Hydroponic systems can be assumed as significant tools for agricultural environments such as plant factories with artificial lighting [4].

Nutrient composition and pH levels are important factors to maintain for enhanced plant factory environments [5]. Therefore, the pH and electrical conductivity (EC) supervision for nutrient status evaluation are common practices in hydroponic solutions used in greenhouse plant production.

On the one hand, hydroponic systems provide a significant opportunity for water saving in agricultural environments as it offers enhanced water efficiency by recycling the irrigation excess water. On the other hand, irrigation water recycling can be assumed as an achievable and significant practice in greenhouses environments which are growing in Mediterranean countries [6].

Hydroponics allows the usage of unsuitable areas for traditional agriculture which is the case of sterile and degraded soil areas. However, the hydroponics systems installation process is expensive and time-consuming [7].

Urban agriculture not only improves the self-sufficiency and resiliency of cities but also provide positive environmental and social advantages [8]. Hydroponic agriculture can be more easily applied to urban environments than open field traditional agriculture.

The basic idea of the Internet of Things (IoT) is the pervasive presence of a variety of objects with interaction and cooperation capabilities among them to reach a common objective [9, 10]. Is expected that the IoT will have a high impact on several aspects of everyday life and this concept will be used in several applications such as domotics, assisted living, e-health and is also an ideal emerging technology to provide new evolving data and computational resources for creating revolutionary software applications [11]. IoT can be assumed as a significant architecture for enhanced agricultural systems. Several IoT applications for environmental supervision using open-source and mobile computing technologies are reported in the literature [12–17]. An intelligent IoT based hydroponic system that incorporates pH, temperature, humidity, level, lighting sensing features and use deep neural network towards providing the appropriate control with an accuracy of 88% is proposed by the authors of [18]. The Titan Smartponics is a low-cost and fully automated hydroponic system that incorporate Arduinos, a Raspberry Pi, open source software for remote monitoring and control [19]. An IoT Hydroponic Farming Ecosystem (HFE) is easy to control and easy to use humidity, nutrient solution temperature, air temperature, pH and EC monitoring system to support non-professional farmers proposed by the authors of [20].

In this document, *iHydroIoT*, an IoT monitoring system for hydroponics is presented. The collected data is stored using Plotly, a data analytics and visualisation library. The solution is composed of a prototype for data collection and an iOS mobile app for data consulting.

The rest of the paper is structured as follows: Sect. 2 is concerned to the methods and materials used in the implementation of the *iHydroIoT* solution; Sect. 3 demonstrates the system operation and experimental results, and the conclusion is presented in Sect. 4.

2 Materials and Methods

The *iHydroIoT* has been developed to be a cost-effective solution that can be easily used by everyone. This solution uses an Arduino Uno as microcontroller unit, a Bluetooth Low Energy (BLE) module as a communication unit and several sensors. The data is collected using the *iHydroIoT* that is connected by serial interface with a Raspberry Pi 2 which is used as a server. The Raspberry Pi is connected to the Internet via Ethernet interface and is responsible for sending the data collected to Plotly service. Plotly is a Python framework for data analytics that can be used as a data visualisation tool for IoT. This framework is an open source JavaScript library for creating graphs and dashboards (Fig. 1).

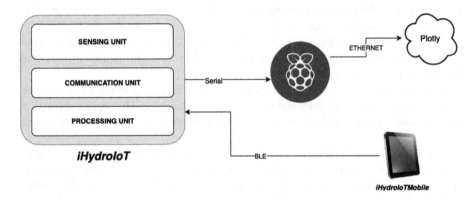

Fig. 1. *iHydroIoT* architecture.

The developed prototype use Arduino as a microcontroller to handle several sensor data such as light, temperature, humidity, CO_2, pH, EC sensors and water level sensors (Fig. 2). The selection of the sensors was made focusing on the cost of the system since the main objective was to test the functional architecture of the solution. Considering that the system is projected to be used in indoor environments where energy is easily accessible, there was no great concern regarding the selection of energy efficient sensors. Other sensors can be added for monitoring specific hydropic parameters. The BLE module is used for wireless data communication with mobile devices.

A brief introduction of the components used in the prototype is shown below:

- **Arduino UNO R3**: is a microcontroller board based on the ATmega328P, support 14 digital input/output pins (6 - PWM), 6 analog inputs, a 16 MHz quartz crystal. The board can be programmable with the Arduino IDE software via USB. The operating voltage is 5 V, the recommended input voltage is 7–12 V, and the limits for input Voltage is 6–20 V.
- **Adafruit NRF8001**: is a BLE module developed by Adafruit that offers wireless communication between an Arduino microcontroller and a compatible iOS or Android (4.3+) device using UART interface.

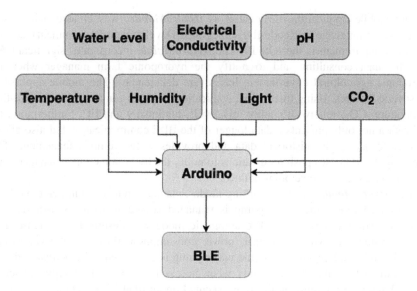

Fig. 2. *iHydroIoT* component diagram.

- **LDR**: is a light-controlled variable resistor that can be assumed for light supervision as the resistance of the sensor decreases when light intensity increase.
- **LM19**: is an analog temperature sensor that operates over a −55 °C to 130 °C temperature range. The temperature error increases linearly and reaches a maximum of ±3.8 °C at the temperature range extremes.
- **HIH-4000**: is a humidity sensor developed by Honeywell. The HIH-4000 support linear voltage output and is suitable for low drain, battery operated systems with a typical current draw of only 200 µA. This sensor has a 3.5% accuracy range and can be operated at a temperature range of −40 to 85 °C.
- **DFRobot CO$_2$ Sensor (MG-811)**: is a CO$_2$ industrial quality module developed by DFRobot. It has MG-811 gas sensor onboard which is not only highly sensitive to CO$_2$ and less sensitive to alcohol and CO but also have low humidity and temperature dependency.
- **DFRobot EC sensor**: is an analog EC meter of the aqueous solution to evaluate the water quality. This module detection range is 0–20 ms/cm, the recommended detection range is 1–15 ms/cm, and the temperature range is 0–40 °C.
- **DFRobot pH sensor**: is an analog 5 V pH sensor, with a measuring range of 0–14 pH with a 0.1 pH accuracy. This sensor can be operated within a temperature range of 0–60 °C and his response time is less than one minute.
- **DFRobot liquid level sensor**: is a photoelectric liquid level sensor that is operates using optical principles without mechanical parts. This sensor's output current is 12 mA, and it can be operated in a temperature range of −25–105 °C. The low-level output is <0.1 and the high-level output is >4.6 V. The sensor's accuracy is ±0.5 mm.
- **DFRobot Water pump**: is an immiscible pump with 4.5–12 V DC power supply, 100–350L/H of capacity and a power range of 0.5–5 W.

The mobile application, designated as *iHydroMobile*, was created using Swift programming language in Xcode IDE (Integrated Development Environment), and the minimum requirement is the iOS 10. The i*HydroMobile* incorporates significant features for data consulting and to notify the hydroponic farm manager when the hydroponic agricultural environment has severe deficiencies. The mobile application has several features. Using this mobile application, the user can access to the real-time temperature, CO_2, relative humidity, pH levels, luminosity and EC. Furthermore, the application not only presents a data logger of the BLE communication but also allows the user to access the historical data and provides water pump management. This mobile application uses push notifications to notify the hydroponic farm manager when a parameter exceeds the setpoint defined.

The i*HydroMobile* was two working modes, manual mode and automatic mode. In the automatic mode, the water pump is managed according to the sensor data for enhanced hydropic agriculture. The automatic mode was designed to increase agriculture productivity while minimises power consumption. Based on the values collected through the built-in sensors the water pump is controlled to be activated during the minimum time possible and only when necessary to promote energy efficiency. Figure 3 shows the mobile application executed on an iPad 3rd generation.

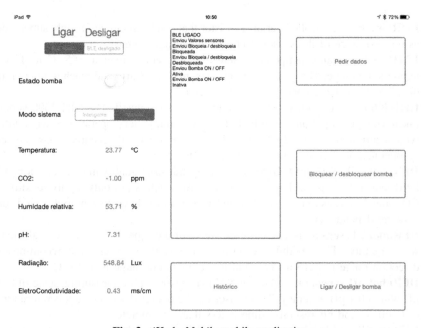

Fig. 3. *iHydroMobile* mobile application.

BLE is a low-power wireless technology for single-hop communication which can be assumed as a significant approach for IoT [21]. Zigbee and BLE were developed for battery-powered. However, BLE outperforms ZigBee in terms of power consumption [22]. BLE uses the 2.4 GHz ISM frequency band with adaptive frequency hopping to reduce interference and includes 24 bit CRC and AES 128 bit encryption technique on all packets to guarantee robustness and authentication [23]. Therefore, this technology was selected to provide data communication between the mobile application and the *iHydroIoT* prototype.

3   Discussion and Results

For testing purposes, the *iHydroIoT* was mounted inside a 21-liter aquarium with the following a dimension: 42 cm in length x 21 in width x 28 in height inside a laboratory of a Portuguese University (Fig. 4). The module is powered using a 230 V-5 V AC-DC 2A power supply. The tests show that under certain conditions the water and environmental quality parameters can be significantly lower than those considered as the high-quality conditions for hydroponics.

Fig. 4. *iHydroIoT* system prototype.

The *iHydroIoT* allows data consulting as graphical or numerical values using the mobile application or Plotly Web framework. A sample of the data collected by the system is shown in Figs. 5, 6 and 7. Figure 5 represents EC in milliSiemens per centimetre (mS/cm), Fig. 6 represents pH levels, and Fig. 7 represents luminosity levels in lux.

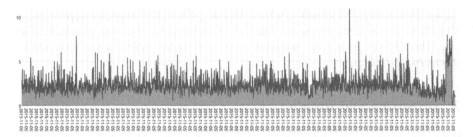

Fig. 5. EC (mS/cm) - data collected in the tests performed.

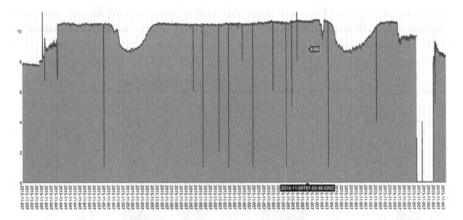

Fig. 6. pH - data collected in the tests performed.

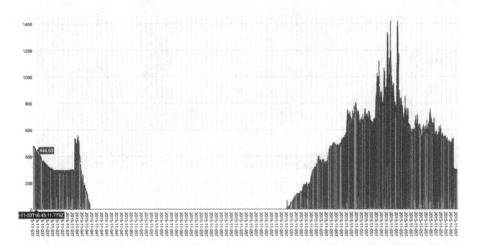

Fig. 7. Luminosity (lux) - data collected in the tests performed

The graphics displaying the hydroponic conditions provide a better perception of the monitored parameters behaviour than the numerical format. On the one hand, the mobile application also provides easy and quick access to collected data and enables a more precise analysis of parameters temporal evolution. Thus, the system is a powerful tool for hydroponics supervision and to support decision making on possible interventions to increase productivity. On the other hand, the proposed IoT approach provide temporal hydroponics data for visualisation and analytics which are particularly relevant to detect unfertile situations and plan interventions to promote a productive hydroponic environment.

The *iHydroIoT* data consulting features are perfectly suited to inspect the historical evolution of the hydroponic parameters and record insalubrious situations for further analysis. Furthermore, the effective productivity results can be compared with the monitored data which is particularly valuable for a correct evaluation and study of the cultivation methods used.

The proposed solution support push notifications feature (Fig. 8) aiming to provide the hydroponic farm manager with timely information and offer them the ability to react in real time to significantly improve hydroponic productivity. Based on values from literature, the maximum and minimum values are predefined by the system, but the user can also change these values for specific purposes. When the parameters exceed the maximum value, the user is alerted to take actions for enhanced hydroponics agriculture. The mobile application stores the history of notifications based on their severity.

Fig. 8. *iHydroMobile* push notification example

The results are very promising as the push notification feature allow the hydroponic farm manager to take action in time for enhanced hydroponics agriculture environments which will improve the productivity of the agricultural plant.

Using *iHydroMobile* is possible to supervise and identify poor water quality conditions and analyse the historical data to discuss possible interventions to improve the hydroponics parameters. The authors believe that the first step is to create reliable quality supervision systems through IoT for enhanced hydroponic agriculture.

Another important advantage of the proposed solution is the scalability associated with the modularity of the system. An installation can start using one station, and new modules can be added, over time, according to the needs of the plant.

Abundant scientific research on infrared thermal imaging (thermography) applied to hydroponics can be found in the literature. Infrared thermography provides an efficient and effective noninvasive method to supervise the surface temperature distribution of a plant for enhanced hydroponics [24]. Infrared thermography (IRT) technology can be assumed as an important method for the precise supervision of plant diseases [25]. Thermography provides temperature visualisation over the leaf surface of plants and the leaf temperature variation shows concerned dehydration, caused by diseases [26]. Therefore, this imaging method should be considered for enhanced hydroponics. In the context of the studies being carried out on the IRT technology applied to the monitoring of trees (Fig. 9), the authors plan to correlate the proposed system results with thermography imaging methods in order to understand how these technologies can be connected for enhanced hydroponics.

Fig. 9. Laboratory tests on IRT applied to plants and trees.

As future work, the main goal is to make technical improvements, including the development of important notifications interfaces such as SMS or e-mail to notify the hydroponic farm manager when a specific parameter reports poor hydroponic conditions. The authors also plan to develop an integrated management Web portal for water quality analytics. This portal should allow the hydroponic farm manager to visualise enhanced dashboards about the water quality metrics for enhanced hydroponic agriculture. In addition, hardware and software improvements have also been planned to connect the developed prototype directly to the Internet using Wi-Fi communication technology. Furthermore, the data collected by the system should be stored using a database management system in order to provide enhanced data visualisations and analysis.

4 Conclusions

In this paper, an IoT light, temperature, humidity, CO_2, pH, EC and water level monitoring system for hydroponics named *iHydroIoT* was proposed. This solution offers a feasibility method to monitor hydroponic agriculture systems in real-time and to guarantee the correct conditions for enhanced productivity.

With the proliferation of IoT and mobile computing technologies, there is a significant perspective to design automatic monitoring systems to improve productivity and quality of hydroponic agriculture.

The results achieved in the study conducted indicate a significant contribution to hydroponic supervision solutions based on IoT and open-source technologies. Using *iHydroIoT*, the collected data can be especially valued to investigate and store the temporal changes of the monitored conditions in order to guarantee that they are established in the course of all the agricultural process.

Compared to existing systems, the *iHydroIoT* supports notifications to alert the hydroponic farm manager in a timely way for enhanced productivity; also, by supporting the history of the environment and water conditions, this system can be used to anticipate technical interventions that avoid unfavourable agricultural conditions.

This solution provides flexibility and expandability as the user can start with only one *iHydroIoT* unity and add more unities if needed regarding the dimension of the greenhouse. Nevertheless, the proposed solution has some limitations. The *iHydroIoT* needs further experimental validation to improve system calibration and accuracy. In addition, quality assurance (QA) and quality control (QC) have also been planned for enhanced product quality traceability. The authors have also planned software and hardware improvements to adapt the system to specific laboratory tests using thermographic experiments applied to wood and trees for enhanced hydroponics.

Acknowledgement. This research is framed in the project "TreeM – Advanced Monitoring & Maintenance of Trees" N. ° 023831, 02/SAICT/2016, co-financed by CENTRO 2020 and FCT, Portugal 2020 and structural funds UE-FEDER.

References

1. Jung, D.-H., Kim, H.-J., Cho, W.-J., Park, S.H., Yang, S.-H.: Validation testing of an ion-specific sensing and control system for precision hydroponic macronutrient management. Comput. Electron. Agric. **156**, 660–668 (2019)
2. Gomiero, T.: Food quality assessment in organic vs. conventional agricultural produce: findings and issues. Appl. Soil. Ecol. **123**, 714–728 (2018)

3. Zanella, A., et al.: Humusica 2, article 17: techno humus systems and global change – three crucial questions. Appl. Soil. Ecol. **122**, 237–253 (2018)
4. Son, J.E., Kim, H.J., Ahn, T.I.: Hydroponic systems. In: Plant Factory, pp. 213–221. Elsevier, Amsterdam (2016)
5. Wada, T.: Theory and technology to control the nutrient solution of hydroponics. In: Plant Factory Using Artificial Light, pp. 5–14. Elsevier (2019)
6. Katsoulas, N., Savvas, D., Kitta, E., Bartzanas, T., Kittas, C.: Extension and evaluation of a model for automatic drainage solution management in tomato crops grown in semi-closed hydroponic systems. Comput. Electron. Agric. **113**, 61–71 (2015)
7. Domingues, D.S., Takahashi, H.W., Camara, C.A.P., Nixdorf, S.L.: Automated system developed to control pH and concentration of nutrient solution evaluated in hydroponic lettuce production. Comput. Electron. Agric. **84**, 53–61 (2012)
8. Romeo, D., Vea, E.B., Thomsen, M.: Environmental impacts of urban hydroponics in europe: a case study in Lyon. Procedia CIRP **69**, 540–545 (2018)
9. Giusto, D. (ed.): The Internet of Things: 20th Tyrrhenian Workshop on Digital Communications. Springer, New York (2010). https://doi.org/10.1007/978-1-4419-1674-7
10. Marques, G.: Ambient assisted living and internet of things. In: Cardoso, P.J.S., Monteiro, J., Semião, J., Rodrigues, J.M.F. (eds.) Harnessing the Internet of Everything (IoE) for Accelerated Innovation Opportunities, pp. 100–115. IGI Global, Hershey (2019)
11. Gubbi, J., Buyya, R., Marusic, S., Palaniswami, M.: Internet of Things (IoT): a vision, architectural elements, and future directions. Futur. Gener. Comput. Syst. **29**(7), 1645–1660 (2013)
12. Marques, G., Pitarma, R.: Agricultural environment monitoring system using wireless sensor networks and IoT. In: 2018 13th Iberian Conference on Information Systems and Technologies (CISTI), Caceres, pp. 1–6 (2018)
13. Pitarma, R., Marques, G., Ferreira, B.R.: Monitoring indoor air quality for enhanced occupational health. J. Med. Syst. **41**(2), 23 (2017)
14. Marques, G., Ferreira, C.R., Pitarma, R.: Indoor air quality assessment using a CO_2 monitoring system based on internet of things. J. Med. Syst. **43**(3), 63 (2019)
15. Marques, G., Roque Ferreira, C., Pitarma, R.: A system based on the internet of things for real-time particle monitoring in buildings. Int. J. Environ. Res. Public Health **15**(4), 821 (2018)
16. Marques, G., Pitarma, R.: An internet of things-based environmental quality management system to supervise the indoor laboratory conditions. Appl. Sci. **9**(3), 438 (2019)
17. Marques, G., Pitarma, R.: A cost-effective air quality supervision solution for enhanced living environments through the internet of things. Electronics **8**(2), 170 (2019)
18. Mehra, M., Saxena, S., Sankaranarayanan, S., Tom, R.J., Veeramanikandan, M.: IoT based hydroponics system using deep neural networks. Comput. Electron. Agric. **155**, 473–486 (2018)
19. Palande, V., Zaheer, A., George, K.: Fully automated hydroponic system for indoor plant growth. Procedia Comput. Sci. **129**, 482–488 (2018)
20. Ruengittinun, S., Phongsamsuan, S., Sureeratanakorn, P.: Applied internet of thing for smart hydroponic farming ecosystem (HFE). In: 2017 10th International Conference on Ubi-media Computing and Workshops (Ubi-Media), Pattaya, Thailand, pp. 1–4 (2017)
21. Gomez, C., Oller, J., Paradells, J.: Overview and evaluation of bluetooth low energy: an emerging low-power wireless technology. Sensors **12**(9), 11734–11753 (2012)
22. Jawad, H., Nordin, R., Gharghan, S., Jawad, A., Ismail, M.: Energy-efficient wireless sensor networks for precision agriculture: a review. Sensors **17**(8), 1781 (2017)
23. Ojha, T., Misra, S., Raghuwanshi, N.S.: Wireless sensor networks for agriculture: the state-of-the-art in practice and future challenges. Comput. Electron. Agric. **118**, 66–84 (2015)

24. James, R.A., Sirault, X.R.R.: Infrared thermography in plant phenotyping for salinity tolerance. In: Shabala, S., Cuin, T.A. (eds.) Plant Salt Tolerance, pp. 173–189. Humana Press, Totowa (2012). https://doi.org/10.1007/978-1-61779-986-0_11
25. Mahlein, A.-K., Oerke, E.-C., Steiner, U., Dehne, H.-W.: Recent advances in sensing plant diseases for precision crop protection. Eur. J. Plant Pathol. **133**(1), 197–209 (2012)
26. Chaerle, L., Hagenbeek, D., Vanrobaeys, X., Van Der Straeten, D.: Early detection of nutrient and biotic stress in Phaseolus vulgaris. Int. J. Remote Sens. **28**(16), 3479–3492 (2007)

Noise Mapping Through Mobile Crowdsourcing for Enhanced Living Environments

Gonçalo Marques$^{(\boxtimes)}$ ⓘ and Rui Pitarma

Unit for Inland Development, Polytechnic Institute of Guarda,
Av. Dr. Francisco Sá Carneiro, nº 50, 6300-559 Guarda, Portugal
goncalosantosmarques@gmail.com, rpitarma@ipg.pt

Abstract. Environmental noise pollution has a significant impact on health. The noise effects on health are related to annoyance, sleep and cognitive performance for both adults and children are reported in the literature. The smart city concept can be assumed as a strategy to mitigate the problems generated by the urban population growth and rapid urbanisation. Noise mapping is an important step for noise pollution reduction. Although, noise maps are particularly time-consuming and costly to create as they are produced with standard methodologies and are based on specific sources such as road traffic, railway traffic, aircraft and industrial. Therefore, the actual noise maps are significantly imperfect because the noise emission models and sources are extremely limited. Smartphones have incredible processing capabilities as well as several powerful sensors such as microphone and GPS. Using the resources present in a smartphone as long with participatory sensing, a crowdsourcing noise mobile application can be used to provide environmental noise supervision for enhanced living environments. Crowdsourcing techniques applied to environmental noise monitoring allow creating reliable noise maps at low-cost. This paper presents a mobile crowdsourcing solution for environmental noise monitoring named *iNoiseMapping*. The environmental noise data is collected through participatory sensing and stored for further analysis. The results obtained can ensure that mobile crowdsourcing offers several enhanced features for environmental noise supervision and analytics. Consequently, this mobile application is a significant decision-making tool to plan interventions for noise pollution reduction.

Keywords: Enhanced living environments · Environmental monitoring · Mobile crowdsourcing · Participatory sensing · Smart city · Smartphones

1 Introduction

The 'environmental noise' can be seen as an unwanted sound produced by human activities that are considered harmful or detrimental to human health and quality of life, while 'noise' was identified as being sound that is 'out of place' or as a form of acoustic pollution as much as carbon oxide (CO) is for air pollution [1].

The concept of the "smart city" has recently been introduced as a strategic device to encompass modern urban production factors in a common framework and, in

© Springer Nature Switzerland AG 2019
J. M. F. Rodrigues et al. (Eds.): ICCS 2019, LNCS 11538, pp. 670–679, 2019.
https://doi.org/10.1007/978-3-030-22744-9_52

particular, to highlight the importance of Information and Communication Technologies (ICTs) in the last 20 years for enhancing the competitive profile of a city as proposed by the authors of [2]. Nowadays cities face interesting challenges and problems to meet socio-economic development and quality of life objectives and the concept of "smart cities" corresponds to answer to these challenges [3]. The smart city is directly related to an emerging strategy to mitigate the problems generated by the urban population growth and rapid urbanisation [4]. The smart city implementation will cause impacts at distinct levels such as impacts on science, technology, competitiveness and on society but also will cause ethical issues as the smart city need to provide correct information access as it becomes crucial when such information is available at a fine spatial scale where individuals can be identified [5].

Noise maps are created with standard methodologies and are based on specific sources such as road traffic, railway traffic, aircraft and industrial. The CNOSSOS-EU was created by European Commission for noise mapping and is a common framework for noise evaluation not only to provide reliable and comparable data on the noise levels personal exposure and the related health implications but also to plan interventions to reduce the exposure to unhealthy noise levels [6]. However, this approach is not reliable because the noise emission models and sources are extremely limited [7].

Several countries such as Germany and the United Kingdom have produced noise maps in the recent past based on statistics. Truth be told, noise evaluation is not an easy task. Environmental noise can be extremely dissimilar on the same location at different hours of the day, but also can vary relatively over small distances at the same space of time. Therefore, noise mapping is a difficult and time-consuming task.

The name crowdsourcing is formed from two words: crowd, referring to the people who participate in the initiatives; and sourcing, which refers to some procurement practices aimed at finding, evaluating and engaging suppliers of goods and services [8]. Crowdsourcing is a concept where the information is collected by a large group of people who submit their data via the Internet, social media and mobile applications. These people can be paid or be done as voluntary work. Crowdsourcing allows not only to avoid costs but also speed and provide the ability to work with people with different skills. On the one hand, this methodology enables building a database of data where people can add new information that in another way was impossible. On the other hand, crowdsourcing uses mobile applications to provide rapid and flexible data collection to everyone as anyone who has the smartphone application can contribute.

Right now, smartphones have incredible processing capabilities as well as a set of very interesting sensors for the study of assisted living, such as GPS, accelerometer, gyroscope, proximity sensor, camera, microphone, NFC and BLE. Smartphones have mobile sensors that turn possible to make activity recognition and detect physical activities such as walking, running, climbing stairs, descending stairs, driving, cycling and being inactive with no additional sensing hardware [9, 10]. The smartphone has a key role in building smart communications architectures for ambient assisted living to know what is happening in the network and detect if older adults need assistance as proposed by the authors of [11]. Therefore, smartphones not only are directly involved in several important applications as they incorporate large potential particularly about sensors, communication technologies and processing power but also face an important adoption for crowdsourcing applications for noise supervision [12].

Due to the well-studied health effects of environmental noise and combining the resources present in a smartphone as long with participatory sensing, a noise mobile application is proposed to provide environmental noise supervision for enhanced living environments. The main objective of the presented application is applying crowd-sourcing techniques to environmental monitoring in order to encourage users to contribute to the data collection process by using their personal smartphone.

In this document, *iNoiseMapping*, a mobile crowdsourcing application for noise pollution is presented. The collected data is stored in a structured database. The solution is composed of an iOS mobile application for environmental noise data collection and data consulting and MySQL database for data storage.

The rest of this paper is structured as follows: Sect. 2 presents the environmental noise health impacts related work; Sect. 3 is concerned to the methods and materials used in the implementation of the *iNoiseMapping* solution; Sect. 4 demonstrates the system operation and experimental results, and the conclusion is presented in Sect. 5.

2   Environmental Noise and Health

The noise effects on health are not only related to annoyance, sleep and cognitive performance for both adults and children but can also be associated with raised blood pressure [13].

Environmental noise pollution may be a novel risk factor for pregnancy-related hypertension, particularly more severe variants of preeclampsia [14]. Long-term exposure to railway and road noise, especially at night, may affect arterial stiffness, a major determinant of cardiovascular disease. Therefore the noise monitoring can be significant to the enhanced understanding of noise-related health symptoms and effects [15]. Poor sleep causes endocrine and metabolic disturbances, several cardio-metabolic and psychiatric problems and anti-social behaviour, both in children and adults. The duration and quality of sleep are risk factors significantly affected by the environment but amenable to modification through awareness, counselling and measures of public health [16].

Pregnancy and childhood exposure to road traffic noise can be associated with a higher risk for childhood overweight has been concluded by the authors of [17]. The World Health Organization (WHO) has recently acknowledged that contrary to the trend for other environmental stressors, noise exposure is increasing in Europe [18]. Therefore, the majority of the developed countries support laws noise regulation at specific hours [19]. Environmental noise must be assumed as a serious public health issue throughout the world. Overall, the evidence suggests that environmental noise should be placed at the forefront of national and international health policies to prevent unnecessary adverse health impacts on the general population [20]. Environmental planning and policy should take both exposures into account when assessing environmental impacts [21]. The variation in noise levels in cities is determined by the cumulative effect of unfavourable or thoughtful city design elements at several scales of a city's general and neighbourhood layout. This is related with the transportation system, the buildings structures, population density, the design of street and building facades, the amount of green space, and the quality of the dwellings concerning sound and vibration features inherent to each city [22]. In the same city, we can find high sound levels at some locations when compared with other places of the same town that are quieter. This is many times related to the city design, especially in cities created a long time ago, not planned and even entirely away from the current mobility needs of the citizens. Currently, environmental noise pollution in cities is based on random sampling. However, these procedures are providing only information relating to a specific sampling and being devoid of details of spatial-temporal variations.

NoiseTube is a participatory sensing solution for noise pollution data collection via mobile crowdsourcing proposed by the authors of [23]. Another approach for noise pollution supervision that implements a crowdsourcing noise pollution monitoring application based on gamification techniques was proposed by [24]. A participatory urban noise mapping system called Ear-Phone that aims to recover the noise map from incomplete and random samples obtained by crowdsourcing data collection has been proposed by the authors of [25]. A middleware solution dedicated to noise monitoring that builds on a refined version of Urban Civics, a platform for urban pollution monitoring currently under development and is focused on the integration of domain-specific sensing applications while applying data assimilation techniques to noise with mobile unplanned was presented by [26]. A ubiquitous crowdsourcing concept where the contributed information is not limited to passively-generated sensor-readings from the device but also includes proactively-generated user's opinions and perspectives, that are processed to offer real-time services to participants was studied by the authors of [27]. A crowdsourcing systems survey which not only provides a better under-standing of crowdsourcing systems but also facilitates future research activities and application developments in the field of crowdsourcing was presented by [28].

3 Materials and Methods

The *iNoiseMapping* has been developed by the authors to be a cost-effective solution that can be easily used by everyone. Using crowdsourcing is possible to create effective noise maps in a less time-consuming and expensive way.

PHP is a fast and multi featured scripting language used for web development. MySQL can be assumed as a relational database management system (RDBMS) database capable of handling a massive number of simultaneous connections. PHP and MySQL technologies offer productivity, scalability, high performance and portability to develop Web applications. The selection of these technologies was first all the fact that are open-source technologies. Furthermore, PHP and MySQL are free develop-ment tools with a large community who contribute every day to add new features and avoid licensing costs. In one hand, PHP is cross-platform, PHP applications are sup-ported by the majority of operating systems such as Solaris, UNIX, Windows and Linux; PHP is an easy language based on the C programming language and provides significantly reduced database configuration time. On the other hand, PHP has a poor quality of handling errors as PHP has a limited amount of debugging tools when compared to other Web development languages. Therefore, this mobile application use PHP Web Services for data communication and MySQL as a database management system. For security proposes the services are not only authenticated but also use HTTPS protocol. Figure 1 shows the *iNoiseMapping* architecture.

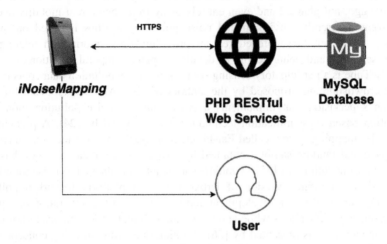

Fig. 1. *iNoiseMapping* architecture.

The *iNoiseMapping* was developed for the iOS operating system as a study that conducts 1472 tests on 100 phones in a reverberation room shows that iOS applications achieve better accuracy than Android applications [29].

This mobile application uses several native frameworks such as AVFoundation for sound recording, CoreAudio for sound analysis and CoreLocation for GPS access (Fig. 2).

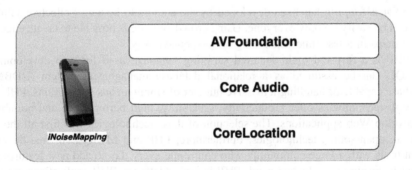

Fig. 2. *iNoiseMapping* frameworks.

AVFoundation is native framework developed by Apple for audiovisual media handling on iOS, macOS, watchOS and tvOS. This framework provides methods for capturing, processing, synthesising, controlling, importing and exporting audiovisual media on Apple platforms. Core Audio is a low-level native Apple framework for sound analysis. Core Location incorporates several methods for device location handling. Using Core Location is possible to get the device's geographic location, altitude, orientation, or position relative to a nearby iBeacon. The framework uses all available onboard hardware, including Wi-Fi, GPS, Bluetooth, magnetometer, barometer, and cellular hardware to gather data.

4 Discussion and Results

The *iNoiseMapping* provides two major features. In the map view scene, the user can consult the participatory data collected. The map view is centred based on the user GPS location. At the sound analysis scene, the user can check the dBA level of the environment and press the "share" button to contribute with real-time noise data from their environment. Figure 3 shows the *iNoiseMapping* features.

Fig. 3. *iNoiseMapping* mobile application developed by the authors.

In order to access the device's GPS location, GPS access must be requested to the user. The *iNoiseMapping* application can access to the device's location data only when the application is used. To enhanced user privacy, the microphone recording feature must be request and the user must accept before audio recording.

The smartphone application allows quick, simple access, intuitive and real-time access to the environmental noise data. Mobile computing in the U.S. has had an exponential growth as adult smartphone device ownership was at 33% in 2011, 56% at the end of 2013 and 64% in early 2015 [30]. In the Netherlands, 70% of global population and 90% of the adolescents own a smartphone [31], in Germany, 40% of the people use a smartphone [32], and 51% of adults owned smartphones in the UK [33]. About 36–40% of smartphone owners use their smartphone 5 min before bed and in the next 5 min after wake up [32]. Actually, smartphones have excellent processing and storage capabilities, and people carry them in their daily lives. For all these reasons a mobile-based crowdsourcing application which uses the resources present in a smartphone and participatory sensing has been developed. On the one hand, this application enables a crowdsourcing approach for environmental noise supervision considering spatio-temporal data for enhanced living environments. On the other hand, the proposed mobile application provides a quick, easy, and intuitive access of the noise monitoring data to the end user. In this way, the user not only can carry the environmental noise data with him for everyday use, but also can contribute by collecting real- time noise data from their environment to a global network which will be used to create efficient and effective noise maps.

Environmental noise pollution control and monitoring policies should be adopted for enhanced living environments. A people's willingness to pay for noise reduction measures study concludes that 80% of respondents were unwilling to pay anything for a noise-control policy [34]. Therefore, cost-effective solutions for noise mapping should be adopted. Participatory sensing is a low-cost approach and can be used to create an effective database with structured spatio-temporal noise data that can be analysed by the city manager to support decision-making in the planning of interventions for noise reduction. Using *iNoiseMapping* is possible to identify specific locations with unhealthy noise conditions and discuss possible interventions for noise reduction. This mobile crowdsourcing approach can provide spatio-temporal sound data which are particularly relevant to plan interventions in order to promote a healthful and productive living environment in smart cities. Consequently, the *iNoiseMapping* is important decision-making tool as this system can not only be used by the city manager authorities to detect unhealthy situations in real-time but also to address behavioural changes to enhance productive environments and wellbeing. On one hand, if several users report unhealthy noise levels in the same area, the city manager can plan inspection activities at the industries and regularly check the sound levels in order to improve the quality of city life. On the other hand, if the noise levels are reported during specific hours of the day and are related to the transportation systems the city manager can plan for changes in traffic flow.

The Internet of Everything (IoE) concept aims to connect people, data, things, and processes in a global network which used intelligently will provide significant enhancements on the daily lives of human beings, particularly in the smart city context [35]. The *iNoiseMapping* allows people to take direct action in the smart city context providing a mobile application to collect environmental sound data and contribute in real-time to the noise mapping creation process for enhanced living environments. Therefore, this mobile application is a useful and cost-effective tool to collect real-time environmental sound data.

As future work, the main goal is to make technical improvements, including the development of important alerts and notifications to inform the city authorities when a specific location is reported with high noise levels. The authors also plan to develop an integrated management Web portal for noise pollution analytics. This portal should allow the municipal authorities to enhanced dashboards about the noise pollution map for enhanced smart cities.

5   Conclusions

Several studies present in the literature advice for noise reduction as this can be assumed as a significant problem for global health. The environmental noise pollution can be related to annoyance, sleep and cognitive performance, raised blood pressure and a risk factor for pregnancy-related hypertension.

Noise maps are created with standard methodologies and are based on specific sources such as road traffic, railway traffic, aircraft and industrial significantly imperfect because the noise emission models and sources are extremely limited.

The resources present in a smartphone as long with participatory sensing allow the creation of a noise mobile application that can be used to provide environmental noise supervision for enhanced living environments in smart cities. The main objective of the presented application is applying crowdsourcing techniques to environmental noise monitoring in order to encourage users to contribute in the data collection process using their personal smartphone. Crowdsourcing techniques applied to environmental noise monitoring can be assumed as a low-cost approach for noise mapping.

The authors conclude that mobile crowdsourcing offers several enhanced features for environmental noise analytics and supervision. The *iNoiseMapping* can be used not only to create reliable noise maps but also to identify specific locations with unhealthy noise conditions and discuss possible interventions for noise reductions. Consequently, the proposed approach provides direct action in the smart city context to people by providing a method to collect environmental sound data in real-time for enhanced living environments. The collected sound data can be analysed to address interventions for enhanced living in smart cities.

The results obtained are encouraging, on behalf of an important contribution to noise mapping methodologies. Nevertheless, the proposed needs further experimental validation and software improvements to adapt the application to specific cases or problems in smart cities such as traffic noise pollution.

Software improvements are planned in order to provide notifications to alert the city authorities when a specific location is reported with high noise levels. The creation of a Web portal for noise pollution supervision and analytics is also planned.

In spite of the influence of environmental noise pollution in daily human activities, systems like this will be significant for enhanced living environments. The authors believe that the first step is to create reliable noise maps through crowdsourcing in order to detect unhealthy noise pollution levels in real time and plan interventions for enhanced living environments.

References

1. Murphy, E., King, E.A.: Principles of environmental noise. In: Environmental Noise Pollution, pp. 9–49. Elsevier (2014)
2. Caragliu, A., Del Bo, C., Nijkamp, P.: Smart cities in Europe. J. Urban Technol. **18**(2), 65–82 (2011)
3. Schaffers, H., Komninos, N., Pallot, M., Trousse, B., Nilsson, M., Oliveira, A.: Smart cities and the future internet: towards cooperation frameworks for open innovation. In: Domingue, J., et al. (eds.) FIA 2011. LNCS, vol. 6656, pp. 431–446. Springer, Heidelberg (2011). https://doi.org/10.1007/978-3-642-20898-0_31
4. Chourabi, H., et al.: Understanding smart cities: an integrative framework. In: 2012 45th Hawaii International Conference on System Sciences, Maui, HI, USA, pp. 2289–2297 (2012)
5. Batty, M., et al.: Smart cities of the future. Eur. Phys. J. Spec. Top. **214**(1), 481–518 (2012)
6. Kephalopoulos, S., Paviotti, M., Anfosso-Lédée, F., Van Maercke, D., Shilton, S., Jones, N.: Advances in the development of common noise assessment methods in Europe: the CNOSSOS-EU framework for strategic environmental noise mapping. Sci. Total Environ. **482–483**, 400–410 (2014)
7. Picaut, J., Fortin, N., Bocher, E., Petit, G., Aumond, P., Guillaume, G.: An open-science crowdsourcing approach for producing community noise maps using smartphones. Build. Environ. **148**, 20–33 (2019)
8. Estellés-Arolas, E., González-Ladrón-de-Guevara, F.: Towards an integrated crowdsourcing definition. J. Inf. Sci. **38**(2), 189–200 (2012)
9. Anjum, A., Ilyas, M.U.: Activity recognition using smartphone sensors. In: 2013 IEEE Consumer Communications and Networking Conference (CCNC), pp. 914–919 (2013)
10. Shoaib, M., Scholten, H., Havinga, P.J.M.: Towards physical activity recognition using smartphone sensors. In: 2013 IEEE 10th International Conference on Ubiquitous Intelligence and Computing, Autonomic and Trusted Computing (UIC/ATC), pp. 80–87 (2013)
11. Lloret, J., Canovas, A., Sendra, S., Parra, L.: A smart communication architecture for ambient assisted living. IEEE Commun. Mag. **53**(1), 26–33 (2015)
12. Burke, J.A., et al.: Participatory sensing. Presented at the Workshop on World-Sensor-Web (WSW): Mobile Device Centric Sensor Networks and Applications (2006)
13. Stansfeld, S.A., Matheson, M.P.: Noise pollution: non-auditory effects on health. Br. Med. Bull. **68**(1), 243–257 (2003)
14. Auger, N., Duplaix, M., Bilodeau-Bertrand, M., Lo, E., Smargiassi, A.: Environmental noise pollution and risk of preeclampsia. Environ. Pollut. **239**, 599–606 (2018)
15. Foraster, M., et al.: Exposure to road, railway, and aircraft noise and arterial stiffness in the SAPALDIA study: annual average noise levels and temporal noise characteristics. Environ. Health Perspect. **125**(9), 097004 (2017)
16. Gupta, A., Gupta, A., Jain, K., Gupta, S.: Noise pollution and impact on children health. Indian J. Pediatr. **85**(4), 300–306 (2018)
17. Christensen, J.S., Hjortebjerg, D., Raaschou-Nielsen, O., Ketzel, M., Sørensen, T.I.A., Sørensen, M.: Pregnancy and childhood exposure to residential traffic noise and overweight at 7 years of age. Environ. Int. **94**, 170–176 (2016)
18. Murphy, E., King, E.A.: An assessment of residential exposure to environmental noise at a shipping port. Environ. Int. **63**, 207–215 (2014)
19. Zanella, A., Bui, N., Castellani, A., Vangelista, L., Zorzi, M.: Internet of things for smart cities. IEEE Internet Things J. **1**(1), 22–32 (2014)

20. Murphy, E., King, E.A.: Environmental noise and health. In: Environmental Noise Pollution, pp. 51–80. Elsevier (2014)
21. Stansfeld, S.: Noise effects on health in the context of air pollution exposure. Int. J. Environ. Res. Public Health **12**(10), 12735–12760 (2015)
22. Lercher, P.: Noise in cities: urban and transport planning determinants and health in cities. In: Nieuwenhuijsen, M., Khreis, H. (eds.) Integrating Human Health into Urban and Transport Planning, pp. 443–481. Springer, Cham (2019). https://doi.org/10.1007/978-3-319-74983-9_22
23. Stevens, M., D'Hondt, E.: Crowdsourcing of pollution data using smartphones. In: Workshop on Ubiquitous Crowdsourcing, held at Ubicomp 2010, Copenhagen, Denmark, 26–29 September 2010 (2010)
24. Martí, I.G., et al.: Mobile application for noise pollution monitoring through gamification techniques. In: Herrlich, M., Malaka, R., Masuch, M. (eds.) ICEC 2012. LNCS, vol. 7522, pp. 562–571. Springer, Heidelberg (2012). https://doi.org/10.1007/978-3-642-33542-6_74
25. Rana, R.K., Chou, C.T., Kanhere, S.S., Bulusu, N., Hu, W.: Ear-phone: an end-to-end participatory urban noise mapping system. In: Proceedings of the 9th ACM/IEEE International Conference on Information Processing in Sensor Networks - IPSN 2010, Stockholm, Sweden, p. 105 (2010)
26. Hachem, S., et al.: Monitoring noise pollution using the urban civics middleware. In: 2015 IEEE First International Conference on Big Data Computing Service and Applications, Redwood City, CA, USA, pp. 52–61 (2015)
27. Mashhadi, A.J., Capra, L.: Quality control for real-time ubiquitous crowdsourcing. In: Proceedings of the 2nd International Workshop on Ubiquitous Crowdsourcing - UbiCrowd 2011, Beijing, China, p. 5 (2011)
28. Yuen, M.-C., King, I., Leung, K.-S.: A survey of crowdsourcing systems. In: 2011 IEEE Third International Conference on Privacy, Security, Risk and Trust and 2011 IEEE Third International Conference on Social Computing, Boston, MA, USA, pp. 766–773 (2011)
29. Murphy, E., King, E.A.: Testing the accuracy of smartphones and sound level meter applications for measuring environmental noise. Appl. Acoust. **106**, 16–22 (2016)
30. Müller, H., Gove, J.L., Webb, J.S., Cheang, A.: Understanding and comparing smartphone and tablet use: insights from a large-scale diary study. In: Proceedings of the Annual Meeting of the Australian Special Interest Group for Computer Human Interaction, pp. 427–436 (2015)
31. van Deursen, A.J.A.M., Bolle, C.L., Hegner, S.M., Kommers, P.A.M.: Modeling habitual and addictive smartphone behavior. Comput. Hum. Behav. **45**, 411–420 (2015)
32. Montag, C., et al.: Smartphone usage in the 21st century: who is active on WhatsApp? BMC Res. Notes **8**(1), 331 (2015)
33. Pearson, C., Hussain, Z.: Smartphone use, addiction, narcissism, and personality: a mixed methods investigation. Int. J. Cyber Behav. Psychol. Learn. **5**(1), 17–32 (2015)
34. Huh, S.-Y., Shin, J.: Economic valuation of noise pollution control policy: does the type of noise matter? Environ. Sci. Pollut. Res. **25**, 30647–30658 (2018)
35. Marques, G.: Ambient assisted living and internet of things. In: Cardoso, P.J.S., Monteiro, J., Semião, J., Rodrigues, J.M.F. (eds.) Harnessing the Internet of Everything (IoE) for Accelerated Innovation Opportunities, pp. 100–115. IGI Global, Hershey (2019)

Environmental Quality Supervision for Enhanced Living Environments and Laboratory Activity Support Using IBM Watson Internet of Things Platform

Gonçalo Marques$^{(\boxtimes)}$ ⓘ and Rui Pitarma

Unit for Inland Development, Polytechnic Institute of Guarda,
Av. Dr. Francisco Sá Carneiro, nº 50, 6300-559 Guarda, Portugal
`goncalosantosmarques@gmail.com`, `rpitarma@ipg.pt`

Abstract. Temperature and humidity are extremely important not only for occupational health and well-being but also for supervising laboratory activities. Laboratories are places characterised by several contamination sources which lead to significant poor indoor quality conditions. Laboratory activities require real-time monitoring supervision. Around 40% of the energy consumed worldwide and around 30% of the carbon dioxide liberated are related to indoor living environments. Further, a substantial amount of this energy is used to provide a satisfactory human perception of the thermal conditions. The IBM Watson IoT Platform provides data integration, security methods, data collection, visualisation, analytics, device management functionalities and allows data to be sent securely to the cloud using MQTT messaging protocol. This document presents a temperature and humidity real-time supervision system based on Internet of Things architecture named *iTemp+*. The system incorporates physical prototype for data acquisition and uses IBM Watson IoT for data storing and consulting. The IBM Watson IoT Platform provides data integration, security methods, data collection, visualisation, analytics, device management, artificial intelligence and blockchain functionalities which are not implemented in the concurrent IoT platforms. The results obtained reveal that IBM Watson IoT platform offers several enhanced features for device management and analytics and can be used as a powerful approach to provide IEQ supervision.

Keywords: Ambient assisted living · Enhanced living environments ·
IBM Watson IoT · IEQ (Indoor Environment Quality) ·
IoT (Internet of Things) · Laboratory environment conditions

1 Introduction

Indoor living environments include several types of spaces, workplaces such as offices, hospitals, public service centres, schools, libraries, leisure spaces and also the cabins of vehicles [1]. In particular, schools are an important place to monitor. Typically, a large number of occupants, the time spent indoors, and the higher density of occupants justify the need to develop automatic supervision systems to provide a healthful and productive workplace for the students, teachers and the school staff [2].

© Springer Nature Switzerland AG 2019
J. M. F. Rodrigues et al. (Eds.): ICCS 2019, LNCS 11538, pp. 680–691, 2019.
https://doi.org/10.1007/978-3-030-22744-9_53

In most higher education establishments in Portugal, laboratories are also used as classrooms and should also be monitored. However, monitoring must ensure different conditions throughout laboratory and teaching activities with reliable data quality. The satisfaction of the indoor thermal conditions is of utmost importance for the occupants, particularly when compared with sound, light and air quality conditions [3]. Typically, the occupants are satisfied with temperature ranges of 17–30 °C. Furthermore, the indoor temperature conditions are affected by several factors. These factors can be classified as physical factors such as humidity, radiant temperature, air temperature and air velocity; and as personal factors such as clothing insulation and metabolic heat [4]. Although, for laboratory experiments, the recommendation is 23 °C (±5 °C) for temperature and <70% for relative humidity. A study on the thermal comfort in a Portuguese school is presented by [5].

Around 40% of the energy consumed worldwide and around 30% of the carbon dioxide liberated are related to indoor living environments. Further, a substantial amount of this energy is used to provide a satisfactory human perception of the thermal conditions. Thermal comfort is a multifaceted subject related with a lot of interconnected parts which are not easy to measure and study. However, the introduction of personalised conditioning techniques is probably the best way to rise the individual thermal comfort acceptance [6].

Indoor environment quality (IEQ) is based on random sampling. Nevertheless, these methods only offer data related to a particular sample and don't provide any details of spatio-temporal changes which are specifically significant in laboratory activities. The IEQ supervision system improves the detection and correction of unhealthy conditions. However, there is a lack of consensus on measuring protocols. Therefore, the same building could have different evaluations which avoid benchmarking [7].

Laboratory ventilation aims to protect the occupants from possible experience to dangerous materials and to offer comfort in an energy efficient manner. Therefore, it is significant to have automatic devices in order to detect any variations during laboratory activity [8]. The indoor microbial levels are influenced by several parameters such as ventilation, temperature, relative humidity and occupants number [9].

IBM Watson IoT Platform is a foundational cloud that offers device connecting and consulting feature as long with an extensive set of built-in and add-on tools for the Internet of Things (IoT) architectures [10]. The IoT concept must be assumed as the ubiquitous presence of physical objects which incorporate sensing capabilities and can cooperate in order to achieve a predefined goal [11, 12]. IoT architectures will bring several effects on actual world and society. The IoT applications continue to enhance several activities of day to day life. IoT has been used as a foundational base to develop several types of applications in numerous contexts such as smart homes and smart cities. Particularly, IoT has been used for ambient assisted living and e-health to create automatic and intelligent solutions which can provide new data and computational methods for enhanced living and occupational health [13].

This document presents an IoT solution for indoor temperature and humidity real-time supervision named *iTemp+*. The system incorporates a hardware prototype for environment data acquisition and use IBM Watson IoT for data storage and consulting (Fig. 1).

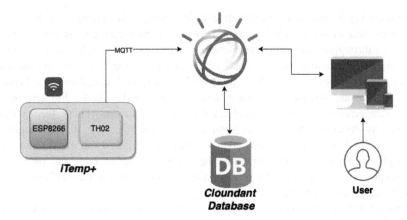

Fig. 1. *iTemp+* architecture.

The *iTemp+* is a totally wireless system which has been developed using open-source technologies. This system has numerous advantages when compared to existing competing systems. The *iTemp+* is a cost-effective solution which incorporates a modular and scalable architecture which lead to an easy installation process. The *iTemp+* use the ESP8266 as microcontroller unit which support built-in Wi-Fi compatibility.

The rest of this document is organized as follows: Sect. 2 describes the related work and Sect. 3 is presents the system architecture of the *iTemp+* solution; Sect. 4 provides the discussion and results, and Sect. 5 concludes the document.

2   Related Work

Numerous systems on indoor quality supervision can be found in the literature. This section presents several IEQ systems that incorporate open-source, and cost-effective technologies are described.

A cost-effective Wireless sensor network (WSN) solution for precision agriculture implemented in a pepper greenhouse which incorporates proper assessment methods for irrigation and facilities monitoring and remote control is being proposed by Ref. [14].

A WSN for temperature distribution supervision system in large-scale indoor space is proposed by Ref. [15]. This system objective is to increase measurement quality wirelessly transmitted, recognize temperature distribution patterns, and improve the distribution of supply air rated flow rate to multiple supply air terminals which are controlled to provide specific temperature conditions.

An IoT solution to control temperature and fire supervision which incorporates Message Queue Telemetry Transportation (MQTT) broker on Amazon Web Services was proposed by Ref. [16].

An Android smartphone application named Orvalho was been developed to evaluate the thermal comfort indicators of animals and individuals by physically inserting meteorological data or by acquiring data from mobile devices using Bluetooth [17].

Various indoor supervision open-source solutions based on IoT which offer data acquisition, processing and transmission features from different locations at the same time and real-time data consulting through Web and mobile applications in proposed by the authors of [18–27].

The *iTemp+* is an automatic temperature and humidity monitoring solution for indoor environments, particularly for the supervision of laboratory activities. Is an entirely wireless solution built on the base of the ESP8266. The IBM Watson IoT platform offer efficient services for device connection using MQTT [28]. In the developed solution, the Watson IoT platform acts as the MQTT broker and is thus responsible for distributing messages to connected *iTemp+* prototype (Fig. 2). The data is stored in a Cloudant DB, a scalable JSON document database for the Web, mobile, IoT and serverless applications.

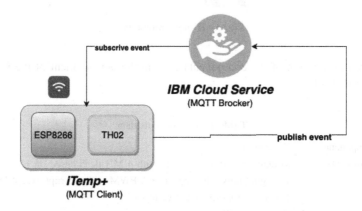

Fig. 2. *iTem+* and IBM Cloud service integration.

3   Materials and Methods

The *iTemp+* was developed to be a cost-effective solution that can be easily connected to Wi-Fi and installed by the final user to avoid installation costs if done by certified professionals. It also avoids privacy issues that come with the home installation. In this section, the hardware and software technical data used in the development of the proposed solution will be discussed in detail.

The temperature and relative humidity sensor used is a cost-effective TH02 sensor (*Grove*). This sensor has reliable data acquisition properties, and the measurement range is 0–70 °C and 0–80% for temperature and relative humidity, respectively. This sensor was selected taking into account the cost-effective approach of the solution but also because the measurement range covers the majority of the scenarios.

The *iTemp+* incorporate a D1 Mini ESP8266 (*Wemos*) module as microcontroller and the temperature and humidity sensor module used is connect via I2C interface. The *iTemp+* hardware is showed in Fig. 3.

Fig. 3. *iTemp+* prototype.

A brief description of the components used in the development of the *iTemp+* is presented in Table 1:

Table 1. Components data.

Component	Description
Wemos D1 mini	Miniaturized Wi-Fi module with 4 MB flash based on ESP-8266
	11 digital pins (all support I2C, PWM and interrupt except D0)
	1 analog input (3.2 max voltage)
	Operating voltage: 3.3 V
	Clock speed: 80 MHz/160 MHz
	Dimensions and height: 34.2 mm × 25.6 mm and 3 g
Grove I2C TH02	Temperature and humidity sensor
	Range: 0%–80% RH; 0–70 °C
	Accuracy: ±4.5% RH; ±0.5 °C
	Operating voltage: 3.3 V–5 V

The *iTemp+* provides an easy Wi-Fi configuration. The *iTemp+* is by default a Wi-Fi client, but if it is unable to connect to any previous configured Wi-Fi network a hotspot will be created with an SSID "iTemp+". The end-user can access this hotspot to configure the Wi-Fi network to which the *iTemp+* is going to connected. Figure 4 represents the Wi-Fi network configuration process.

Fig. 4. *iTemp+* easy installation process.

4 Discussion and Results

The system test has been conducted inside a laboratory located in a Portuguese university using one *iTemp+* station. The monitored laboratory is naturally ventilated by manually open of windows or doors and doesn't have dedicated ventilation slots. The temperature and humidity have been supervised on the context of thermography experiments to test the supervisory solution in the course of laboratory activities but also in real-time environmental supervision of teaching activities for enhanced living environments and well-being (Fig. 5).

Fig. 5. Thermography experiments supported by the *iTemp+* prototype.

The tests performed indicates the *iTemp+* ability to provide real-time supervision not only for laboratory experiments but also for enhanced living environments and occupational health using IBM Watson IoT platform. The collected data can be stored for further analysis and correlation among laboratory experiments to study the effect of the indoor conditions on the tests conducted. This IoT platform allows the user to consult the collected data history through a browser-compatible device. Figures 6 and 7 represent temperature (°C), and relative humidity (%) data collected respectively.

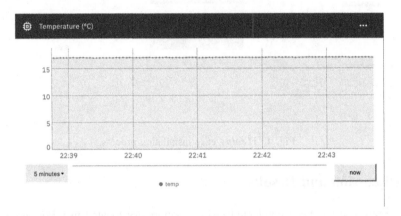

Fig. 6. Temperature monitoring data (°C) collected in the tests performed.

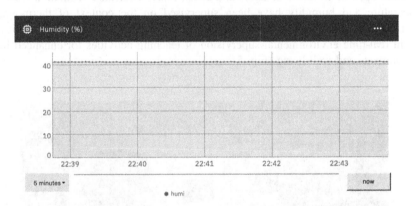

Fig. 7. Relative humidity monitoring data (%) collected in the tests performed.

The IBM Watson also allows the user to create a custom dashboard for a detailed examination of the complete temporal variation. Consequently, the *iTemp+* solution is a significant method for temperature and humidity data assessment for environmental monitoring. The proposed solution incorporates several advantages when compared with the majority of the competing systems such as miniaturized size, scalability, and modularity with lead to an easy installation.

The IBM Watson IoT offers an integrated solution for device management and analysis. Using this platform, the developer can easily configure and manage his devices and as well as consult the state of the device. Additionally, he can access a full detailed log of the device activity (Fig. 8).

Fig. 8. IBM Watson IoT device management

The IBM Watson IoT incorporates another significant advantage for IoT architectures; using this platform is possible to access a full detailed network usage of the IoT devices. A custom dashboard for device management created by the authors is shown in Fig. 9.

Fig. 9. IBM Watson IoT network usage management dashboard

The IBM Watson IoT Platform provides data integration, security methods, data collection, visualisation, analytics and device management functionalities which are not implemented in the concurrent IoT platforms (Table 2). Moreover, the IBM Watson IoT Platform provides artificial intelligence and blockchain integration.

Table 2. Summary of similar IoT platforms.

IoT platform	Integration	Security	Data collection	Visualization	Analytics	Device management
IBM Watson	REST API	√	HTTPS, MQTT	√	√	√
Ericsson platform	REST API	√	CoAP	×	×	√
Xively	REST API	×	HTTP, HTTPS, Sockets, MQTT	√	×	√
ParStream	R, UDX API	×	MQTT	√	√	×

People's considerations of temperature conditions are supported by their habits and behaviour. On the one hand, prolonged exposure to the high temperature levels increase the occupants' temperature requirements and turn the difficulty to get a comfortable exposure to environments with neutral temperatures. On the other hand, uncomfortable thermals experiences can bring the lead to an easier thermal adaptation [29].

A personal comfort model is a novel methodology for thermal comfort. That methodology calculates people thermal comfort behaviour, instead of considering a standard value extracted from a large population. IoT and artificial intelligence algorithms can "learn" the occupant thermal comfort by analysing data collected from the environment where the person is [30]. The IBM Watson IoT platform offers tools not only to create machine learning and identification of patterns but also to supervisor and enhance diverse machine learning and deep learning models by performing automatic tests and matching results. Therefore, this platform can be assumed as a unique platform to develop new and personal IEQ supervision systems.

Furthermore, the IBM Watson IoT platform's built-in blockchain allows users to add selected IoT data to private blockchain ledgers that can be included in shared transactions for enhanced IoT security applications [31]. The IBM Watson IoT platform can be assumed as one of the most powerful artificial intelligence systems and is already used in the healthcare field to support medical diagnostics [32].

As future work, the principal objective will be to perform technical enhancements including the devolvement of a mobile and smartphone application, using the IBM Cloud services to provide the configuration of setpoints to notify the user when the IEQ is poor.

5 Conclusions

This document had proposed an IoT solution for indoor temperature and humidity real-time supervision named *iTemp+*. IEQ is affected by several factors such as noise, light, air and temperature conditions. Poor IEQ significantly affects well-being and occupational health, especially of children and older adults who are most fragile.

Technological enhancements and the increase of low-cost controller and sensors lead to the proliferation of IoT systems which incorporate the important potential for the development of smart solutions for smart cities in general and indoor monitoring in particular. Real-time monitoring systems turn possible not only to quantity the IEQ index in distinctive locations but also to offer reliable information to identify insalubrious conditions automatically.

The *iTemp+* has several advantages when compared to the majority of existing solutions. The proposed solution is scalable and modular which lead to an easy installation. Further, the *iTemp+* offer accurate data collection and consulting in real-time, providing important progress in the current quality assessment methods. The *iTemp+* is a totally wireless system which provides easy installation and configuration features.

IBM Watson IoT can be used as an IoT platform to supervise the environmental quality conditions of locations where technological activities are performed such as laboratories. For example, it can be implemented in laboratories where this IoT platform can provide an accurate association between the verified results and the environmental conditions. We conclude that IBM Watson IoT offers several advantages for connecting, device management, analytics and security. This platform not only supports easy and secure methods for device connection but also provides a configurable dashboard to manage the data storage, data transformation actions and third-party services integration. The IBM Watson IoT allows network and storage supervision and to define rules to trigger automatic actions that include alerts, email, IFTTT, Node-RED flows, and external services. When compared to several existing IoT platforms, the IBM Watson incorporates built-in artificial intelligence and blockchain support which are not implemented in similar platforms. In the future, the authors plan to use those resources for enhanced environmental quality prediction and security respectively. Technical enhancements to adapt the proposed solution to specific domains such as schools, laboratories and industry have been planned as future work. Due to the negative and significant effect IEQ in human behaviour, solutions as the *iTemp+* will lead to enhanced living environments and occupational health.

Acknowledgement. This research is framed in the project "TreeM – Advanced Monitoring & Maintenance of Trees" N. ° 023831, 02/SAICT/2016, co-financed by CENTRO 2020 and FCT, Portugal 2020 and structural funds UE-FEDER.

References

1. de Gennaro, G., et al.: Indoor air quality in schools. Environ. Chem. Lett. **12**(4), 467–482 (2014)
2. Madureira, J., et al.: Indoor air quality in schools and its relationship with children's respiratory symptoms. Atmos. Environ. **118**, 145–156 (2015)
3. Yang, L., Yan, H., Lam, J.C.: Thermal comfort and building energy consumption implications – a review. Appl. Energy **115**, 164–173 (2014)
4. Havenith, G., Holmér, I., Parsons, K.: Personal factors in thermal comfort assessment: clothing properties and metabolic heat production. Energy Build. **34**(6), 581–591 (2002)
5. Pereira, L.D., Cardoso, E., da Silva, M.G.: Indoor air quality audit and evaluation on thermal comfort in a school in Portugal. Indoor Built Environ. **24**(2), 256–268 (2015)
6. Rupp, R.F., Vásquez, N.G., Lamberts, R.: A review of human thermal comfort in the built environment. Energy Build. **105**, 178–205 (2015)
7. Heinzerling, D., Schiavon, S., Webster, T., Arens, E.: Indoor environmental quality assessment models: a literature review and a proposed weighting and classification scheme. Build. Environ. **70**, 210–222 (2013)
8. DiBerardinis, L., Greenley, P., Labosky, M.: Laboratory air changes: what is all the hot air about? J. Chem. Health Saf. **16**(5), 7–13 (2009)
9. Hwang, S.H., Seo, S., Yoo, Y., Kim, K.Y., Choung, J.T., Park, W.M.: Indoor air quality of daycare centers in Seoul, Korea. Build. Environ. **124**, 186–193 (2017)
10. Akinsiku, A., Jadav, D.: BeaSmart: a beacon enabled smarter workplace. In: NOMS 2016 - 2016 IEEE/IFIP Network Operations and Management Symposium, Istanbul, Turkey, pp. 1269–1272 (2016)
11. Giusto, D. (ed.): The Internet of Things: 20th Tyrrhenian Workshop on Digital Communications. Springer, New York (2010). https://doi.org/10.1007/978-1-4419-1674-7
12. Marques, G.: Ambient assisted living and internet of things. In: Cardoso, P.J.S., Monteiro, J., Semião, J., Rodrigues, J.M.F. (eds.) Harnessing the Internet of Everything (IoE) for Accelerated Innovation Opportunities, pp. 100–115. IGI Global, Hershey (2019)
13. Gubbi, J., Buyya, R., Marusic, S., Palaniswami, M.: Internet of Things (IoT): a vision, architectural elements, and future directions. Future Gener. Comput. Syst. **29**(7), 1645–1660 (2013)
14. Srbinovska, M., Gavrovski, C., Dimcev, V., Krkoleva, A., Borozan, V.: Environmental parameters monitoring in precision agriculture using wireless sensor networks. J. Clean. Prod. **88**, 297–307 (2015)
15. Zhou, P., Huang, G., Zhang, L., Tsang, K.-F.: Wireless sensor network based monitoring system for a large-scale indoor space: data process and supply air allocation optimization. Energy Build. **103**, 365–374 (2015)
16. Kang, D.-H., et al.: Room temperature control and fire alarm/suppression IoT service using MQTT on AWS. In: 2017 International Conference on Platform Technology and Service (PlatCon), Busan, South Korea, pp. 1–5 (2017)
17. de Oliveira Júnior, A.J., et al.: Development of an android APP to calculate thermal comfort indexes on animals and people. Comput. Electron. Agric. **151**, 175–184 (2018)
18. Marques, G., Pitarma, R.: IAQ evaluation using an IoT CO_2 monitoring system for enhanced living environments. In: Rocha, Á., Adeli, H., Reis, L.P., Costanzo, S. (eds.) WorldCIST'18 2018. AISC, vol. 746, pp. 1169–1177. Springer, Cham (2018). https://doi.org/10.1007/978-3-319-77712-2_112
19. Pitarma, R., Marques, G., Ferreira, B.R.: Monitoring indoor air quality for enhanced occupational health". J. Med. Syst. **41**(2), 23 (2017)

20. Marques, G., Pitarma, R.: Smartwatch-based application for enhanced healthy lifestyle in indoor environments. In: Omar, S., Haji Suhaili, W.S., Phon-Amnuaisuk, S. (eds.) CIIS 2018. AISC, vol. 888, pp. 168–177. Springer, Cham (2019). https://doi.org/10.1007/978-3-030-03302-6_15

21. Marques, G., Pitarma, R.: Using IoT and social networks for enhanced healthy practices in buildings. In: Rocha, Á., Serrhini, M. (eds.) EMENA-ISTL 2018. SIST, vol. 111, pp. 424–432. Springer, Cham (2019). https://doi.org/10.1007/978-3-030-03577-8_47

22. Marques, G., Pitarma, R.: Monitoring health factors in indoor living environments using internet of things. In: Rocha, Á., Correia, A.M., Adeli, H., Reis, L.P., Costanzo, S. (eds.) WorldCIST 2017. AISC, vol. 570, pp. 785–794. Springer, Cham (2017). https://doi.org/10.1007/978-3-319-56538-5_79

23. Marques, G., Roque Ferreira, C.: Pitarma,: A system based on the internet of things for real-time particle monitoring in buildings. Int. J. Environ. Res. Public Health 15(4), 821 (2018)

24. Salamone, F., Belussi, L., Danza, L., Galanos, T., Ghellere, M., Meroni, I.: Design and development of a nearable wireless system to control indoor air quality and indoor lighting quality. Sensors 17(5), 1021 (2017)

25. Akkaya, K., Guvenc, I., Aygun, R., Pala, N., Kadri, A.: IoT-based occupancy monitoring techniques for energy-efficient smart buildings. In: 2015 IEEE Wireless Communications and Networking Conference Workshops (WCNCW), New Orleans, LA, USA, pp. 58–63 (2015)

26. Marques, G., Pitarma, R.: A cost-effective air quality supervision solution for enhanced living environments through the internet of things. Electronics 8(2), 170 (2019)

27. Marques, G., Ferreira, C.R., Pitarma, R.: Indoor air quality assessment using a CO_2 monitoring system based on internet of things. J. Med. Syst. 43(3), 67 (2019)

28. da Silva, A.F., Ohta, R.L., dos Santos, M.N., Binotto, A.P.D.: A cloud-based architecture for the internet of things targeting industrial devices remote monitoring and control. IFAC-PapersOnLine 49(30), 108–113 (2016)

29. Luo, M., Wang, Z., Brager, G., Cao, B., Zhu, Y.: Indoor climate experience, migration, and thermal comfort expectation in buildings. Build. Environ. 141, 262–272 (2018)

30. Kim, J., Schiavon, S., Brager, G.: Personal comfort models – a new paradigm in thermal comfort for occupant-centric environmental control. Build. Environ. 132, 114–124 (2018)

31. Kshetri, N.: Can blockchain strengthen the internet of things? IT Prof. 19(4), 68–72 (2017)

32. Branger, J., Pang, Z.: From automated home to sustainable, healthy and manufacturing home: a new story enabled by the internet-of-things and industry 4.0. J. Manag. Anal. 2(4), 314–332 (2015)

Combining Data from Fitness Trackers with Meteorological Sensor Measurements for Enhanced Monitoring of Sports Performance

Anna Wachowicz⑩, Bożena Małysiak-Mrozek$^{(\boxtimes)}$⑩, and Dariusz Mrozek⑩

Institute of Informatics, Silesian University of Technology,
ul. Akademicka 16, 44-100 Gliwice, Poland
anna.m.wachowicz@gmail.com,
{bozena.malysiak-mrozek,dariusz.mrozek}@polsl.pl

Abstract. Systematic analysis of training data has become an inherent element of building a sports condition and preparation for sports competitions. Today's progress in the development of various fitness trackers, smart wearables and IoT devices allows monitoring the level of development of athletic abilities not only for professional athletes but also for sports enthusiasts and sports amateurs. Meteorological conditions prevailing on a given day can significantly affect the effectiveness of training and abilities of a person during sports competitions. However, in order to properly analyze particular sports achievements and the effectiveness of sports efforts, the training data from body sensors should be appropriately combined with weather sensor data. In this paper, we show that, due to approximate nature, this process can be implemented by using the fuzzy join technique.

Keywords: Fuzzy sets · Fuzzy logic · Data analysis · Smart devices ·
Human activity monitoring · IoT · Sensors

1 Introduction

In recent years, monitoring of physical activity gained popularity among both sports enthusiasts and cutting-edge technologies fans. As the IoT devices market extends, smartwatches and smart bands are no more out of reach for an average person. They allow monitoring a wide range of activities and parameters – from simply walking and counting the number of steps to controlling advanced

This work was supported by Microsoft Research within Microsoft Azure for Research Award grant, pro-quality grant for highly scored publications or issued patents of the Rector of the Silesian University of Technology, Gliwice, Poland (grant No. 02/020/RGJ19/0166 and 02/020/RGJ19/0167), and partially, by Statutory Research funds of Institute of Informatics, Silesian University of Technology, Gliwice, Poland (grant No. BK/204/RAU2/2019).

© Springer Nature Switzerland AG 2019
J. M. F. Rodrigues et al. (Eds.): ICCS 2019, LNCS 11538, pp. 692–705, 2019.
https://doi.org/10.1007/978-3-030-22744-9_54

workout with tracking position with GPS signal, recognizing the type of activity and measuring the level of body hydration. People who do sport professionally use wearables as one of many work tools. Sport amateurs also want to monitor and improve their training results, however, they do not have as many devices as professionals nor advice from a sports coach. Wearables, along with corresponding mobile web applications, help amateurs to control the progress of training quality. These applications also provide many useful reports and allow to share the results of the training with other people in social media or online forums for sports enthusiasts.

As online communities gain more and more popularity, people use them for exchanging experiences. They describe the observations and ask various questions. The popular topic is how the weather affects the training performance. Although this issue is covered in many scientific papers, the results usually apply to the professionals. Moreover, from the technical point of view, the answer to this question necessitates joining the training data from wearable devices with weather parameters. Weather parameters can be retrieved from online Web services. However, they are usually provided hourly, while the training may start at any time. This requires the implementation of some flexibility while joining both types of data. This flexibility can be modeled by fuzzy sets.

In this paper, we apply a fuzzy join technique while flexibly combining training data with weather conditions. The fuzzy join is performed as a preparation step for further data analysis allowing to investigate the impact of weather conditions on the training performance. The training data are gathered with the use of smartwatch used for monitoring the running training. With the use of the fuzzy join algorithm, data are combined with the most accurate weather measurements retrieved from an online weather service. The output is then loaded to the Microsoft Azure cloud data center where the analysis can be conducted. In this paper, we present the outcome of the analysis of the impact of weather features such as temperature, air humidity, and pressure or wind speed on the effectiveness of amateurs' running. We also show the duration of particular operations related to data pre-processing performed by the fuzzy join module.

2   Related Works

IoT technologies and wearable devices are frequently applied in monitoring various physiological parameters in terms of people's health state and the activities they perform. World literature gives many examples of such applications. For example, in [14] Yamato shows the platform for the analysis of posture and fatigue on the basis of data gathered from electrocardiograph and accelerator. The analysis can be performed in a smartphone or in the Cloud. Lara and Labrador [8] presented the survey on the usage areas of the human activity recognition (HAR) systems. The authors mention the types of activities recognized by *state-of-the-art* HAR systems which may cover areas such as daily activities, fitness, military, ambulation or transportation. They widely explain the feature extraction step distinguishing the techniques used for measured attributes that are grouped into three sets: *time-domain, frequency-domain, and others*. The most popular methods for the time-domain group are mean, standard deviation, variance, mean absolute deviation and correlation between axes, whereas the techniques for the frequency-domain set are Fourier Transform and Discrete Cosine Transform. A review of various types of wearable devices and inner sensors used for observing human activities is shown in [10]. In the paper, Mukhopadhyay describes also the architecture of monitoring systems and networks containing various sensors and devices. In [12], Toh et al. draw attention to the design issues for wearable devices. As rightly pointed out by authors, in order to meet all users' requirements smart wearables should be light and low-energy consuming. Devices should be able to send data to the paired application, so that users can access the history of performed activity, like training, or be informed that there is something wrong with a monitored person, regardless of whether it is a sports amateur or an elderly that wears the smart band. In [11], authors describe wearable and implantable sensors for distributed mobile computing. They also present the difficulties and complications that may happen while using wearables. Similarly to the system that we have built and present in Sect. 4, the system presented in [11] aims to analyze the impact of weather conditions on the running training parameters.

The relationship between weather conditions and the training performance is not only intuitive but also confirmed by published research. In [4,5,13], the authors study the impact of the temperature and other atmospheric measurements on the performance of marathon runners. All of these articles describe the optimal temperatures for the best running performance among men and women which are between $10\,^{\circ}\mathrm{C}$–$15\,^{\circ}\mathrm{C}$. One of the most recent studies that investigate the relationship between various physical exercises (including the running) and weather conditions are presented in [3]. The author analyzes seasonal upper-body strength resistance and running endurance performance, and studies if there are any relationships between the efficiency of these activities and weather conditions. Comparing to previous papers, the running distance is 5 km and the research shows that participants of the conducted experiment gained better results in summer and spring, which are hotter seasons. In our paper, we not only investigate these relationships in the data analysis step, but also focus on the preparation of data with the use of fuzzy join technique before the main analysis begins.

The term *fuzzy join* is widely used in scientific literature but may have different meanings and applications. Many of published papers, including [1,2,7,15], use the term while combining data sets on the basis of flexible character data matching and string similarity with the use of various distance functions, like the Hamming distance. Meanwhile, articles [6] by Khorasani et al. and [9] by Małysiak-Mrozek et al. show how to flexibly combine big data sets with the use of fuzzy join operation by applying the fuzzy sets-based techniques on the numerical attributes. In our paper, we show how we utilize this idea on the numerical values of the time attribute while combining sensor data from a wearable device (a smartwatch) with meteorological data from weather sensors.

3 Fuzzy Sets for Flexibility

Fuzzification of selected attributes of sensor readings may introduce some flexibility while joining various data collections. Fuzzy sets can play a particular role here. The fuzzy sets theory assumes that the membership degree $\mu(x)$ of an object to the set A may be represented with countless values within the unit interval $[0, 1]$ [16]. This stays in contrast to the classical set theory that assumes that membership of an object to a set is bivalent – the object either belongs to the set or does not belong to it.

Assuming that X is the universe of points (objects) and x is an element of X, the fuzzy set A in X is defined as an ordered collection of pairs:

$$A = \{(x, \mu_A(x)) | x \in X\}, \tag{1}$$

where μ_A is the *membership function* defining the set A, and $\mu_A(x)$ is the *membership degree* of the element x to the set A, which takes a value from 0 to 1.

Graphically, the *membership function* is usually represented as a triangular or trapezoidal function. Triangular function is used when there is only one situation such as the value of membership is equal to 1. Figure 1a shows sample fuzzy set *training time around 9:00 AM* defined with the use of the triangular membership function. This type of characteristic function is defined by three parameters a, b and c, where $a \leq b \leq c$, as follows:

$$\mu_A(x; a, b, c) = \begin{cases} 0, & x \leq a \\ \frac{x-a}{b-a}, & a < x \leq b \\ \frac{c-x}{c-b}, & b < x \leq c \\ 0, & c < x \end{cases} \tag{2}$$

On the other hand, the trapezoidal characteristic function is described by four parameters a, b, c and d, where $a \leq b \leq c \leq d$, and is defined as follows:

$$\mu_A(x; a, b, c, d) = \begin{cases} 0, & x \leq a \\ \frac{x-a}{b-a}, & a < x \leq b \\ 1, & b < x \leq c \\ \frac{d-x}{d-c}, & c < x \leq d \\ 0, & d < x \end{cases} \tag{3}$$

Figure 1b shows a sample fuzzy set *morning time* defined with the trapezoidal membership function, together with the calculation of the membership degree for the beginning of sample sports trainings.

4 Cloud-Based Monitoring and Data Analysis System

Training data processing and further analysis are performed in the Cloud-based system presented in Fig. 2. Runners are equipped with the Garmin smartwatch, which gathers training parameters with the use of various sensors. In

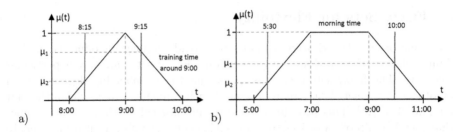

Fig. 1. Sample fuzzy sets defined for the training time: (a) *training time around 9:00 AM* defined with the use of the triangular membership function, (b) *morning time* defined with the trapezoidal membership function, and calculation of the membership degree for the beginning of sample trainings.

our research, we focused on the effectiveness of running/jogging for a selected group of 15 sports amateurs jogging systematically within a period of one year.

The Garmin smartwatch allowed to collect the following data on the performed training:

- date and time of the training,
- training duration,
- heart rate,
- distance,
- calories burned,
- raw GPS data for determining the route,
- running cadence (number of steps a runner takes per minute),
- average speed,
- training type.

Data produced by the smartwatch during the training are collected as training data files stored in the *.tcx* format. *.tcx* is the acronym for Training Center XML introduced by Garmin Company. The format enables the exchange of GPS tracks as an activity with parameters of monitored training, including running, biking, and other forms of activity. The data produced by the smartwatch create a collection of data *at rest*, i.e., these are not constantly monitored data streams, but historical, offline data that are sent to the Cloud after the training is finished.

The Edge gateway module is responsible for transmitting the data to the Cloud. However, before it happens, the training data are combined with weather conditions for the day of performed training. To this purpose, we invoke a URL request to the Dark Sky Web service by using appropriate API (Application Programming Interface). The Web service accepts the date, time and coordinates on the input and returns the following parameters describing the weather conditions on the output:

- temperature – real and apparent,
- air humidity,

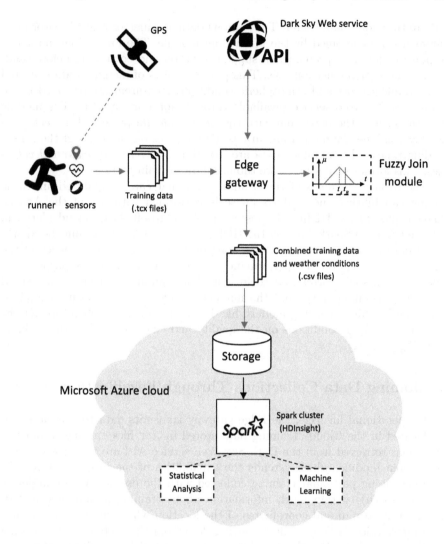

Fig. 2. Architecture of the Cloud-based system for training data analysis.

- air pressure,
- wind speed,
- dew point,
- cloud cover,
- UV index.

Data collected by the smartwatch are supplemented by the most appropriate meteorological data from weather sensors provided by the Dark Sky Web service in the Edge gateway. The gateway is a device, like an electronic unit, mobile phone or a field computer, which pre-process the data and transmits the data to the Cloud data center. The training data and meteorological sensor measurements are combined by the Fuzzy Join module. The Fuzzy Join module is responsible for data preparation, supplementation and combining before sending the data for further analysis. This phase consists of merging data collected by wearable sensors and atmospheric conditions measurements provided by an online weather Web service (available through appropriate API). The module uses the idea of the fuzzy join with the *fuzzy umbrella* presented in Sect. 5, to flexibly combine the data from various data sources on the basis of the sensor reading times. The combined data are stored as *.csv (comma-separated) values* files. This format enables to store tabular data in plain text.

Due to large volumes of the training data that can be analyzed and wide scaling capabilities, the analysis phase is performed with use of the Apache Spark engine in the HDInsight cluster in the Microsoft Azure cloud platform. We used Apache Spark 2.3.0 on the HDInsight cluster 3.6.0. Within the Spark-based data analysis we can perform statistical analysis of the influence of the weather conditions on the performance of the training. For this purpose, we calculate Pearson's correlation coefficient. The training efficiency is measured by average running speed and the number of calories burnt during exercises. With the Machine Learning models, like Linear Regression, we can predict the impact of weather conditions on the quality and efficiency during the training (running/jogging).

5   Joining Data Collections Through Fuzzification

In its operational lifecycle, the Edge gateway transmits data to the data center located in the Cloud. Training data stored in *.tcx* files and meteorological conditions retrieved from the Dark Sky Web service API are joined within the Fuzzy Join module, which extends the capabilities of the Edge gateway. The data describing particular training collected with the use of a wearable device (a smartwatch) contain much information about training parameters, including the time stamp and coordinates of the location where the training begins. This information is used while retrieving data from the Web service providing weather conditions. Weather conditions are retrieved by specifying the date, time, and coordinates of the training. They are delivered as *.json* objects containing hour-by-hour daily measurements (air temperature (real and apparent), the percentage level of air humidity, dew point, wind speed, air pressure, and others). The aim of the fuzzy join is to find the most accurate weather conditions based on the time the training begins.

The fuzzy join algorithm (Algorithm 1) implemented in the Fuzzy Join module calculates the value of a membership degree of the time of training to the fuzzy sets U created for each full hour of weather measurement t_{m_j}:

$$\forall_{t_{m_j}} \quad U_j = \{(t, \mu_U(t)) | t \in T\}, \tag{4}$$

where $t \in T$ represents all possible time points on the timeline T, and:

$$Supp(U_j) = (t_{m_j} - 1\,\mathrm{h}, t_{m_j} + 1\,\mathrm{h}), \tag{5}$$

$$Core(U_j) = \{t_{m_j}\}, \tag{6}$$

where $Supp(U_j) = \{t \in T, \mu_U(t) > 0\}$ is the support of the fuzzy set U_j, and $Core(U_j) = \{t \in T, \mu_U(t) = 1\}$ is the core of the fuzzy set U_j. Such defined fuzzy sets U_j are called *fuzzy umbrellas* and they may cover various training times $t_i \in T$. The fuzzy join algorithm for combining training data with weather conditions is presented Algorithm 1.

Algorithm 1. Combining training data with weather conditions with the use of Fuzzy join algorithm

---

1: **procedure** FuzzyJoin(t_i, t_{m_j}) ▷ Meteorological conditions are read at full hours
2:   **for each** training time t_i **do**
3:     Convert the beginning of training (t_i) into seconds
4:     Convert the time of weather conditions measurement t_{m_j} into seconds
5:     Find neighbouring times of meteorological measurements $t_{m_{j-1}}, t_{m_{j+1}}$ for the time t_{m_j} and convert them into seconds
6:     $\mu_{U_{j-1}} \leftarrow 1 - \frac{|t_{m_{j-1}} - t_i|}{3600}$
7:     $\mu_{U_j} \leftarrow 1 - \frac{|t_{m_j} - t_i|}{3600}$
8:     $\mu_{U_{j+1}} \leftarrow 1 - \frac{|t_{m_{j+1}} - t_i|}{3600}$
9:     **for each** $\mu \in \{\mu_{U_{j-1}}, \mu_{U_j}, \mu_{U_{j+1}}\}$ **do**
10:       **if** $\mu \notin \langle 0; 1 \rangle$ **then**
11:         Skip μ
12:       **end if**
13:     **end for**
14:     Join training data for time t_i with meteorological sensor readings from the time $t_m \in \{t_{m_{j-1}}, t_{m_j}, t_{m_{j+1}}\}$ for which $\mu_{t_m} = \max\{\mu_{U_{j-1}}, \mu_{U_j}, \mu_{U_{j+1}}\}$
15:   **end for**
16: **end procedure**

For each training (starting at time t_i), the algorithm converts the time of weather conditions measurement t_{m_j} (Fig. 3) and the beginning of training t_i into seconds (e.g., 10 AM for weather conditions measurement is converted to 36000 s). In the next step, it finds times of meteorological measurements neighbouring to the t_{m_j} ($t_{m_{j-1}}, t_{m_{j+1}}$) and convert them into seconds. Then, it computes the differences between the weather conditions measurement hours ($t_{m_{j-1}}$, t_{m_j}, $t_{m_{j+1}}$) and the beginning of the training (t_i) (values in seconds). It takes the absolute values of the results and computes the ratio of the values to the number of seconds in one hour (3600 s). For each of the times of weather conditions measurement ($t_{m_{j-1}}$, t_{m_j}, $t_{m_{j+1}}$) the value of membership function (μ_U) is calculated as the difference of 1 and the calculated ratio. If the calculated value is not between 0 and 1 it has to be rejected, as it is out of the range of the membership degree. Finally, the algorithm takes the maximum of all values of the calculated membership degree and combines training data with those meteorological conditions for which the membership degree is the highest.

The use of fuzzy join allows finding the best matching of the weather conditions to the training as they are chosen by the nearest hour of measurement. The concept of the fuzzy umbrella is presented in Fig. 3. On the timeline, there are hours of weather conditions measurements from sensors in meteorological stations. On the μ axis, there are placed values of membership function computed for each hour of weather measurement.

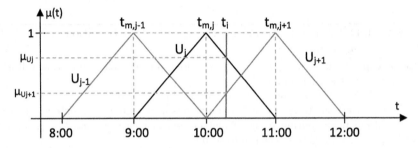

Fig. 3. Fuzzy umbrellas defined for the time of weather sensors measurements $(t_{m_{j-1}}, t_{m_j}, t_{m_{j+1}})$ and calculation of membership degree for training time t_i.

6   Experimental Results

Our experiments covered (1) the analysis of correlations between training parameters like distance, time, average speed, etc. and meteorological measurements, and (2) verification of the performance of the fuzzy join operation. We decided to conduct the experiments for each person independently as everyone had different running habits and experiences. The training data that we collected concerned running on an average distance of 10 km. The runners were amateurs with different level of running experience, various running habits and frequency. Among them, there were men and women, at the age between 20 and 55. The standard deviation of the distance was approximately 1 km whereas for the about 1-h training standard deviation value was near to 7 min. The running parameters had normal distribution. Weather conditions, in which the trainings were performed also varied. For example, the range of measured temperatures was from $-15\,°C$ up to $32\,°C$, while the air humidity was between 26% and 100%.

The results for each of the analyzed set of data (workout) varied. The average values of correlations are shown in Table 1. The correlation measure was presented with the absolute value (without direction) of the Pearson correlation coefficient defined as:

$$\rho_{X,Y} = \frac{cov(X,Y)}{\sigma_X \sigma_Y} \tag{7}$$

where X, Y are random variables, $cov(X, Y)$ is the covariance, and σ_X, σ_Y are the values of standard deviations of X and Y.

Table 1. Ranges of linear correlations between running parameters and meteorological conditions

ρ	Duration	Distance	Calories	Average speed
Temperature	0.022–0.224	0.053–0.224	0–0.196	0.046–0.286
Air humidity	0.045–0.386	0.046–0.205	0.014–0.174	0.023–0.313
Air pressure	0.006–0.132	0.021–0.228	0.014–0.350	0.026–0.255
Wind speed	0.019–0.500	0.021–0.364	0.019–0.292	0.012–0.646

The results of the analyzed correlations are diversified for each of the runners. The widest range may be observed for the interactions between training parameters and the wind speed – from only 1%, which means no correlation at all, up to 65%. Large values of the Pearson coefficient were noticed for the air humidity – between this feature and the duration of the training, the correlation for one of the persons reached nearly 39%. For the temperature, opposite to what could be expected, the results were not so satisfying, however, the maximum value of the correlation between the temperature and average speed exceeded 28%. The summary of the average results for all of the runners is presented in Table 2.

Table 2. Average correlations between running parameters and meteorological conditions

ρ	Duration	Distance	Calories	Avg speed
Temperature	0.099	0.144	0.115	0.138
Air humidity	0.153	0.112	0.073	0.144
Air pressure	0.063	0.088	0.117	0.106
Wind speed	0.205	0.119	0.125	0.170

On average, the results of our experiments show that there are correlations between trainings parameters and weather conditions. Usually, the average value of the Pearson coefficient is between 10% and 20%. Although the value of the correlation seems to be low, particular values for individual athletes are strongly differentiated. This shows that some of them are able to perform the training assumptions regardless of the prevailing weather conditions. The strongest correlations may be noticed between running parameters and the wind speed while the weakest correlation is for the air pressure. Despite the results seem satisfying, it is important to remember that for every runner the correlations were different – for ones they were stronger, for others – negligible. Moreover, there

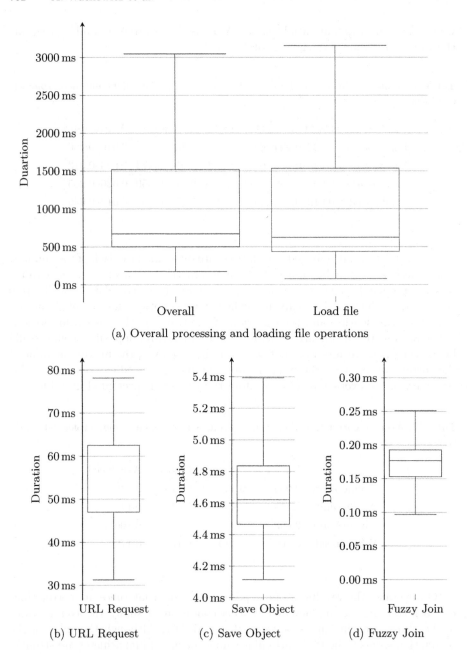

(a) Overall processing and loading file operations

(b) URL Request   (c) Save Object   (d) Fuzzy Join

Fig. 4. Duration of particular operations on one set of the training data.

are also many other factors that may have an impact on the results but were not
analyzed during our experiments – like the quality and time of the meal before
training, the overall condition of the runner or even their health state.

In the second series of experiments, we tested how the fuzzy join operation affects the performance of the data processing on the Edge gateway (Fig. 2). The Edge gateway and the Fuzzy Join module combine data before sending the data to the Cloud for further analysis. The performance tests were conducted on PC station with 8 GB RAM and processor Intel(R) Core(TM) i7-3537U CPU @ 2.00 GHz, controlled by the 64-bit Windows operating system. To obtain the most reliable results, during experiments no other applications were running on the machine.

We tested the performance of the fuzzy join on the data set consisting of 850 files (*.tcx*) with data from real training (running/jogging). Results of our experiments presented in Figs. 4 prove that the execution time of the Fuzzy Join algorithm (Fig. 4d) is negligible in comparison to other operations. On average, it takes less than 0.20 ms which is 0.0015% of overall time spent on processing a particular (one out of 850) data file.

During the operation named *Load file* (Fig. 4a) the *.tcx* training data file is opened and all of the needed data are retrieved and prepared for further analysis. The time of this operation is related to the size of the file which depends on the number of measurements. Duration of the URL request (Fig. 4b) to the Web service is determined by the network efficiency and traffic. However, the average time of this step is much shorter than the time of the previous operation. In Fig. 4c we also presented the duration of saving combined training and weather data to an output file performed by the Fuzzy Join module. The time of this operation is on average about 5 ms, which is about 0.44% of the overall time needed for processing a single file. Although the maximum time attracts attention due to its high value, the values of other measures point that it is rather a singular outlying result than a frequent issue.

7   Conclusion

Application of fuzzy sets while joining sensor data on the Edge gateway allows not only to flexibly combine training data collected with a smartwatch with weather conditions but also delivers information on the compatibility of combined time moments for both data sets. This can be an important factor while analyzing the correlation between the performance of the training and the weather parameters, and planning tactics for future training activities.

The fuzzy join algorithm operates on numerical representation of time stamps, similar to the solution presented in [9] that operates on numbers in Big Data cloud environments, and in contrast to works [1,2,15] that operate on strings. Moreover, likewise it was presented in works by Yamato [14], Revathi Pulichintha Harshitha et al. [11], Małysiak-Mrozek et al. [9] the whole solution is built upon the cloud infrastructure, which allows for scalable data analysis. However, the fuzzy join is performed on the Edge device, which reduces the amount of the processing work performed in the Cloud.

The data processing performed on the Edge gateway consists of many operations, out of which the fuzzy join takes the least time. This shows that the operation will not introduce any significant delays in data pre-processing. Meanwhile, it is very important, since it supplements the existing training data with additional information that may shed new light on the analyzed data.

During our preliminary data analysis on the Spark cluster in the Cloud, we could notice the existence of correlations between some weather measurements and running efficiency. Some of them were strong and some of them were weak – they largely depend on a person who does the activity. Other factors like talent or psychical strength, which are also mentioned in [13] and [5] but were out of our analysis, could contribute to the results. These elements are difficult to determine. An important issue, also hard to classify, is the level of runner's experience and sports form during a particular workout. Still, we believe that the presence of correlations between meteorological measurements and training parameters is an interesting matter and worth further studies in our future works.

References

1. Afrati, F.N., Sarma, A.D., Menestrina, D., Parameswaran, A., Ullman, J.D.: Fuzzy joins using MapReduce. In: 2012 IEEE 28th International Conference on Data Engineering, pp. 498–509, April 2012. https://doi.org/10.1109/ICDE.2012.66
2. Deng, D., Li, G., Hao, S., Wang, J., Feng, J.: MassJoin: a MapReduce - based method for scalable string similarity joins. In: 2014 IEEE 30th International Conference on Data Engineering, pp. 340–351, March 2014
3. Dhahbi, W.: Seasonal weather conditions affect training program efficiency and physical performance among special forces trainees: a long-term follow-up study. PLoS One **13**(10) (2018). https://doi.org/10.1371/journal.pone.0206088
4. El Helou, N., et al.: Impact of environmental parameters on marathon running performance. PLoS One (2012). https://doi.org/10.1371/journal.pone.0037407
5. Ely, M.R., Cheuvront, S.N., Roberts, W.O., Montain, S.J.: Impact of weather on marathon-running performance. Med. Sci. Sports Exerc. **39**(3), 487–493 (2007)
6. Khorasani, E.S., Cremeens, M., Zhao, Z.: Implementation of scalable fuzzy relational operations in MapReduce. Soft Comput. **22**(9), 3061–3075 (2018). https://doi.org/10.1007/s00500-017-2561-3
7. Kimmett, B., Srinivasan, V., Thomo, A.: Fuzzy joins in MapReduce: an experimental study. Proc. VLDB Endow. **8**, 1514–1517 (2015). https://doi.org/10.14778/2824032.2824049
8. Lara, O.D., Labrador, M.A.: A survey on human activity recognition using wearable sensors. IEEE Commun. Surv. Tutor. **15**(3), 1192–1209 (2013)
9. Małysiak-Mrozek, B., Lipińska, A., Mrozek, D.: Fuzzy join for flexible combining big data lakes in cyber-physical systems. IEEE Access **6**, 69545–69558 (2018). https://doi.org/10.1109/ACCESS.2018.2879829
10. Mukhopadhyay, S.C.: Wearable sensors for human activity monitoring: a review. IEEE Sens. J. **15**(3), 1321–1327 (2015)
11. Revathi Pulichintha Harshitha, S., Narramneni, P., Raghavee, N.S.: Body sensor using internet of things (IoT). ARPN J. Eng. Appl. Sci. **13**(8), 2916–2922 (2018)
12. Toh, W.Y., Tan, Y.K., Koh, W.S., Siek, L.: Autonomous wearable sensor nodes with flexible energy harvesting. IEEE Sens. J. **14**, 2299–2306 (2014)
13. Vihma, T.: Effects of weather on the performance of marathon runners. Int. J. Biometeorol. **54**(3), 297–306 (2010). https://doi.org/10.1007/s00484-009-0280-x

14. Yamato, Y.: Proposal of vital data analysis platform using wearable sensor. In: Proceedings of the 5th IIAE International Conference on Industrial Application Engineering (2017)
15. Yan, C., Zhao, X., Zhang, Q., Huang, Y.: Efficient string similarity join in multi-core and distributed systems. PLoS One **12**(3), 1–16 (2017)
16. Zadeh, L.A.: Fuzzy sets. Inf. Control **8**, 338–353 (1965)

Collaborative Learning Agents (CLA) for Swarm Intelligence and Applications to Health Monitoring of System of Systems

Ying Zhao[1]([✉])[iD] and Charles C. Zhou[2][iD]

[1] Naval Postgraduate School, Monterey, CA 93943, USA
yzhao@nps.edu
[2] Quantum Intelligence, Inc., Monterey, CA 93943, USA
charles.zhou@quantumii.com
http://www.quantumii.com

Abstract. The system of systems is the perspective of multiple systems as part of a larger, more complex system. A system of systems usually includes highly interacting, interrelated and interdependent sub-systems that form a complex and unified system. Maintaining the health of such a system of systems requires constant collection and analysis of the big data from sensors installed in the sub-systems. The statistical significance for machine learning (ML) and artificial intelligence (AI) applications improves purely due to the increasing big data size. This positive impact can be a great advantage. However, other challenges arise for processing and learning from big data. Traditional data sciences, ML and AI used in small- or moderate-sized analysis typically require tight coupling of the computations, where such an algorithm often executes in a single machine or job and reads all the data at once. Making a generic case of parallel and distributed computing for a ML/AI algorithm using big data proves a difficult task. In this paper, we described a novel infrastructure, namely collaborative learning agents (CLA) and the application in an operational environment, namely swarm intelligence, where a swarm agent is implemented using a CLA. This infrastructure enables a collection of swarms working together for fusing heterogeneous big data sources in a parallel and distributed fashion as if they are as in a single agent. The infrastructure is especially feasible for analyzing data from internet of things (IoT) or broadly defined system of systems to maintain its well-being or health. As a use case, we described a data set from the Hack the Machine event, where data sciences and ML/AI work together to better understand Navy's engines, ships and system of systems. The sensors installed in a distributed environment collect heterogeneous big data. We show how CLA and swarm intelligence used to analyze data from system of systems and quickly examine the health and maintenance issues across multiple sensors. The methodology can be applied to a wide range of system of systems that leverage collaborative, distributed learning agents and AI for automation.

Keywords: Collaborative learning agents · Swarm intelligence · System of systems · Health monitoring · Lexical link analysis ·

J. M. F. Rodrigues et al. (Eds.): ICCS 2019, LNCS 11538, pp. 706–718, 2019.
https://doi.org/10.1007/978-3-030-22744-9_55

Distributed computing · Parallel computing · Data fusion ·
Machine learning · Artificial intelligence

1 Introduction

The system of systems is the viewing of multiple systems as part of a larger, more complex system. For example, a Navy ship is a system of systems. The internet of things (IoT) is a system of systems. A system of systems usually includes highly interacting, interrelated and interdependent sub-systems that form a complex and unified system. Maintaining the health of such a system of systems requires constant collection and analysis of the big data from sensors. The data for a system of systems are often collected in a distributed fashion from the sensors installed in the sub-systems. Fusing and analyzing the data from heterogeneous sensors in a holistic approach the key to successfully detecting problems, monitoring and maintaining the health of a system of systems.

In a separate perspective, as the size of data increases for data analytics such as machine learning (ML) and artificial intelligence (AI), the statistical significance for these methods often improves purely due to the increased data size. This positive impact of big data drives proliferated considerations of ML/AI applications.

However, other challenges arise. For example, the computational concept map/reduce - an analytic programming paradigm for big data, which consists of two tasks: (1) the "map" task, where an input data set is converted into key/value pairs; and (2) the "reduce" task, where outputs of the "map" task are combined to a reduced key-value pairs, serves as the cornerstone of many big data algorithms and their variations. The paradigm typically include computers used in parallel computations (e.g., hadoop clusters) to be physically clustered in the same location.

Traditional data sciences used in small- or moderate-sized analysis typically require tight coupling of the computations of the "map" and "reduce" steps in a typical big data algorithm. Such an algorithm often executes in a single machine or job and reads all the data at once. How can these algorithms be modified so they can be executed in parallel? If the data is processed in parallel and parsed into subsets, how to leverage the art and science of fusing the results as phrased in the "reduce" step? Making a generic case for a ML/AI algorithm running in a parallel environment proves to be a difficult task. Furthermore, running such an algorithm in a distributed environment is even more challenging, for example, using an agent to compute part of the analysis separately in sub-systems of a system, and then combing the results.

In this paper, we describe a novel infrastructure called collaborative learning agents (CLA) and the application in an operational environment, namely swarm intelligence, where a swarm agent is implemented using a CLA. This infrastructure enables a collection of swarms working together, not only for fusing heterogeneous big data sources in a parallel and distributed fashion, but also for effectively performing customized analytics such as ML/AI algorithms as if they are as a single agent. We show a use case to use CLA for monitoring the health of a system of systems.

2   Collaborative Learning Agents (CLA)

Our previous work [8] shows the architecture of CLA. A single agent representing a single system capable of ingesting and analyzing data sources while employing a process (i.e., an unsupervised learning process) that separates patterns and anomalies within the data. Multiple agents can work collaboratively in a network. This collaboration is achieved through a peer list defined within each agent, through which each agent passes shared information to its peers. Each agent initially analyzes its own input or content data separately and then fuses the results with its peers'.

In detail shown in Fig. 1, an agent CLA j includes an analytic engine with an algorithm for data fusion and one for ML/AI that can be customized externally. The fusion algorithm integrates the local knowledge base $b(t,j)$ with an input knowledge base $B(t-1,i)$ from its peers i and forms a new knowledge base $B(t,j)$. $B(t-1,i)$ represents all knowledge from $i's$ network up to point $t-1$. The ML/AI algorithm can be an anomaly detection algorithm, for example, such an algorithm like lexical link analysis (LLA) that assesses the total value of the agent j by separating the new knowledge base $B(t,j)$ into the categories of patterns, emerging and anomalous themes and computes a total value $V(t,j)$ [1,2]. LLA functions as both an fusion and ML/AI (unsupervised learning) algorithms (see Sect. 4). A knowledge base $B(t,j)$ contains two components: The first component is an association list which contains pairwise correlations or associations between two word features for structured data or bi-gram word pairs for unstructured data. The second component is a context/concept list, which essentially the same set of context points such as timestamps, geo-locations or file names used in the fusion step to fuse with data from multiple agents.

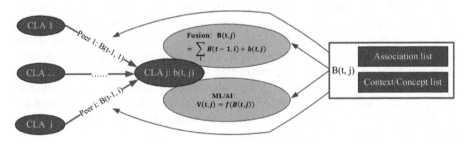

Fig. 1. CLA detail: each agent contains a fusion and ML/AI engine. The fusion is represented as an additive term here as a special case of Step 1 in Sect. 5. The function forms of the fusion and ML/AI can be customized.

3   Swarm Intelligence (SI)

The CLA concept has an analogue in nature. As human being often ponder: What is the mechanism behind that flocking swarms successfully achieve collective goals such as looking for food or going to places in an optimized fashion using only local and simple communications as shown in Fig. 2 (left) [12]. Often swarms can maximize a total value, e.g., get to a food target in a shortest distance. Swarms find an optimal solution using the pheromone, or the chemical substances produced and released into the environment by a mammal or an insect, which affects the behavior or physiology of others. The concept is simulated in work in AI, i.e., swarm intelligence (SI). SI is the collective behavior of natural or artificial, decentralized and self-organized systems. The expression was introduced in the context of cellular robotic systems as shown in Fig. 2 (right) [13].

Fig. 2. Left: natural flocking swarm behaviors [12]. Right: swarm intelligence has been simulated in the context of cellular robotic systems. It has been used for design armed forces, wireless communications, cellular automata, peer-to-peer networks where the whole system has stronger collective intelligence than individual systems [13]

4   Lexical Link Analysis (LLA) and CLA

4.1   LLA as a Text Analysis Tool for Unstructured Data

In a LLA, a complex system can be expressed in a list of attributes or features with specific vocabularies or lexicon terms to describe its characteristics. LLA is a data-driven text analysis. For example, word pairs or bi-grams as lexical terms can be extracted and learned from a document repository. LLA automatically discovers word pairs, clusters of word pairs and displays them as word pair networks. LLA is related to but significantly different from so called bag-of-words (BOW) methods such as Latent Semantic Analysis (LSA [3], Probabilistic Latent Semantic Analysis (PLSA) [4], WordNet [5], Automap [10], and Latent Dirichlet

Allocation (LDA) [6]. LDA uses a bag of single words (e.g., associations are computed at the word level) to extract concepts and topics. LLA uses bi-gram word pairs as the basis to form word networks and therefore network theory and methods can be readily applied here.

4.2   Extending LLA to Structured Data

Bi-gram also allows LLA to be extended to numerical or categorical data. For example, for structured data such as attributes from databases, we discretize and then categorize attributes and their values to word-like features. The word pair model can further be extended to a context-concept-cluster model [8]. In this model, a context is a word or word feature shared by multiple data sources. A concept is a specific word feature. A context can represent a location, a timestamp or an object (e.g. file name) shared across data sources. In the use case in Sect. 7, a timestamp is the context.

4.3   Three Categories of High-Value Information and Value Metrics

The word pairs in LLA are divided into groups or themes. Each theme is assigned to one of the three categories based on the number of connected word pairs (edges) within a cluster (intra-cluster) and the number of edges between the themes (inter-cluster):

- Authoritative or popular (P) themes: These themes resemble the current search engines ranking measures where information containing the dominant eigenvectors rank high because the elements of the dominant eigenvectors tend to not only connect to each other but also connect to the elements outside a group. They represent the main topics in a data set and are insightful information in three folds: (1) These word pairs are more likely to be shared or cross-validated across multiple diversified domains, so they are considered authoritative; (2) These themes could be less interesting because they are already in the public consensus and awareness, so they are considered popular; (3) The records associated with these themes are considered normal. A popular theme has the largest number of inter-connected word pairs. The content associated with popular themes disseminate faster.
- Emerging (E) themes: These themes tend to become popular or authoritative over time. An emerging theme has the intermediate number of inter-connected word pairs.
- Anomalous (A) themes: These themes may not seem to belong to the data domain as compared to others. They are interesting and could be high-value for further investigation.

Community detection algorithms have been illustrated in Newman [9,10], a quality function (or Q-value), as specifically defined as the "modularity" measure, i.e., the fraction of edges that fall within communities, minus the expected value of the same quantity if edges fall at random without regard for the community structure, is optimized using a "dendrogram" like greedy algorithm. The

Q-value for modularity is normalized between 0 and 1 with 1 to be the best and can be compared across data sets. Formation of the modularity matrix is closely analogous to the covariance matrix whose eigenvectors are the basis for Principal Component Analysis (PCA) [10]. Modularity optimization can be regarded as a PCA for networks. Related methods also include Laplacian matrix of the graph or the admittance matrix and spectral clustering [7]. Newman's modularity assumes a subgraph deviates substantially from its expected total number of edges to be considered anomalous and interesting, therefore, all the clusters or communities (i.e., popular, emerging and anomalous themes regardless) found by the community detection algorithms are considered to be interesting. However, this anomalousness metric does not consider the difference among the communities or clusters.

In LLA, we improve the modularity metric by considering a game-theoretic framework: In a nutshell and in a social network, the most connected nodes are typically considered the most important nodes. However, in LLA, we consider emerging and anomalous information are more interesting and correlated to high-value information. Also, for a piece of information, the combination of popular, emerging and anomalous contributes to the total *value* of the information. Therefore, we define a value metric as follows:

Let the popular, emerging and anomalous value of the information i be $P(i)$, $E(i)$ and $A(i)$ computed from LLA respectively, the total value $V(i)$ for i is defined as in (1).

$$V(i) = P(i) + E(i) + A(i) \tag{1}$$

In the use case in Sect. 7, we show that the value metrics are correlated with high-value information, e.g., anomalous profiles of Navy engine data.

5 Recursive Learning in CLA and SI

The key advantage of using CLAs relies on using a collection of agents or artificial swarms to perform a task difficult to perform by individual agents. Assume each swarm consists of a CLA and processes part of the total sensor data.

- An agent j represents one sensor or part of the total sensors, operates on its own like a decentralized data analyzer. A single agent does not communicate with all other sensors but only with the ones that are its peers. A peer list is specified by the agent, for example, in Fig. 7, there are three agents in total, CLA 1, CLA 2 and CLA 3. CLA 1 has two peers CLA 2 and CLA 3; CLA 2 has one peer CLA 1; and CLA 3 has one peer CLA 1.
- An agent j collects, analyzes from its domain specific data knowledge base $b(t, j)$. For example, $b(t, j)$ may represent the statistically significant features and associations based on the data observed only by agent j.
- An agent j also includes an analytic engine with two algorithms (i.e., a fusion and ML/AI algorithm) that can be customized externally. We use the two algorithm LLA1 (fusion) and LLA2 (ML/AI) in the implementation of LLA to illustrate the process. The fusion algorithm (LLA1) integrates the local

knowledge base $b(t,j)$ and the global knowledge base $B(t-1,j)$ into a new knowledge base $B(t,j)$. The ML/AI algorithm (LLA2) assesses the total value of the agent j by separating the total knowledge base into the categories of patterns, emerging and anomalous themes based on the total knowledge base $B(t,j)$ and generates a total value $V(t,j)$. The whole process is displayed as follows:

- Step 1: $B(t,j) = LLA1(B(t-1,p(j)), b(t,j))$;
- Step 2: $V(t,j) = LLA2(B(t,j))$.

Where $p(j)$ represents the peer list of agent j.

– The total value V(t,j) is used in the global sorting and ranking of relevant information.

In this recursive data fusion, the knowledge bases and total values are completely data-driven and automatically discovered and unsupervised-learned from the data. Each agent consists the exact same code, yet collects and analyzes its own data apart from other agents. This agent design has the advantages of decentralized and distributed models: performing learning and fusing simultaneously and in parallel.

6 Fusion and Context Learning in CLA and SI

In a CLA's fusion step, if agent j's local model $b(t,j)$ shares vocabulary or word features with the knowledge bases that its peers pass to, i.e., $B(t-1,i)$, the fusion part is simply to modify and update the association list to reflect the new data. Meanwhile, a so-called context learning is performed at each agent using the context/concept list if the agents do not share common word features or vocabulary as follows:

– Step 1: Each agent loops through Peer i in its peer list, and list all contexts and associations from its peers and local data $b(t,j)$.
– Step 2: For each concept (word feature) i_c in $B(t-1,i)$, check if agent j's local data $b(t,j)$ and concept j_c to see if it has the same context. If yes, concept i_c and concept j_c in agent i and agent j is linked and the association is added to the knowledge base $B(t,j)$.

Figure 1 also shows the update algorithm and context learning algorithm. Both are part of the whole fusion algorithm (LLA1). The LLA2 refers the part of the LLA algorithm that categorize word features into popular, emerging and anomalous ones.

7 Use Case

At the heart of the US Navy are thousands of machines that drive the ships and submarines. The U.S. Navy mission is to maintain, train and equip combat ready naval forces capable of winning wars, deterring aggression and maintaining freedom of the seas. The US Navy needs to harness big data, data sciences, and

ML/AI to better understand these machines as system of systems. A test data set was culled from the Navy's engine rooms around the world, link data sciences, ML/AI. The data was used in the Hack The Machine event in Cambridge, Massachusetts in September, 2017 organized by the US Navy [11].

The data set is a typical health maintenance use case where multiple sensors are used to monitor a system of systems (e.g., a ship) to see if it is operated normal or if there are any "health" issues. The sensor data collected can be in a variety of heterogeneous formats such as numerical values, image and text etc. The correlations and associations of the multiple sensor data are not necessarily known before the data collection. The sensors can also be installed in a distributed fashion, for example, in different ships or in the same ship but different ship subsystems.

Figure 3 is a ML/AI paradigm to learn from historical data of system of systems (e.g., ship 1 and ship 2) and then apply the knowledge patterns learned and discovered to the new data (e.g., ship-x). The CLAs in a swarm intelligence can reside in the systems or subsystems in a distributed fashion in this case.

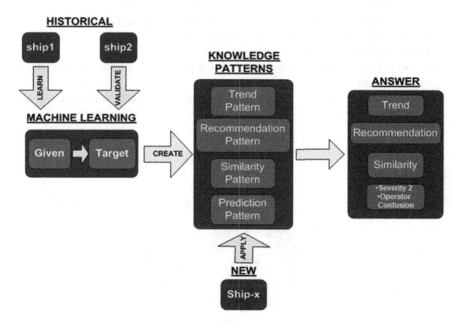

Fig. 3. Data sciences, ML/AI meet to check the health of a system of systems

Figure 4 shows an sample of the original data set with about 50 variables with timestamps over a period of a year (7/2016 to 6/2017). Each numerical variable is discretized into LLA word features based on the initial statistics such as means and standard deviations. The total LLA features $n = 160$. For example, $bearing\_temp\_aft\_bt\_107.97\_136.70$ represents a feature from the original sensor measurement variable $bearing\_temp\_aft$ with its value between 107.97

and 136.70, where 107.97 and 136.70 is the mean and the mean plus one standard deviation. The values were generated automatically and initially within a CLA. If only one CLA is used with the fusion of a set of agents (swarms), in order to perform the ML/AI in the second step, the association list $B(t, j)$ needs to be computed with the amount of the computation $O(n^2 m)$, where m is the number of contexts (i.e., timestamps) that can link these word features.

DateTime	VOLTA	TURBI	TURB	SW PL	STATC	STATC	STATC	STATC	STATC	STATC	STATC	POWE	POWE	LUBE	(LO SU	HP AI	GEN S	GEN L	GEN L	GEN L	GEN L
7/22/2016 1:44	5.5	84	0	30	103	98	98.4	98.8	97.3	96.5	102.5		0	1	1	2900	1.5	25.1	1	8	9
7/22/2016 1:44	7	84	0	30	102	96.9	98.6	98.8	97.5	96.9	102.3		0	1	1	2900	1.5	25.8	1	8	9
7/22/2016 1:44	11.7	85	0	30	101	101.2	96.5	99.4	97.1	94.5	100.2		0	1	1	2900	1.5	15.5	1	8	9
7/22/2016 1:44	20.5	85.9	0	30	105	97.7	104.9	103.5	97.9	90.4	99.6		0	1	1	2900	1.5	15.4	1	8	9
7/22/2016 1:45	33.7	102.5	0	30	108	107.8	105.3	103.5	98.6	88.9	96.7		0	1	1	2900	1.5	14.8	1	8	9
7/22/2016 2:12	49.5	351.6	0	30	137	113.7	129.7	121.5	119.5	137.3	120.9		-3.9	1	1	2900	1.5	12.7	1	8	9
7/22/2016 2:12	36	343.8	0	30	126	126	119.9	124.8	119.7	93.4	108.2		0	1	1	2900	1.5	12.5	1	8	9
7/22/2016 2:12	26.7	336.9	0	30	123	123.4	119.7	112.1	118.6	111.5	123.4		0	1	1	2900	1.5	12.2	1	8	9
7/22/2016 2:12	21.7	334	0	30	127	117.4	114.8	115.2	116	119.9	127.3		0	1	1	2900	1.5	12.1	1	8	9
7/22/2016 2:12	16.1	329.1	0	30	125	118	118.4	124.8	117.4	112.5	120.3		2	1	1	2900	1.5	11.9	1	8	9
7/23/2016 4:47	47.2	346.7	0	30	130	129.9	102.5	112.1	114.1	117.2	120.9		-17.6	1	1	2900	1.5	12.6	1	8	9

DateTime	GEN B	GEN A	GEN A	GEN A	GEAR	FUEL (	FUEL (	FUEL F	FUEL F	FREQ	EXCTR	ENGIN	ENGIN	ENGIN	ENG L	ENG L	ENG L	ENCL	CURR	COAL	BLEED	BEA
7/22/2016 1:44	86.2	94	16	17	65	61	82	210	1	0.1	85.9	0.03	2205	8.2	86	9	12	84.8	0	2	0	8
7/22/2016 1:44	86.3	94	16	17	65	61	82	210	1	0.1	85.9	0.05	2380	4.6	86	9	12	84.4	0	2	0	8
7/22/2016 1:44	86.7	94	16	17	65	61	82	210	1	0.1	86.3	0.03	2705	7.4	86	9	12	84.8	0	2	0	8
7/22/2016 1:44	86.7	94	16	17	65	61	82	210	1	0.1	86.7	0.06	3069	15	86	9	12	84.8	0	2	0	8
7/22/2016 1:45	93.4	94	16	17	65	61	82	210	1	0.1	92.4	0.06	2549	14.6	84	9	12	85	0	2	0	9
7/22/2016 2:12	157.6	111	16	17	65	61	82	210	1	0.1	144.4	0.04	2898	8.9	152	9	12	109.2	0	2	0	15
7/22/2016 2:12	157.2	111	16	17	65	61	82	210	1	0.1	143.6	0.06	2800	7.5	152	9	12	109.8	0	2	0	15
7/22/2016 2:12	156.4	111	16	17	65	61	82	210	1	0.1	142.6	0.03	2351	6.3	152	9	12	109.8	0	2	0	15
7/22/2016 2:12	155.3	111	16	17	65	61	82	210	1	0.1	142.1	0.03	2212	5.7	152	9	12	110	0	2	0	15
7/22/2016 2:12	154.7	111	16	17	65	61	82	210	1	0.1	141.4	0.03	2022	5	152	9	12	109.8	0	2	0	15
7/23/2016 4:47	166.4	103	16	17	65	61	82	210	1	0.1	154.8	0.05	2878	10.6	119	9	12	129.5	0	2	0	16

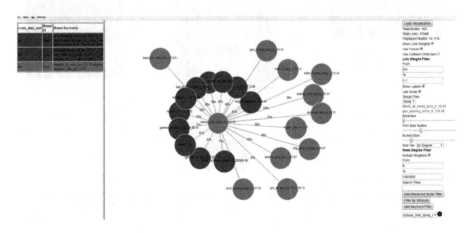

Fig. 4. An example from the original sensor data set with about 50 variables with timestamps over a period of a year (7/2016 to 6/2017)

Three results were discovered by a single CLA as follows:

- Three clusters were discovered by a single CLA as shown in Fig. 5.
  - Two green clusters represent normal running conditions.
  - Blue clusters represents outliers findings (anomalies).
- Characteristics of the anomaly cluster 1: Class A, Ship #1 with all gas turbine generator data.
- If Turbine Inlet Temp. > 409.60 °F, then the blue parts/units have a high likelihood to fail (anomalies) in near future and should be checked.

Fig. 5. Two green clusters represent normal running conditions (Color figure online)

For this maintenance sensor data, the CLA generated 160 word features from the 50 sensors. The identified features as shown highlighted in blue in Fig. 8 out of the total 160 ones are more sensitive to the engine operating performance.

Figure 6 shows all the time series for selected variables in the anomaly group distributed along time points when all the sensor data are processed together, i.e., in one CLA. The anomaly time points are shown in blue dots which have higher emerging scores in the y-axis.

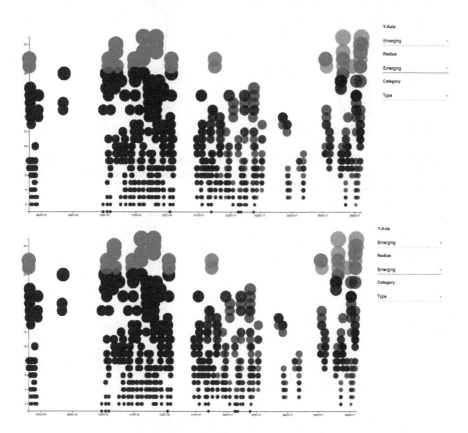

Fig. 6. Top: all three groups shown in a time series relationship with the anomaly time points when all the sensor data are together, i.e., in one CLA. Bottom: All three groups shown in a time series relationship with the anomaly time points in three separate CLAs and then fused together (Color figure online)

To illustrate the use of SI, we divided the 160 features into three groups and each set of the features and associated data are processed separately in three CLAs as shown in Fig. 7 (left). Figure 7 (right) shows the peer lists for three agents for themselves. The agents do not have to fully connected to each other. Each agent periodically performs the algorithms LLA1 and LLA2, and the whole system converges to an equilibrium state where every agent acquires the same global knowledge base. Each agent can also decide not to publish some of its knowledge base $B(t, j)$. In this case, we may call the agent possesses private information or retains expertise for itself.

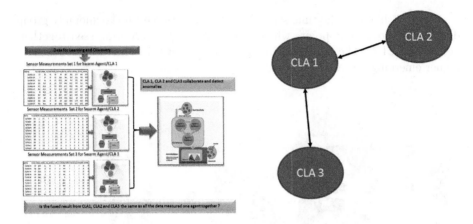

Fig. 7. Left: SI is shown in three CLAs. Right: three CLAs do not have to be fully connected. The knowledge bases "spread out" over many iterations of the fusion and ML/AI algorithms LLA1 and LLA2

CLA1 word features	Percentage of records	CLA2 word features	Percentage of records	CLA3 word features	Percentage of records
lube_oil_filter_dp_0.5	0%	gen_sw_strnr_dp_lt_0.27	0%	current_mt_2.15	0%
lube_oil_filter_dp_7	1%	hp_air_pres_mt_2975.81	0%	bearing_temp_fwd_mt_168.11	1%
turbine_inlet_temp_bt_171.42_290.51	1%	gen_bearing_temp_mt_168.14	1%	bleed_air_head_pres_bt_18.81_52.87	1%
turbine_inlet_temp_mt_408.60	2%	exctr_bearing_temp_mt_165.42	2%	bearing_temp_aft_mt_165.43	2%
lube_oil_filter_dp_6	2%	exctr_bearing_temp_bt_107.98_136.70	2%	bearing_temp_aft_bt_107.97_136.70	2%
power_mt_1.77	2%	gen_bearing_temp_bt_110.38_139.26	3%	bearing_temp_fwd_bt_110.41_139.26	3%
power_factor_mt_620629.36	2%	gen_air_temp_mt_114.22	4%	bleed_air_head_pres_mt_86.93	4%
lube_oil_filter_dp_5	6%	fuel_oil_head_pres_mt_73.81	6%	eng_lube_oil_temp_bt_89.97_109.97	9%
sw_pump_dis_pres_mt_43.76	13%	fuel_filter_dp_lt_0.54	7%	engine_lo_pres_lt_6.47	11%
stator_3_temp_mt_117.90	16%	gen_lo_clr_out_pres_bt_3.21_7.64	9%	eng_lo_clr_out_pres_mt_10.88	11%
stator_6_temp_mt_120.74	17%	gen_air_clr_out_p_bt_9.29_16.58	9%	eng_lube_oil_temp_mt_129.97	12%
stator_1_temp_mt_124.48	17%	fuel_oil_head_pres_lt_45.03	9%	coalescer_diff_pres_lt_0.57	13%
stator_highest_temp_mt_130.35	17%	gen_lo_pres_lt_9.79	9%	engine_lo_pres_mt_14.13	13%
stator_2_temp_mt_124.13	17%	gen_lo_clr_in_pres_mt_15.23	10%	engine_vibs_mt_0.07	16%
voltage_lt_14.70	17%	gearbox_oil_pres_bt_65.58_67.36	11%	eng_lo_clr_in_pres_bt_5.22_11.26	16%
stator_4_temp_mt_117.06	19%	fuel_oil_head__temp_mt_97.46	12%	coalescer_diff_pres_mt_9.05	17%
stator_4_temp_lt_84.64	19%	gen_air_clr_in_p_mt_26.50	12%	current_lt_0	21%
power_factor_lt_0	19%	gen_air_clr_out_p_mt_23.87	12%	eng_lo_clr_out_pres_lt_3.22	21%
stator_5_temp_mt_120.61	19%	gen_lo_clr_out_pres_mt_12.07	12%	bleed_air_head_pres_bt_52.87_86.93	22%
stator_5_temp_lt_94.19	21%	fuel_manifold_pres_lt_194.27	17%	eng_lo_clr_out_pres_bt_3.22_7.05	23%
stator_3_temp_lt_94.14	21%	gearbox_oil_pres_mt_69.14	17%	enclosure_temp_mt_153.41	23%
stator_6_temp_lt_92.22	21%	gen_air_clr_in_p_lt_13.38	17%	eng_lo_clr_in_pres_mt_11.26	26%
stator_1_temp_lt_96.92	21%	fuel_filter_dp_mt_3.42	18%	enclosure_temp_lt_92.31	27%
stator_2_temp_lt_94.97	21%	frequency_lt_0	18%	eng_lube_oil_temp_lt_89.97	27%
lube_oil_filter_dp_1	23%	fuel_oil_head__temp_lt_86.94	18%	bearing_temp_fwd_lt_110.41	27%
stator_highest_temp_lt_103.49	23%	gen_lo_pres_bt_13.45_17.11	19%	bearing_temp_aft_lt_107.97	28%
lube_oil_filter_dp_3	24%	fuel_manifold_pres_mt_256.41	19%	engine_speed_lt_2218.24	28%
voltage_mt_41.14	24%	gen_lo_clr_in_pres_bt_5.19_10.21	20%	engine_speed_mt_2830.48	28%
turbine_inlet_temp_lt_171.42	25%	gen_sw_strnr_dp_mt_1.65	21%	enclosure_temp_bt_92.31_122.86	28%
sw_pump_dis_pres_lt_31.26	26%	hp_air_pres_lt_2263.07	23%	engine_vibs_bt_0.03_0.05	30%
sw_pump_dis_pres_bt_37.51_43.76	27%	gen_air_temp_lt_76.10	23%	engine_vibs_lt_0.03	31%
lube_oil_filter_dp_4	29%	hp_air_pres_bt_2263.07_2619.44	27%	engine_speed_bt_2524.36_2830.48	33%
voltage_bt_27.92_41.14	30%	gen_lo_pres_mt_17.11	28%	engine_speed_bt_2218.24_2524.36	33%
stator_highest_temp_bt_103.49_116.92	32%	frequency_bt_0.02_0.12	29%	engine_lo_pres_bt_10.30_14.13	33%
stator_3_temp_bt_94.14_106.02	33%	gen_bearing_temp_lt_110.38	30%	coalescer_diff_pres_bt_4.81_9.05	38%
stator_1_temp_bt_96.92_110.70	35%	exctr_bearing_temp_lt_107.98	31%	enclosure_temp_bt_122.86_153.41	43%
stator_2_temp_bt_94.97_109.55	35%	gen_lo_clr_in_pres_lt_5.19	33%	engine_vibs_bt_0.05_0.07	45%
lube_oil_filter_dp_2	36%	gen_lo_clr_out_pres_lt_3.21	34%	coalescer_diff_pres_bt_0.57_4.81	54%
stator_6_temp_bt_92.22_106.48	37%	gearbox_oil_pres_lt_65.58	34%	engine_lo_pres_bt_6.47_10.30	64%
stator_5_temp_bt_94.19_107.40	39%	gen_air_temp_bt_76.10_95.16	34%	eng_lo_clr_out_pres_bt_7.05_10.88	67%
stator_4_temp_bt_84.64_100.85	40%	gen_sw_strnr_dp_bt_0.96_1.65	36%	eng_lube_oil_temp_bt_109.97_129.97	74%
stator_5_temp_bt_107.40_120.61	43%	gen_air_clr_out_p_lt_9.29	39%	eng_lo_clr_in_pres_bt_0_5.22	80%
stator_4_temp_bt_100.85_117.06	44%	fuel_manifold_pres_bt_194.27_225.34	39%	bearing_temp_aft_bt_136.70_165.43	90%
stator_6_temp_bt_106.48_120.74	46%	fuel_oil_head__temp_bt_86.94_92.20	45%	bearing_temp_fwd_bt_139.26_168.11	91%
stator_2_temp_bt_109.55_124.13	48%	fuel_filter_dp_bt_1.98_3.42	52%	bleed_air_head_pres_lt_18.81	95%
stator_1_temp_bt_110.70_124.48	48%	gen_air_clr_in_p_bt_19.94_26.50	52%	current_bt_-1.77_2.15	100%
stator_highest_temp_bt_116.92_130.35	49%	gen_air_clr_in_p_bt_13.38_19.94	54%		
voltage_bt_14.70_27.92	50%	fuel_oil_head_pres_bt_45.03_59.42	58%		
stator_3_temp_bt_106.02_117.90	52%	fuel_filter_dp_bt_0.54_1.98	59%		
sw_pump_dis_pres_bt_31.26_37.51	55%	fuel_oil_head__temp_bt_92.20_97.46	60%		
power_lt_0	58%	fuel_manifold_pres_bt_225.34_256.41	60%		
power_bt_-6.82_1.77	62%	fuel_oil_head_pres_bt_59.42_73.81	61%		
turbine_inlet_temp_bt_290.51_408.60	94%	gen_lo_clr_in_pres_bt_10.21_15.23	72%		
power_factor_bt_-399489.21_620629.36	100%	gearbox_oil_pres_bt_67.36_69.14	73%		
		gen_air_temp_bt_95.16_114.22	74%		
		gen_air_clr_out_p_bt_16.58_23.87	75%		
		gen_sw_strnr_dp_bt_0.27_0.96	78%		
		gen_lo_pres_bt_9.79_13.45	79%		
		gen_lo_clr_out_pres_bt_7.64_12.07	80%		

Fig. 8. Variables in all three swarms and highlighted variables are more important and sensitive in the emerging groups (Color figure online)

As shown in Fig. 6, the swarm CLAs (bottom) generate the identical results and time series visualization as if all the sensors in one swarm (top). The swarm agents can compute the exact same fusion and ML/AI computation as if the data is collected and processed in a single system. Therefore, the total correlation computations $O(n^2 m)$ is distributed and decentralized among three agents.

8 Conclusion

In this paper, we showed how collaborative learning agents and swarm intelligence are used to analyze data from a system of systems. We showed an application and a use case of quickly examining the health and maintenance issues of a sample Navy ship which might be used to generate early warning and recommendations.

A single agent/swarm is able to identify features that are anomalous and more sensitive to the engine performance. Multiple agents/swarms collaborate to distribute and decentralize the computation as if the data and computation are collected and analyzed all together in a single system.

The mechanism described in this paper is not a simple map/reduce mechanism or a collection of parallel processes because the collective behavior of swarm agents are iterated and converge towards to a stable state as if all the big data are processed in a single swarm. Each swarm actually follows a game-theoretic dual process of finding an equilibrium for itself and meanwhile achieving the maximum social welfare for the whole system. The final state of swarms is decided not only by the data collected individually but also by how the data of different agents interact, correlate, and associate with each other, just like a community of swarms, social animals and humans. An agent has one or multiple types of expertise as special data for itself. Agents share knowledge and collaborate when applying different types of expertise.

Each agent possesses the exact same code but analyzes different (sensor) data from each other. This agent design has the advantages for decentralized and distributed computing, performing learning and fusion simultaneously and in parallel as in the internet of things (IoT). Swarm intelligence is an important aspect of IoT systems or broadly defined system of systems.

Acknowledgements. Thanks to the "Hack The Machine" event (https://www. hackthemachine.ai) at the Massachusetts Institute of Technology (MIT), Cambridge, MA, organized and sponsored by the US Navy. The views and conclusions contained in this presentation are those of the authors and should not be interpreted as representing the official policies, either expressed or implied of the U.S. Government.

References

1. Zhao, Y., Gallup, S.P., MacKinnon, D.J.: System self-awareness and related methods for improving the use and understanding of data within DoD. Softw. Qual. Prof. **13**(4), 19–31 (2011). Accessed http://asq.org/pub/sqp/

2. Zhao, Y., Mackinnon, D.J., Gallup, S.P.: Big data and deep learning for understanding DoD data. J. Defense Softw. Eng. 4–10 (2015). Special Issue: Data Mining and Metrics. http://www.crosstalkonline.org/storage/flipbooks/2015/201507/index.html

3. Dumais, S.T., Furnas, G.W., Landauer, T.K., Deerwester, S.: Using latent semantic analysis to improve information retrieval. In: Proceedings of CHI 1988: Conference on Human Factors in Computing, pp. 281–285 (1988)

4. Hofmann, T.: Probabilistic latent semantic analysis. In: Proceedings of the Fifteenth Conference on Uncertainty in Artificial Intelligence, Stockholm, Sweden (1999)

5. Miller, G.A.: WordNet: a lexical database for English. Commun. ACM **38**(11), 39–41 (1995)

6. Blei, D., Ng, A., Jordan, M.: Latent Dirichlet allocation. J. Mach. Learn. Res. **3**, 993–1022 (2003). Accessed http://jmlr.csail.mit.edu/papers/volume3/blei03a/blei03a.pdf

7. Ng, A., Jordan, M., Weiss, Y.: On spectral clustering: analysis and an algorithm. In: Dietterich, T., Becker, S., Ghahramani, Z. (eds.) Advances in Neural Information Processing Systems, vol. 14, pp. 849–856. MIT Press (2002). Accessed http://ai.stanford.edu/~ang/papers/nips01-spectral.pdf

8. US patent 8,903,756: System and method for knowledge pattern search from networked agents (2014). Accessed https://www.google.com/patents/US8903756

9. Newman, M.E.J.: Fast algorithm for detecting community structure in networks (2003). Accessed http://arxiv.org/pdf/cond-mat/0309508.pdf

10. Newman, M.E.J.: Finding community structure in networks using the eigenvectors of matrices. Phys. Rev. E **74**, 036104 (2006)

11. HackTheMachine (2017). https://www.hackthemachine.ai/level-project/

12. Perretto, M., Lopes, H.S.: Reconstruction of phylogenetic trees using the ant colony optimization paradigm. Genet. Mol. Res. **4**(3), 581–589 (2005). Accessed http://www.funpecrp.com.br/gmr/year2005/vol3-4/wob09_full_text.htm

13. Beni, G., Wang, J.: Swarm intelligence in cellular robotic systems. In: Dario, P., Sandini, G., Aebischer, P. (eds.) Robots and Biological Systems: Towards a New Bionics?. NATO ASI, vol. 102, pp. 703–712. Springer, Heidelberg (1993). https://doi.org/10.1007/978-3-642-58069-7_38

Computationally Efficient Classification of Audio Events Using Binary Masked Cochleagrams

Tomasz Maka$^{(\boxtimes)}$ (ID)

West Pomeranian University of Technology,
Szczecin Zolnierska 49, 71-210 Szczecin, Poland
tmaka@wi.zut.edu.pl

Abstract. In this work, a computationally efficient technique for acoustic events classification is presented. The approach is based on cochleagram structure by identification of dominant time-frequency units. The input signal is splitting into frames, then cochleagram is calculated and masked by the set of masks to determine the most probable audio class. The mask for the given class is calculated using a training set of time aligned events by selecting dominant energy parts in the time–frequency plane. The process of binary mask estimation exploits the thresholding of consecutive cochleagrams, computing the sum, and then final thresholding is applied to the result giving the representation for a particular class. All available masks for all classes are checked in sequence to determine the highest probability of the considered audio event. The proposed technique was verified on a small database of acoustic events specific to the surveillance systems. The results show that such an approach can be used in systems with limited computational resources giving satisfying classification results.

Keywords: Audio events · Cochleagram mask · Time–frequency units

1 Introduction

As a part of auditory scene analysis, acoustic events detection plays an essential role in the machine listening systems. The number of events and its occurrence in time creates a specific structure of the acoustic environment. Acoustic events detection is a well-studied problem. Recently, due to the popularity of deep learning paradigm, many more robust solutions have been proposed [8,12,17]. However, these systems require a lot of data to create robust models. The requirement of memory and computational resources, in this case, can be significant and cause difficulties in using them into low–power systems. Moreover, such solutions are sensitive to the varying real acoustic conditions which cause performance deterioration. The event detection system in many practical applications has to be run continuously to detect the specific events and perform actions according

© Springer Nature Switzerland AG 2019
J. M. F. Rodrigues et al. (Eds.): ICCS 2019, LNCS 11538, pp. 719–728, 2019.
https://doi.org/10.1007/978-3-030-22744-9_56

to the type of event. Additionally, such a system is often organised as a set of separate and cooperated modules powered by a battery which become more and more popular in IoT (Internet of Things) systems [2,13]. Therefore, the overall computational cost needs to be minimised primarily if the system is dedicated to acoustic surveillance. Moreover, such distributed system gathering the information from many modules placed in various locations may require a mechanism to fuse information of detected events [16] which can be used to track moving sound sources in the monitored area. Using many IoT devices, it is possible to balance the calculations and to map a collection of events to several IoT modules according to the characteristics of acoustic scenes. A set of such devices with acoustic sensors can also be used for tracking sound sources by a dedicated spatial configuration of modules. Additionally, the detection of acoustic events in many cases needs a quick reaction which requires reliable and secure communication [1]. Event in the acoustic scene is often characterized by an abrupt change in energy and frequency properties in various bands of an audio stream. The process of audio event identification involves a comparison of a current signal frame with a time-frequency template. The situation of overlapping events makes it harder to detect due to the shared data in time and frequency domains. The model of events can be represented in various forms, and the detection stage may exploit a matched filtering [9], supervised learning [15] or deep learning [4] approaches. The audio event detection process depends on the many factors and a lot of techniques is applied in the analysis chain. Furthermore, such systems are rarely considered in the context of low memory requirements and computational expenditures.

The selected representation of audio signal based on the time and frequency domains plays an important role in the detection accuracy. Thus, various sets of features, its dependencies and different configurations are used in the analysis. For example, authors in [7] proposed a hierarchical structure with different feature sets with SVM classifier and found for 7 event classes that only MFCC features and their derivatives are more useful for the event classification. A method of using various audio features with a bag-of-features concept for sound events detection with low computational cost has been presented in [3]. The proposed system uses soft quantisation, supervised cookbook learning, and temporal modelling. The feature set includes MFCC, GFCC, loudness and temporal index attributes and the detection stage exploits the SVM classifier with a sliding window approach. The joint properties in time and frequency domains have been used in the work [19]. For overlapping sound event detection, a nonnegative matrix factor 2-D deconvolution and RUSBoost techniques were used. The method exploits spectral and temporal transition characteristics of the audio signal using features calculated from activations obtained from Mel spectrogram. In [10], an analysis of robust sound event recognition in adverse conditions was presented. The proposed technique uses missing feature cepstral coefficients, and ESTI Advanced Front End feature to detect the events in four different types of additive noise.

In this study, a generation and analysis of acoustic events models in time–frequency (TF) plane are presented. We proposed a simple scheme to determine a set of TF units for a given number of acoustic events by preserving its parts with the highest energy. The obtained binary masks for a specific event are then used in the process of classification. The paper is organised as follows. In the next section a process of time–frequency representation calculation, binary mask estimation an classification of acoustic events is introduced. Subsequently, in Sect. 3 an experimental evaluation is described including the thresholding analysis used in binary mask generation and an event detection evaluation process for an example database of audio events. Finally, a short discussion of the proposed technique and obtained results is presented.

2   Audio Event Classification

The process of audio event classification has to identify which time–frequency units explicitly belongs to an acoustic event. For this reason, various audio representations are exploited in existing systems. The distribution of energy in the signal representation depends on the type of sound source and the acoustic conditions like background noise, reverberation and others. The selection in such circumstances require a lot of computational power due to requirement of adaptative mechanisms.

2.1   Peripheral Auditory Representation

As a basic description of audio events we have selected cochleagram due to its importance in machine hearing [11]. The cochleagram is a model that reflects basilar membrane mechanics in the inner ear and is calculated by using gammatone filters which cover the cochlea frequencies range. Also, such representation is more robust to noise in comparison to the spectrogram representation [14]. The audio signal is converted into cochleagram in the following steps [18]:

- Bandpass filtering by a set of gammatone filters in the selected frequency range (e.g. from 50 Hz to 8 kHz).
- Calculation of the time–domain envelopes using half–wave rectification of signals at the outputs of the filter bank.
- Applying a static nonlinearity function (e.g. square root).

The obtained time–frequency representation has different frequency resolution compared to the spectrogram. The impulse response of gammatone bandpass filters can be expressed in the following form [6]:

$$g_t(t) = t^{n-1} \cdot e^{-2\pi \cdot b(f_0) \cdot t} \cdot \cos(2\pi \cdot f_0 \cdot t), \qquad t \geq 0,$$

where n is the order of the filter, f_0 denotes the filter centre frequency [Hz] and $b(f_0)$ is the bandwidth for a given f_0 frequency. The bands and the centre frequencies of the filters used in gammatone filter bank are estimated according to the equivalent rectangular bandwidth (ERB) of human auditory filters. In our experiments, we have used 4th order ($n = 4$) bandpass filters, and their bandwidth can be approximated with the formula:

$$b(f) \approx 1.019 \cdot (24.7 + 0.108 \cdot f).$$

In the filter bank, the centre frequencies of the filters f_0 are located across frequency proportionally to their bandwidths $b(f_0)$. An example set of the gammatone bandpass filters is depicted in Fig. 1.

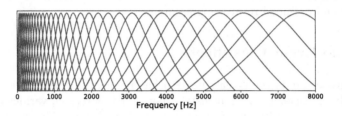

Fig. 1. A part of the gammatone filter bank (32 out of 128 band-pass filters are shown for clarity) used in the cochleagram calculation.

2.2   Binary Mask Computation

To extract TF units from cochleagram specific for a defined group of events, we decided to use a masking scheme. The binary mask is computed in the training phase where for a set of events the energy of TF units are amplified by increasing the shared values of consecutive events. All input training data is aligned and optionally interpolated in the time domain for the unification of its size. Then every cochleagram is thresholded using L_1 value and added up to binary mask template. The resulting temporary mask is eventually thresholded using level L_2, and then after binarisation, the final representation is obtained. This process converts every TF unit in the mask to value 1 when source value is greater than zero else, it replaces with value 0. The whole described scheme is illustrated in Fig. 2.

The thresholding operation is described by the following formula for both modules where $k = 1, 2$:

$$H_k(x) = x \cdot \left\lfloor \frac{\operatorname{sgn}(x - L_k) + 1}{2} \right\rfloor. \tag{1}$$

The motivation behind this scheme is the selection of the TF units with the dominant energy in the events. The TF units selected below the threshold are removed from the mask template. An essential assumption in this scheme is that all events used in the binary estimation process have to be time aligned according to their onsets. The final mask depends on the number of input audio items in the training set. For example, in Fig. 3, an evolution of the mask structure depending

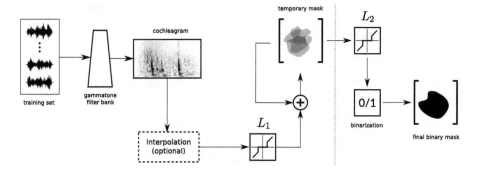

Fig. 2. The computational scheme for binary mask generation.

on the number of input events is shown. The obtained masks can be efficiently coded and stored with a small memory footprint due to their sparsity and binary representation.

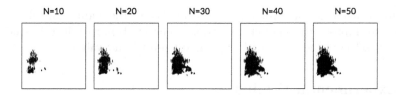

Fig. 3. An evolution of binary mask for different number of training signals.

Because the thresholding operation removes TF units, the analysis window may be reduced after mask estimation. Initially, the window has the size equal to the longest event in the set. In Fig. 4 the final width of the binary mask is presented for 5 example events. The duration of the event's mask is dependent on the type of sound source and acoustic conditions of recorded audio templates.

2.3   Events Classification

Let's consider a set of event classes $\Theta = \{C_1, C_2, C_3 \ldots\}$. The event classification in our approach is performed by multiply the cochleagram with a binary mask for successive classes. The class presence probability can be expressed as follows:

$$P(\theta) = \frac{\mathbf{e}^{\mathrm{T}} \cdot (\mathbf{A} \circ \mathbf{B}_\theta) \cdot \mathbf{e}}{\mathbf{e}^{\mathrm{T}} \cdot \mathbf{B}_\theta \cdot \mathbf{e}}, \tag{2}$$

where $\mathbf{A}$ denotes input cochleagram, $\mathbf{B}_\theta$ is the final mask for class θ, $\mathbf{e}$ is the all-ones vector, and $\circ$ is the Hadamard product operator.

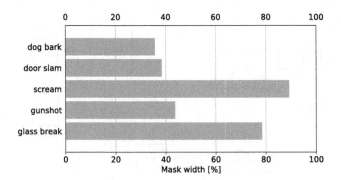

Fig. 4. The width of the final mask in the analysis window.

The final result is determined by the selection of the class with the highest probability:

$$\tilde{\theta} = \arg\max_{\theta \in \Theta} \left[P(\theta) \right]. \tag{3}$$

After calculating the probabilities for all classes, the additional rules can be applied to improve the final classification accuracy. However, in this study, we have just selected the event with the highest probability value.

3 Experimental Evaluation

The performance of the proposed technique was evaluated by using a set of acoustic events recorded in clean conditions with one channel and 44.1 kHz sampling rate. The dataset we used in the experiments contains five different acoustic events occurring in acoustic surveillance situations. The events include *'screaming'*, *'dog bark'*, *'gunshot'*, *'door slam'* and *'glass break'* sounds. Every event is in isolated form and is aligned in the class to the time onsets. The total number of items in the set contains 250 individual recordings with 70/30 data split to use as training and testing sets.

3.1 Thresholding Analysis

In the proposed approach two parameters have a direct influence on the mask generation. At each iteration of mask creation the input TF plane is thresholded using L_1 level, then the final mask is additionally thresholded using level L_2. In this way, the performance of the system can be tuned to the acoustic environment. The values are determined as a percentage value of the whole dynamic range of the cochleagram. To verify how the thresholds affect the effectiveness of classification, we have generated binary masks for all the combinations of both L_1 and L_2 values with step in subsequent attempts equal to 10%. The classification results for five classes are depicted in Fig. 5.

To illustrate the results we have decided to use violin plots [5] as it additionally shows local density estimates. For the analysed dataset the best accuracy has been achieved with $L_1 = 30\%$ and $L_2 = 70\%$. Selection of these parameters to obtain the best results should be performed whenever a new dataset will be used.

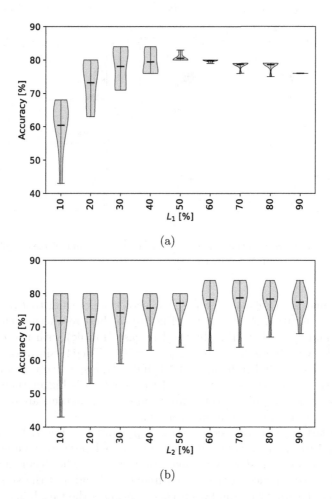

Fig. 5. The influence of thresholding on the classification accuracy for 5 classes with gradual changes of level L_1 (a), and L_2 (b).

3.2 Classification

We evaluated the performance of our technique for three different sets of randomly selected events contained 3 (*'dog bark'*, *'gunshot'*, *'glass break'*), 4 (*'dog bark'*, *'gunshot'*, *'door slam'*, *'glass break'*) and 5 (*'screaming'*, *'dog bark'*, *'gunshot'*, *'door slam'*, *'glass break'*) classes. Then for each case a binary mask was estimated as is shown in Fig. 6.

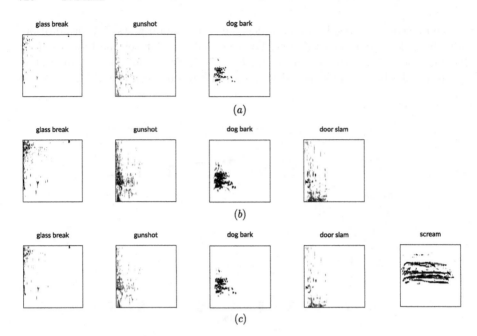

Fig. 6. The final representations of binary masks for 3, 4 and 5 classes obtained for the best thresholds L_1 and L_2 in each case.

For the first set (Fig. 6a), the best thresholds were equal to $L_1 = 60\%$ and $L_2 = 80\%$. In situation of four classes (Fig. 6b) the best result was achieved with $L_1 = 40\%$ and $L_2 = 60\%$. Finally, the last case (Fig. 6c) with five classes was obtained with thresholds $L_1 = 30\%$ and $L_2 = 70\%$. The classification accuracies for each case were equal to 95.6%, 83.3% and 84% respectively. It's interesting to observe that in every case the second threshold L_2 is bigger than L_1 which suggest that more TF units selected from source cochleagrams are omitted than in the final thresholding before binarisation.

For the last case, a confusion matrix is presented in Fig. 7. It follows that events 'dog bark' and 'glass break' were recognized perfectly, while the most mistakes occur for 'gunshot' and 'door slam' classes. The occurring mistakes are related to similarities in the shared frequency band and the similarities in the duration. The main reason for misclassification is the variability of physical properties of sound sources. The changes are rather small, but they have a direct impact on the computed mask. Moreover, the range of frequencies in cochleagrams calculated in our study was limited to 50–8000 Hz range what could have been influenced the final representation of the mask. Finally, it is difficult to indicate unambiguously the way to adjust the parameters of the proposed system to maximise the classification accuracy. As always it is a kind of the trade–off between the efficiency and the computational cost. Despite the low computational expenditures, the proposed approach has to be adapted to the application taking into account the events recorded in the target acoustic conditions.

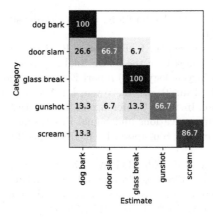

Fig. 7. The confusion matrix for 5 classes.

4 Conclusion

A computationally effective and straightforward approach to acoustic event classification is presented. The proposed method uses binary masks to determine the discriminative TF units for the joint structure of the same type of acoustic events. The fast parametrisation and classification stages along with a very low memory requirement make the proposed approach attractive to low–resource and low–power applications. Such a solution may be implemented as an auxiliary module with low computational resources to introduce additional information in multimodal systems used in smart environments. As an example application, we have selected a simple surveillance system with five specific acoustic events. The performed experiments show how to configure the mechanism of binary mask estimation. The achieved classification accuracy is acceptable in situations where the number of events is limited to a few. The presented scheme can be easily adapted to real-time analysis using the sliding window approach. In future work, the robustness analysis to background noise and the influence of the overlapping level between events will be investigated.

References

1. Ali, M.I., et al.: Real-time data analytics and event detection for IoT-enabled communication systems. J. Web Semant. **42**, 19–37 (2017)
2. Antonini, M., Vecchio, M., Antonelli, F., Ducange, P., Perera, C.: Smart audio sensors in the internet of things edge for anomaly detection. IEEE Access **6**, 67594–67610 (2018)
3. Grzeszick, R., Plinge, A., Fink, G.A.: Bag-of-features methods for acoustic event detection and classification. IEEE/ACM Trans. Audio Speech Lang. Process. **25**(6), 1242–1252 (2017)
4. Hertel, L., Phan, H., Mertins, A.: Comparing time and frequency domain for audio event recognition using deep learning. In: International Joint Conference on Neural Networks, IJCNN 2016, Vancouver, Canada, 24–29 July 2016 (2016)

5. Hintze, J.L., Nelson, R.D.: Violin plots: a box plot-density trace synergism. Am. Stat. **52**(2), 181–184 (1998)
6. Holdsworth, J., Nimmo-Smith, I., Patterson, R., Rice, P.: Implementing a gammatone filter bank. Annex C of the SVOS final report (part a: the auditory filter bank) APU (applied psychology unit) report 2341, Cambridge, UK, February 1988
7. Huang, W., Lau, S., Tan, T., Li, L., Wyse, L.: Audio events classification using hierarchical structure. In: Joint Conference of the Fourth International Conference on Information, Communications and Signal Processing, and Fourth Pacific Rim Conference on Multimedia, Singapore, 15–18 December 2003, vol. 3, pp. 1299–1303 (2003)
8. Jansen, A., Gemmeke, J.F., Ellis, D.P.W., Liu, X., Lawrence, W., Freedman, D.: Large-scale audio event discovery in one million YouTube videos. In: 42th IEEE International Conference on Acoustics, Speech and Signal Processing, ICASSP 2017, New Orleans, USA, 5–9 March 2017 (2017)
9. Kintzley, K., Jansen, A., Hermansky, H.: Event selection from phone posteriorgrams using matched filters. In: 12th Annual Conference of the International Speech Communication Association, INTERSPEECH 2011, Florence, Italy, 27–31 August 2011, pp. 1905–1908 (2011)
10. Leng, Y.R., Tran, H.D.: Using blob detection in missing feature linear-frequency cepstral coefficients for robust sound event recognition. In: 13th Annual Conference of the International Speech Communication Association, INTERSPEECH 2012 (2012)
11. Lyon, R.F.: Human and Machine Hearing. Cambridge University Press, Cambridge (2017)
12. McFee, B., Salamon, J., Bello, J.P.: Adaptive pooling operators for weakly labeled sound event detection. IEEE Trans. Audio Speech Lang. Process. **26**(11), 2180–2193 (2018)
13. Navarro, J., Vidaa-Vila, E., Alsina-Pags, R.M., Hervs, M.: Real-time distributed architecture for remote acoustic elderly monitoring in residential-scale ambient assisted living scenarios. Sensors **18**(8), 2492 (2018)
14. Sharan, R.V., Moir, T.J.: Cochleagram image feature for improved robustness in sound recognition. In: IEEE International Conference on Digital Signal Processing, DSP 2015, Singapore, 21–24 July 2015, pp. 441–444. IEEE (2015)
15. Sharma, A., Kaul, S.: Two-stage supervised learning-based method to detect screams and cries in urban environments. IEEE/ACM Trans. Audio Speech Lang. Process. **24**(2), 290–299 (2016)
16. Siantikos, G., Sgouropoulos, D., Giannakopoulos, T., Spyrou, E.: Fusing multiple audio sensors for acoustic event detection. In: 9th International Symposium on Image and Signal Processing and Analysis, ISPA 2015, pp. 265–269, September 2015
17. Takahashi, N., Gygli, M., Pfister, B., Gool, L.V.: Deep convolutional neural networks and data augmentation for acoustic event detection. In: 17th Annual Conference of the International Speech Communication Association, INTERSPEECH 2016, San Francisco, USA, 8–12 September 2016 (2016)
18. Wang, D., Brown, G.J.: Computational Auditory Scene Analysis: Principles, Algorithms, and Applications. IEEE Press/Wiley-Interscience, Hoboken (2006)
19. Yang, W., Krishnan, S.: Sound event detection in real-life audio using joint spectral and temporal features. Sig. Image Video Process. **12**(7), 1345 (2018)

Author Index